THE CHEMICAL FORMULARY

THE
CHEMICAL FORMULARY

A CONDENSED COLLECTION OF VALUABLE, TIMELY,
PRACTICAL FORMULAE FOR MAKING THOUSANDS
OF PRODUCTS IN ALL FIELDS OF INDUSTRY

VOLUME II

Editor-in-Chief

H. BENNETT

1935
CHEMICAL PUBLISHING CO., INC.
200 Park Ave. South   New York, N. Y. 10003

The Chemical Formulary, Volume II

ISBN: 978-0-8206-0260-8

Chemical Publishing Company:
www.chemical-publishing.com
www.chemicalpublishing.net

First Edition:

© **Chemical Publishing Company, Inc.** - New York 1935-2011

Second Impression:

Chemical Publishing Company, Inc. - 2011

Printed in the United States of America

PREFACE TO VOLUME II

The gratifying reception accorded Volume I of the Chemical Formulary together with the helpful and constructive criticisms received from reviewers and chemists have manifestly proved the need for a book of this type covering modern formulation in commercial chemistry.

While Volume I is complete in itself, the Editors felt it was impossible within the scope of one book to include all the formulae compiled for the numerous subject headings in the book. Volume II therefore is not a duplication or revision of Volume I but an entirely new work giving further formulae on the subjects treated in the first volume as well as more detailed information on processes and fundamental principles involved.

It will be noticed that all patented formulae have the patent number included. A helpful article on what is patentable in chemical compounding: infringements, licensing, etc., is another important addition to the book. It must be borne in mind in this connection that patented formulae cannot be used in the manufacture of commercial products unless prior arrangements have been made with the patentee.

The Editorial Board has been considerably enlarged and consequently it has been possible to include formulae hitherto unavailable.

A certain amount of criticism was directed toward the use of trade-names in Volume I. It was contended by the critics that formulae containing trade-names should be eliminated regardless of their value. Considerable thought was given to this contention and it was felt that, inasmuch as chemical trade-name products are being used in an ever-increasing number of formulae in every class of chemical manufacturing, these formulae should be included unless the application was exceptionally limited.

A second subject of criticism was the non-uniformity of systems of weights and measures used in the book. Since there is no uniformity in such systems in commercial practice and since the main purpose of the book is to familiarize the reader with commercial practice it was thought best not to attempt to standardize these systems.

In the Preface to Volume I, it was emphasized that the chemistry taught in schools and colleges is rightly confined to synthesis, analysis and engineering whereas in commercial manufacture many of the products so made are not synthetic or definite chemical materials but consist of mixtures, blends or highly complex compounds,

Because of the paucity or antiquity of the literature in this field and because of the difficulty encountered even by experienced chemists on entering new fields a definite need has existed for a modern compilation of formulae for chemical compounding and treatment.

In addition to an Editorial Board composed of chemists and engineers in many industries, publications, laboratories, manufacturers and individuals have been consulted to obtain the latest and best information in the numerous fields covered in the book.

It is important to remember that repeated experiments may be necessary to get the best results, especially when the field is intricate or unfamiliar. Again, although many of the formulae are being used commercially, some of them have been taken from patent specifications and the literature. Since these sources are subject to various errors and omissions, due regard must be given to this factor.

Formulae must be considered chiefly as starting points, variations have to be made to meet individual requirements and specifications. In cases of doubt or difficulty it is advisable at all times to consult other chemists or technical workers familiar with the particular field. This applies particularly in the case of the layman, as while a certain expense is involved this is more than compensated for by the saving of time, money and material.

As mentioned in Volume I it is hoped that those who have found a work of this kind helpful, will bring to our attention any errors they come across and will feel free at all times to make any constructive criticisms or suggestions.

TABLE OF CONTENTS

ADHESIVES

Adhesive for Aluminum

Dextrin (very soluble type)	16 lb.
Sodium Metasilicate	4 lb.
Glucose	2 lb.
Water	25 lb.

Thermoplastic Adhesive

Shellac	45 lb.
Gum Copal	55 lb.
Castor Oil	20 lb.

dissolved in

Alcohol	65 lb.
Toluol	15 lb.

This solution which is of fairly high viscosity is spread by suitable means, the solvent allowed to evaporate and the coating dried thoroughly. It can be softened after application, by heat.

Adhesive for Laminating Satin (or Other Fabric) to Paper

Casein	10 lb.
Ammonia (concentrated)	1.5 lb.
Water	40 lb.

Heat until casein is completely dissolved then add it slowly and with constant stirring to

Latex (concentrated)	100 lb.

Self-Sticking Adhesive (Non-Drying)

Nitro-cotton (20 sec.)	5 lb.

dissolve in

Ethyl Acetate	4 lb.
Toluol	2 lb.

Add the following solution which has been previously prepared

Ester Gum	4.5 lb.
Castor Oil	3 lb.
Dibutyl Phthalate	3 lb.
Ethyl Acetate	2 lb.
Toluol	2 lb.

Low Alkali Flour Adhesive
U. S. Patent 1,904,619

A glue having a moderate alkali content and a thinning or "cutting" agent is prepared at less than 43° C. as follows:

Tapioca Flour	100 parts
Potassium Bichromate	0.25–2.0 parts
Calcium Oxide (approx.)	1 part

Caustic Soda	5 parts
Water (approx.)	200 parts

Adhesive for Bands of Transmission Belting

Syrian Asphalt	1 part
Rosin	1 part
Gutta Percha	4 parts
Gasoline	6 parts
Carbon Disulphide	15 parts

Allow to stand until dissolved then stir and cover. The surfaces to be glued are rubbed and the adhesive applied, the bands being pressed between two hot metal plates.

Non-Permanent Adhesive
British Patent 404,589

A composition, which is adhesive when moist and disintegrates and is non-adhesive when dry, is formed by heating to 100° C., glucose 40, lactose 8, sucrose 8 and water 40 and, after cooling, adding French chalk 10 and flour 80.

Quick Dissolving Casein Adhesive
Formula 1
U. S. Patent 1,959,185

21 parts of the ground crude casein are soaked for a few minutes in a solution of from 1 to 2 parts of di-isopropyl, or dibutyl, naphthalene sulphonic sodium salt; a mixture of 3 parts of calcium hydroxide, 6 parts of aqueous caustic soda of 40° Bé. strength and 35 parts of water are then added, an adhesive ready for use being obtained after short stirring and homogenizing.

The aforesaid sulphonic acid salts may be replaced by sulphonation products or mineral or tar oil fractions the constituents of which have a high molecular weight, such as fractions boiling about 180° to 200° C., by sulphuric esters or sulphonic acids of palmitic, stearic, myristic, oleic, ricinoleic and like acids of the fatty acid series or of esters, amides, hydroxyalkyl amides of the said sulphonation products or, preferably, by the alkali metal salts of the aforesaid compounds.

Formula 2
U. S. Patent 1,959,185

42 parts of ground crude casein are intimately mixed with from 2 to 4 parts

of di-isopropyl, or di-butyl, naphthalene sulphonic sodium salt, 6 parts of calcium hydroxide and 4.2 parts of ground solid caustic soda. The mixture may be stored and furnishes on stirring with water an adhesive of any desired consistency which is entirely free from knots or lumps and which possesses a very high adhesive power.

Acid or Neutral Casein Solution
U. S. Patent 1,893,608

Casein (55 lb.) is soaked in water (30 gal.) at 65–70° containing sodium fluoride (5 lb.) and ammonium chloride (2 lb.) until swelling ceases, and crystalline borax (6 lb.) is added to produce a homogeneous viscous dispersion of pH 4.6–4.7, boric acid being added if necessary to produce the desired pH.

Casein "Solution"

Casein	5	lb.
Water	4½	gal.
Ammonia	11	oz.
Phenol	1½	oz.

Mix casein with hot water, add ammonia, stir until smooth and add phenol.

Casein Binder for Cork

Casein	32.7	parts
Glycerine	37.15	parts
Latex	12.75	parts
Sodium Silicate	12.4	parts
Triethanolamine	5.0	parts

Casein Adhesive
U. S. Patent 1,919,158

An intimate dry mixture of calcium hydroxide 10, sodium fluoride 4.2, and either peanut meal 84 and sodium phosphate 3, or peanut meal 76, wood flour 10, and sodium carbonate 1%, is stirred for 15 minutes with 3.8 parts of water to give a thin paste which is thickened by warming and stirring with 5–6 parts of 40% sodium hydroxide for 2 hours.

Liquid Casein Glue

Water	2,000	lb.
Casein	425	lb.
Borax (as dry crystals)	55¼	lb.
Phenol	81	lb.
Terpineol	10	oz.

Moisture-Proof "Cellophane" Adhesive
U. S. Patent 1,948,334

An adhesive composition comprised by weight from 70 to 29 parts of G rosin, 29 to 70 parts Venetian turpentine, and 1 to 20% of diethylene glycol monobutyl ether.

The method of preparing a composition for causing moisture-proof cellophane to adhere to similar or other materials, which comprises heating solid rosin to approximately 120° C. to liquefy the same, adding thereto Venetian turpentine, cooling the mixture to approximately 100° C., and then adding thereto a glycol ether, such as ethylene glycol monobutyl ether.

Moisture-Proof "Cellophane" Adhesive
U. S. Patent 1,953,104

1. Aqueous dispersion of rubber (45% rubber)	20	parts
Water-soluble agglutinant	50	parts
Condensed glycerol	10	parts
2. Aqueous dispersion of rubber (45% rubber)	5	parts
Water-soluble agglutinant	15	parts
Triethanolamine	2	parts
Ethyl lactate	5	parts
3. Aqueous dispersion of rubber (45% rubber)	5	parts
Water-soluble agglutinant	15	parts
Glycerol	3	parts
Ethyl lactate	5	parts

As a water-soluble agglutinant any of a large number of known compositions may be used, but to clearly define the above examples there is hereinafter set forth certain specific examples which have given successful results:

1. Starch, such as		
Cornstarch	10	parts
Dextrin	2	parts
Water	63	parts
2. Starch, such as		
Cornstarch	3	parts
Dextrin	1	part
Gum Arabic	4	parts
Water	22	parts

"Cellophane" Adhesive
U. S. Patent 1,929,013

An adhesive suitable for use on cellophane and the like contains gum 16.5, glyptal resin 17.9, acetone 16.6, ethyl acetate 9.0, castor oil 8.3, a 30% aqueous gelatin solution 14.5 and a 20% aqueous casein solution 17.2 parts.

Adhesive for Celluloid to Wood

Glue	10	parts
Water	10	parts
Glycol Bori-Borate	10	parts

Adhesive for Cellulose Acetate
U. S. Patent 1,950,954

Formula 1

Diethyl Phthalate	5-100 parts
Ethyl Lactate	25-100 parts
Methyl Acetate	5- 50 parts

Formula 2

Dibutyl Phthalate	100 parts
Butyl Acetate	50 parts
Ethyl Acetate	50 parts

Formula 3

Tricresyl Phosphate	5 parts
Monoethyl Ester of Ethylene Glycol	100 parts
Acetone	5 parts

Formula 4

Tricresyl Phosphate	25 parts
Ethyl Oxybutyrate	100 parts
Ethyl Alcohol	25 parts

Formula 5

Diethyl Phthalate	50 parts
Dioxan	100 parts
Ethyl Alcohol	50 parts

Formula 6

Triacetin	100 parts
Methyl Oxybutyrate	25 parts
Methyl Acetate	25 parts

Formula 7

Dibutyl Tartrate	100 parts
Ethyl Lactate, Class B	20 parts
Diacetone Alcohol, Class B	20 parts
Ethylene Dichloride	15 parts

Formula 8

Dimethyl Phthalate	5 parts
Monomethyl Ether of Ethylene Glycol	90 parts
Methyl Acetate	5 parts

Formula 9

Monomethyl Xylene Sulphonamide	100 parts
Ethyl Lactate	50 parts
Ethyl Alcohol	50 parts

Formula 10

Paraethyltoluol Sulphonamide	25 parts
Dioxan	100 parts
Acetone	10 parts

While these formulae are applicable to cellulosic plastics in general, Formulae 2, 3, 4, and 5 are especially suited for cellulose nitrate plastics, whereas Formulae 6, 7, 8, 9, 10, are especially suitable for cellulose acetate plastics.

Electrical Conducting Cement

Lampblack	1 part
Litharge	2 parts
Glycerol	3-8 parts

Electrical Lamp Circuit Paste

A device for maintaining the circuit when the filament of one of a number of series-connected lamps fails consists of a paste, containing powdered tin, carbon or other conductive substance, used to secure the cap to the bulb. As an example the paste consists of powdered conductive substance one part, lead oxide two parts and sufficient glycerol to form a paste that sets hard.

Electrical Heater Unit Cement

This method covers the process to be followed in the manufacture of cement for coffee percolator heater units (expressed in percentages by weight).

Powdered Silica	63 %
Pyrolusite (200 mesh)	9.5%
Sodium Silicate (1 part) Sp. Gr. 1.25	
Potassium Silicate (2 parts) Sp. Gr. 1.25	27.5%

Reduce the specific gravity of the potassium and sodium silicates to 1.25, using water not over 60° C. as a thinner. Then mix the two liquids. Place the silica and pyrolusite together in a mixing machine, then add the liquid. Continue to mix until a smooth uniform cement is obtained.

Make this cement only in such quantities that will be used within a week or 10 days.

Metal to Glass Cement

Plaster Paris	56%
Fine Clean Sand	20%
Rosin	20%
Drier	4%

Sufficient linseed oil is added to form a stiff paste.

Adhesive to Join Metal to Glass or Wood
Formula 1

Black Pitch	3 parts
Rosin	10 parts
Yellow Wax	5 parts
Powdered Brick	3 parts

Formula 2

Tar Resin	85 parts
Plaster	10 parts
Venice Turpentine	8 parts

Adhesive to Join Metal Letters to Glass

Marine Glue	12 parts
Slaked Lime	25 parts
Venice Turpentine	10 parts
Oil of Turpentine	5 parts
Copal Oil Varnish	35 parts
Boiled Linseed Oil	15 parts

Adhesive for Photographic and Optical Lenses

These are used hot.

Formula 1

Black Pitch	85 parts
Tallow	10 parts
Powdered Wood Charcoal	5 parts

Formula 2

Black Pitch	10 parts
Tallow	10 parts
Powdered Wood Charcoal	10 parts

Optical Glass Cement
U. S. Patent 1,921,948

Polyvinyl Chloride	33 parts
Polymerized Vinyl Acetate	42 parts
Canada Balsam	25 parts

Adhesives for Glass and Porcelain

Surfaces must be heated before application.

Formula 1

Burgundy Pitch	8 parts
Shellac, Orange	2 parts
Gum Mastic	4 parts
Gum Elemi	4 parts
Sulphur	12 parts
Kaolin	10 parts

Formula 2

Black Pitch	1 part
Tallow	1 part
Slaked Lime	Sufficient

Cement for Wood, Porcelain, etc.

Melt together equal parts of shellac, gutta-percha, and rosin. Apply to the article, which must be dry and warm.

Grafting Adhesives

The following formulae are for protecting the cuts made by grafting, from the action of air, sun and humidity. The adhesives used hot are more economical than those used cold but somewhat more difficult to apply.

Formula 1

Black Pitch	125 parts
Rosin	500 parts
Yellow Wax	120 parts
Tallow	110 parts
Sifted Wood Ash	Sufficient

Formula 2

Black Pitch	150 parts
Rosin	150 parts
Yellow Wax	200 parts
Tallow	60 parts
Sifted Wood Ash	40 parts

Formula 3

Black Pitch	280 parts
Burgundy Pitch	280 parts
Yellow Wax	160 parts
Tallow	140 parts
Sifted Wood Ash	140 parts

Formula 4
(National Horticultural School, Versailles)

Black Pitch	12 parts
Burgundy Pitch	50 parts
Rosin	12 parts
Yellow Wax	10 parts
Tallow	6 parts

This is applied hot.

Formula 5
Applied hot.

Burgundy Pitch	75 parts
Rosin	25 parts
Tallow	20 parts
Ochre	30 parts

Formula 6
This does not flow in the sun and can be applied in all weathers.

Black Pitch	50 parts
Yellow Wax	10 parts
Linseed Oil	20 parts

Formula 7
Applied cold.

Black Pitch	60 parts
Burgundy Pitch	50 parts
Yellow Wax	25 parts
Tallow	50 parts
Denatured Alcohol	25 parts

Formula 8

Coal Tar	10 parts
Black Pitch	10 parts
Burgundy Pitch	10 parts
Yellow Wax	5 parts
Whiting	10 parts
Denatured Alcohol	8 parts
Essence of Turpentine	5 parts

Formula 9
Applied hot.

Black Pitch	20 parts
Burgundy Pitch	20 parts
Rosin	5 parts
Yellow Wax	5 parts
Tallow	3 parts

Formula 10
Can be thickened by addition of wood soot.

Coal Tar	15 parts
Yellow Wax	35 parts
Tallow	15 parts
Linseed Oil	35 parts

These can be used also for covering the crevices, fissures and bruises produced on vegetation by insects, and which it is desired to heal quickly.

Formula 11
This is recommended for combating the canker of apple trees. It is spread

on the wound which has previously been carefully rubbed.

Black Pitch	30 parts
Rosin	50 parts
Yellow Wax	5 parts
Tallow	10 parts
Denatured Alcohol	100 parts

Latex Adhesive (Patented)

To 100 parts of rubber as creamed latex add 20–40 parts of rosin oil as emulsion; 30–200 parts of carbon tetrachloride emulsion (carbon tetrachloride 60, water 38, glue 1, sodium oleate 1); 2–5 parts of oil of wintergreen. The use of the oil of wintergreen is optional. If a filler is desired, e.g. whiting, then 50–100 parts of this may be added to the above formula. Latex can be creamed with any available creaming agents such as pectin, Karaya gum, alginates, Irish moss, Iceland moss, and it may be creamed more than once if one creaming is unsatisfactory. Instead of rosin oil, one may use pine tar, white pine pitch or cumar, or any softener for rubber. In place of CCl_4, one may use either benzol, high flash coal tar, naphtha or solvent naphtha, in emulsion form. These water insoluble organic liquids, by reason of their volatility, assist in the removal of water during the drying of the adhesive. Water dispersions in general do not dry as rapidly as rubber cements made with volatile organic solvents, and by the above described addition, the adhesive of the present invention is caused to dry more rapidly. Instead of oil of wintergreen, one may use either terpineol, anisic aldehyde, phenyl ethyl alcohol, methyl salicylate, oil of lavender.

Sticky Latex Adhesive

60% concentrated latex is diluted to a concentration of 45% with water and stabilized by the addition of a small quantity of casein dissolved in dilute ammonia (at the rate of 1 gram casein per 100 cc. of the original concentrated latex). The latex mix is then freed of ammonia as rapidly as possible by drawing air through it. The next step is to place 150 cc. of the ammonia-free latex in a large vessel, as considerable frothing occurs, together with 12.5 cc. of 20-volume hydrogen peroxide. The liquid is then slowly but efficiently stirred and warmed on a steam bath until frothing subsides. This takes 20 to 30 minutes. The whole is then cooled, and a further 12.5 cc. of the 20-volume peroxide added. Finally the mix is heated for 3 to 4 hours on the steam bath, with constant stirring. If any tendency to clotting appears, a little dilute ammonia should be added. After cooling, the oxidized latex is strained, and a small quantity of ammonia added as a preservative.

Vulcanizable Latex Cement
U. S. Patent 1,745,084

90 parts sulphur: 15 parts bentonite: 590 parts water, are intimately mixed. 3% saponia may be added. To this is added 462 parts latex containing 38% solids. The bentonite keeps the sulphur from settling out. This makes a fluid cement free from volatile and inflammable solvents. Used for setting bristle brushes.

Liquid Glue.
U. S. Patent 1,950,483

Animal glue is mixed with about 20–50% of urea or biuret to form a product which is liquid at ordinary temperatures and which is hygroscopic when dried and suitable for use on articles which are to be exposed to low temperatures.

Liquid Glue

Liquefied Glue (conc.)	100 parts
Lactic Acid, 85% U.S.P.	10 parts
Alcohol (den.)	6 parts
Glycerin	2 parts
Boric Acid	1 part
Benzoic Acid	1 part
Oil of Sassafras	sufficient to cover glue odor

Glue Defoaming
U. S. Patent 1,957,513

Forty parts paraffin wax and ten parts aluminum stearate are heated until the aluminum stearate is dissolved. Fifty parts of sulphonated tallow are added and the mixture is heated to about 75° C. for 15 minutes under constant stirring and allowed to cool.

As an example of the use of the defoaming agent, produced as above, determine the amount of dry glue in the glue solution to be treated, and add to the glue solution an amount of the defoaming agent equal to two per cent of the amount of dry glue. This mixture should be made at a temperature of 60° C. and when thoroughly mixed, cooled and dried.

Glue Defoaming
U. S. Patent 1,927,927

In order to produce a smooth-working glue and to prevent foaming when water is added to a stale dry glue base such as one containing casein, alkaline earth hydroxide and alkali metal compounds serving as casein solvents, a small proportion of pine oil is added to the water used, separately from the glue base.

Glue Binder for Cork Composition

Granulated Glue	1 part
Glycerin (95%)	1 part

Warm the glycerin in a steam bath, add the glue and stir until thoroughly mixed. Let stand for twenty-four hours. The product is a tough, rubbery, heavy material.

Glue for Sticking Labels to Tin

Dissolve copal gum in acetone. The best way is to cover the gum with acetone and to allow to stand for a day or two, then stir and thin to required consistency with acetone.

Glue and Gelatin Preservative
U. S. Patent 1,932,338

The addition of 0.1% ortho cyanphenol to gelatin solutions or 0.25% para cyanphenol to glue solutions prevents fungus growths.

Hardening of Gelatin
British Patent 408,085

Gelatin is hardened by addition (0.2–5% by weight) of an acid derivative of glyoxal (bisodium bisulphite or tetraacetate).

Preserving Glue
U. S. Patent 1,925,819

A small proportion (suitably about 0.1–5.0%) of ortho phenylphenol or its sodium salt is added to animal glue or similar material as a preservative.

Making Glue from Bones
U. S. Patent 1,951,260

1000 kilograms of bones are macerated for about 24 hours with 2000 litres of 7½% hydrochloric acid, whereas the complete maceration of the same quantity of bones would require 5000 litres of 7½% acid. Upon the expiration of this period the salt solution formed is run off and the bones are repeatedly washed in fresh water until the remaining acid has been completely removed.

The washed bones are then treated several times with boiling water until the largest part of the glue contained in the bones has gone into solution, although only about 40% of the amount of hydrochloric acid theoretically necessary was used.

The glue still remaining in the bones can be obtained by the pressure method. Only one or two pressings are necessary for this purpose.

"Waterless" Glue
U. S. Patent 1,955,075

Defibrinated blood is acidified at 40° C. with 0.5% of lactic acid (the percentages refer to the dry condition of the blood) and then 2.3% of ammonium sulphate is added in dissolved condition. The mixture is continually stirred at 35–40° C. and is neutralized with alkalies after 1 to 3 hours, until an alkaline reaction just occurs. 8 to 12% of concentrated glycerin and 5% of alum are then added, intimately mixed and dried according to the form of the dry product on vacuum cylinders or gelatinizing machine.

The method of using the dry glue is so carried out, that the dry glue in suitable form is introduced between the bodies to be glued. The whole is then heated to a temperature of 100 to 120° C. and pressed together. By the hygroscopic substances water is added to the dry glue in such a quantity, that sufficient moisture is present for rendering soluble the glue corresponding exactly to the temperatures employed, so that the detrimental influence effected by the excess water in other gluing methods is avoided and moistening of the bodies to be glued is not necessary. By the addition of easily fusible salts or of salts, which have a high percentage of crystal water, the melting temperature of the dry glue is kept in practice very low, so that even a heating up to 100° C. suffices.

Waterproof Albumen Glue

Add 1 part of albumen to 1½ parts slightly tempered water (to make for better receptivity). Pour the albumen into the water, allowing it to precipitate at will. What does not readily precipitate, press under, but *do not stir*. Let it soak if need be overnight, after which the swelling of the flakes will have been consummated and stirring will bring it to a fine solution. Do not strain, since the small percentage of fibrinogen forms a very valuable matter from the standpoint of waterproofing.

After the solution is ready, prepare on the side, by creaming in a little cold water, 15% paraformaldehyde based on dry weight of albumen, and stir it in rapidly. Follow immediately with 5½%

ammonium hydroxide, also slightly diluted for better dispersion, stirring very rapidly. Then, allow the solution to set up into a firm jellylike mass, which, after standing 1 hour, will return to solution easily worked. This is the solution to employ and will last for approximately 8 hours to its inversion.

Insoluble Albumen Coatings

Dissolve 1 kilogram albumen in 8 litres water.

Add about 800 cc. water in which have been dissolved 40 to 50 cc. formaldehyde (40%).

Upon the addition of formaldehyde, the albumen solution becomes jellylike and hard, the time of setting being delayed by the addition of a small amount of ammonium hydroxide to the solution. Therefore, it is imperative to make immediate use of this jelly as a coating, since setting takes place rapidly, and only what one might be prepared to use in a short time should be made up. After 24 hours, the coating becomes insoluble in water.

Padding Glue

Glue (6–8 grade Nat.		
Assoc.)	33	lb.
Glycerin	33	lb.
Water	4 gal. 2	qt.
Titanium Dioxide	4	lb.
Beta Naphthol	1¼	oz.
Methyl Salicylate	3	oz.

After passing the titanium dioxide through a 20 mesh sieve mix it with the glycerin, add the beta naphthol and water then stir in the glue. Let soak overnight and then melt; mix the methyl salicylate in thoroughly and store in tin pails.

Glue Deodorizer

Glue can be almost completely deodorized by adding 3.0% by weight of potassium nitrate to the glue when starting to cook it.

Marble or Onyx Adhesive or Cement

Granular Marble	67%
Zinc Oxide	15%
Whiting	18%

The above ingredients are mixed dry and kept that way; for use the powdered material is mixed with sodium silicate until a paste condition is obtained. Cements that are smoother can be obtained by mixing whiting or fine silica with sodium silicate until a smooth paste is obtained. There are several such mixtures on the market at present.

Adhesive or "Masking" Tape
U. S. Patent 1,933,026

Tape, made as below and coated, remains substantially permanently tacky and does not require the use of heat, a solvent, or the application of any agent other than mechanical pressure. The tape sticks with great tenacity to a highly polished or glazed surface. When it is removed it does not leave any stain. The coating further prevents permeation of the alcohol, cellulose acetate or other solvent with which the varnish or lacquer is mixed.

The tape is unimpaired after a single use and may be removed and used over again and again. It may be rolled on a core without the use of slip sheeting or otherwise treating to prevent adhesion of successive laps in the roll. The adhesive is colorless, tasteless, odorless, non-inflammable, and non-poisonous.

As a suitable reagent for treating kraft paper, first prepare a 20% aqueous solution of ammonium monophosphate (acid). Mix this solution with a 40% solution of commercial formaldehyde on the basis of 3 parts of the ammonium monophosphate solution by volume with 1 part of the formaldehyde solution.

During the mixing operation noted, there is a rise in temperature due to the breaking down of the formaldehyde and ammonia with the formation of hexamethylene tetramine. In order to prevent overheating the solutions may be mixed in a water jacketed mixer.

Next mix 1 part of ethylene glycol monoethyl ether acetate which acts as a check or modifier of the water content, to 2 parts of diethylene glycol by volume. In this solution the diethylene glycol is used as a hygroscopic agent to hold a desired water content. Triethanolamine is added in varying proportions to control the adhesive properties, thereby producing an adhesion of the desired degree. A mixture of 3 parts of the latter reagent with 1 part of the former produces a desired degree for the purposes specified.

The mixture referred to in the immediately preceding paragraph is mixed with the mixture resulting from the first two described reagents on the basis of 3 parts of the former with 1 part of the latter.

The final mixture or impregnating reagent may be placed in a bath and the paper webbing passed through the bath so that the fiber of the paper will absorb the solution. The webbing is then conducted through squeeze rolls to remove the excess of the solution.

While the web is still damp or moist, however, latex solution is applied to one surface of the tape, preferably by spraying. The web is then dried preferably by the application of heat, and in about 3 minutes it is dry enough to form into rolls. The impregnating solution is absorbed by the latex, increasing the tackiness of the latter, and acts as a binder to cause the latex coating to become firmly bonded to the fiber of the paper and tends to vulcanize and limit oxidation or decomposition of the latex solution. Ordinarily the solution is applied to a full roll of paper and the individual tapes are slit from the roll after the coating operation is completed.

Masking Tape Adhesive

Pale Crepe	50 parts
Tube Reclaim	50 parts
Cumar RS	30 parts
Bardol	10 parts
Zinc Oxide	25 parts
Lithopone	75 parts

Mill together—cut with solvent (benzol or naphtha) to the desired consistency and spread.

Adhesive for Metal
U. S. Patent 1,631,265

Bentonite	20%
Sodium Silicate (65% solution)	58%
Rubber Solution	7%
Water	15%

The rubber solution may vary from 2% to 25% and may be latex or a solution of rubber in benzol, carbon tetrachloride, etc. For glass or wood, 2% to 10% of whiting or slaked lime is added.

Newspaper Agglutinant

Water	120 gal.
Fish Glue	90 gal.
Acetic Acid	30 gal.
Glycerin	10 gal.
White Medium Cooking Dextrin	560 lb.

The dextrin is stirred into the water and heated up gently. The glue is liquefied and the acetic acid added to it and this mix pumped to the tank containing the dextrin mix. After a uniform mix has been obtained the glycerin is incorporated and the liquid then drawn off through a straining-cloth.

Non-Slip Compound for Rugs (Patented)

Water	8 lb.
Glue	¾ lb.

Glycerin	1 oz.
Latex	8 lb.
Formalin	½ oz.

Dissolve glue in water, add glycerin and then latex. Allow to stand one day and add formalin.

Enamel to Paper Adhesive
U. S. Patent 1,936,152

Clay 27 parts, zinc oxide 6.5 parts and starch 6.5 parts are used in preparing a paste which is suitable for uniting paper to enameled surfaces.

Adhesives
French Patent 748,441

An adhesive suitable for waterproofing paper, etc., is made by dissolving a vegetable gum (Dammar, Manila) in a volatile solvent (benzine, alcohol, acetone) and adding 3–6% of a vegetable fatty oil (linseed oil).

Adhesive for Bonding Paper to Moistureproof "Cellophane"

This preparation successfully sticks paper to moistureproof "Cellophane," "Sylphwrap" and similar types of transparent sheeting.

Solids

Cellulose Nitrate dry basis (High Viscosity) by weight	7%
Rosin	50%
Ester Gum	40%
Castor Oil	3%

Solvents

Benzol	48%
Methanol	32%
Butyl Acetate	20%

Solids 45%   Solvents 55%

The solvents are mixed and then the dry cellulose nitrate added. If the cellulose nitrate is already in solution, an equivalent amount of solvent is deducted from the above before adding the solvent portion. Clean, colorless, transparent celluloid scrap may be used instead of the cellulose nitrate. The rosin and ester gum are powdered and added and stirred until solution is complete. The castor oil is then added.

Wax Paper Adhesive
U. S. Patent 1,964,380

Chicle	10	parts
Dammar	10	parts
Mineral Oil	2½	parts

Paper-Hanging Paste

An excellent adhesive paste of this type may be made by heating up, with constant stirring, the following:

10 lb. best tapioca starch, 80 lb. water, and 1¾ oz. of phthalic anhydride. By increasing the phthalic anhydride content to five times its value and adding the water in the proportion of 5:1 of starch an extremely sticky paste is obtained that is valuable for many purposes. The replacement of phthalic anhydride by oxalic acid gives a less smooth paste that is not nearly so transparent on drying, but still has other very useful features, and can be used for hand-labelling of bottles or as a paper board lining paste. Other poster gums have the following composition: 220 lb. farina, 4 lb. alum, 3 lb. paraformaldehyde; 184 lb. potato starch, 539 lb. of tapioca starch, 2¾ lb. alum, and 2 lb. paraformaldehyde; 980 lb. of wheat flour, 108 lb. sago, 15 lb. of alum, and 5 lb. of formalin. The pastes all serve for various types of bill-posting work.

Jelly gums made from starch find wide applications and can be used for veneering. A gum of this type may be made by heating together at 45° C. 1000 lb. of sago, 200 lb. of tapioca starch, 3000 lb. of water, and 360 lb. of 25% caustic soda solution. After 4 or 5 hours' digestion the mass is neutralized with nitric acid.

Cement for Paper and Fabrics

Cut Celluloid	560 g.
Collodion	600 cc.
Acetone	120 cc.
Ester Gum	5 g.

Thin as desired with acetone. This cement can be thinned with acetone and used as a clear semi-flexible water resistant coating.

Low-Temperature Curing Rubber Cement

	A	B
	parts	parts
Rubber	100	100
XX zinc oxide	10	10
ZBX (zinc butyl xanthate)	6	..
DBA (dibenzyl amine)	..	4
Sulphur	..	6

In these products zinc oxide is essential, but may be varied. Also, colors and standard compounding ingredients may be used as desired. The ratio of ZBX, DBA, and sulphur to the rubber in the above is correct.

The 4 pounds of DBA may be replaced by 6 pounds of aniline if a reduction in cost is important. Aniline, however, is very volatile, and some danger exists of the fumes entering the ZBX cement and causing it to semi-cure. There is also the well-known toxic property of aniline to be considered. DBA is a very stable and comparatively non-volatile, non-toxic amine especially made for this work and well adapted to the job.

ZBX should be kept in a cool place. It is relatively unstable and will decompose easily at summer heat and above. Under proper storage conditions it is stable for a period of 3 to 6 months.

Separate cements A and B are stable for a comparatively long period of time. Blended cement should be used the day it is blended.

The products A and B are made up as 2 separate cements, preferably with benzol as solvent. Naphtha of the proper volatility may be used. It is preferable to use 2 cement churns; but, if only one is used, it must be thoroughly cleaned after mixing the one in order not to contaminate the other. The cements may be stored in separate containers in the tube splicing room. The cement operators draw about equal amounts of each cement into their hand pails, stir well, and use. Tubes spliced with this cement will cure in 1 to 2 hours in a warm room at 175° F. Steam cured splices have been made in 3 minutes at 25 pounds' steam.

Rubber Cement

Smoked Sheets	100 parts
Cumar V3	100 parts

Solvent—Equal parts of Benzol and 72° Naphtha.

The rubber is broken down on the rubber mill with about 10 parts of Cumar. The mixture is dissolved in the rubber churn and the remainder of the Cumar added.

Adhesive for Vulcanized Rubber
U. S. Patent 1,937,861
Formula 1

As a specific example of this invention 100 parts by weight of reclaimed inner tubes are mixed with 100 parts of rosin and 5 parts of powdered quick-lime. The mixture is thinned with 300 parts of gasoline. The resulting cement is an excellent adhesive, adhering permanently to wood, metal, glass, fabric, leather, rubber, etc. It is particularly useful for glueing vulcanized rubber matting to floors, etc., since the adhesion is not appreciably affected by moisture nor by the heat of the summer sun. Similar results are secured when magnesia or litharge is substituted for the lime.

Formula 2

A mixture is prepared containing 100 parts by weight of reclaimed rubber, 75 parts of rosin and 25 parts of calcium resinate, thinned with 300 parts of gasoline. The product may be employed for the same purposes as that of Formula 1.

Rubber to Metal Cement

For cementing rubber or gutta-percha to metal: Pulverized shellac is dissolved in ten times its weight of pure ammonium hydroxide. In three days the mixture will be of required consistency. The ammonia penetrates the rubber, and enables the shellac to take a firm hold, but as it all evaporates in time, the rubber is immovably fastened to the metal, and neither gas nor water will remove it.

Bonding Rubber to Metal
Formula 1
U. S. Patent 1,931,879

70 parts by weight of the semi-fluid resin obtained by extracting crude balata with a light petroleum solvent is mixed with 30 parts by weight of masticated smoked sheet rubber and maintained at 100° C. until a solution results. The solution of rubber in the balata resin is then mixed with 10% to 20% of phenol sulphonic acid, and after stirring well at 100° C., is heated to 120° to 130° C. in an air oven. After approximately 30 minutes a vigorous reaction takes place accompanied by frothing. The mixture is then removed from the oven and stirred well. When the frothing begins to subside the mass is quickly cooled in cold water and the roughly crushed material is washed to remove excess acid.

The dried product is dissolved in a suitable organic solvent such as toluene or carbon tetrachloride and applied in this form as an adhesive for rubber to metal.

A metal plate cleaned by sand blasting, pickling, or any other suitable manner is coated with the adhesive solution so that a thin even layer is obtained. After the solvent has evaporated a layer of uncured rubber is attached by pressing into contact and the rubber layer is then vulcanized by heat in the usual manner.

During the vulcanization it is desirable to hold the rubber in position by clamps in order to ensure good contact, but a very slight pressure suffices. After vulcanization the rubber is firmly attached to the metal, and the degree of adhesion even on heating to 110° is only very slightly impaired, the resistance of the rubber and metal to separation being still very good.

If the rubber covered plate is heated subsequently, e.g., at 100° C. for several hours, the degree of adhesion is further enhanced.

Formula 2
U. S. Patent 1,931,879

5 grams masticated smoked sheet rubber is dissolved in 50 grams of the semi-fluid resin obtained by extracting crude balata with a light petrolatum solvent by maintaining the mixture at 100° C. with occasional stirring until the rubber gives a smooth viscous solution. To the solution 5.5 grams of phenol sulphonic acid is added and stirred in at 100° C. The mixture is then heated to 120° C. and a vigorous reaction takes place accompanied by a copious evolution of gases containing sulphur dioxide. When the reaction has subsided the dark colored hard reaction product is dissolved in benzene and the solution applied to a steel plate as in Formula 1. After applying a sheet of compounded rubber and vulcanizing by suitable treatment (viz. heating for one hour with steam under a pressure of 20 pounds) and cooling, the rubber sheet is very firmly attached to the metal surface.

Formula 3
U. S. Patent 1,931,879

To 70 parts of the semi-fluid resin obtained by extracting crude balata with a light petroleum solvent is added 50 parts by weight of pure toluene and then 30 parts of masticated smoked sheet rubber thinly sheeted. The mixture is maintained at 100° C. for about 3 hours and well stirred from time to time until a smooth solution is obtained which is then heated to 110° C. and 4 parts of weight of hot phenolsulphonic acid (at 110° C.) added and well stirred in.

The reaction mixture is then maintained at 110° C. for about 5 to 10 hours, the maintenance of constancy of temperature of the mass being assisted by the evaporation of the toluene as the temperature tends to rise. After heating the product forms a viscous thermoplastic mass which is dissolved in toluene to give about a 20% solution which may be used for the attachment of rubber to metal in the manner described in Example 1.

Formula 4
U. S. Patent 1,931,879

25 parts of masticated smoked sheet rubber are incorporated into 75 parts of the semi-fluid resin obtained by extract-

ing crude balata with a light petroleum solvent, by heating at 100° C. and occasionally stirring during about 3 hours. When a smooth homogeneous solution has been obtained the temperature of the mass is raised to 110° C. and 3.5 parts of phenolsulphonic acid are well stirred in and the temperature maintained between 110-115° C. for 30 to 60 hours and the mass well mixed from time to time. Gases are evolved which contain sulphur dioxide. At the end of the treatment the heating is discontinued and the mass well homogenized by working in a steam heated internal mixer. This adhesive is soluble in organic solvents such as toluene and carbon tetrachloride and forms an excellent bond for rubber to metal when used in the manner described in Formula 1. The solid adhesive may also be used for attaching glass to wood or metal as for example in fixing lenses or mirrors in camera viewfinders. A solution of the adhesive is also valuable for attaching cork particles to metal plates as well as for general adhesive purposes.

In each of the above formulas resins extracted from rubber or from gutta-percha may be substituted for the resins extracted from balata and in substantially the same proportions or mixtures of resins of two or more sources of the above type may be used. A composition containing 70 parts of gutta-percha resin, 30 parts of rubber and 20 parts of phenolsulphonic acid may for example be used.

Rubber to Steel Adhesive

Raw Rubber	30	parts
Zinc Oxide	50	parts
Iron Oxide	15	parts
Sulphur	5	parts
Lime	1¼	parts

Make a thick solution with benzol and apply with brush to steel plate. When dry apply prepared rubber sheet and vulcanize.

Rubber to Steel Adhesive

Tung Oil	100	parts
Sulphuric Acid Sp.Gr. 1.84	6	parts

Add the acid to the oil slowly while stirring and then allow to stand for twenty four hours. The mixture is then ready to be applied to the steel surface which can be accomplished with a brush, after which the vulcanizable rubber is placed against the steel plate and cured.

Linseed, castor, or rosin oil can be substituted for the tung oil.

Rubber to Steel Adhesive

Tung Oil	100 parts
Phosphorus Oxychloride	50 parts

These substances are mixed and then heated for 24 hrs. at 110 deg. Centigrade and then applied to the steel plate when curable rubber is placed against the steel plate and cured.

Rubber Adhesive

A rubber bond can also be made if raw rubber such as smoked sheet is milled until plastic and then gas black or zinc oxide is added and well mixed in by milling. This milled rubber is then placed in a container and benzol and tin tetrachloride added. This mixture is then heated at the boiling temperature of benzol for two hours when it will be ready to apply to the steel plate.

Rubber to Glass Adhesive

Gutta Percha (Dental)	4	oz.
Rosin	2½	oz.
Burgundy Pitch	½	oz.
Gum Sandarac	½	oz.
India Rubber	¼	oz
Chloroform to make	1	lb.

The material is made by dissolving the gutta percha in chloroform, adding the rosin, pitch and sandarac in the order named and shaking until dissolved. The India rubber is dispersed in benzene, as little as possible and added to the chloroform solution.

Non-Drying Adhesive
French Patent 754,446

White crepe rubber 32-6, factis 36, elemi resin 4, benzene 13, methylhexalin acid ester of adipic acid 51.6 and chalk 108.1 parts.

Plastic Adhesive
U. S. Patent 1,892,123

A composition remaining plastic after prolonged weathering contains resin gum (100) dispersed in reclaimed rubber (50-200), the viscosity being reduced by volatile solvents.

Universal Adhesive
U. S. Patent 1,956,899

A glue adhering to all kinds of materials comprising a composition of:

White Crepe Rubber	3.265 lb.
Factis	3.605 lb.
Resin	0.400 lb.
Benzene	1.300 lb.

Methylhexalinhydrogenester
of the adipic acid   5.167 lb.
Chalk   10.815 lb.

Syndetickon (Adhesive)

Sugar	6	oz.
Water	18	oz.
Slaked Lime, sifted	1½	oz.
Gelatin	6	oz.

Dissolve the sugar in the water by boiling and add the lime with constant stirring. Set aside for a few days to settle and decant the clear solution. In this clear solution soak the gelatin for 24 hours, and then heat on a water bath until dissolved.

Flexible Sealing Composition
U. S. Patent 1,906,749

A composition that remains flexible comprises : oxidized and polymerised castor oil 60, alumina 5, mica 15, and asbestos less than 15 parts. The temperature of mixing should be 150° C. during at least the incorporation of alumina.

Lead Seal for Pipe Fittings

Powdered Lead (325 mesh) 90 parts
Cup Grease   10 parts

This gives a perfect seal, and has an advantage over red lead in that it does not harden. It allows of pipe lines being taken down without splitting pipe or breaking fittings.

Sticking Labels to Tin

Gummed labels will adhere to tin, if some tincture of Benzoin is first applied to the tin.

Tin Adhesive
U. S. Patent 1,627,278

A water emulsion of latex, containing about 35% rubber solids, is mixed with a colloidal suspension of bentonite. The final proportions are—10 lbs. of 35% rubber latex emulsion: 18 lbs bentonite: 100 lbs. water. A preservative may be added.

Adhesive for Lacquered Tins

Wheat Flour	2 to 4	oz.
Corn Starch	¼	oz.
Alum	¼	oz.
Sodium Salicylate	¼	oz.
Honey	1 to 2	oz.
Venice Turpentine	2	oz.
Water	8	oz.
Glycol Bori-Borate	¼	oz.

Mix the solids with the water and heat the mixture on a water bath until a stiff paste results. Then add the sodium salicylate and the honey, and after mixing well, slowly pour in the Venice turpentine and stir until it is thoroughly distributed.

Tin Paste

Ten pounds of cornstarch is thickened in six gallons of boiling water, and then one quart of glucose is added. To this mixture add one pint of glue and four to eight ounces of phosphoric acid. Mix well.

Tile Adhesive
British Patent 397,678

Mix 15 parts of 75% rubber latex, 3–12 parts of silica flour and 0.5–1 part of gum. Coloring matter may be incorporated or painted on the tile back.

Transfer Adhesive

1. Gum Arabic	2	lb.
2. Water	1	gal.

Soak overnight and decant clear liquid.

3. British Gum	1	lb.
4. Fish Glue (liquid)	2	lb.
5. Glycol Bori-Borate	2	lb.
6. Methyl Cellosolve	2	lb.

Dissolve 5 in 6 by slow agitation and then add 3 and 4 to it, mixing well. Mix with gum solution warming to 60° C.; stir until uniform.

Veneer Glue
French Patent 750,403

A glue suitable for veneering is made by dipping a light cloth into a bath containing agar-agar 23, sugar 12, formaldehyde 5, and water 9.3%. It is dried and dipped into a bath containing phenol 0.5, glycerol 1.5, oxalic acid 1.5, kaolin 5%, and glutin, and again dried. The required amount is cut off and gluing is obtained by applying heat and pressure.

Vacuum Preserve Jar Ring Adhesive
British Patent 397,374

Soft manila gum 34, kaolin 33, methylated spirits and castor oil 5 parts. The adhesive may be thinned by aqueous ammonia and glycerol.

Veneer Glues

Make up a solution of blood albumen with calcium hydroxide $(Ca(OH)_2)$, which will produce a jellylike mass, which adheres to the rollers of a spreading machine, if such is used, and is not absorbed too much by the wood. Use the following proportions:

1 kilo albumen dissolved in 7½ litres filtered water.

1 kilo slacked lime mixed with enough water to give the solution a sp.gr. of 9° Baumé. To obtain this, about 4 litres of water are necessary.

7½ litres albumen solution to 1 litre calcium mixture and 1½ litres of water. After about 9 minutes, this solution approaches a consistency like honey and after 15 hours is a firm gelatin.

This prepared glue is then spread on the fibre, wood or veneer, which is immediately pressed between heated plates in a hydraulic press (or high pressure press) at about 85° C. Through the application of heat, the albumen becomes insoluble and binds evenly.

If lime injures the fibre, it is necessary to employ a smaller amount of water. For example,

1 kilo albumen to 2 litres of water will yield a very heavy glue. If the cemented joints are treated in a kettle with formaldehyde fumes, the albumen becomes insoluble in water.

White Veneer Glue

White Tapioca Flour	100 lb.
Barium Peroxide	9 oz.

Mix the barium peroxide with the tapioca flour, preferably in a rotary mixer, and leave in the mixer for approximately one hour, or until the barium peroxide is thoroughly blended into the tapioca flour. In use, one part of vegetable glue is added to two and one-fourth parts of cold water and agitated until a creamy paste is formed. This is heated, with constant stirring, to a temperature of 165° F.; within about one-half hour the mixture becomes transparent, viscous and tacky. It is then ready for use. This product is particularly desirable for wood veneers as it is pure white in color, practically neutral in reaction, and hence will not stain or discolor thin veneers. Sheer blocks and panels prepared with this glue, according to the standard methods in use at the Forest Products Laboratory, indicate sheering strengths of over 2600 pounds per square inch, when used on high-grade hard marble blocks.

Cabinetmakers Glue (waterproof)

Formula 1

Add 3 drams potassium bichromate dissolved in smallest possible quantity hot water to 1 quart hot thick glue immediately before using. Keep out of sunlight.

Formula 2

Swell 1 part good grade glue in equal amount of water. After glue has absorbed as much water as possible (without losing semi-solid form), add 1 part linseed oil, heat, and stir to even consistency.

Formula 3

Melt and stir to even consistency, 12 parts glue in 15 parts water. Stir in 2 parts rosin. When thoroughly mixed, stir in 4 parts turpentine. Apply hot.

Water-Resistant Silicate Adhesive
U. S. Patent 1,949,914

Formula 1

45.8 parts of an aqueous solution of copperamino-sulphate as is obtained by mixing 14.4 parts of blue vitriol with 16.5 parts of water and 14.9 parts of aqua ammonia of sp. gr. 0.90, is added slowly to 200 parts of a 42.5° Bé. sodium silicate solution of ratio 1: 3.2. The solution is diluted with 14.6 parts of water to bring its viscosity to the viscosity of the original sodium silicate solution. The modified silicate has good adhesive properties and a more rapid rate of set than the original silicate. The water resistivity of its film as an adhesive for paper is illustrated by the following:

Chip-board joints formed by application of the silicate and aged for seven days at room temperature had a life of four days to complete failure under heavy tension in water, compared with twenty minutes' life for the straight sodium silicate controlled points. Bond writing paper joints made and tested as above had a life of three days in water, compared with five minutes for the straight sodium silicate controlled joints. The modified silicate solution was perfectly stable after eight and a half months of age.

Formula 2
U. S. Patent 1,949,914

23.4 parts of a solution of a complex copperamino-sulphate obtained by dissolving 7.2 parts of blue vitriol, 7.7 parts of water and 8.5 parts of aqua ammonia sp. gr. 0.90, is added slowly to 200 parts 35° Bé. sodium silicate solution of ratio 1: 3.2, with good stirring and diluted with 17.4 parts of water. An additional amount of copperamino obtained from 36.6 parts of a 20% aqueous solution of blue vitriol is then added and the viscosity adjusted by dilution to that of the original silicate. This solution is stable for at least six weeks and shows an even better water resistance than films obtained from the modified silicate of Formula 1.

Films produced from such modified sodium silicate solutions when produced for instance, on glass, show a considerably greater resistance to water than straight sodium silicate films. Such films, when broken away from the surfaces on which they are formed, are only slightly dissolved after two months' sojourn in water whereas when made from straight sodium silicate they would dissolve in a few minutes.

Formula 3
U. S. Patent 1,949,914

12.7 parts of soy bean meal is dispersed by grinding in 22.3 parts of water and 65 parts of a 38.5% solution of sodium silicate having a ratio of $1Na_2O$ to $3.2SiO_2$. When the soy bean meal is well dispersed, 194 parts of the silicate solution, 5.1 parts of wood flour, and 7.6 parts of water are added. The solution is next modified with a solution of cupricamino-sulphate consisting of 14.3 parts of blue vitriol, 15 parts of aqua ammonia of 0.90 sp. gr., and 41.3 parts of water. The cupricamino-solution is added slowly with thorough agitation to the modified silicate solution. Then 13.1 parts of blown China-wood oil containing 0.13 part of the sodium salt of petroleum sulfonic acids is added slowly with thorough agitation of the solution.

The high water resistance of an adhesive having the above composition is illustrated by the following data: Chip-board joints glued with the composition and aged for seven days at room temperature has a life of forty-seven days to complete failure under tension in water, compared with twenty minutes' life for the joints which had been glued with unmodified sodium silicate.

Formula 4
U. S. Patent 1,949,914

45 parts by weight of a solution of the zincamino-sulphate (10 parts zinc sulphate heptahydrate, 20 parts water, and 15 parts aqua ammonia, 0.90 sp. gr.) is added to 200 parts of sodium silicate (42.5° Bé., ratio 1: 3.25) diluted with 10 parts of water, with high speed stirring. Then 10 parts of water is added to bring its viscosity to the same value as the original silicate solution.

This modified silicate solution has good adhesive properties, and a more rapid rate of set than the original silicate. For example, a film on glass of this modified silicate dried to tack-free state in three

minutes, compared with eight minutes for the original silicate. The so obtained film when flaked from glass and placed in water is not completely dissolved after four months of contact. Chip-board joints of this adhesive has a life in water of 2½ days, compared with 30 minutes for control joints of the original silicate. This modified silicate solution has shown perfect stability for four months.

At normal water content the above composition has a paste-like consistency. As an example, a sodium silicate-copper-amino-paraffin composition of paste-like consistency is prepared by adding to the composition described under Formula 1 a 50% water emulsion of paraffin in an amount equal to 50% paraffin based on the silicate solids. The film of the composition on glass is very water repellent, continuous and does not crack or peel and is particularly adapted in the field of coatings and sizes. The composition can, of course, be modified within wide limits, both as to the water repellent and the thickening agent and as to the composition of the modified silicate. The composition can likewise be modified further by the addition of pigments which may serve both as thickening agents or dispersing agents for the paraffin, etc., and impart to the compositions a distinct color.

The composition can, for instance, be used as a substitute for asphalt in the paper board industry.

Instead of the water repellent substances, or in addition to them, drying oils, such as blown China-wood oil, blown linseed oil, rubber latex, particularly in the concentrated form, and many other water insoluble adhesives may be added. Here again thickening agents which in addition act as stabilizing agents, for the adhesive besides the China-wood and linseed oils, substances, such as wheat flour, starch, pectin, salts of alginic acid, kelp flour, etc., may be added.

Waterproof Adhesive

Cumar Resin (light)	75 lb.
Shellac	15 lb.
Aroclor 1262	10 lb.
Ethyl Acetate	30 lb.
Toluol	20 lb.

Machine Labeling Paste

660 lb. of water is run into a tank and heated to 50° C. and the hydrogen ion concentration adjusted to 4.6, using bromcresol green as the indicator, and sulphuric acid as the acidifying medium. 600 lb. of farina are now added and

stirred in for 15 minutes the pH again being adjusted to 4.6 after stirring. 14 lb. of malt is mixed with 36 lb. of water and 20 lb. of this mixture added to the tank and the contents heated to 55° C.; 15 minutes later a further 10 lb. of the malt extract solution is added and the temperature raised to 65° C. After the lapse of another 15 minutes, 5 lb. of the malt extract solution is added and the temperature raised to 73° C. and maintained at this level for 80 minutes, then raised to 95° C., cooled to 70° C., and the rest of the malt added. The malt is killed with 4 lb. of caustic soda solution (25 per cent) when the required viscosity is reached.

Bag Sealing Adhesive

For this purpose 1,400 lb. of the above conversion are prepared and to this is added 1,980 lb. of a yellow dextrin and 1,760 lb. of a light yellow dextrin that has not so far been converted. Phenol to the extent of 10 lb. is added as a preservative.

Spot Gumming Paste for Use on Platen Printing Presses

Gum Acacia	2	lb.
Glycerine	3	oz.
Strained Honey	3	oz.
Water	6	lb.
Sodium Benzoate	1½	oz.

Sprinkle the acacia over the water. Do not attempt to mix, but allow to set undisturbed until dissolved, which requires about twelve hours. When dissolved stir thoroughly and mix in the sodium benzoate followed by the honey and glycerine. Use as you would a printing ink.

Printers Tableting Composition

To a 70% ammoniated rubber latex mix in 20% by volume of any good fast drying liquid glue. To color the composition add any aniline dye which is water soluble and fast to alkali. The coloring is best done if a definite weight of aniline dye be dissolved in a given weight and strength of ammonium hydroxide and a previously determined amount of this stock coloring solution mixed into the latex mixture.

This material has the advantage of never becoming tacky even on contact with water, being water insoluble after it dries. It has better strength, is much more flexible and dries within ten to fifteen minutes. Only one coat is required.

Glucose-Glycol Paste

Water	50 parts
Diethylene Glycol	10 parts
Glucose (43° Bé.)	20 parts
Beta Naphthol	0.1 part
Ammonia Alum	0.3 part
Flour (soft winter wheat)	19.6 parts

Heat in double boiler while stirring until a smooth paste results. The above formula is used in the U. S. Government Printing Office.

Inexpensive Sealing Wax

Rosin	70%
Orange Shellac T. N.	10%
"Celite" (Infusorial Earth)	20%

Melt the rosin and add the shellac. Stir until evenly mixed. Add the Celite and mix thoroughly. Remove from the fire and pour into the moulds. If it is to be colored any desired colored pigment can be mixed with the Celite. The formula may be varied if desired. If a harder wax is wanted increase the percentage of Celite.

Druggists' Paste

Mix four ounces of powdered yellow dextrin with enough boiling water to make a thin paste. Boil this until clear and add a few drops of methyl salicylate as a preservative. This paste is simple, inexpensive and permanent.

Paste for Vitreous Enamel
U. S. Patent 1,936,152

For pasting labels to vitreous enamel and porcelain.

Corn or Wheat Starch	5 parts
Zinc Oxide	1 part

Mix and grind together in ball mill. Add sufficient water before using to make a paste of the consistency desired.

Photo-Library Paste

White dextrin, 10 oz.; potato starch, 10 oz.; water, 7 pints. Make into a paste with cold water, then add rest of water, and heat in double boiler till a smooth solution is made. Stir constantly. Add 3 ounces of glycerin, then 30 drops carbolic acid and ⅛ oz. formaldehyde, 10 drops oil of sassafras, and stir in thoroughly. Pour into jars while warm.

Card Paste

British Gum white	42 lb.
Corn starch	5 lb.
Glucose	15 lb.
Water enough to make	25 gal.

Paste the gum and starch with cold water. Raise to a boil using total quantity of water and boil 1 hour. Cool down and add the glucose.

Office Mucilage

Gum arabic, 8 oz.; water, 20 oz. Dissolve in warm water, let stand several days. Do not heat. When dissolved, bring to a gentle boil, add glucose, 32 oz. Mix well. Let cool and add acetic acid, 1 oz. and carbolic acid ¼ oz.; with stirring.

Mucilage for Labels

Macerate five parts good glue in eighteen to twenty parts of water for a day, and to the liquid add nine parts of rock candy, and three parts of gum arabic. The mixture can be brushed upon paper while lukewarm; it keeps well, does not stick together, and when moistened adheres firmly to bottles. For labels of bottles, it is well to prepare a paste of good rye flour and glue, to which linseed oil, varnish and turpentine have been added in the proportion of half an ounce of each to the pound. Labels prepared in the latter way do not fall off in damp cellars.

Glazing, Calking and Pointing Compositions
British Patent 398,057

A composition which retains its plasticity for long periods comprises aluminium 1–30, oil, e.g. soy-bean, linseed, perilla, cottonseed, olive, rapeseed, corn, etc., about 30, mineral filler, e.g. china clay, titanium white, ground shale or slate, zinc white, silica, barium sulphate, etc., 1–50, fibrous material, e.g. asbestos, shredded or ground rope or rags, 1–20% and the remainder a fluid vehicle, e.g. mineral spirits. The preferred composition is whiting 12, magnesium silicate 17, asbestos 5.45, soybean oil 30.6, varnish 16.2, mineral spirits 9, and powdered aluminium 9 parts.

Putty Hot and Cold Glazing

Putty Mixture	275 lb.
Raw Lindeed Oil	3⅜ gal.
Japan Drier	⅛ gal.
Mineral Spirits	⅜ gal.

For cold glazing it may be used as is, or a little more oily. For hot glazing the user may want it a little stiffer. The putty mixture is a dry blend of approximately 73% of 200 mesh domestic whiting and 27% imported Belgium whiting; so that the formula above would be equivalently 200 lb. of domestic whiting and 75 lb. of Belgium whiting. This may also be varied to say 225 lb. of the former and 50 lb. of the latter.

Primeless Sash Putty

Domestic Whiting	91 lb.
Belgium Whiting	26 lb.
Silica (200 mesh)	55 lb.
Boiled Linseed Oil (containing drier)	2¾ gal.
Mineral Thinner	⅛ gal.

Skylight Putty

5% leaded Zinc Oxide	24 lb.
Borate of Manganese	2 lb.
Lamp Black	2 oz.
H. B. Boiled Linseed	4 gal.
Heavy Gloss Oil	4 gal.
Long-Oil Spar Varnish	4 gal.
Soapstone Talc	16 lb.
No. 60 Asbestos Fiber	16 lb.

The heavy bodied linseed is regular material obtainable from all the large oil mills; the heavy gloss oil is 8 lb. of rosin to the gallon of thinner; the waterproof spar varnish should be long-oil to an extent not less than 45 gallons of oil to the cwt. of resin, and have true body—i.e., the amount of thinner should not exceed 1½ times the oil; the No. 60 asbestos fiber is a Johns-Manville grade.

The product is not ground; a pony mixer is used. The zinc, with borate and lamp black, is first worked into a paste with the oil, subsequently adding the talc and sufficient vehicle, and finally feeding in the fibrous asbestos. This latter should be added in portions to insure wetting and thorough pulling apart to avoid masses of dry fiber.

For new work and for all embedding, the knifing consistency as formulated should be used. The lights should be embedded; if large dimension and too much play between the glass and the frame, oakum may be caulked in at intervals along the edges to support a uniform spacing within the frame. For re-glazing or faceglazing of skylights, the putty may be produced in special consistency (suggest a normal gloss oil instead of the heavy body) or it may slightly be thinned but only as necessary for using in a hand-pressure glazing gun.

Glass roofing, as on art buildings and auditoriums, is constructed of long panes of wire-glass, the end of each upper pane overlapping the one below same as in laying shingles. Driving rains will force the water up beyond the lower lap unless well sealed. Some roofs have melted together 2 parts of rosin and 1 part of tallow, adding a little linseed oil; this is brushed onto 2-inch wide strips of heavy muslin and laid over the joint, pressing down to exclude air pockets.

Steel Sash Putty

Steel sash putty dries very firm, even to the point of complete hardening, whereas skylight putty should not harden because maximum expansibility is required continuously for effectual sealing against leakage. Steel sash putty is generally furnished in dark color, gray predominating. A natural gray color is supplied by the formulation herewith:

Domestic Whiting	190 lb.
Belgium Whiting	80 lb.
Sublimed Blue Lead	25 lb.
Reduced Fish Oil	4¾ gal.
Bodied Linseed Oil	1¼ gal.
White Oil Drier	⅛ gal.

This product has proven very satisfactory to steel frame fabricators. The blue lead is not only essential for color but it imparts solidity and some measure of adhesion. The treated fish oil is 100 gallons of clear light sardine oil heated to 575° F. (not to exceed 600°), pulled off and allowed to cool in the kettle overnight, and reduced cold the next day with 45 gal. of mineral thinner. This produces an exceptionally pale liquid a little lighter body than raw linseed oil, drying properties also similar, and it imparts some water repellence Sp. Gr. 0.912. The bodied linseed oil is 50 gal. of V.M. Oil run to 600° F. and held 5 hours; allowed to cool in the kettle. Sp. Gr. 0.972. This oil gives good wetting of the pigments and also of the surface to which the putty is applied. The drier carries 2 gal. of wood oil and 4 gal. of linseed to the cwt. of hardened rosin; also 6 lb. of lead acetate incorporated with that base quantity of oil. The weight per gallon of the product is 7.55 lb.

Litharge Putty

Finely ground Litharge (80 lb.), red lead (8 lb.), flock asbestos (10 lb.) and boiled linseed oil (1½ gal.), mixed by mechanical means.

General purpose sealing for bell-and-spigot joints on earthenware, stoneware, fused silica and glass; also for jointing purposes.

Becomes moderately hard in 7 days. Resistant to nitric acid (30 to 56%), but not to be used for nitric acid vapors which are very hot, unless pipe joints are plugged with asbestos rope, as there is a tendency to catch fire.

Black Putty

Boiled tar (or soft pitch) and raw linseed oil in equal quantities, mixed with dry china-clay to consistency of glazier's putty.

Flexible sealing for bell-and-spigot joints on hydrochloric acid plant.

Good acid and weather resisting qualities. Does not harden, crack or shrink when in use; being permanently soft it is easily removed from pipe joints when necessary. Pipe joints should be plugged with asbestos. If made with anthracene oil (sp. gr. 1.12) the quality is said to be greatly improved.

Soft (Asbestos) Putty

White asbestos powder (40 parts), blue asbestos fiber (8 parts), tallow (2½ parts), whiting or china-clay (10 parts) and boiled linseed oil (21 parts), mixed by mechanical means (preferably in edge-runner mill.

Flexible sealing for bell-and-spigot joints and other connections in contact with nitric acid (hot or cold) and nitric acid vapors.

Can be made in quantity and stored ready for use. Asbestos fiber should be well "tensed out" to the nature of fluff. Surface of joints should be finished off with Portland cement (or silicate cement) and painted with tar or weather-proof coating.

Litharge Cement

Finely ground litharge and glycerine mixed to consistency of stiff paste.

Permanent metal-to-metal joints; also applicable for rigid joints on earthenware, stoneware, fused silica and glass; very satisfactory for fixing acid-resisting tiles.

Sets in about 30 minutes, giving a hard stone-like mass. Setting can be delayed by diluting the glycerine with water or by adding china-clay to the litharge. Should not be used for the joints of metal tanks subject to wide temperature fluctuations. Resistant to steam and oil up to 400° F., and to weak acids (sulphurous acid).

Baryta Cement

Precipitated barium sulphate (free from carbonate) and sodium silicate solution sp. gr. 1.30, mixed to consistency of mortar.

Rigid and acid-resisting joints in warm positions.

Sets to hard mass under moderate degree of heat. Speed of setting can be increased by using potassium silicate solution. Should not be made with finely ground natural barytes.

Silicate Cement

Coarse white asbestos powder and sodium silicate solution sp. gr. 1.25, mixed to consistency of fairly stiff dough and used immediately.

Permanent rigid joints of all types; top coating for soft putty joints.

Sets to hard mass, but it is advisable to coat the joints with tar when exposed to weather.

Acid-Resisting Building Cement
British Patent 110,258

Ground stoneware 30 mesh (8 parts), fine Leighton sand (7 parts), ground Staffordshire blue-brick 60 mesh (2 parts), plaster of Paris (0.12 part) and sodium silicate solution sp. gr. 1.30 (3 parts), mixed by mechanical means (preferably in edge-runner mill) and used within one hour.

For setting acid-resisting brick in the building of chimneys, flues, for acid gases, absorption towers and tanks, suitable for acid-resisting floors not subject to large amounts of water.

Sets quickly enough to carry the weight of the bricks. Water or steam should not be allowed in contact with the finished work until thoroughly dry; drying can be hastened by raising temperature to 210° F. for several days. Unaffected by mineral acids over a wide temperature range, by sulphur dioxide and trioxide, nitrous acid fumes, chlorine and sulphuretted hydrogen.

Heat-Resisting Cement

a. Graphite (2 parts), black oxide of manganese (2 parts), common salt (1 part), borax (1 part), iron filings free from oil (4 parts), dry powdered clay (8 parts) and water, mixed to consistency of mortar for immediate use.

Retort work generally; furnace brickwork; flue pipes in hot positions.

Heat-resisting qualities are procured to the greatest degree by slowly raising the temperature (when set and dry) to "white heat."

b. Powdered pumice (9 parts), asbestos powder (1 part) and sodium silicate solution sp. gr. 1.38, mixed to consistency of mortar.

Silicated Fire-Clay Cement

Common fire-clay mixed with weak sodium silicate solution sp. gr. 1.02.

For use in very hot positions on furnace brickwork, gas producers, flues, etc.

Sets to a harder mass than usual (in warm positions) if made unduly wet.

Rust Joint Cements

Quick-setting: Cast-iron filings free from oil (80 to 100 parts), sublimed sulphur (2 parts) and salammoniac (1 part), well mixed and merely made damp to touch with water, for immediate use.

Slow-setting: Cast-iron filings (200 parts), sulphur (1 part) and salammoniac (2 parts).

Permanent iron to iron joints; gasworks retorts, superheater pipes, etc.; also applicable for fixing iron in earthenware or stoneware.

Iron filings should pass 1/16-inch mesh. Mixture is tamped into the joint with the aid of steel chisels. Gives hard gas-tight joints which do not leak when heated, but should not be exposed to acid fumes. The slow-setting mixture gives the strongest joint.

Rusting Portland Cement

Cast-iron filings free from oil (40 parts), black oxide of manganese or sublimed sulphur (10 parts), salammoniac (1 part) and Portland cement (20 to 40 parts), mixed with water to consistency of mortar.

Permanent iron to iron joints; very useful for sealing cracks and for general repairs to iron vessels.

Portland cement is added to reduce the risk of undue expansion with consequent bulging from cracks and crevices.

Rubber Cements

a. Masticated raw rubber and raw linseed oil in equal quantities, heated together and worked into a stiff putty with flock asbestos.

Flexible sealing for bell-and-spigot joints on hydrochloric acid plant.

Does not harden, crack or shrink. Resistant to hydrochloric acid gas, but not suitable for high temperatures.

b. Masticated raw rubber (1 part), hot raw linseed oil (2 parts) and pipeclay (3 parts).

Flexible sealing for bell-and-spigot joints on hydrochloric acid plant.

Remains permanently soft and very impervious, but not suitable for high temperatures.

Rubberized Pitch Cement

Masticated raw rubber (1 part) and shellac (2 parts), dissolved in the smallest quantity of turpentine and mixed with hot pitch (3 parts).

Useful for making gas-tight joints between glass and metal.

Becomes firm, but not brittle, when cold.

Red Oxide Jointing Putties

a. Red oxide of iron free from silica (5 parts) and boiled linseed oil (1 part), mixed to stiff paste.

Jointing putty for iron to iron, iron to lead, and lead to lead.

Resistant to hydrochloric acid and sulphuric acid and suitable for temperatures up to 500° F.

b. Red oxide of iron free from silica (2½ parts) and boiled linseed oil (1 part), mixed to consistency of dough.

Jointing putty for iron to iron, iron to lead, and lead to lead.

Resistant to hydrochloric acid and sulphuric acid and suitable for temperatures up to 500° F.

c. Red oxide of iron and pine tar.

Jointing putty for iron to iron, iron to lead, and lead to lead.

Does not harden so soon as red lead, but very adhesive on flanges under pressure.

Red Lead Jointing Putty

Red lead added to white lead ground in linseed oil to form stiff paste, which is then "pounded" (adding more red lead) until it no longer softens and sticks to the fingers.

Jointing putty for steam, water (hot or cold) and general pipe-work to resist fluid pressure.

For superheated steam use red lead, white lead and graphite in equal quantities, mixed with boiled linseed oil.

Red Lead Cement

Red lead and glycerine mixed to stiff paste.

Permanent joints for acid resisting cast-irons.

Sets to hard joint which is resistant to acids and acid fumes.

Sulphur Cements

a. Fine dry sand (6 to 7 parts) and sulphur (4 to 3 parts) heated to 300° F., well stirred and used in semi-plastic or molten state.

For setting acid-resisting brick in the building of acid towers and tanks; rigid joints on brick, earthenware and stoneware.

Tensile strength nearly 400 lb. per sq. inch. Resistant to acids and acid fumes up to 200° F. Bell-and-spigot joints should be plugged with asbestos rope before pouring the cement.

b. Pitch and sulphur melted together and used hot.

For setting acid-resisting brick in the building of acid towers and tanks; rigid joints on brick, earthenware and stoneware.

Litharge-Asbestos Cement

Finely ground litharge (1 part), very fine sand (7 parts), asbestos powder (2 parts) and sodium silicate solution sp. gr. 1.38, mixed to consistency of putty.

General sealing purposes against acid fumes at high temperature.

Ultimate hardness and rate of setting can be altered by varying the amount of sand used.

Oil-Resisting Cement

Common rosin boiled with 16–17 per cent caustic soda, the residue which separates on cooling being mixed with plaster of Paris.

Proves very satisfactory against paraffin oil and hot mineral jelly (vaseline).

Retort and Crucible Lutes

a. Soft pitch mixed with starch.

This meets the requirements of an "all carbon" mixture.

b. Powdered glass or fine sand and sodium silicate solution.

Difficult to remove.

c. Common clay, protected externally with Portland cement.

Easily removed. Resistant to nitric acid vapors.

Retort Patching Cements

a. Alumina (1 part), fine sand (4 parts), slaked lime (1 part), borax (½ part) and water, mixed to stiff paste.

Suitable for fire-clay retorts and crucibles.

b. Iron filings (10 parts), brick clay (20 parts), fire-clay (15 parts), common salt (8 parts), and water, mixed to stiff paste.

Suitable for iron retorts and crucibles.

Brick Bedding Plastic Cements

a. Dry fire-clay and anhydrous tar, mixed hot to give workable plastic mass which hardens on cooling.

For impregnating and bedding the bricks of acid-resisting floors, dumps, etc.

Bricks should be warmed on hot plate and dipped before bedding. Joints to be as narrow as possible, sealed by forcing in hot plastic mass with the aid of heated irons of suitable shape.

b. Pitch (11 parts) and ozokerite (1 part), melted together and poured hot or applied in warm plastic state.

For impregnating and bedding the bricks of acid-resisting floors, dumps, etc.

This mixture can be poured into the joints of brick floors. In the case of new floors, bricks should be dipped before laying. Floors in situ should be warmed by charcoal fire before pouring to fill joints and crevices.

Weather-Proof Coating

Fire-clay (56 pounds) and tar (1 gallon), well mixed by mechinical means and applied hot.

For coating bell-and-spigot joints to provide weather resistance.

Can be used in warm positions.

Iron to Iron Cements

a. Red oxide jointing putty.

Suitable for temperatures up to 500° F.

b. Red lead and litharge in equal quantities, mixed with glycerine.

Resists the action of alkalies.

c. White lead (6 parts), sublimed sulphur (6 parts) and borax (1 part), moistened with strong sulphuric acid and used immediately.

Should be allowed 7 days to harden.

Cast-Iron to Cast-Iron Cement

Rust joints.

Iron to Lead Cement

Red oxide jointing putty.

Suitable for temperatures up to 500° F.

Lead to Lead Cements

a. Powdered brick (2 parts) mixed with molten black rosin (1 part) and applied hot.

Not suitable for temperatures above 160° F.

b. Red oxide jointing putty.

Suitable for temperatures up to 500° F.

Acid-Resisting Cast-Iron Cement

Red lead cement.

Glass to Metal Cements

a. Litharge cement.

Resistant to steam and oil up to 400° F.

b. Litharge (2 parts), white lead (1 part), copal varnish (1 part) and boiled linseed oil (3 parts).

c. Chlorinated derivations of diphenyl.

Said to give extremely good glass to metal joints.

d. Rosin (5 parts), beeswax (1 part) and red oxide of iron (1 part), heated together and applied hot.

Not suitable for temperatures above 90° F.

e. Glass is coated with copper mirror and dipped into molten lead; the adherent of lead is then joined to metal surface by soldering.

Glass to Glass Cement

Asbestos powder (3 parts), precipitated barium sulphate (1 part) and sodium silicate solution sp. gr. 1.52.

Resistant to sulphuric acid.

Brick to Lead Cement

Silex (finely crushed flint) mixed with sodium silicate solution sp. gr. 1.38.

Resistant to sulphuric acid.

Brick to Brick Cements

a. Calcined magnesite and water mixed to consistency of mortar.

Bricks should be heated to 150° F. and laid with thin joints.

Resistant to fused alkalies.

b. Powdered pumice and sodium silicate solution sp. gr. 1.38.

Resistant to sulphuric acid.

c. Silex 20 mesh (1 part), silex 100 mesh (1 part) and sodium silicate solution sp. gr. 1.38, mixed to consistency of mortar.

Bricks should be heated to 150° F. and laid with thin joints.

Resistant to acid fumes up to 350° F.

d. Sulphur cement.

Not suitable for temperatures above 200° F.

Earthenware (or Stoneware) to Iron Cements

a. Silex or fine sand (20 parts), litharge (2 parts), unslaked lime (1 part) and boiled linseed oil, mixed to consistency of mortar.

Useful for fixing steel shafts in stoneware impellers of acid pumps and acid gas exhausters.

b. Cast-iron filings free from oil (20 parts), plaster of Paris (50 parts), and salammoniac (3 parts), mixed with weak vinegar to thin paste.

c. Sulphur and pitch in equal quantities, melted together and applied hot.

Earthenware and Stoneware Cements

a. Litharge cement.

Resistant to weak acids, such as sulphurous acid.

b. Silex, china-clay or fire-clay mixed with boiled tar.

c. Silex and sodium silicate solution sp. gr. 1.38, mixed to consistency of dough and applied by pressing into the joints.

Unaffected by mineral acids, except hydrofluoric acid, but difficult to remove.

d. Powdered pumice and sodium silicate solution sp. gr. 1.38.

Resistant to sulphuric acid.

Fused Silica ("Vitreosil") Cements

a. Powdered glass (1 part), powdered feldspar (1 part), kaolin (2 parts) and asbestos powder (1 part), mixed with sodium silicate (10 parts) and water (13½ parts).

Suitable for luting purposes on pipes at 600° C.

b. Powdered barytes (or china-clay) and asbestos powder, mixed with sodium silicate solution to form stiff paste.

Suitable for bell-and-spigot joints in contact with acids and salts up to 600° C.

Acid Resisting Cements for Joining Ceramic Tiles to Brick or Concrete

a. Litharge cement.

Tiles must be perfectly dry when laid and joints should be as narrow as possible. Allow 3 days to harden.

b. Asbestos powder (2 parts), precipitated barium sulphate (3 parts) and sodium silicate solution sp. gr. 1.38.

Hardened, after setting, by preliminary treatment with dilute sulphuric acid.

Fireback, Fire-Clay Crucibles, etc. Cements

a. Silicated fire-clay.

Surface of fire-brick should be thoroughly damp in order to obtain good adhesion. Openings of considerable size are sealed by incorporating broken fire-brick.

b. Litharge (1 part), borax (2 parts), slaked lime (2 parts) and water, mixed to stiff paste.

Soft Rubber, Hard Rubber, etc. Cements

a. Masticated raw rubber (1 part), boiled linseed oil (4 parts) and fire-clay (6 parts).

Resistant to hydrochloric acid.

b. Chlorinated rubber.

Resistant to nitric acid (42 per cent).

Chlorine Resistant Cement

Finely powdered glass and Portland cement in equal quantities, mixed with sodium silicate solution sp. gr. 1.38.

Setting can be delayed by diluting the silicate with water.

Nitric Acid Vapor Resistant Cements

a. Soft asbestos putty.

b. Short asbestos fiber, sodium silicate and mineral oil (5%).

c. Silex (20 to 30 lb.) and flock asbestos (10 lb.), mixed with boiled linseed oil (1½ gal.).

d. Finely powdered glass, precipitated barium sulphate, china-clay and sodium silicate solution sp. gr. 1.26, mixed to consistency of putty.

Nitric Acid (Hot or Cold)
Resisting Cements
a. Soft asbestos putty.
b. Silex or fine sand (1 part) and
flock asbestos (1 part), mixed with sodium silicate solution sp. gr. 1.30 (2
parts).
c. Litharge putty.

Hydrochloric Acid Gas
Resisting Cements
a. Fire-clay (2 parts) incorporated
into molten mixture of rosin (1 part)
and sulphur (1 part) and applied hot.
Said to be very satisfactory for electrolytic cells.
b. China-clay (1 part by volume) and
fine white sand or silex and sand (2
parts by volume), mixed with sodium
silicate solution.
Mass can be toughened by the addition of fine dry casein powder to the
extent of 5 per cent of weight of silicate used.
c. Black putty.

Sulphuric Acid and Nitric Acid
Resisting Cements
a. Asbestos powder (2 parts) and precipitated barium sulphate (1 part),
mixed with sodium silicate dilution sp.
gr. 1.30 (2 parts).
b. Powdered pumice and potassium
silicate solution sp. gr. 1.26.

Acid Fumes Resisting Cements
a. Silex or fire-clay mixed with bitumen and boiled linseed oil.
For normal or moderately warm temperatures.
b. Litharge-asbestos cement.
Suitable for high temperatures.

Petroleum Oil Resisting Cements
U. S. Patent 1,497,782
a. Glycerine (9 parts by volume) and
water (1 part by volume) made into
stiff putty with litharge (9 parts by
weight) and red lead (1 part by
weight).
Sets in 24 hours. Formerly used for
fixing gauge glasses on low-pressure
steam boilers.
b. Whiting or china-clay mixed with
sodium silicate solution sp. gr. 1.26.
c. Shellac (30 parts), alcohol (33
parts), rosin (20 parts), gypsum (2
parts) and iron oxide (15 parts).
For use as jointing putty.

Steam (Low Pressure)
Resisting Cement
Linseed meal made into stiff paste
with water.
The addition of sour milk, lime water
or weak glue increases the strength of
the joint.

Steam (Superheated)
Resisting Cements
a. Cast-iron filings (88 parts), asbestos fiber (11 parts) and salammoniac
(1 part), made damp with water and
used immediately.
This particular formula for rust
joints has been found very satisfactory
for super-heater pipes.
b. Silex or fine sand mixed with 10
per cent magnesium chloride solution
and applied as a putty; the surface of
the joint is then well soaked with 30
per cent sodium silicate solution.

Gas Leak Cement
Plaster of Paris mixed with a thick
solution of Scotch glue, applied hot.
Sets to a tough gas-tight mass on
cooling, but is not waterproof; surface
should be coated with tar.

Oil Leak Cement
Scotch glue (2 parts) softened by
water (7 parts) and heated, glycerine
(1 part) being added.
With increased amount of glycerine
this mixture becomes softer and of a
more rubbery nature on cooling. Not
suitable for steam leaks.

Steam Leak Cement
Black oxide of manganese mixed with
raw linseed oil.
Steam pressure must be released, but
the pipe should be kept warm to dry the
oil. Becomes thoroughly hard in 24
hours.

Jointing, Sealing Putty and Cement
for Earthenware and Stoneware
Plant
Joints on earthenware, stoneware, or
fused silica should be capable of being
"broken" (when necessary) without
damage to sockets or flanges. The surface of the joint exposed to the acid
gases or liquid should be reduced to the
minimum. When a setting cement is
used for sealing bell-and-spigot joints
it must not be tamped too hard, as most
cements of this type expand on setting

and may easily fracture the bell of the pipe. The use of a non-setting putty obviates this danger; it also gives the structure a slight degree of flexibility which is often desirable.

Tower Sections (Horizontal Joints) Cement

Flannel strips soaked for 24 hours in mixture of red lead (2 parts), precipitated barium sulphate (3 parts) and boiled linseed oil.

Foundations for base of tower should be prepared with asphalt, "boiled down" tar and sand, or pitch and sulphur.

Pipelines (Flanged Joints) Cements

a. Rubber washers.

b. Asbestos millboard (2 mm. thick) soaked in boiled oil or tar and applied hot.

Resistant to strong nitric acid.

c. Asbestos millboard soaked in sodium silicate solution and smeared with silicate cement.

Resistant to strong nitric acid.

d. Asbestos millboard soaked in boiling mineral jelly (vaseline) for 30 minutes.

Tower Sections and Pipelines (Bell-and-Spigot Joints) Cements

a. Soft asbestos putty.

Flexible joint. Suitable for nitric acid (hot or cold) and nitric acid vapors.

b. Molten sulphur.

Rigid joint. For temperatures up to 250° F.

c. Sulphur cement.

d. Asbestos fiber, china-clay, precipitated barium sulphate, boiled tar and pitch.

For temperatures up to 200° F.

e. Caustic soda (2 lb.), water (2 lb.), linseed oil (4 lb.) and sodium silicate (50 lb.), mixed in order specified and heated, china-clay (60 lb.) being incorporated.

For temperatures up to 150° F.

f. Black putty.

Suitable for hydrochloric acid.

g. Rosin (1 part), sulphur (1 part) and finely crushed stoneware (2 parts), heated and used in semi-plastic state.

Suitable for hydrochloric acid and nitric acid.

h. China-clay (3 parts), finely crushed stoneware (1 part) and hot tar, mixed to consistency of thick paste.

Suitable for hydrochloric acid and nitric acid.

j. Crude vaseline (40 parts), crude paraffin (10 parts), asbestos fibre (12 parts), powdered stoneware or other silicious material (30 parts), mixed hot but used cold.

Suitable for cold nitric acid and cold sulphuric acid. Asbestos fibre must be well "teased out" to nature of fluff and slowly incorporated as last ingredient.

k. Asbestos fibre reduced to fluff and sodium silicate solution sp. gr. 1.30, mixed to plastic mass capable of being worked between the fingers.

Surface of joint should be painted with concentrated sulphuric acid before sealing. When hardened the exterior of the joint should also be treated likewise.

l. Dry china-clay mixed with heavy mineral oil to consistency of putty.

Easily removable.

Repairs to Cracks, Cements for

a. Litharge cement.

Resistant to sulphuric acid.

b. Litharge (40 parts), asbestos fibre (5 parts) and boiled linseed oil (3 parts).

Resistant to nitric acid.

c. China-clay mixed with hot tar.

Resistant to hydrochloric acid.

d. Asbestos powder (2 parts), precipitated barium sulphate (3 parts) at sodium silicate solution sp. gr. 1.38.

Hardened after setting by treatment with strong sulphuric acid.

Tile Linings (Setting) Cement

Litharge cement.

Resistant to sulphuric acid.

Putties

Where appearance is only a minor consideration, construction companies use a bituminous (black) plastic made with non-drying oils and long asbestos fiber. This is somewhat smeary in application and does not permit a smooth neat job, but it remains soft and elastic and withstands weathering remarkably long. It is made as follows:

"D" Asphaltum (soft)	400 lb.
Gilsonite	100 lb.
Black Fish Oil	7 gal.
Crude Black Oil	7 gal.
Stove Distillate	70 gal.

Melt the asphaltum and gilsonite at 550° F. and hold until in complete solu-

GASKETS AND LUTES RESISTANT TO SOLVENTS.

Nature of Jointing.	Acetone.	Alcohol, Ether, etc.	Benzol, etc.	Creosote.	Petroleum Oils.	Tar (hot).	Ammonia.	Alkaline Lyes.	Sulphuric Acid.	Steam.	Water (cold).
Manilla board coated with stiff paste of whiting and vaseline.	×					×					×
Manilla board coated with magnesite paste *†.	:	×	:	:	:	:	:	:	:	:	×
Manilla board coated with asbestos paste *†.	:	:	:	×	×	:	×	:	×	:	
Asbestos millboard coated with graphite paste *†.	:	:	×	:	:	:	:	:	:	:	
Asbestos millboard soaked in mixture of fire-clay and water.	:	:	×	:	×	:	×	:	:	:	×
Asbestos millboard soaked in sodium silicate solution (sp. gr. 1.38) and applied wet.	:	:	:	:	:	:	×	:	:	:	
Compressed fibre coated with old shellac varnish.	:	:	:	:	:	:	:	:	:	:	
Stiff brown paper coated with old shellac varnish.	:	:	:	:	:	:	:	:	:	:	
Sheet lead coated with asbestos paste *†.	:	:	:	:	:	:	:	:	:	:	
Sheet rubber coated with white lead paste *†.	:	×	:	:	:	×	×	:	:	:	
Compressed cork soaked in mixture of glue and glycerine and applied wet and hot.	:	:	×	:	:	:	:	:	:	:	
Compressed cork soaked in equal quantities of vaseline (or edible paraffin) and paraffin wax (or ozokerite).	:	:	:	:	:	:	:	×	:	:	
Flannel coated with asbestos paste *†.	:	:	:	:	:	×	:	:	:	:	
Red lead (with or without white lead) and boiled linseed oil †‡.	:	:	×	:	:	:	:	:	:	×	
Graphite and boiled linseed oil †.	:	:	×	:	×	:	:	:	:	:	
Litharge and glycerine mixed to stiff paste ‡.	:	:	×	:	×	:	:	:	:	:	
Gum shellac made into paste with 10% ammonia water §.	:	:	:	:	:	:	×	:	:	:	
Asbestos fibre mixed with soft soap.	:	:	:	:	×	:	:	:	:	:	
Asbestos fibre mixed with fire-clay and water.	:	:	:	:	×	:	:	:	:	:	
Asbestos fibre soaked in sodium silicate solution (sp. gr. 1.38) and applied wet.	:	:	:	:	:	:	:	:	:	:	

* Made with boiled linseed oil.
† Compositions containing linseed oil must be given time to harden before pipes are put into service.
‡ Permanent metal-to-metal joints, if the faces of the flanges are not faulty.
§ Screw threads *must* be freed from dirt and oil before application.

tion, then add both oils and heat to 575° F. Cool to 450° F. and reduce.

The black fish oil is a very dark crude and cheap oil, unfiltered and full of stearines.

For overglazing where the lights of glass overlap, a semi-liquid coating is made by mixing into the base vehicles while hot ¾ lb. of long-fiber asbestos to each gallon.

For plastic putty for cementing the glass to the frame, the following mixture is made in a regular pony chaser:

Base Vehicle (above)	5	gal.
Stove Distillate	1½	gal.
"Asbestine"	50	lb.
Long-fiber Asbestos	5	lb.

This product is stiff and must be applied by knifing or with a small trowel.

Joint Cement for Roofing Slabs

In the East and South, cement slabs called cementiles are quite commonly used in constructing factory roofs. The joints of these tiles are first partly filled in with a non-shrinkable cement, and above this flush with the tile surface is run a waterproof expansive plastic for protection. An Eastern manufacturer of cementile roofing slabs also makes the joint cement or putty. They buy large quantities of paint skins from paint manufacturers, and use this as the base material, cooking same with an addition of fish oil, subsequently churning it with such filling material as asbestine or whiting, short asbestos fiber, and red oxide for color. The final protective is a well-known commodity. Its salient features are: a soft but firm plasticity; a condition of slime for easy slip in troweling; slow setting during manipulation, but later becomes surface-set out of dust and dirt; the power to retain its softness and cohesiveness within the joint, indefinitely. These features have been very well reproduced in the following formulation:

5% Leaded Zinc Oxide	24	lb.
Borate of Manganese	½	lb.
Spanish Red Oxide	8	lb.
Treated China Wood Oil	4	gal.
Sulfurized Fish Oil	4	gal.
Medium Body Gloss Oil	4	gal.
"C" Asbestos Fiber	32	lb.

The compounding of this would be similar to compounding a skylight putty. The prepared wood oil is composed of 40 lb. of limed rosin and 20 gal. of wood oil heated to 425° F. and held there about 2 hours until very heavy—but no stringing; then reduced immediately with 50 gal. kerosene.

The above plastic is run into the tile joints with a hand-pressure caulking and glazing gun, fitted with either the standard or the extra large caulking nozzle.

Litharge-Glycerine Cement

A product somewhat akin to putty is that compound familiarly known as Litharge-Glycerine Cement, which is valuable for a number of purposes for which ordinary cement and putty would be neither practicable nor desirable. Probably many may feel that they know how to mix this cement for usage, but those who merely combine these two ingredients really would not be doing it efficiently, for best results. The cement is correctly produced by adding to a mixture of 5 parts of 95% pure glycerine and 3 parts of water, sufficient finely ground litharge to form a plastic of any required consistency. Variation in the amount of water will influence the time of setting and to some extent the general characteristics, but all modifications within the range of 1 to 3 parts of water with 5 or 6 parts of glycerine will attain satisfactory hardness. Its normal hardening time is about 10 minutes, but it may be made to remain soft for a longer period by an addition of 10% of inert material such as silica, iron oxide, or Fuller's earth. Such admixtures do not detract from the ultimate hardening or strength, but also are beneficial in preventing possible cracking. Litharge-Glycerine Cement will withstand a high degree of combined heat and moisture. A very common usage is for forming water-tight connections between iron pipes and porcelain fittings; and for cementing glass aquariums, etc. Its most conspicuous feature is its resistance to practically all acids not of full strength. It is used to good advantage in temporarily sealing leaks at seams, around the bottoms, and around flanges, etc., of storage tanks filled with varnish; these temporary repairs hold until the contents of the tanks are used when a permanent repair can be made.

Marine Putty

This type of putty, which hardens under water, may be made as follows:

Hydraulic Cement	30	lb.
Plaster of Paris	7½	lb.
Litharge	10	lb.
Belgium Whiting	20	lb.
Lead Carbonate (dry)	10	lb.
Boiled Linseed Oil	3	gal.

Hydraulic Cement

On the seaboard, hydraulic cement is better known as sea-water cement. This type differs from regular Portland cement for land construction in being darker color and containing a minimum of tri-calcium aluminate, the constituent in cement which is rapidly attacked by sea water. Whereas regular cement contains 10–15% tri-calcium aluminate, this is minimized to 2% in sea-water cement.

Painters' Lead Putties

These varieties of putty, which are also termed *Hard Putty* and *Carriage Putty*, will vary in lead content from almost straight lead to approximately 75% and 50%; the admixtures being whiting and/or silica.

Dry White Lead	50 lb.
White Lead in Oil	20 lb.
Whiting	25 lb.
Silex	6 lb.
Boiled Linseed Oil	$\frac{3}{16}$ gal.
Gold Size Japan	$\frac{3}{16}$ gal.
Rubbing Varnish	$1\frac{1}{8}$ gal.

These mixtures are allowed to stand 72 hours to thoroughly wet down and sweat, and then are kneaded into putty. The silex used is the live quartz silica mainly used for the making of paste wood fillers.

The composition of the solids in a representative painters' hard putty with lower lead content would be 50% dry white lead, 35% whiting, and 15% silica.

Non-Shrinkable Putty

A non-shrinkable type of putty containing about 20% of lead in the pigment is the following:

Whiting	125	lb.
White Lead, dry	$37\frac{1}{2}$	lb.
Silica	$12\frac{1}{2}$	lb.
Raw Linseed Oil	$3\frac{1}{2}$	gal.
Flour Paste	$10\frac{1}{4}$	lb.

The flour paste is 2 lb. of wheat flour beaten up in about 1 quart of cold water and then poured into 3 quarts of boiling water, and boiled 5 minutes. Yield $10\frac{1}{4}$ lb. net.

Swedish Putty

The foregoing non-shrinkable putty is very similar to a product called Swedish Putty, purported to be excellent for wood, iron or stone. Another type of Swedish Putty without lead is made as follows:

Part I

Rye Flour	2	lb.
Water	$1\frac{1}{2}$	gal.
Whiting	20	lb.

Mix the flour with ½ gal. of cold water, then pour the mixture into the remainder of the water (1 gal.) at the boiling point. After the mixture has cooled stir in 20 lb. of whiting.

Part II

Whiting	50 lb.
Gold Size Japan	2 gal.
Raw Linseed Oil	1 gal.

Grind the ingredients of Part II in a paint mill. Then combine the two parts in a pony chaser, and thicken with more whiting to the required plasticity for knifing.

Metal Furniture Baking Putty

A product which will not crack, peel, shrink, or run with heating, is made as follows:

Bolted Whiting	5 lb.
Boiled Linseed Oil	1 pt.
Flour Paste	1 pt.

Mix all very thoroughly, first mixing the oil and whiting, and adding the flour paste. The flour paste is as given for non-shrinkable putty. In all cases of preparing flour pastes, the flour and cold water should be beaten until entirely free from lumpiness; and during the subsequent cooking, should be continually stirred.

Stopping Putty

This is a dry mixture of 2 lb. of "Alabastine," 1 lb. of wheat flour, and 1 lb. of Portland cement. When ready to use, 1 lb. of this mixed powder should be thoroughly worked up to a stiff putty with 8 fluid ounces (½ pint) of cold water. This putty sticks to stone, wood, brick, etc.; it is used for filling knot holes, cracks, etc. Keep the dry powder in an air-tight jar.

Tesso Duro

This product is Italian hard plaster used in making bas-relief casts. When dried, it becomes very hard and durable. An American putty-like plastic of this kind is sold by art goods dealers, department stores, and wherever paints and unpainted furniture are sold. It is used alike by the amateur and professional trade, by art classes and in manual training departments of public schools, for the decoration of furniture in the white, picture frames, bric-a-brac, floor lamps, etc.; and is adaptable to any type of finish particularly to producing a polychrome effect on lamp stands.

This product, per formula below, remains soft and easily-worked for quite a period of time, using a small trowel, spatula, or by forming with the hands:

Fish Glue	1⅝ gal.
Water	½ gal.
Oil of Sassafras	3 fl. oz.
Boiled Linseed Oil	5 pt.
Belgian Whiting	37½ lb.
Domestic Whiting	50 lb.

Decorative Compound

This product is practically the same as the above molding clay, but is a semi-paste which can be applied with a stiff brush if preferred to troweling or knifing.

Fish Glue	4 gal.
Water	1 gal.
Oil of Lavender	6 fl. oz.
Raw Linseed Oil	1 gal.
Bolted Danish Whiting	50 lb.
Rubbing Varnish	1 gal.
Bolted Danish Whiting	20 lb.

(Colors in oil may be added, if shading is desired.)

Wood Dough or Plastic Putty

In the past few years there has come into more or less extensive use several products of this type for general repair work in the home and also in the shop, for obliterating blemishes, open joints, screw holes, etc., in cabinet work. One very popular and efficient product is marketed by a well-known brand name, and enjoys a wide sale. The formula herewith is not that identical material, but it does produce a highly satisfactory composition of the same type:

Gum Solution	1 gal.
Glycerin	3 pt.
Butyl Alcohol	3 pt.
Whiting	8 lb.
Wood Flour	24 lb.
Dope (Solution)	8 gal.

The "gum" solution is 16 lb. of gum rosin (WW Rosin) cold-cut (dissolved) in 1 gal. of methyl acetone; the "dope" is another cold-cut solution, basis of 1 lb. of "movie"-film scrap to each gallon of methyl acetone. The picture-film scrap should be desilvered by washing in hot water to remove its gelatin coating and then laid out in the sun and air to dry; but preferably it is obtainable cleaned and ready for cutting from dealers.

The above wood dough product is a soft workable putty easily applied to all kinds of depressions to be surfaced. The work or job should not be left in a rough state because the putty dries and hardens very rapidly; the ultimate sanding down later is rather difficult unless the putty has been smoothly applied.

Onyx Cement

This putty is used in fair quantity for special purposes. This is termed *Onyx Cement* because its specific utility is for bonding slabs of onyx, marble, glass, and their imitations, to the walls of public buildings. It is necessarily of rather firm plasticity because of the weight it must partially support. Uniform handfuls of the putty are attached to the wall foundation at intervals about 18 to 24 inches apart; the slabs mentioned are then stood upright on their bases, and are pressed back steadily and firmly into the mounds of putty. Suction, and the adhesive strength of the putty, securely hold the marble and glass permanently in place. The same material, plain or colored, is embedded in the joints between the slabs. The composition of this putty follows:

Domestic Whiting, 350 mesh	100	lb.
Domestic Whiting, 200 mesh	100	lb.
"Super-Sublimed" White Lead	40	lb.
White Oil Drier	1¼	gal.
Bodied Linseed Oil	1¼	gal.
Bodied Linseed Oil	2½	gal.

The two kinds of whiting have been described before. The white oil drier and bodied linseed oil are the same as specified in steel sash putty and the bodied oil is the commercial linseed oil with drier but not bodied.

Black Packing Compound

This specialty is required by makers of corrugated iron culverts. They are the aqueducts for streams crossing the highways and for surface-sewers under driveways in rural districts, etc. There is first applied hot a thoroughly-tested bituminous mastic pavement along the line of flow where erosion is greatest, approximately the lower one-quarter or one-third of the inside circumference. This coating practically fills the valleys of the corrugations and to the extent of building up a thickness of perhaps ¼ inch over the rises.

For this purpose the culvert manufacturer supplies a plastic for cold application. The composition is 3 parts by weight of sawdust and 1 part asbestos fiber, thoroughly churned together with enough coal tar solution to form a putty that may be applied by hand to the abraded spots in the paved section of the culvert.

Acid Proof Cement

Asbestine	3 parts
Barium Sulphate	4 parts
Potassium Silicate	3 parts

Make into a paste when desired for use. If a slow setting cement is desired, use sodium silicate.

Acid-Proof Cements
British Patent 386,045

A cement of the kind comprises finely divided silica and sodium silicate solution and contains also 1–5% sodium fluosilicate or sodium fluoborate and, optionally 5–20% of a lubricant such as graphite. The sodium silicate should have a ratio of SiO_2 to Na_2O of not less than 3.5 : 1.

Aquarium Cement

Litharge	10 parts
Plaster of Paris	10 parts
Powdered Resin	1 part
Dry White Sand	10 parts
Boiled Linseed Oil	Sufficient

Mix all together in the dry state, and make into a stiff putty with the oil when wanted for use.

Do not use the tank for three days after cementing. This cement hardens under water, and will stick to wood, stone, metal, or glass, and, as it resists the action of sea-water, it is useful for marine aquaria. The linseed oil may have an addition of drier to the putty made up four or five hours before use, but after standing fifteen hours, however, it loses its strength when in the mass.

Can Sealing Cement
U. S. Patent 1,765,134

14% bentonite is dispersed in 86% water. 12 parts of this dispersion is added to 70 parts of latex and gently mixed. To this is added 18 parts of gum karaya solution, prepared by making a water suspension containing 3% of gum, and adding 2% sodium carbonate.

Cement, for Celluloid or Movie Films

This quick drying adhesive is very effective in bonding celluloid to itself or connecting lengths of motion picture films.

Cellulose Nitrate (high viscosity) or clean transparent celluloid scrap—by weight	10%
Acetone	8%
Methanol	27%
Benzol	45%
Methyl Cellosolve	10%

For use with movie films, the gelatin emulsion surface is removed with steel wool, sand paper or scraped off with a pen knife. Other celluloid surfaces should preferably be cleaned and slightly roughened. The above liquid is applied with a brush and the moistened surfaces pressed together.

Electrode Cement
German Patent 583,347

A cement, particularly useful for securing the middle electrodes of spark plugs, is prepared by mixing with water glass solution a composition comprising alumina 50–53, aluminium hydroxide 10, powdered quartz 15, and powdered lead glass 2–5 parts.

Cementing (Cellulose Acetate to Copper)
U. S. Patent 1,908,601

A mixture (equal volumes) of ethyl alcohol and lactic acid is applied to the clean dry metal and the cellulose acetate is pressed in contact therewith.

Cements for Correcting Faults of Foundry Pieces

Black Pitch	1 part
Rosin	1 part

This is incorporated into finely sifted iron filings to obtain a paste-like product. The piece to be repaired is heated and some of the above cement is placed cold on the defective parts and an iron heated to a dark red heat is passed over it.

Counter Cement

Latex (40% solids)	20 gal.
Karaya Solution (2% Gum)	40 gal.
Resin Emulsion (5% Resin)	4 gal.
Water	25 gal.

Counter Cement—Quick Drying

Latex (40% solids)	20 gal.
Karaya Solution (2% Gum)	20 gal.
Resin Emulsion (5% Resin)	4 gal.
Water	50 gal.

DeKhotinsky Cement
Formula 1

U. S. Army specification No. 3–125 does not stipulate a desired composition; however, the material is approximately 60% shellac; 40% pine tar. The pine tar is heated, then the shellac added and while stirring constantly the temperature is maintained for 1 hour at 100° C., avoiding heating over 125° C.

Formula 2

An excellent general cement for sticking glass to metal or making similar airtight joints has the following composition:

Pine Tar	80 lb.
Solid Shellac	150 lb.

Digest on a water bath and after stirring thoroughly let cool, preferably in molds in the form of sticks about a half-inch square.

In using, warm both articles to be cemented and then apply the cement. Do not overheat.

Cement for Glassware

The above product is solid cement for laboratory and other glassware; it is furnished in 2 ounce sticks. A shellac creosote cement developed in the laboratories of the British Scientific Instrument Research Association is reputed to be more adhesive and less brittle, and does not become infusible by repeated heating. The composition is reported to be:

Superfine Shellac	50 lb.
Wood Creosote	5 lb.
Terpineol	2 lb.
Ammonia (sp. gr. 0.88)	1 lb.

The ammonia, terpineol, and creosote are added to the shellac and the mixture gently heated until fused, stirring well; after which it may be molded into sticks. For making joints between metal and glass or similar materials, the surfaces to be joined should be clean, and should be warmed before the cement is applied. The cement must be fused as needed.

Dental Cement

This cement consists of a solid and a liquid. To make the solid, fuse zinc oxide in fire clay pots. Cool and break out. Grind to 400 mesh. (Pebble or ball mills.) This mass is colored to desired shade. Final results are often improved if an admixture of from 10 to 15% of Portland cement is made.

The liquid consists of

Zinc Phosphate	1 lb.
Glacial Phosphoric Acid	20 oz.
Water (Distilled)	10 oz.

Filter carefully through glass wool. Keep air and moisture tight. To use, combine solid and liquid. If set is too fast, evaporate the liquid until desired speed is attained. If too slow, add small amounts of water until satisfied. When mixing, add very small amounts of powder to the liquid on a slab or plate, mix well and add more. If amalgam, gold or tin is added to the cement during the mixing process, the cement takes the name of the admixture, as tin cement, etc.

Dental Model Plaster
British Patent 385,896

Dental models are made from a composition comprising at least 75% of calcium sulphate, ½ water, the weight by volume of which "loosely put in" is above 0.9 gram per cubic centimeter and "shaken in" above 1.45 grams per cubic centimeter, the "strewed in quantity" of which, until a paste that can be poured is attained, being at least 230 grams per 100 cubic centimeter and its "thickest consistency" 300 grams per 100 cubic centimeter. In an example 95 parts by weight of a plaster of Paris having the above properties is mixed with borax 0.05, potassium sulphate 0.5, pulverized marble 1.5 and pulverized quartz 3 parts, 330 grams of the mixture are mixed with 100 cubic centimeters of water, setting in ½ hour and having a Brinell hardness of about 7 kilograms per square millimeter after 1 hour.

Furnace Cements

In boiler settings the firebricks are often set in a cement made simply of ground firebrick, raw fireclay and water. Some shrinkage takes place as the joints dry out. The strength of such cement is developed only by the sintering which takes place when the furnace is fired. Parts of the brickwork are never reached by the firing, and these, if bonded simply with fireclay and water, never have much strength.

One way of improving the cement is to add a small amount of "N" or "O" Brand or "S" Brand silicate of soda to the mix. Up to a quart per gallon of water may be used. This makes the cement much more sticky. As it air dries, it does not shrink away from the brick, but the wall remains gas-tight. The strength of the cement previous to firing is greatly increased and the bond is amply strong in the portions of the furnace which the higher heats do not reach. "S" Brand gives a somewhat more refractory result than "N" or "O." A more alkaline silicate makes a stickier but somewhat less refractory cement.

Still better results are obtained when specially prepared cements are used. These have the advantage of quick set, great strength and tightness, and high refractoriness. Numerous combinations of refractories are used, of which the following are examples. The amounts of water depends somewhat upon the fineness of the ingredients. In mixing the cements start with a little less than the amounts of water given in the formula and add more if necessary, to get the right consistency.

Formula 1

Silicon Carbide	77 lb.
Raw Fireclay	23 lb.
"U" Brand Silicate of Soda	9 lb.
Water, about	8 lb.

Formula 2

Silicon Carbide	94 lb.
Raw Fireclay	6 lb.
"U" Brand Silicate of Soda	11 lb.
Water, about	11 lb.

"D" Brand may be used instead of "U," in formulae No. 1 and No. 2. Formula 2 has more silicate than No. 1, because it has less clay, and without a larger amount of silicate it might not be sufficiently plastic. If silicon carbide firesand is not available, crystalline magnesium oxide or chrome ore is suggested as an alternative.

Dry Refractory Cement

Silicon Carbide Firesand	50 lb.
Raw Fireclay	50 lb.
"G" Brand Powdered Silicate of Soda	17.5 lb.

This is an excellent dry cement, to be mixed with water on the job. The refractory ingredients may be varied greatly according to convenience. The "G" Brand powdered silicate of soda is sufficiently soluble to act as the necessary binder when the dry cement is stirred up with cold water and then allowed to set by air drying. If less solubility is desired varying amounts of "SS–65 Powdered" may be combined with the "G" Brand.

These formulae are satisfactory also for the linings of furnaces for heating brass crucibles, and for other similar refractory purposes.

Furnace Paint

An excellent furnace paint for coating furnaces may be made by the following formula:

Calcined Fireclay	75 lb.
Raw Fireclay	25 lb.
"U" Brand Silicate of Soda	18 lb.

Add water enough to make a thick paint; apply two or three coats with a broom or stiff brush.

This paint is of special value in the arch and bridge walls of boiler furnaces subjected to high heat, in potters' kilns, and in welding and malleable iron furnaces. On oil furnaces it will protect the brick from the pitting and eating action of the high pressure flame. This paint also does good service in protecting the metal zone of foundry cupolas.

Refractory Linings

Excellent refractory linings for Rockwell and other direct-melting furnaces are made of several formulae with silicate of soda. The following are examples:

Formula 1

Silicon carbide firesand	70 lb.
Ground quartz, 40 mesh and finer	16 lb.
Ground mica, 40 mesh and finer	4 lb.
Raw fireclay or kaolin	10 lb.
"U" Brand silicate of soda	4 lb.
Water, about	10 lb.

Formula 2

Silicon carbide firesand	100 lb.
"U" Brand silicate of soda	13.5 lb.
Water, about	8 lb.

Mix the dry ingredients thoroughly. Dilute the silicate and stir into the dry material. Mix thoroughly until free from lumps. The consistency should be about that of moulding sand.

For making the lining, a wooden form is set inside the shell as in pouring concrete, having a space of from four to six inches between the form and the outer shell. The cement is rammed in tight between the form and the shell. It is allowed to dry somewhat, and a wood fire is built in the furnace to help set the cement. The furnace is heated up slowly to permit gradual drying out and final set.

Pointers

To prevent the closing of the furnace openings by the expansion of the lining, cut back the lining from the edge of each opening at least one-quarter of an inch.

Ram the linings by hand, it gives best results. A twisting movement places the material better than a direct blow.

Use the right amount of silicate. The formulae have been worked out with care and should be followed.

Use the "U" Brand silicate. Other brands should not be substituted for it.

Use plenty of water, but use it with care. Many foundrymen have used, with good results, as much as ten parts of water to every one part of silicate. The water is used only to make the mixture workable. For this it may be used freely.

Dry slowly. Too much heat at first will form steam in the silicate and weaken the cement by making it porous.

Patching and Stove Cements

For local jobs of patching firebrick linings, or filling cracks in furnace walls, or setling the fireboxes of ordinary domestic heaters, ranges and waterbacks, cements can be made up with almost any convenient refractory material, mixed up to a thick paste with silicate of soda and sometimes a little water. Where high heats are to be expected use:

| Silicon carbide firesand | 50 parts |
| Raw fireclay | 50 parts |

Make up with undiluted "U," "D" or "N" Brand to a suitable consistency. Where the temperatures are not so high, raw and calcined clay, with asbestos if desired, may be used. For household ranges a quantity of finely sifted coal ashes with "N" or "U" Brand silicate, or with egg preserving "water-glass," will make a very satisfactory home-made cement.

Coke Oven Cements

In preparing the charging holes in coke ovens, it is necessary to use a cement which can be applied to the surfaces while they are hot. For this purpose the following formula is recommended:

Calcined clay	12 parts
Raw clay	8 parts
"GC" silicate	2 parts
Water	11 parts

This is applied by means of a cement gun. The "GC" is sufficiently soluble to dissolve and form the bond before the heat dries out the water. The relative amounts of calcined and raw clay may be varied, and other refractory materials included, but in any case about 10% of "GC" should be used.

Sagger Mending

Very satisfactory cements for mending saggers can be made from a mixture of raw and calcined clay, with "U," "K," "N" or "O" Brand silicate. Usually about two parts of silicate to one of clay gives a good result, with sometimes a little water added to make a cement of smooth, thick creamy consistency. It is best to wet both broken surfaces of the sagger with water, taking care to apply enough to prevent a premature set due to absorption of water from the wet cement. Enough should be used to allow the cement to stay sticky for five minutes. The broken parts are pressed together, so that a little cement is squeezed out, and allowed to stand undisturbed over night.

Chimney Cements

Industrial brick chimneys often need to be acid resistant as well as heat resistant. The best cement for the acidproof brick in lining such chimneys is one made of ground quartz and "S" Brand silicate as next described in the section on Acid Tower Cements.

Acid Tower Cements

In concentrators for sulphuric acid, and in various other applications in the chemical industries, acid resistance is obtained by means of specal acidproof brick set in suitable acidproof cements. These cements consist chiefly of ground quartz or silica and "S" Brand silicate of soda.

Quartz is used as the filler because it is not attacked by the acid. Any quartz that is practically pure silica will be satisfactory. It is best to use graded sizes, one-half to be about 20 mesh quartz and one-half about 100 mesh including fines. This gives the densest possible cement.

"S" Brand silicate is used because it contains the least amount of alkali. Also, it is the quickest setting. The relative amounts of silicate and ground quartz will usually be:

Ground quartz, 20 mesh	50 lb.
Ground quartz, 100 mesh and fines	50 lb.
"S" Brand silicate	70–100 lb.

The amount of "S" Brand will depend upon the grading of the quartz. If a large proportion of fines is used the mass is increased in volume and requires more silicate to wet it. Enough silicate should be used to give a cement of the consistency of thick cream.

The brick should be laid by dipping, with the thinnest possible joints. The bricklayers on acidproof jobs should be impressed with the fact that in this work thin joints are more important than straight cement lines. Usually it is best not to build over six feet per day. If more is attempted it is likely that the pressure will squeeze out some of the partly set cement from the lower course.

The cement sets by drying. After it is thoroughly dried, when the acid touches the cement it will take up soda from the surface, but will leave the cement strongly bound by the silica of the "S" Brand. At the same time the cement is made entirely waterproof by the acid.

For good results this drying must be thorough and carefully carried out. It is best to allow air drying for thirty days before putting into service, but if drying must be speeded up by heating, the temperature must be held below the boiling point for some tome to allow most of the water to evaporate. Sudden application of high heat would convert the water into steam with subsequent puffing and disruption of the bond.

Brushing the cement lines with dilute acid before the cement is completely dry will give a quicker set and immediate water resistance, but this will be at the expense of ultimate strength.

Some proprietary cements include a small quantity of a dry acidic material with the dry filler. This is dissolved by the water of the silicate and reacts with the silicate, forming a bond more quickly than does the regular "S" Brand-quartz mixture. Other special ingredients are sometimes included.

Digester Linings

In lining sulphite digesters, used in converting chipped wood to sulphite pulp for paper making, the usual practice is to have a course of acidproof brick set with acid-proof cement next the shell, a second course with about a two-inch space from the first, and the space between filled with an acidproof packing. Several formulae are used with "U," "D" or "C" Brand silicate of soda. One formula for the brickwork cement is as follows:

Ground quartz, about 20 mesh	2 parts
Portland cement	1 part
"U" Brand silicate	Sufficient

The amount of "U" Brand to use should be just enough to make a good working consistency. It is often diluted somewhat, up to 10 per cent. This hastens the set. The cement is made up in small units. While the bricklayer is placing each brick his helper mixes up the cement for the next one. Here, as in the acidproof cements, careful drying is a very important factor in obtaining a bond which is strong and resistant.

For packing between the courses, four parts of ground quartz are used to one part of Portland cement, and larger quartz may be used.

Other cements are made up with ground firebrick or ground acidproof brick, in place of some or all of the ground quartz. Also in some cases "C" Brand silicate is preferred to "U" Brand.

One mixture which gives good acid and water resistance is prepared by mixing 20 parts pulverized silica, 12 parts English Ball Clay, 4 parts "G" Brand silicate (powdered) and 9 parts water.

A very fine acidproof cement, though an expensive one, is prepared from litharge, glycerin and silicate. Eight parts of litharge with four parts "U" Brand silicate diluted to 30° Baumé and one part glycerin will set firm in three minutes. Additional glycerin may be added to extend the working time. With two parts glycerin, the setting time is 6 minutes, and with four parts 25 minutes. Using more concentrated "U" Brand speeds up the reaction. Such mixtures are sometimes used for pointing a silicate-Portland cement.

Mixtures such as these resist for long periods the hot calcium bisulfite liquors to which they would be exposed in normal service. Of course, they must be applied to the brick while the mixture is wet enough to spread well. No movement of the brick is permissible after the initial set.

Tunnel Cements

In brick-lined railroad tunnels it is necessary to have the brick set with a mortar that will not be destroyed by the combustion fumes from the locomotives. Acidproof cements made from ground quartz and "S" Brand silicate may be used. "N" and "O" Brands also have given good results and the cement may include some fireclay with the quartz.

Abrasive Wheel Binder

This is a special application of a silicate of soda cement. "J" Brand silicate is used, and a little powdered plastic clay to bind the abrasive grain. A typical formula is:

Abrasive grain	100	lb.
Powdered clay	12½	lb.
"J" Brand silicate of soda	12½	lb.

"J" Brand is a very viscous silicate. It should be used without any addition of water, even though it seems impossible to get a complete mix with such thick material. A slow power mixer is used. The abrasive grain is first weighed into the mixer. Then the right amount of clay is weighed, and spread out on the scale pan. The thick "J" Brand is then weighed on the clay, and the clay and silicate slid together into the mixer. Otherwise, it is difficult to get the right amount of silicate, as some will stick tight to the weighing pan. When the mixing is complete, it should be found that each abrasive grain is covered with a sticky film of the silicate. The material is then tamped into moulds. This is an art requiring some little skill to produce good wheels. The wheels in the moulds are then air-dried, below 212° F., until hard enough to stand handling. They are then baked at around 400°–500° F. until thoroughly hardened and until the silicate and clay have reacted to form an insoluble bond.

In some special conditions silicates slightly different from "J" Brand have been found desirable and are available.

Also when the abrasive used is silicon carbide (such as "Carborundum," or "Crystalon"), "J-Special" silicate should be used instead of "J" Brand. Some reaction takes place between silicon carbide and "J" Brand silicate, with the liberation of hydrogen gas. This interferes with the formation of the proper bond. "J-Special" is a silicate which takes care of any hydrogen that is liberated and thus makes a strong bond with the silicon carbide.

Both these brands, "J" and "J-Special," have been in regular use by silicate wheel makers for many years.

Wood Substitute

Non-shrinking moulded products can be made from wood flour by a binder of Portland cement and "GC" powdered silicate of soda, with a little water.

Wood flour	10 parts
Portland cement	10 parts
"GC" silicate of soda	10 parts
Water	3½ parts

Mix thoroughly, add water as above to dampen the mass, and mould under heavy pressure. Such products are durable and water resistant, though they are not refractory. Similar products can be made with part or all mineral filler instead of wood flour, but the proper amount of such heavier material should be much greater than of the flour.

A mixture which sets up to form a hard product capable of being tooled like hard wood is prepared by mixing wood-flour and fine ball clay in proportion of about three to two, adding water, a little at a time, until the mass is of moulding consistency and then adding one part of "SSC" Powdered silicate. This mixture is lighter and less rock-like than the product of the above formula, but it is not water resistant. The addition of some zinc oxide would improve it in this respect.

Refractory Soapstone Products

To make a moulded product which shows no permanent change in size or shape if heated to redness, the following formula is recommended:

Ground soapstone	100 parts
"U" Brand Silicate of Soda	15 parts
Water	25 parts

Dilute the "U" Brand and mix thoroughly with the ground soapstone. Mould to the shapes desired and allow to dry. When air-dry throughout, bake with a gradually rising temperature, which should be held for some time at just under the boiling point of water. This product will then withstand temperatures up to red heat without warping or cracking. It is not, however, entirely waterproof.

Lining for Petroleum Cracking Chambers

A recent development is the use of "N" or "O" Brand silicate of soda in making a cement coating for the steel reaction chambers used in certain processes in cracking petroleum. This is noted here as typical of the kind of special results sometimes obtainable from silicate of soda. The metal surface is first completely cleaned and somewhat roughened by sand blasting with coarse, sharp, heated sand. The coating is as follows:

Commercial furnace cement (silicate binder)	60 lb.
White silica foundry sand	30 lb.
"N" or "O" Brand silicate of soda	10 lb.
Asbestos	1 lb.
Water	about 1¼ lb.

This is sprayed on with a cement gun, using about 80 lb. pressure to form a layer ⅛ to ¼ inch thick, 100 pounds of mixture to 100 square feet of surface.

The sprayed coat is then treated lightly with a brushing solution, richer in silicate, five pounds of furnace cement, 11.8 pounds of silicate of soda and about 3 pounds of water. The coating is then very carefully dried or cured. A time schedule is worked out by which the temperature is slowly raised, taking 24 hours to get it up to 900° F. At the critical point, that at which the water boils, from 200° to 225° F., the temperature is allowed to increase only five degrees per hour for five hours.

China, Glass and Metal Cements

Porcelain to Metal: Two parts powdered fluorspar and one part powdered glass mixed with sufficient "N" Brand to form a workable mass sets rapidly and forms a strong bond. Mixtures of whiting and silicate are also useful for this purpose.

Glass to Metal: Where it is desired to bond two non-porous surfaces and ordinary cements are unsatisfactory, a sheet of paper soaked in a silicious silicate and placed between the two is often satisfactory. A glass window in an aluminum box was thus prepared by using a paper gasket soaked in "S" Brand.

Sealing a glass tube into a brass fitting, 7 parts of "N" Brand with 3 parts of whiting are used. A more water resistant mixture is the silicate-litharge-glycerin cement described in the section on Digester Linings. In a case where the joint has to withstand a vacuum, asbestos thread or paper soaked in "N" Brand is wrapped around the inner tube and slipped into the outer tube. After 20 minutes of drying, it is ready for use. It is necessary in such cases to use fairly long-fibred asbestos as mixtures with short-fibred material have a tendency to shrink.

China and Glass: "E," "O" and "U" Brands are sometimes used for this purpose. They are fairly satisfactory for articles that are kept dry, but they will not stand much washing. Spread the silicate with a sliver of wood along both edges of the break, press together at once before the silicate has time to set and hold together with string or other means of pressure. Apply so that little or none is squeezed out as wiping off while wet may disturb the joints, and if left until set, it will be hard to get off at all. Be sure that the joint does not slip or shift while setting. If there are several pieces, it may be best to build up one at a time. This mending should be sufficient for ornaments, or a lamp body that is not much handled, but it will not stand soaking in water.

Spark Plug Cements

Gas-tight cements for sealing metallic electrodes in porcelain spark plug bodies made of 14 parts of 40° silicate ("N" Brand) with 30 parts of kaolin give a cement which, after heating to 1000° C., is strong, hard and not porous.

To obtain the greatest possible density it is advisable to use various particle sizes; a fairly large proportion should pass through bolting cloth (350 mesh). A small amount of plastic clay, perhaps 10%, would also increase the density.

Any considerable dilution of the silicate would tend to reduce the strength, but by adding a very small amount of water the silicate viscosity is greatly decreased and more filler can be added. A denser mixture naturally results.

Water-Resistant Cements

The silicates which are used as binders for refractory cements are not of themselves waterproof, and the ordinary silicate cements, therefore, are not entirely water-resistant. As noted in the data regarding acidproof cements, a cement made with "S" Brand, after being thoroughly dried, may be made insoluble on the surface by brushing with some acid material. Salammoniac has been found very satisfactory in certain cases.

Another formula that gives a water-resistant product consists of "C" Brand or "D" Brand silicate, diluted to 40° Bé. and used as a binder with ground limestone, whiting, or some other form of calcium carbonate. At first such a cement behaves no differently from other formulae of silicate and mineral filler, but in the course of time certain reactions take place which result in an insoluble cement. This result is usually noticeable within a week. The addition of a small amount of zinc oxide is often effective to increase water resistance.

Some forms of whiting are much more reactive than others. This is true also of clays.

Gasket Cement

One type of cement containing shellac is made by dissolving 12 parts white glue in 6 parts acetic acid, and then adding this solution to 2 parts gelatin in 16 parts water. After mixing, add 2 parts shellac varnish.

Gasket Sealing Cement (Non-setting)

Graphite	32 parts
Cup Grease	2 parts
Heavy Fuel Oil	8 parts
Bodied Linseed Oil	2 parts
Adheso Wax	1 part
Rosin Oil	1 part
Lithopone	1 part

This formula may be made thicker or thinner by altering the proportion of the fuel oil.

Insole Cement

Latex (40% Solids)	70 gal.
Casein "Solution"	9 gal.
Resin Emulsion	6 gal.
Water	15 gal.

Insulating Cement
U. S. Patent 1,933,271

Slag wool	73 %
Long fibered asbestos	13 %
Bentonite	6 %
Medium fat clay	7 %
Anhydrous sodium carbonate	1 %

The constituents of the cement are mixed in the dry state by tumbling or pugging and the product in this form can be kept indefinitely. For application it is mixed with enough water to form a thin putty and is thrown or trowled onto the surface to be insulated. When the surface to which it is to be applied is hot it has been found beneficial to use more water than for a cold surface.

One of the main uses of a cement of this kind is to cover the surface of brickwork in certain parts of furnaces and ovens and to stop up openings in the brickwork. Since much of the brickwork in certain types of furnaces is of a relatively temporary character, it may result in considerable economy if the cement applied to same can be recovered when the brickwork is torn down. This cement (unless it has been subjected to a sintering temperature) can be restored to practically its original plasticity and adhesiveness by simply adding water and kneading it.

Iron Cement ("Smooth-on" Type)

Precipitated or Reduced Iron (Powdered)	60 lb.
Ammonium Chloride (Powdered)	2 lb.
Sulphur (Powdered)	1 lb.

Mix all the ingredients thoroughly until a homogeneous powder is attained. To use: Stir water into the mixture to form a light paste. Press or pour the material into the crack blowhole or the like to be repaired, and allow the mass to solidify. This forms a strong union with the underlying steel or iron mass and, when solidified, will withstand fairly high pressures and strains.

Cement for Laminated Products
U. S. Patent 1,935,434

Thirty pounds of blood albumin are dissolved in fifty pounds of cold water. One pound of black oxide of manganese is stirred up in one pound of cold water and twelve ounces of cobalt oxide is stirred up in twelve ounces of cold water. The solutions, or suspensions, of each of these substances, as the case may be, are then thoroughly mixed with a pint of formaldehyde and a pint of ammonia This mixture thoroughly stirred is then ready for use.

The solution may be applied with a brush, spreader spray or other suitable means to layers of wood, cloth, paper, etc., which are to be cemented together. After application, the parts to be united are allowed to dry. When dry, they are subjected to heat and pressure, which may vary according to the character of the laminæ used. Use a temperature of about 140° C. and a pressure of from 400 to 500 pounds per square inch for about 1¼ hours on a mass of layers approximately ¾ inch thick. Materials made up in this way and having this cement as a binder have been subjected to very strenuous service without breaking down. For example, the life of drop hammer boards (friction bearing devices) made from laminations of wood cemented with this cementing composition has been doubled.

The addition of cobalt oxide to the cement is important. It is found that the addition of the cobalt oxide causes the firm adhesion of the laminæ, makes the material waterproof and capable of withstanding heat. It is also found that the cobalt oxide makes the material very hard. By varying the proportions of cobalt oxide, the cemented laminated material may be varied from a flexible to a very hard material. The greater the proportion of cobalt oxide added the harder the material will be. In place of cobalt oxide uranium oxide or chromium oxide may be used.

The oxide of manganese may, if desired, be omitted, but its use imparts added adhesion to the laminations.

Laboratory Cement

This product is excellent for gas-tight joints and other uses where it is desirable to stick two objects together.

Asphalt	1 part
Rosin	6 parts
Rubber	3 parts
Turpentine	1 part

An old inner tube from an auto serves as a satisfactory source of rubber. Cut it into strips two or three millimeters wide. Mix the rosin and asphalt and heat with a small flame until melted and then add the rubber. Heat the mass at as low a temperature as possible and still keep it fluid, until the rubber is nearly all melted. This should require two or three hours. The turpentine is then added and the mixture is heated two or three hours more; or until the mass becomes homogeneous. It should be stirred at intervals of fifteen to thirty minutes. It may then be cast into sticks. The cement is applied by holding the stick in a small gas flame until it liquefies, or catches fire, when it is applied directly to the objects to be cemented. A harder and less pliant cement may be made by increasing the amount of rosin.

Cement, Waterproof, for Leather Belts and Shoe Soles

This composition is a quick drying cement for building up and joining cement for building up and joining leather belts and for putting on "Compo" soles on shoes. It is made up as follows:

	% by weight
Cellulose Nitrate High Viscosity (dry basis)	10
Tricresyl Phosphate	4
Acetone	26
Benzol	25
Methanol	15
Ethyl Acetate	20

The cellulose nitrate may be purchased in solution and the solvent content deducted from the solvents used above. The solvents are mixed and the cellulose nitrate brought into solution with frequent stirring.

This solution is applied with a brush or spatula to cleaned roughened surface of the leather faces which are to be bonded. The coat is applied to each face and allowed to dry while apart. Then a second coat is spread over the first and the faces tightly pressed together while still wet and allowed to dry in position.

U. S. Navy Linoleum Cement

Alcohol	4⅝ gal.
Gum Shellac	17⅙ lb.
Gasoline	9/16 gal.
Crude Rubber	1½ oz.
Whiting	56 lb.

The rubber should be cut up small and allowed to soak, swell, and dissolve in the gasoline overnight; the shellac should be dissolved in the alcohol by agitation. The two solutions should be combined on a pony mixer pan, and the whiting should be sifted and incorporated into the vehicle by thorough mixing until the cement is smooth and entirely free from aggregations of the pigments.

Litharge Cement No. 1

Litharge	3¾ lb.
Glycerine	1 lb.

When the space into which this cement is to be poured is narrow—3/16 inch or smaller—as, for instance, in cementing paper tubes into porcelain insulators, more glycerine may be used, but in no case shall it be made thinner than 2¾ lb. of litharge to 1 lb. of glycerine. The cement should always be used as thick as it can be successfully poured, as the less glycerine used the harder and stronger the cement. Cement mixed with a greater amount of glycerine than specified will never set hard and may not set at all. If the cement sets too quickly, the litharge should be dried out in oven at about 110° C., as moisture hastens the setting.

The cement shall be thoroughly mixed and all small lumps broken up.

As soon as cement is mixed it shall be poured into place and left until thoroughly set before the piece is moved or jarred.

Litharge Cement No. 2

To 3½ lb. of litharge add 1 lb. of high gravity (Dynamite) glycerine.

For slow-setting, use No. 40 litharge.
For rapid-setting, use No. 30 litharge.

Litharge Cement No. 3

Litharge	50 parts
Glycerine	30 parts

Mix well and apply where desired. Let stand for twelve hours. Useful for mending broken china, glass, etc., and for cementing glass or china to metals.

Litharge-Glycerin Cement

Cement composed of a mixture of litharge and glycerin has a wide application in industry. It is extensively used in the manufacture and installation of electrical equipment, gasoline pumps and chemical equipment, for setting acid-resistant brick, for fastening metal to glass and for various other purposes.

The mixture depends to some extent upon the purpose and the user. For a very quick drying cement, which will set hard in about 10 minutes, fine grained litharge is added to a mixture of two parts glycerin and five parts water. Dry litharge is used in an amount which makes a paste of a consistency suitable to the user. For a slower drying cement the water may be omitted and litharge added to 95 per cent grade glycerin, and for still slower drying 98 per cent (dynamite) glycerin may be use. The rate of setting is faster the finer the litharge is, seeming to be a direct function of the specific-surface of the litharge particles per unit weight. A mixture often used consists of 3¼ parts litharge to 1 part 98 per cent glycerin, which will set hard in from 30 to 90 minutes, depending upon the kind of litharge. Litharge should be added to the vehicle immediately before using or the mixture will harden before it can be applied, so that it can not be used.

Litharge and Glycerine Mortar

Litharge	25	parts
Cement	10	parts
Sand	30	parts
Glycerine	9.5	parts

The glycerine is a 64% by weight solution, made by diluting pure glycerol with water.

Magnesium Oxy-Chloride Cement

This product is excellent for joining metal to metal or metal to porcelain or glass.

Calcined Powdered Magnesite 25 g.
Magnesium Chloride Solution
(24° Bé.) 10 cc.

The magnesia powder is thoroughly mixed to a smooth paste with the chloride solution. After use the cement dries hard when air dried overnight.

Mercury Arc Rectifier Cement

This cement is intended for molding into the openings of the glass bulbs used with mercury arc rectifiers.

Portland Cement	
(P.D. Spec. 1573)	4 lb.
Shellac No. 3	1 lb.

Mix and knead thoroughly and use at once. The cement hardens and sets on application of heat, becoming somewhat porous. If used in the mercury arc rectifiers one hour's run at full load will be sufficient to harden the cement.

If the cement expands under the heat, trim off with a coarse file or some other convenient way.

Pettman's Cement

This product is used by the British for munition work, and has the following composition:

Orange Shellac	7½ lb.
Denatured Alcohol	8 lb.
Stockholm Tar	5 lb.
Venetian Red Oxide	20¾ lb.

Pitch Cements for Transmission Belts

These preparations are used to assure the adherence of the belts on the pulleys and prevents slipping. The Bourdais Formula for this product is as follows:

Pitch	5 parts
Rosin	3 parts
Yellow Wax	4 parts
Whale Oil	13 parts

In winter the whale oil is increased to 15 parts. Neatsfoot oil can be substituted for the whale oil. The belt is rubbed with this mixture or it can be spread on the pulleys.

Crystal Cement for Porcelain

This product is readily made by pouring acetic acid (about 20 per cent) over pieces of clear glue, sufficient to cover, and heating until a homogeneous clear thick-flowing mass is formed. It is used by heating the edges of broken articles and applying the previously melted cement by means of a brush; the surfaces of the fracture being pressed firmly together and the mended object allowed to remain undisturbed for twelve to twenty-four hours.

Cement for Porcelain, Metals or Stoneware

Dried casein soaked in an equal weight of water for two hours. At the end of that time sodium silicate and lime are stirred in and the cement is ready for use.

Refractory Cement
U. S. Patent 1,952,119

A refractory, slag-resistant cement contains substantially by weight dead-burned, finely ground magnesite 50 parts; electrically fused, finely ground magnesia 15 parts; refined zircon sand (60 mesh) 25 parts; and refined zircon milled (300 mesh) 10 parts, bonded with sodium silicate solution of specific gravity of about 1.3, wherein magnesite shrinkage is counteracted by expansion of magnesia-zircon compounds formed when said cement is fired at temperatures up to 2900° F. to volatilize the binder and coalesce the particles thereof.

Transparent Cement

White caoutchouc (in sheets)	10 parts
Chloroform	10 parts
Dissolve cold, then add	
Mastic	2 parts

Allow to stand till dissolved, shaking at intervals to hasten solution. An excellent cement for glass.

Tar Cements for Roofs
Formula 1

This material is heated for application.

Rosin	35 g.
Slate Powder	35 g.
Mica	30 g.
Coal Tar	sufficient

Formula 2
German Patent 90,094

This cement is especially suitable for sticking pasteboard in manufacture of double roofs.

Coal Tar	75 g.
Chalk	100 g.

Formula 3
German Patent 98,071

This product used for roofs also for coating metal angles and joints and protecting them from rust.

Gas tar	100 g.
Cement	500 g.

Transformer Lead Cement

This is a resinous cement intended to be applied well melted. It is suitable for cementing leads in transformer cases, etc.

It is made as follows (stated in parts by weight):

Cimmerian Gum	13 parts
Powdered kaolin	18 parts
Powdered Rosin	29 parts
Powdered Brick Dust (50 mesh)	40 parts

Powdered Rosin, brick dust and kaolin are well mixed while cold. They are put in an iron vessel and heated until the rosin melts, stirring well during the heating. When the rosin is well melted, the impregnating gum—previously melted in another vessel, is poured into the mixture, and the whole stirred until thoroughly mixed. After the cement is thoroughly mixed, it is poured into the molds.

The brick dust should pass through No. 50 mesh screen before mixing.

In melting the rosin, the pot should be heated gradually, care being taken to keep the rosin from sticking to the pot.

If bubbling occurs at the surface of the liquid, this indicates the presence of moisture. The cement should be kept well melted until bubbling ceases.

Waterproof Cement

Glue	10	parts
Water	15	parts
Latex	5	parts
Formalin	3	parts
Alpha-naphthol	.5	part

Wood Heel Cover (Folding Cement)

Latex (40% Solids)	36 gal.
Resin emulsion (5% Resin)	15 gal.*
Gum karaya solution (2% Gum)	25 gal.†
Water	15 gal.

Flange Cement & Filler
U. S. Patent 1,919,037

This composition comprises vermiculite 29, corn syrup 61, and, if desired, powdered coal 9 parts by weight.

Masonry Joint Filler
U. S. Patent 1,931,643

A mixture of approximately 80% Portland cement 12% pulverized iron and 8% of lime or other coloring material (all by volume) to which is added enough water to make the composition plastic. Take a stiff brush and apply the composition along the masonry joints without taking any particular pains to avoid having the material contact with the faces of the bricks or other masonry material. After the caulking material has taken an initial set, (about one hour) the entire surface of the building or other structure may be washed with water to remove the surplus material on the faces of the joints and on the surrounding portions thereof. The caulking material, however, will have found its way into any openings in the masonry joints and that portion thereof will remain in place.

Wood Crack Filler

There has been on the market for years a wood crack filler (water putty) in dry powder form for mixing when ready to use. This preparation is for checks, cracks, and holes in the wood of interior trim and similar work, and for cracks and seams in old and new floors. The dry powder is an intimate mixture or blending of the following:

Quartz Silica	10	parts
Plaster of Paris	2	parts
Dextrine	1½	parts

The quartz silica is the same character of silica as used in the so-called "silex" paste wood fillers for furniture finishing. In this mixture, dextrine is the binder (pulverized gum arabic could be substituted); dextrine is the cheaper material, now almost universally used for adhesives. Make a thick paste of the above with water. The wood surface ought to be slightly moistened before applying this crack filler to assure best results. Apply with pressure, then smooth off the top with a putty knife. When dry carefully sand off flush with the surrounding surface.

AGRICULTURAL AND GARDEN SPECIALTIES

Agricultural Oil Sprays

Agricultural oil sprays are usually divided into two groupings, those used in the summer and those used in the winter, and also there is a separation between sprays used on deciduous and citrus. On deciduous during the dormant season heavy oils are used. On the Pacific Coast these oils are generally classed as Brown Neutral and Lemon Neutral, and 100 Pale, all oils with the following approximate general specifications:

Gravity °API	24.5
Flash °F. Min.	300
Fire °F. Min.	340–360
Visc. Sayb. @ 100° F.	100–120
Pour Point °F	–30
Color A.S.T.M.	2–4
Unsulphonated Res.	73%

These are prepared tank-mix style by the growers by the use of 12 ounces up to one pound of a mixture called Blood Albumen Spreader, which consists of 3 parts of fullers earth and one part of dried blood albumen to the 300 or 400 gallon spray tank in the field. Ten or 15 gallons of water are introduced, then the emulsifier and then the oil at the desired strength for which it is to be used, usually 3% or 4%, and the tank is filled with water.

These same products are prepared commercially two ways. The paste emulsions contain approximately 83% oil and are emulsified by pumps or colloidal mills, with a mixture of casein, soap and ammonia in varying amounts.

One common formula to make a stock solution is—

Oil	100 gal.
Water	33 gal.
Casein (finely powdered)	3 lb.
Ammonia (28%)	¼ gal.

There are many varieties of this emulsifier used by different commercial companies. Some introduce colloidal clay with the casein. Others use colloidal clay and albumen, the albumen dissolved in a rosin soap.

The same oil is also used in miscible oil form. These usually run from 80 to 85% mineral oil, the rest as emulsifier which is a cresol soap, the fat in the soap being either generally fish oil or whale oil. The finished product is a liquid and shows generally from 1½% to 4% phenols, and from 3% to 5% soap.

Agricultural Spray Emulsion: (Stock Emulsion)

Formula No. 1

Highly Refined White Min. Oil (Spray Stock "A")	87 lb.
White Olein	8.8 lb.
Triethanolamine	3.5 lb.

Formula No. 2

Highly Refined White Min. Oil	95 lb.
Diglycol Oleate	5 lb.

Formula No. 3

Highly Refined White Min. Oil (Spray Stock "A")	65 lb.
Cocoanut Oil Soap (40%)	2 lb.
Water	33 lb.

Cherry Leaf Beetle Spray

Since the larvae of the cherry leaf-beetle do not thrive on the foliage of the cultivated cherry but apparently feed entirely on the wild bird cherry, no means of combating the pest during this stage is known. The most practical method that has been devised for protecting cherries and peaches is to spray the trees as soon as the beetles appear. On cherry trees, applications of 8 pounds of paste arsenate of lead with 100 gallons either of water or of bordeaux mixture (8–8–100) have proven very efficient. In spraying great care should be exercised to cover the foliage thoroughly, both on the upper and lower surfaces, with the material. Paste arsenate of lime in the proportions of 6 pounds to 100 gallons of water and with bordeaux mixture as above, proves effective. This arsenical should be used experimentally as there is some doubt as to its safeness on foliage of fruit trees.

Nicotine sulphate (40%), one-half pint in 60 to 80 gallons of water, is an effective contact spray. In using it on

CHERRY SPRAYS

Time of Application	Spray Mixtures	Enemy	Dust Mixtures
When bud scales separate and expose green blossom buds. (For sweet cherries only.)	Lime-sulphur, 11 gal. Nicotine sulphate, ¾ pint. Water to make 100 gal. or Nicotine sulphate, ¾ pint. Soap, 5 or 6 lb. Water to make 100 gal.	Scale Aphids Aphids	No satisfactory dust for scale. Control of aphis by 90–10 sulphur lead-arsenate dust with 2% nicotine not yet demonstrated. Thorough dusting with 2% nicotine dust should reduce number of insects.
Just before blossoms open.	Lime-sulphur, 2½ gal. Water to make 100 gal.	Brown-rot Blossom blight	Apply 95–5 sulphur-lead-arsenate dust.
When petals fall.	Lime-sulphur, 2½ gal. (Sweet cherries, 2 gal.) Arsenate of lead, 2½ lb.* Water to make 100 gal.	Leaf-spot Brown-rot Curculio	Apply 90–10 sulphur-lead-arsenate dust, or if curculio is abundant 80–20 sulfur-lead-arsenate dust.
10 days after petals fall or when shucks are off.	Lime-sulphur, 2½ gal. (Sweet cherries, 2 gal.) Arsenate of lead, 2½ lb.* Water to make 100 gal.	Leaf-spot Brown-rot Curculio	Apply 90–10 sulphur-lead-arsenate dust, or if curculio is abundant 80–20 sulphur-lead-arsenate dust.
As Early Richmond cherries show red on one side.	Lime-sulphur, 2½ gal. (Sweet cherries, 2 gal.) Arsenate of lead, 2½ gal.* Water to make 100 gal.	Maggot Leaf-spot Brown-rot Curculio	Apply 90–10 sulphur-lead-arsenate dust.
As Montmorency cherries show red on one side.	Lime-sulphur, 2½ gal. (Sweet cherries, 2 gal.) Arsenate of lead, 2½ lb.* Water to make 100 gal.	Maggot Leaf-spot Brown-rot	Apply 90–10 sulphur-lead-arsenate dust.
After picking.	Lime-sulphur, 2½ gal. (Sweet cherries, 2 gal.) Arsenate of lead, 1 to 2 lb.* Water to make 100 gal.	Leaf-spot	Apply 95–5 sulphur-lead-arsenate dust.

* The amount of arsenate of lead is given for powder form; if paste form is used, twice as much is required.
If heavy rains are of frequent occurrence, apply dust mixtures about one week after the first treatment and make a third application one week after the second treatment.

should first direct the treatment to the foliage and then thoroughly spray all beetles which have dropped to the ground. On account of its safe properties nicotine solutions may be employed without danger of injury to peach foliage. Arsenate of lead sometimes causes injury to peach leaves, and in its use on this fruit great care should be exercised.

Dormant and Summer Sprays

An emulsion made according to the formula petroleum oil 100 gallons, water 33 gallons, finely powd. casein 3 lb., NH_3 soln. (28%). 1 quart has been successfully used on a large scale for insect control in the Pacific Northwest. Emulsification is effected in the spray tank, oils of about 105 seconds viscosity (Saybolt) and about 60% unsulfonated residue being used for dormant sprays, and 70-75 seconds viscosity and 85-90% unsulfonated residue for summer sprays.

A Homemade Colloidal Copper Spray

A concentrate is prepared as follows: 1 pound of cupric sulphate is dissolved in 2 quarts of water, and 1 pint blackstrap molasses is added. The mixture is stirred well and made slightly alkaline by adding 0.4 pound of lye dissolved in 1 quart of water. The concentrate is allowed to stand until it turns yellow and is used at a rate equivalent to 4 pounds of cupric sulphate to 50 gallons of water for the control of late blight of potatoes.

Derris Emulsion Spray
U. S. Patent 1,934,057

Extract 2000 grams ground derris roots with 4 liters of benzol. Distil solvent ''in vacuo.'' A reddish brown resin results. Into a colloid mill feed 80 grams of this resin with 750 grams of oil soluble sodium sulfonates dissolved in 2½ litres of white mineral oil. Run thru again mixing with 1½ liters water. A thick emulsion results. Dilute with water (fifty times its weight) to use as insecticidal spray for trees.

Preparation of Derris Extract Sprays
(Patented)

In preparing sprays from commercial acetone extracts of derris containing 3-5% rotenone and 16-18% total extractives, large quantities of a precipitate are formed when the extract is added to 250-500 or more parts of water, aqueous solutions of soap, or various other wetting agents. This precipitate clogs spray nozzles, and as it occludes active constituents, the spray is rendered less effective. If diluted 1:1 with acetone before the water or aqueous solutions are added, the formation of this precipitate is prevented. Residues from sprays prepared in this way with sulfonated castor oil (1:400) retain their toxicity to thrips much longer than when potassium coconut oil soap (1:300) is the wetting agent. The spray possesses good wetting properties.

Fly Spray for Outside Use

Where kerosene odors are not objectionable, a cheap, yet effective insect spray can be prepared by adding 1 pound of powdered pyrethrum flowers to 3 gallons of kerosene. Stir occasionally, and then allow to settle until clear enough to use in sprayer.

Fruit Tree Self Roller Spray

	No. 1	No. 2
	6% Oil	8% Oil
Oil, Lubricating	6 gal.	8 gal.
Water	3 gal.	4 gal.
Potash fish-oil soap	6 lb.	8 lb.
Water—enough to make	100 gal.	100 gal.

Place the soap and water in an iron kettle and heat until the soap is dissolved, then add the oil and heat until the contents come to a boil. After boiling has commenced remove from the fire and emulsify by pumping the liquid twice back into itself while still very hot. The resulting emulsion is quite stable and can be kept for several months.

If steam is available the cooking can be done in the spray rig using the pump to emulsify the material. When steam is used allowance must be made for condensation, and the amount of water used reduced accordingly.

Insecticide Emulsion Sprays
U. S. Patent 1,949,798

Formula No. 1

Glyceryl mono-oleate	0.10- 2.0%
Spray oil	99.90-98.0%

Glyceryl mono-oleate and spray oil are completely miscible and form very stable mixtures which can be emulsified as desired by agitating with the desired quantities of water. In general, 99-97% by volume of water to 1-3% by volume of spray oil-glyceryl mono-oleate gives a very satisfactory insecticide.

Formula No. 2

Glyceryl di-oleate	0.20– 2.0%
Nitrobenzene	2.50– 5.0%
Spray oil	97.30–93.0%

Formula No. 3

Glycol mono-oleate	0.20– 2.0%
Butyl esters of whale oil fatty acids	5.00–10.0%
Spray oil	94.80–88.0%

The nitrobenzene and butyl esters of whale oil fatty acids of Formulas No. 2 and No. 3 serve as insect poisons where a higher degree of toxicity is required in the emulsions made from these compositions.

Formula No. 4

Glyceryl mono-acetyl ricinoleate	0.20– 2.0%
Butyl ricinoleate	1.00–10.0%
Spray oil	98.80–88 %

The butyl ricinoleate in this case serves a double function, namely, as a bonding agent and as an insect poison. As previously indicated some of the emulsifying agents herein disclosed are relatively insoluble in spray oil and hence in order to get a desirable type of emulsion it is necessary to incorporate the emulsifying agent in a material which is completely miscible with the spray oil.

	No. 1 6 per cent oil	No. 2 8 per cent oil
Oil	6 gal.	8 gal.
Water	3 gal.	4 gal.
Emulsifier { Kayso (a Casein product or Bordeux Mixture Copper Sulphate Quicklime *	1 lb.	1 lb.
	1 lb.	1 lb.
	1 lb.	1 lb.
Water to make	100 gal.	100 gal.

* If hydrated lime is used, double the quantity.

If Kayso is used for the emulsifier sift it into the water in the spray tank and mix thoroly. If Bordeaux is used, prepare copper sulfate solution and lime separately, each in half the required amount of water, and pour these two liquids together into the spray tank and mix thoroly. Next add the oil and emulsify with the spray-gun blowing the mixture back into itself in the tank. When emulsified, which will take but a few minutes, dilute with water to the required concentration and, with the agitator still running, drive to the orchard.

Spray for Insect Control in Empty Grain Bins

A spray consisting of 8% miscible petroleum oil and 3 pounds caustic soda

Formula No. 5

Glyceryl mono-ricinole-ate	0.20– 2.0%
Triolein	1.00–10.0%
Spray oil	98.80–88 %

The triolein here serves as a bonding agent between the oil and emulsifying agent.

Emulsions prepared with water and compositions such as those disclosed above are particularly effective against red spiders, plant lice, the red scale of citrus fruits, and other pests which infest fruit trees. When thoroughly agitated with water and sprayed onto the trees until the foliage of the latter is dripping wet, effective destruction of the insects has consistently resulted without any injurious effect on the trees.

* Spray oils are mineral oils of 70–300 seconds Saybolt viscosity at 100° F. and from 85–100% in unsulfonated residue.

Cold-Mixed Emulsions

The cold-mixed lubrication oil emulsions are less stable than the cooked type and should be used immediately after making. They are, however, easier to prepare and just as effective against the leaf roller. The following formula is used in their preparation:

in 100 gallons water will destroy grain insects in empty bins.

Vermin Fluid

Mineral Spirits	90 gal.
Turpentine	5 gal.
Myrbane oil	2 gal.
Carbolic acid	3 gal.

Control of Thrips

Fumigation with 1 ounce of naphthalene (grade 16) per 100 cubic feet killed 95% of the insects. Mixtures of ½ ounce of powdered naphthalene and 0.025 fluid ounce of nicotine killed 95% of the insects.

Formulae for Treatment of Insect Infestation in Stored Grain

Formula No. 1
"80–20" Mixture

Carbon tetrachloride	80%
Carbon bisulphide	20%

Use 2 gallons per 1000 bushels.

Formula No. 2
Dep't. of Agriculture Mixture

Ethylene dichloride	75%
Carbon tetrachloride	25%

Use 2 gallons per 1000 bushels.

Formula No. 3
Commercial Weevil Killer

Carbon bisulphide	50 gal.
Carbon tetrachloride	100 gal.
Denatured Alcohol	1 gal.

Mix Carbon tetrachloride with alcohol, add carbon bisulphide, and saturate mixture with sulphur dioxide. Use 1½ gallons per 1000 bushels.

Formula No. 4
Dri-ice Dope

Mix well and scatter in grain for each 1000 bushels:

Ethylene oxide	2 lb.
Solid carbon dioxide	20 lb.

(Apt to effect germination.)

Formula No. 5
Carbon Dioxide Insecticide

Liquid Carbon dioxide	90%
Methyl Formate	10%

Vapor applied to moving grain (as it enters bin).

Formula No. 6

The only grain fumigant that kills eggs and larvae, as well as adult insects, is Chlorpicrin. While the initial cost is somewhat greater, Chlorpicrin (Tear-gas fluid) used 2 pounds per 1000 bushels is the best available grain fumigant.

Peach Cottony Scale Sprays

Formula No. 1

Red engine oil	2 gal.
Water	1 gal.
Potash fish-oil soap	2 lb.

Formula No. 2

Red engine oil	2 gal.
Water	1 gal.
Calcium caseinate (Kayso)	4 to 8 oz.

Pear Midge Spray

Formula No. 1

The following formula is recommended for the control of the pear midge and pear psylla:

Nicotine Sulphate	¾ to	pint
Lime-sulphur	11	gal.
Water to make	100	gal.

Formula No. 2

If a second treatment is necessary the following mixture should be used to avoid an excess of lime-sulphur on the foliage:

Nicotine sulphate	¾ to	1 pint
Lime		10 lb.
Water		100 gal.

Formula No. 3

The use of a late-dormant application of lubricating oil emulsion for the control of pear psylla is popular with some growers. If an orchard has received the late-dormant oil treatment, then the mixture to use for the control of pear midge (first application) would be the following:

Nicotine Sulphate	¾ to	pint
Lime		10 lb.
Copper Sulphate		2 lb.
Water		100 gal.

If a second application is necessary use a mixture prepared according to formula No. 2.

It is to be noted that in the systems of treatment given it has been assumed that the grower desires to include a fungicide. Occasionally a grower has an orchard where psylla is absent, but a fungicide is needed in the midge application. Under such conditions, formula 1 should be modified to contain only 2½ gallons of lime-sulphur, or a spray mixture can be prepared after formula No. 3. If either mixture be used, a second application, if necessary, should consist of a mixture made according to formula No. 2.

Pyrethrum Spray

Potash Coconut Oil Soap (40%)	1.25	gal.
Pyrethrum Extract (20 fold)	13	fl. oz.
Water	100	gal.

Dissolve the soap in the water; add the extract and stir. The material is then ready to use. This dilute mixture must not be allowed to stand overnight, or much of its effectiveness will disappear.

Satisfactory results are likely to follow that type of spraying which wets all the insects thoroughly.

Squash-Borer Sprays

Equipment	Ingredients	Amount on acre basis
Power sprayer	Nicotine sulfate (10%)	3 pints
	Soap	6 lb.
	Water	100 gal.
Hand sprayer	Nicotine sulfate (40%)	1 pint
	Soap	12 oz.
	Water	12½ gal.
Power or hand sprayer	Calcium arsenate	5 lb.
	Flour paste	6 lb.
	or	
	Hydrated lime	12 lb.
	Water	100 gal.
	Calcium arsenate	5 lb.
	Bordeaux mixture†	100 gal.
Power sprayer	Lead arsenate	5 lb.
	Fish oil	1 qt.
	Water	100 gal.

Blue Tobacco Mold Spray

Dissolve one pound blue octriol in 2 quarts water; add 1 pint of molasses and then make slightly alkaline by slowly adding a solution of 5 ounces of caustic soda in one quart of water. This stock solution should be stored for at least 1 week before use. For spraying purposes, 1 part of the stock solution is diluted with 30 parts of water, and 0.5% by weight of potash soft soap is added as a spreader.

Bordeaux-Arsenate Spray

For plants such as potatoes, tomatoes, or egg plants, the best protection is a spray of bordeaux mixture, 4–6–50, to which calcium arsenate has been added at the rate of 5 pounds per 100 gallons of spray. In restricted areas, as in cold frames, gardens, or small fields, the spray may be suitably applied by a hand sprayer. In larger fields, where the growth of the crop permits, the potato sprayer may be used advantageously. If the coverage at spraying seems incomplete or if flea beetles are exceptionally abundant, the affected plants should be double sprayed. One or two applications should follow the first at weekly intervals to maintain coverage on the foliage against both the spring and summer broods. In cases of cold frames, or where plants are growing rapidly, this interval may require shortening to meet the needs of the situation.

Plant Insects Controlled with Pyrethrum

While pyrethrum is primarily considered as a contact poison,—to kill sucking insects, such as aphis, squash bugs, leaf hoppers, and thrips,—many chewing insects, such as blister beetles, cucumber beetles, Mexican bean beetles, Japanese beetles, and some caterpillars are also killed when thoroughly sprayed with it. A good wetting and spreading agent is essential, in order to cover the insect and plant. Liquid soaps, such as cocoanut oil soap, potassium oleate, and similar soaps offer the cheapest and possibly the best spreaders. A spray solution containing from 0.25 to 0.5 per cent actual soap will wet most, if not all, kinds of plants and insects.

The great objection against pyrethrum insecticides is that they readily decompose and lose their toxic properties within 2 or 3 days when exposed to sunlight and other atmospheric factors. For this reason pyrethrum sprays should be applied only for immediate kill of the insect pests present on the plant. In order to keep the plant free from insects all the time, frequent spraying is necessary.

Plant Killer
U. S. Patent 1,913,141

Sodium thiosulphate pentahydrate (11 parts) is heated at 110° with cuprous cyanide (1 part) to form a complex tetrathiosulphatocyanocuprite which is highly toxic to plant life and may be used for removing plants from roads, paths, etc.

Mexican Bean Beetle Spray

Magnesium Arsenate	1½ oz.
Water	1 gal.

Stir well and spray on under side of bean leaf while the arsenate is in suspension. Excellent for lima beans, string beans, and wax beans.

CAUTION. Do not spray on string beans after the pod has formed.

Mexican Bean Beetle Control

Three dust mixtures applied at the rate of 30 pounds per acre have given good results. These are as follows:

Formula No. 1

	On an acre basis	For the small garden
Calcium arsenate	15 lb.	1 lb.
Dehydrated copper sulfate	15 lb.	1 lb.
Hydrated lime	70 lb.	5 lb.

Formula No. 2

Magnesium arsenate	20 lb.	1 lb.
Hydrated lime	80 lb.	4 lb.

Formula No. 3

Barium fluosilicate	20 lb.	1 lb.
Hydrated lime	40 lb.	2 lb.

Following are three spray mixtures which have given good results when applied at the average rate of 100 gallons per acre:

Formula No. 1

	On an acre basis		For the small garden	
Calcium arsenate	3	lb.	2½	oz.
Bordeaux mixture	100	gal.	5	gal.

Formula No. 2

Calcium arsenate	3	lb.	1	oz.
Hydrated lime	9	lb.	3	oz.
Water	100	gal.	2	gal.

Formula No. 3

Magnesium arsenate	3	lb.	1	oz.
Kayso or flour paste	3	lb.	1	oz.
or				
Skim milk	3	qt.	2	oz.
Hydrated lime	½	lb.	½	oz.
Water	100	gal.	2	gal.

Fungicide

U. S. Patent 1,910,223

A suspension of Schweinfurth green (1) (1 kilogram per litre) in water is mixed with a solution of sodium palmitate (500 grams in 3 litres of water and a little ethylalcohol; after gentle warming, green plates of $(C_{15}H_{31} \cdot CO_2)_2Cu$, 0.3 $Cu(AsO_2)_2$ (11) separate. When powdered, (11) forms a highly poisonous fungicide and insecticide for dusting on plants; unlike (1) it has no injurious effects on the leaves.

Sulphur-Arsenate Fungicide and Insecticide

Sulphur, finely ground	85 lb.
Manganese Arsenate	10 lb.
Aluminum Hydroxide	5 lb.

Weigh together and thoroughly mix. This dust is an efficient fungicide and insecticide, particularly adapted for use on roses for the control of mildrew and black-spot. Application may be made with a hand duster.

Copper Lime Arsenate Dust

Copper Sulphate Mono-hydrated	20 lb.
Lime, Hydrated	70 lb.
Calcium Arsenate	10 lb.

After thoro mixing the dust is ready for use. It is best to prepare in quantities required for immediate use, yet may be kept for a time in air-tight containers. When exposed to the air the mixture becomes lumpy. Application may be made with a hand or power duster. This dust is useful upon potatoes for the control of blights, hopperburn, and such insects as flea beetles, potato bugs and leaf hoppers. It may be substituted for bordeaux mixture in nearly all cases when a copper spray is recommended.

Termite Insecticide

Ortho-di-chlor-benzene, preferably used pure, though often diluted with kerosene or similar material.

Control for Damping-off of Seedlings

Charcoal, Dried Muck or Kieselguhr	85 parts
Formaldehyde, 40%	15 parts

The formaldehyde must be well mixed with the dust carrier. Lumps can be most easily broken up by passing through a coarse sieve. If stored in an air-tight container the mixture will keep indefinitely. In use 8 ounces of the formaldehyde dust are added to a bushel of soil and thoroly distributed by shoveling over several times. The treated soil is then used for filling beds or flats, the seed planted and well watered.

Control for Scab of Gladiolus Bulbs

Prepare a suspension of one ounce of finely pulverized mercurous chloride or calomel in one gallon of water and soak bulbs for 5 minutes. In extremely bad cases increase the amount of calomel to 6 ounces for each gallon of water.

Dust Control for Oats Smut

Diatomaceous earth or Kieselguhr	4½ lb.
Formaldehyde	½ lb.

Mix thoroughly, taking care to break up lumps. Finally pass through a sieve. Store in an air-tight container. Finely ground charcoal or dried, sifted muck may be substituted for the diatomaceous earth. Place grain to be treated in a barrel or some container that can be rotated, and sift the dust over the seed at the rate of 3 ounces or 3 heaping tablespoonfuls per bushel. Mix for 2 or 3 minutes, then sack. After standing over night seed will be ready to sow, yet may be kept indefinitely without injury to germination. This treatment is not poisonous and grain not required for seeding may be fed to stock.

Control for Brown Rot of Peaches and Plums

Sulphur, finely ground	6 lb.
Lime, hydrated	3 lb.
Glue, granulated	3 oz.
Water	50 gal.

Mix the lime and sulphur. Dissolve glue in 3 to 4 gallons of boiling water (removed from flame). Then stir in the lime and sulphur mixture. Allow to stand for an hour, then dilute to 50 gallons in spray tank and apply. A good covering of this spray on the fruit at harvest time is particularly desirable to prevent rot in market.

Fire Blight of Pear and Apple Canker Treatment

Zinc chloride	9 lb.
Muriatic acid (Conc.)	3 oz.
Water	1 qt.
Alcohol (denatured)	7 pts.

Use an enameled kettle. Add acid to water and dissolve the zinc chloride in the acid solution using heat if necessary. After cooling pour into the denatured alcohol. Store in glass bottles. Scrape loose bark from cankers, then apply the solution with a paint brush. The liquid penetrates the dead bark and wood rapidly and sterilizes by killing the fire blight bacteria, thus preventing spread of infection. This treatment is particularly useful for eradicating infections upon large limbs, where is is not desirable to cut out the dead wood of the canker because of impairing the strength of the limb. Brushes will last longer if the acid is neutralized with a solution of baking soda, after use.

Control for Stinking Smut or Bunt of Wheat

Thoroughly coat seed with finely ground copper carbonate. The dust is used at the rate of 2½ to 3 ounces per bushel. Treatment may be made by spreading the grain upon a floor, sifting the powder over it and shoveling until thoroughly mixed; or better, in some form of rotatory container such as a barrel or cement mixer. Treated grain may be kept indefinitely without injury to germination.

Control of Bermuda Grass

Two applications, each of 200 pounds per acre, of either calcium chlorate or sodium chlorate with a 6-weeks interval, are necessary to obtain effective results.

Control of Black Pecan Aphid

The aphid is effectively controlled by adding nicotine sulfate (to give a 1:4000 concentration) to the usual summer sprays of Bordeaux mixture (3–4–50) and following these, if necessary, by early fall spraying the summer-oil emulsion (0.5–100) also containing nicotine sulfate.

Rose Chafer Control

Spray as soon as the beetles appear using, on grapes, arsenate of lead (dry), 5 pounds; molasses, 2 gallons (or glucose 25 pounds); and water, 100 gallons. On cherries use the same formula, but use glucose instead of molasses. Spray a second time, if necessary, from four days to one week after the first application, using the same mixture as first applied. Since the addition of molasses or glucose destroys the adhesiveness of the arsenate of lead, care should be exercised to avoid applying the material just previous to a rain. If a rain occurs within 24 hours after spraying, it may be necessary to repeat the application as soon as weather conditions permit.

Rose Leaf-Hopper Control

Although parasites render valuable service in the control of this pest, it frequently becomes necessary to supplement their work with artificial control measures. If treated at the proper time, this leaf-hopper is not difficult to hold in check. Affected roses should be thoroughly sprayed while the insects are in the nymphal stages, using a solution of nicotine and soap in the proportions of three-fourths pint of nicotine sulphate and five pounds of soap to one hundred gallons of water. Where only a few plants are to be treated, the mixture may be prepared at the rate of one teaspoon level full (four cubic centimeters) of nicotine sulphate, 40 per cent, and two ounces of soap to one gallon of water. The spray is effective only while the insect is in the immature stages, for the adults fly so quickly that they cannot be reached with the spray.

Combating Coccids on Citrus Plants and Tea Bushes

The application of petroleum emulsions in the summer is more effective and causes less damage to foliage than does lime sulfur. Vegetable oil soaps can successfully be replaced with petroleum soaps or soaps obtained from oxidation products of petroleum. Petroleum soaps are very effective for controlling scale on citrus trees and tea bushes. Tea bushes should be sprayed in summer preferably with lubricating oil and kerosene emulsions so as not to affect the flavor of the tea. Emulsions prepared of oxidation products of mineral oils (concd. solns.) as well as of green soap but with a higher oil content (jellies) are unsuitable for summer spraying of tea bushes and tangerine trees because of their burning effect on the foliage. They may be used in a more dilute form together with nicotine or pyrethrum. Machine oil emulsion with green soap as emulsifier is very effective without injury to the foliage. Lubricating oils are more effective than illuminating oils, machine oil No. 2 being best. A good summer spray consists of 3% dolphin soap, 1% machine oil emulsion (32% of oil and 40% of soap) and 6% of a soap-kerosene emulsion (66% kerosene and 20% soap).

Potato Flea-Beetle Spray

	On an Acre Basis	For the Small Garden
Calcium Arsenate *	5 lb. (3 lb.)†	1 oz. (½ oz.)
Bordeaux Mixture ‡	100 gal.	1 gal.

Potato-Flea Beetle Dust

	On an Acre Basis	For the Small Garden
Tobacco Dust	50 lb.	1 lb.
Hydrated Lime	50 lb.	1 lb.
Calcium Arsenate *	20 lb. (15 lb.)	1 lb.
Monohydrated Copper Sulphate	20 lb. (15 lb.)	1 lb.
Hydrated Lime	60 lb. (70 lb.)	5 lb.

* Lead arsenate may be substituted for calcium arsenate except in the case of beans.
† Numbers in brackets denote the proportions to be used for beans.
‡ A home-made mixture may be prepared as follows: Dissolve 1 ounce of copper sulfate in 2 quarts of water, add 2 ounces of hydrated lime to 2 quarts of water, pour together stirring; finally, add 1 ounce of calcium arsenate. Use glass, wooden or earthenware containers.

Seed Disinfectant
U. S. Patent 1,934,804

Phenyl Mercury Hydroxide	3%
Talc or Diatomaceous earth	97%

Seed Disinfectant
U. S. Patent 1,934,803

Alpha chlormercurithiophene	1.5 %
Calcium Carbonate (powder.)	97.5
Charcoal (powder.)	1.0 %

Mix together and dust on seeds.

Fertilizer
U. S. Patent 1,922,909

A fertilizer which is suitable for lawns comprises dried activated sewage sludge 40–75, lead arsenate 15–25, water 5–20 and an organic colloid such as glue 1–6%.

Non-Caking Fertilizers
U. S. Patent 1,939,165

As new articles of manufacture, fertilizers are made comprising a homogeneous salt containing 3 molecular proportions of ammonium nitrate and 1 molecular proportion of ammonium sulphate.

Based Ammonium Sulphate Fertilizer
U. S. Patent 1,918,454

A mixture of ammonium sulphate (75), calcium superphosphate (2.5), peanut-hull meal or similar vegetable material (3.75), and dolomite (18.75%) is intimately ground to a dry, free-running powder.

Poison Baits for Wood Lice

One part by weight of Paris green to 56 parts of dried blood is the most effective poison found.

Poison Bait for Army Worms and Cutworms

Addition of ½ pound casein to the bait (bagasse 10, molasses 20, arsenic trioxide 1 lb. and water 2 quarts) greatly improves the coherence of the particles composing the mixture and adds considerably to its life by preventing breakage, scattering and washing away when beaten upon by drenching showers of rain.

Wire Worm Insecticide
Canadian Patent 338,803

A compn. consisting of lead arsenate 3, bluestone 2, copper carbonate 2, pitch 1, salt 1, and sugar 1 pound may be used dry or wet for the destruction of wire worms and other insects.

Worm Killer on Lawns

Chlorinated lime. 1 pound to 20 gallons of water.

Weed Killer Solution

This weed killer is principally a 30% solution of crude ammonium sulphocyanate.

Directions for use:

In practice it is difficult to distribute a small amount of concentrated solution uniformly with an ordinary sprinkling can. It is desirable to add 2 gallons of fresh water to each gallon of liquid as received, and stir well. Four gallons of this mix is sufficient to kill all the weeds on a square rod (16½ ft. square) when dealing with such stubborn weeds as quack grass, thistles of all kinds, poison ivy, etc. The ground should be gone over twice to insure good coverage with a sprinkling can. When a high pressure type of sprayer is used, one coverage should be sufficient. For such weeds as mustard, kale, plantain, mullen, ragweed, burdock, etc., ½ the above amount or 2 gallons of the mix per square rod should be enough.

Thorough coverage is very important.

Weed Killer
U. S. Patent 1,946,462

A composition of matter, for use in preventing the growth of vegetation and in the destruction of weeds, comprising a mixture of all of the distillates of coal tar which are driven off therefrom up to a temperature of substantially from 800° F. to 1,000° F.

Weed Killer

Grass and weeds along walls, etc., may be killed for a whole year by spraying with a 5 to 10% solution of sodium arsenite. The spray should be directed to the base of the plant. Foliage so sprayed is poisonous to animals.

Weed Killer

Aluminum Potassium Sulphate 1 lb.
Ferrous Sulphate 1 lb.

Mix the above and dissolve in 1 gallon water.

Zinc Chlorate Weed-Killer
U. S. Patent 1,924,107

Claim is made for a mixture of zinc chloride (3 pts.) and sodium chlorate (4 pts.) with a small quantity of an acid salt, e.g., sodium bisulphate. The mixture deliquesces at normal temp. and humidity, has only a slight corrosive action on iron, and is repellent to cattle.

Chemical Weed Killers

Sodium Chlorate used at the rate of 10 pounds per square rod has proved very effective as a weed killer. A less expensive substitute is Calcium Chloride powder or flakes. The calcium chloride is used by applying the powder or flakes around the stems of the weeds.

Tree Banding Composition

Banding compositions are usually made from mineral or vegetable oils and fats or rosin. The disadvantage of applying the banding directly to the tree is that it usually has to be renewed twice during the winter in order to maintain its efficiency. The majority of horticulturists prefer the paper bands. The following are formulae for banding composition:

Formula No. 1

Rosin, powdered	25 lb.
Boiled linseed oil	5 gal.

Formula No. 2

Yellow wax	12 lb.
Tallow	56 lb.
Rosin	12 lb.
Chalk	20 lb.
Castor oil	Sufficient

Melt the wax, tallow, and rosin; stir in the chalk, and thin down with castor oil to the required consistency.

Tree Insect Bands

These are applied very hot on rolls of paper or thin cardboard which are wrapped around the trunk first covered with hemp and then fastened tightly with cord.

Formula No. 1

Rosin	1 g.
Linseed Oil	1 g.

Melt together and stir.

Formula No. 2

Burgundy Pitch	15 g.
Pine Turpentine	5 g.
Linseed Oil	5 g.
Rape Seed Oil	5 g.

Melt together.

Formula No. 3

Burgundy Pitch	10–20 g.
Pine Turpentine	5 g.
Linseed Oil	5 g.
Olive Oil	6 g.

Melt together.

Formula No. 4

Wood Tar	10 g.
Fish Oil	4 g.
Light Petroleum Oil	4 g.

Formula No. 5

Coal Tar	1 g.
Fish Oil	1 g.

Formula No. 6

Wood Tar	10 g.
Coal Tar	20 g.
Heavy Petroleum Tar	5 g.

Formula No. 7

Rosin	10 g.
Fish Oil	4 g.
Degras	4 g.

Formula No. 8

Wood Tar	1 g.
Black Pitch	1 g.
Fish Oil	1 g.
Light Petroleum Oil	1 g.

Formula No. 9

Wood Tar	20 g.
Fish Oil	5 g.
Light Petroleum Oil	5 g.

Ant Poisons

Mix 9 pounds of granulated sugar, 6 grams of tartaric acid (crystallized), and 8.4 grams of benzoate of soda. Boil slowly for 30 minutes in 9 pints of water. Allow to cool. Dissolve 15 grams of sodium arsenite (C.P.) in ½ pint of hot water. Cool. Add poison solution to syrup and stir well. To the poisoned syrup add 1¼ pounds of honey. Mix thoroughly.

Mix 1 part of tartar emetic with 20 parts of extracted honey. Mix 1 part of tartar emetic with 20 parts of bacon drippings or grease. Rub tartar emetic on raw or cooked meat and bacon rind.

Mix 1 ounce of Paris green with 16 ounces of brown sugar.

ANIMAL PREPARATIONS

Sheep Dips (Non-Poisonous)

	No. 1	No. 2	No. 3
Creosote	10	10	10 gal.
Yorkshire wool grease	42	42	42 lb.
Coconut oil	14	14	—
Caustic soda lye, 68° Tw.	5	5	5 gal.
Common rosin	56	112	112 lb.
Stearin fat			100 lb.
Caustic soda lye, 30° Tw.			15 gal.

Melt the fat and saponify with the lye. Melt in separate pan 100 lb. of rosin and add 20 gal. caustic lye, 18° to 20° Tw. Add contents of the two pans together while boiling, draw the fire, cool down, and add 6 to 8 gal. creosote oil. Stir well, add as much creosote as the oil will carry. Sample after each gallon of creosote has been added by dropping a little on a glass, and if it sets hard it is finished. Stop adding creosote as soon as the batch goes flabby. Melt the rosin and add the fat and coconut oil: then add the soda lye and stir in the creosote or dead oil.

Blowfly Dressing

A paste prepared from ½ ounce of Paris green, 5½ ounces of kaolin and 18 ounces of soft soap solution (1–2%) and applied directly to the wounds was effective in killing blowfly maggots in sheep. The treatment reduced the number of fly re-strikes occurring in 6–20 days.

Treatment of Stomach Worms in Sheep

Carbon tetrachloride, tetrachloroethylene and cupric sulphate and mustard, administered orally, have little if any effect on tapeworms, Moniezia sp., in lambs. Good control seems to be obtained by the administration of an aqueous solution of 2.33 grains arsenic trioxide and 105 grains magnesium sulphate. A mixture of sodium arsenite and cupric sulphate was only slightly efficient.

Mixture for Fowl Ailments

A useful mixture of sulphate of copper and iron for tonic and worm deterrent purposes is: Copper sulphate, 3 ounces; iron sulphate, 1 ounce; vinegar, 1 quart. Dose: 1–2 ounces added to each gallon of the drinking water. This is suitable for all ages of poultry, as the dose automatically becomes regulated by the amount of water taken. It is commonly employed in cases of catarrh, roup, and fowl-pox, also in worm infestations, and in anaemic conditions.

Cough Electuary for Horses

The following formula is suggested:

Powdered camphor	½ oz.
Powdered myrrh	½ oz.
Potassium chlorate	1 oz.
Honey	4 oz.
Glycerine	4 oz.

Dose: One tablespoonful three times a day.

Horse Conditioning Powder

Gentian Root Powder	4 oz.
Sulfur Powder	4 oz.
Potassium Nitrate	1 oz.
Glaubers Salts	2 oz.
Fenugreek Powder	1 oz.
Licorice Root Powder	4 oz.

A tablespoonful at meals is a usual dose.

Dog Vermifuge

Oil of chenopodium	16	minims
Turpentine	2	minims
Oil of aniseed	10	minims
Castor oil	3½	fl. drachms
Olive oil	3	fl. drachms

For a full-size or medium-size puppy under six weeks old give ½ drachm in a teaspoonful of milk. Between six and eight weeks the dose is 1 drachm, and at eight weeks 1 drachm, to be repeated in an hour. If the bowels do not act within an hour, give ½ to 1 teaspoonful of castor oil. For small puppies reduce the doses to one-half, and for the toy breeds to one-quarter. If no worms be expelled, the mixture may be repeated in a few days. The only really satisfactory method of treating animals is to dose each one separately, otherwise there is no way of insuring that every animal obtains a correct dose.

Blister Salve (Veterinarian)

(Poisonous internally)

Powdered cantharides	2 oz.
Red iodide mercury	3 oz.
Powdered euphorium	2 oz.

Mix to even consistency with sufficient petrolatum to make salve.

Warble Fly Cure

Derris powder is deadly to the grub of the warble fly, but difficulty is met in getting the powder into contact with the grub, which lives beneath the hide of a cow's back. An effective method of making this contact is, according to the United States Department of Agriculture, to prepare a mixture of 4 parts of derris powder, 3 of gum arabic, 1 of glue, and 1 of tannic acid, with enough water to make a stiff paste, which is to be rolled into slender rods and dried. The treatment consists of inserting the rod into the hole in the hide, and breaking it off at the surface of the skin.

It would seem, too, that a little benzocaine would relieve the pain to the cow that may be caused by the bot and by the hard roll of poison.

Warble Fly Wash

Powdered derris root is used as a wash for dressing cattle infected by the warble fly, in the following manner: A mixture is made of derris root, 1 lb.; soft soap, 1 lb.; in water, 1 gallon. The mixture must be freshly prepared, and is used by scrubbing the warbled area with a stiff cane brush during each of four monthly dressings. One gallon of solution is sufficient to dress 50–60 cattle once.

Fly Chaser for Cattle

Crude Carbolic Acid	10 parts
Tanners Oil	10 parts
Crude Petroleum	20 parts

Use this as a spray.

Lice and Mite Killer

Crude Carbolic Acid	10 parts
Waste Crank Case Oil	10 parts
Coal Oil	10 parts

Spray chickens with this as often as needed or saturate sawdust with it and dust in nests.

Cattle Spray

Pyrethrum Extract (20 lbs. of Flowers per gal.)	1 gal.
Steam Distilled Pine Oil	1 gal.
50/75 Viscosity Neutral Oil	7 gal.
Light Mineral Oil	11 gal.

CLEANERS AND SOAPS

Coloring Soap and Other Products

Color lends a greater sales appeal to practically every product. The purpose of coloring soap and soap products, therefore, is to give them greater sales appeal, and not to cover any defects, as many people think. The soap maker, today, not only manufactures soaps and shampoos, but also bath salts, deodorizing blocks, sweeping compounds, floor oils, floor waxes, and emulsions. In this chapter it is intended to consider all these products and present those dyes which have been found best suited for the particular purposes.

Before the discovery of aniline dyes and their development during the past forty years, the coloring and dyeing of all materials was done by vegetable dyes or mineral colors. With the advent of the aniline dyes, the vegetable dyes and the mineral colors were for the most part discarded. Today, there are very few vegetable colors used in any industry, let alone soap. There will be discussed in this article the few vegetable dyes and mineral colors that are still being used and also the coal tar dyes, their advantages and their disadvantages.

The vegetable colors will first be considered. The one which enjoys the greatest use is chlorophyll. Chlorophyll is a green color and is extracted from the green leaves of plants such as stinging nettles or spinach. It comes in three types—water soluble, alcohol soluble, and oil soluble. The commercial product in each case is a liquid or a paste. Of the three, the oil soluble chlorophyll enjoys the greatest sale. It is used particularly for the coloring of olive oil soaps such as silk boil-off, and automobile soaps. It is also used in the coloring of cosmetic creams, paraffine oils and waxes. Contrary to general belief, chlorophyll is not very fast. Liquid soap colored with chlorophyll will not hold its color when packed in tin containers. The light fastness of chlorophyll in certain solutions is poor. The coloring cost, when compared to a good aniline dye is excessive. There is a coal tar substitute which gives the exact shade of chlorophyll, is faster than chlorophyll, and has greater coloring power. It comes in a water, alcohol, or oil soluble form.

Alkinet and Bixin, yellow and red vegetable colors, are still being used somewhat for the coloring of cheese and butter, but they find no use in the coloring of soaps due to their low strength.

Carmine may be mentioned at this point although it is not a vegetable color. It is made from the dried bodies of the female insect coccus cacti, an insect which lives on certain cactus plants in Mexico, Central America, and South America. At one time, carmine was used extensively. Its use today as a red color is limited.

Of the mineral colors, the two most important are ultramarine blue and vermilion. Ultramarine blue is used in the manufacture of blue mottled laundry soap. Vermilion red is used in the manufacture of a red mottled soap. Chrome green has been used for the coloring of soap but it has been largely replaced by aniline dyes. Spanish oxide or iron oxide is used in some cases for sweeping compounds.

Next to be considered are the coal tar derivatives or what are commonly called aniline dyes. The analine dyes of commerce are generally powders or crystals. When considered in light of their solvents, they may be divided into these three groups: 1—water soluble, 2—spirit soluble, 3—oil and fat soluble.

Not all of the aniline dyes are fast. For most soaps and sanitary products, it is essential to have a dye or color that is fast to alkali and light. For some products, alkali fastness is essential and light fastness may be disregarded. As a general rule, the water soluble and oil soluble colors have good fastness to both alkali and light. The spirit soluble colors, with one or two exceptions, do not have good fastness to alkali and light.

It has been mentioned that water, alcohol, and oil soluble colors are used. It might be advisable, therefore, to consider the preparation of each of these types of colors for use.

Water Soluble Colors: Dissolve the color in the proportion of two to three ounces of color to a gallon of hot water. Do not use a tin or an iron container as a chemical reaction will set up and will tend to decrease or destroy the coloring power. Filter to insure that all the color is dissolved. Undissolved particles of color cause spots and blotches. It is not necessary to make fresh color solutions each time. The color must be stirred if it has not been used in some time as some dyes have a tendency to settle out of the color solution.

Alcohol Soluble Colors: Disssolve from two to three ounces of color to a gallon of alcohol. Filter to insure the absence of undissolved color. Alcohol soluble colors are also soluble in acetone, ethyl acetate, and some lacquers. They are also soluble in perfume oils.

Oil Soluble Colors: These are soluble in all vegetable and mineral oils and waxes, fats, oleic and stearic acid, fatty acids of all kinds, paradichlorbenzol, perfume oils, ethyl acetate, toluol and lacquers. When dissolving the color in oils, waxes, or fatty acids, heat should be employed to get the full solution of the color.

It is important to remember that soap and soap products are merely colored and not dyed, i.e. the color does not combine chemically. Consequently, if too much color is used, the color will bleed out of the product.

Now, the various types of materials to be colored will be considered in regard to the dyes best suited for the purposes.

Milled Soaps

The primary requisite for colors here is light fastness. Water, alcohol, and even oil soluble colors are used. The water soluble colors are recommended as best for general use on milled soaps. The alcohol soluble colors have a tendency to blister. The oil soluble colors are incorporated in very few types of milled soaps. The recommended dyes are:

Pink	—Rhodamine B Extra
Salmon Pink	—Rhodamine 6G Extra
Green	—Cyanine Green
Golden Yellow	—Metanil
Blue	—Alizarine Blue
Red	—Cloth Red
Amber	—Bismarck Brown
Lemon	—Fluorescin
Canary Yellow	—Fast Light Yellow
Heliotrope	—Violamine
Violet	—Acid Violet

These colors are all water soluble and, with the exception of the Rhodamine and the Bismarck Brown, will mix with one another to give any shade desired. They will have good fastness. For a two hundred pound batch of soap, you will require one-sixteenth of an ounce of Rhodamine B Extra; one-half ounce of Fluorescein; one-half ounce of Violamine. All the other colors require one ounce per two hundred pounds of soap. Where very delicate shades are required, these proportions can be cut down one-third. The color, in liquid form, is best added in the amalgamator. If spots or blotches form, it is a sign that some of the color was undissolved. It is, therefore, advisable to make sure that the solution is clear.

Cold, Half-Boiled, and Boiled Soaps

The requirements for a color for these types of soaps are fastness to light, alkali, and heat. Water or oil soluble colors are used. Where a water soluble color is used, the color is dissolved in water. Where the oil soluble color is used, the color is dissolved in some of the oil or fat. In the cold soaps, the water soluble color is added in liquid form after saponification has started. In figged soaps, the color is crutched in after saponification is completed. Never add dry dye to the soap mass or to the lye. Dry dye causes spots and blotches. The water soluble colors recommended are:

Pink	—Rhodamine B Extra
Salmon Pink	—Rhodamine 6G Extra
Green	—Cyanine Green
Golden Yellow	—Metanil
Blue	—Alizarine Blue
Red	—Cloth Red
Amber	—Bismarck Brown
Lemon	—Fluorescein
Canary Yellow	—Fast Light Yellow
Heliotrope	—Violamine
Violet	—Acid Violet

The oil soluble colors recommended are:

Green	—Alizarine Green Oil soluble plus Azo Yellow
Amber	—Azo Amber

Soap Bases and Liquid Soaps

The primary requisites here are fastness to light, alkali, and contact with tin. Water soluble colors are used for the liquid soaps. In the soap bases, the oil soluble colors may be used. The same colors listed for milled soaps and cold soaps can be used here. One exception, however, is Naphthol Green. Naphthol Green is used by many soap makers to get a leaf green shade. In many cases where liquid soap or soap bases have been colored with Naphthol Green and packed in tin containers, the color has faded out due to the chemical action resulting between the dye, the soap, and the tin. A recommended substitute consists of Cyanine Green and Tartrazine. The most popular shades and the colors used to obtain them are:

Pink	—Rhodamine B
Yellow	—Metanil Yellow
Blue	—Alizarine Blue
Amber	—Bismarck Brown
Opal	—Fluorescein
Strawberry	—Rhodamine B and Bismarck Brown

Boiled Automobile and Silk Boil-Off Soaps

These products are colored either with chlorophyll or with a water or an oil soluble aniline dye. The replacement of the chlorophyll with aniline dye due to the excessive coloring cost of the chlorophyll is recommended. The dye is added after saponification is completed and before the soap is settled. A pound colors 4500 pounds of soap.

Laundry Soap

Where it is desired to give a laundry soap a deeper tone or a more brownish cast, an oil-soluble amber is recommended, as it will not stain the clothing. For laundry powders which are already manufactured and which must be made darker in tone, or browner, a water soluble dye can be used. Do not use more than one pound of color for 16,000 pounds of soap, otherwise the dye will stain the clothing.

Medicated Soaps

These are generally colored red. Either a water soluble or an oil soluble red is used. The water soluble dye is a Rhodamine derivative. The oil dye is an Azo compound. The oil soluble dye is soluble in the cresylic acid. A pound of dye colors 2500 pounds of soap. These two colors can be used in milled, cold, or semicold medicated soaps.

Bath Salts

The requisites for colors for bath salts are fastness to alkali and light. There are two ways of coloring bath salts. One is to get the color and odor combined and use the proportions recommended by the manufacturer, generally a pound to one hundred to two hundred pounds of bath salts. For the small manufacturer this is the most practical and most convenient method. The other method is to use water or alcohol soluble colors, and add the perfume afterwards. When water soluble colors are used, the solution is made as concentrated as possible. Color some of the salt very heavily and then mix this up with the rest of the salt. This will minimize the water used. Add the perfume and then tumble or mix. The colors recommended are the water colors given at the end of this article.

Emulsions

There are two kinds of emulsions. 1—oil in water emulsions, 2—water in oil emulsions. The oil in water emulsions are best colored with water soluble dyes. The water which is used in the emulsions is first colored. If the emulsion is to be colored after completion, the color is dissolved in as little water as possible and the concentrated dye solution is added to the emulsion and stirred vigorously. The following colors are recommended. The proportions are anywhere from one pound to four hundred gallons, to one pound to twelve hundred gallons, depending upon the depth of shade desired.

Pink	—Rhodamine B Extra
Green	—Cyanine Green
Golden Yellow	—Metanil Yellow
Canary Yellow	—Tartrazine
Blue	—Alizarine Blue
Red	—Cloth Red
Heliotrope	—Violamine
Opal	—Fluorescein
Black	—Nigrosine

Water in oil emulsions are best colored with oil soluble colors. The colors are dissolved in the oils before emulsification. The colors recommended are:

Yellow	—Azo Oil Yellow
Red	—Azo Oil Red
Black	—Oil Black
Orange	—Azo Oil Orange
Blue	—Alizarine Oil Blue
Violet	—Alizarine Oil Violet
Green	—Oil Green

With the exception of the black, one pound colors 200 gallons. One pound of the black colors 50 gallons.

Wax Emulsions

Wax emulsions, wax polishes and wax pastes are colored with oil soluble colors. The same colors as for the water in oil emulsions are recommended. In the case of shoe polish pastes, both the water and oil soluble colors are used. The water soluble color is dissolved in the water and the oil soluble color is dissolved in the wax before emulsification takes place. In the case of the wax floor pastes, the color is added to the solvent for the wax.

Cosmetic Creams and Lotions
See Emulsions.

Nail Polishes

Either basic dyes soluble in acetone and ethyl acetate are used, or oil soluble dyes. The basic dyes generally used are Rhodamine B (Pink), Safranine Y (Red). They are used alone or mixed with Auramine (Yellow) or Chrysoidine (Orange) to give all desired shades. The oil reds given below are used, or shaded with oil yellow.

Deodorizing Blocks

These products are colored in the same manner as bath salts. Buy combinations of colors and perfumes and use the proportions recommended, or use the oil soluble colors mentioned at the end of the article. If the blocks are molded, dissolve the color in the molten paradichlorbenzene. If the blocks are pressed, dissolve the color in the perfume oils and spray.

Mineral Oils

Brilliantines and mineral oils are colored with oil soluble colors given at the end of this article. A pound of color generally colors 1600 gallons.

Washing Powders

Dish washing and cleaning compounds made from trisodium phosphate modified soda, soda ash, or combinations of same are generally colored with water soluble Fluorescein. The proportion used is one pound of color to 1250 pounds of compound. The color of the dyed compound is peach. When dissolved in water, a greenish fluorescence is given. The use of this color in the cleaning compounds is covered by patents. For coloring washing compounds made from the above chemicals where a fluorescence is not desired, use any of the water colors given at the end of this article.

Sweeping Compounds

Water or oil soluble red or green is used. The water soluble green is Malachite Green. The water soluble red is Croceine Scarlet. The best oil soluble green is an Alizarine Oil Green, the red is an Azo Oil Red. Where the water soluble colors are used, the color is dissolved in water and the sawdust is colored first. Then the oil and sand are added afterwards. If the oil soluble colors are used, the oil is colored first and then mixed thoroughly with the sawdust and the sand

is added afterwards. The whole mass is then thoroughly mixed.

Washing and Bluing Powders or Tablets

These tablets may be made of a soap or cleansing agent like Trisodium Phosphate Bluing. Three kinds of bluing may be used. 1—Ultramarine Blue, 2—Soluble Prussian Blue, 3—Aniline Blue. The Aniline Blue is the best to use as it gives a more attractive finish. The Prussian Blue may cause rust spots when the laundered material is ironed. The proportion of the Aniline Blue is one pound to 2000 pounds of compound.

Glycerin Anti-Freeze Mixtures

These products are colored with basic colors. The shades used are a scarlet which is a mixture of Safranine and Phosphine, a green which is a mixture of Malachite Green and Auramine, a blue which is Methylene Blue zinc free, or Victoria Blue. The proportions are one pound to 1500 to 3000 gallons depending upon the depth of shade desired.

Silicate of Soda Compounds

Three coloring media are used. Two are dyes and one is an indicator. The indicator is Phenolphthalein. It gives a color from a pale pink to a deep wine color depending upon the amount of indicator added. The others are Fluorescein which gives a yellow with a greenish fluorescence, and Eosine which gives a red with a yellowish fluorescence. In both cases, the water soluble Fluorescein and Eosine should be used. The silicate will act as a solvent. Stir thoroughly to insure perfect solution.

In closing, we list the colors recommended for quick references.

Water Soluble Dyes

These can be used for coloring milled, cold, semi-boiled soaps, liquid soaps and bases, shampoos, toilet waters, bath salts and emulsions.

Pink	—Rhodamine B Extra
Salmon Pink	—Rhodamine 6G Extra
Green	—Cyanine Green
Golden Yellow	—Metanil
Blue	—Alizarine Blue
Red	—Cloth Red
Amber	—Bismarck Brown
Lemon	—Fluorescein
Canary Yellow	—Tartrazine
Heliotrope	—Violamine
Violet	—Alizarine Violet

Alcohol Soluble Dyes

These can be used for coloring nail polishes, anti-freeze glycerin, denatured alcohol, and shellac.

Pink —Rhodamine B
Red —Safranine Y
Blue —Methylene Blue ZF
Violet —Methyl Violet
Green —Malachite Green
Yellow —Auramine
Black —Nigrosine
Orange —Chrysoidine
Brown —Bismarck Brown

Oil Soluble Dyes

They can be used for coloring emulsions, nail polishes, waxes, wax pastes, oleic and stearic acid, lacquers, acetone, toluol, cello-solve, creams, mineral oils, petrolatum, paradichlorbenzene and orthodichlorbenzene.

Red —Azo Oil Red
Yellow —Azo Oil Yellow
Orange —Azo Oil Orange
Black —Oil Black
Blue —Alizarine Oil Blue
Violet —Alizarine Oil Violet
Green —Oil Green
Brown }
Amber } —mixtures of the above

Alpine Soap

Cocoanut Oil	25 kg.
Castor Oil	5 kg.
Caustic Soda (38° Bé.)	15 kg.
Oil of Lemon	25 g.
Oil of Peppermint	55 g.
Oil of Rosemary	44 g.
Oil of Carroway	30 g.

The oils are saponified with the caustic soda and then the essential oils are added. The soap should be colored a very light green.

Camphor Soap No. 1

Cocoanut Oil	10 kg.
Caustic Soda (40° Bé.)	5 kg.
Camphor	1½ kg.
Alcohol	Sufficient

After the saponification the camphor is introduced dissolved in the smallest quantity of alcohol possible.

Camphor Soap No. 2

White Soap	20 kg.
Camphor	1 kg.
Oil of Sweet Almonds	300 g.
Oil of Rosemary	1 kg.

Carbolic Acid Soap

Cocoanut Oil	10 kg.
Caustic Soda (40° Bé.)	5 kg.
Carbolic Acid	½ kg.

The carbolic acid is added after the oil is saponified.

Sulphur Soap

Cocoanut Oil	10 kg.
Caustic Soda (40° Bé.)	5 kg.
Sulphur Flowers	2 kg.

Sulphur Soap
U. S. Patent 1,957,918

A substantially odorless sulphur soap comprise approximately one part sulphur, one part resin soap, one part of a two to one mixture of iodized starch and an oily or waxy body such as vaseline, and about ten parts of an alkali soap.

Sulphur-Tar Soap

Cocoanut Oil	10 kg.
Coal Tar	2 kg.
Caustic Soda (40° Bé.)	5½ kg.
Sulphur Flowers	2 kg.

The cocoanut oil and the coal tar are first melted together. After cooling, the mixture is saponified with the caustic soda and then the sulphur is added with constant stirring.

Vaseline Tar Soap (Patented)

Cocoanut Oil	10 kg.
Vaseline	1½ kg.
Coal Tar	2 kg.
Caustic Soda (40° Bé.)	5½ kg.

The cocoanut oil, vaseline and coal tar are melted together and after cooling are saponified with the caustic soda.

Tar Soap

Cocoanut Oil	10 kg.
Coal Tar	2 kg.
Caustic Soda (40° Bé.)	5½ kg.

The cocoanut oil and the coal tar are heated together and the mixture is saponified with caustic soda when the temperature reaches 38° C.

Iodine Soap

Cocoanut Oil	10 kg.
Caustic Soda (40° Bé.)	5 kg.
Potassium Iodide	1½ kg.
Water	2 l.

The potassium iodide, dissolved in the water, is added with constant stirring after the oil and soda had been saponified.

White's Hand Soap

Powdered Pumice Stone	4½	lb.
Green Soap	1¼	lb.
Potassium Carbonate	280	gr.
Glycerin	2	fl. oz.
Water	26	fl. oz.

Dissolve the soap and potassium carbonate in the water by the aid of heat; add the glycerin, and rub this solution up well in a mortar with the pumice stone until it is of a paste-like consistency.

Pumice Soap

Cocoanut Oil	10 kg.
Caustic Soda (28° Bé.)	10 kg.
Powdered Pumice	15 kg.
Oil of Lavender	50 g.
Oil of Thyme	20 g.

The cocoanut oil and caustic soda are first saponified and then the other ingredients are added.

Thum's Grit Soap for Surgeons

Cottonseed Oil	500 cc.
Stearic Acid	500 g.
Sodium Hydroxide	150 g.
Alcohol	150 cc.
Sodium Chloride Sol. 20%	sufficient
Distilled Water	sufficient
Powdered Pumice	300 g.

Heat together the cottonseed oil and stearic acid until the latter is completely dissolved. Then add the sodium hydroxide, dissolved in a liter of distilled water, and heat for fifteen minutes with constant stirring. Next add the alcohol and stir until saponification is effected. This is shown by the mixture becoming homogeneous in a few minutes. Then add one liter of a 20% aqueous solution of sodium chloride, and stir vigorously. Allow this to stand until the soap is hardened. The alkaline liquid which remains at the bottom of the container is then drained out through a hole punched in the soap mass on one side. The mass is then washed two or three times with distilled water, melted, and while still on the fire the powdered pumice is thoroughly incorporated. While still hot it is poured into suitable molds. In 24 hours the soap is hard enough to use.

Antiseptic Soap

"Titrol" is described as a transparent, non-poisonous, glycerine soap having high germicidal potency by reason of its "Titrol" content. Manufacture of the soap is simple, being carried out in a jacketed pan, to the following formula:

Tallow	305 lb.
Resin	35 lb.
Cocoanut Oil	400 lb.
Spirit	48 gal.
Caustic Soda at 38° Bé.	415 lb.
Glycerine	120 lb.
"Ti Tree" Oil	3 %

Method of working is to heat the tallow and resin in the pan until the latter is completely dissolved, then add the cocoanut oil and glycerine, and bring the temperature to 140° F. Add the caustic solution (which must be cold), crutching meanwhile, and continue the crutching from ten to fifteen minutes. Next add the spirit, when saponification will proceed rapidly. Continue the crutching until saponification is complete, which should take thirty or more minutes.

At this stage the soap should be fairly strong in alkali in order to counteract the effects of the "Ti-Tree Oil," which should not be added to the pan. After the addition of this the soap should be tested and the amount of free alkali brought down to not more than the .05% by the addition of the necessary amount of cocoanut oil. The addition of a certain quantity of water will now probably be necessary in order to make the soap transparent and free from cloudiness. Water must be added gradually—not more than three gallons at a time to this charge. Test soap by cooling a portion on a metal plate.

Antiseptic Soap

Sixty kilograms cocoanut oil, 15 kilograms peanut oil, 25 kilograms light colored red oil, saponified with approximately 46 kilograms caustic potash solution, 50° Bé. neutralized with 3 kilograms Turkey red oil, (or possibly still more red oil) dissolved in 300 kilograms water, or more, but not over 400 kilograms.

Let stand a few days and filter. Then add 10 kilograms glycerine, and 1 kilogram potash thoroughly stirred and dissolved in 10 kilograms of distilled water; but before filtering, add approximately 3 kilograms soap perfume oil, mixing thoroughly.

Add 2.60 kilograms benzyl-sodium paraoxybenzoate dissolved in 1.30 kilograms of warm water, 1.40 kilograms ethyl-sodium paraoxybenzoate dissolved in 0.70 kilograms warm water, 2 kilograms methyl paraoxybenzoate dissolved in 10 kilograms alcohol.

Add each solution separately.

The disinfectant value amounts to about 40 parts carbolic acid per 60 parts soap. The quantities added may therefore be smaller, if desired.

An antiseptic cream soap is obtained by adding to the finished soap cream the required amounts of esters, in solution, and working same in thoroughly.

A solid antiseptic soap is obtained by drying an ordinary primary toilet soap rather more than it would be dried for ordinary soaps (say with from 80 to 83–85% fatty acid), and then milling the paraoxy benzoic ester solution (or solutions) in with it. After mixing, and after passing once through the mill, further additions may be milled in. Now, in order for the solutions to be distributed evenly throughout the soap in all cases, the ester solutions should be treated by themselves. In certain cases, where' the operations are carefully and skillfully carried out, the ester solution and the other added materials can, perhaps, be worked in at one and the same time.

Davis' Liquid Antiseptic Soap

Cottonseed Oil	300.0 g.
Alcohol	200.0 cc.
Water	450.0 cc.
Sodium Hydroxide	45.0 g.
Potassium Carbonate	10.0 g.
Ether	15.0 cc.
Liquefied Phenol	25.0 cc.

Add a mixture of 100 cc. water and 200 cc. alcohol to the oil, mix thoroughly. Dissolve the sodium hydroxide and the potassium carbonate in 325 cc. of water; add this solution to the oil mixture. Lastly add the ether and phenol.

Disinfectant Soaps
British Patent 407,039
Formula No. 1

Sulphonated Palmitic Acid	99 parts
Mercuric Chloride	1 part
Soda Ash to neutralize	Sufficient

Formula No. 2
British Patent 407,309

Hydroxyethane sulphonic acid oleic acid ester	48 parts
Soda ash to neutralize	Sufficient
Ammonium sulfate	20 parts
Water	19 parts
Mercuric Chloride	1 parts

Hexalin Soaps

A few of the better recipes for Hexalin soaps are as follows: 1.25 kilograms of linseed or soy bean oil, 10 kilograms of caustic potash 50° Bé., 12 kilograms of hexalin, 53 kilograms of water. This soap remains liquid.

I—25 kilograms of crude lauric acid, 13 kilograms of caustic potash 50° Bé., 25–40 kilograms of hexalin, 35–25 kilograms of water. By adding silicate, calcified soda ash, and caustic soda, syruplike soaps may be prepared. More caustic soda must be added to give products resembling soft soap.

II—40 kilograms of crude lauric acid, 20 kilograms of caustic potash, 8 kilograms of hexalin, 32 kilograms of benzine or similar solvent.

III—50 kilograms of coconut oil, 25 kilograms of caustic soda 38° Bé., 6 kilograms of hexalin, 20 kilograms of quartz sand, 100 grams of ultramarine blue. Add bitter almond oil or terpineol for scent.

Palm Oil Soap
Belgium Patent 391,389

Soap is made from palm oil by cooking the same at a temperature of 150 degrees C. and then allowing the mass to stand, and mixing with an equal volume of 18 degrees Bé. sodium hydroxide at a temperature of 60 degrees C. Sodium hydroxide solution, which contains ultramarine, is added to color the product.

Castor Oil Soap

Castor oil soap is readily prepared as follows: Solution of potassium hydroxide (80 per cent), 2 fl. oz; industrial alcohol, 1 fl. oz.; castor oil, 3½ oz. by weight; mix in a conical beaker and set aside in a warm place; in about ten minutes a yellow transparent jelly of castor oil soap is produced.

Liquor Cresol Saponatus

The official formula contains a great excess of soap, and a good preparation can be made by shaking 5 fluid ounces of cresol with the above quantity of castor oil soap and adding sufficient water to produce to 10 fluid ounces Liquid Castor Oil Soap. A syrupy liquid, which contains about 66 per cent of soap, and remains liquid on keeping, can be made by mixing the above quantity of castor oil soap with sufficient water to produce 7½ fluid ounces "Ether Soap."—Liquid castor oil soap 4 fluid ounces, Industrial alcohol 1 fluid ounce, ether, 5 fluid ounces. Liniment of Turpentine, Modified.—Liquid castor oil soap 2½ fluid ounces, camphor ½ ounce, oil of turpentine 13 fluid ounces, water to make 20 fluid ounces. Mix the liquid soap with an equal volume of water, add 1 fluid ounce of the turpentine and shake to emulsify. Dissolve the camphor in the rest of the turpentine, and add this, 1 or 2 fluid ounces at a time, to the primary emulsion, shaking well after each addition. Add sufficient water to produce 20 fluid ounces.

Translucent Cocoanut Oil Soap

Cocoanut Oil	20	kg.
Caustic Soda (36° Bé.)	10	kg.
Caustic Potash (30° Bé.)	1½	kg.
Oil of Anise	60	g.
Oil of Peppermint	40	g.

The cocoanut oil is first saponified with the caustic soda and then the caustic potash is added with constant stirring. The mixture is heated gently until clear. After about an hour the kettle is uncovered and the essential oils are incorporated.

Castile Soap

For 1 ton of olive oil some 350 or 360 gallons of lye is required. Put the oil into the pan with 150 gallons lye at 18° Tw., and boil with free steam for two or three hours, and then throw in sufficient quantity of salt to separate the spent lye. Turn off steam and let it rest for four or five hours. Draw off the spent lye and boil up again, at the same time adding 100 gal. lye at 32° Tw. Light the fire and boil four or five hours, taking care the soap does not boil over the pan. This done, again separate with salt. Allow to settle four or five hours; draw off the spent lye and boil again with 110 gallons of lye at 42° Tw.

Castile Soap

	Formula No. 1	Formula No. 2
Olive oil	40 parts	30 parts
Ground nut oil	30 parts	— parts
Cottonseed oil	— parts	30 parts
Tallow oil	30 parts	40 parts

Formula No. 3

Olive Oil	30 parts
Lard	30 parts
Palm kernel oil	40 parts

Formula No. 4

Palm oil (bleached)	50 parts
Sesame oil	20 parts
Tallow	30 parts

Lye as above in proportion.

Static Destroying Soap

22 pounds of white curd soap (containing 70 per cent of fat) are dissolved in water, magnesium chloride or sulphate being added to the solution so long as a precipitate of magnesium oleate (Magnesia soap) continues to form. This latter is purified with boiling water, dried, melted at 130° C., treated with 15 pounds of cold petroleum, and dissolved in 20 gallons of benzine. One pound of the product is sufficient to protect 7,000 gallons of benzine from spontaneous ignition by electrical excitation.

Substantially Anhydrous Persalt Soap
U. S. Patent 1,940,570

A product readily soluble in water is prepared by mixing about 100 parts of highly split, saponifiable and saturated acids of nondrying oils such as coconut and tallow oils with 44 parts of calcined soda, permitting saponification by the spontaneous heat of reaction, stirring the mass until it becomes stiff, adding 15 parts sodium perborate and shaping the material as desired before it becomes cold and brittle.

Transparent Soap
British Patent 342,400
Formula No. 1

A batch of fatty matter consisting of 61 kilograms of tallow, 18 kilograms of coconut oil, 11 kilograms of castor oil and 10 kilograms of resin is after careful purification saponified with 45.8 kilograms of purified soda lye of 39° Bé. The transparent, hot, almost neutral soap paste is thereupon applied to cooled rollers in such a manner, that the temperature in the course of few (2–3) seconds is reduced from 90–100° C. to about 20° C. The ribbons or plates are scented, milled once or twice on cold rollers and are thereupon with a fat content of about 71 per cent passed through a slowly operating press or plodder, the head of which is cooled. The string of soap is cut up and pieces of suitable size and shape are formed from the same. The soap obtained is extremely transparent and is immediately in transportable state. When stored the soap will dry without altering its shape. The soap has a final fat content of about 73–75 per cent.

British Patent 342,400
Formula No. 2

A batch of fatty material consisting of 92 kilograms of crude palm oil and 8 kilograms of resin is purified, completely saponified with 42.9 kilograms of soda lye of 38° Bé. and subjected to a further treatment as described above.

The transparent soaps obtained are non-filled, milled toilet soaps of high quality and glass-like transparency. On account of their high percentage of fat, usually amounting to 71–75 per cent, they are very economical in use and are not spent as quickly as the transparent soaps hitherto known. The novel soaps are agreeable to the skin and may be produced at low cost. The advantages of the novel process reside in the shortening of the time required for the production and in the fact that no storing time is needed. The soaps may, therefore, be brought on the market immediately after the same have been manufactured. In contrast thereto, the hitherto known low-percentage transparent soaps have to be stored for a considerable time before the desired high degree of transparency is obtained. The novel soaps may be scented in the same manner as ordinary toilet soaps.

Composition for Prevention of Rancidity in Soap

I.	Anhydrous Lanolin	690 g.
	Paraffin wax 52/54 degrees m.p	314 g.
	Beeswax, preferably white	200 g.
	White oil or paraffin oil	390 g.
II.	Boiling water	200 g.
	Pulverized borax	16 g.
III.	Sodium thiosulphate crystals	600 g.
	Water	200 g.
IV.	Water glass (Sodium silicate)	390 g.

In making the preparation, the mixture of fats (I) is first melted at a temperature of 85 degrees C. While the molten mixture is kept at this temperature, the hot solution (II) is allowed to flow into the mixture in a thin stream and the mixture is then allowed to boil for a few minutes. Care must be taken in this operation, as the mass increases markedly in volume and will overflow the vessel if the latter is not large enough. The mixture is then allowed to stand for some time and agitated now and then. When the mass has cooled down to a temperature of approximately 50 degrees C., the warm solution (III) which has a temperature of approximately 55 to 60 degrees C. is added and mixed thoroughly with the other ingredients, until a homogeneous solution is obtained. Then solution (IV) containing the sodium silicate is added and also stirred with the other ingredients. The final product is a salve-like mass, perfectly homogeneous, which is added to the soap in the proportion of one per cent, that is one kilogram of the mixture is added per hundred kilograms of soap.

Hard Coconut Oil Soap

French Patent 749,751

The soap contains water 280 liters, coconut oil 550, soda 100 and talc 5 parts.

Persalt Soaps

British Patent 381,211

Stable and readily soluble soaps containing persalts are made by saponifying fatty acids at a low temperature with approximately but no more than double the amount of calcined soda necessary, and adding the persalt before the saponification is completed. Hydrocarbons insensitive to oxygen, and borax, sodium triphosphate, pyrophosphate, bicarbonate, or silicate may be added at the same time. Part of the soda may be replaced by potash or potassium carbonate to reduce brittleness, or liqued grain soap may be added to the product. The products may be powdered, be extruded to form threads, filaments, wool, or tiny needles, or be rolled to platelets or to films which are scraped off the rollers by knives or needles.

Examples: (1) 70 kilograms of split coconut oil fatty acid and 30 kilograms of split tallow oil fatty acid are saponified with 30 kilograms of calcined soda and 5 kilograms of potash or potassium carbonate at 35° C., just before the mass becomes viscous 12–13 kilograms of sodium perborate are added; when nearly cold the mass is rolled to platelets or strips. (2) 100 kilograms of a mixture of split fatty acids from hardened palm kernel and ground nut oils saponified at 35° C. with 33 kilograms of calcined soda and 10 kilograms of potash or potassium carbonate; just before the mass becomes solid a doughy mixture of 14 kilograms of sodium perborate with 14 kilograms of sodium silicate is added; when saponification is almost complete the mass is rolled to form strips and then pressed to rods from which platelets are scraped off by rotating knives. (3) 100 kilograms of distilled fatty acid mixed with 22 kilograms of benzine, xylene, or hydrogenated hydrocarbons are saponified as in example (1); 16 kilograms of sodium perborate are added.

Toilet Soap Powder
U. S. Patent 1,943,253

A toilet soap powder comprises finely divided high grade soap and pyrophyllite in the proportions of about 25% of soap to 75% of pyrophyllite, and both ingredients being in a finely subdivided state, the particles of pyrophyllite for the most part or entirely passing through a 100 mesh screen and the soap particles being of similar fineness and the particles of pyrophyllite and soap being thoroughly intermixed with each other, said soap particles dissolving almost instantaneously when the powder is added to water.

Transparent Milled Soap Chips

A batch of fatty matter consisting of 61 kilograms of tallow, 18 kilograms of coconut oil, 11 kilograms of castor oil and 10 kilograms of resin is after careful purification saponified with 45.8 kilograms of purified soda lye of 39° Bé. The transparent, hot, almost neutral soap paste is thereupon applied to cooled rollers in such a manner, that the temperature in the course of few (2-3) seconds is reduced from 90–100° C. to about 20° C. The ribbons or plates are scented, milled once or twice on cold rollers and are thereupon with a fat content of about 71 per cent passed through a slowly operating press or plodder, the head of which is cooled. The string of soap is cut up and pieces of suitable size and shape are formed from the same. The soap obtained is extremely transparent and is immediately in transportable state. When stored the soap will dry without altering its shape. The soap has a final fat content of about 73–75 per cent.

A batch of fatty material consisting of 92 kilograms of crude palm oil and 8 kilograms of resin is purified, completely saponified with 42.9 kilograms of soda lye of 38° Bé. and subjected to a further treatment as described above.

Soap Bubble Liquid

Castile Soap	28 g.
Glycerine	118 cc.
Distilled Water	236 cc.

Dissolve the soap in the water. Add the glycerine, and mix thoroughly. Allow to stand until the liquid clears at the bottom. Siphon off this liquid for use.

Cotton Scouring Soap

Edible Tallow	245 lb.
Caustic Soda Solution (30%)	10 gal.
Water	1,100 lb.

Melt the tallow, enter the caustic soda and boil 5 hours. Make up the volume with water. Used as a cheap soap for scouring cotton cloth.

Persil Soap Powder

Powdered Soap	20 parts
Sodium Carbonate	30 parts
Water Glass	7 parts
Sodium Perborate	10 parts

Soap Powder
British Patent 378,973

A solution of 1 pound of crystallized magnesium sulphate in a small quantity of water, 10 pounds of 75° Tw. sodium silicate, and 22.5 pounds of anhydrous sodium carbonate are well mixed into 50 pounds of soap in a pasty or molten condition; when the temperature is below 50° C., 9.5 pounds of sodium perborate are added, and the mixture reduced to powder.

Laundry Washing Aid
U. S. Patent 1,943,519

In the use of washing compounds the detergent properties are greatly increased if solid matter in a finely divided state be present in the composition in order that the same may assist in the mechanical removal of the dirt from the fabrics.

Any of the following different mixtures added to the laundry water give satisfactory results:

Formula No. 1

Bentonite	50 to 70 parts
Soda ash	30 to 50 parts

Formula No. 2

Bentonite	75 to 85 parts
Caustic soda	15 to 25 parts

Formula No. 3

Bentonite	75 to 90 parts
Tri-sodium phosphate	10 to 30 parts

Formula No. 4

Bentonite	75 to 85 parts
Sodium borate	15 to 25 parts

Sodium silicate may also be used in proportions ranging from 20% to 40%.

Laundry Soap

Manila Type Coconut Oil, or Coconut Oil Fatty Acids,	50 lb.
Stearic Acid, (single pressed)	450 lb.
Caustic Soda, (solid or flake)	75 lb.

The caustic soda is to be dissolved in sufficient water to make a solution of 15 degrees Baumé (1.116 specific gravity). The fatty material is placed in the soap kettle, the caustic soda solution is run in slowly, steam being turned into the coils or jacket of the kettle and the entire mass boiled vigorously. As the caustic soda solution is taken up by the fat, more and more is added until saponification is complete. The soap is then grained out with dry salt, re-boiled and finished as usual. If the resultant product is too hard it can be made softer by increasing the proportionate amount of coconut oil.

This formula is based upon production of about 800 pounds of finished soap containing 30% moisture. The amounts can be varied proportionately for any capacity charge.

Laundry Mixes
Formula 1
Dry Soap and Metasilicate Mixture

Where one mixture is used for break and suds:

Make up:

Soap	3 lb.
Sodium Metasilicate	2 lb.

Where different mixtures are used for break and suds:

Make up:

Break—	
Soap	3 lb.
Sodium Metasilicate	4 lb.
Suds—	
Soap	2 lb.
Sodium Metasilicate	1 lb.

Formula 2
Dry Soap and Metasilicate Used Separately

Break—use ⅓ pound Sodium Metasilicate per 100 pounds goods.

Suds—use a total of ⅙ pound Sodium Metasilicate per 100 pounds goods, using most Sodium Metasilicate the first suds and gradually diminish.

Formula 3
Soap and Sodium Metasilicate in Solution

Where one solution is used both for break and suds:

Make up:

Water	100 gal.
Soap	30 lb.
Sodium Metasilicate	15 lb.

Where different solutions are used for break and suds:

Make up:
Break—

Water	100 gal.
Soap	22½ lb.
Sodium Metasilicate	30 lb.
*Suds—	
Water	100 gal.
Soap	30 lb.
Sodium Metasilicate	15 lb.

*Where this solution is used for the break, ⅛ pound additional Sodium Metasilicate per 100 pounds goods should be used.

Laundry Bluing Composition
U. S. Patent 1,921,635

Level bluing of textile materials is obtained by using a solution of an ammonium salt, e.g., ammonium sulphate impregnated with about 0.8–2.0% of an acid blue dye (e.g., No. 707, Color Index).

Water Softener

Sodium metaphosphate and sodium metaborate prevent the formation of soap curds, and will even dissolve calcium and magnesium soaps. For this reason they may be used in laundries to turn out whiter clothes, with a saving in soap. They are excellent for rinsing the hair after it has been washed to remove any remaining soap, thus replacing the time-honored use of lemon juice for this purpose.

Silicates in Soap

The animal and vegetable fats and oils used in soap-making are combinations between fatty acids and glycerin. Soaps are essentially combinations between these fatty acids and alkali, either soda or potash. In boiled soaps the glycerin which is separated by the alkali is drawn off, and in cold process soaps it is retained in the soap. The kind of stock affects the character of the soap, and many different mixtures of grease stock are in use. Household soaps often contain rosin, which also combines with alkalies. All soaps contain water and carry certain amounts of other substances.

If a suitable amount of silicate of soda is properly incorporated in soap while it is being made, the silicate will readily mix in, and as the soap cools and stiffens the silicate stiffens up with it. The water which the silicate contains is merged with the water which the soap contains and as the soap dries out the whole mass becomes uniformly hardened. The silicate in the course of drying tends to become harder than the soap would naturally become, so that the use of silicate, as already stated, makes the soap firmer. This action makes it possible to produce soap with satisfactory working qualities from fats or formulae that would otherwise make too soft a soap.

Mixtures of soap and "N" Brand silicate, unless some additional alkali is provided, may become grainy. This additional alkali may be supplied in some cases by having the grained soap from the kettles finished "strong." Usually, however, a suitable amount of caustic soda lye is added to the silicate. This should be done at least 24 hours before the silicate is mixed with the soap. The amount of caustic thus added varies with the character of the soap stock and the judgment of the soapmaker. For example, one may add 1.85 pounds of solid caustic (containing 76% Na_2O), per hundred pounds of "N" Brand. This is equivalent to 1.41 parts of Na_2O and makes a silicate solution of a ratio approximately 1:2.8. This would be similar to diluting the "K" Brand already mentioned. In some kinds, more caustic is used, but it is best not to exceed 5.85 pounds per hundred pounds of "N" Brand. This amount is equivalent to 4.45 pounds of Na_2O, making a silicate of a ratio about 1:2.2. The caustic is added to the silicate in the form of a lye solution varying from 30° to 36° Baumé; the figures above are for the solid caustic.

The use of a more alkaline silicate than "N" Brand not only avoids the labor involved in mixing the caustic with the silicate, but insures that the proportions are correct. In addition, the chance that free caustic may get into the soap is eliminated. In such cases soapmakers may find it advantageous to use a silicate which is even more alkaline, that is to say, contains more sodium oxide per unit of silica, than those given above. In such cases there is greater danger that the caustic will damage things on which the soap is used unless an alkaline silicate is added direct.

In making cold-process and semi-boiled soaps, a suitable amount of additional caustic is provided in the formulae for the batches and none need be added to the silicate before using. If a more alkaline silicate is used, there should be a proportionate decrease in the amount of caustic.

Silicates in Boiled Soaps

The raw soap is made in the usual way, by boiling with weak lye ("killing change"), graining with brine and settling, boiling up with strong lye ("strengthening change"), graining with additional strong lye, and settling.

Rosin can be included if desired ("rosin change").

The details of the formulae used and of the process of handling vary greatly according to the stocks that are available and the results desired. The following gives one formula and process, which can be taken as typical:

Tallow	100 parts
Cottonseed Oil	30 parts

Boil with weak lye (8° to 10° Baumé) until the soap will not take up any more. Open with salt or brine. Settle and run off the settlings to be worked for glycerin. If rosin is to be used, next run in lye of 18° to 20° Baumé and boil up, adding the rosin and additional lye as may be required. Open with brine and run off the lye.

After the rosin change is run off, or, if no rosin is used, after the stock change, add lye for the strengthening change, at 13° to 14° Baumé if closed steam is used or 20° to 22° if open steam. Boil, adding fresh lye as required. A "head" or foam will rise on the surface of the soap; continue boiling until the foam disappears and the soap settles to a smaller space in the kettle. The soap should be in a pea grain, and lye thrown up by the boil should settle down quickly through it. When the soap reaches this point it has taken all the strength that it will combine with. Add water and boil with open steam; the soap will thin out. Boil until the soap is of uniform consistency, adding water gradually. When it reaches the point that the soap is thin and the lye will separate on a paddle it is done. Settle overnight and run off the lye under the soap into a tank for future use. Finish the soap in the usual way, taking care to leave enough free alkali to combine with the amount of "N" Brand silicate that is to be added. The soap at this stage should contain about 31% water.

The soap at a temperature of 185° to 200° F. is run into the crutchers. While crutching, from 20 to 25 pounds of "N" Brand silicate per 100 pounds of this raw soap usually can be mixed in. Larger amounts can sometimes be used. The silicate should be at a temperature of about 85° to 110° F. The soap is run from the crutchers to the frames at about 140° F. Frames are usually stripped after about forty-eight hours and the soap allowed to stand three days before cutting.

As previously mentioned, this is only one method of making boiled soap. Various factors, such as the kinds of fats and oils used, the conditions under which the caustic is added to and allowed to react with the silicate, the temperature of the soap from the kettle, the temperature of the silicate, and other details, affect the amount of silicate that can be incorporated in the soap. By careful manipulation, the maximum amount is much higher than the above-mentioned figures.

Silicates in Cold Process Soap

In the cold process of soap making, instead of prolonged boiling of the fats with weak lye, settling, reboiling, etc., as already outlined, the whole process is carried out in a crutcher, in a single operation and in a very short time. The exact quantities of fats, strong lye and other ingredients required are carefully weighed or measured out, and everything that goes into the crutcher remains in the soap. There is no spent lye to drain off and the glycerin is retained in the soap. The equipment required for the cold process is much less than for boiling, as no kettles are needed and no steam. It is hard to make a good, uniform soap by the cold process. Unless the mixing is very thorough, there will be spots containing excess alkali and others containing excess fats or oils.

Many formulae are in regular use, but the following may be taken as typical:

Formula 1

Tallow	75 lb.
Cocoanut Oil	25 lb.
Caustic Soda Lye (35.5° Baumé made of 76% Caustic)	75 lb.
"N" Silicate of Soda	125 lb.
Pearl Ash Lye (36° Baumé)	20 lb.

Formula 2

Tallow	75 lb.
Cocoanut Oil	25 lb.
Caustic Soda Lye (35.5° Baumé made of 76% Caustic)	70 lb.
"N" Silicate of Soda	100 lb.
Pearl Ash Lye (36° Baumé)	17 lb.

Refined cottonseed oil up to 30% can be substituted for an equal weight of tallow, if it is hard. If the tallow is soft or mixed with grease, use less oil. The soap will not be quite so hard and will take longer to harden, but it will be a good washing soap. In these formulae the amounts of caustic are calculated to include the proper excess for the silicate to take up.

Cold Process

Three weighing tanks are usually arranged; one to supply the exact amount of grease stock, another for the exact amount of lye, and the third for the silicate. With the silicate is mixed the pearl ash lye.

The whole amount of grease stock is first run into the crutcher. Its temperature should be about 145° to 150° F. in cold weather and 125° to 130° F. in summer. The crutcher is started and then the whole amount of the lye is quickly run into the grease and the mixture crutched rapidly until it begins to thicken up. This marks the beginning of the reaction between the fat and the alkali, and is accompanied by considerable heat. The whole amount of the mixed silicate and pearl ash lye is then run in rapidly. As this mixes with the soap the whole will thin out. The crutching is continued and in a few minutes the whole mass will gradually turn creamy. The whole process is a quick one, taking from ten to fifteen minutes. When the soap is thick enough for a mark made on it to remain, it is quickly dropped into a frame and the frame moved immediately to the spot where it is to stand to cool. The formation of the soap goes on to some extent in the frame while standing, and it is particularly important that the frame should not be moved or shaken until the soap is cold.

Semi-Boiled Process

This is another method by which soaps can be made. It is similar to the cold process just described, except that in this process, a higher temperature is reached. The various ingredients may be run into the mixer or crutcher at around 140° F. and the reactions will cause the temperature to rise nearly or quite to the boiling point. Or a steam-jacketed crutcher may be used. The ingredients are run in at the same temperature as for cold soaps, but when the reactions begin the temperature is raised by means of steam to about the boiling point and kept there until the soap is uniform.

This process has many of the advantages of the cold process and makes a better soap. The fats are more completely saponified, and the soap is more uniform. The time of operation is longer than for the cold soap. A typical formula should be as follows:

Tallow	315 lb.
Cocoanut Oil	55 lb.
Soda Lye, 35° Baumé	280 lb.
"N" Brand Silicate of Soda	185 lb.
Pearl Ash Lye, 32° Baumé	30 lb.

Warm the stock to 140° F. and mix in the other ingredients as described for cold process soaps. The mixture is then allowed to stand for one to one and a half hours, until it is observed to become heated. The crutching machine is then started slowly. When the materials have combined into a homogeneous mass, the soap is run into the frame. It should then be crutched or stirred by hand for fifteen to twenty minutes to avoid the formation of streaks.

Soaps Made from Fatty Acids

These are made by essentially the same processes as the other soaps. In making boiled soap from fatty acids, there is no separation of glycerin, as the glycerin has already been removed. But otherwise, the process is similar to that already described. Soda ash may be used instead of caustic soda lye, though care must then be taken to guard against excessive foaming from the carbon dioxide of the soda ash.

The details of fatty acid soap-making are varied, depending upon the stocks and processes, and the kind of soap to be made.

Liquid Soaps and Shampoos

Liquid soap for use in offices, etc. (14 per cent. fatty acid content) may be of similar composition to the following:

Coconut Oil (free fatty acid less than ½ per cent.)	126 lb.
Caustic Potash Liquor (38° Bé.)	90 lb.
Glycerin	17 lb.
Water	560 lb.

A liquid soap for workshop use is made by the following formula:

Coconut oil	220 lb.
Caustic Potash at 38° Bé.	157 lb.
Glycerin	26 lb.
Water	418 lb.

A good quality shampoo is made as follows:

Resin (finest W.W. grade)	1 lb.	3 oz.
Coconut Oil (refined)	2 lb.	8 oz.
Glycerin	16 lb.	4 oz.
Olive Oil	13 lb.	0 oz.
Caustic Potash Liquor (38° Bé.)	9 lb.	10 oz.
Water	57 lb.	8 oz.

This shampoo may be mentholated by adding the required amount of menthol dissolved in spirit, as before detailed.

A pine tar type shampoo is obtained by adding 2 per cent. of tar water, obtained as follows:

Take 250 grams wood tar with 15 grams sodium bicarbonate and 1,000 grams of water. Place in a vessel and keep at 35° C. to 40° C. for three hours, during which time the vessel is shaken frequently. The mixture is stored for three days and then filtered.

A brown tint is imparted to the shampoo by means of a suitable water-soluble dye.

The preparation of shampoos and liquid soaps is as follows:

The manufacture of shampoos and liquid soaps is best carried out in enamelled jacketed pots. Thirty pounds steam pressure is adequate. Stirring devices are not really essential. A clean wooden paddle will suffice.

The pot is charged with the required amounts of the selected oils and the contents raised to 160° F. to 180° F., when the saponifying agent (which is always caustic potash standardized at 38° Bé.) is added, whilst the contents are gently stirred. From twenty to thirty minutes suffice to complete saponification, then a sample is taken and tested carefully. It is essential that there shall be at least 0.05 per cent. free alkali present, expressed as Na_2O. Failure to observe this condition will result in imperfect clarification of the finished product and impair its lathering properties.

Upon the result of the test an addition of caustic potash liquor is added, if indicated. Should the free alkali lie between 0.05 and 0.15 per cent., it will be unnecessary to correct with oil.

The base will now be of a translucent pale yellow appearance of a fairly stiff consistency; this latter factor will be variable according to the amount of olive or palm oil used.

Distilled water at as high a temperature as possible is now added in the desired proportion, and the contents of the pot are gently stirred till solution is obtained. The glycerin is then added.

The product is now allowed to cool and the perfume and/or medicament incorporated with the rest.

The clarification of the shampoo or liquid soap is effected in a small diameter deep cylindrical enamelled vessel with a deep conical base. Fitted at the junction of the straight side of the cone is a tap to facilitate the withdrawal of the clarified liquid.

Too much importance cannot be placed on the need for effective clarification at as low a temperature as possible, and the deep conical bottom influence materially the rate of settlement of the suspended matter.

An addition of potassium carbonate of the order of one-half of 1 per cent. dissolved in distilled water will be found invaluable in preventing subsequent formation of cloudiness.

For large-scale production at least two clarification vessels should be employed, so that orders may be filled from the vessels in turn.

Automobile Cleaner

Kerosene	40 parts
Turkelene	10 parts

Mix together and while stirring with high speed stirrer, run in slowly

Water	100 parts

Where an abrasive is desirable add some infusorial earth or air-floated silica.

Cleaning Paste

Stearic Acid	100 lb.
Caustic soda solution 30° Bé.	54 lb.
Soda Ash	10 lb.
Water	836 lb.

Paint Cleaner

Kerosene	40 parts
Diglycol Stearate	10 parts
Water	60 parts

Heat together to 60–70° C. and stir with high speed mixer until emulsified. Continue stirring until temperature falls to 35° C. If a stronger cleaner is needed dissolve 5 parts trisodium phosphate in water used.

Paint Brush Cleaner

Kerosene	2 pints	Mix (1)
Oleic Acid	1 pint	
Strong liquid ammonia, 28%	¼ pint	Mix (2)
Denatured Alcohol	¼ pint	

Slowly stir 2 into 1 till a smooth mixture results. Directions: To clean brushes, pour into a can and stand the brushes in it over night. In the morning wash out with warm water.

Stain Remover, Mercurochrome

Syrupy Phosphoric Acid	5 parts
Ethyl Alcohol	95 parts

Iodine Stain Remover

Sodium Thiosulphate	10 parts
Water	90 parts

Wash the stained portion of fabric in the solution until the color has disappeared. Rinse thoroughly in water. Dry in the sun. This treatment will remove iodine stains from cotton, linen, rayon, silk, and woolen fabrics.

Dry Cleaning Composition
U. S. Patent 1,944,859

A composition useful for cleaning fabrics, clothing and the like comprising the following ingredients in the indicated proportions:

Tetrachlorethane	1%
Dichloroethyl Ether	3–5%
Carbon Tetrachloride	4–6%
Stoddard Solvent	90%

A Non-Inflammable Type Wash

	by volume
Naphtha (Distillation Range 200–300° F.)	40%
Carbon Tetrachloride	60%

Cleaning Fluid
U. S. Patent 1,921,054

A solution suitable for removing grease and dirt from cloth or leather, etc., comprises tertiary amyl alcohol 4%, ligroin (boiling point about 80–120°) 16% and carbon tetrachloride 80%.

Removing Perspiration Stains

Perspiration stains are removed from woolen and cotton goods with sodium hyposulphite, or a perborate of sodium solution, and subsequent washing with water; from silk and satin, also with dilute sodium hyposulphite solution, or, from silk with strong salt water in which the article is allowed to remain 3 to 4 hours. From worsted and cheviot garments perspiration stains are removed by brushing with benzine, and finally by washing with soap and water. If, as is frequently the case, the ground color of colored goods is injured by this process, it has to be remedied by re-dyeing.

Fabric-Cleaning Composition
U. S. Patent 1,943,519

Bentonite 50–70 is dispersed in water together with sodium carbonate 30–50 parts for maintaining the bentonite in a "highly fluid" condition beyond the maximum gel effect on the bentonite, and a small amount of soap also is used with this composition.

Composition for "Dry Cleaning"
U. S. Patent 1,944,859

Tetrachlorethane 1%, dichloroethyl ether 3–5%, carbon tetrachloride 4–6%

and ''Stoddard solvent'' 90% are used together.

To Remove Acriflavine Stains

Fresh stains may be removed from the skin quite readily by washing with soap and warm water. If stains are deep or have become dry, rub with a cloth dipped in a solution containing 1% of potassium permanganate and 1% of hydrochloric acid. After thorough washing with water the brown color remaining is removed by rinsing with acidulated 3% hydrogen peroxide solution. This treatment is, of course, irritating if there are any wounds or abrasions.

Ordinary laundering will remove the stains from wash goods. For non-wash goods, especially woolens, the permanganate method described above is suggested. The cloth should be rinsed thoroughly after dipping in each solution. Some dyes are altered or bleached. Therefore, the method should be first tried on an unexposed part of the cloth.

Dry-Cleaning or Dyeing of Fabrics, Etc.
British Patent 396,434

The original handle and appearance are preserved by using a solvent consisting of, e.g., paraffin wax, olive oil, or other non-volatile vegetable or mineral oil (one pound) dissolved in Stoddard solvent (gasoline, flash point higher than 38°) or carbon tetrachloride, trichlorethylene, etc. (5 gallons).

Cleaning Fluid (for Degreasing Textiles and Leather)
U. S. Patent 1,921,054

A mixture of tertiary-amyl alcohol 4 per cent by volume, ligroin (boiling point 80–120°) 16 per cent by volume, and carbon tetrachloride 80 per cent by volume is used.

Removing Iron Stains from Celanese
Method No. 1

The goods are soaked in a 20% solution of common salt for 1½ hours at 85° C. when the flattened fibers will swell out and the surface of the garment will regain a more uniform appearance.

Method No. 2

The goods are soaked in 2% acetic acid for ½ hour. Extract lightly and steam, brushing the proper direction of the fabric with a soft brush.

Rust Spot Remover for Silk and Other Fine Materials

Water	1 gal.
Gum Tragacanth	3 oz.
Potassium Bifluoride	7 oz.

Let gum and about 1 pint of the water soak overnight in an aluminum pot. Stir out to a homogeneous mass and add the remaining water in small portions, stirring each portion properly. Heat the mass on water-bath to about 150° F. and add the finely powdered potassium bifluoride. Stir until dissolved. This material should preferably be kept in aluminum tubes or receptacles. For removal of rust spots apply a few drops, rub lightly with the finger and the spot will disappear in a few minutes. Rinse in lukewarm water.

Cleansing Composition
German Patent 584,477

A composition for removing grease from household articles, etc., contains an alkali or alkaline earth acetate and a subordinate amount of silica or water glass with or without other usual components. A typical composition contains sodium acetate 25%, borax 20%, crystallized soda 50%, and water glass 5%.

Grease Remover

Carbon Tetrachloride	100 parts
Benzine	10 parts

Mixed Solvent for Removing Grease Spots
Formula 1

Ammonia Water	100 g.
Ether	300 g.
Alcohol	300 g.

Formula 2

Benzol	100 g.
Alcohol	60 g.
Ammonia	20 g.
Ether	100 g.
Oil of Turpentine	400 g.

Formula 3

Benzine	100 g.
Carbon Tetrachloride	100 g.

Cleaning Composition
Canadian Patent 335,693

A cleaning composition contains ethyl ether 16 fluid ounces, carbon tetrachloride 16 fluid ounces, finely ground pumice stone, talc or kieselguhr 1 pound, 28% ammonia water 8 fluid ounces, soap 5–10 ounces and water 29 gallons. This composition is used for cleaning rugs, carpets and the like.

Cleaning Composition
Swedish Patent 78,917

Add 420 grams of sodium carbonate, 210 grams of sodium bicarbonate, and 4.9 kilograms of soap cuttings to 22.5 liters of water. This is emulsified with 80 grams of petroleum and 80 grams of lubricating oil containing 10.2 kilograms of pulverized pumice in suspension. To this may be added 30 grams of bitter-almond oil and 10 grams of carmine.

Cleaning Composition
U. S. Patent 1,940,558

A non-saponaceous hand-cleaning composition, comprising a liquid mixture of toluol, butanol, butyl acetate, ethyl acetate, and one-half to one pound of lanolin per gallon of such liquid.

Soapless Hand Cleaner
U. S. Patent 1,940,558

Laquer Thinner	1 gal.
Lanolin	8–16 oz.

This removes lacquer from hands and leaves skin soft and pliable.

Dairy Utensil Cleaner
U. S. Patent 1,937,229

A mixture of 562 grams of sodium hypochlorite solution of specific gravity 1.125 and 250 grams of caustic soda solution of specific gravity 1.383 (about 11%) is prepared. 500 grams of this solution are admixed with 300 grams of sodium waterglass of 38–40° Bé. This forms a concentrated solution which can be diluted for use in cleaning or disinfecting.

For the purpose of disinfection the concentrated solution is employed in suitable dilution. For obtaining the best results, solutions which contain $\frac{1}{4}$ to 2% of the concentrated liquid can be employed.

Dairy Chlorine Rinse Calculations

To prepare 10 gallons of chlorine rinse containing 100 parts per million from a stock solution testing 2.5 per cent chlorine:

$$\frac{100 \text{ (parts per million)}}{2.5 \text{ (per cent)}} \times \frac{10 \text{ gallons}}{2.64 \text{ (factor)}} = 151.5 \text{ cubic centimeters of the stock solution to use}$$

The formula can be used to prepare different strengths of chlorine rinse and also varying amounts. The strength of the stock solution must be known.

Household Cleaner

Soap Powder	2%
Soda Ash	3%
Trisodium Phosphate	40%
Finely Ground Silica	55%

Mix well and put up in the usual containers.

Household Cleaner

Soap Flakes	4	oz.
Water	1½	qt.
Powdered Borax	3	oz.
Ammonia (Household)	4	oz.
Ether	1	oz.

Dissolve the soap flakes in boiling water then add the borax. Allow to cool, then add the ammonia and the ether. Set aside over night.

Wood work—use without water—apply with a soft rag.

Carpets and rugs—mix a pint of the cleaner with one gallon of slightly warm water and apply with a cheap scrubbing brush.

General Household Cleaning Mixtures

Dissolve one-half cupful of good white soap flakes in one-half cupful of boiling water. When cool add ten tablespoonfuls each of alcohol and glycerine, and 4 tablespoonfuls of chloroform or carbon tetrachloride. Use two tablespoonfuls of the mixture to a pint of water.

Window Cleaning Fluid

Strong Ammonia	
Water	2 tablespoonfuls
Whiting	4 tablespoonfuls
Alcohol (95% or	
denatured)	4 tablespoonfuls
Water to make one pint	

Apply very thinly to the glass. Allow to dry and rub off with a soft paper or cloth. It may be poured onto a wet rag and applied after most of the dirt has been washed off.

"Nitro" Solvent for Cleaning Guns

Amyl Acetate	4 oz.
Benzol	4 oz.
Motor Oil (S.A.E. 50)	8 oz.

Apply the cloth and pull through gun barrel, repeating with fresh cloth until the cloth comes through unstained. This leaves a thin coating of oil as a protective film in the gun barrel.

Cleaning Combs and Hairbrushes

Dip into cold water containing a tablespoonful of strong ammonia water per quart of water.

Toilet Bowl Cleaner

Potassium acid sulphate is one of the substances sold commercially for this purpose.

Solvent for Grease in Drain Pipes

Potassium Hydroxide, Dry Flake	99 parts
Aluminum, Fine Powder	1 part

Mix well and keep dry.

Drain Pipe Cleaner
U. S. Patent 1,938,560

In preparing the solvent a dry granular or flaked caustic soda, aluminum shavings, and dry finely divided aromatic organic materials are intermixed in desired proportions. The proportions of the several ingredients are subject to considerable variation, but highly satisfactory results are obtained by employing the following proportions by weight and which proportions are preferably used:

Caustic Soda	1 lb.
Aluminum Shavings	½ oz.
Aromatic Organic Materials	½ to 1 oz.

The several ingredients are intermixed dry and are suitably packaged to exclude moisture. The aromatic organic substance preferably employed comprises powdered spices, such for example as ginger, cinnamon, cloves and the like, which spices may be used separately or in various combinations.

By adding the aromatic organic matter of the kinds above specified to the caustic soda and aluminum, the chemical reactions set up on mixing the ingredients with water results in the ebullition or agitation and resultant foaming of the liquid being greatly augmented, and the volume of generated gases greatly increased, thereby increasing the effectiveness of the solvent in its action in cleaning out drain pipes, traps and the like; the ebullition and foaming imparting an effective physical aid to the solvent and heat actions on accumulations within a drain pipe or trap thereby quickening the action of the material in removing such accumulations.

An important result attained by the employment of the aromatic organic ingredient resides in the fact that the irritating, noxious and lethal fumes ordinarily resulting from the reaction of the caustic or aluminum will be rendered non-irritating, inoffensive to the sense of smell and non-poisonous.

Wall Paper Cleaner

Water	1	gal.
Salt	5	lb.
Aluminum Sulphate	4	oz.
Kerosene	4½	fl. oz.
Flour	9¾	lb.
Color and Perfume	Sufficient	

Use first grade, clear winter-wheat flour. Into a vessel that will hold three times the amount of water called for, put the water, salt, aluminum sulphate, and heat with occasional stirring to 180° F. Set off, and add the kerosene and scenting oil. Stir thoroughly. When it has cooled down to about 170° F. add the flour slowly, stirring steadily as it is sifted in to keep the mixture from lumping. If after adding the kerosene and scent, the temperature drops below 170° reheat to this point before adding the flour. In cold weather this last temperature should be 175° to overcome the coolness of the flour. If the batch turns out too short and is crumbly, increase the heat until the desired consistency is obtained. As flours differ in dryness, it may be necessary in some cases to use a little more water.

For coloring either of the above cleaners, you can use the following mixture for pink; Red aniline, 30 grains; glycerine, 1 dram and enough denatured alcohol to make 1 fluid ounce; for blue, use laundry bluing.

Cleansing composition for wallpaper, painted surfaces, etc. Ger. 584,312. Coarse rye flour is treated with two-thirds of its wt. of hot dil. aq. water glass soln.

Rug Cleaner

Trisodium Phosphate	1 oz.
Soft Water	1 gal.

The rug or upholstery should be sponged lightly with the liquid but should not be saturated. It is a very effective cleaner.

Cleaning Silver
German Patent 569,473

Silver articles are placed in an aluminum container with the following solution:

Sodium Bicarbonate	9.25 parts
Soap	0.50 part
Glucose	0.25 part
Water	1000.00 parts

Aluminum Cleaner
British Patent 379,152

Cleansing agents not attacking aluminum, tin, or their alloys are obtained by dissolving water-insoluble silica in a solution or melt of trisodium phosphate, and drying or solidifying the product, preferably by atomization to obtain it finely divided. Quartz, kieselguhr, and silicic acid gel are specified forms of silica. For example, 1,000 kilograms of trisodium phosphate is melted in its water of crystallization, and 100 kilograms of quartz or kieselguhr is dissolved therein with stirring; the product is atomized.

Aluminum Cleaner
U. S. Patent 1,935,834

Aluminum Sulphate	15-20%
Trisodium Phosphate	85-80%

Aluminum Cleaner
British Patent 390,751

Calcined Sodium Phosphate	80–70%
Sodium Metasilicate	20–30%

1–5% solutions of the above will not attack aluminum at 90° C.

Cleaning Composition (For Aluminum)
U. S. Patent 1,890,214

A water-soluble cleaning powder comprises tartaric, citric, or other solid organic fruit acid, 99%, and an alkali fluoride (sodium fluoride), 1%.

Cleaner for Chemical Glassware

Pulverize a 10 ounce cake of good grade cleaning and polishing grit cake soap, such as ''Bon Ami.'' Cut a 12 ounce cake of good grade of rosin laundry soap, such as ''Octagon'' into thin slices and add just enough water to cover the mass. Slowly heat on a hot plate until the soap his dissolved in the water and a clear solution results. Add this liquid soap mixture to the powdered grit cake in a beaker or earthen jar, stirring the mixture well. Allow to stand over night or until the resulting mixture has solidified into a soft mass. The mixture can then be easily applied to the wet glassware in the usual manner with a brush or the hands, scrubbing thoroughly and finally rinsed in running water. It is only necessary from time to time to add small quantities of water to keep the mixture at a proper consistency.

Metal Cleaner
U. S. Patent 1,935,911

A concentrated metal cleaning composition which readily dissolves in water forming a solution having comparatively low surface tension characteristics and which when applied to metal surfaces forms a uniform and continuous film may be produced by combining the following ingredients in the proportions, by weight, specified:

Orthophosphoric Acid 75%	69.5%
Mono-Butyl Ether of Ethylene Glycol	17. %
Oleic Acid	00.5%
Saponin	1.0%
Water	12. %

A sufficient quantity of a sugary substance such as dextrine and molasses may be substituted in place of part of the water to build up the viscosity to the desired degree.

The saponin and desired amount of dextrine and molasses are dissolved in the water and are then added to the phosphoric acid and thoroughly mixed. The mono-butyl ether of ethylene glycol and oleic acid are then added to the mixture.

Part of the mono-butyl ether of ethylene glycol and part of the water may be replaced by mono-ethyl ether of ethylene glycol or the two ether derivatives may be used together in any desired proportion that will leave sufficient grease solvent to do the work required. The mono-ethyl ether of ethylene glycol, aside from being a cheaper material, affords the advantage of being more miscible in water and is equally as good a solvent for vegetable oils as is the mono-butyl ether of ethylene glycol. Under certain conditions, as where the oils to be removed from metal surfaces or dissolved in order that the rust removing acid and rust inhibitor may act on the metal, are mainly vegetable in nature; it is advantageous to replace, at least a part of, the mono-butyl ether of ethylene glycol with the mono-ethyl ether of ethylene glycol. In practice, two forms of the cleaning composition may be employed; one of which is to be washed from the body that is prepared for painting; and the other is adapted to be wiped off with dry cloths. It is preferred to employ, as the oil solvent, the mono-butyl ether of ethylene glycol in the first form, whereas the mono-ethyl ether of ethylene glycol may be employed, at least in part in the second form. Part or all of the mono-butyl ether of ethylene glycol may be replaced with ethyl methyl ketone. Monobasic acids, such as linoleic acid ricinoleic, acrylic and crotonic acids, may be employed in cleaning compositions of this character as surface tension reducing agents and they may be used either alone or in conjunction with saponin.

Saponin, which is a soap-like compound obtained from licorice bark or soap tree bark, may be replaced by other organic soaps of vegetable derivation, such as sarsaponin which is obtained from sarsaparilla bark.

Butcher's Tool Cleaner
Austrian Patent 135,349

Sodium Borate	15 parts
Sodium Bicarbonate	70 parts
Alum	15 parts

This composition deodorizes and disinfects as well as cleans.

Plastic for Cleaning Typewriter Keys

Hydroresin	30. g.
Zinc Oxide	25. g.
Latex (72%)	5. g.
Triethanolamine	1.5 g.
Stearic Acid	3.5 g.
Clay	5.0 g.
Carbon Black	3.0 g.

The hydroresin and zinc oxide are heated together and when thoroughly mixed, the clay, triethanolamine, stearic acid, and carbon black are added at a temperature of 250° F. The latex is then added and the mixture stirred until fairly cool.

The consistency of the product may be changed by varying the quantities of latex and triethanolamine.

Machinery Cleaning Compounds

The following are for use as heavy duty cleaners for degreasing machinery. They are to be used hot and the concentration depends primarily on the amount and type of work done.

	Formula No. 1	Formula No. 2
Tri-Sodium Phosphate	15 parts	25 parts
Sodium Meta-Silicate	10 parts	20 parts
Sodium Hydroxide	5 parts	2 parts
Soda Ash	10 parts	33 parts
Soda Soap	60 parts	20 parts

General Cleaner

Dissolve 20 pounds of soap flakes in 55 gallons of water. To this solution add enough of a mixture of one part

sodium metasilicate, and 9 parts silica (200–400 mesh), so that the resulting product forms a stiff paste.

This formula has been used very satisfactorily in keeping college and other institutional buildings in attractive condition.

Rubber Blanket Cleaner

Naphtha having a maximum heavy ends of 0.5% and a distillation range of 200–300° is used for cleaning rubber blankets.

Perborate Cleanser
German Patent 581,464

A protective covering is provided in cleansing agents containing perborate or other bleaches to prevent the particles of the same from coming into contact with moisture, carbon dioxide and alkalies. The protective covering consists of water-soluble waxes, water-soluble paraffin or its derivatives, hard fatty acids, hydrogenated soft fatty acids, resinic acids or mixtures. The protective agent is incorporated in the soft soap base of the cleanser but in spite of this fact, the coating dissolves first in the wash liquor, because it does not form a solution, strictly speaking, but an emulsion. Hard fatty acids and hydrogenated soft fatty acids dissolve directly in the lukewarm, wash liquor, forming soaps with the alkali in the liquor. However, this reaction is stated not to take place in the moist, soft soap. The bleaching mixture, which is added to the soft soap, may be made by mixing 75 to 90 parts of a suitable bleaching agent with approximately 10 to 25 parts of glycol stearate or similar product, and forming granules from the mixture in the usual manner. Or, three parts of anhydrous stearic acid are mixed with two parts of sodium perborate or the like and the mixture formed into a solid mass and then granulated. After the granules are sprayed with a solution of water-soluble wax, they are mixed to soft soap to give the finished cleansing compound.

Detergent and Emulsifier
German Patent 568,703

An effective emulsifier and detergent is made from soft soap and starch by the following process: Place 7 parts of potato flour in a mixer, add 43 parts of soft soap and mix at ordinary temperatures. Then add 50 parts of caustic soda in small portions. After sufficient stirring, draw off the dry, loose powder. To shorten the time of stirring, the mixer may be cooled so that the reaction takes place at 40° C.

"High Gelatinating" Bentonite
U. S. Patent 1,934,267

Bentonite of the Wyoming variety is used with about 2–5% of its weight of an alkaline agent for increasing the "gelling value" of the bentonite, this agent being formed of calcium sulphate 20–33 and magnesium oxide 80–67%.

Cleaner for Gelatin Films and Matrices
U. S. Patent 1,924,892

Alcohol	95 parts
Diethylamine	5 parts

Straw Hat Bleach

It is composed of a mixture of equal parts sulphur and tartaric acid. A teaspoonful of this mixture is used in a saucer of water and the hat scrubbed with the mixture, after which it is rinsed and allowed to dry in the sun.

Bleaching Powder
Formula No. 1

Borax	50 parts
Soda Ash	45 parts
Sodium Thiosulphate	5 parts

To use, dissolve the finely pulverized mixture in one liter of boiling water and dip the clothes in this solution.

Formula No. 2

Soap Powder	90 parts
Sodium Perborate	10 parts

Javelle Water

Dissolve 3 pounds of washing soda (dry) or 6 pounds of the crystals in 1 gallon of water. Mix 1½ pounds chloride of lime in 1½ gallons of water. Mix the two liquids and stir well. Let stand several hours in a covered crock, then filter clear, adding enough more water to make 2½ gallons of the finished product. Use soft water in the making.

Bleaching Liquid for Tobacco Stains
Solution No. 1

Sodium Bisulphite	5%
Water	95%

Solution No. 2

Potassium Permanganate	3%
Water	97%

Apply Solution No. 2 first, then Solution No. 1. Wash well with water. Repeat if necessary.

Dirt and Rust Remover for Automobile Radiators

Rust, dirt and foreign matter can be removed to increase the cooling efficiency of automobile radiators by dissolving one pound or less of sodium bisulphate in the water before adding to the radiator. The engine should be run until the solution is hot and remains so for about one hour. Then drain out and flush out well with water.

Mechanic's Powdered Hand Cleaner

Olive Oil Soap	25 lb.
Borax, Powdered	25 lb.
Tri-Sodium Phosphate	17½ lb.
Tripoli, Double-Ground	125 lb.

Mix the above in a powder mixing and sifting machine but do not run the materials through the sifter mechanism. If it is desired to scent the material use any desired perfume oil which can be sprayed into the mixing material in a fine atomized spray while the material is being mixed. One pound of perfume oil per 100 pounds of cleaner is usually sufficient.

The tripoli should be Missouri tripoli as this grade produces a very velvety feel not possible with other abrasives, yet it does not cut the skin. The particles are round, softer than the skin, and cleans by absorption as well as abrasion.

The above hand cleaner is also ideal for children's knees and elbows. It can be used by women without harm. Only about two grams are required to clean the dirtiest pair of hands.

Below is a formula for another Mechanic's Hand Cleaner which can be made at a lower cost. The same directions regarding mixing should be followed as for that above.

Soap	10 lb.
Tri-Sodium Phosphate	10 lb.
Borax, Powdered	5 lb.
Tripoli, Single-Ground	50 lb.
Bentonite Clay (200 mesh)	20 lb.

Mechanic's Grit Hand Paste Soap

Cocoanut Oil Soap	1 lb.
Potassium Carbonate	2 oz.
Tripoli, Double Ground	8 oz.
Water	3 qt.
Mineral Seal Oil	8 oz.

Cocoanut oil soap can be prepared by formulas found in this volume, or it may be substituted by using Kirk's flake laundry soap for small quantities. Dissolve the soap in hot water and add the potassium carbonate and stir until dissolved. Add the mineral seal oil and agitate vigorously until cold, finally adding the tripoli which is stirred until an even homogeneous mass is obtained. Let set over night before packaging. The ratio of water to soap in the formula is variable and may be changed as desired within reasonable limits.

Waterless Hand Cleaner

Cocoanut Oil Soap	1 lb.
Potassium Carbonate	2 oz.
Tri-Sodium Phosphate	1 oz.
Water	4 qt.
Fine Asbestos Powder	1 lb.

Heat the water and dissolve the soap, potassium carbonate and tri-sodium phosphate in it, stirring thoroughly. Allow to set undisturbed until cold then work in the powdered asbestos fibre and allow to set about twelve hours or over night before packaging.

Chemical Washing and "Antiquing" of Oriental Carpets and Rugs

The subdued color of the old carpets is now faked by dulling and fading the colors, and their luster imitated by a caustic soda or bleaching powder "chemical wash." The color of dyed yarn is saddened by boiling for 2 hours in a bath containing 0.25–1.5 ounces of sumac extract or 0.5–3 ounces of sumac leaves per 10 gallons of water, and drying without rinsing. If this is not sufficient, it is followed by treatment with a cold dilute copperas solution. One or 2 pounds Igepon T or A per 100 gallons of water for 15–30 minutes is recommended for washing. The color is faded and the wool lustered by treating for 5–10 minutes with a 0.125–0.5° Tw. bleaching powder solution either by immersion or by flooding the rug on a floor and vigorously brushing in the direction the pile lies. After rinsing with cold water until free of chlorine odor, it is acidified with a solution containing 3% of acetic acid; it is brushed well and then squeezed or centrifuged. In some cases the chlorine treatment is preceded by a 15–30 minute treatment at 30–40° with a solution containing 1 pound of caustic soda or sodium phosphate per 50 to 80 gal. of water. Where luster rather than fading of the color is desired, the rug is sometimes treated with a cold 80° Tw. caustic soda solution. This attacks the wool only very slowly and is rinsed off. Dilute caustic solution is also sometimes painted on the edges of the rug to remove some of the pile and simulate wear. A bath not longer than 30 to 1 and containing 3% of Eulan NK, on the weight of the rug, is recommended for mothproofing, working cold for 45 minutes. A solution containing 1–1.25 ounces of Eulan NK per gallon can be sprayed on the rug. Chlorinated carpets lose luster when sprayed in this way so they are best immersed in a bath. Various solutions of Turkey-red oil, sugar and glycerol; shellac in borax or ammonia; dextrin in glycerol; tartaric acid with glue or ox gall; decoctions of Panama wood and glue; oil or wax emulsions, etc., are often applied for finishing.

Alberene Laboratory Table Dressing

Treat lightly with thin paraffin oil. This penetrates surface and is non-drying. Varnishes or non-drying oils should not be used for this purpose.

Alberene Stone Top Dressing

Paraffin	1,000 g.
Kerosene	700 c.c.
Gasoline	600 c.c.

Melt paraffin over a water bath and cool to about 65° C. Pour in kerosene slowly with vigorous stirring. Allow to cool to about 50° C. and add gasoline. Thoroughly mix. Keep in stoppered bottle, away from flame.

The above mixture should be heated over a water bath until it becomes a liquid, and applied to the stone by means of a cloth. The surface of the stone must not be moist and not too cool. After the thin layer of the dressing has set, polish with dry soft cloth.

Coating Composition
French Patent 756,481

A composition for application to metals, wood, stone or paper is made by mixing tar and coal-tar pitch with the addition of chlorinated aliphatic hydrocarbons and a mixture of homologs and derivatives of benzene, the mass being energetically agitated and submitted to pressure. One example contains tar 21, pitch 13, trichloroethylene 3 and homologs and derivatives of benzene 11 kilograms.

Waterproof Coating Compound
U. S. Patent 1,957,179

Petroleum asphalt 27%; gilsonite 12%; blown bean oil 10%; asbestos fiber 16%; pigment 1%; and benzine 34%; the mixture of substantially liquid consistency and the fiber remains in suspension therein.

Waterproof Bituminous Compositions
U. S. Patent 1,949,229

Bituminous material such as pitch 15–45 is mixed with clay 15–40, sand 15–50, fiber such as asbestos 17 parts and water to form a plastic mass, which is shaped and dried and then heated to above the melting point of the bituminous material but below 425°.

Aluminum Sealer for Creosoted Wood

Formerly, several coatings of ordinary products were required to seal up the bleeding through of both tarry and creosotic wood preservatives with which the butts of the poles are either coated or impregnated. Now with the present type of paint one coat only effectively seals in the bleeding material so that it becomes the finishing coat itself. The paint is applied preferably by spraying, the section of the poles so finished being about six or eight feet of the treated butts above ground. The composition of the paint is per the following formulation:

Cumar Mixing Varnish	37½ gal.
Tung-Ester Varnish (80 gal.)	10 gal.
Heavy-body Gloss Oil	37½ gal.
Mineral Spirits	7½ gal.
Aluminum Powder (dry)	185 lb.

The cumar varnish is 28-gal. length of oil (Tung), the resinous constituent being 80% cumar resin and 20% fused lead resinate.

Tung-Ester Gum Varnish (80-gal.)	72½ gal.
Bodied Linseed Oil	10 gal.
Mineral Spirits	7 gal.
Lead-Manganese Concentrated Drier	½ gal.
Paste Aluminum	135 lb.

It will be noticed that only 1½ pounds of aluminum paste was used to each gallon of vehicle, but the priming of both redwood and white pine was in excellent condition of coating unprotected during a full year, and the same primer with only one coat full body of pure lead-zinc-titanox white paint over it appeared in perfect condition over the same period of time.

Coating to Render Asphalt Non-Sticky
U. S. Patent 1,916,970

A mixture of 90–98% of talc and 10–2% of bentonite is agitated and mixed with water and applied as a thin coat.

Soluble Gilsonite
U. S. Patent 1,925,085

Gilsonite is heated at 163–205° for 30 minutes, whereby it becomes, on cooling, readily soluble in petrol or benzene.

Corrosion Resisting Coating
U. S. Patent 1,932,156

Distilled Castor Oil	40 parts
Undecylenic Acid	10 parts
Copper Oxide	1 part
Iron Oxide or Colcothar	1 part
Petroleum Spirit	24 parts
Practical Anhydrous Ethyl Alcohol	24 parts

According to the different cases of employment, the above composition may be added with for instance 8 to 20 per cent pulverulent graphite, or 8 to 20 per cent hydrated magnesia silicate.

The product thus prepared is applied on the surface of the metal to be protected which is always more or less oxidized, without previously cleaning said surface, by dipping or by means of a brush, pistol or any known means.

The metal thus prepared is then heated, preferably in a furnace or stove or by means of a flame. During heating, complicated chemical reactions takes place.

The metallic oxides coating the surface metal to be protected are first dissolved. Then the undecylenic acid attacks the surface of the metal itself, and finally a chemical combination takes place, which gives as result a solid layer integral with the metal to be protected.

During heating and the subsequent chemical reactions, a smoke gas escapes. Heating must be progressive until there is no longer escape of smoke. The chemical reaction is then terminated. With the formula given above, the temperature should rise progressively up to 250 centigrade.

According to the thickness of the protecting coating desired, a second or third layer of the same product may be applied successively, or of one of the variants of the product, while operating in the same manner.

The successive layers combine to each other and form a resulting homogeneous layer. The latter affords the following improvements in the protection of metals:

It stops all the previous corrosions.
It has remarkable resistance to oxidizing agents, such as salt air, dampness, etc., chemical agents, acids or bases. It is insoluble in hydrocarbons, alcohois or esters.

It is hard, harder than red copper or aluminum. It is flexible and endures several bendings. Its adherence to the metal enables it to support shocks. Its electrical resistance is very high.

Protection of Lead Against Corrosion
British Patent 407,276

Lead cables or pipes for laying underground are smeared with a mixture of vaseline 55%, rosin 40%, and sulphur 5%, whereby an adherent, protective film of lead sulphide is formed on the surface.

Flexible Gelatin Coating

Gelatin	10	parts
Water	100	parts
Sulfo Turk C	25	parts
Butyl Cellosolve	2.5	parts

Leather "Dope"

The usual coating composition for the prime or first coat may be as follows:

Cellulose Nitrate	40 lb.
Vegetable Oil (Castor)	55 lb.
Wood Alcohol	12 gal.
Amyl Acetate	8 gal.
Refined Fusel Oil	2 gal.
Benzine 62° Bé	7 gal.
Benzine 71° Bé	13 gal.

Pigment sufficient to produce the desired shade.

Spraying Lacquer for Leather

Cotton Wet (½″ R. S.)	140 parts
Tricresylphosphate	20 parts
Amyl Acetate	91 parts
Butyl Acetate	81 parts
Butanol	132 parts
Ester Gum*	25 parts
Kettle Bodied Linseed Oil*	25 parts
Xylol	10 parts
Kettle Bodied Linseed Oil	100 parts

* Fused together @ 300° C.

Protecting Metallic Surfaces from Sulphur Compounds
U. S. Patent 1,896,141

The surfaces are coated with sodium silicate to protect them from sulphur.

Washable Protective Coating for Oil Tanks
British Patent 404,874

A composition consisting of approximately equal weights of water-glass and finely-divided barium sulphate is used.

Coating Composition for Artificial Leather
U. S. Patent 1,898,540

A mixture of dry pyroxylin (1 part), boiled linseed oil (1 part), solvent mixture, e.g., ethyl alcohol, ethyl acetate, butyl acetate, and toluol (10 parts), drier (0.002 part), and crude gluten (0.1 part).

Coating Lacquer for Producing Artificial Leather, etc., on Coating Machine

Cotton, Wet (½″ R. S.)	28	parts
Benzol	80	parts
Ethyl Acetate	50	parts
Blown Soya Oil	15	parts
Blown Linseed Oil	15	parts
Pyrol Black	1.25	parts
Paris X Black	1.13	parts
Steel Blue	.12	part

Impregnation Composition for Stiffening Shoe Tips

Glue	250 g.
Water	250 g.
Portland Cement	300 g.
Ceresin	100 g.
Coal-tar pitch	100 g.

Compositions for Lining Pipes
German Patent 593,284

Pipes for conveying hot waste gases are lined with a composition comprising powdered chamotte 20, clay 30, comminuted cast iron 32, asbestos wool 1 and caustic soda or caustic potash 1 part. The composition is made into a paste with water, applied to the pipe and dried at a gradually rising temperature, preferably until the free surface of the lining sinters.

Protective Coating for Steel
U. S. Patent 1,936,533

Linseed oil fatty acids	9.30 lb.
Triethanolamine	1.50 lb.

Heat to 350° F., then with agitation heat as rapidly as foaming permits to 480° F. and hold until 1 c.c. does not mix with 5 c.c. of 95% alcohol, but leaves the alcohol layer nearly colorless.

To .90 pound of the cooled product add 1.05 pounds of mineral spirits and .15 pound of orthophosphoric acid with vigorous stirring. (The product should be entirely clear when a sample is diluted with 10 parts of mineral spirits.)

The composition made in accordance with the above example will give a very thin and very adhesive primer coat, when used alone on metal and baked at 450° F.

When incorporating this product into paint products it may be added in any amount from the smallest having any appreciable effect, which will be found to be about 3%, up.

A satisfactory baking schedule for these products is 10 minutes at 450° F., about 10 minutes being used to attain this temperature, and another 10 minutes to cool the finished article. Lower temperatures will greatly increase the time required to dry the film thoroughly, but when the film has been dried the phosphoric acid will show its characteristic effect. 300° F. to 500° F. will usually cover the range of temperature which would be considered practical for using the materials described herein. The special condition for the use in part products, in the ordinary course of paint development work, will be indicated to those skilled in the paint art.

A typical example of the use of the material in a paint product is as follows:

Gilsonite, 1 part by weight; prepared oil (usually containing iron or manganese drier), 1 part by weight; black pigments, .16 part by weight; mineral spirits, 2 parts by weight; ester phosphate compound (above), .4 part by weight; 30% citric acid solution in alcohol, .04 part by weight.

The gilsonite is fluxed with the oil, and the black pigments ground in accordance to usual paint practice. The phosphate compound and citric acid solution are stirred into the finished material.

A hydroxy polybasic aliphatic acid such as citric or tartaric acid is included in the above composition to overcome the tendency of these compounds, when added to some paint and varnish materials, though not when used alone, to give a peculiar alligatored appearance to the force-dried film. If citric or a similar acid is also added to these mixtures, the alligatoring effect is eliminated, and, consequently, the use of such an acid, where necessary, constitutes an integral part of this invention.

The product made according to the first example above will also give a very thin and very adhesive primer coat, or protective coat, when used alone on metal and baked at 450° F.

In some cases it is only partly soluble in alcohol at the start and may, or may not, pass through a soluble stage and eventually become insoluble, depending on the proportion of ingredients and heat treatment accorded them. This is explained by the following table of solubility in alcohol:

Bodied Tri-Esters, insoluble.
Tri-Esters, insoluble.
Bodied Mono- and Di-Esters, partly soluble.
Mono- and Di-Esters, soluble to partly soluble.
Triethanolamine, Glycerine, Oil Acids, soluble.

The use of oils instead of oil-acids in these compositions, permits the use of heat-bodied material when it is desired to increase the viscosity, or reduce the crawling tendencies of the product when coated on steel and baked at high temperatures. It is, however, sometimes feasible to body the compound made from oil-acids after its formation, especially in the case of the wood oil derivative.

An effective composition for use alone on steel is made as follows:

Linseed Oil, Bodied Two
Hours at 625° F.   5.00 lb.
Triethanolamine   1.60 lb.

Heat one hour at 480° F. in covered kettle with slow stirring. Cool to 400° F. and add:

Kerosene   8.00 lb.
High Boiling Coal-Tar
Naphtha   12.00 lb.

To the thinned material add in a slow steam while vigorously stirring:

85% Commercial Phos-
phoric Acid   1.00 lb.

This composition yields a very thin hard film when flowed on steel and heated 10 minutes at 450° F.

Tennis Racket String Coating
U. S. Patent 1,956,441

Boil one-half an ounce of starch in sixteen ounces of water and a suitable gum arabic solution may be made by dissolving six and three-quarter ounces of the gum in sixteen ounces of water.

To the mixture of equal parts cooked starch solution and gum arabic solution there is added an approximately equal quantity of a solution of any of the bichromates having the property of hardening colloids upon exposure to light, preferably the sodium, potassium or ammonium salt. The bichromate solution used may be of anywhere from 1 to 4% strength. The proportions, therefore, of the compound are approximately one part starch solution, one part gum arabic solution, all of the solutions being made, of course, by mixing the substances with water.

As a result of actual experiment it is found that the present compound, when brushed on tennis racket strings and exposed to sunlight, more nearly approaches the texture and qualities of real sheep gut than any other compound now used, such as shellac, varnish or glue, and that it is able to take the impact and punishment of stroking the ball and adds from 25 to 100% to the life of the strings.

Translucent Glass Coating

Heat China wood oil under a hood for twenty minutes at a temperature of 235° C. Allow to cool, paint on window glass and allow to dry. This simple process gives a very good frosted glass effect to any ordinary glass surface.

Glazing Solution
U. S. Patent 1,892,980

Casein 100, borax 7–15, sodium phosphate 7–15, and hexamethylene tetramine 0.5–8 pounds are ground dry and sprayed with sassafras oil 1 ounce and a non-drying oil 1–5 ounces; the whole is diluted with water.

Synthetic Resin for Paints
British Patent 378,094

Para-hydroxydiphenyl (100 parts), reacts with commercial formaldehyde (100 parts) by heating to form a resinoid; this is then heated with China wood oil (100 parts) at about 210° C. until a sample on cooling remains clear and can be diluted with an equal weight of cold linseed oil without clouding. Linseed oil is then added slowly, keeping the temperature about 200° C., and the temperature is gradually raised and heating continued until the desired viscosity has been attained. Further amounts of linseed oil and small quantities of driers can then be added. For example, 800 parts of linseed oil, 14 parts of litharge, and 2 parts of cobalt acetate may be used. When cool, the composition is ready for mixing with pigments and volatile thinners to produce rapid-drying paints of very high resistance to weathering, moisture, and weak alkalies. If the heating of the composition is continued, either before or after adding the driers, the viscosity increases, and a tough, rubbery solid suitable as a binding medium in plastic compositions is obtained. Another method of making such compositions is to heat a mixture of commercial cresol (100 parts), China wood oil (200 parts), and hexamethylenetetramine (25 parts) at 200° C. until foaming ceases; linseed oil (400 parts) is then added and the mass is heated at 300° C. until it has acquired the necessary viscosity, whereupon 300 parts of linseed oil and 20 parts of litharge are added. In a further modification, 100 parts of phenol are digested with 100 parts of linseed oil and 1 part of phosphoric acid at boiling. temperature, under reflux, for 8 hours; when the product has cooled to 100° C., 30 parts of trioxymethylene are added and refluxing is continued until the mass has thickened and the odor of formaldehyde has disappeared. The mass is mixed with 500 parts of linseed oil, heated to 260° C. until of the desired viscosity, cooled to 220° C., and mixed with the appropriate driers. The use of China wood oil is generally desirable, but it is not essential, and for compositions where lower viscosity is desired it may be omitted. On the other hand, a vehicle suitable for paints may be obtained by cooking the phenolic resinoid with tung oil alone; for example, a paint of good brushing and covering properties, and drying within two hours to a highly glossy, waterproof film, is obtained by mixing as little as 5 per cent of phenolic resinoid with 95 per cent of tung oil and cooking at about 315° C. until the sample shows gas-proofness.

According to a variation of the process, which forms the subject of a specific claim, 37.9 parts of xylenol, 8.14 parts of paraformaldehyde, 1.22 parts of triethanolamine, and 6.1 parts of China wood oil are heated together at 110° to 115° C. until a sample cures in 4 to 6 seconds at 200° C., another 6.1 parts of the oil are added, with stirring until the mixture becomes clear, and then the remainder (20.3 parts) of the oil is added, with stirring and heating. When the product is clear, solvent naphtha (20.3 parts) is stirred in. The product is stated to be particularly suitable for coating the coils of motors or other electrical apparatus, since it gives films which are flexible, tough, and strong, and have good electrical insulating properties. It is also suitable for making laminated stock, moulded articles, etc.

High Tension Cable Lacquers

No. 1

Nitrocellulose (15–20 sec. R.S.)	32–34%
Tricresyl Phosphate	40–38%
Aroclor 1242	28%

No. 2

Benzol	62%
Alcohol	22%
Ethyl Acetate	16%

Dissolve 1 part of No. 1 in 3 parts of No. 2.

"Lacquer" for Food Containers

U. S. Patent 1,945,584

The lacquer or paint has the following advantages:

•1. It is not brittle, thus doing away with the objectionable feature mentioned.
2. When dry, it remains brilliant, thus giving the container an attractive appearance.
3. It dries more rapidly than any other paint at present in use.
4. It is perfectly adhesive and waterproof.
5. Containers having a ring or strip of rubber for sealing, do not suffer any damage as the protective preparation does not attack rubber.

The protective paint is prepared as follows:

Denatured Alcohol

This alcohol must be prepared by adding 3% of shellac and 3% of rosin and allowing them to dissolve gradually by suspending them on a perforated stockinette lined tray in the alcohol.

Shellac Stock Solution

To be prepared in a wooden cask of 800 liters capacity equipped with a removable steam coil. 600 liters of denatured alcohol should be lowered into the cask and 22.2 kilos of shellac placed in the suspended tray. By next morning the shellac will be found to have dissolved to a turbid solution. Remove the empty tray, place the steam coil in the bottom of the cask and heat the solution to 50° C. (122° F.), keeping this cask covered with a tight fitting cover. Once this temperature has been reached, remove the steam coil and allow the solution to settle out for 24 hours, after which pass it without disturbing the settlings to another cask which acts as the holding vat for the finished shellac stock solution. The settlings to be filtered through a thick layer of cloth and the filtrate to be poured into the stock solution.

Rosin Stock Solution

To be prepared in exactly the same way as described for the shellac stock solution, using the same quantities, i.e. 600 liters of denatured alcohol and 22.2 kilos of rosin broken up so that none of the pieces measure one to one and a half inches square.

Mixing Paint
1. Gold Lacquer

Transfer 300 liters of the shellac stock solution and 300 liters of the rosin stock solution to an 800 liter wooden cask and mix well. Draw about 15 liters of the mixture off into a bucket and stir 1 kilo 950 grammes of Sudan yellow RR aniline into the liquid until no lumps remain. Pour the paste into the cask and rinse the bucket three times with the paint from the cask adding 3 kilos 900 grammes of castor oil to the second rinse and stirring well to dissolve. Mix the contents of the cask well. Keep the cask well covered.

2. Blue Lacquer

Transfer 300 liters of shellac stock solution and 300 liters of rosin stock solution to an 800 liter wooden cask and mix well. Draw about 15 liters of the mixture off into a bucket and stir 2 kilos 880 grammes of Victoria blue B base aniline into the liquid until no lumps remain. Pour the paste into the cask and rinse the bucket three times with the paint from the cask adding 3 kilos 900 grammes of castor oil to the second rinsing and stirring well to dissolve. To the paint in the cask add 60 liters of gold lacquer and mix well. Keep the cask well covered.

Crystallizing Lacquer
U. S. Patent 1,880,419

Compositions of cellulose acetate (100 parts) and 2:4:6–tribromoanisole (25–150 parts) are used.

Cellulose Acetate-Mastic Lacquer

A cellulose acetate-mastic lacquer can be prepared by mixing separate solutions of the following ingredients: 20 parts mastic in 90 cubic centimeters methylene chloride and 10 cubic centimeters alcohol; 20 parts cellulose acetate (containing the equivalent of 62 per cent acetic acid) in 100 cubic centimeters methylene chloride. To incorporate "mowilith" in a primary cellulose acetate lacquer, a solution is prepared of 20 parts of the synthetic resin in 90 parts tetrachlorethane and 10 parts alcohol. The solution is then mixed with 20 parts cellulose acetate (62 per cent acetic acid) in 100 cubic centimeters of the same solvent mixture.

Cellulose Fatty Ester Film and Lacquer
British Patent 399,814

An aryl phosphate, e.g., triphenyl, tritolyl, monotolyl diphenyl or trinaphthyl phosphate, is used as a plasticizer, for higher cellulose esters, e.g., stearate, laurate, oleate, palmitate or for cellulose phenylacetate or crotonate. The material may include also other cellulose esters, cellulose ethers, fats, waxes, oils, gums, resins, etc. A solution for making films contains cellulose stearate 100, benzene 300–500 and triphenyl phosphate (I) 10–50 parts. A solution for coating or for making filaments contains cellulose oleate 100, acetone 300–500 and (I) 10–50 parts. The composition may be used in lacquers or varnishes or in a molding powder.

Lacquer, Airplane Dope, Non-Inflammable

The following formula is used by the U. S. Government for coating the fabric of the wings of airplanes. While it may be applied by brush, it is preferably sprayed on. It is given in percentage by weight:

Cellulose Acetate	7.5%
Triphenyl Phosphate	2.5%
Acetone	30.0%
Benzol	30.0%
Methanol	20.0%
Di-Acetone Alcohol	10.0%

The solvents are mixed in the above proportions and the cellulose acetate added with continued stirring to avoid formation of lumps. When solution is complete, the triphenyl phosphate is added and again stirred. The mass is allowed to settle at room temperature and then decanted, or if need for immediate use may be readily filtered by passing same through a piece of finely woven fabric.

Aeroplane Fabric "Dope"

Dissolve and mix one part celluloid in one part acetone. Add 2 parts 95% alcohol and mix. Add 1/30 part castor oil and mix. 5-7 coats make good weatherproof fuselage covering. Will shrink paper or canvas drumhead tight.

Automobile Lacquers

In the manufacture of automobile lacquers the following formula gives very satisfactory results on metal finishing. It may be found necessary to slightly change the physical characteristics of the film produced and this can be effected by modification of formula, changing the oil ratio or if desirable resins, of the type usually used in lacquer work, may be incorporated. The formula is given in parts by weight:

½" Wet R. S. Cotton	56.0 parts
Butyl Acetate	37.5 parts
Butanol	37.5 parts
Ethyl Acetate	25.0 parts
Ansol	25.0 parts
Toluol	125.0 parts
A.D.M. No. 100 Oil	46.8 parts

Base Colors

The following quantities of pigment to the above formula:

White
Titanox B or C	40 parts
Zinc Oxide	80 parts

Blue
Steel Blue	8 parts
Alcohol	5 parts

Green
Chrome Green	20 parts

Scarlet Red
Scarlet Toner	12 parts

Yellow
Chrome Yellow	20 parts

Black
Carbon Black	4 parts
Alcohol	5 parts

Tinted Colors
Dark Blue
95 parts base blue lacquer and 4 parts white lacquer mixed.

Light Blue
100 parts base blue lacquer and 100 parts white lacquer mixed.

Pale Blue
786 parts base white lacquer and 25 parts base blue lacquer mixed.

Light Green
300 parts base white and 100 parts base green lacquer mixed.

These lacquers should be thinned approximately 2 parts lacquer to 1 part thinner by volume for spraying.

Lacquers Containing Linseed Oil

Lacquers containing linseed oil possess peculiar advantage over some of the more conventional types. They do not sweat, as castor oil lacquer does; since linseed oil can be bodied to a high degree of viscosity, lacquers can be made with it containing a substantially higher total solids content than is usually met with, without sacrificing gloss or flexibility. Owing to the peculiar properties of linseed oil, certain conditions are essential to its successful use in lacquers, the important ones of which are these:

Owing to its property of "drying" it should never be used in the limpid state. It should always be heavy bodied. Two types of bodied oil are available: oxidized oil, produced by blowing air through the oil at an elevated temperature; and polymerized oil, produced by heating oil to a high temperature under conditions which minimize oxidation so that the bodying effect is the result entirely or nearly so, of polymerization. Such oils are the common "kettle bodied" oils of the varnish maker, and also the more modern "specially treated" polymerized oils now available on the market, which can be obtained in all degrees of viscosity and of beautifully pale color.

The heavier the body, the more suitable linseed oil is in lacquers. For this reason blown oils are less suitable than polymerized oils. Blown oils however mix more readily with nitrocellulose solutions. When polymerized oils are used (they are referred to in the formulas as "Kettle Bodied" Oils), special solvents are necessary. Blown oils can be used in any of the formulas to replace polymerized oils, but the quality of the lacquer will be diminished. Polymerized oils cannot be used in the formulas calling for blown oil.

Except when otherwise specified, the gums and oil indicated should be melted together and heated to 300° C. When the mixture is sufficiently cool, the xylol (when called for) is added and stirred in, and any loss by evaporating compensated for by further additions. The final mixture is a varnish which can be more readily combined with the nitrocellulose solution than the oil and gums could be separately.

The preferred procedure is as follows:

Weigh cotton first and wet down with solvents which are not nitrocellulose solvents (alcohol, benzol, toluol, butyl alcohol, etc.). Then add remainder of solvents, and stir. When cotton is completely dissolved, add gums and oil, previously combined as described above (unless otherwise specified). If the lacquer is to be pigmented, the pigment may be ground in a paint mill, using the gum-oil-xylol combination as the vehicle, thinned if necessary with a portion of the higher boiling solvents; or the whole mixture may be ground in a ball mill, adding the pigments after all the other ingredients are thoroughly combined.

Lacquers should always be thinned with the same solvents, in the same proportions, as are used in making the lacquer.

Explanation of Terms Used in the Formulas

R. S. Cotton—"Regular" Soluble Nitrocellulose.

A. S. Cotton—Alcohol Soluble Nitrocellulose.

"Wet" Cotton — Nitrocellulose is shipped wet with 40% alcohol. In the laboratory where relatively small stocks are kept, part of the alcohol will be lost in storage, and this must be allowed for in making up a lacquer. "Wet" Cotton refers to the original standard cotton containing 40 pounds of alcohol for every 100 lb. of cotton.

"Dry" Cotton—Nitrocellulose from which the original alcohol has been removed by evaporation.

Automobile Spraying Laquer

½" R.S. Cotton, Wet	14 parts
Tricresylphosphate	3 parts
Butyl Acetate	12 parts
Butanol	8 parts
Toluol	20 parts
Ester Gum ⎱ Cold	3 parts
Toluol ⎰ Cut	3 parts
Kettle Bodied Linseed Oil	4 parts

For Base Colors use pigments in the following proportions:

Bone Black	4 parts
Steel Blue	2 parts
Chrome Yellow	4 parts
Chrome Green	4 parts
Toluidine Red	2 parts
Chrome Orange	4 parts
Titanox White	7 parts

Automobile Spraying Lacquer, High Gloss

½" R.S. Cotton, Wet	14. parts
Tricresylphosphate	2. parts
Amyl Acetate	9.1 parts
Butyl Acetate	8.1 parts
Butanol	13.2 parts
Toluol	23.7 parts
Ester Gum *	7.5 parts
Kettle Bodied Linseed Oil *	7.5 parts
Xylol	3. parts

* Fused together at 300° C.

For colors use pigments in the following respective proportions:

Timonox White	4 parts
Chrome Yellow	2 parts
Paris X Black	1 part
Scarlet Toner	1 part

Flat Lacquers

The formulas listed below are indicative of the type flat lacquer that can be produced, using A. D. M. No. 100 Oil. (Archer-Daniels-Midland)

In formula 1 the carnauba wax is incorporated by first pulverizing the wax and then grinding in the vehicle, using a pebble mill.

Formula 2 is made in the same manner as No. 1 only substituting aluminum stearate for carnauba wax.

Formula 1

Wet ¼" Cotton	56 parts
Toluol	80 parts
Butanol	20 parts
Ethyl Acetate	106 parts
Ansol	44 parts
Ethyl Lactate	10 parts
A. D. M. No. 100 Oil	35 parts
Carnauba Wax	20 parts

This lacquer is at spraying consistency.

Formula 2

Wet ¼" Cotton	56 parts
Toluol	80 parts
Butanol	20 parts
Ethyl Acetate	106 parts
Ansol	44 parts
Ethyl Lactate	10 parts
A. D. M. No. 100 Oil	47.4 parts
Aluminum Stearate	12 parts

This lacquer is at spraying consistency

Clear Furniture Lacquers

The first five formulas listed below indicate the difference in types of clear

lacquer that can be effected in using
A. D. M. (Archer-Daniels-Midland) No.
100 Oil. Formulas 1 and 2 are practically the same, the only change being
in the amount of A. D. M. No. 100 Oil
used. However, in formulas 3 and 4
changes in type of cotton used should be
noted. The use of these lower viscosity
type cotton permits of increasing the
solids contents at spraying consistency.

Formula 5 indicates the flowing properties of a lacquer in which A. D. M. No.
100 Oil is used. This formula is representative of a lacquer which flows sufficiently so that it is not necessary to rub
with an abrasive to remove ''orange
peel'' before polishing.

Formula 1

Wet ½" Cotton	56	parts
Toluol	80	parts
Butanol	20	parts
Ethyl Acetate	106	parts
Ansol	37.2	parts
Ethyl Lactate	10	parts
A. D. M. No. 100 Oil	46.8	parts

Thin approximately 2 lacquer to 1
thinner for spraying.

Formula 2

Wet ½" Cotton	56	parts
Toluol	80	parts
Butanol	20	parts
Ethyl Acetate	106	parts
Ansol	37.2	parts
Ethyl Lactate	10	parts
A. D. M. No. 100 Oil	71	parts

Thin approximately 2 lacquer to 1 thinner for spraying.

Formula 3

Wet ¼" Cotton	56	parts
Toluol	80	parts
Butanol	20	parts
Ethyl Acetate	106	parts
Ansol	44	parts
Ethyl Lactate	10	parts
A. D. M. No. 100 Oil	47	parts

This lacquer is at spraying consistency.

Formula 4

Wet 23 C.P. Cotton	56	parts
Toluol	67	parts
Butanol	17	parts
Ethyl Acetate	87	parts
Ansol	37	parts
Ethyl Lactate	9	parts
A. D. M. No. 100 Oil	47	parts

This lacquer is at spraying consistency.

Formula 5

Wet ¼" Cotton	56	parts
Cellosolve	46	parts
Butanol	33	parts

Butyl Acetate	33	parts
Ethyl Acetate	33	parts
Lacquer Gasoline	103	parts
Xylene	103	parts
Alcohol	64	parts
A. D. M. No. 100 Oil	59	parts

This lacquer is at spraying consistency.

Formula 6

½" R. S. Cotton, Wet	14.	parts
Tricresylphosphate	3.	parts
Amyl Acetate	7.23	parts
Butyl Acetate	6.44	parts
Butanol	10.47	parts
Toluol	18.8	parts
Ester Gum	1.66	parts
Blown Linseed Oil	3.32	parts
Xylol	1.0	part

Formula 7

½" R. S. Cotton, Wet	462.	parts
Amyl Acetate	446.	parts
Butyl Acetate	396.	parts
Butanol	644.	parts
Toluol	1155.	parts
Ester Gum*	247.5	parts
Kettle Bodied Linseed Oil*	247.5	parts
Xylol	99.	parts
Tricresylphosphate	66.	parts

* Fused together at 300° C.

A high quality lacquer which can be
polished to a deep lustrous piano-finish.

Formula 8

½" R. S. Cotton, Wet	112	parts
Toluol	160	parts
Butanol	40	parts
Ethyl Lactate	20	parts
Tricresylphosphate	14	parts
Ansol	88	parts
Ethyl Lactate	212	parts
Ester Gum	30	parts
Blown Linseed Oil	30	parts

A cheap lacquer for producing a quick
high gloss finish.

Formula 9

½" R. S. Cotton, Wet	14.	parts
Tricresylphosphate	3.	parts
Amyl Acetate	7.23	parts
Butyl Acetate	6.44	parts
Butanol	10.47	parts
Toluol	18.8	parts
Ester Gum*	2.92	parts
Kettle Bodied Oil*	5.83	parts
Xylol	1.75	parts

* Fused together at 300" C.

Formula 10

½" R. S. Cotton, Wet	28.	parts
Toluol	40.	parts
Butanol	10.	parts
Ethyl Lactate	5.	parts

Tricresylphosphate	3.5 parts
Anhydrous Alcohol	22. parts
Ethyl Acetate	53. parts
Ester Gum	10. parts
Blown Linseed Oil	20. parts

Designed especially for use on soft woods, which by reason of their excessive expansion and contraction, require a more flexible lacquer than hard woods.

Formula 11

(High Gloss, Colorless, good filler. For spraying).

½″ R. S. Cotton, Wet	56. parts
Toluol	80. parts
Butanol	20. parts
Ethyl Lactate	10. parts
Tricresylphosphate	7. parts
Alcohol	40. parts
Ethyl Acetate	110. parts
Ester Gum	20. parts
Blown Linseed Oil	10. parts

An inexpensive lacquer for general purposes.

Flat Lacquer for Furniture, Clear
(For Spraying)

½″ R. S. Cotton, Wet	56. parts
Toluol	80. parts
Butanol	20. parts
Ethyl Lactate	10. parts
Tricresylphosphate	7. parts
Anhydrous Alcohol	44. parts
Ethyl Acetate	106. parts
Ester Gum	15. parts
Blown Linseed Oil	15. parts
Corn Starch*	32.7 parts

* Ground in Oil and T C P.

Furniture Lacquer, Spraying

½″ R. S. Cotton, Wet	56 parts
Toluol	80 parts
Butanol	20 parts
Ethyl Lactate	10 parts
Tricresylphosphate	7 parts
Anhydrous Alcohol	44 parts
Ethyl Acetate	106 parts
Ester Gum	10 parts
Blown Linseed Oil	40 parts

Furniture Lacquer, Pigmented
(For Spraying)

½″ R. S. Cotton, Wet	14 parts
Tricresylphosphate	3 parts
Butyl Acetate	24 parts
Butyl Alcohol	16 parts
Toluol	40 parts
Kettle Bodied Linseed Oil	7 parts
Ester Gum	3 parts
Xylol	2 parts
Pigment	Sufficient

Furniture Enamel-Lacquer

½″ R. S. Cotton, Wet	56 parts
Toluol	80 parts
Butanol	20 parts
Ethyl Lactate	10 parts
Tricresylphosphate	7 parts
Ansol	43 parts
Ethyl Acetate	110 parts
Elemi Gum	30 parts
Blown Linseed Oil	20 parts
Pigment	Sufficient

One-Coat Furniture Lacquer, White

½″ R. S. Cotton, Wet	14. parts
Tricresylphosphate	2. parts
Amyl Acetate	9.1 parts
Butyl Acetate	8.1 parts
Butanol	13.2 parts
Toluol	23.7 parts
Ester Gum*	10. parts
Kettle Bodied Linseed Oil*	20. parts
Xylol	6. parts
Titanium Oxide	19. parts
Zinc Oxide	37. parts

This lacquer is designed for heavy pigmentation, to give sufficient covering in one coat. When other colors are substituted for white, care should be exercised in selecting pigments having relatively high opacity.

* Fused together at 300° C.

Furniture Lacquer, Colored

½″ R. S. Cotton, Wet	7. parts
4″ R. S. Cotton, Wet	7. parts
Dibutylphthalate	3. parts
No. 10 Thinner	75. parts
Ester Gum*	4. parts
Blown Linseed Oil*	8. parts
Cobalt Linoleate	.02 part
Xylol	2. parts

Thinner

Butyl Acetate	15. parts
Butanol	17. parts
Ethyl Acetate	8. parts
Alcohol	10. parts
Toluol	50. parts

* Fused together at 300° C.

For base colors, use previous formula with the folowing pigments:

White	
Zinc Oxide	12 parts
Timonox	10 parts
Black	
Paris X	2 parts
Yellow	
Chrome Yellow	6 parts
Red	
Scarlet Toner	4 parts

Piano Lacquer, Clear

½" R. S. Cotton, Wet	14. parts
Tricresylphosphate	2. parts
Kettle Bodied Linseed Oil	20. parts
Butyl Acetate	24. parts
Butanol	16. parts
Toluol	40. parts
Hercusol 80	6. parts
Turpentine	6. parts

Spraying Lacquer for Farm Machinery

½" R. S. Cotton Wet	14. parts
Tricresylphosphate	2. parts
Butyl Acetate 12 ⎫ Butanol 8 ⎬ Toluol 20 ⎭	55. parts
Ester Gum*	5. parts
Kettle Bodied Linseed Oil*	10. parts
Xylol	3. parts

*Fused together at 300° C.

Yellow

Chrome Yellow	15. parts

Dark Orange

Basic Red Chromate	15. parts

Outdoor White Lacquer
(For Trucks, etc.)

½" R. S. Cotton, Wet	14. parts
Tricresylphosphate	3. parts
Amyl Acetate	7.23 parts
Butyl Acetate	6.44 parts
Butanol	10.47 parts
Toluol	18.8 parts
45% Dammar in alcohol	8. parts
Kettle Bodied Linseed Oil	4. parts
Zinc Oxide	6. parts
Antimony Oxide	5. parts

Flat White Surfacing Lacquer, Outdoor
(For Wood)

½"R. S. Cotton Wet	1.47 parts
4" R. S. Cotton Dry	.63 part
20" R. S. Cotton Dry	.42 part
Alcohol	.42 part
Kettle Bodied Linseed Oil*	1.67 parts
Ester Gum*	1.05 parts
Butyl Acetate	7.14 parts
Butanol	4.76 parts
Toluol	11.90 parts
Tricresylphosphate	1.68 parts
Zinc Oxide	12.60 parts

* Fused together at 300° C.

Outdoor White Gloss Lacquer
(For Wood)

½" R. S. Cotton, Wet	14. parts
Amyl Acetate	9.1 parts

Butyl Acetate	8.1 parts
Butanol	13.2 parts
Toluol	23.7 parts
Tricresylphosphate	2.0 parts
Ester Gum	2.5 parts
Blown Linseed Oil	4.7 parts
Kettle Bodied Linseed Oil	7.8 parts
Titanox	10.0 parts

Outdoor Spraying Lacquer
(Black)

½" R. S. Cotton, Wet	14. parts
Butyl Acetate	16.5 parts
Butanol	11. parts
Toluol	27.5 parts
Ester Gum*	5. parts
Kettle Bodied Linseed Oil*	10. parts
Xylol	3. parts
Paris X Black	3. parts
Tricresylphosphate	2. parts

* Fused together at 300° C.

White

Zinc Oxide	10 parts
Titanox	10 parts

Red

Scarlet Toner	5 parts

Gutta-Resin Lacquer

1. Solution mixture: benzol, 3 parts; toluol, 2 parts.
2. Solution mixture: ethyl acetate, 6 parts; butyl alcohol, 4 parts; butyl propionate, 3 parts; butyl acetate, 2 parts.
3. Solids: (½ sec.) dry pyroxylin, 1 part; (5 to 10 sec.) dry pyroxylin, 1 part; hard gutta percha resin, 2 parts.
Mix in the following proportions: 10 parts solution 1; 7½ parts solution 2; 2 parts solids.
The gutta resins are good softeners for rubber.

Lacquer Bronzing Liquid

Wet Nitrocellulose (15 sec.)	10 lb.
Toluol	50 lb.
Ethyl Acetate	17 lb.
Butyl Acetate	12 lb.
Blown Castor Oil	1 lb.
50% Ester Gum Solution in Toluol	10 lb.

Bronzing Liquid

½" R. S. Cotton, Wet	21 parts
Toluol	50 parts
Alcohol	53 parts
Butyl Acetate	50 parts
Tricresylphosphate	5 parts
Ester Gum	7 parts
Blown Linseed Oil	8 parts

For use with aluminum, gold, and colored bronze powders. May also be used as a clear colored lacquer, the color being obtained with oil-soluble aniline dyes.

Flexible Electrical Cable Lacquer
Canadian Patent 338,769

A coating composition comprises nitrocellulose 10, tritolyl phosphate 5–15, a nonvolatile mineral oil 0.2–1.0 parts by weight. The composition is used for coating flexible electric cables and similar products.

Cabinet Green Primer for Metal
Formula 1

½" R. S. Cotton, Dry	100.	parts
Alcohol	40.	parts
Butanol	104.7	parts
Toluol	188.	parts
Amyl Acetate	72.3	parts
Butyl Acetate	64.4	parts
Tricresylphosphate	30.	parts
Ester Gum*	25.	parts
Kettle Bodied Linseed Oil*	75.	parts
Xylol	10.	parts
Titanox	21.	parts
Chrome Yellow	5.6	parts
Paris Black	.7	part
Indian Red	1.0	part

* Fused together at 300° C.

Formula 2

½" R. S. Cotton, Dry	100.	parts
Alcohol	40.	parts
Butanol	104.7	parts
Toluol	188.	parts
Amyl Acetate	72.3	parts
Butyl Acetate	64.4	parts
Tricresylphosphate	30,	parts
Ester Gum*	29.8	parts
Kettle Bodied Linseed Oil*	29.8	parts
Xylol	12.	parts
Titanox	21.	parts
Chrome Yellow	5.6	parts
Indian Red	1.0	part
Paris Black	.7	part

*Fused together at 300° C.

Low Viscosity Lacquer
U. S. Patent 1,909,935

Ethyl-cellulose (viscosity 1.0) 10–15 parts, resin or gum (ester gum or phenol-formaldehyde resin prepared with an acid catalyst) 10 parts, and softener or plasticizer (such as diethyl phthalate) 1–3 parts; 20 parts of this mixture are dissolved in 100 parts of a solvent made up of ethyl acetone 67 parts, acetone 22 parts, toluene 11 parts. Coating compositions prepared in the above way may also be used for producing films suitable for photographic purposes.

Clear Metal Lacquer

Use the following percentages by weight:

Nitro Cellulose (½ sec.)	10%
Flexoresin DA-1	10%
Damar Gum Dewaxed (6 lb. cut)	10%
Ethyl Acetate	8%
Butyl Acetate	12%
Butyl Alcohol	6%
Toluol	36%
Xylol	8%

This lacquer has good adhesion to brass, steel, tin, silver, bronze, and gold. It should be thinned with one part of lacquer to one part of thinner for spraying. However, if one wishes to dip, it should be thinned; 70 parts of lacquer to 30 parts of thinner.

Cabinet Green Surfacer for Metal
Formula 1

½" Cotton, Dry	10.	parts
Alcohol	4.	parts
Tricresylphosphate	2.	parts
Amyl Acetate	9.	parts
Butyl Acetate	9.	parts
Butanol	9.	parts
Ethyl Acetate	14.	parts
Toluol	39.	parts
Ester Gum*	5.	parts
Kettle Bodied Linseed Oil*	15.	parts
Xylol	4.	parts
Chrome Yellow	8.2	parts
Paris Black	1.	part
Titanox	24.	parts
Indian Red	1.8	part
Asbestine	15.	parts

*Fused together at 300° C.

Formula 2

½" Cotton, Dry	10.	parts
Alcohol	4.	parts
Thicresylphosphate	2.	parts
Amyl Acetate	9.	parts
Butyl Acetate	9.	parts
Butanol	9.	parts
Ethyl Acetate	14.	parts
Toluol	39.	parts
Ester Gum*	5.	parts
Kettle Bodied Linseed Oil*	15.	parts
Xylol	4.	parts
Chrome Yellow	16.5	parts
Paris Black	2.2	parts
Titanox	48.	parts

Indian Red	3.65	parts
Asbestine	30.	parts

˙ Fused together at 300° C.

Cabinet Green Finishing Lacquer for Metal

Formula 1

½″ Cotton, Wet	600.	parts
Tricresylphosphate	86.	parts
Amyl Acetate	390.	parts
Butyl Acetate	347.	parts
Butanol	566.	parts
Toluol	1015.	parts
Ester Gum*	321.5	parts
Kettle Bodied Linseed Oil*	321.5	parts
Xylol	86.	parts
Paris Black	16.8	parts
Chrome Yellow	231.2	parts
Titanox	448.8	parts
Indian Red	111.2	parts
Prussian Blue	15.2	parts
Ferrite Yellow	36.	parts
Cosmic Black	57.6	parts

*Fused together at 300° C.

Formula 2

½″ R. S. Cotton, Dry	436.	parts
Alcohol	174.	parts
Amyl Acetate	398.	parts
Butyl Acetate	353.	parts
Butanol	575.	parts
Toluol	1035.	parts
Tricresylphosphate	87.	parts
Chrome Yellow	101.8	parts
Paris Black	16.3	parts
Titanox	320.	parts
Indian Red	23.8	parts
Ester Gum	327.	parts
Blown Linseed Oil*	327.	parts
Xylol	90.	parts

* Fused together at 300° C.

White Primer for Metal

½″ R. S. Cotton, Wet	14.	parts
Amyl Acetate	7.23	parts
Butyl Acetate	6.44	parts
Butanol	10.47	parts
Toluol	16.30	parts
Ester Gum ⎫ cold	3.	parts
Toluol ⎭ cut	3.	parts
Blown Linseed Oil	3.	parts
Zinc Oxide	3.	parts
Tricresylphosphate	3.	parts

White Surfacer for Metal

½″ R. S. Cotton, Wet	56.	parts
Tricresylphosphate	8.	parts
Amyl Acetate	36.	parts

Butyl Acetate	36.	parts
Butanol	36.	parts
Ethyl Acetate	56.	parts
Toluol	156.	parts
Ester Gum	20.	parts
Kettle Bodied Linseed Oil	20.	parts
Blown Linseed Oil	40.	parts
Zinc Oxide	75.	parts
Lithopone	345.	parts
Ultramarine Blue		Sufficient

White Finishing Lacquer for Metal

½″ R. S. Cotton, Wet	140.	parts
Tricresylphosphate	30.	parts
Amyl Acetate	98.	parts
Butyl Acetate	85.	parts
Butanol	140.	parts
Toluol	217.	parts
Ester Gum ⎫ cold	50.	parts
Toluol ⎭ cut	50.	parts
Blown Linseed Oil	50.	parts
Titanox	225.	parts
Ultramarine Blue		Sufficient

Colored Finishing Lacquer for Metal
Base Lacquer

½″ R. S. Cotton, Wet	14.	parts
Tricresylphosphate	3.	parts
Amyl Acetate	7.23	parts
Butyl Acetate	6.44	parts
Butanol	10.47	parts
Toluol	18.8	parts
Ester Gum*	2.5	parts
Kettle Bodied Linseed Oil*	7.5	parts
Xylol	2.	parts

* Fused together at 300° C.

Pigments for Base Colors

Ferrite Yellow	8	parts
Indian Red	8	parts
Paris X	1	part
Chrome Green	6	parts
Titanox	8	parts

Outdoor Lacquer for Metal
(Black)

20″ R. S. Cotton, Wet	14.	parts
Dibutylphthalate	3.	parts
Ester Gum*	3.	parts
Blown Linseed Oil*	7.	parts
Cobalt Linoleate	.02	part
Xylol	2.	parts
Butyl Acetate	12.	parts
Butanol	13.	parts
Ethyl Acetate	6.	parts
Alcohol	7.5	parts
Toluol	37.5	parts

| Cosmic Black | 4. | parts |
| Paris X Black | 1. | part |

An inexpensive lacquer for machinery, metal fences, etc.

* Fused together at 300° C.

Brass Lacquer, Clear
Formula 1

Film Scrap	15.	parts
Xylol	60.	parts
Benzol	40.	parts
Tricresylphosphate	3.	parts
Ethyl Acetate	17.	parts
Butyl Acetate	25.	parts
Blown Linseed Oil	4.	parts
Ester Gum	4.	parts
Alcohol	26.	parts

Formula 2

4″ R. S. Cotton, Wet	28.	parts
Xylol	80.	parts
Benzol	40.	parts
Tricresylphosphate	10.	parts
Ethyl Acetate	17.	parts
Butyl Acetate	25.	parts

Formula 3

20″ R. S. Cotton, Wet	21.	parts
Xylol	60.	parts
Benzol	40.	parts
Tricresylphosphate	7.	parts
Ethyl Acetate	17.	parts
Butyl Acetate	25.	parts
Blown Linseed Oil	8.	parts
Alcohol	20.	parts

Formula 4

Film Scrap	15.	parts
Xylol	60.	parts
Benzol	40.	parts
Ethyl Acetate	17.	parts
Butyl Acetate	25.	parts
Blown Linseed Oil	8.	parts
Alcohol	26.	parts

Formula 5

Film Scrap	18	parts
Xylol	60.	parts
Benzol	40.	parts
Tricresylphosphate	3.	parts
Ethyl Acetate	17.	parts
Butyl Acetate	25.	parts
Blown Linseed Oil	3.	parts
Ester Gum	4.	parts
Alcohol	30.	parts

Brass Dipping Lacquer, Clear

½″ R. S. Cotton, Wet	10.5	parts
4″ R. S. Cotton, Wet	10.5	parts
Alcohol	55.	parts
Toluol	55.	parts
Butyl Acetate	50.	parts
Tricresylphosphate	5.	parts

Ester Gum	7.	parts
Blown Linseed Oil	8.	parts
Ethyl Acetate	5.	parts

Thinner for Brass Dipping Lacquer

Alcohol	55.	parts
Toluol	55.	parts
Butyl Acetate	50.	parts
Ethyl Acetate	5.	parts

Indoor Wall Lacquer, Cream

½″ A. S. Cotton, Wet	105	parts
Butyl Propionate	180	parts
Tricresylphosphate	20	parts
Ethyl Acetate	135	parts
Butyl Acetate	135	parts
Naphtha	195	parts
Ester Gum	90	parts
Blown Linseed Oil	60	parts
Zinc Oxide	82	parts
Titanium Oxide	41	parts
Chrome Yellow	6.74	parts
Prussian Blue	Sufficient	

Outdoor Lacquer, White

½″ R. S. Cotton, Wet	14	parts
Tricresylphosphate	3	parts
Butyl Propionate	15.3	parts
Ethyl Acetate	11.3	parts
Butyl Acetate	11.3	parts
Naphtha	16.2	parts
Ester Gum	5	parts
Blown Linseed Oil	10	parts
Zinc Oxide	14	parts
Titanium Oxide	7	parts

Outdoor Clear Spraying Lacquer

½″ R. S. Cotton, Wet	14	parts
Butyl Acetate	16.5	parts
Butanol	11	parts
Toluol	27.5	parts
Ester Gum*	5	parts
Kettle Bodied Linseed Oil*	10	parts
Xylol	3	parts
Tricresylphosphate	2	parts

* Fused together at 300° C.

Outdoor Aluminum Spraying Lacquer
Use 14.25 parts aluminum powder in the above formula.

Brushing Lacquer, Clear

½″ A. S. Cotton, Dry	300	parts
Alcohol	120	parts
Anhydrous Alcohol	900	parts
Butyl Propionate	720	parts
Tricresylphosphate	90	parts
Butanol	180	parts

Ethyl Lactate	180 parts
Ethyl Acetate	300 parts
Ester Gum	225 parts
Blown Linseed Oil	225 parts

Brushing Lacquer

½" A. S. Cotton, Wet	14 parts
Tricresylphosphate	2 parts
Ester Gum*	7.5 parts
Kettle Bodied Linseed Oil*	1.5 parts
Xylol	3 parts
Butyl Acetate	24 parts
Butanol	16 parts
Toluol	40 parts

* Fused together at 300° C.

Pigment with sufficient pure pigment color to give desired covering.

Gloss Brushing Lacquer

This formula is suggested for making a gloss brushing lacquer which can be pigmented by incorporating the usual pigments now used in the industry. Grinding procedure shown previously to be used.

Wet ½" A. S. Cotton	35 parts
Butyl Cellosolve	35 parts
Cellosolve	30 parts
Butyl Propionate	30 parts
Alcohol	48 parts
Lacquer Gasoline	96 parts
*A. D. M. No. 100 Oil	35.1 parts
Ester Gum	25 parts

* Archer-Daniels-Midland.

Outdoor Brushing Lacquer, White

½" R. S. Cotton, Wet	14 parts
Tricresylphosphate	3 parts
Butyl Propionate	15.3 parts
Ethyl Acetate	11.3 parts
Butyl Acetate	11.3 parts
Naphtha	16.2 parts
Ester Gum	5 parts
Blown Linseed Oil	10 parts
Zinc Oxide	14 parts
Titanium Oxide	7 parts

Outdoor Clear Brushing Lacquer

½" R. S. Cotton, Wet	14 parts
Tricresylphosphate	2 parts
Amyl Acetate	26.5 parts
Butanol	19.5 parts
Toluol	35 parts
Ester Gum*	7.5 parts
Blown Linseed Oil*	7.5 parts
Xylol	3 parts

* Fused together at 300° C.

Quick-Drying Outside Lacquer-Paint

½" A. S. Cotton, Wet	28 parts
Ansol	150 parts
Butyl Propionate	40 parts
Tricresylphosphate	6 parts
Butanol	12 parts
Blown Linseed Oil	40 parts
Naphtha	14 parts
Gum Elemi	36 parts
Titanium Oxide	42 parts
Zinc Oxide	84 parts

Dipping Lacquer, Clear

½" R. S. Cotton, Wet	14 parts
4" R. S. Cotton, Wet	14 parts
Ethyl Acetate	45 parts
Amyl Acetate	5 parts
Butanol	8 parts
Butyl Propionate	20 parts
Toluol	25 parts
Benzol	20 parts
Ester Gum	9 parts
Blown Linseed Oil	9 parts
Tricresylphosphate	7 parts

Dipping Lacquer

The following is a suggested formula for the manufacture of a dipping lacquer.

Wet ½" R. S. Cotton	14 parts
Wet 4" R. S. Cotton	14 parts
Ethyl Acetate	45 parts
Amyl Acetate	5 parts
Butanol	8 parts
Butyl Propionate	20 parts
Toluol	45 parts
A. D. M. No. 100 Oil	29.2 parts
Ester Gum	10 parts

Dipping Lacquer, Clear, for Wood
Formula No. 1

4" R. S. Cotton, Wet	28 parts
½" R. S. Cotton, Wet	28 parts
Toluol	80 parts
Butanol	20 parts
Ethyl Lactate	10 parts
Tricresylphosphate	7 parts
Anhydrous Alcohol	44 parts
Ethyl Acetate	106 parts
Ester Gum	10 parts
Blown Linseed Oil	40 parts

Formula No. 2

½" R. S. Cotton, Wet	14 parts
4" R. S. Cotton, Wet	14 parts
Tricresylphosphate	6 parts
Camphor	2 parts
Blown Linseed Oil	20 parts
Ester Gum	5 parts
Benzol	10 parts
Toluol	32 parts
Butanol	9 parts

Anhydrous Alcohol   11 parts
Ethyl Acetate   49 parts
Butyl Propionate   15 parts
For tool handles, toys, etc.

Dipping Lacquer, for Metal, Red, Clear

½" R. S. Cotton, Wet   28 parts
Tricresylphosphate   7 parts
Toluol   50 parts
Ethyl Acetate   60 parts
Butyl Propionate   5 parts
Ethyl Lactate   5 parts
Ester Gum   15 parts
Blown Linseed Oil   5 parts
National Oil Red No. 0   Sufficient

May be used with any oil-soluble dye to produce other colors.

Dipping Lacquer, White, for Wood

Formula No. 1

½" R. S. Cotton, Wet   14   parts
Tricresylphosphate   4   parts
Amyl Acetate   7.23 parts
Butyl Acetate   6.44 parts
Butyl Alcohol   10.47 parts
Toluol   16.3 parts
Ester Gum ⎱ cold   3   parts
Toluol   ⎰ cut.   3   parts
Blown Linseed Oil   9   parts
Zinc Oxide   10   parts
Titanium Oxide   5   parts

For the handles of kitchen utensils, toys, etc.

Formula No. 2

½" R. S. Cotton, Wet   14   parts
Tricresylphosphate   4   parts
Amyl Acetate   7.23 parts
Butyl Acetate   6.44 parts
Butanol   10.47 parts
Toluol   16.3 parts
45% Dammar in Alcohol   6.66 parts
Blown Linseed Oil   9   parts
Zinc Oxide   10   parts
Titanium Oxide   5   parts

For the handles of kitchen utensils, toys, etc.

Dipping Lacquer for Metal (Green)

½" R. S. Cotton, Wet   14 parts
4" R. S. Cotton, Wet   14 parts
Tricresylphosphate   7 parts
Butanol   8 parts
Benzol   20 parts
Toluol   25 parts
Amyl Acetate   5 parts
Ethyl Acetate   45 parts
Butyl Propionate   20 parts
Ester Gum   9 parts

Blown Linseed Oil   9 parts
Chrome Green   12 parts
For Black, use
Super Spectra Black   .79 part

Watch Dial Enamel, for Spraying or Dipping

½" R. S. Cotton, Wet   56 parts
Ethyl Acetate   40 parts
Butyl Acetate   60 parts
Butanol   16 parts
Butyl Propionate   40 parts
Toluol   50 parts
Benzol   40 parts

Dammar Sol. ⎰ Ethyl Acet. 6.3 ⎱
           ⎱ Acetone   6.3 ⎰
             Benzol   29.4 ⎫ 111 parts
             Alcohol   33.0 ⎭
             Dammar   36.0

Blown Linseed Oil   18 parts
Tricresylphosphate   10 parts
Titanium Oxide   15 parts
Zinc Oxide   30 parts

An excellent high gloss lacquer for general purposes.

Coating Lacquer, Black (For Coating Machine)

20" R. S. Cotton, Wet   14 parts
Ethyl Acetate   40 parts
Alcohol (F. 30)   20 parts
Benzol   20 parts
Blown Linseed Oil   30 parts
Paris X Black   1 part

Golf Ball Lacquer

4" R. S. Cotton, Wet   14   parts
Dibutylphthalate   3   parts
Butanol   19.5
Amyl Acetate   13.5 ⎱
Butyl Acetate 12.0 ⎰ 50   parts
Toluol   35.0
45% Dammar in Alcohol   5.55 parts
Kettle Bodied Linseed
  Oil   5   parts
Titanox   10   parts

Pure White Lacquer

½" R. S. Cotton, Wet   7 parts
4" R. S. Cotton, Wet   7 parts
Dibutylphthalate   3 parts
Butyl Acetate   6 parts
Butanol   6 parts
Ethyl Acetate   12 parts
Alcohol   13 parts
Toluol   40 parts
45% Dammar in Alcohol   8 parts
Blown Soya Oil   7 parts
Zinc Oxide   6 parts

Titanox	5 parts
Ultramarine Blue	Sufficient

Wall Paper Lacquer, Clear (Glossy, Colorless)

½" R. S. Cotton, Wet	28 parts
Ethyl Acetate	22 parts
Butyl Acetate	32 parts
Butanol	9 parts
Butyl Propionate	21 parts
Toluol	27 parts
Benzol	20 parts
Dammar Sol.:	
Ethyl Acetate	6.3
Acetone	6.3
Benzol	29.4 ⎫ 60 parts
Alcohol	33.0
Dammar	36.0
Blown Linseed Oil	10 parts
Tricresylphosphate	5 parts

Floor Lacquer

½" R. S. Cotton, Wet	14 parts
Dibutylphthalate	5 parts
Ester Gum	5 parts
Ethyl Acetate	17 parts
Butyl Acetate	17 parts
Naphtha	31 parts

Playing-Card Lacquer

Film Scrap	9 parts
Xylol	40 parts
Benzol	60 parts
Ethyl Acetate	25 parts
Butyl Acetate	17 parts
Alcohol	30 parts

Also suitable for coating silverware and similar metal articles.

Ebony Finish for Wood

½" R. S. Cotton, Wet	14 parts
Dibutylphthalate	2 parts
Butyl Acetate	21 parts
Butanol	14 parts
Toluol	35 parts
Ester Gum	5 parts
Kettle Bodied Linseed Oil	7 parts
Xylol	2.4 parts
Amyl Black No. 15827	.5 part

Linoleum Lacquer, Hard Finish
Formula No. 1

20" R. S. Cotton, Wet	188 parts
Blown Soya Oil	67 parts
Blown Linseed Oil	67 parts
Ethyl Acetate	395 parts
Benzol	885 parts

Formula No. 2

½" R. S. Cotton, Wet	56 parts
Toluol	80 parts
Butanol	20 parts
Ethyl Lactate	10 parts
Tricresylphosphate	7 parts
Ansol	44 parts
Ethyl Acetate	106 parts
Ester Gum	10 parts
Blown Linseed Oil	10 parts
Gum Elemi	10 parts

Linoleum Lacquer, Flexible Finish
Formula No. 1

20" R. S. Cotton, Wet	188 parts
Blown Soya Oil	134 parts
Blown Linseed Oil	134 parts
Ethyl Acetate	395 parts
Benzol	885 parts

Formula No. 2

½" R. S. Cotton, Wet	56 parts
Toluol	80 parts
Butanol	20 parts
Ethyl Lactate	10 parts
Tricresylphosphate	7 parts
Ansol	44 parts
Ethyl Acetate	106 parts
Ester Gum	7.5 parts
Blown Linseed Oil	15 parts
Gum Elemi	7.5 parts

Crackle Lacquer (Green)

½" R. S. Cotton, Wet	59 parts
Ethyl Acetate	68 parts
Butyl Acetate	99 parts
Toluol	235.6 parts
Wax-free Dammar	33.5 parts
Alcohol	219.9 parts
Benzol	78 parts
Chrome Green	226 parts
Asbestine	85 parts
Magnesium Carbonate	51 parts

This lacquer when sprayed over a contrasting previous coat of lacquer will produce an ornamental cracked effect. The first coat must be heavy, and allowed to dry before spraying the crackle lacquer. Any color may be used to replace the green, but the high pigmentation must be maintained.

"Aroclor" Lacquers
Formula No. 1

A dammar lacquer for sanding and polishing; of excellent durability. This formula is expressed in parts by weight.

½ Sec. Nitrocellulose	
(Dry Basis)	100.0 parts
Dammar	80.2 parts
Aroclor 1262	20 to 39.5 parts
Dibutyl Phthalate	20 to 0 parts

Formula No. 2

An ester gum lacquer for sanding and polishing; of excellent durability especially in air near sea-coast. This formula is expressed in parts by weight.

½ Sec. Nitrocellulose	
(Dry Basis)	100 parts
Ester Gum	80 parts
Aroclor 1262	20 parts
Dibutyl Phthalate	20 parts

Formula No. 3

A softer lacquer, very flexible, not suitable for sanding, of good weathering qualities. This formula is typical of the lacquers in which the Aroclor performs the function of the resin. The clear lacquer made according to this formula is practically colorless. It is especially useful for white enamels in those cases where unusual flexibility is desired. This formula is expressed in parts by weight.

½ Sec. Nitrocellulose	
(Dry Basis)	100 parts
Arocolor 1262	80 to 70 parts
Lindol (Tricresyl Phosphate)	39 to 70 parts

Pigments:

For white enamels either French Process Zinc Oxide or Titanox may be used. As examples of the amount of pigment, the white pigments may be used in the proportion of 147 parts by weight on the basis of the above formulas or 40% of the weight of total solids including the pigment. For the colored enamels the pigments may be used in the usual proportions down to the carbon blacks which may be used in the proportion of 15 parts by weight on the basis of the above formulas or 6.4% of the total solids including the pigment.

Solvents:

The mixed solvent should be one in which the high boiling constituent is amyl or butyl acetate. Other types of mixed solvents in which the high boiler is of an entirely different type are well known to the industry, but consistently good results have not been obtained with such solvents for lacquers containing the Aroclors. We, therefore, strongly recommend the use of a balanced solvent in which the least volatile constituent is amyl or butyl acetate.

Nitrocellulose Lacquer Formulae

Water Proof Cement

This formula is expressed in per cent by weight.

Nitrocellulose (15–20 sec.)	13.3%
Camphor	1.7%
Benzol	45.0%
Acetone	35.0%
Diacetone Alcohol	5.0%

Silk or Cloth Lacquer

This formula is expressed in per cent by weight.

Nitrocellulose (½ sec.)	10.0%
Lindol	6.5%
Ethyl Acetate	10.0%
Butyl Acetate	10.0%
C. D. Alcohol	10.0%
Toluol	53.5%

Lacquer Plasticizer
U. S. Patent 1,955,348

A mixture of 2,000 parts of castor oil and 2 parts of selenium powder is heated for three hours at 250° C. The mixture at this point appears to be the very same as castor oil. The mixture is allowed to cool and to stand several days. It changes over into a pasty mass containing about 37% of a white solid and about 63% of a liquid very similar to castor oil. The separation of this mixture is accomplished by treating 10 parts of the mixture with 25 parts of petroleum ether or other aliphatic hydrocarbon solvent. The liquid portion dissolves the petroleum ether, and is thereby thinned. The solid portion can be removed by filtering and further purified by a similar washing with petroleum ether. The liquid portion which has been diluted with petroleum ether is heated until no more petroleum ether is evolved (about 2 hours at 150° C. is sufficient). In the treatment of castor oil with selenium, the amount of solid portion decreases as the length of time of heating is increased.

A typical example of a nitrocellulose composition containing these improved softeners for use in clear lacquers follows (stated in parts by weight):

Pyroxylin	12	parts
Dammar Gum	3	parts
Dibutyl Phthalate	4	parts
Liquid Portion of Selenium Treated Castor Oil	2.6	parts

Rubbing Lacquer

R. S. Nitrocellulose		
(½ sec.)	56	parts
Toluol	80	parts
Butanol	20	parts
Ethyl Acetate	106	parts
Anhydrous Alcohol	37.2	parts
Ethyl Lactate	10	parts
Lacquer Grade Linseed Oil	46.8	parts

Lacquer Containing Wax

U. S. Patent 1,942,902

Wax is reduced by first crushing and coarse grinding the large lump of wax to approximately 20 mesh to the inch size. The resulting coarse powder is then placed in a closed pebble mill and ground in the presence of a solvent such as butyl acetate, ethyl acetate, etc., to which may be added a quantity of nitrocellulose. With 150 pounds of carnauba wax, and 38 gallons butyl acetate use 15 pounds of dry nitrocellulose. (The addition of nitro cellulose, while desirable, is not, however, essential to the process.) The grinding in the pebble mill, as stated, is continued for the purpose of reducing the particle size of the wax until such time as a sample taken from the batch indicates that the product is smooth enough for use. This generally requires from 24 to 60 hours, depending upon the type of mill employed.

At the end of this grinding operation, there is a resulting mass composed of approximately 33.8 per cent by weight of finely divided carnauba wax, approximately 62.8 per cent by weight of butyl acetate, and approximately 3.4 per cent by weight of nitrocellulose.

The resulting product is mixed and blended by agitation with the desired quantity of conventional cellulose ester lacquer and is thereupon ready for application by brush or spray gun.

Another method of preparation which may be followed is as follows. The wax is first disintegrated by melting and while in melted condition is poured into any organic solvent which is compatible with a cellulose ester lacquer. The solvent may, however, be poured into the melted wax. The mass is agitated continuously until it cools to room temperature. If not sufficiently smooth for use at this stage of the process, the mass is placed in a mill, preferably a pebble mill, and the particles are reduced, either with or without the addition of nitrocellulose as in the cold process, until the particle size of the wax is sufficiently small to produce the desired blend. In carrying out this second mentioned process, a very satisfactory breaking up or reduction of the wax particles may be accomplished during the so-called agitation step by causing the mass, while cooling, to circulate rapidly, e.g., through rotary or centrifugal pumps. The mechanical operation of the pumps appears to exert a pronounced smoothing action on the wax and a very satisfactory reduction of particle size may be accomplished in this manner.

The mechanical circulation and agitation may be practiced in both the cold and hot processes and in the latter the circulation of the mass may be continued after the mass has cooled to room temperature and until such time as the desired and necessary smoothness is obtained. In any event, the wax, after being reduced and smoothed as stated is added to the conventional cellulose ester lacquer and is then ready for use.

Lacquer made according to the present invention as herein described has an opaque, cloudy appearance, due to the dispersion of the carnauba wax in solid particle form and the further fact that the carnauba wax is substantially insoluble in the other constituents of the composition.

Lacquer made according to this invention, as described, is a thoroughly commercial product. It is readily applied to the surface to be finished in the same manner as other lacquers and dries with a hard wax finish, free from lustre and possessing qualities superior to those obtained from applying a conventional wax and then polishing the same. Furthermore, the superior finish is obtained economically and expeditiously, without the expense and drudgery required to apply wax finishes as formerly.

Lacquer Bottle Seal, Iridescent

This is a very attractive coating composition for use in sealing corked bottles and giving the effect of being covered with a tin foil cap. It may also be used as a brushing lacquer, and if more solvents are used may be applied from a spray gun. It is composed of the following, in percentages by weight:

Cellulose Nitrate (½ sec.)	10%
Ester Gum	5%
Dibutyl Phthalate	5%
Acetone	10%
Methanol	12%
Benzol	20%
Ethyl Acetate	25%
Butyl Alcohol	5%
Butyl Acetate	8%

The solvents are mixed and the cellulose nitrate added and stirred until solution is complete. The ester gum is powdered and then added. This is a quick-drying lacquer and of the proper body for dipping. The cork is driven flush with the head of the neck of the bottle. Same is turned neck down, dipped into the lacquer to the desired depth, removed and allowed to dry in this inverted position,

Reducing the Body of Scrap Celluloid for Use in Lacquers

Scrap celluloid, such as clean clear sheeting from movie film, x-ray film, camera films, etc., may have the viscosity or body of its solution reduced by the following method.

The cellulose nitrate is dissolved in the solvent mixture suitable for the purpose for which it is to be ultimately used. An amount of crystalline sodium acetate equal to from 2% to 4% of the weight of the dry cellulose nitrate used is dissolved in a minimum amount of 95% ethyl or methyl alcohol and added to the above solution of cellulose nitrate. This mass is heated to about 50° to 60° in a closed vessel, or one equipped with a reflux condenser and maintained at this temperature until the solution has had its viscosity reduced to the desired body.

The smaller amount of sodium acetate used, the higher the temperature or the longer the time required to reduce the body to the same degree. Other substances may be used such as a solution of ammonia in methanol, dimethylamine in methanol, etc., but while these substances are more active and can be used in smaller amounts, they usually darken the ultimate solution to a greater extent than does sodium acetate.

Lacquers Containing Synthetic Resins

The working up of the urea type of resin with cellulose esters is not always a very simple matter. The amount of cellulose ester used must be double that of the artificial resin. If this ratio is not adhered to the finished masses exhibit great brittleness, poor adhesive power, and diminished stability towards water.

Following are some useful mixtures for lacquers of this type:

Formula 1

Nitrocellulose	25 parts
Castor Oil	20 parts
Normal Butyl Alcohol	125 parts
Isobutyl Acetate	125 parts
Urea Resin	25 parts
Isobutyl Alcohol	50 parts

The whole is well mixed together. The lacquer can be applied by spraying. To produce a product of suitable viscosity for this purpose the following may be added:

Ethyl Alcohol	75 parts
Isobutyl Alcohol	125 parts
Ethyl Acetate	100 parts
Isobutyl Acetate	100 parts
Benzol	100 parts

Formula 2

Another lacquer incorporating an artificial resin and a natural resin has the following composition:

Nitrocellulose	300 parts
Castor Oil (Oxidized)	450 parts
Kauri Copal	150 parts
Isobutyl Alcohol	2,250 parts
Normal Butyl Acetate	2,250 parts
Urea Resin	450 parts
Normal Butyl Alcohol	900 parts

As diluent the following mixture is suitable:

Ethyl Alcohol	225 parts
Isobutyl Alcohol	375 parts
Ethyl Acetate	300 parts
Isobutyl Acetate	300 parts
Benzol	300 parts

The lacquer described above, together with the diluent, will give a highly glossy coating which adheres well to the surface to which it is applied.

In place of the castor oil, animal oils such as cod oil, shark oil, sardine oil, etc., may be used. Esters of phosphoric acid may be included as additional softening agents. A useful formula is as follows:

Nitrocellulose	60 parts
Shark Oil (Oxidized)	45 parts
Ethyl Alcohol	300 parts
Ethyl Acetate	200 parts
Butyl Acetate	100 parts
Toluol	300 parts
Urea Resin	60 parts
Isobutyl Alcohol	80 parts

This lacquer does not require the addition of a diluent since its consistency is such that it can be easily applied directly.

Formula 3

The preparaton of the following lacquer is even more complicated. Nitrocellulose (15 parts) is heated with cod oil (10 parts) in presence of a nickel catalyst at 160–170° C. for 10 minutes, the catalyst is removed, and air is blown through the product heated at 160–170° C. during four hours to produce a highly viscous, brown product. Obviously this product can only be obtained if a stable nitrocellulose is used; otherwise decomposition will ensue. The product is dissolved in the following solvent:

Normal Butyl Alcohol	75 parts
Glycol Monoacetate	50 parts
Butyl Acetate	25 parts
Toluol	75 parts

A solution of 15 parts of urea resin

in 20 parts of normal butyl alcohol is added to the above lacquer, whereby a product giving coatings of good adhesive power and high elasticity is obtained.

Formula 4

The use of sardine oil is illustrated is the following preparation. Firstly, a solution is prepared from:

Nitrocellulose	20.0 parts
Sardine Oil	7.5 parts
Tricresylphosphate	5.0 parts
Normal Butyl Alcohol	75.0 parts
Glycol Monoacetate	37.5 parts
Butyl Acetate	25.0 parts
Toluol	75.0 parts

To this solution is now added the artificial resin mixture, made from urea resin 16 parts and isobutyl alcohol 19 parts. Glossy coatings are obtained with this lacquer.

Formula 5

As a final example, the following lacquer containing oxidized linseed oil may be quoted:

Nitrocellulose	30 parts
Tricresylphosphate	40 parts
Urea Resin	30 parts
Ethyl Alcohol	20 parts

This mass is preferably worked up in a kneading machine at 60–70° C. After thoroughly mixing, castor oil (20 parts) and ethyl alcohol (30 parts) are added. The whole is then further kneaded at the same temperature as before and finally at 100° C. in order to expel the alcohol used.

Pearl Essence, Synthetic

The particles which give the characteristic sheen to certain fish may be produced artificially by the following method.

A nearly saturated solution of barium chloride is prepared and allowed to clarify by standing, or may be filtered for immediate use.

A more dilute solution of sodium hyposulphite is also prepared and filtered. This solution is then heated to about 80° C. and added to the barium chloride solution with slow stirring until precipitation stops.

These small plate crystals are then washed by decantation until free from sodium chloride and excess sodium hyposulphite by testing with silver nitrate. The crystals are then dried.

They are then mixed with such suitable vehicles as nitrocellulose lacquers, varnishes and the like. Amounts as small as 5% of the weight of the vehicle produce the Pearl Essence or Fish Scale effect. Very attractive color effects may also be produced by the addition of small amounts of spirit soluble, or oils soluble dyestuffs to the vehicle.

Paste Filler (Lacquer Undercoat)

Linseed Oil	5– 7%
Lithopone and Black Aluminum Lake	70–80%
White Spirit (Petroleum)	25–13%

Grind together until uniformly smooth. This is thinned down with white spirit and sprayed.

Lacquer Spray Pressures

Many finishers neglect one of the most important points of their art—the air pressure they use in spraying. This is more apparent when they change from one finishing material to another, than it is where the same material is used continuously.

Because the difference in specific gravity and vicosity of finishing materials affects the proper spraying to a great degree, it is essential that the air pressure be regulated to suit each different material so that it will properly atomize, since each material will require a different air pressure.

The following pressures, have been found from experience, to be satisfactory.

Heavy-Bodied Materials	10 to 20–lb
Low-Viscosity Lacquers	30 to 40–lb
Clear Furniture Lacquers	50 to 75–lb
Lacquer Enamels	60 to 80–lb
Primers	60 to 90–lb

The style of the spray gun and the type of equipment have a great deal to do with the best actual air pressure required but the above figures will serve as a basic standard in most cases.

Lacquers having solvents with high boiling points, which are not so volatile as others, may be successfully sprayed at 100 lb. pressure

Many spray operators set their air gauges at about 60 lb. pressure and leave them there for all classes of work and all types of materials. Reasonable attention to air pressure will repay any efforts in the better finishing which it produces.

Non-Inflammable Lacquer

Cellulose Acetate	20 parts
Chlorinated Rosin	10 parts
Dibutyl Phthalate	4 parts

Acetone	33 parts
Ethyl Acetate	10 parts
Ethyl Lactate	10 parts
Toluol	10 parts

Lacquer Remover

B B' Dichlorethylether	4 parts
Acetone	5 parts
Methanol	5 parts
Benzol	2 parts

Quick Drying Lacquer

Archer-Daniels-Midland

No. 100 Oil	1.04 lb.
Nitrocellulose (¼" dry)	1.56 lb.
Thinner	1. gal.

The thinner used with this formula has the following composition:

Toluol	61%
Butyl Acetate	26%
Butanol	13%

This lacquer will dry in about five minutes and can be sanded in about one-quarter to one-half hours, compared to one-half to two hours for other coatings.

Removal of Intermediate Layer of Cinema Film

The removal of the top gelatin layer from scrap cinema film, for use in the manufacture of quick-drying lacquers and adhesives, is comparatively easy, but the intermediate layer, which is generally of a protein character, and which is to be found on most film on the European market, adheres very tenaciously to the celluloid, and cannot be removed by the usual solvents such as hot water and ferments. It has recently been found that such agents as hydrogen peroxide and hot concentrated phosphoric acid will overcome this difficulty and the following example is given: Wash 30 kilograms of scrap film in hot water to remove the emulsion coat, and then treat for one hour at 80 to 100° C. with 0–6 kilograms of 40% hydrogen peroxide in 200 to 300 kilograms of water. Another method is to immerse washed cinema film in 90 per cent of phosphoric acid at about 80° C.

Brush Lacquer and Variation for Book Lacquer

½ Sec. R.S. Nitrocellulose (Wet)*	860 g.
Cellosolve	480 cc.
Butyl Cellosolve	240 cc.
Ethyl Acetate	240 cc.
Dibutyl Phthalate	180 cc.

Add 1400 cc. of A or B or 480 cc. of C.

A.

Bodied Tung Oil	120 g.
Ester Gum	120 g.
Benzene	600 cc.
Toluene	600 cc.

B.

Rosin	1500 g.
Tung Oil	1500 g.
Glycerine	300 g.

Heat to 200° C. then blow air through mixture, keeping it at 200° for 1 hour, test to see if it remains clear on cooling, if not continue operation until it does, then cool to about 100° and add:

| Benzene | 2700 cc. |
| Toluene | 2700 cc. |

A covered kettle with the necessary outlets should be used.

The above mixture can be thinned to desired viscosity with Ethyl Acetate.

To make a good Book Lacquer add 140 cc. of dibutyl phthalate, 50 cc. creosote and 5 g. of quinine sulfate.

The creosote and quinine sulfate keep roaches from bothering the lacquer and the dibutyl phthalate increases the flexibility.

* If nitrocellulose is dry, use 600 g.

Grinding Procedure for Pigmented Lacquers

Grinding the steel blue, chrome green and carbon black in the lacquer vehicle in the oall mill is recommended. The other pigments can be conveniently ground in the oil through a burr stone mill.

However, the black, green and blue can be ground in the oil through the burr stone mill if certain precautions are taken. To grind the blue in the oil dilute the 35.1 parts of oil with 5 parts of denatured alcohol, mix in 8 parts of steel blue and grind through burr stone mill until pigment is sufficiently fine. It it absolutely necessary that the mill be kept cool during the grinding of this pigment, otherwise it will liver with the oil. As soon as the paste has been ground mix it with an equal weight of the cotton solution. When mixed with the cotton solution it can be stored for an indefinite period. Use the same precautions and procedure for grinding the chrome green. However, it is necessary to dilute the oil with alcohol before grinding.

The carbon black pigments are amorphous and highly oil absorbing. They make pastes which are rather hard to break up. We, therefore, recommend grinding in the ball mill, using the lacquer as a vehicle.

"Non-Caking" Pigmented Lacquer
U. S. Patent 1,863,834

A method of preparing non-caking cellulose lacquers consists of grinding a mixture of the pigment and rubber solution with a solvent mixture and a gum, and adding this to a cellulosic material mixed with an ester, a gum, and a solvent, the mixture being adjusted to brushing viscosity by addition of a solvent. For example, 55 parts of titanox are ground in a pebble mill for 12 hours with a mixture of 5 per cent solution of rubber in gasoline 11 parts, ethyl alcohol 6.1 parts, butyl acetate 1.3 parts, toluol 2.3 parts, gum dammar 6.1 parts, and the mixture is added to a combination of butyl acetate 16, ethylene glycol ethyl ether 10, butyl alcohol 10, ethyl alcohol 7.9, gasoline, 3.4 toluol 4.2, gum dammar 3.9, half-second nitrocellulose 10, dibutyl phthalate 4 parts (all by weight). If desired, the pigment may be incorporated with the protective agent (rubber) by dry-milling without the use of a solvent or diluent, and the mass then mixed with the other constituents.

Homogeneous Pigmented Lacquer
U. S. Patent 1,911,104

The process consists in first preparing a dispersion of the pigment with dry cellulose ester in a mixture of liquids which is capable of causing the cellulose ester to swell but has no solvent effect on it. This may be done, for example, by slowly grinding a mixture of 5 parts of low-viscosity nitrocellulose, 5 parts of titanium oxide, ¼ part of ethyl acetate, ¼ part of blown castor oil, and 7 parts of a very low-boiling petroleum naphtha, preferably one having a boiling point of 70° to 175° C. The grinding may be effected by placing the mixture in a pebble mill and allowing it to rotate therein for 24 hours or more. When the required homogeneous dispersion has been obtained, the excess of petroleum naphtha is strained off, and the residue is dissolved in a suitable nitrocellulose solvent, whereby a completely dispersed pigmented solution is obtained. As an example of a finished lacquer of great opacity and covering power is mentioned the product obtained by mixing 2 "parts" of the above nitrocellulose pigment mixture with 1 gallon of varnish such as spar varnish, and ½–⅛ gallon of solvent, such as butyl lactate.

Clear Gloss Wood Lacquer

The quantities in these formulas are expressed in percentage by weight.

Formula No. 1
(High Grade)

¼ sec. Nitrocotton	13 %
Ester Gum 8 lb. cut *	16 %
Dammar Gum 6 lb. cut (dewaxed)	10 %
Blown Castor Oil	1.6%
Dibutyl Phthalate	1.6%
Butanol	9 %
Ethyl Acetate	10 %
Toluol	13.8%
Xylol	25 %

Formula No. 2
(Cheap Grade)

¼ sec. Nitrocotton	13 %
Ester Gum 8 lb.*	23 %
Blown Castor Oil	1.6%
Dibutyl Phthalate	0.8%
C.D. Alcohol No. 1	9 %
Butanol	4.6%
Ethyl Acetate	15 %
Butyl Acetate	7 %
Toluol	26 %

* The Ester gum cut referred to above is made by dissolving 8 pounds of gum in one gallon of toluol. Likewise the Dammar cut is made by dissolving the gum in toluol and later adding C. D. Alcohol to precipitate the wax. The flat lacquer is usually placed in a pebble mill and allowed to grind for about 10 hours. All of the above lacquers can be sprayed without thinning.

Clear Flat Wood Lacquer

¼ sec. Nitrocotton	12.5%
Ester Gum *	18.5%
Carnauba Wax	2.3%
Blown Castor Oil.	1.5%
Dibutyl Phthalate	0.7%
C.D. Alcohol	8.5%
Butanol	4 %
Ethyl Acetate	14 %
Butyl Acetate	8 %
Toluol	30 %

Durable Automotive Enamel Base

This formula gives quantities in percentage by weight.

½ sec. Nitrocotton	23%
Dammar Gum 6 lb. (Dewaxed)	19%
Blown Castor Oil	4%
Dibutyl Phthalate	4%
Butanol	7%
Ethyl Acetate	6%
Butyl Acetate	16%
Toluol	12%
Xylol	9%

Leather Enamel Base

This formula gives quantities in percentage by weight:

15°20 sec. Nitrocotton	13 %
Blown Castor Oil	9.5%

Dibutyl Phthalate	4.5%
C.D. Alcohol No. 1	10 %
Ethyl Acetate	14 %
Butyl Acetate	7 %
Toluol	42 %

Pigmented pastes may be added to both of the above lacquer bases. They both require thinning of one part of lacquer to one part of thinner in order to use in a spray gun.

Shellac Lacquer

A shellac lacquer of beautiful clarity, excellent toughness, perfect leveling, and exceptionally quick drying, may be made on the following actual working formula:

De-waxed Bleached Shellac	2 lb.
Solvent { 80% Fusel Oil / 20% Methanol }	1 gal.

The wax-free shellac must be very dry for best results, so spread it out onto pans or otherwise to evaporate any moisture, then cut it in the solvent by agitation of the mixture as in making ordinary shellac varnish. For a small lot as above, divide the quantity into two 1-gallon glass bottles and repeatedly shake by hand. Finally, filter through filter paper.

Shellac Filler

Ground Shellac	1 part
China Clay	2 parts
Denatured Alcohol	1 part

Grind to a uniform paste.

Treatment of Shellac with Sulphur, Shellac Varnish with Sulphur Monochloride

Varnish films, prepared from shellac which has been heated with 3–4% of sulphur at 180° until the cooled product is green, possess improved elasticity, adhesion and resistance to water and abrasion. The vulcanized shellac can be colored by relatively small amounts of pigments or dye and, when pressed, the sulphur does not darken the product or damage the mould.

Addition of 0.5–10% of a 10% solution of sulphur monochloride in carbon tetrachloride to a shellac varnish improves the water resistance and mechanical properties and reduces its color.

Formulas for Pastes

No. 100 Oil	Pigment	Alcohol	Added	Grinds
35.1	Titanox	40	..	3
35.1	Steel Blue	8	5	6–10
35.1	Chrome Green	20	..	5
35.1	Chrome Yellow	20	..	5
35.1	Carbon Black	4	5	3
35.1	Scarlet Toner	12	..	4

Paste A

A. D. M. No. 100 Oil	500 parts
Paris Black	25 parts
Chrome Yellow	150 parts
Titanium Dioxide	50 parts
Zinc Oxide	50 parts
Chrome Green	10 parts
Raw Sienna	15 parts

Paste ground through burr stone mill four times or until pigments are sufficiently fine.

Vehicles for Some Luminous Pigments

Satisfactory vehicles for calcium sulphate (2 parts) were: (a) Resoglaz resin (2 parts), diphenyl phthalate (1 part), and toluol (10 parts); (b) 30% dammar solution in turpentine (8 parts); (c) 20% chlorinated rubber solution in PhMe (7 parts) and 50% chlorodiphenyl resin solution in turpentine (3 parts), silicon and cellulose ester media gel and evolve hydrogen sulphide; linseed oil and oleoresins reduce the luminosity. Grinding the pigment into the vehicle has the same effect.

Phosphorescent Pigments

A method for producing phosphorescent pigments from alkaline earth sulphides have been developed. This is stated to be superior to the usual method, which uses a zinc sulphide base, in that commercially available materials can be used. The following formulas are suggested:

Formula No. 1
(Greenish-Blue Phosphorescence)

Strontium Hydroxide	20.7 g.
Sulphur	8.0 g.
Lithium Sulphate	1.0 g.
Colloidal Bismuth (0.3% Aqueous Suspension)	6.0 cc.

Heat the mixture for 40 minutes in a porcelain crucible, to the point of incandescence. Then allow the mass to cool slowly.

Formula No. 2
(Red Phosphorescence)

Barium Oxide	40.0 g.
Sulphur	9.0 g.
Lithium Phosphate	0.7 g.
Cupric Nitrate Solution (0.4% Solution in Ethyl Alcohol)	3.5 cc.

The lithium phosphate may be replaced by a mixture of magnesium phosphate with the carbonate or sulphate of lithium.

Formula No. 3
(Red Phosphorescence)

Magnesium Phosphate	40.0 g.
Lithium Sulphate	0.7–1.0 g.
Cupric Nitrate Solution (0.4% Solution in Ethyl Alcohol)	3.5 cc.

Ultraviolet Paints

The following formulae produce a strong fluorescence when exposed to ultra-violet light:

Blue-violet—Dissolve five parts of vaseline and 12 parts of white paraffin wax of melting point 140° F. in 175 parts of benzene and add, by stirring in, five parts of finely powdered calcium salicylate. Alternatively, five parts of aesculin may be substituted for the calcium salicylate.

Apple-green—Repeat the above mixture and substitute anthracene for the calcium salicylate.

Brilliant Green—Dissolve 20 parts of cellulose acetate in 300 parts by weight of chloroform into which one part of vaseline to 15 to 37 parts of chloroform are dissolved. Mix the solutions thoroughly and introduce 10 to 30 parts of finely powdered potassium uranyl sulphate.

Orange Yellow—Substitute zinc sulphide containing about one part in 1000 of manganese for the potassium uranyl sulphate in the above formula.

Red—Mix 100 parts of zinc sulphide with 200 parts of cadmium sulphate and incorporate in gum. The consistency of the mixture determines the brilliance of resulting hue.

Violet Luminous Paint

Twenty parts, by weight, of calcium oxide (burnt lime) free from iron; 6 parts by weight of sulphur; 2 parts by weight of starch; 1 part by weight of a 0.5 per cent solution of bismuth nitrate; 0.15 part by weight of potassium chloride; 0.15 part by weight of sodium chloride. The materials are mixed, dried and heated to 1300° C. (2373° F.). The product gives a violet light. To make this effective, it is exposed for a time to direct sunlight, or a mercury lamp may be used. Powerful incandescent gaslight also does well, but requires more time.

Fabric Paint

½ Second Nitrocotton	20 parts
Tricresyl Phosphate	24 parts
Cellosolve	20 parts
Toluol	90 parts
Basic Dye	5 parts

Metallic Paint Vehicle for Fabrics

70 Second Nitrocotton	5 parts
Tricresyl Phosphate	6 parts
Blown Rapeseed Oil	4 parts
Cellosolve	20 parts
Toluol	50 parts

Cold Water Paint in Powder Form

Precipitated Whiting or Chalk	230 parts
China Clay	50 parts
Slaked Lime	20 parts
Casein	50 parts
Lithopone	50 parts

Plastic Paint

Mica (250 Mesh)	30 parts
Molding Plaster	10 parts
Asbestos	10 parts
Asbestine	30 parts
Casein	10 parts
Soda Ash	1 part
Lithopone	9 parts

Plastic Paint Powder

Whiting	1000 parts
Clay	520 parts
Glue	60 parts
Gypsum	80 parts

Outside White Paint

The following formula has given a paint which wears splendidly and has unusual luster.

Titanium Oxide—Paste	1 gal.
Linseed Oil—Boiled	1 gal.
China Wood Oil (Spar) Varnish	¼ gal.

Gloss Paint
U. S. Patent 1,879,045

Heat 1000 parts of stand oil to 110°–120° C., add 10 parts 2, 3 hydroxynaphthoic acid, and stir to aid solution. Let cool. Grind 500 parts of zinc oxide in 330 parts of the treated oil, and reduce the paste with 110 parts of thinner (turpentine and mineral spirits) and sufficient cobalt drier.

Stone Cement and Anti-Slip Paint

Zirconium Silicate	9 parts
W T G Clay	1 part

Add 50% solution of water glass until of right consistency.

U. S. Navy Ship Bottom Paint

Formula No. 1
Anti-Corrosive

Alcohol	80	gal.
Gum Shellac	85	gal.
Pine Oil	6¾	gal.
Zinc Dust (dry)	110	lb.
Zinc Oxide (dry)	305	lb.

Formula No. 2
Anti-Fouling

Alcohol	76	gal.
Gum Shellac	163	lb.
Pine Oil	12½	gal.
Zinc Oxide (dry)	165	lb.
Indian Red (dry)	165	lb.
Oxide of Mercury	75	lb.

Formula No. 3
Copper Paint for Wooden Vessels

Alcohol	76	gal.
Gum Shellac	163	lb.
Pine Oil	12½	gal.
Zinc Oxide (dry)	165	lb.
Indian Red (dry)	165	lb.
Cuprous Oxide	75	lb.

Swedish Railway Paint

Red Iron Oxide	15.3%
Ferrous Sulfate	3.6%
Rye Flour	5.1%
Water	76.0%

Dissolve the ferrous sulphate in the boiling water. Stir the rye flour into the solution and boil for 15 minutes. Stir in the oxide, mix thoroughly and boil for another 15 minutes. When cold it is ready to use.

Swedish Farm Paint

Red Iron Oxide	13.0%
Ferrous Sulfate	3.2%
Rye Flour	3.8%
Water	76.0%
Linseed Oil	4.0%

Stir the rye flour into most of the cold water and boil 20 minutes. Dissolve the ferrous sulphate in a small portion of the water and add the solution to the rye flour gruel. Add the oxide, mixing thoroughly. Boil 15 minutes. Just before removing from the fire add an amount of linseed oil equal to 4% of the weight of the other ingredients. When cold it is ready to use.

Paint Primer
German Patent 591,084

A priming composition, on which water paints or oil paints or both in turn can be applied, comprises an unboiled vegetable oil 40–65, glycerol or like alcohol 30–55, and spermaceti about 5%, together with an optional amount of a mineral filler other than calcium carbonate.

Non-Thickening Zinc Oxide Paints with Good Hiding Power

Concentrated zinc oxide-linseed oil pastes do not thicken appreciably and any soaps formed lose their thickening properties when aged. Hence, to prepare non-thickening zinc oxide paints, a paste is first prepared with 85 parts zinc oxide and 15 parts low-acid-number linseed oil. After aging, this paste may be thinned to 72–75% pigment with ordinary oil without any danger of thickening.

One Coat Fast Drying Paint
(Brushing)

Where it is desired to produce a quick drying, dense covering brushing paint we suggest the following formula. The high percentage of oil used in this formula permits the use of large quantities of pigments which produces exceptional covering qualities.

Wet ¼" "Cotton"	28 parts
Xylol	28 parts
Butyl Cellosolve	68 parts
Oleum Spirits	82 parts
Ethyl Acetate	20 parts
Butyl Lactate	20 parts
Paste A	94 parts

Paste A

Cotton	29.0%
Oil (A. D. M. No. 100)	71.0%

One Coat Fast Drying Paint
(Spraying)

The following formula indicates the large quantity of diluent that it is possible to use in Archer Daniels Midland No. 100 lacquers. This formula contains 75% toluol and 25% cotton solvents. Because of the high percentage of oil used in this formula it is possible to incorporate larger quantities of pigments, thereby making exceptionally dense covering products.

Wet ¼" "Cotton"	56 parts
Butyl Acetate	25 parts
Butanol	25 parts
Ethyl Acetate	50 parts
Toluol	300 parts
Paste A	188 parts

Thin 15–20% for spraying.

Quick-Drying Paint, White

½" R. S. Cotton, Wet	140 parts
Tricresylphosphate	30 parts

Blown Linseed Oil *	203 parts
Rosin *	177 parts
Xylol	6 parts
Zinc Oxide	500 parts
Non-Bronzing Blue	30 parts
Butyl Acetate	364 parts
Butanol	216 parts
Toluol	464 parts

* Fused together at 300° C.

Quick-Drying Paint
German Patent 568,693

A paste is made containing:

Zinc Oxide	100 lb.
Hexamethylene Tetramine	3 lb.
Linseed Stand Oil	100 lb.

To the paste add

| Cobalt drier (cobalt resinate dissolved in an equal weight of mineral spirits) | 4 lb. |

and mineral spirits to give brushing consistency.

Quick Drying Coating

Resinlac solution is a 3½ pound cut of shellac rosin broken up small, viz:

| Shellac Rosin | 140 lb. |
| Denatured Alcohol | 40 gal. |

Cut in a churn same as for regular shellac, then strain through fine cloth into a paint agitator. The above paint may be merely an intimate mixture of the components, but preferably the pigment should be run once through a roller mill to effect complete wetting of it with the solution and thereby obviate the tendency otherwise to caking in the containers after filling into gallon pails.

This quick-drying paint is used by automobile manufacturers and other machinery makers. It is brushed or sprayed onto the inside surface of crank and transmission cases to bind foundry sand or other grit adhering to the castings and which otherwise would be abrasive in the bearings and gears when assembled. The paint must be oil- and grease-proof. Lacquer-type product has supplanted the above in some plants. For certain machinery, white is used instead of the red.

Synthetic Resin House Paint
Pigment:

Titanox B	49%
Lead Carbonate	49%
Litharge	2%

Vehicle:

No. 1319 Beckosol Solution	44%
Raw Linseed Oil	55%
Cobalt Drier (2% Cobalt)	1%

Synthetic Resin House Paints
Formula No. 1

Grind 6 pounds of Cryptone (high strength lithopone containing 50% zinc sulphide and 50% barium sulphate) into one gallon of the following varnish:

Durez Resin No. 500	100 lb.
China Wood Oil	45 gal.
Bodied Linseed Oil	5 gal.

Heat the above together to 460° F. and hold a few minutes for body. Reduce at 400° F. with:

| Mineral Spirits | 75 gal. |
| Drier | 1 gal. |

Formula No. 2

The same varnish may be used in the following formulation:

| Pigment | 64% |
| Vehicle | 36% |

Pigment:

Leaded Zinc Oxide	35%
Cryptone	45%
Asbestine	10%
Silica	10%

Vehicle:

Durez Varnish	40%
Refined Linseed Oil	43%
Turpentine	15%
Drier	2%

Dry Water Paints
U. S. Patent 1,947,498

As an example of a preferred formula for an improved paint composition, the following is given:

Titanium Pigment	15.0%
Mineral Filler	19.0%
Special Velvet Filler	39.8%
Casein	12.0%
Mica	5.0%
Borax Glass	1.1%
Sodium Fluoride	1.0%
Irish Moss	.1%
Lime	7.0%

Ten to thirteen gallons of water are required for 100 pounds of dry powder to make a paint ready for use.

The titanium pigment preferred is an intimately wet mixed mixture of titanium dioxide and barium sulphate, containing about 25% of the titanium dioxide and 75% barium sulphate. Other titanium pigments may be used, such as pure titanium oxide, or various mixtures of pure titanium oxide with barium sulphate, calcium sulphate, or other inert fillers. The titanium pigment adds greatly to the opacity of the paint and increases the ability to hide or to obscure the surface to which it is applied. The titanium dioxide pigments, because of their inertness toward most chemical reagents ordinarily encountered, are particularly adapted for use in water paints of this nature. The amount of titanium pigment present may vary between 0 to 70%, a preferred range being 5%–70%.

The mineral filler preferred is a material sold under the trade name "Metronite," and consists of a very white filler, which is composed of a mixture of tremolite, dolomite, and smaller amounts of talc, silica, calcite, etc. Metronite carries the other ingredients and at the same time, imparts considerable hiding power or opacity. It adds to the brushing and leveling qualities. It may be replaced in whole or in part by other inert fillers, such as calcium, magnesium, or barium carbonates by silicates such as kaolin, talc, or mica, by silica, by barium sulphate or calcium sulphate in their various forms, or by other fillers or similar materials, or by mixtures of any of these. The percentage of mineral filled may vary from 0 to 40%.

Special velvet filler is made from a naturally occurring mixture of a number of secondary magnesium minerals, including talc, Brucite, magnesite, and chlorite. Its exceptional fineness and plasticity makes it a valuable constituent of the paint. It imparts brushing and leveling qualities and aids materially in keeping the ready-for-use paint in suspension. It has good hiding power and permits a decrease in the percentage of the more expensive pigments, such as titanium pigment. In other words, the use of the velvet filler produces hiding property at low cost. The percentage of velvet filler may vary from 0 to 75%.

Casein acts as the binder for the paint and imparts hardness, durability, water-resistance, and sealing qualities. It is preferably coated with mineral oil or other water-repellent substances to slow up the rate of solution and lessen any tendency toward lumping. This coating of the casein is accomplished by simple mixing of mineral oil with the casein at the time of manufacture. The preferred screen analysis of the casein is:

All through 40 mesh.
Through 40 on a 100 mesh 30–45%.
Through 100 mesh 55–70%.

The proportions may be varied between 10–15%.

The mica lessens the tendency of the dried paint film to check-crack where it is applied too heavily. It decreases any pulling action, which would cause the dried paint to lift off weak bases to which it has been applied. It increases the continuity and decreases the porosity of the dried paint film so that the surface is better sealed for the application of succeeding coats of the paint itself, or of lacquers or oil paints. The mica also improves the application qualities, making the paint work more easily under the brush. The mica must be very fine, all passing a 200 mesh, and at least 97% passing a 235 mesh screen. A water-ground product is preferred, but a dry-ground mica may be used. The proportions may vary between 0% and 12%.

The borax glass is substantially anhydrous sodium borate ($Na_2B_4O_7$), prepared by driving off the water of crystallization from borax by fusing. It is very important that the borax be anhydrous, as any water of crystallization may cause certain of the ingredients present to react together in the dry mix and form gritty particles, which show up in an objectionable manner in the dried paint film. It should all pass a 200 mesh screen. Its most important function is to prevent mold growth. It also helps in the solution of the casein and in preventing the formation of lumps. The borax glass may vary from 0.5–2%. If the paint is to be marketed in the paste form the equivalent amount of borax, crystallized sodium borate ($Na_2B_4O_7.10H_2O$) may be substituted for the anhydrous salt if desired.

The main function of the sodium fluoride is to prevent the formation of lumps and of foam in the process of mixing the dried powder paint with water. Its action in this respect is very marked as it enables the user to mix the paint easily without an objectionable lumping or foaming. Sodium fluoride also assists in the solution of the casein and is a preservative and mold-preventative. Potassium fluoride may also be used. The proportions of sodium fluoride may vary from 0% to 2%.

Irish moss or chondrus prevents settling of the mixed paint. It may be replaced by pectin, agar-agar, or similar gelling agents. The amount of Irish moss used may vary from 0 to 0.5%.

The hydrated lime may be from either high calcium or dolomitic lime, a white product being preferred. It may vary from 0 to 12%. The lime acts both as a solvent for the casein, and as an insolublizer. A high calcium lime is preferred as there is less danger of its aging in the package.

It should be understood that suitable lime-proof colors or pigments may be added to the above formula to secure the desired tint or color. To increase durability and weather-resistance, drying oils, such as linseed or china wood oil, may be added to the water-paint mixture, the oils, easily forming an emulsion. The amount and kind of oil used depend on the degree of weather-resistance desired, and to some extent, on the nature of the pigments and fillers used. The paint may be marketed as a dry powder to be mixed with water, or water and oil, prior to use. It also may be sold in semi-paste form with water alone, or with water and oil. The semi-paste may be formed by mixing the dry powder with the liquid, or if desired, by grinding through some type of paint mill after mixing.

In developing very brightly colored paint, there can be used an alternate formula, which appears to have considerable merit, not only for interior, but also for exterior use. An example of a preferred formula of a colored paint of this nature is as follows:

Pigments and Fillers	74.4%
Mica	10.0%
Casein	12.0%
Borax Glass	1.1%
Sodium Fluoride	1.0%
Potassium Dichromate	1.5%

Water is added to the above powdered mixture in the proportion of 10 to 18 gallons of water per 100 pounds of dry powder. Drying oil may also be added to the paint-water mixture, as in the first named formula.

The lime in this formula has been entirely eliminated, and the solution of the casein is brought about by the borax and sodium fluoride. This provision leaves the mixture only slightly alkaline and makes it possible to use colors which are not fast to lime or strong alkalis. In fact, pigments like chrome yellow (lead chromate), and iron blues (ferric ferrocyanide) have been used without any bad results apparent after standing in the wet mix for weeks.

The potassium dichromate ($K_2Cr_2O_7$) produces a washability and resistance to weather, which appears to be equal, if not superior, to the results produced by lime. In fact, on exposure to weather, there appears to be less tendency to check crack than when lime was used. The equivalent amount of chromate (K_2CrO_4) works just as well as the dichromate, and other bases such as sodium or ammonium may be substituted for potassium. Other less oxidized chromium compounds, such as chromic sulphate ($Cr_2(SO_4)_3$), or chromic chloride ($CrCl_3$) may also be used.

For pigments, many of the pigments commonly used in paint may be used such as zinc sulphides, zinc oxides, and titanium dioxide pigments for whites, iron oxides, lead chromates, and cadmium sulphides for yellows, oranges and reds; ferro-cyanides, for blues; chromium oxides and various mixtures of blues and yellows for greens; for blacks, iron oxides and carbon pigments such as gas blacks, lamp blacks, and bone blacks may be used. Many of the organic colors, both lakes and toners, may be used to give a wide variety of brilliant shades.

The fillers, my consist of any of the fillers commonly used in paints, such as barytes, whiting, clay, talc, etc., depending upon the nature of the coloring pigment.

In the modified formula, the mica may vary from 3-15%; the casein from 6-18%; the borax glass from 0.5-2%; the sodium fluoride from 0.5-2% and the potassium dichromate from 0.25-5%. The sum of the pigments and fillers would in each case comprise the balance of the formula. The percentage range of pigment would depend upon the nature of the pigment and the amount of covering and hiding power desired. For instance, a strong pigment like a gas black or an iron blue might require 5-10%, while a white pigment, such as lithopone, might require in the neighborhood of 75%. Conversely, the fillers would vary from 0-70% depending on the nature of the pigments and of the fillers themselves.

In either the preferred formula or the modified formula it may be found desirable to add about 0.2% of tribromophenol to the dry paint mixture. The addition of the tribromophenol entirely prevents, in most cases, the growth of molds of any color on the surface of the dried paint film. If any mold does grow, it is usually white and is not noticeable. Except in very severe humidity conditions, the interaction between the tribromophenol and the borax entirely prevents the formation of the molds. Instead of tribromophenol, other halogen substitution products of phenol or other compounds containing the phenol hydroxyl group, such as cresols or napthols may be used, such as sodium trichlorphenate or parachlormetacresol.

High Grade Whitewash

Slack a bushel of a good grade quicklime and remove the lumps by pouring through a screen. Add sufficient water to produce a good whitewash, which is usually forty gallons

Then add 20 pounds of whiting, 17 pounds of rock salt, and 12 pounds brown sugar in the order stated. Stir thoroughly and if necessary add some more water until the desired fluidity is attained.

Apply with a brush. Two coats are needed on wood, and three on brick or stone surfaces. This produces a coat that does well in preserving the surface, looks well, and does not readily rub or wash off. The sugar serves as a binder.

White Distemper

Casein	1000 parts
Urea	340 parts
Hexamethylenetetramine	210 parts
Lithopone	6950 parts
Zinc Oxide	1000 parts
Lime	500 parts

Hot Water Calcimine

Water-Floated Chalk Whiting	86%
China Clay	10%
Hot Water Animal Glue (High Grade)	4%
Zinc sulphate (Preservative)	1 part to 300.

Cold Water Plastic Paint
U. S. Patent 1,954,291

About 60 pounds of glue are mixed with approximately 125 pounds of water and 2 pounds of a preservative, such as zinc sulphate, alum or the like, preferably zinc sulphate, and allowed to stand for several hours. In preparing the "tinting material" 1000 pounds of whiting, free from strong alkali reaction, is charged into a kettle or retort heated to substantially 75° F. and 520 pounds of china clay and 60 pounds of the prepared glue is gradually added to the whiting and clay in the hot kettle where the mixture is heated for substantially two hours. Substantially 80 pounds of gypsum is then added and the mixture heated to 215 to 230° F. for a limited period of time, leaving approximately 4% moisture in the mass. To approximately 184 pounds of this mixture of prepared "tinting material" is then added approximately 100 pounds of plaster of Paris, 4 pounds of retarder, preferably fruit sugar, although other known retarders are available, 4 pounds of oil cake and 36½ pounds of mica, and the whole is intimately mixed and ground to such a fineness that substantially 90% will pass through a 150 mesh screen. The fineness of the material may vary within substantial limits.

The material is then ready for packaging and shipment in the form of a dry powder, requiring only the addition of cold water to make it ready for application to wall or ceiling surfaces.

Water Paint
U. S. Patent 1,947,497

A paint suitable for use on interior walls, etc., after admixture with water and linseed oil comprises a mineral filler such as a tremolite and dolomite mixture, 59.7 parts, casein 6 parts, a titanium pigment 20 parts, zinc oxide 3 parts, Irish moss 0.1 part, hydrated lime 9 parts, borax 2 parts and tribromophenol 0.2 parts.

Water Rubber Paint
British Patent 391,973

Thirty parts of commercial sodium silicate dissolved in an equal amount of water is added to 150 parts of a 60 per cent rubber latex obtained by centrifugal action. The mixture is diluted with water to a concentration of 5 per cent and to it is added a 5 per cent solution of 24 parts of aluminium sulphate in water. The precipitate in aqueous suspension is allowed to remain for several hours on a filter-cloth, after which time a thick, irreversible, creamy precipitate remains. A white, water-resistant distemper can then be prepared as follows: 3.75 parts of gum acacia and 3.75 parts of gelatin are dissolved in 7.5 parts of 0.880 ammonia and 985 parts of water, and this is mixed in an end runner mill with 500 parts of zinc oxide and 500 parts of china clay; 900 parts of the above rubber precipitate (16 per cent concentration) are added and mixing is continued until the paint has the necessary smoothness.

Rubber Paint
British Patent 399,394

A viscose composition that can be shaped, brushed, or sprayed consists of an aqueous rubber dispersion, ammonia, water glass, and zinc or lead salts of one or more weak acids, e.g., carbonic or boric acid. Vulcanizing agents and fillers may be added. For example, the following is a representative formula:

Sulphur	200 g.
Zinc Carbonate	10 g.
Water Glass (specific gravity 1.25)	50 cc.
Ammonium Hydroxide (25%)	15 cc.
Latex (60%)	1000 cc.

Coating Composition
Canadian Patent 336,932

A coating composition is manufactured by dissolving in a thinner, an oil-modified glyceryl phthalate resin in the ratio of about 90.3 to 107.2 parts by weight of thinner to 1 part by weight of resin and thereby obtaining a solution having a viscosity between 10 and 20 poises. Drier is incorporated into the solution which is diluted to a specific gravity of about 0.9. Flaked aluminum is incorporated into the solution in the ratio of 1.5-2.6 pounds of aluminum flake to 1 gallon of solution.

Rubber Paint
British Patent 402,865

Four parts by volume of casein is saturated with 6 parts by volume of water. Twenty-four parts by volume of normal rubber latex (or the equivalent amounts of concentrated latex and water) and ⅓ part by volume of ammonium hydroxide (specific gravity 0.88) are added. Vulcanizing agents, accelerators and cellulose lacquer (1¼ parts by volume) may be incorporated if desired.

Rubber Paints

Flat Paints.—Flat paints as is well known are used either as a finishing coat where a matt surface is required, or as an undercoat for a gloss paint, and it is now generally accepted that the production of a good gloss finish is largely dependent on the quality of the undercoat. The drawbacks of the usual flat paint are the lack of flow—the brush marks giving the finish a "liney" or striated appearance, as already stated—and the settling out of the pigments when the tins of paint are allowed to stand for any length of time.

The addition of a proportion of rubber modified, materially improves the quality of flat paints in both these respects.

Formulae have been worked out on these lines, replacing part of the oil-content by rubber in various recognized formulae for flat paints, such as linseed oil thinned with white spirit, gold size and turpentine, and so forth. As the result of experiments, a combination of rubber and stand oil give the most promising results. The following is a typical formula expressed in parts by weight:

Pigment	100–150 parts
Milled Crepe Rubber	10 parts
Lead Linoleate	1 part
Stand Oil	10 parts
Volatile Thinner	90–100 parts

The proportions of pigment and thinner will vary with the type of pigment used, a heavy pigment such as white lead requiring less oil than one of low specific gravity such as titanium white.

It is possible to prepare a flat paint with the modified rubber solution alone, but although this has excellent flowing properties, it lacks adhesion, and the combination of rubber and oil is far more satisfactory.

A rubber solution containing cobalt linoleate can also be used. Any of the usual pigments and fillers can be incorporated. Among those which have been tried are: lithopone, zinc oxide, titanium white, white lead, antimony oxide, Paris white, china clay, barytes, red ochre, yellow ochre, burnt umber, chrome yellow, emerald green, Brunswick green, Prussian blue ultra-marine and bronze and aluminum powders.

Ready-Mixed Gloss Paints.—The ordinary type of ready-mixed paint prepared by grinding the pigments in refined linseed oil, or a combination of this and boiled oil, and thinning with a relatively small proportion of turpentine, does not flow sufficiently to eliminate brush marks, and the dried film, although possessing a fair gloss, has a striated surface. The addition of a proportion of rubber has the same effect here as in the case of the flat paints, improving the flow and eliminating the brush marks, with the result that the gloss is also enhanced.

The addition of rubber to various proprietary brands of gloss paint showed that a proportion in the order of 4 per cent is sufficient to eliminate brush marks and produce almost an enamel gloss.

It is preferable to incorporate the rubber with the pigments in the preparation of the paint, following the procedure described under flat paints. But in this case, the use of a rubber solution containing lead linoleate is unsatisfactory, and it is preferable to use cobalt linoleate as the catalyst.

The proportion of rubber to oil may range from 15 to 30 per cent.

The following are typical examples of formulation expressed in parts by weight:

Formula No. 1

Zinc Oxide	100 parts
Pale Boiled Oil	55 parts
Rubber Solution	22 parts
Terebene	2 parts
White Spirit	11 parts

Formula No. 2

Lithopone	150 parts
Pale Boiled Oil	32 parts
Refined Linseed Oil	32 parts
Rubber Solution	32 parts
White Spirit	15 parts

The rubber solution in each case is a 50 per cent solution of milled crepe rubber in white spirit, with the addition of 2.5 per cent of cobalt linoleate.

As an illustration of this effect of replacing a portion of the oil in a gloss paint with rubber, a specimen was exhibited which had been painted in two sections with a red paint, made as follows:

	A	B	
Venetian Red	100	100	parts
Milled Crepe (+2½ per cent Cobalt Linoleate)	—	6.5	parts
Refined Linseed Oil	22	22	parts
Boiled Oil	6	9.5	parts
Terebene	2	2	parts
White Spirit	4	9	parts

Although section B contained a materially higher proportion of volatile thinner, it had much better gloss, and was free from brush marks.

As such paints are used as finishing coats, it is important to ascertain whether the addition of rubber affects the durability in any way. Tests have been carried out in the usual manner by exposing panels to natural weathering and also to accelerated weathering by means of the weatherometer apparatus. These clearly indicate that whilst no improvement in durability can be claimed for paints made with rubber, the addition of rubber to an oil gloss paint certainly does not in any way impair its resistance to decay.

Enamels.—The addition of rubber to ready-mixed paint as described above brings ordinary gloss oil-paint nearer to an enamel appearance, giving a perfect flow and thereby enhancing the gloss. The addition of rubber to enamels does not give such marked advantage, as owing to the low pigmentation and the use of highly polymerized oils, the flowing properties of enamels are in general excellent. Various forms of modified rubber have, however, been tried to replace the resins which are at present used in the manufacture of enamel.

Of these some attention has been given to molten rubber. The molten rubber is prepared by heating raw rubber at a temperature of up to 300° C. until it is liquefied. Linseed oil and driers can be incorporated in the raw rubber, and the whole heated in the usual manner of preparing standard varnishes, and then thinners added to obtain the usual consistency for varnish.

Molten rubber alone will not dry, and when sufficient driers are added to give the requisite drying properties, the film formed contracts on exposure, causing severe cracking. To correct this the addition of oil is necessary.

A good example is as follows:

Molten Rubber	100 parts
White Spirit	100 parts
Terebene	12 parts
Cobalt Drier	12 parts
Red Ochre (Venetian Red)	100 parts

The defects of molten rubber as a paint vehicle may be obviated by using it in conjunction with oil. That is to say, the varnish is made up partly of molten rubber and partly of linseed oil. A paint containing these in the proportion of 50/50, and made by "cooking up" the ingredients in the presence of the drier, appears to have good aging properties, and to yield a film which does not readily crack.

Frosting Varnish.—The addition of rubber solution to China wood oil gives a frosting varnish which will give the desired effect in a more regular manner than when China wood oil is used alone.

The rubber solution used contains cobalt linoleate, and the following is an example of a frosting varnish:

Rubber solution	20.25	parts
(Containing milled crepe 10 parts, cobalt linoleate ¼ part, and white spirit 10 parts)		
China wood oil	10	parts
Terebene	1	part
White Spirit	10	parts

Washable Distempers.—A water-paint which will brush out satisfactorily can readily be made with a mixture of casein and latex, but this is unsuitable owing to the tendency of the casein solution to harden up in the tin on standing. To obviate this, part of the casein is replaced by glue. Such a paint will wash down well, and show very little discoloration on aging.

The following is a good example:

Glue solution	25 parts
Casein solution	25 parts
Latex	30 parts
Lithopone	100 parts

Drying oils can, if desired, be incorporated with the above, and for some purposes are an advantage, but tend to discolor the paint more rapidly.

Distempers can also be satisfactorily prepared by using a rubber solution (as used for the oil-paints). The solution readily emulsifies with a glue solution, to which the pigments can be incorporated. The following is an example of this type of distemper:

Glue solution	20	parts
Rubber solution	16.2	parts
(Containing milled crepe 8 parts, cobalt linoleate 0.2 parts, and white spirit 8 parts)		
Water	25	parts
Lithopone	100	parts

Anti-Rusting Paint

10 parts of coal tar pitch (free of ammonia) are boiled together with 2 parts of graphite and one part of red oxide of lead. One part of sulphur and 2 parts of ground pumice are then added, the product being diluted with suitable thinners. If charcoal tar pitch is used instead of coal tar pitch, this must be free from acetic acid if perfect anti-rusting qualities are to be obtained.

Protective Paint for Rafters

75 parts of coal tar are melted together with 20 parts of rosin pitch and 5 parts of rosin. This mixture is applied hot, and white sand sprinkled over the film.

Aluminum Tar Paint

50 parts of coal tar is mixed with 20 parts of crude benzene, 20 parts of toluol, 10 parts of xylol and suitable quantities of aluminium powder.

Bag-Printing Paint

90 parts of bitumen (mineral pitch) are mixed with 90 parts of Canadian balsam (balm), 120 parts of oil of turpentine and suitable quantities of bone black.

Sealing Material for Paint Barrels, Cans, Etc.

Formula No. 1

Ten parts of Burgundy pitch, and 10 parts of rosin are melted together with 25 parts of Japanese wax, 15 parts of carnauba wax, 35 parts of paraffin and 5 parts of tallow-oil.

Formula No. 2

50 parts of Burgundy pitch (a white resin), 40 parts of rosin and 10 parts of yellow wax are melted together and mixed with 5 to 10 parts of pigments.

Formula No. 3

73.5 parts of rosin pitch are melted together with 12 parts of rosin, 51 parts of Japanese wax (obtained from the berries of Rhus succedana), 49 parts of carnauba wax and 6 parts of red oxide of lead.

Formula No. 4

24.5 parts of rosin pitch, 12 parts of Japanese wax, 12 parts of ochre.

Steel Rust Prevention Paint

Polyvinyl acetal	20 parts
Alcohol	60 parts
Toluene	10 parts
Naphtha	10 parts
Red lead	90 parts

Rust Proof Black Paint for Metal

25 parts coal tar are thinned by 20 parts of benzol, then 15 parts of silica black, grade A* are added with vigorous stirring. Both the benzol and the silica black act as dryers, and the coal tar as a binder for the pigment. This paint is not only more rust proof, but more acid resistant than asphaltum.

* Note: Silica black is a new black pigment patented under U. S. No. 1,940,352, entitled ''A Process for Treating Powdered Coal.''

Black Paint for Wood

This is a superior paint for preserving wood under all sorts of weather conditions. 10 parts of raw linseed oil are thinned by 3 parts of turpentine and 2 parts of Japan drier, then 5 parts of silica black, grade A, are added with vigorous stirring. This paint will also give good results when applied to composition roofing, automobile tops, canvas and tarpaulin.

Paint for Moderately Heated Metal Surfaces

Formula No. 1

10 parts coal tar are thinned by 3 parts benzol, then 5 parts of silica black, grade A, are added with vigorous stirring. This paint stood a sizzling heat for 52 hours without showing signs of cracking, peeling or burning off.

Formula No. 2

The following formula was found to be almost as good as Formula No. 1, and both of them were several times better than six commercial black paints used for comparison.

10 parts coal tar are thinned by 5 parts furfural and 5 parts formaldehyde to which are added with vigorous stirring 5 parts of silica black, grade A.

Fire-Resistant Paint
U. S. Patent 1,706,733

This patent involves the use of a large percentage of borax which, upon being heated, forms a protective glaze.

Example:

Finely Powdered Borax	5 lb.
White Lead	1 lb.
Zinc Oxide	1 lb.
Asbestine	5 oz.
Barytes	3 oz.
Lead Borate	1 oz.
Zinc Borate	1 oz.
Linseed Oil (Boiled)	3 pts.
Treated Tung Oil	1 pt.
Japan Drier	¼ pt.
Varnish	¾ pt.

Fire-Proof Preparation for Roofs

Quick lime should be slaked in a closed box to prevent the escape of steam. The lumps should be removed from the resulting creamy liquid by pouring through a sieve. To every six gallons of this lime add twenty pounds of rock salt and four gallons of water. After this thoroughly boil and skim.

The next step is to add slowly two pounds of sodium hydroxide and thirty-five pounds of fine lake sand. A suitable pigment may be added if desired. Heat and stir until the mixture is homogeneous. Apply with a paint or whitewash brush. In addition to rendering a roof incombustible to falling sparks this liquid stops small leaks, and prevents the growth of fungi.

Waterproof Aluminum Paint

Spar Varnish	1 gal.
Aluminum Powder	2 lb.

Thin to desired consistency with good grade of mineral spirits. Avoid the use of turpentine.

Traffic Line Paint

"Neville" Hard Resin	550 lb.
Light Naphtha (Gasoline or V.M.P.)	350 lb. (56 gal. approx.)
Commercial Xylol	100 lb. (14 gal. approx.)

Dissolve cold in tumbling drum or agitator tank. This results in a thin bodied vehicle, fast-dyeing (ten minutes) in which a high percentage of pigments may be incorporated without losing gloss. A small amount (2 to 5%) of bodied Linseed Oil is sometimes added. Any desired pigment combination may be used dependent on the color desired.

Waterproof Paint

Take eighty-two parts of ochre, nine parts of carbon black and make a paste with sufficient boiled linseed oil. Then add a soap solution composed of two parts of yellow soap dissolved in seven parts of water. Thoroughly mix, and if necessary reduce the consistency with boiled oil until it can be applied with a brush. Best results are obtained by applying one coat, then after an interval of three days, applying another.

Finally three days later a finishing coat of varnish should be applied. This is best prepared by making a paste of carbon black and boiled linseed oil. This preferably should be thoroughly ground, if not it should be extremely well mixed. Sufficient boiled oil must then be used to reduce the paste to the consistency of a thick varnish. This combination gives a very satisfactory waterproof paint.

Waterproofing Paint

Bakelite No. 254 Varnish (40 gals. in Length of China Wood Oil and with a Total Solid Content of 50%)	1	gal.
Zinc Chromate	1½	lb.
Zinc Dust	10	lb.

The pigments should preferably be ground in the vehicle. When stored over longer periods, cans should be provided with air-vents, as hydrogen gas sometimes is generated in small amounts in the material.

This paint may be applied on wood or metal to give permanent waterproof coatings.

Asphalt Emulsion Paint and Waterproofing
U. S. Patent 1,940,431

In the case of Montan wax-asphalt dispersions, it is desirable from a cost standpoint to use as little Montan wax as possible, since even the crude Montan wax is much more expensive than most types of asphalt. As little as about 10% to 15% of crude Montan wax need be used to furnish sufficient protective colloid in the product.

Assuming that the asphalt is air-blown and has a melting point of 150° F. (ball and ring test) and the crude Montan wax has a melting point of 175° F., the mixture is melted and heated to about 300° F., at which temperature it flows readily. To approximately 35 parts of the molten mixture is added approximately 65 parts of a ¾% solution of caustic soda at a temperature of 150° F. The melted mixture and solution are preferably maintained under rapid agitation as they are mixed, so as to effect a rapid and uniform dissemination of the solution throughout the mass and thus to promote uniform chemical reaction and dispersion. Inasmuch as the specific heat of the molten mixture is much less than that of water, little, if any, liberation of steam takes place at the final temperature produced, which is above the melting point of the thermoplastic materials but below the boiling point of water. The resulting dispersion is of a creamy consistency, so that water may be added and readily distributed there through in large quantities. It is composed of uniformly fine particles in the order of magnitude of 1/5000 to 1/10000 of an inch in diameter and is characterized by its stability even at as low as 1% solids content. Such a dispersion is suitable for use as a paint, at a solids content of about 30%, under which conditions it has remarkably good covering power and is comparable in this respect to the so-called asphalt paints prepared by "dissolving" asphalts in organic solvents. Such paints may be applied with the usual paint brushes and may set under usual room temperature conditions, say, 70° F., as continuous films in about an hour. The dispersion is suitable for use in the impregnation of felts, papers, yarns, textile fabrics, and the like, to render them waterproof. It is an excellent size for paper pulp intended more especially for the production of waterproof papers, as the resulting papers are free from the unsightly asphalt specks appearing when the usual asphalt dispersions are employed. Evidently the usual asphalt dispersions when added to a dilute aqueous pulp suspension undergo premature precipitation of a kind in which the fine particles of asphalt unite into particles of microscopic size, whereas in the case of the product of the present invention no such premature reaction takes place. The dispersed particles may, however, be fixed on the fibers by the use of paper-makers' alum or other suitable salts or acids. in which case a loose precipitate, quite different from that incident to dilution of some dispersions with water is produced.

To Prevent Bleeding of Asphaltic Molded Products When Lacquering

Dip the articles in a water shellac solution and allow to thoroughly dry. Shellac solution is prepared as follows:

Orange or White Shellac (Dry)	100 gs.
Ammonium Hydroxide (28%)	40 cc.
Water	10 cc.
Glycerine	1 g.

Mix the ammonia, water and glycerine. Put the shellac in this solution and allow to soak in a covered container over night. Stir thoroughly in the morning and place on a water bath and heat to 125° F., stirring occasionally until dissolved. Solution may be used either hot or cold. The concentration of the shellac may be increased by using more or less water

Mold Paint for Molds Made of Plaster of Paris for Production of Plaster of Paris Ornaments

Dissolve scrap celluloid in a solvent mixture composed of 50% acetone and 50% amyl acetate. This will require about 36 hours, during which time it should be shaken occasionally. When thoroughly dissolved add sufficient solvent made as above to bring it to the consistency of very thin syrup. Before applying to the mold surface give the mold a coat of very thin orange shellac made by dissolving one part of ordinary 4 pound cut orange shellac in three volumes of pure alcohol. Do not use radiator alcohol. After about three hours of drying apply the celluloid solution allowing about one hour between coats. Give as many coats as may be necessary to completely fill the pores of the plaster and make a smooth glossy finish. Additional coats beyond this point should not be given however. Keep the molds out of direct rays of the sun and away from heat higher than ordinary room temperature.

To Give Plaster of Paris Products a Marble-Like Finish

To the mixing water used to mix the plaster dissolve one ounce of aluminum potassium sulphate to each gallon of water.

First Coat Material

This product is used to save paint and

give smooth job on very old, very rough, or very porous material.

Ten pounds whiting mixed to stiff paste with water are thoroughly incorporated with 25 pounds white lead mixed to stiff paste with linseed oil. Add ½ gallon sweet- or butter-milk and mix well. Then add ½ gallon linseed oil and mix well. Apply with short bristled brush.

Metallic Pigment Protective Paint
U. S. Patent 1,953,508

500 to 600 grams of aluminum-silicon alloy, very finely ground and subsequently air-sifted, said alloy being obtained by the liquation of an alloy lower in silicon, and containing about 32% of metallic aluminum and about 53% of metallic silicon (the remainder consisting of oxides and other compounds of these two metals) are stirred to a workable paint with 500 to 400 grams of a mixture of a suitable binding medium (such as for example boiled linseed oil) with a small quantity of copal and with a diluent (such as turpentine oil). Thus, for instance, the mass is brought into distributable condition with the aid of a mixing apparatus and a colloid mill.

The paint is applied with a brush, or a paint-spraying gun, and furnishes a gray to dark gray coating.

According to the purpose in view, the boiled linseed oil may be replaced by lacquer varnish and the like.

In all cases the most suitable proportion of pigment to varnish depends, on the one hand, on the composition of the pigment and, on the other, the purpose in view (whether brushed or sprayed paint, linseed oil or lacquer paints, zapon lacquers, etc.). This proportion can easily be ascertained, in each case, by preliminary experiment.

The usual commercial driers and diluents can also be employed with these paints.

Prevention of Paint Livering
U. S. Patent 1,836,265

The addition of 0.2–1% citric or tartaric acid prevents livering.

Bakelite Type Insulating Varnish
British Patent 382,861

Xylenol		
(211–232° C.B.P.)	45	parts
Paraformaldehyde	10.75	parts
Triethanolamine	1.5	parts
Chinawood Oil	47.75	parts
Solvent Naphtha	8	parts
Mineral Spirits	16	parts

Heat the first four ingredients to 110–115° C. and hold until a sample ''cures'' at 200° C. in 10–15 seconds; turn off heat and add the other two quickly.

Paint Remover
Formula No. 1

Paraffin wax	3 parts
Wood spirit	30 parts
Acetone	25 parts
Benzol	20 parts
Carbon tetrachloride	15 parts
Xylol	10 parts

The paraffin wax is dissolved in the benzol-carbon tetrachloride-xylol mixture and the wood spirit-acetone mixture added. This precipitates the wax and forms a paste when it is set.

Formula No. 2

Paraffin wax	4 parts
Benzol	8 parts
Wood spirit	7 parts

The above two types of paint removers are very quick acting and enable the old paint to be easily scraped off. It is necessary that all the wax should be removed from the surface before painting is commenced, so it is recommended to go over the surface with turpentine or petrol.

Formula No. 3

Benzene or Toluene	1 part
Ethyl Acetate	1 part
Ethyl Alcohol	1 part

Add paraffin and nitrobenzene enough to make a fairly viscous mixture.

Formula No. 4

Lye	2⅔	lb.
Water	2	qt.
Aqua Ammonia	2	qt.
Sodium Silicate	2½	gal.

Paint Remover
U. S. Patent 1,838,908

The following mixture is suitable for removing paint from surfaces to be coated with pyroxylin lacquers:

Benzol	35 parts
Alcohol	
(Ethyl or Denatured)	30 parts
Acetone	20 parts
Ethyl Acetate	14 parts
Hard Paraffin Wax	1 part

Paint Remover and Floor Cleaner

Lye	3 lb.
Sodium Bicarbonate	1 lb.

Dissolve in water, apply to paint and let stand. If only a little water is added, the remover should not be allowed to stand too long or it will injure the fiber of the wood beneath the paint.

Paint and Varnish Remover
U. S. Patent 1,938,714
Formula No. 1

Ortho-dichlorobenzene	7 parts
Propylene Dichloride	6 parts
Benzene	1 part
Carbon Tetrachloride	1 part

Formula No. 2

Ortho-dichlorobenzene	1 part
Propylene Dichloride	3 parts
Benzene	1 part
Carbon Tetrachloride	½ part
Acetone	½ part

The mixture shown in Formula No. 1 may advantageously be used where the paint or varnish layer to be removed is thick and where a long time of contact therewith of the solvent action is required. This applies especially in the case of wood or fiber composition surfaces. Formula No. 2 is best adapted to use where the paint or varnish covering is thin and may be removed more quickly, as on metal or glazed surfaces.

Paint and Varnish Remover

Amylene Dichloride	80 parts
Alcohol	40 parts
Naphtha	20 parts
Diglycol Stearate	2 parts

Paint, Varnish, Lacquer, Shellac and Enamel Remover

Soda Ash	100 g.
Infusorial Earth	50 g.
Toluene	900 cc.
Ethyl Acetate	150 cc.
Benzene	900 cc.
Acetone	900 cc.
Methanol	600 cc.
Dioxan	50 cc.
Paraffin	20 g.
Methyl Salicylate	20 cc.

Melt paraffin by putting solid cut in fine chips in mixture and setting on steam bath until wax is melted, then shake to mix well.

Leave on 15 to 60 minutes depending on speed of action.

TABLE I
Composition and Viscosities of Varnishes (Independent Work)

No.	Resin	Linseed Oil Kind	%	China Wood Oil Kind	%	Viscosity Wiesbaden	G-H Standards Philadelphia
A-1	Albertol 209 L	American Stand Oil	30%	Raw	70%	B (65 cp)	—
D-1	Albertol 209 L	Ger. Var. Oil Ger. Stand Oil	15% 15%	Stand	70%	B (65 cp)	—
A-2	Amberol F-7	American Stand Oil	30%	Raw	70%	G (165 cp)	E + (130 cp)
D-2	Amberol F-7	Ger. Var. Oil Ger. Stand Oil	15% 15%	Stand	70%	B (65 cp)	E + (130 cp)
A-3	Albertol 111 L	American Stand Oil	15%	Raw	85%	C (85 cp)	—
D-3	Albertol 111 L	Ger. Stand Oil	15%	Stand	85%	A (50 cp)	—
A-4	Amberol BS/₁	American Stand Oil	15%	Raw	85%	C (85 cp)	E (125 cp)
D-4	Amberol BS/₁	Ger. Stand Oil	15%	Stand	85%	B (65 cp)	D (100 cp)

Numbers A-1, A-2, A-3, and A-4 are American method varnishes and D-1, D-2, D-3 and D-4 are German method varnishes. The above data refer to the varnishes made independently at Wiesbaden. The varnishes made at Philadelphia, the viscosities of which are given in the last column, have the same composition as the corresponding varnishes made at Wiesbaden, except that in D-2 and D-4, American linseed oil was made.

Formulas and Cooking Procedures

The following formulas and methods were decided upon at Wiesbaden for the later independent work at Wiesbaden and Philadelphia. These were followed except for the modifications as described.

American method (for all grades of modified phenol formaldehyde resins): [Raw China Wood Oil + Resin + Heat] + Linseed Stand Oil.

Varnishes A-1 and A-2

Albertol 209L (or Amberol F-7)	100 grs.
Raw China Wood Oil	210 grs.

Heat together to 302° C. (575° F.) in about 30 minutes. Add

Am. open kettle— Stand Linseed Oil	60 grs.
Am. open kettle— Stand Linseed Oil	30 grs.

Cool to 270° C. (520° F.) and hold a few minutes for body. Then add

Mineral Spirits	400 grs.

Cool to 245° C. (475° F.) and thin with

Varnishes A-3 and A-4

Raw China Wood Oil	255 grs.

Heat to 260° C. (500° F.) in 18 to 20 minutes. Add

Albertol 111 L (or Amberol BS/$_1$)	100 grs.

Heat to 302° C. (575° F.) in 15 minutes. Add

Am. open kettle—Stand Linseed Oil	45 grs.

Hold at 235° C. (450° F.) a few minutes for body.

Mineral Spirits	400 grs.

Wiesbaden—All four varnishes made using prescribed methods and materials.

Philadelphia—Varnishes A-2 and A-4 were made as prescribed.

German method (for the harder types of modified phenol formaldehyde resins): [Linseed Oil (Varnish and stand) + Resin + Heat] + Stand China Wood Oil + Heat.

Varnishes D-1 and D-2

Albertol 209L (or Amberol F-7)	100 grs.
Varnish Linseed Oil	45 grs.

Heat to 260° C. (500° F.) in about 10 minutes. Add

German Polymerized Linseed Stand Oil	45 grs.

Heat to 260° C. (500° F.) and hold about 5 minutes or until clear.

Stand China Wood Oil	210 grs.

Add in small portions keeping the temperature at or above 240° C. (465° F.), then raise to 500° F. if necessary to secure a clear solution. Cool to 450° F. and add

Mineral Spirits	400 grs.

Wiesbaden—In making D-1 it was necessary to heat the Albertol 209 L and the varnish linseed together to 280° C. for 20 minutes, then 10 minutes longer at the same temperature with the stand oil to secure a clear varnish. D-2 worked more easily, only 15 minutes heating of Amberol F-7 with German varnish linseed oil and 5 minutes with German stand linseed oil, being required.

Philadelphia—D-2 made according to the prescribed formula and method except that American open kettle stand linseed oil was used. Less difficulty was experienced than at Wiesbaden in making

a clear varnish, again indicating the superior solvent properties of the American open kettle stand oil over the German polymerized closed kettle stand oil.

German method (for the softer types of modified phenol formaldehyde resins): Stand Oils (Linseed and China Wood) + Resin + Heat.

Varnishes D-3 and D-4

Stand China Wool Oil	255 grs.
German Polymerized Linseed Stand Oil	45 grs.

Heated together to 260° C. (500° F.) in 17 to 20 minutes and held a few minutes for combination.

Albertol 111 L (or Amberol BS/$_1$)	100 grs.
Mineral Spirits	400 grs.

Wiesbaden — Varnishes D-3 and D-4 made according to prescribed formula and method, the stand China Wood oil being made very viscous, as above described, in order to secure viscosities approaching those of A-3 and A-4.

Philadelphia — Varnish D-4 made according to formula and method except that American open kettle stand linseed oil was used. Stand China Wood oil was made as heavy as possible in order to secure same viscosity as A-4.

An amount of concentrated liquid drier, equivalent to .02% Cobalt and .4% lead on the weight of the oil, was added to each varnish Additional mineral spirits was added to the cooled varnishes to replace that lost by vaporization during the thinning.

Water Varnish

British Patent 395,299

A varnish comprises a colloidal aqueous solution of wood oil stand oil, e.g., wood oil stand oil 75, cobalt resinate 2, and turpentine oil 25 parts, may be mechanically dispersed in 100 parts water to which 1 part of an emulsifying agent, e.g., potassium resinate, has been added.

Varnish Composition and Shellac Substitute

U. S. Patent 1,942,413

A mixture such as Batu gum 18–20, rosin 10–20, lime 1–2 and Chinawood oil 20–40 parts, which has been heated to over 260° is dissolved in a varnish thinner such as naphtha and benzine.

Shellac-Oil Varnish

These formulas are stated in parts by weight.

Formula No. 1

Shellac	20 parts
Linseed Oil	40 parts
Calcium Oxide	1 part

The lime is added at 250° C. (482° F.), and when the oil clears the temperature is 290-300° C. and then the shellac added. With a larger proportion of lime lower incorporation temperatures can be used.

Formula No. 2

Glycerine	4 parts
Linseed Oil	10 parts
Shellac	6 parts

The glycerine and oil are heated to 270° C. (518° F.) and the shellac added slowly. A small addition of litharge, sodium carbonate, etc., enables less glycerine to effect a satisfactory solution.

Formula No. 3

Rosin	9 parts
Linseed Oil	10 parts
Shellac	9 parts

The shellac is added at 250-270° C.

Formula No. 4

Albertol	12 parts
Linseed Oil	10 parts
Shellac	9 parts

The shellac is added at 250-270° C.

Formula No. 5

Certain high-boiling solvents, e.g., tricresyl phosphate (b.p. 300° C.), triacetin (b.p. 259° C.), etc., also are used as incorporating agents. A mixture of about 10 parts of linseed oil, 8 parts of solvent, and 8 parts of shellac is heated until the solvent has evaporated and the shellac remains in solution in the oil.

Formula No. 6

Modified shellacs which are oil soluble, are prepared from rosin with glycerine, lime, magnesia, etc., as hardening agents, e.g.:

Rosin	20 parts
Shellac	16 parts
Calcium Oxide	1 part

are heated together to 280° C. (536° F.). The rosin dissolved in oil previously heated to 250-270° C. Similarly, a mixture of rosin and shellac is esterified with glycerine, and the resulting ester dissolved in oil heated to 270° C.

In the preparation of a shellac and oil solution when the shellac separates and cures due to incorrect conditions, the whole has to be immediately cooled and the cured shellac separated out. This material then can be reconditioned by the following process: The cured shellac is powdered and slowly added to rosin heated to 270° C. Five parts of rosin dissolves about three parts of this shellac; the resulting ester dissolves in linseed oil heated to 270° C.

Formulae of Varnishes

Formula No. 1

Limed Rosin (5%)	100 lb.
Fused Lead Resinate	5 lb.
Cobalt Linoleate (5%)	8 oz.
China Wood Oil	40 gal.
Kettle Bleached Linseed	5 gal.
Petroleum Thinner (48° Bé.)	80 gal.

Body oil and rosin at 560° to heavy string, check with linseed, hold at 480° F. and stir in driers, thin at 400° F.

Formula No. 2

Limed Rosin (5%)	100 lb.
Fused Lead Resinate	5 lb.
Cobalt Linoleate (5%)	8 oz.
China Wood Oil	30 gal.
Kettle Bleached Linseed	5 gal.
Petroleum Thinner (48° Bé.)	60 gal.

Body oil and rosin at 560° to heavy string, check with linseed, hold at 480° F. and stir in driers, thin at 400° F.

Formula No. 3

Lined Rosin (5%)	100 lb.
Fused Lead Resinate	5 lb.
Cobalt Linoleate 5%	4 oz.
China Wood Oil	18 gal.
Kettle Bleached Linseed	2 gal.
Petroleum Thinner (48° Bé.)	36 gal.

Body wood oil and 50 lbs. rosin at 560°, check with remainder of rosin crushed and linseed, hold at 480° and stir in driers, thin at 400°.

Formula No. 4

Clear Amber Congo	90 lb.
Limed Rosin (5%)	10 lbs.
Fused Lead Resinate	5 lb.
Cobalt Linoleate (5%)	4 oz.
China Wood Oil	18 gal.
Kettle Bleached Linseed	2 gal.
Petroleum Thinner (48° Bé.)	24 gal.
Turpentine	12 gal.

Fuse Congo at 625°; add wood oil heated to 350° with rosin, body at 560°, check with linseed and hose, hold at 480° and stir in driers, thin at 400°.

Formula No. 5

Four-Hour Resin	100 lb.
Fused Lead Resinate	5 lb.
Cobalt Linoleate (5%)	4 oz.
China Wood Oil	30 gal.
Kettle Bleached Linseed	5 gal.
Petroleum Thinner (48° Bé.)	64 gal.

Body resin and wood oil at 560°, check with linseed, hold at 480°, and stir in driers, thin at 400°.

Formula No. 6

Four-Hour Resin	100 lb.
Fused Lead Resinate	5 lb.
Cobalt Linoleate	4 oz.
China Wood Oil	20 gal.
Kettle Bleached Linseed	5 gal.
Petroleum Thinner (48° Bé.)	48 gal.

Cook same as No. 5.

Formula No. 7

Ester Gum, Acid No. 6	100 lb.
Fused Lead Resinate	5 lb.
Cobalt Linoleate	4 oz.
China Wood Oil	30 gal.
Kettle Bleached Linseed	5 gal.
Petroleum Thinner (48° Bé.)	56 gal.

Cook same as No. 5.

Formula No. 8

Ester Gum, Acid No. 6	100 lb.
Fused Lead Resinate	5 lb.
Cobalt Linoleate	4 oz.
China Wood Oil	25 gal.
Kettle Bleached Linseed	5 gal.
Petroleum Thinner (48° Bé.)	46 gal.

Cook same as No. 5, except that 40 pounds ester gum are held out of the kettle and added with the check oil.

Fused Lead Resinate

Rosin "N"	100 lb.
Lead Acetate	40 lb.

Kettle Bleached Linseed

A good grade of varnish oil run one hour at 580° F.

Four-Hour Type Nevindene Varnish

Nevindene	100 lb.
China Wood Oil	25 gal.
Mineral Spirits	40 gal.
Liquid Drier "A"	10 lb.
Lead Metal*	0.045 %
Cobalt Metal*	0.045 %
Manganese Metal*	0.00045%

The China wood oil and one-half of the Nevindene are heated to 550° F. for 30 minutes and removed from the fire and the rest of the Nevindene is added with stirring. The kettle is replaced on the fire when the temperature of the batch has dropped to 500° F. and held at this temperature until the proper body is reached, the pill is very heavy when tested on a cold surface. Do not hold until a string forms from the stirrer. Water cool with hose to 450° F. and thin with mineral spirits. The liquid drier is added any time after the thinner. This varnish is recommended for floors, window sills and furniture.

This varnish dries dust-free in one hour and can be recoated after four hours. It has a high gloss and depth of finish. It is highly water and alkali resistant. The film has some yellowing tendency and is not gas-proof in an open flame oven. Samples did not show evidence of skinning or precipitation of drier during aging of one week.

* Percentages of lead, etc., are based on oil content.

Four-Hour Type Nevindene and Rosin Ester Varnish

Nevindene	20 lb.
Rosin Ester	80 lb.
China Wood Oil	25 gal.
Mineral Spirits	40 gal.
Liquid Drier "A"	12 lb.
Cobalt Metal*	0.054 %
Lead Metal*	0.054 %
Manganese Metal*	0.00054%

The china wood oil and 60 lbs. of the rosin ester are heated to 575° F. in 30 minutes and removed from the fire; the rest of rosin ester and the Nevindene are added with stirring. The kettle is replaced on the fire and held at 500° F. until the proper body is reached. Water cool and thin with mineral spirits at 450° F. The liquid drier is added after the thinner.

This varnish dries dust-free in one hour and can be recoated after four hours. The film is almost free from yellowing and has a tendency to gas check in an open flame oven. Samples skinned in 48 hours but showed no evidence of drier precipitation. The film has a high gloss and very slight silking. This varnish is recommended for interior coatings.

* Percentages of cobalt, etc., are based on oil content.

Phenol-Nevindene Type Varnish—Quick Drying

Nevindene	100 lb.
China Wood Oil	25 gal.
Tertiary Amyl Phenol	10 lb.
Mineral Spirits	40 gal.
Liquid Drier "A"	12 lb.
Lead Metal*	0.054 %
Cobalt Metal*	0.054 %
Manganese Metal*	0.00054%

* Percentages of lead, etc., are based on oil content.

The China wood oil, one-half of the Nevindene and the tertiary amyl phenol are heated to 580° F. in 30 minutes and removed from the fire and the rest of Nevindene is added with stirring; the kettle is replaced on the fire and held at

520° F. until heavy body is reached. It is then water cooled and thinned with mineral spirits at 450° F. The liquid drier is added at room temperature.

This varnish dries dust-free in one and one-half hours and can be recoated after 4 hours. It is water and alkali proof. The film has a yellowing tendency and gas checks in an open flame oven.

Nevindene Rubbing Type Varnish

Nevindene Resin	100 lb.
China Wood Oil	12 gal.
Mineral Spirits	28 gal.
Liquid Drier ''A''	8 lb.
Lead Metal*	0.075 %
Cobalt Metal*	0.075 %
Manganese Metal*	0.00075%

The chnia wood oil and 50 lb. of Nevindene are heated to 590° F. in 35 minutes, removed from the fire and the rest of the Nevindene added with stirring. When the temperature has dropped to 500° F. replace the kettle on the fire and hold until a very heavy pill is reached, water cool and thin with mineral spirits at 450° F. The liquid drier is added after the thinner.

This varnish dries dust-free in one hour and can be rubbed after 6 to 8 hours. It is alkali and water resistant. The varnish does not skin or precipitate any drier. It is gas-proof and has considerable after-yellowing. The film has a high gloss and very slight silking. Recommended for water, acid and alkali resistance.

* Percentages of lead, etc., are based on oil content.

Nevindene Floor and Deck Varnish Type

Nevindene	100 lb.
China Wood Oil	17 gal.
Bodied Linseed Oil	3 gal.
Mineral Spirits	34 gal.
Liquid Drier ''A''	10 lb.
Lead Metal*	0.056 %
Cobalt Metal*	0.056 %
Manganese Metal*	0.00056%

The China wood oil and 70 lb. of Nevindene are heated to 590° F. in 30 minutes, removed from the fire and the linseed oil and rest of the Nevindene are added with stirring. When the temperature has dropped to 500° F., replace the kettle on fire and hold until a heavy body is reached. Water cool to 450° F. and thin with mineral spirits. The liquid drier is added after the thinner.

This varnish dries dust-free in one hour and gives a hard film in 24 hours. It is alkali and water resistant. The film does

not gas check and has considerable after-yellowing. Samples show no tendency to skin or precipitate the drier on aging. It has a high gloss and very slight silking.

* Percentages of lead, etc., are based on oil content.

Nevindene-Rosin Medium Length Type Mixing Varnish

Nevindene	85 lb.
W. W. Rosin	15 lb.
China Wood Oil	17 gal.
Bodied Linseed Oil	8 gal.
Mineral Spirits	40 gal.
Liquid Drier ''A''	24 lb.
Lead Metal*	0.108 %
Cobalt Metal*	0.108 %
Manganese Metal*	0.00108%

The china wood oil, Nevindene and one-half of the linseed oil are heated to 575° F. in 30 minutes, remove from fire, add rosin and rest of linseed oil with stirring. When the temperature has dropped to 500° F., replace kettle on fire and hold at 500° F. for heavy pill; water cool and add mineral spirits at 450° F. The liquid drier is added after the thinner.

This varnish dries dust-free in one hour. It has high gloss and excellent flowing. The film is gas-proof and has considerable yellowing tendency. Samples showed no skinning or precipitation of drier. Recommended as grinding liquid for quick-drying enamels.

* Percentages of lead, etc., are based on oil content.

Nevindene-Lead Resinate-Rosin Medium Lenth Four-Hour Type

Nevindene	75 lb.
Lead Resinate (Fused)	
(Pb 16%)	10 lb.
W. W. Rosin	15 lb.
China Wood Oil	25 gal.
Mineral Spirits	40 gal.
Liquid Drier ''A''	24 lb.
Lead Metal*	0.1 %
Cobalt Metal*	0.1 %
Manganese Metal*	0.001%

The china wood oil and Nevindene are heated to 575° F. in 25 minutes, removed from the fire and the lead resinate and rosin are added with stirring. When the temperature has dropped to 500° F., replace the kettle on the fire and hold at 500° F. for heavy pill; water cool and thin at 450° F. Add liquid drier after the thinner.

This varnish dries dust-free in one

hour. It has high gloss and excellent flowing. The film is gas-proof and has slight tendency to yellow. Samples showed no precipitation of drier, but slight tendency to skin. Recommended for interior use.

*Percentages of lead, etc., are based on oil content.

Nevindene-Beckacite Medium Length Four-Hour Type

Nevindene	75 lb.
Beckacite No. 1001	25 lb.
China Wood Oil	25 gal.
Mineral Spirits	40 gal.
Liquid Drier ''A''	24 lb.
Lead Metal*	0.1 %
Cobalt Metal*	0.1 %
Manganese Metal*	0.001%

The chinawood oil, Beckacite and one-half of the Nevindene are heated to 525° F. in 25 minutes, removed from the fire and the rest of the Nevindene added with stirring. Replace kettle on the fire and hold at 500° F. for heavy pill, water cool and thin at 450° F. Add liquid drier.

This varnish dries dust-free in one hour and has considerable tendency to yellow. It is not gas-proof in an open flame oven. Samples showed no skinning or precipitation of driers. Recommended for furniture and floors.

*Percentages of lead, etc., are based on oil content.

Nevindene-Bakelite Medium Length Four-Hour Type

Nevindene	75 lb.
Bakelite No. 254	25 lb.
China Wood Oil	25 gal.
Mineral Spirits	40 gal.
Liquid Drier ''A''	12 gal.
Lead Metal*	0.05 %
Cobalt Metal*	0.05 %
Manganese Metal*	0.0005%

The chinawood oil, Bakelite and one-half of the Nevindene are heated to 520° F. in 25 minutes, removed from the fire and the rest of the Nevindene is added with stirring; replace on fire and hold at 490° F. for heavy pill, water cool and thin at 450° F. Add the liquid drier after the thinner.

This varnish dries dust-free in one hour and is gas-proof. It has a tendency to skin and show after-yellowing. Recommended for floors and furniture.

*Percentages of lead, etc., are based on oil content.

Nevindene-Fused Kauri Medium Length Four-Hour Type

Nevindene	75 lb.
Fused Kauri	25 lb.
China Wood Oil	25 gal.
Mineral Spirits	40 gal.
Liquid Drier ''A''	30 lb.
Lead Metal*	0.135 %
Cobalt Metal*	0.135 %
Manganese Metal*	0.00135%

The chinawood oil, Kauri and one-half of the Nevindene are heated to 575° F. in 30 minutes, removed from the fire and the rest of the Nevindene is added with stirring. When the temperature has dropped to 500° F., replace kettle on fire and hold at 500° F. for heavy pill; water cool and thin at 450° F. Add liquid drier after the thinner.

This varnish dries dust-free in one and one-half hours and extra hard in 24 hours. It is slightly dark and gas checks in an open flame oven. Samples showed no skinning or precipitation of drier. It has considerable yellowing.

*Percentages of lead, etc., are based on oil content.

Nevindene-Fused Congo Medium Length Four-Hour Type

Nevindene	75 lb.
Fused Congo	25 lb.
China Wood Oil	25 gal.
Mineral Spirits	40 gal.
Liquid Drier ''A''	24 gal.
Lead Metal*	0.10 %
Cobalt Metal*	0.10 %
Manganese Metal*	0.001%

The chinawood oil, Congo and one-half of the Nevindene are heated to 575° F. in 30 minutes, removed from the fire and the rest of the Nevindene is added with stirring. When the temperature has dropped to 500° F., replace kettle on fire and hold at 500° F. for heavy pill; water cool and thin at 450° F. Add liquid drier after the thinner.

This varnish dries dust-free in one and one-quarter hours and fairly hard in 24 hours. It is fairly pale but has tendency to yellow. Samples showed no skinning.

*Percentages of lead, etc., are based on oil content.

Nevindene-Calcium Zinc Resinate Mixing Type Varnish

Nevindene	60 lb.
Calcium and Zinc Resinate (Ca 3%, Zn 2%)	40 lb.
China Wood Oil	20 gal.
Mineral Spirits	30 gal.

Liquid Drier "A"	20 lb.
Lead Metal*	0.11 %
Cobalt Metal*	0.11 %
Manganese Metal*	0.0011%

The china wood oil, Nevindene and one-half of calcium-zinc resinate are heated to 580° F. in 30 minutes, removed from the fire and the rest of the calcium-zinc recinate is added with stirring. When the temperature has dropped to 500° F., replace kettle on the fire and hold at 500° F. for heavy pill, water cool and thin at 450° F. Add liquid drier after the thinner.

This varnish dries dust-free in one hour and has high gloss and good flowing characteristics. It is gas-proof and showed no skinning or precipitation of drier. It has considerable yellowing tendency. Recommended as a liquid for quick drying paints and enamels.

* Percentages of lead, etc., are based on oil content.

Nevindene-Litharge Short Spar Type

Nevindene	100	lb.
China Wood Oil	30	gal.
Litharge	7½	lb.
Cobalt Acetate	½	lb.
Mineral Spirits	45	gal.
Lead Metal*	0.28%	
Cobalt Metal*	0.05%	

The china wood oil and one-half of the Nevindene are heated to 580° F. in 30 minutes, remove the kettle from the fire and add the litharge with stirring, then add the rest of the Nevindene with stirring. When the temperature has dropped to 500° F., replace kettle on the fire and hold at 500° F. for heavy pill, then add cobalt acetate, water cool and thin at 450° F.

Varnish has cloudy precipitate of lead soap. The film dries dust-free in one hour and is slightly dark. It is almost gas-proof, shows considerable yellowing and no skinning.

* Percentages of lead, etc., are based on oil content.

Nevindene-Limed Rosin High Gloss Mixing Type

Nevindene	50 lb.
Limed Rosin (6% Lime)	50 lb.
Mineral Spirits	45 gal.
Liquid Drier "A"	26 lb.
Lead Metal*	0.1 %
Cobalt Metal*	0.1 %
Manganese Metal*	0.001%

The china wood oil and limed rosin and one-half of the Nevindene are heated to 580° F. in 30 minutes, removed from the fire, and the rest of the Nevindene is added with stirring. When the temperature has dropped to 500° F., replace kettle on the fire and hold for heavy pill. Water cool and thin at 450° F. Add liquid drier after the thinner.

This varnish is quick drying. It has a high gloss and good flowing. It has slight tendency to yellow and is not gas-proof. Samples showed no skinning. Recommended as a liquid for high gloss enamels and paints.

* Percentages of lead, etc., are based on oil content.

Nevindene-Calcium-Zinc Resinate High Gloss Mixing Type

Nevindene	50 lb.
Calcium-Zinc Resinate	
(Ca 3%—Zn 2%)	50 lb.
China Wood Oil	30 gal.
Mineral Spirits	45 gal.
Liquid Drier "A"	40 lb.
Lead Metal*	0.15 %
Cobalt Metal*	0.15 %
Manganese Metal*	0.0015%

The china wood oil, one-half of the Nevindene and Calcium-Zinc Resinate are heated to 580° F. in 30 minutes, removed from the fire and the rest of the Nevindene added with stirring. When the temperature has dropped to 500° F., replace kettle on the fire and hold for heavy pill; water cool and thin at 450° F. Add liquid drier after the thinner.

This varnish is quick drying. The gloss and flowing are good. It is gas-proof and shows no skinning. It has a slight tendency to yellow. Recommended for high gloss paint and enamels.

* Percentages of lead, etc., are based on oil content.

Nevindene-Ester Spar Varnish Type

Nevindene	70 lb.
Rosin Ester	30 lb.
China Wood Oil	35 gal.
Bodied Linseed Oil	5 gal.
Mineral Spirit	50 gal.
Liquid Drier "A"	16 lb.
Lead Metal*	0.045 %
Cobalt Metal*	0.045 %
Manganese Metal*	0.00045%

The china wood oil and Nevindene and one-half of the ester are heated to 565° F. in 30 minutes, removed from the fire and the linseed oil and the rest of the rosin ester added with stirring. When the temperature has dropped to 500° F., the kettle is replaced on the fire and held at 480° F. until a long string is formed when the pill is applied to a cold surface; water cooled and thinned at 450° F. Add the liquid drier after the thinner.

This varnish is quick drying and shows very little after-yellowing. Samples showed skinning after 24 hours, and slight cloudiness when the varnish was first tested. It is not gas-proof. Recommended as a varnish for outside surfaces.

*Percentages of lead, etc., are based on oil content.

Nevindene-Calcium-Zinc Resinate Spar Mixing Type

Nevindene	50 lb.
Calcium and Zinc Resinate (Ca 3%—Zn 2%)	50 lb.
China Wood Oil	50 gal.
Mineral Spirits	50 gal.
Liquid Drier ''A''	32 lb.
Lead Metal*	0.09 %
Cobalt Metal*	0.09 %
Manganese Metal*	0.0009%

The China wood oil, Nevindene and one-half of the calcium-zinc recinate are heated to 575° F. in 30 minutes, removed from the fire and the rest of the calcium-zinc resinate is added with stirring. When the temperature has dropped to 500° F., replace the kettle on the fire and hold at 500° F. until a long string forms when the pill is tested on a cold surface. Water cool and thin at 450° F. Add the liquid drier after the thinner.

This varnish is quick drying and has very slight after-yellowing. Sample showed no skinning or precipitation of drier. It is not gas-proof in an open flame oven. Recommended for paints and enamels.

*Percentages of lead, etc., are based on oil content.

Nevindene-Bakelite Spar Varnish Type

Nevindene	75 lb.
Bakelite No. 254	25 lb.
China Wood Oil	34 gal.
Bodied Linseed Oil	10 gal.
Mineral Spirits	55 gal.
Liquid Drier ''A''	25 lb.
Lead Metal*	0.065 %
Cobalt Metal*	0.065 %
Manganese Metal*	0.00065%

The china wood oil, Bakelite and Nevindene are heated to 525° F. in 30 minutes, removed from the fire and the linseed oil added with slight stirring. Replace kettle on fire and hold at 480° F. for heavy pill, indicated by long string when the pill is tested on a cold surface.

Water cool and thin at 450° F. Add liquid drier after the thinner.

This varnish is quick drying and gas-proof. It has slight after-yellowing but no skinning or precipitation of drier. The film is very hard in 24 hours. Recommended for outside surfaces.

*Percentages of lead, etc., are based on oil content.

Nevindene-Congo Spar Varnish Type

Nevindene	75 lb.
Fused Congo	25 lb.
China Wood Oil	34 gal.
Bodied Linseed Oil	10 gal.
Mineral Spirits	55 gal.
Liquid Drier ''A''	34 lb.
Lead Metal*	0.088 %
Cobalt Metal*	0.088 %
Manganese Metal*	0.00088%

The china wood oil, Congo and two-thirds of the Nevindene are heated to 585° F. in 25 minutes, removed from the fire and the linseed oil and the rest of the Nevindene are added with stirring. When the temperature has dropped to 500° F., replace kettle on the fire and hold at 500° F. for heavy pill, water cool and thin at 450° F. Add the liquid drier after the thinner.

This varnish dries dust-free in one and one-half hours and fairly hard in 24 hours. The film is gas-proof but considerably dark in color and has considerable after-yellowing. No skinning or precipitation of drier was noticed. Recommended for outside surfaces.

*Percentages of lead, etc., are based on oil content.

Nevindene Long Spar Mixing Type

Nevindene	100 lb.
China Wood Oil	40 gal.
Bodied Linseed Oil	10 gal.
Mineral Spirits	50 gal.
Liquid Drier ''A''	18 gal.
Lead Metal*	0.04 %
Cobalt Metal*	0.04 %
Manganese Metal*	0.0004%

The china wood oil and Nevindene are heated to 540° F. in 25 minutes, removed from the fire and the linseed oil added with slight stirring. Replace kettle on fire and hold at 490° F. for heavy body. Water cool and thin at 450° F. Add liquid drier after the thinner.

This varnish is quick drying. It has a tendency to skin and gas check, but no evidence of precipitation of drier was noticed. The film has a high gloss and slight yellowing tendency. This varnish will make an excellent vehicle for quick drying enamels for outside exposure. Recommended as a liquid for outside paints and enamels.

* Percentages of lead, etc., are based on oil content.

Special Varnishes

In the following formulae heat the resins in a non-ferrous pot until the resins become fluid. Add the oil (to this) which has been previously heated to about 500° F., a little at a time with stirring. Keep heating until a test portion placed on a cold surface assumes a milky appearance. Continue heating until a sample when chilled becomes crystal clear. Allow batch to cool somewhat and add the volatile solvent. The drier is best mixed in cold.

Finest Finishing Body Varnish

Best Pale Kauri	40 lb.
Sierra Leone Copal.	20 lb.
Old Tanked Linseed Oil	12 gal.
American Turpentine	12 gal.

After the varnish is made, the driers are churned in the following proportions to every 600 gal.:

Powdered Litharge	84 lb.
Zinc Sulphate	84 lb.

The usual time for churning is about six hours. After the process is completed, the varnish is allowed to settle for twelve to eighteen months and is then finally finished off with lime (2 ounces) and manganese borate (2 ounces) to every 10 gallons.

Best Pale Wearing Body Varnish

Best Pale Kauri	45 lb.
Sierra Leone Copal	20 lb.
Linseed Oil	13 gal.
Turpentine	13 gal.
To every 600 gal. when churning—	
Litharge	84 lb.
Zinc Sulphate	84 lb.

Finest Pale Hard Body Varnish

Pale Kauri	85 lb.
Sierra Leone Copal	35 lb.
Linseed Oil	9 gal.
Turpentine	13 gal.
To every 600 gal. when churning—	
Litharge	84 lb.
Zinc Sulphate	126 lb.

Finest Pale Elastic Carriage Varnish

Best Kauri	60 lb.
Linseed Oil	12 gal.
Turpentine	12 gal.
To every 600 gal. when churning—	
Litharge	84 lb.
Zinc Sulphate	84 lb.

Best Pale Hard Carriage Varnish

Best Kauri	60 lb.
Linseed Oil	8 gal.
Turpentine	12 gal.
To every 600 gal. when churning—	
Litharge	84 lb.
Zinc Sulphate	126 lb.

Pale Flatting Varnish

Kauri	60 lb.
Chips	15 lb.
Linseed Oil	8 gal.
Turpentine	14 gal.
To every 600 gal. when churning—	
Litharge	84 lb.
Zinc Sulphate	126 lb.

Coach Makers' Gold Size

Best Kauri	75 lb.
Linseed Oil	7 gal.
Powdered Litharge	12 lb.
Manganese Dioxide	½ lb.
Turpentine	50 gal.

Best Pale Terebene Varnish

Best Kauri	80 lb.
Linseed Oil	15 gal.
Powdered Litharge	10 lb.
Manganese Dioxide	½ lb.
Turpentine	50 gal.

Pale French Oil Varnish

Sierra Leone Copal	20 lb.
Pale Kauri	10 lb.
Linseed Oil	6 gal.
Litharge	½ lb.
Manganese Borate	1 oz.

Turpentine at discretion—about 8 to 10 gal.

Special Outside Oak Varnish

Kauri	60 lb.
Linseed Oil	12 gal.
Turpentine	12 gal.
To every 600 gal. when churning—	
Litharge	84 lb.
Zinc Sulphate	84 lb.

Special Inside Oak Varnish

Kauri	70 lb.
Linseed Oil	10 gal.
Turpentine	12 gal.

To every 600 gal. when churning—

Litharge	84 lb.
Zinc Sulphate	126 lb.

Pale Copal Varnish

Special Outside Oak Varnish	½ part
Special Inside Oak Varnish	½ part

Hard-Drying Copal Varnish

Special Inside Oak Varnish	9 parts
Pale Gold Size	1 part

Pale Outside Elastic Oak Varnish

Kauri	65 lb.
Linseed Oil	13 lb.
Turpentine	13 lb.

To every 600 gal. when churning—

Litharge	96 lb.
Zinc Sulphate	96 lb.

Pale Quick-Rubbing Varnish

No. 1 Kauri	150 lb.
Borate Oil	9 gal.
Turpentine	20 gal.
Heavy Naphtha	18 gal.

Popular Inside House Varnish

No. 2 Pontianiac	150 lb.
Flake Litharge	2 lb.
Borate Oil	18 gal.
Turpentine	19 gal.
Heavy Naphtha	19 gal.

Borate Oil

Linseed Oil	150 gal.
Manganese Borate	8 lb.
Flake Litharge	4 lb.

The oil is heated in a portable kettle to 565° F., then removed from the fire, and 8 lb. of manganese borate slowly added. After all the manganese is in, the mass is stirred slightly and then allowed to cool to 550° F., when the litharge is added while stirring. After all the driers are in, the whole is kept stirred for three-quarters of an hour. This oil will dry in six hours, and is frequently used in making varnishes from recipes on similar lines to the example given above.

Linseed oil for varnish making should be free from "foots" and mucilage, hence the use of old tanked oil. Oil, however, which has not undergone long tanking can be used if specially treated and prepared. In order to do this a special liquid is necessary, known as the "Doctor," which is made as follows:

Linseed Oil	6 gal.
Powdered Oxide of Manganese	12 lb.
Turpentine	12 gal.
Benzene (68–70%)	22 gal.
Acetate of Lead	6 lb.

Six gallons of linseed oil are placed in a varnish kettle and the oxide of manganese added. The kettle is then placed on the fire and well stirred with an iron stirrer. In a short time the mixture will commence to bubble, and the bubbling must be kept up hard by regulating the fire, so that in a short time the mixture commences to head over. The liquid now commences to make its own heat and rise in the pot, and at this stage it may be necessary to use a whisk to keep the head down.

At the end of thirty to forty-five minutes a drop should be taken out and examined. If the drop hardens quickly, and can be crushed up to a fine powder under the palette-knife, and is brownish pink in color, the kettle can be removed from the fire and allowed to cool down. After it has cooled, the turpentine can be added. This should be done slowly at first, more being added as the liquid cools. When adding the turpentine the liquid is well stirred. Next the benzene is added. If up to this point the preparation has been correctly made, the liquid should be quite thin like water. The lead acetate is now mixed with some refined linseed oil to the consistency of a thick liquid and passed through a triple roller mill to reduce it to a fine condition, and is then added to the turpentine-benzene mixture in the pot and well stirred. The whole is then allowed to stand for two days, after which the prepared "doctor" can be added at the rate of 10 gallons to 500 gallons of linseed oil.

The oil should then rest in a tank for two months, when it will be found that all the "foot" and mucilage will have settled out at the bottom of the storage tank and a bright clear oil equal to old tanked oil produced which will not "break" on being heated when used for varnish making. Such oil is especially useful for making varnishes requiring a minimum of tanking after being made. Such varnishes are made as follows:

The resin is run in the usual way, and the requisite amount of doctored linseed oil added. In a separate kettle the required amount of driers, consisting of lead linoleate and zinc linoleate, or cobalt linoleate and zinc linoleate, is melted and naphthaline added and the whole thinned with turpentine, about half the amount required, the remaining turpentine being added in the ordinary way. The varnish is then allowed to stand till bright, and is then transferred to a tank and air blown in by a blower for six hours. The varnish is then allowed to settle for a week, and finally passed through a centrifugal clarifier straight into cans. Varnishes so made compare very favorably with varnishes of long tanking and storage, and are quite suitable for both outside and interior decoration.

Pale Outside Elastic Oak

Pontianiac Resin	65	lb.
Doctored Oil	117	lb.
Turpentine	117	lb.
Lead Linoleate	7.5	lb.
Zinc Linoleate	7.5	lb.
Naphthalene	1.5	lb.

Pale Inside Oak Varnish

Pontianiac Resin	70	lb.
Doctored Oil	90	lb.
Turpentine	127	lb.
Lead Linoleate	7.5	lb.
Zinc Linoleate	7.5	lb.
Naphthalene	1.5	lb.

Best Pale Hard Carriage Varnish

Best Pale Kauri	77	lb.
Doctored Oil	90	lb.
Turpentine	135	lb.
Lead Linoleate	7.5	lb.
Zinc Linoleate	7.5	lb.
Naphthalene	1.5	lb.

Scratch-proof Floor Varnish

Amberol F-7 (Extra Light)	80	lb.
Amberol BS-1 (Extra Light)	20	lb.
China Wood Oil (Extra Light)	20	gal.
Linseed Oil (Body Q)	2	gal.
Varsol	50	gal.
Turpentine	5	gal.
Lead Resinate	4	lb.
Cobalt Drier (4% Co.)	2	lb.

Heat Amberol F-7 and china wood oil as fast as possible to 565° F. Gain to 575° F. Introduce BS-1 and linseed oil and stir until incorporated. Add lead resinate. Then cool with 4–5 gallons of water, cut with varsol and turpentine. Add cobalt drier at about 300° F.

This varnish may be used as is or cut 25% with varsol or turpentine for floor brushing.

High Lustre Aluminum Finish

Amberol No. 801 Varnish, 10 gallons in length of bodied linseed oil, containing 25% bodied soy bean oil acid with a total solid content of 50%	½	gal.
Amberol No. 226 Varnish, 44 gallons in length of china wood oil containing 10% of linseed oil and with a total solid content of 50%	½	gal.
Cobalt drier (4% Co)	¼	gal.
Fine Aluminum Lining	1¼	lb.
Varsol	¼	gal.

The aluminum powder should be suspended in the varsol before addition of the varnishes for a proper distribution and webbing of the powder.

The combinations dry in less than 3 hours with a very high, metallic lustre.

Altenburg Varnish

Powdered Glass (Very Fine)	200	g.
Mastic, Refined	375	g.
Sandarac, Refined	180	g.
Venice Turpentine	188	g.
Alcohol (96%)	1875	g.

The first three ingredients are first thoroughly incorporated with the Venice Turpentine. The alcohol is then added and the mixture is heated on a water bath.

Coal Tar Varnish
French Patent 751,305

A varnish is made from coal-tar pitch 65%, phenol 5%, and benzine 30% by weight.

Water-Lac Varnish

This product is used for tempera painting (with distemper colors), show card work, and as a size for under same, and is prepared by this formula:

Borax	1	lb.
Water	2	gal.
Bleached Shellac	5	lb.

First, dissolve the borax in the water, then add the shellac and heat nearly to boiling . . . but do not boil. It should be strained through fine cheese cloth.

Aqua-Spirit Lac
Part 1

Denatured Alcohol	1	gal.
Bleached Shellac	4	gal.

Part 2

Water	½	gal.

Pulv. Borax	4 oz.
Glycerine	4 fl. oz.

First "cut" the shellac, then prepare Part 2 by dissolving the borax in the water and adding the glycerine and finally combine by stirring Part 1 constantly while adding Part 2. As in the preceding case, this solution should be strained when cold, through fine cheesecloth to remove all insoluble particles.

Picture-Film Varnish

De-waxed Shellac	3.70 lb.
Butyl Alcohol	27.57 lb.
90% Benzol	43.78 lb.
70-second Cotton	3.49 lb.
Butyl Acetate	5.84 lb.
Ethyl Acetate	14.59 lb.
A.D.M.–100 Oil	1.03 lb.

The shellac should be cut in butyl alcohol, strained, then added to the other mixture

Colored Varnishes (Light-Fast)
U. S. Patent 1,491,058
Formula No. 1

In 100 parts of warm commercial spirit varnish (containing as essential part a resin, for instance shellac) there are dissolved 0.25 parts of Victoria blue B, highly concentrated, whereupon 0.5 part of uranylnitrate is added. The varnish is colored blue fast to light.

Formula No. 2

10 parts of crystal violet 6B are mixed with 20 parts of thorium nitrate. 0.75 part of this mixture is dissolved in 100 parts of commercial spirit varnish with application of heat. The varnish is colored fast to light.

Non-Slip Floor Polish

Shellac	5 oz.
Gum copal	1 oz.
Venice turpentine	1 oz.
Denatured Alcohol to make	26 oz.

Light Fast Colored Spirit Varnish
U. S. Patent 1,925,208

Spirit varnishes are colored with Malachite green or Victoria blue B and 0.5% perchloric acid is added to render the color fast to light.

Fine Automobile Varnish

Pale Hardened Rosin	50 lb.
XXXX Ester	25 lb.
China Wood Oil	30 gal.
Stand Oil	30 gal.
Gum Turpentine	75 gal.
Dammar Varnish	30 gal.
Turpentine Drier	1 pint

The hardened rosin is neutralized to acid value 78.62 with lime and glycerin, and contains lead and manganese linoleates. The stand oil is 88% linseed and 12% wood oil held 3 hours at 575° F. The dammar varnish is a cold cut solution proportioned 7 lb. Batavia gum to each gallon of turpentine; this was added to the hot batch immediately after thinning, so as to amalgamate thoroughly by heat but not hot enough to "throw" the color.

Air Drying Varnishes
Formula 1
Thin Air-Drying Varnish

Gilsonite	38 parts
Linseed Oil Varnish	3 parts
Benzine	59 parts

This is used on machinery, lamps, tools, etc.

Formula 2

Coal tar pitch	55 parts
Slaked lime	0.5 part
Benzine	44.5 parts

This is a useful varnish for iron-work. Varnishes containing coal-tar pitch cannot be incorporated with linseed oil or benzine.

Formula 3
Matt Varnish

Gilsonite varnish (1) above	76 parts
Lampblack	8 parts
Benzine	16 parts

Formula 4
Matt Varnish for Groundwork

Gilsonite Varnish (1) above	65 parts
Lampblack	16.5 parts
Benzine	18.5 parts

Stoving Varnishes
Formula 1
Bright Thin Varnish Stoving at 60° C.

Gilsonite Varnish (Air-Drying Varnish No. 1 above)	90 parts
Linseed Oil varnish	5 parts
Lampblack	0.5 part
Bone black	0.5 part
Benzine	4 parts

Formula 2
Bright Semi-Stoving Varnish

Gilsonite Varnish (Air-Drying Varnish No. 1 above)	69 parts
Blue Oil	23 parts
Lead Manganese Drier	1.5 parts
Benzine	6.5 parts

Formula 3
Semi-Stoving Matt Varnish

Gilsonite Varnish (Air-Drying Varnish No. 1 above)	55 parts
Blue Oil	15 parts
Lead Manganese Drier	1 part
Benzine	15 parts
Lampblack	14 parts

Formula 4

The best bituminous varnishes are made with the aid of natural or artificial copals, and the following is an example to be stoved at 120° C.

Gilsonite varnish (Air-Drying Varnish No. 1 above)	34 parts
Congo copal varnish	40 parts
Blue Oil	20 parts
Lead Manganese drier	1.5 part
Benzine	4.5 parts

This gives a fine japan with a high gloss.

Flexible Synthetic Resin Varnish

Albertol 1.16Q	100 parts
Wood Oil	80 parts
Stout Linseed Stand Oil	220 parts
Lead Manganese Drier (4% Pb, 2% Mn.)	5 parts
Cobalt Drier (1.5% Co.)	2 parts
White Spirit	250 parts

Heat the two thickened oils to 240° C. and maintain till the viscosity is high. Dissolve the albertol as the temperature drops to 180°, and add the driers and diluents.

Synthetic Varnish

A clear homogeneous product is obtained, for example, by treating a mixture of 500 parts, heat-polymerized linseed oil, 250 parts vinyl chloride, 1 part sodium perborate and 5 parts of acetic anhydride at a temperature of 95° C. for 15 hours in an autoclave. A product with similar properties is obtained by autoclave treatment of a mixture of equal parts of asymmetrical dichlorethylene and thickened linseed oil for 10 hours at 125° C.

Glyptal Type Varnish
British Patent 391,508

A mixture of glycerol (17 parts), tung oil (108 parts) and colophony (100 parts) is heated at 200° C. for 45 minutes, phthalic anhydride (37 parts) is added, and the heating is continued at 200° C. for 40 minutes.

Rubbery Varnish Composition

A satisfactory product may be obtained by gradually heating 45 kilograms of pitch-like residue from the distillation of fatty oils with 4 kilograms of a mixture of carnaubyl and ceryl alcohol to a temperature of 260° C., when esterification is completed in about 30 minutes. Gradual addition is then made of 4 kilograms of litharge previously ground to a paste with a mixture of 6 kilograms of linseed stand oil and 4 kilograms of boiled linseed oil. Complete solution is gradually effected on maintaining at a minimum temperature of 250° C., at which stage the mass is cooled to 130 to 140° C. and thinned with 90 kilograms of solvent naphtha. Suitable solvents for the final dark, tough, rubber-like mass are turpentine, naphthal and teralin, and the solution can be worked up both to oil varnishes and nitrocellulose-oil combination lacquers.

Violin Varnish

Sandarac	160 parts
Mastic	80 parts
Alcohol	21 parts
Turpentine Varnish	750 parts

Mix and set aside in a warm place, agitating occasionally until solution is complete; then strain.

Special Violin Rosins

By combining any grade of ordinary untreated pine rosin with various percentages of London Rosin Oil or second run rosin oil, violin rosins can be made that will give any desired result such as extra loud music for dancing, etc. Sometimes it is advisable to incorporate up to 1 per cent of high melting paraffin wax in the mixture to prevent accumulation on the bow.

Flexible Insulating Varnish
British Patent 393,034

	No. 1 parts	No. 2 parts
Phthalic anhydride	213	235
Glycerol	129	118
Succinic acid	456	235
Ethylene glycol	242	230
Rosin	200	200
China wood oil	80	80

The ingredients are mixed and heated to 250° C. as quickly as possible, and kept at that temperature until the resin has become clear and has an acid value less than 25 milligrams of sodium hydroxide per gram. The temperature used may vary from 230° to 270° C. Such resins are stated to be unaffected by heating at 100° C.–105° C. for two weeks or more. Resin No. 1, when in reasonably thin films, will cure in about four hours at 135°–140° C. No. 2 is a somewhat faster curing resin and gives harder and tougher films, but is slightly less flexible than No. 1.

Insulating Varnish

Bakelite varnish	24 gal.
Naphtha 140–220° F.	6 gal.
Ethylene dichloride	12 gal.

Paper Varnish

Sandarac	5 parts
Venice Turpentine	3 parts
Denatured Alcohol	15 parts

Waterproofing Varnish for Fishing Lines

This formula is stated in parts by weight.

Pyroxylin	100 parts
Castor Oil	250 parts
Amyl acetate	400 parts
Magnesium carbonate	2 parts
Methyl Alcohol	600 parts

Wrinkle Finish Varnish
U. S. Patent 1,934,034

A base for a wrinkle finish varnish, comprising raw Chinawood oil and a hard fusible rosin in proportion of not less than 5% and not more than 10% of the weight of the oil, the compound having been subjected to heat of 350 to 550 degrees Fahrenheit for a period of four to eight hours.

Coach Japan
Formula No. 1

Raw Linseed Oil	50 gal.
Red Lead	200 lb.
Granulated Manganese Dioxide	25 lb.
TN Shellac	50 lb.
Pure Turpentine	110 gal.

Directions

Heat the oil to 200° F., add part of the red lead and all the manganese dioxide, continue heating to 320–330° F., shut off the fire, and constant stirring will run the temperature to 450° F.; hold at 450° for 3 hours. (The balance of the red lead is to be added slowly after the batch has raised in the kettle; if all put in at once, the batch would rise and flow over.) After holding 3 hrs. at 450°, add the shellac and hold at 400° for 2 hours, then cool to 310° and reduce.

This product has always been made in a set-kettle of ample size, with large fume stack overhead; heating was by gas burner underneath, although the rotating oil-burner also could be used. The bricked-in kettle was intentionally used for several reasons: the preponderance of active metallic driers to be incorporated demands a round bottom to enable thorough stirring and surety in scraping the bottom of the kettle to prevent the salts adhering and burning; it offers a wider expanse of surface for sprinkling in the driers and likewise freer exit of the gases evolved; and then there is the factor of safety—obviously running goods of this type the foaming from the driers and the shellac is very great and requires the ability of a good and strong kettle man because the rising of the batch must be under complete control by constant whipping. Should the batch get away and overflow, the enclosed burner is easily turned off and there is no fire pit for the batch to run into as might occur in the regular varnish stacks. Would advise that the varnish maker be attended by an assistant when making these goods.

The red lead used was 80% Pb_3O_4, 20% PbO. The granulated manganese dioxide contained approximately 76% pure MnO_2. Destructively-distilled turpentine cannot be used.

Formula No. 2

Identical in composition, purpose, and characteristics except that the diluent is equal parts of steam-distilled wood turpentine and petroleum naphtha. To this extent the product is not quite equal in certain features to the No. 1, and is intended for use where it may be necessary to economize.

Both of these products have been in use over forty years, and not surpassed in essential quality and characteristics by any others. Being composed of a large percentage of oil, they serve as excellent grinding mediums. The next requirement is that colors ground in a medium of this type, when thinned with turpentine, must have good flowing, flatting, drying, and binding properties. A standard test is this: Grind 4 ounces of bone black (drop black) in 4½ avoirdupois oz. of the japan; reduce 2 oz. of the paste with 1 oz. avoirdupois of pure turpentine. Apply this Black with a camel-hair brush onto a panel already coated in with primer and filler; it should dry in 45–50 minutes so same will not rub up when varnished over.

Prevention of "Skinning" of Varnish

If the following is used as the solvent, or thinner in a cooked wood oil resin varnish skinning is prevented.

Dipentene	75 parts
Petroleum Spirits	25 parts

The addition of one-eighth–one-half ounce per gallon of crude (wood distillation) guiacol will also prevent skinning and often retards gelling.

Shock-Resistant Varnish
U. S. Patent 1,950,820

A varnish consisting, for example, of chlorinated rubber and benzene gives, on application upon smooth surfaces, transparent coatings which are very beautiful in themselves but which tend to flake off, film being formed; this tendency is the more marked, the smoother the surface. Such coatings become fully resistant to shock, when hard substances such as quartz-meal, carborundum and the like of grain size 40–100μ are added thereto in accordance with the invention. The coatings produced with these mixtures are characterized by extraordinarily secure attachment and do not flake or split off even when struck or dropped.

Varnish Remover

Acetone	1 part	by volume
Chloroform	1 part	by volume
Denatured Alcohol	4 parts	by volume
Benzene	6 parts	by volume
Diglycol Stearate	⅛ part	by weight

White Priming Coat for Cypress

White Lead	100	lb.
Linseed Oil	2½	gal.
Benzol	2	gal.
Turpentine	1½	gal.

Ungelled China Wood Oil
U. S. Patent 1,903,666

China wood oil is heated rapidly to temperatures above 600° F., held for a sufficient number of seconds, then cooled rapidly to 200° F., without jellying.

The method consists of forcing the china wood oil at pressures of 15 to 35 pounds, through externally heated spiral or helical coils. The ungelled treated oil is suitable for varnish. Higher temperatures or long periods of treatment produce oils without drying properties, suitable as plasticizers.

Oxidized Tung Oil

Tung oil is kept at 100° C. for six or more hours, while a stream of air is blown through it. It will then dry to a clear film and may be used in making varnish.

"Scarlet" Chrome Yellow Pigment
Solution A

Lead Acetate	200 lb.
Hot Water	100 gal.

Solution B

Normal Potassium Chromate	77 lb.
Hot Water	100 gal.

Solution C

Caustic Soda (100%)	30 gal.
Hot Water	20 gal.

Solution B is run into A at a moderate speed with good stirring, then allowed to stand one hour and the top liquor run off as far as possible. Solution C is then run in rapidly with good stirring, and the mixture boiled with open steam for 15 minutes. The color must be thoroughly washed before drying. A bright shade and a soft product are thus obtained.

Manganese Resinate Drier

148 pounds of rosin (preferably F grade) are melted and raised to 250° during 40 minutes, 9 pounds of dry manganese dioxide are sifted in during ½ hour, and the temperature then held at 310–315° until reaction is complete (40–60 minutes) when the melt is rapidly cooled, e.g., by pouring into a shallow pan. A rosinate solution in mineral oil and containing 4.5–5% manganese is obtained.

Liquid Drier

Lead Nuodex (16% Lead)	5.0 lb.
Cobalt Nuodex (4% Cobalt)	20.0 lb.
Manganese Nuodex (4% Manganese)	0.2 lb.
Mineral Spirits	62.8 lb.

The resulting metal content of this drier is: lead 0.9%, cobalt 0.9%, and manganese 0.009%.

Bodied Linseed Oil

A good grade of varnish linseed oil is heated at 590° F. for four hours.

Wicker Varnish Stain

Soft Gum Sandarac	250 parts
Methanol	675 parts
Aniline Dye	28 parts

Black Acid Proof Stain for Wood

Solution No. 1

Iron Sulphate	4 parts
Copper Sulphate	4 parts
Potassium Permanganate	8 parts
Water to make	100 parts

Solution No. 2

Aniline	12 parts
Hydrochloric Acid	18 parts
Water to make	100 parts

This formula is stated in parts by weight.

Apply two coats of No. 1 hot; the second coat as soon as the first is dry. Rub off excess of last coat when dry and apply two coats of No. 2. Dry thoroughly and apply a coat of linseed oil, using a cloth instead of a brush to obtain a thinner coat. The oil may be thinned with turpentine if desired. The color develops to ebony black in a few hours. The surface may be washed with soap and water and is not easily affected by acids or alkalies. 100 cubic centimeters of No. 1 mixture will cover approximately 20 square feet of surface. 50 cubic centimeters of No. 2 mixture will cover approximately 20 square feet also.

Wood Finishes

Furniture, caskets, toys, pianos, etc., that are made from raw wood are given certain types of stains and finishes that beautify them and protect them against wear.

The general procedure is first to apply stain, then a wood filler and finally one or more coats of lacquer or varnish.

Stains

Stains may be divided into four main groups, benzol stains, water stains, oil stains and finally lacquer or varnish stains.

Benzol stains are oil soluble dyes dissolved in benzol and are blended to give the desired shades, such as walnut, mahogany, oak, maple, etc. The concentration of dye will vary from perhaps one ounce up to one pound per gallon of benzol depending largely upon the strength of the dyes and upon the depth of the shade desired.

These solutions are applied by either brushing, spraying or dipping and are then permitted to dry. The wood is then filled with a pigment paste filler. A coat of sanding lacquer is then applied over the filler and when it is dry, it is sanded to give a smooth surface. It is then followed by a coat of shellac to prevent the dyes from bleeding through subsequent applications of lacquer and varnish. Over the shellac one or more coats of either lacquer or varnish are applied in a gloss, semi-gloss or flat finish depending upon the effect desired. The final coat may be rubbed and polished if a hard rubbing finish is used.

The wood filler is a paste that is first thinned with turpentine or substitute turpentine in the proportion of about ten to fifteen pounds of filler to one gallon of thinner. It is then brushed on the wood and permitted to set up for about twenty to thirty minutes and then the excess wiped off. This is done by wiping with a soft cloth against the grain of the wood so as not to remove the pigment particles from the wood pores.

Wood fillers are often colored with pigments to impart not only a filling but also a staining quality. A typical filler is as follows:

Japan Drier	3 gal.
Turpentine	5 gal.
Linseed Oil	10 gal.
Silica	250 gal.
Asbestine	150 gal.

The above may be colored by adding to or partially substituting umbers, siennas, carbon black and other colored pigments for the asbestine. This may be used to stain and fill at the same time.

Water Stains

Water stains are similar to benzol stains except that water soluble dyes are used instead of oil soluble dyes and water is the vehicle instead of benzol. Also, no shellac is necessary to seal these stains as these do not bleed as the benzol stains do. Water stains are also more resistant to fading than are the benzol types. One must consider all of these factors in choosing the type of stain for one's work.

Oil Stains

These stains are made by grinding colored pigments in oil and then diluting rather excessively with turpentine or substitute turpentine. The combination of pigments that are used depends of course upon the color desired. For example, one shade of walnut stain would be:

Burnt Umber	80	lb.
Linseed Oil	7½	gal.
(Grind and add)		
Drier	1	gal.
Turpentine	24	gal.

This is applied the same as the other stains, permitted to dry overnight and then finished with lacquer or varnish. No filler is used with this stain as the pigment serves to both fill and stain at the same time. Also no shellac is used as in the case of benzol stains since no bleeding can take place. This stain is employed only in cheaper work where it is desired to eliminate the extra operations of filling. The grain of the wood is not so clearly revealed as by the other stains.

Varnish and Lacquer Stains

For very cheap work, particularly for cheap toys, merely one application of a varnish or lacquer in which dyes are dissolved is used. This stains and finishes in one operation. Of course the result is not comparable to the finish obtained by the more complete methods, but it is satisfactory for the purpose for which it is used.

Lacquer and Varnish Finish

After staining and filling a sanding lacquer is applied. This is a lacquer that is designed to sand easily and to seal the imperfection in the wood. When dry it is sanded with very fine sandpaper to give a smooth surface. Over the sanding surface a coat of clear wood lacquer is sprayed. For the best type of finish a rubbing lacquer is used high in solid content with a high ratio of nitro-cellulose. (See Vol. 1, Page 227.) This is first water sanded with fine sandpaper then rubbed with a rubbing compound and then waxed. This gives the highest possible type of finish of glass-like hardness and smoothness. Where the labor of rubbing and polishing is to be eliminated a clear lacquer high in solid content and lower in nitro-cellulose is used. This may be either in a high gloss, semi-gloss or flat finish. (See Vol. 1, page 227.)

Varnishes are also employed in finishes of varying degrees of glass. With varnish it is possible to obtain a higher gloss than with lacquer. They cannot, however, be made so hard as lacquer and will not rub to as smooth a finish as lacquer. Also, all varnishes will impart a yellow tinge to the surfaces and cannot be made so clear and water white as lacquer. This discoloration is particularly true of the bakelite types which otherwise are the most suitable varnishes for this type of work. For formulae, see page 238, Vol. 1, Four Hour Varnish and Forty gallon Phenolformaldehyde type of Varnish.

These varnishes can be converted into flat finishes by grinding aluminum or zinc stearate and magnesium carbonate into them. For example:

Aluminum Stearate	4 oz.
Magnesium Carbonate	4 oz.
Turpentine	1¼ gal.

Acid Proofing of Wooden Shelves

The above preparation is not suitable for direct application on wooden surfaces that have been oiled or painted, since it readily peels off. First cut strips of cloth to fit the shelves and soak these strips in the dressing, after it has been warmed as directed above. After the cloth has been saturated with the warmed dressing, press out excess liquid and lay strips on shelves. Allow to set for several hours, then take soft cloth and polish. The treated strips will lay flat on the surfaces and protect the wood.

Non-Drying Stain for Wood
U. S. Patent 1,896,662

The pigment is ground into about ⅕ of its weight of a drying oil, naphtha (one part or somewhat less) is added, and the mixture stirred into a paraffin oil (12 parts) at 60° having distillation range 250–300° and a density of approximately 0.86.

Mahogany Stain
U. S. Patent 1,925,749

A ''fadeless'' mahogany stain is prepared by mixing with the steam-extracted water-insoluble extract of quebracho wood sufficient hot concentrated alkali solution such as sodium carbonate or sodium hydroxide to give a pH of about 11 or 12, and digesting with sufficient added hot water to give a pH of about 7.0 to 8.5 in the final product.

Wood Stain to Duplicate Antique, Weathered, or Burned Finishes

Allow steel wool to set in dilute acetic acid (or vinegar) a few days. When desired depth of color is obtained, remove steel wool (all of it) and the stain will keep indefinitely. Any shade from weathered gray to black may be obtained.

Antique Finish for Wood

Chrome Green (Dark)	3 tspf.
Van Dyke Brown	2 tspf.
Lampblack	2 tspf.
Turpentine	1 pint
Linseed Oil (Boiled)	1 pint
Japan Drier	a few drops

This should be coated over the entire moulding; when it starts to set, wipe off the high places with a clean rag. Go over the rest of the moulding and take off all excess glaze that is not needed.

COSMETICS

Face Powders and Talcs

The most important consideration in the manufacture of face powders and the various talcum powders is in the selection of the raw materials. First, and most important, is the talc employed which should be judged on the basis of slip and smoothness, grit, color, mica content, fineness, acid soluble materials, and specific gravity both actual and apparent.

These properties should be carefully considered in the selection of the talc and more so after a selection has been made in checking of subsequent shipments from the raw material source. For the better products the Italian and Manchurian talcs are to be recommended. French talcs find their use in the medium grade products, and particularly in compacts. Californian talcs are also suitable for medium grade products, while the various other talcs are employed in low-priced products.

Talcum Powders

Dusting Powder No. 1

Talc	95 lb.
Boric Acid	2 lb.
Magnesium Carbonate	3 lb.
Perfume	4–8 oz.

Dusting Powder No. 2

Talc	85 lb.
Magnesium Carbonate	10 lb.
Boric Acid	2 lb.
Zinc Stearate	3 lb.
Perfume	4–8 oz.

After-Bath Powder No. 1

Talc	80 lb.
Zinc Stearate	10 lb.
Boric Acid	3 lb.
Magnesium Carbonate	7 lb.

After-Bath Powder No. 2

Talc	85 lb.
Magnesium Carbonate	7 lb.
Zinc Stearate	7 lb.
Boric Acid	1 lb.

The zinc stearate is used for the adhesiveness and softness which it imparts to the powder. The boric acid is used for its antiseptic action. The magnesium carbonate is used for securing lightness and fluffiness. Substitutions, such as the use of magnesium stearate in place of zinc stearate, the use of a light precipitated chalk in place of magnesium carbonate can be made. The incorporation of other antiseptic bodies, such as methyl-para-hydroxy-benzoic-acid, tertiary-chlor-butanol, chlor-meta-xylenol, is effected by melting these materials into the perfume oil (with the addition of a small amount of alcohol if desired). The perfume is incorporated in the usual manner.

The procedure in the manufacture of these talcum products is as follows:

Dry materials are mixed (usually in a horizontal type enclosed mixer) for a period of time. The perfume is added to a quantity of magnesium carbonate or of the mixed powder, equivalent to twenty (20) times the weight of perfume oil, mixed and then brushed through a forty (40) mesh wire screen, and then through at least a ninety (90) mesh silk screen. The perfume mixture is then added to the full batch of dry materials and bolted through a silk screen of at least one-hundred (100) mesh. A mesh of two-hundred (200) should be used for the highest quality product. At times a pebble mill is used for a simultaneous mixing and grinding operation, preparatory to sifting. The powders may be tinted slightly by using the same color as used in a face powder to secure a rachel, peach, flesh or any other desired shade. The amount of color usually used is about 20% that used in the equivalent face powder shade. However, in talcums for men, the colors are about equivalent in intensity to regular face powder shades.

Face Powders

Formula No. 1—Heavy

Talc	40 parts
Magnesium Carbonate	5 parts
Zinc Oxide	10 parts
Zinc Stearate	5 parts
Rice Starch	10 parts
Kaolin	30 parts
Color—See Coloring	
Perfume	6–14 oz.

Formula No. 2—Medium

Talc	50 parts
Zinc Oxide	15 parts

Zinc Stearate	10 parts
Kaolin	20 parts
Precipitated Chalk	5 parts
Color—See Coloring	
Perfume	6–14 oz.

Formula No. 3—Light

Talc	65 parts
Zinc Oxide	10 parts
Zinc Stearate	10 parts
Kaolin	10 parts
Precipitated Chalk	5 parts
Color—See Coloring	
Perfume	6–14 oz.

Coloring for Face Powders and Talcs

The colors necessary to secure most shades of face powder are yellow ocher, geranium lake, persian orange lake, orange lake, burnt umber, burnt sienna, ultramarine blue, violet lake and green lake. These colors are diluted to make color bases as follows:

Rachel Color Base

Yellow Ocher	1 part
Talc	4 parts

Flesh Color Base

Geranium Lake	1 part
Talc	9 parts

Peach Color Base

Persian Orange Lake	1 part
Talc	3 parts

Orange Color Base

Orange Lake	1 part
Talc	9 parts

Grey Color Base

Burnt Umber	1 part
Talc	5 parts

Tan Color Base

Burnt Sienna	1 part
Talc	3 parts

Blue Color Base

Ultramarine Blue	1 part
Talc	9 parts

Lavender Color Base

Violet Lake	1 part
Talc	10 parts

Green Color Base

Green Lake	1 part
Talc	10 parts

To secure the various shades of face powder, the following are used in conjunction with the above bases.

Light Cream

Base	100 lb.
Rachel Color Base	24 oz.

Rachel

Base	100 lb.
Rachel Color Base	48 oz.

Medium Rachel

Base	100 lb.
Rachel Color Base	64 oz.
Tan Color Base	8 oz.

Dark Rachel

Base	100 lb.
Rachel Color Base	48 oz.
Tan Color Base	16 oz.
Lavender Color Base	4 oz.
Gray Color Base	8 oz.

Various intermediate shades can be secured by combining two or more of the above formulae, or by increasing or decreasing the components of these formulae, or by the addition or deletion of items from these formulae. The procedure in the manufacture of these powders is: mix the dry materials, including the color base, in a horizontal type enclosed mixer, or else in a pebble mill, until all the materials are thoroughly distributed. The perfume is worked into about twenty (20) times its weight of either magnesium carbonate or powder base, and sifted through a wire screen and then a silk cloth as mentioned under talcum powders. The perfume mass is then added to the dry materials and the entire mixture is brushed through a sixty (60) mesh wire screen and then bolted through at least a one-hundred and twenty (120) mesh silk, with the finer mesh screens being more advisable. If it is so desired, the powder may be ground through a hammer mill or similar apparatus.

In the formulation of the powder, these factors are considered:

Talc is used for slip and its lubricating effect.

Kaolin—there are two (2) types available—the osmo or colloidal kaolin, and the bolted kaolin.

The *osmo kaolin* is much whiter and drier than the bolted, and has its main use in bodying the powder, giving it slip and coverage, and for its powers of absorption, especially in regard to perspiration and moisture.

The *bolted kaolin* is greener and much more moist than the osmo kaolin, and it is used to secure additional slip and a creaminess that can not be secured otherwise.

Rich Starch is used for the smoothness it imparts. Its absorption of moisture or perspiration with a subsequent swelling of the particles, resulting in enlarged pores, is a question of reasonable doubt, and can not be answered with entire satisfaction.

Magnesium Carbonate is used for securing lightness and bulkiness of the powder, and also as an absorbent of perfumes.

Zinc Oxide is used for its tinting covering powers.

Titanium Oxide is used for its covering powers which it possesses to a much greater degree than zinc, oxide, but its tendency to pack and "ball" a powder in which it is used makes it a product not usually found in face powders; zinc oxide still being considered more favorable for securing cover. However, if care is taken in the formulation of a product, excellent powders can be made with titanium oxide as a component of the powder.

Zinc Stearate is used in securing adhesiveness of the powder, as well as softness and lightness.

Magnesium Stearate has properties similar to zinc stearate but it is heavier. It finds its main use to replace zinc stearate where the use of zinc stearate is prohibited.

Precipitated Chalk is used in securing the smoothness that rice starch imparts without running into the difficulties that might be encountered in the use of rice starch by reason of its property of swelling.

Compact Powders

Compact powders are made in the same manner as are compact rouges, except that coloring is done with color bases, the same as are used in face powders. About 50% of the amount of color used in face powders is usually sufficient to give the same intensity of color.

Liquid Face Powder

Liquid face powders are suspension of chalk and zinc oxide in a water, alcohol and glycerin solution. The coloring is done with the color bases that are used in face powders.

Liquid Powder

Zinc Oxide	3 lb.
Precipitated Chalk	3 lb.
Diethylene Glycol	1 pint
Alcohol	4 pints
Perfume	4 oz.
Water	4 gals.

Color

(See Face Powder)

Rachel—Yellow Ochre Base	1	oz.
Tan—Burnt Sienna Base	1	oz.
Flesh—Geranium Base	1	oz.
Peach—Persian Orange Base	½	oz.

Further ideas for coloring may be had by referring to the various shades and the combinations necessary to secure them.

The zinc oxide may be replaced in whole or in part with titanium oxide.

Diethylene glycol may be used in place of glycerin.

Another type product is:

Liquid White (for Skin)

A lotion for hand and arms contains 2,500 parts witch hazel extract, 5,000 parts rose water, 1,000 parts alcohol, 1,800 parts glycerin, 100 parts tallow, 100 parts magnesium carbonate, 50 parts magnesium stearate and 1,000 parts antipyrine. First, the antipyrine is dissolved in the witch hazel extract and rose water. Then glycerin is added. The perfume used is absorbed by the magnesium carbonate, magnesium stearate and tallow. Then alcohol is added. This suspension is strongly shaken for two days. The milk is filtered through coarse filter paper. The two preparations are united with vigorous stirring and decanted. This preparation is applied with cotton. The skin is rubbed and the preparation is allowed to dry. The skin remains white the entire evening. The advantage of this preparation over ordinary liquid powder is that a dull white effect is obtained, lasting 4 to 6 hours.

Dry Rouges

Dry rouges were originally made by mixing talc and carmine with a mucilage of tragacanth, placing the plastic mass in metal trays or cups and allowing them to harden. These products were then sold in these trays or cups.

The manufacturing technique was then developed to the point wherein with a slight modification of the formula, the plastic mass was placed on discs, allowed to dry and then turned into shape by a cutting tool.

The present day manufacture of rouges is done in several different ways. The most simple is a mixture of the following:

Talc	40 lb.
Kaolin	35 lb.
Zinc Oxide	15 lb.
Precipitated Chalk	10 lb.

Mix the above with sufficient dry color to give the desired shade (see following), grind in a ball mill for a period of time sufficient to distribute all color particles throughout the entire mass. The material is then bolted through at least a 140 mesh silk, and is then moistened with a tragacanth solution of a strength of ¼ oz. gum tragacanth and ¼ oz. boric acid to 1 gallon water. This solution is added to the powder mass at the rate of 1 oz. to every pound of dry material. The perfume oil is also incorporated at this stage.

The slightly wetted powder is brushed through at least a 30 mesh wire screen and bolted through at least a 60 mesh silk screen. The rouge material is then ready for pressing.

The type press used for this particular formula is a foot press which has a fast downward and upward stroke. The metal disc or cup onto or into which the rouge is to be pressed is moistened with a tragacanth solution of a strength of 1 oz. of gum tragacanth to 1 gallon of water.

Colors for Rouges
To Be Added to a 100 Pound Batch
Orange

Scarlet Lake	7 lb.
Yellow Ocher	8 oz.
Indian Red	4 oz.

Light No. 1

Indian Red	3 lb.
Burnt Sienna	1 lb.
Brilliant Lake	2 lb.
Maroon Lake	8 oz.

Light No. 2

Scarlet Lake	4 lb.
Brilliant Lake	2 lb.
Yellow Ocher	8 oz.
Indian Red	8 oz.
Geranium Lake	4 oz.

Geranimum No. 1

Geranium Lake	5 lb.
Scarlet Lake	7 lb.
Brilliant Lake	3 lb.

Geranium No. 2

Geranium Lake	12 lb.
Orange Toner	4 oz.
Crimson Lake	8 oz.

Medium No. 1

Brilliant Lake	18 lb.
Scarlet Lake	2 lb.
Maroon Lake	1 lb.

Medium No. 2

Brilliant Lake	12 lb.
Geranium Lake	4 lb.
Maroon Lake	3 lb.
Indian Red	3 lb.

Lip Rouge—Indelible
Formula No. 1

Castor Oil	7 lb.
Lanolin	1 lb.
Beeswax	1 lb.
Bromo Acid	12 oz.
(tetrabrom fluorescein)	
Lake Color	12 oz.
Perfume	as desired

Formula No. 2

Castor Oil	6 lb.
Cetyl Alcohol	1 lb.
Stearic Acid	4 oz.
Lanolin	1 lb.
Glycerol Mono Stearate	1 lb.
Bromo Acid	8 oz.
(tetrabrom fluorescein)	
Lake Color	4 oz.
Perfume	as desired

The lake color mixtures used in the previous formulae may be used to secure the various shades. The procedure is the same as in the above formulae.

Lip Pomade

Mineral Oil	6 lb.
Petrolatum	2 lb.
Paraffin	2 lb.
Ozokerite	¼ lb.
Beeswax	¾ lb.
Perfume	½ oz.

The materials are melted together and poured into a suitable mold. These sticks are intended for use in softening of lips and preventing chapping of the lips.

Materials such as: lanolin, absorption base, olive oil, cocoa butter, may also be introduced into the formula. From 2-4 oz. of zinc oxide ground into the product will give a whiter stick and will aid as a healing agent.

Materials such as: menthol, camphor, thymol and similar medicants, may be used in small quantities and most conveniently introduced into the product by molding them into the perfume and then adding the mixture to the melted oils and waxes.

Lipstick (Non-Indelible)
For Theatrical Use

Petrolatum	4 lb.
Paraffin	2 lb.
Mineral Oil	1 lb.
Carnauba Wax	6 oz.
Lanolin	8 oz.
Lake Color	1 lb.
Perfume	as desired

The same lake color mixtures as are used in the greasy cream rouges are suggested to secure the various shades. The procedure is the same.

Lipstick—Indelible
Formula No. 1

Castor Oil	4	lb.
Cocoa Butter	2	lb.
Stearic Acid	1½	lb.
Paraffin	2	lb.
Beeswax	1¾	lb.
Carnauba Wax	2	oz.
Lanolin	8	oz.
Bromo Acid	12	oz.
(tetrabrom fluorescein)		
Lake Color	12	oz.
(see cream rouge)		
Propyl-para-hydroxy-benzoate	1	oz.
Perfume	2	oz.

Formula No. 2

Castor Oil	6	lb.
Glyceryl-mono-stearate	2	lb.
Stearic Acid	½	lb.
Cetyl Alcohol	½	lb.
Bromo Acid	10	oz.
(tetrabrom fluorescein)		
Erythrosine	¼	oz.
Oil Soluble Red	16	grains
Perfume	1½	oz.

The oils and waxes are heated together and the oil soluble color is dissolved therein. The bromo acid and erythrosine are then added and the entire mass is ground through an ointment mill. For this particular type lipstick, various shades may be secured by the use of various of the oil soluble colors, such as: yellow and orange in combination with the red, and the use of other water soluble dyestuffs in place of the erythrosine, such as: tartrazin, ponceau, etc.

Formula No. 3—Changeable Orange

Castor Oil	5	lb.
Cocoa Butter	3	lb.
Ceresin	3	lb.
Beeswax	2	lb.
Bromo Acid	2	oz.
(tetrabrom fluorescein)		
Propyl-para-hydroxy-benzoate	2	oz.
Perfume	2	oz.

The oils and waxes are melted. The bromo acid and the benzoate are then added and the entire mixture is filtered hot.

Mascara—Soapless Type—Poured

Triethanolamine	14 lb.
Stearic Acid	20 lb.
Oleic Acid	5 lb.
Ricinoleic Acid	5 lb.
Carnauba Wax	30 lb.
Ozokerite	15 lb.
Petrolatum	6 lb.
Perfume	1 lb.

These materials are all melted together.

The following colors are ground into the molten mass to secure the various shades:

Black	
Charcoal Black	5 lb.
Brown	
Burnt Umber	10 lb.
Burnt Sienna	1 lb.
Indian Red	3 lb.
Blue	
Ultramarine Blue	12 lb.
Titanium Oxide	2 lb.

Creams

Cold creams are the most basic and still the most important creams that are sold. Cold creams are usually formulated using mineral oil as a softening and cleansing agent, and emulsifying with water by the action of borax on beeswax.

A rather soft but exceptionally smooth cream is made as follows:

Mineral Oil	1 gal.
Beeswax	1¾ lb.

Heat the above to 160° F. Dissolve 1½ oz. of borax in 5 pints of water, heat to 160° F. and add this solution to the oil and wax with rapid stirring. When the temperature drops to 140°, add 1 oz. of perfume oil and pour the cream at about 120°.

This basic formula may be modified by replacing up to half of the beeswax with paraffin, ceresin, ozokerite or spermaceti.

The oil may be replaced in part by petrolatum or by the vegetable oils. If vegetable oils are used, a preservative should be employed.

Materials such as lanolin and absorption base may be introduced in small quantities.

Cold Cream

Mineral Oil	54 %
White Wax	18 %
Absorption Base	5.5%
Borax	1 %
Water	21 %
Perfume	.5%

Melt the white wax, add the mineral oil. Dissolve borax in part of water with heat. Add to melted fats. Heat rest of water, stir in absorption base until smooth and mix with fats. Agitate thoroughly and when just above solidifying point, add perfume.

Cleansing Cream

A second type of cold cream is based on the action of triethanolamine on stearic acid.

The following are examples of this procedure:

Cleansing Cream

	Mineral Oil	76 lb.
	White Wax	5 lb.
1.	Spermaceti	26 lb.
	Trihydroxyethylamine Stearate	20 lb.
2.	Perfume	1 lb.
3.	Glycerin	4 lb.
	Water	92 lb.

Heat Nos. 1 and 3 separately to 200° F.; then add Nos. 1 to 2 slowly, stirring thoroughly. When the cream begins to set, the perfume is added and stirred in. Allow to stand over night. Stir thoroughly the next morning and package. This cream will not sweat oil during hot weather and will maintain its consistency.

A third type of cream is that in which the emulsifying agent is either glyceryl monostearate or glycosterin.

These creams are emulsions of oil in water and for that reason evaporate quickly, and produces a cooling effect. They are much more water soluble than the beeswax type creams. These creams should be packed in air tight jars as there is a tendency for a small amount of water to separate from them.

The following are examples of this type product:

Cold Cream (Non-Greasy)

1.	Glyceryl Monostearate	22 lb.
2.	Petrolatum (Vaseline)	16 lb.
3.	Paraffin Wax	12 lb.
4.	Mineral Oil	30 lb.
5.	Water	98 lb.

Heat first four ingredients to 170° F. and stir together. Then slowly with stirring pour in the water which has been heated to the same temperature. Stir thoroughly and then allow to stand (hot) until air bubbles are gone. Add perfume and stir and pour at 110–130° F. Cover jars as soon as possible.

Neutral Cleansing Cream

1.	Mineral Oil	80 lb.
2.	Spermaceti	30 lb.
3.	Glyceryl Monostearate	24 lb.
4.	Water	90 lb.
5.	Glycerin	10 lb.
6.	Perfume to suit.	

Heat 1, 2 and 3 to 140° F. and stir into it slowly 4 and 5 heated to same temperature. Add perfume, at 105° F., stir slowly until cold; after allowing to stand for 5 minutes stir until smooth and pack.

A four purpose cream that cleans, nourishes, stimulates and acts as a powder base is made as follows:

Mineral Oil	3 pints
Petrolatum (white) (heat to 140° F.)	½ lb.
Water	4½ pints
Glycerin	5 oz.
Preservative	½ oz.

Heat to 140° F. and add slowly with stirring to oil mixture. As the temperature falls, a gelatinous mass forms at 120° F. 1 oz. perfume oil is added while stirring and the gelatinous mass changes to a white cream. Slow stirring is continued until cold. This cream may be packed either in tubes or jars.

This cream can be modified by various coloring agents and perfume as under cold cream to obtain specialty creams. Since it is neutral there may be incorporated in it viosterol, or gland or hormone extracts.

Liquefying Cleansing Creams

This type of cream is composed of approximately 50% mineral oil together with petrolatum to give sufficient viscosity so that when the cream liquefies on the skin, it suspends the dirt which is removed from the pores.

The following formulae give excellent results:

Formula No. 1—Soft Translucent

Mineral Oil (light or medium)	56 parts
Paraffin	25 parts
Petrolatum (white)	19 parts

Formula No. 2—Medium Translucent

Mineral Oil (light or medium)	50 parts
Paraffin	18 parts
Petrolatum (white)	23 parts
Spermaceti	9 parts

Formula No. 3—Medium Opaque

Mineral Oil (light or medium)	50 parts
Paraffin	30 parts
Petrolatum (white)	20 parts

Formula No. 4—Hard Opaque

Mineral Oil (light or medium)	45 parts
Paraffin	25 parts
Petrolatum (white)	20 parts
Spermaceti	10 parts

The ingredients are melted and stirred together on a water-bath and 4 oz. of perfume is added per 100 lb. These creams are poured at the lowest possible temperature and allowed to stand undisturbed until solid.

This cream possesses exceptional penetrating powers and is absorbed very readily by the skin.

Vanishing Creams

Vanishing creams are greaseless creams, essentially stearic acid soaps with excess suspended stearic acid dispersed in water. Pearliness is the effect produced by the crystalline stearic acid in suspension.

For a soft type vanishing cream, triethanolamine is used as a saponifying agent.

| Stearic Acid (heated to 170° F.) | 24 lb. |

Heat the following to 170° F.:

Triethanolamine	1 lb.
Glycerin	8 lb.
Water	8 gal.

and add to the melted stearic acid slowly, while stirring rapidly for a few minutes until emulsification is completed.

When the temperature falls to 135° F. add 4 oz. perfume oil and stir intermittently at slow speed until cold.

Allow to stand for a few days stirring slowly at least once a day for a few minutes.

For harder products, potassium carbonate or hydrate is used as a saponification agent. For example:

| Stearic Acid (heat to 170° F.) | 24 lb. |

Then heat, also to 170° F., the following:

Potassium Carbonate	5 oz.
Glycerin	8 lb.
Water	12 gal.

and add to the melted stearic acid. The procedure is the same as above. Use 5 oz. of perfume.

Lotions (Astringent)

Astringent lotions are usually based on alcohol as the active ingredient or on an alcohol containing product, such as witch hazel. For example;

Astringent Lotion—Mild

Menthol	4 oz.
Zinc Phenol Sulphonate	5 lb.
Camphor	4 oz.
Perfume	8 oz.
Alcohol	10 gal.

All of the above are dissolved together and 90 gallons of witch hazel is added to it. The product may be colored slightly by the use of a water soluble color.

Astringent Lotion—Medium Strength

Alcohol	35 gal.
Borax	2 oz.
Zinc Phenol Sulphonate	4 lb.
Perfume	2 lb.
Gum Camphor	8 oz.
Glycerin	3 gal.

All of the above are dissolved together and water sufficient to bring the volume to 100 gallons is added.

Astringent Lotion—Strong

Alcohol	50 gal.
Ethyl Amino Benzoic Acid	8 oz.
Parachlor Meta Xylenol	8 oz.
Menthol	8 oz.
Thymol	4 oz.
Lavender Oil	3 lb.
Glycerin	5 gal.
Vanillin	8 oz.

All of the above are dissolved together. Water is added to make 100 gallons.

Other materials, such as: resin, benzoin, peru or styrax may be used in small quantities.

Propylene glycol, or diethylene glycol may be used in place or the glycerin.

Lotions
Face Lotions

Skin milks and milky face lotions are made with trihydoxyethylamine stearate or else with mucilages of the various gums, as for example:

Liquid Cleansing Cream

The following milky cream is stable and an effective cleanser. It will even remove indelible lipstick and rouge from the skin in addition to the usual grime and dirt. It leaves the skin clean, fresh and stimulated and serves as a perfect powder base without any harmful effects.

Stearic Acid	3 lb.
Mineral Oil (Heat to 170° F.)	2 gal.
Water	3 gal.
Triethanolamine	1 lb.
Diethyleneglycol	3 lb.
Diethyleneglycol Ethyl Ether	2 lb.

Heat to 170° F. and add slowly with rapid stirring to the melted stearic acid and mineral oil. Continue stirring until temperature falls to 150° F. when 2 oz. of perfume is added. Stir until cool. A thicker cream is made by replacing part of the mineral oil by petrolatum.

Pearly Finishing Astringent Lotion

Gum Tragacanth	½ oz.
Water (Warm)	5 pints

Allow to stand for a day and stir into it

Alcohol	3 pints

To 12 pounds of soft vanishing cream (see Vanishing Cream formulae) which has stood long enough to develop a pearly sheen, there is added slowly a gallon or more, if desired, of the above gum solution. Stirring must be slow but thorough. Filter through cheesecloth and bottle.

This lotion may be colored if desired. This lotion is shimmering and pearly. It is quick drying and leaves the skin in a fresh, soft condition ready for the application of powder.

Skin Milks

Milky preparations for use on skin can be made with lanolin, cucumber milk and almond milk. The following are some examples:

Formula No. 1

Lanolin	50 parts
Pure Castile Soap	3 parts
Glycerin	20 parts
Rose Water	300 parts
Tincture of Benzoin	5 parts
Perfume Bouquet	10 parts
Water	612 parts

Formula No. 2

Lanolin (Melt on water bath)	30 parts

Then add the following mixture, gradually:

Warm Rose Water	200 parts
Potash Soap (pure)	10 parts
Glycerin (in solution)	20 parts

Then add the following:

Perfume Composition	10 parts
Tincture of Benzoin	30 parts

Remove the entire mixture from water bath and mix with

Cucumber Juice	700 parts

Freshly percolated, warmed

The mixture is then agitated until it cools off.

Formula No. 3

Shelled almonds, 70 parts, are crushed with addition of sufficient rose water to give stiff paste. Then the following mixture is added:

Tincture of Benzoin	20 parts
Benzaldehyde	2 parts
Rose Oil	1 part
Borax	7 parts

and 50 parts glycerin in sufficient rose water to give total of 1,000 parts.

Mixture is allowed to stand several days and then filtered through hair sieve.

Liquid Cleansing Cream (Non-Greasy)

1.	Beeswax	1.5 parts
2.	Spermaceti	6.5 parts
3.	Cherry Kernel Oil	6.0 parts
4.	Glycosterin	4.0 parts
5.	Water	122.0 parts
6.	Alcohol	3.0 parts
7.	Galagum	1.0 part
8.	Borax	3.0 parts
9.	Perfume	3.0 parts
10.	Glycerin	4.0 parts

Melt together 1, 2 and 3. Heat while stirring 4, 5, 7 and 8 together until uniform. Mix these two solutions stirring until uniform. Stir in 6, 9 and 10 and mix until uniform.

Almond Cream Liquid

Oil Sweet Almonds	1 lb.
Spermaceti	2 lb.
Beeswax	2 lb.
Castile Soap Powdered	3 lb.
Borax	2 lb.
Quince Jelly	1 lb.
Alcohol	1 pint
Water	4 pints

Melt the spermaceti and wax together. Dissolve the soap and borax in hot water. Mix these together and add balance of ingredients. Stir and filter through cloth.

Lemon Juice Lotion

Pectin	2.5 parts
Lemon Juice	9.5 parts
Water	88 parts
Preservative	0.15 part

Skin Smoothener

Boric Acid	3 drams
Tragacanth	8 grams
Glycerin	3 drams
Distilled Water	16 oz.

Boil—stir until a clear jelly is obtained.

Hand Lotions
Hand Cleanser and Conditioner

1. Mineral Oil	70 lb.
2. Olive Oil	8 lb.
3. Trihydroxyethylamine Stearate	14 lb.
4. Water	70 lb.
5. Perfume	2 lb.

Heat Nos. 1, 2 and 3 together to 140° F. and stir until homogeneous. Add No. 4 slowly while stirring and then stir in the perfume. Continue stirring until cool. By varying the amount of water a thicker or thinner preparation will be formed. The thicker preparations are put up in tubes and are now carried by men and women, especially motorists, who, when water is not available, merely put a little of this cleaner on their hands, rub it in and then wipe off with it the grease, oil, paint or dirt present. Not only is this an excellent detergent but it leaves the skin smooth, and produces a cooling sensation and prevents chapping during cold weather.

Hand Lotion

Boric Acid	1 dram
Glycerin	6 drams

Dissolve by heat and mix with

Lanolin	6 drams
Petrolatum	1 oz.

The borated glycerin should be cooled before mixing. Add any perfume desired.

Hair "Restorers"

These products are sold and used on faith. Some are helped by the accompanying massage and later washing. There is nothing known which actually will grow hair where it is wanted.

Preparations for Baldness
Ointment

Pilocarpine Hydrochloride	20 oz.
Precipitated Sulphur	120 oz.
Parachol	60 oz.
Balsam of Peru	60 oz.
Resorcinol Monoacetate	30 oz.
Petrolatum	900 oz.
Water	60 oz.
Perfume to suit.	

Dissolve the pilocarpine in water and mix with absorption base. Mill the sulphur and the monoacetate with part of the petrolatum. Melt the rest and stir in the absorption base and add finally the sulphur mass. Mix thoroughly.

Brilliantines and Pomades
Dressing for "Kinky" Hair

Beefsuet	16 oz.
Yellow Beeswax	2 oz.
Castor Oil	2 oz.
Benzoic Acid	10 grains
Perfume	sufficient

Melt the suet and wax, add the castor oil and acid, allow to cool, and add perfume.

Hair Preparations
Hair Tonic

Tannic Acid U.S.P.	0.5
Salicylic Acid U.S.P.	1.0
Castor Oil U.S.P.	24.5
Resorcinol Monoacetate	5.0
Alcohol	69.0
Perfume	sufficient

Eau de Quinine Hair Tonic

Tincture of Cantharidin	6 oz.
Quinine Hydrochloride	1 oz.
Tincture of Capsicum	2 oz.
Glycerin	3 oz.
Bay Rum	6 gal.
Tincture of Cudbear	sufficient to color

Shampoos
Milky Hair Wash (Kerosene)

Trihydroxyethylamine Stearate	10 lb.
Kerosene	150 lb.
Pine Oil	6 lb.
Water	250 lb.

Heat the trihydroxyethylamine stearate and the kerosene to a temperature of 140° Fahrenheit, and stir them together until dissolved; then pour in the pine oil with continued stirring. Now allow the water to run in slowly while stirring. If the pine oil is objectionable, however, any other oil may be substituted for it. It may be colored beautifully by means of any water-soluble dye free from salt.

Soapless Shampoo

Turkey Red Oil	10 lb.
Mineral Oil	10 lb.
White Oleic Acid	10 lb.
Alcohol	2–10 lb.
Perfume	4 oz.

Mix the above materials in the order given. If desired, the cost can be reduced by adding an additional amount of water. The water should be added carefully with stirring. The addition of water should be stopped just before a cloudiness appears.

These shampoos are used by pouring a little into the hand and rubbing to a creamy consistency with water and then applying to the hair which must be wet.

Shampoo

Oleic Acid	55 lb.
Cocoanut Fatty Acids	40 lb.
Triethanolamine	50 lb.
Diethylene Glycol Monoethyl Ether	55 lb.
Perfume	1 lb.

The product prepared in this way is a liquid soap of a clear red color, which can be diluted with water to any desired consistency or concentration. Glycerin and/or alcohol may replace in whole or in part the diethylene glycol ethyl ether.

Lemon Rinse

1. Lemon Oil	3	oz.
2. Alcohol	14	lb.
3. Citric Acid	3½	lb.
4. Tartaric Acid	4½	lb.
5. Water	16	lb.

Dissolve 1 in 2 and add to it slowly with stirring 3 and 4 which have been dissolved in 5.

Hair Fixative

Water	20 gal.
Gum Tragacanth	1 lb.
Boric Acid	1 lb.
Preservative	sufficient

Allow to stand overnight and stir until uniform: then stir in

Perfume Oil	4 oz.
Color	to suit

Hair Setting Preparation

Psyllium Seed	1 oz.
Distilled Water	5 gal.
Water soluble Perfume	

Prepared by boiling for five minutes, straining and mixing with an equal bulk of alcohol.

Permanent Wave Solution

Borax	3.75 parts
Sodium Bicarbonate	3.50 parts
Linseed Oil	0.17 part
Starch	0.40 part
Water	99.00 parts

Finger Wave Lotion

Borax	600 g.
Acacia	80 g.
Boiling Water	18 liters
When cold add:	
Spirit of Camphor	75 cc.
Perfume	as desired

Artificial Sunburn Liquids

Formula No. 1

Powdered Cudbear	20 lb.
Powdered Henna	4 lb.
Peanut or Almond Oil	32 lb.

Macerate at 120° F. for 3 hours and filter.

Formula No. 2

Quinine Sulfate	2 lb.
Witch Hazel	5 lb.
Lanolin	10 lb.
Peanut Oil	92 lb.

Formula No. 3

Peanut Oil	60 lb.
Olive Oil	35 lb.
Bergamot Oil	1 lb.
Laurel Berry Oil	3 lb.
Chlorophyll	1 lb.

Formulae B and C above require exposure of skin to sun.

Anti-Perspiration Cream

1. Lanolin Hydrous	1	part
2. Benzoinated Lard	90	parts
3. Zinc Oxide	6.5	parts
4. Salicylic Acid	1.2	parts
5. Benzoic Acid	0.9	part
6. Perfume Oil	0.4	part

Dissolve (4) and (5) in small amount of alcohol; mix into (1) and then work into (2). Grind in (3) until smooth and then work in (6).

Freckle "Removers"

Two grams of zinc sulphophenylate, 30 grams of distilled water, 2 grams of ichthyol, 30 grams each of anhydrous lanolin and petroleum jelly and 2 grams of lemon oil or other suitable perfume, will give good results.

Preparations with a bleaching action are made containing 1500 grams of wool grease, 530 grams of almond oil, 110 grams of beeswax, 150 grams of bora:., 150 grams of hydrogen peroxide (100% by volume) and 10 grams of yellow petrolatum.

Beauty Masks and Clays
Face Clay

Clay	100 lb.
Water (Cold)	20 gal.
Tincture of Benzoin	3 pt.
Perfume	3 oz.

Add the water to the clay and grind till smooth. Evaporate until 150 lb. remain. Run through mill to smooth clumped particles; cool and mix in the benzoin and perfume. Fill in collapsible pure tin tubes.

Beauty Pack

Tragacanth	25 parts
Alcohol	40 parts
Calamine	80 parts
Zinc Oxide	30 parts
Zinc Stearate	50 parts
Glycerin	60 parts
Lime Water	1000 parts

Mix the tragacanth in the alcohol. Then add to the lime water. Rub up zinc stearate, zinc oxide and calamine with glycerin. Add tragacanth, alcohol, lime water mixture to calamine, zinc oxide, zinc stearate and glycerin mixture.

Leg and Arm Blemish Covering

Stearic Acid	4 lb.
Diethylene Glycol	16 lb.

Heat to 180° F. and to this add while stirring the following solution heated to 140° F.

Caustic Potash	4 oz.
Water	16 pints

When uniform work in following:

Zinc Oxide	15 lb.
Yellow Lake	12 oz.
Persian Lake	4 oz.
Perfumed Oil	4 oz.

The colors may be varied to give more suitable shades.

Mole and Blotch Covering

Collodion	1 gal.
Zinc Oxide	1 lb.
Geranium Lake	½ oz.
Yellow Ocher Lake	1½ oz.

Mosquito Preparations

The following application is suggested as a means of preventing insect bites:

Cedar Oil	2 drams
Citronella Oil	4 drams
Spirits of Camphor	add 1 oz.

This should be smeared on the skin of the exposed parts as often as is necessary. Cod-liver oil used in the same way has been highly recommended, and in combination with quinine it makes an effective "sunburn and midge cream," a formula being as follows:

Quinine Acid Hydrochloride	5 parts
Cod-liver Oil	20 parts
Anhydrous Wool Fat	75 parts
Oil of Lavender (or geranium)	a sufficiency

The irritation of a mosquito or fly bite may be allayed by gently rubbing the puncture with a moist cake of soap, or by applying a 1 per cent alcoholic solution of methol, or 1–20 aqueous carbolic lotion. Hydrogen peroxide or weak ammonia solution dabbed on is also useful. If the bite shows signs of sepsis, constantly renewed hot boric fomentations should be applied, or if a limb is implicated, hot saline arm or leg baths.

Mosquito Cream

Good results can be secured from a composition containing 5 parts powdered wheat starch, 10 parts water, 45 parts glycerin 28° Bé., 30 parts lanolin and 5 to 10 parts oil of clove. Starch is rubbed into smooth paste with water; glycerin is mixed in and mass converted into jelly-like consistency by heating and agitating; it is then allowed to cool.

Bath Preparations
Liquid Toilet Ammonia
(For Bath)

Ammonium Stearate (Paste)	8 oz.
Ammonia 28°	6 oz.
Water	50 oz.
Glycerin	2 oz.

Perfume to suit, avoiding the use of aldehydes and unstable esters.

Borated Bathing Solution

Boric Acid	10.0 g.
Alum powdered	2.5 g.
Camphor	1.5 g.
Alcohol	120.0 cc.
Water enough to make	500.0 cc.

Perfume to suit, dissolved in Alcohol.

Bath Salts and Water Softeners

The most widely sold bath salts are products that are based on sodium-sesqui-carbonate. Sodium Bicarbonate and Sodium Chloride are also used.

Formula No. 1

To 100 pounds of sodium sesqui-carbonate is added a mixture of

Dye Stuff	1 oz.
(to give desired shade)	
Perfume Oil	12 oz.
Alcohol	1 pint

The entire mass is mixed until the color and perfume are thoroughly dispersed.

Other bases are as follows:

Sodium Bicarbonate	100 lb.
Sodium Bicarbonate	50 lb.
or Sodium-Sesqui-Carbonate	50 lb.
Sodium Chloride	100 lb.

All of the above bases are colored and perfumed as shown previously. Care should be taken in the selection of the crystals as regards crystal size, appearance and uniformity. In securing dye stuffs, it is necessary that the type base for which they are intended be specified.

Pine Needle Concentrate
(For Bath)

Many pine-needle oil preparations now marketed, do not take into account that when they are put into water the oil floats on top and only makes contact with a very small portion of the body. By using the following formula the oil is emulsified and spreads uniformly through the bath, giving the entire body the benefit of the pine needle oil.

1.	Pine Needle Oil	10 lb.
2.	Sodium Sulforicinoleate	10 lb.
3.	Water	5 lb.
4.	Fluorescein	To Suit
5.	Oleic Acid	½ lb.

Mix 1, 2 and 3 until dissolved. Add 3 slowly with stirring. Add 4 and stir until dissolved.

The above formula when thrown into water disperses uniformly to give a milky green solution. Other oils may be substituted for Pine Needle Oil. If a lower cost is desired, part of the pine oil may be replaced by mineral, olive or cottonseed oil and a larger amount of water may be added.

Pine Needle Milk

The first step in the process is to prepare a 5% solution of 80% soda soap in 95% alcohol.

Then, titurate the following items:

Pulverized White Gum Tragacanth (Finest)	5 parts
Soap Solution	100 parts

Then, add the following mixture to the paste:

Pine Needle Oil	45 parts
Juniper Oil	5 parts
Alcohol 95%	125 parts

Add the following next:

Water at 30° C.	550 parts

and the entire mixture is agitated for a long time. A thick emulsion is formed, resembling a cod liver oil emulsion. It is ready for use and can be added directly to the bath.

Astringent substances, such as oak bark extract, may be added to the emulsion, but this must be done during the manufacturing process.

Removing of Tattoo Marks
I.

Pepsin and papain have been proposed as applications to remove the epidermis. A glycerol solution of either is tattooed into the skin over the disfigured part; and it is said that the operation has proved successful. Papain, 5; water, 25; glycerol, 75; diluted hydrochloric acid, 1. Rub the papain with the water and hydrochloric acid, allow the mixture to stand for an hour, add the glycerin, let it stand for three hours and filter.

II.

Apply a highly concentrated tannin solution to the tattooed places and treat them with a tattooing needle as the tattooer does. Next vigorously rub the places with a lunar caustic stick and allow the silver nitrate to act for some time until the tattooed portions have turned entirely black. Then take off by dabbing. At first a silver tannate forms on the upper layers of the skin, which dyes the tattooing black; with slight symptoms of inflammation a scurf ensues, which comes off after fourteen or sixteen days leaving behind a reddish scar. The latter assumes the natural color of the skin after some time. The process is said to have good results.

Obviously such treatments are heroic and carry along with them the risk of permanent scarring. It is therefore a job for a trained dermatologist rather than for a layman.

Perfume Bases

Unless the amount of perfume base used in manufacturing is considerable, it is not advisable to make them on a small scale. It will probably be cheaper to buy small amounts of these bases from reputable manufacturers than to make them. To make them it is necessary to make a considerable investment in numerous materials. The slightest error in compounding will result in expensive spoilage. The beginner cannot attempt to try out new compounding without excessive waste of time and materials.

Perfumes

The compounding of perfumes and perfume oils is rather complex. These products are made from mixtures of natural oils together with synthetic aromatic chemicals and natural isolates, as well as certain animal derivatives.

Certain of the aromatic chemicals are necessary to secure reproductions of certain of the natural flower odors, and they, when blended properly with the natural flower oils, give products of the desired character.

In the preparation of extracts, an oil is added to alcohol at anywhere from 8 to 16 ounces per gallon of alcohol, although in certain cases up to 20 ounces are used.

For pre-fixing alcohol, small amounts of the natural resins or gums, or small amounts of the animal derivatives, such as: ambergris, civet or castorium, are allowed to stand in alcohol for at least a month before it is used. The addition of small amounts of water to an alcoholic extract will reduce the tendency towards the alcoholic sharpness.

Toilet Waters

Toilet waters are made in a similar fashion to the perfume extracts, excepting that a 60–70% alcoholic concentration is used, and from 3–6 ounces of oil are used per gallon of 60–70% alcohol.

Shaving Creams

Shaving creams are special types of soap.

A shaving cream must

1. Lather freely and rapidly.
2. Lather in hot or cold water.
3. Be dense and firm.
4. Be capable of being worked into a dense and voluminous lather.
5. Must not form too soluble a lather which would wash off with excess water.
6. Lather must not dry rapidly but should remain moist for some time.
7. Must be a powerful emulsifying agent, cut surface tension and have good degreasing properties.
8. Must be stable in tube or jar and

not dry out or turn hard and gummy and maintain the same consistency for all reasonable temperatures.

The problem of the shaving soap is a problem of balance, so as to obtain a combination which most nearly gives the desired result.

The addition of a sufficient amount of glycerin will help keep the lather moist. The amount generally used is about 10% of the finished cream.

Analysis of the average shaving cream will generally show as follows:

Actual soap content	40%
Water	50%
Glycerin	10%

For the rapid lather a very "soluble" soap is required. If the cream consists entirely of rapid lathering soap, it will be too soluble and will wash away in hot water or on vigorous rubbing, therefore, a large quantity of the "less-soluble" soap is required. The more soluble soaps are made from the more soluble oils. These are represented by coconut oil and palm kernel oil. Because of their solubility, they will give a rapid lather, will lather up in cold water or in hard water, but will wash away in hot water or on vigorous rubbing. Because both coconut and palm kernel contain lower molecular weight acids, they will irritate the face if used in too high concentrations. They are generally limited to about 15% or less of the total fat content. While both are satisfactory, coconut is the more widely used, since the odor of palm kernel is more likely to occur in the finished soap. However, a type of deodorized palm kernel has recently been made available.

The soap required to give a more lasting lather, which will retain its body in hot water, must contain a soap such as tallow, stearic acid or palm oil. If a very dense, persistent lather is required, fats containing large amounts of behenic acid may be used.

The consistency desired is obtained not only by a balancing of soaps according to the fatty acids contained, but also by the proper balancing of sodium and potassium soaps. Too much sodium soap cannot be used because of its hardness.

The proper blends of soaps, glycerin and water, is all that a shaving cream consists of. Some contain borax and other fillers. A typical shaving cream formula would be as follows:

Coconut Oil	9
Tallow	3
Stearic Acid	28
Sodium Hydroxide	1

Potassium Hydroxide	7
Glycerin	10
Water	45

Sodium Hydroxide is prepared as a 20° Bé. solution, using part of the water, in the formula.

Potassium Hydroxide is prepared as a 35° Bé. solution.

Glycerin, coconut oil and tallow are melted in the tank. The sodium hydroxide is run in slowly making sure that saponification is complete.

The excess fat is now saponified with potash, ½ the potash is added to the tank and the mass agitated until saponification appears to be complete. The stearic acid is melted and added and finally the remainder of the potash solution. The mass is stirred until neutralization is complete, and then adjusted to the amount of free stearic acid desired. Three per cent excess stearic acid is commonly used.

This soap when made will be very thick while hot, but will soften on cooling. It is possible to keep the soap thin while hot as by finishing with a large excess of stearic acid which may be later neutralized by adding the appropriate amount of potash solution to the cold soap with suitable agitation.

Liquid Shaving Creams

Stearic Acid	200 g.
Triethanolamine	10 g.
Water	800 g.

Thicker Creams

Stearic Acid	200 g.
Triethanolamine	10 g.
Anhydrous Sodium Carbonate	10 g.
Water	800 g.

After Shaving Preparations
Almond Cream for After Shaving

Formula No. 1

| Potassium Carbonate | 1 oz. 130 grains |
| Distilled Water | 15 oz. |

Dissolve Potassium Carbonate in water, filter.

Formula No. 2

Gum Tragacanth	175 grains
Glycerin	10 oz.
Borax	1 oz.
Distilled Water	64 oz.

In 20 oz. hot water dissolve Borax then add Gum Tragacanth and Glycerin. Allow to stand 12 hours, stirring frequently. When gum has formed mucilage add the remaining 44 oz. of water while stirring and strain through muslin.

Formula No. 3

Stearic Acid triple pressed	5 oz. 260 grains
Oil Sweet Almond	3 oz.
Ethyl Amino Benzoate	½ oz.

Melt acid and oil together and add Ethyl Amino Benzoate. Stir until dissolved and adjust temperature to 70° C.

After Shave Lotion

Menthol	1 dram
Boric Acid	2½ oz.
Glycerin	5 oz.
Alcohol	5 quarts
Water, to make	5 gal.
Perfume	

Dissolve menthol in alcohol. Add boric acid, perfume, and glycerin. Stir thoroughly until everything is dissolved. Add water. Filter. This preparation may be colored by adding enough color to give shade desired.

Styptics
Styptic Pencils

The following are the methods adopted for the manufacture of alum pencils: White: Liquefy 100 gm. of potassium alum crystals by the aid of heat. Remove any scum and avoid overheating, particularly of the sides of the vessel in which liquefaction is being carried out. The molten liquid should be perfectly clear. Triturate a mixture of French chalk in fine powder, 5 gm., glycerin 5 gm. to a paste, incorporate with the liquefied alum and pour into suitable molds. A white appearance can be imparted to the resulting pencils by the addition of more French chalk. Clear: Carefully liquefy potassium alum crystals so as to avoid loss of water of crystallization, adding a small amount of glycerin and water (about 5 per cent) until a clear liquid is obtained. This is poured, whilst hot, into suitable molds, previously smeared with fat. The solidified pencils are rendered smooth by rubbing them with a moistened piece of cloth.

Styptic Powder

An excellent styptic powder results from the mixture of 50% powdered talc and 50% phthalyl peroxide. The latter often contains up to 40% of its weight as phthalic acid; this is beneficial and acts as a stabilizer. The mixture is antiseptic.

Nail Polish

The formulation of a suitable nail polish presents problems peculiar in itself.

The properties desired in the finished product are:

1. Ease of application
2. Drying time
3. Appearance of dry film
4. Permanency

Ease of application is essential. If the polish is too thin, it will tend to flow too readily when applied to the nail, and will give difficulty in securing a smooth even coat. If the polish is too thick, a lumpy, streaky finish will result. In other words, the viscosity of the polish should be such that it will allow an even film to be brushed upon the nail. The drying time should be such that when the nails of the second hand are finished, the coat on those of the first hand should be sufficiently dry to permit the second application. Naturally, this applies only to the so-called "2 coat polishes."

The dry film should present an even appearance, any ridges, streaks, or even pinholes being absent. Finally, a good nail polish should remain on the finger nails for at least 5–7 days with little diminution of its original brilliance, and should show no signs of cracking and peeling.

True solvents, such as acetone, butyl acetate, amyl acetate, etc., give free flowing solutions whose viscosity can be influenced by increased concentration of low viscosity cotton, or by the addition of non-solvents, such as toluene, xylene, etc. Commercial nitrocellulose is manufactured in various viscosities, ½ second, 4 seconds, 15–20 seconds, 40 seconds, etc. However, ½ second regular soluble nitrocellulose generally furnishes the basis of nail polishes. This permits the incorporation of a sufficient solid content, whereas the higher viscosity cottons, even in small quantities, give a much too viscous product.

"Regular Soluble Cotton" is nitrocellulose soluble in acetone, amyl acetate, etc., but not in ethyl alcohol. There is another type of nitrocellulose produced, the alcohol soluble type. This type of cotton is sometimes used in formulating nail polishes where a high alcohol content is desired. However, the film of this type of cotton is not as strong as that of an equivalent amount of Regular Soluble Cotton. Where the incorporation of a large percentage of low boiling solvent is desired, the use of R. S. Cotton and ethyl acetate is preferable.

The solvents most commonly used in nail polishes are: ethyl acetate, absolute denatured ethyl alcohol, butyl acetate, normal butyl alcohol, amyl acetate, glycol ethers (cellosolve, cellosolve acetate, butyl and methyl cellosolve) and acetone oil. The non-solvents are toluene, benzol and VMP Naphtha. Most polishes contain little or none of these non-solvents as they have a disagreeable odor which is objectionable in the finished product. The evaporation rate of solvents is related, in most cases, to their boiling points.

In formulating, it is necessary to make sure that at all times there is sufficient true solvent for the cotton present. Otherwise, although the polish may be clear, the resultant film deposited may be cloudy due to the "throwing out" of solution of the nitrocellulose. The presence of resins further complicates this problem, as the solvents must also be balanced to insure sufficient solvents being present to prevent the resins from being thrown out.

The solvents boiling below 100° C. generally constitute 50% more of the total solvents of a nail polish. This insures a sufficiently rapid evaporation rate. If the solvents are very volatile and the air humid, the rapid evaporation cools the air about the film to below the dew point and the condensing moisture whitens and makes the film opaque. This is commonly termed "blushing." A film that has "blushed" quickly peels. This condition is alleviated by incorporation of small amounts of high boiling solvents that have the property of absorbing the condensed moisture, preventing the precipitation of cotton and resins, and causing the water to evaporate with the constituents of the polish. These compounds are the glycol ethers and acetone oil.

The manner in which nitrocellulose is deposited from solution depends upon the solvent used. In many cases, the resultant film is ridged and rippled. Certain solvents have the ability to "flat" the film and markedly alleviate the above condition. Such solvents are: ethyl alcohol, butanol, cellosolve and methyl cellosolve.

Pyroxylin solutions have, to a pronounced degree, the property of contracting upon drying, and this causes it to buckle away from the surface to which it has been applied. To prevent this, substances called "plasticizers" are incorporated. These plasticizers are high boiling organic solvents which very slowly evaporate from the film. But small amounts of these substances are used, as their too liberal use would retard the setting of the film. The commonly used ones are castor oil, tricresyl phosphate, dibutyl phthalate, butyl stearate and camphor. The one that gives more plasticizing value, ounce for ounce, than any of the others is tricresyl phosphate. It tends to discolor and blacken with age, changing the color of the polish.

Castor Oil is widely used, but in slight excess it softens the film. Camphor is objectionable because of its odor and the fact that the luster of polishes containing camphor rapidly diminishes and in some cases the surface of the film soon presents a dull pitted appearance. The best of the lot is dibutyl phthalate, as it gives a good plasticizing effect, is stable and relatively odorless.

A number of resins can be used in pyroxylin lacquers and it is here that the formulator has his evident choice as well as one of his greatest troubles. The resins are two types, natural and synthetic.

Many of the natural resins and gums must be treated before incorporation in pyroxylin lacquers, as they contain waxes and other constituents which are incompatible with nitrocellulose. Each has its own treatment, to remove the insoluble matter.

The synthetic resins need no previous treatment before incorporation in lacquers, but in the main they have the drawback that they are colored compounds, yielding lacquers that are suited only for dark colored polishes.

All resins should be used as stock solutions in appropriate solvents, and the solutions assayed for resin strength from time to time thus insuring the proper percentage of this ingredient in the finished product.

All components should be combined on a weight basis as this alleviates any errors due to expansion or contraction. The total solids (cotton, resin and plasticizers) constitute from 10–30% of the polish, depending upon the desired thickness of final film. Plasticizers are added in the ratios of 20–30% of the weight of dry cotton, or 5% of the resin content. Actual mixing is done in glass or tin lined containers. All motors or shafting should be grounded and adequate ventilation provided.

Formula No. 1

½ Sec. R.S. Wet Cotton	24 oz.
Ethyl Acetate	25 oz.
Butanol	5 oz.
Toluene	48 oz.
Dammar Solution	19 oz.
Cellosolve Acetate	4 oz.
Dibutyl Phthalate	2 oz.
Tricresyl Phosphate	2 oz.
Butyl Acetate	25 oz.

Formula No. 2

Dry Alcohol Soluble Cotton	12 oz.
Shellac	1 oz.
Castor Oil	1 oz.
Ethyl Alcohol	50 oz.
Ethyl Acetate	20 oz.
Butanol	5 oz.
Amyl Alcohol	6 oz.
Acetone Oil	5 oz.

In coloring polishes, it is best to secure dye stuffs for this particular purpose, from dye stuff houses.

Nose and Throat Spray
Formula No. 1

White Mineral Oil	4 ounces
Menthol	5 grains
Camphor	10 grains
Eucalyptol	5 grains

Formula No. 2

White Mineral Oil	99%
Ephredine	1%

Sore Throat Relief

Mix tincture of ferric chloride, 30 gm., alcohol, 30 gm., and potassium chlorate, 60 gm., with sufficient water to make 250 cc.

Antiseptic Inhalant

Eucalyptol	20.0 cc.
Menthol	7.5 gr.
Oil of Rosemary	10.0 cc.
Oil of Pine Needles	10.0 cc.
Oil of Lavender	3.0 cc.
Oil of Jack Rose Comp.	2.0 cc.
Brilliant Green	trace
Ethyl Alcohol (S.D.) q.s.	100.0 cc.

Dissolve the menthol in the oils. Make a strong solution of brilliant green in alcohol. Use enough to give finished product a green tint. Add the remaining alcohol to make 100 cc.

Antiseptic for Telephone Mouthpiece
Formula No. 1

Stearic Acid	6.00 parts
Alcohol	20.00 parts

Formula No. 2

Sodium Hydroxide	1.35 parts
Alcohol	10.00 parts
Water	5.00 parts

Hand Lotions
Formula No. 1

Stearic Acid	3.15%
Glycerin	6.00%
Potassium Hydroxide	.15%
Water	80.30%
Alcohol	8.50%
Perfume	.50%
Quince Seed	1.25%
Preservative	.15%

Dissolve the potassium hydroxide in one-third of the water, bring the rest of the water to a temperature of 80° C.; add the quince seed and soak for six hours. Melt the stearic acid in the glycerin, dissolve the perfume in the alcohol. Add the potassium hydroxide solution to the melted fatty substance and boil for a minute. Allow the temperature to drop to about 70° C. and then stir it into the quince seed mucilage. Stir occasionally until cool; then slowly add the alcohol, perfume and preservative.

Formula No. 2

Stearic Acid	3.00%
Alcohol	5.00%
Mineral Oil	5.00%
Triethanolamine	2.00%
Glycerin	5.00%
Water	75.85%
Quince Seed	3.50%
Perfume	.50%
Preservative	.15%

Melt the stearic acid in the mineral oil and glycerin. Bring the temperature to about 70° C.; boil the triethanolamine in the water and add the melted fats. Stir vigorously until the temperature drops to about 50° C., then slowly stir in the quince seed mucilage and then slowly add the perfume and preservative dissolved in alcohol.

Formula No. 3

Stearic Acid	1.50%
Soap (Powdered Neutral)	1.00%
Alcohol	2.00%
Glycerin	4.90%
Boric Acid	2.50%
Borax	2.50%
Quince Seed	2.50%
Water	82.45%
Perfume	.50%
Preservative	.15%

Put the quince seed into half the water to which the preservative has been added. Allow to soak 24 hours and strain through muslin. Heat the rest of the water and dissolve in it the borax. Add the melted stearic acid and stir vigorously. Add the soap, boric acid and glycerin. Agitate until cool. Add the quince seed mucilage and the perfume dissolved in alcohol.

Formula No. 4

Beeswax, White	1.00%
Quince Seed	2.75%
Stearic Acid	1.65%
Borax	2.50%
Glycerin	3.00%
Water	85.45%

Perfume	.50%
Alcohol	3.00%
Preservative	.15%

Melt stearic and beeswax in glycerin. Add borax dissolved in hot water. Add mucilage and other ingredients as before.

Formula No. 5

Powdered Tragacanth	1.75%
Pulverized Neutral White Soap	3.00%
Glycerin	7.00%
Alcohol	8.00%
Borax	.50%
Preservative	.15%
Boric Acid	1.00%
Water	78.10%
Perfume	.50%

Dissolve the preservative in half the water and add the tragacanth and allow it to stand until dissolved, stirring it occasionally. Heat the remainder of the water; add the glycerin, borax and boric acid, and when dissolved, add the soap and stir until cool, then add the mucilage. Stir gently and add the perfume dissolved in the alcohol, a little at a time.

Formula No. 6

Spermaceti	2.00%
White Beeswax	.25%
Glycerin	6.00%
Soap	3.00%
Borax	.50%
Sodium Benzoate	.20%
Quince Seed	3.40%
Alcohol	4.00%
Water	80.00%
Perfume	.50%
Preservative	.15%

Add the quince seed to two-thirds of the water in which the preservative has first been dissolved. Allow to soak for 24 hours and strain. Add the glycerin, borax and sodium benzoate to the remainder of the water and stir until dissolved. Melt and add the spermaceti and white wax, stir rapidly for one or two hours then add the soap. Add the mucilage next and finally add the perfume dissolved in the alcohol, in a very thin stream.

Formula No. 7

Spermaceti	3.75%
White Beeswax	.75%
Glycerin	2.00%
Pulverized Neutral White Soap	1.50%
Borax	.35%
Almond Oil	3.00%
Quince Seed	2.50%

Alcohol	1.50%
Water	84.00%
Preservative	.15%
Perfume	.50%

Proceed as in Formula No. 6.

Formula No. 8

White Beeswax	4.125%
Glycerin	2.000%
Pulverized Neutral White Soap	3.375%
Borax	.375%
Almond Oil	3.000%
Honey	1.250%
Quince Seed	1.500%
Alcohol	1.500%
Water	80.550%
Witch Hazel	1.500%
Boric Acid	.175%
Perfume	.500%
Preservative	.150%

Add the quince seed to two-thirds of the water in which the preservative has been dissolved; allow to stand 24 hours and strain through muslin. Boil the remainder of the water, add the soap, boric acid, witch hazel, glycerin, borax. Melt and add the beeswax, spermaceti, oil of almond, and the honey. Add the quince seed solution quickly. Agitate slowly until cool, then add the perfume and alcohol.

Formula No. 9

Citric Acid	1.00%
Glycerin	5.00%
Alcohol	6.00%
Boric Acid	.50%
Lemon Oil	1.00%
Benzoate of Soda	.50%
Quince Seed	2.50%
Water	83.50%

Dissolve the benzoate of soda in two-thirds of water, add the quince seed and allow to soak 24 hours, then strain through muslin. Heat the remainder of the water and dissolve the boric and citric acids in it. Add the glycerin; then add the quince seed mucilage. Mix the oil of lemon in the alcohol and add it slowly with continuous stirring.

Formula No. 10

Bay Rum	2 oz.
Alcohol	2 oz.
Witch Hazel	2 oz.
Glycerin	3 oz.
Gum Tragacanth	⅛ oz.
Water	1 qt.

Soak the gum in the water for 24 hours, then stir until thoroughly dispersed, add remaining ingredients, scent as desired.

Formula No. 11

Water	100 lb.
Di-Glycol Stearate	5.6 lb.
Preservative	¼ lb.

Heat to 160° F.; turn off heat and begin stirring; when temperature drops to 140° F. add

Water	50 lb.
Alcohol	7 lb.
Perfume	Sufficient

Hand Lotion

Witch Hazel	58 parts
Bay Rum	58 parts
Spirits of Camphor	20 parts
Tincture of Benzoin	4 parts
Carbitol	180 parts

Tragacanth Hand Lotion

Gum Tragacanth (Ribbon No. 1)	25 g.
Glycerin, U.S.P.	500 c.c.
Water, Distilled	3500 c.c.
Benzoic Acid, U.S.P.	3 g.
Salicylic Acid, U.S.P.	1 g.
Perfume	Sufficient
Coloring (if desired)	Sufficient

Dissolve the benzoic and salicylic acids in the glycerin at low heat, stirring well.

The gum tragacanth used should be of finest grade, Aleppo type, No. 1. Stir gum into the water, allow to stand for a minimum of 24 hours, with occasional stirring. Add color in solution, if desired. Mix until uniform. Pass through double-thickness of fine-mesh cheesecloth. After straining stir the glycerine-acid mixture into the gum-water mass. Mix with mechanical mixer for 30 minutes. Strain through single-thickness cheesecloth. Add perfume, and mix until mass is well blended. Bottle.

This lotion is stable, and will not separate. It will keep indefinitely when corked.

Hand Cream

Glyceryl Monostearate	10%
Cocoa Butter	2%
Glycerin	5%
Anhydrous Lanolin	2%
Stearic Acid	1%
Water	80%

Add the ingredients to the water, turn on the heat and bring to the boiling point with constant stirring. When all the ingredients are melted and a clear emulsion is formed, shut off the heat and continue the mixing until the product is cold.

Mechanic's Protective Hand Cream

Glyceryl Monostearate	8%
Magnesium Stearate	14%
Beeswax	3%
Petrolatum	10%
Mineral Oil	5%
Water	60%

Heat together to 70° C. and stir until cool.

"Honey and Almond" Cream

Cold Cream	½	oz.
Almond Oil	½	oz.
Glycerin	½	oz.
Boric Acid	1	oz.
Solution of Soda, B.P.	1½	oz.
Quince Mucilage	5	oz.
Water to make	5	pints

Stir the cold cream, almond oil, and solution of soda together until a uniform soapy emulsion is obtained. Dissolve the boric acid in 60 ounces of warm water; to this add the glycerin and quince mucilage, and add the mixture slowly, and with constant stirring, to the mortar contents. Perfume with spirits of almonds and rose when cold, and make up.

Dry Skin Nourishing Cream

Absorption Base	25 g.
Di-Glycol Stearate	15 g.
Olive Oil	15 g.
Preservative	1.3 g.
Water	75 g.

Melt absorption base, di-glycol stearate, and olive oil and add to water which has been heated to same temperature and contains preservative. Mix well and stir slowly till cool. Incorporate perfume and package.

Skin Smoothener

Boric Acid	2 parts
Gum Tragacanth	8 parts
Carbitol	3 parts
Moldex	1 part
Distilled Water	250 parts
Perfume and Color	Sufficient

Soak the gum in cold water for one hour, then heat to boiling, add the boric acid and the moldex, and cool. Add the perfume and color to the carbitol, and pour this into the mixture. Then add the balance of the distilled water.

Cleansing Cream
Formula No. 1

U.S.P. Flaked White Beeswax	10 parts
Ceresin Wax (Melting Point 138° F.)	5 parts
Mineral Oil (Viscosity 68/75)	65 parts
Water	40 parts
Borax	½ part
Perfume	Sufficient

This formula is stated in parts by weight.

The beeswax, ceresin and a portion of the mineral oil are placed in a steam jacketed kettle and the temperature in melting the waxes should be at no time in excess of 80° C. (The emulsifying power of beeswax is retarded or broken down when the temperature is above 80° C.) When the waxes have melted, the steam should be so regulated (this can be accomplished only by actual experiment on the part of the cream maker) that when the balance of the mineral oil is added cold, the temperature will be approximately 57° C. It will take a little experimenting to gain the knack of properly controlling the kettle temperature in order to speed production.

When the melted waxes and mineral oil are at a temperature of 57° C., the perfume oil should be added and the agitator started. The agitator should have only one propeller at the bottom, causing a complete vortex when agitating the melted waxes and oil, and forming only a slight vortex when all the water and borax have been added.

In another kettle, the water and borax should be kept at exactly the same temperature, 57° C. It is important that agitation be continued while the water and borax are being added and while the curds form until the cream begins to thin down and becomes a homogeneous mass. As soon as this occurs, stirring should be continued for a few minutes only. When the cream must stand before passing into the filling machine, it might be necessary to agitate it to destroy any crust caused by surface cooling.

Formula No. 2

Liquid Cleansing Cream

Glyceryl Monostearate	6.0%
Water	75.5%
Glycerin	10.0%
Stearic Acid	1.0%
Oleic Acid	2.0%

Mineral Oil	5.0%
Perfume	.5%

Put all the ingredients into a kettle excepting the perfume and heat until melted and mass is slimy. Agitate continuously until mass is cool. Then perfume.

Formula No. 3
Colorless Liquid Skin Cleanser

Glycerin	22.0%
Alcohol	10.0%
Water	64.5%
Tribasic Sodium Phosphate	3.0%
Perfume	.5%

Dissolve the phosphate in the water, add the glycerin. Mix the alcohol and perfume and add to the first solution. Mix and filter.

Formula No. 4
Quick Melting, Mineral Jelly

Mineral Oil	50.0%
Petrolatum	33.5%
Spermaceti	10.0%
Stearic Acid	6.0%
Perfume	.5%

Melt waxes and petrolatum in oil, add perfume when cool.

Formula No. 5
Cold Cream Type

Beeswax	15.0%
Petrolatum	14.0%
Mineral Oil	50.0%
Water	20.0%
Borax	.5%
Perfume	.5%

Melt wax and petrolatum. Add oil. Dissolve borax in hot water. Add to above with stirring. Perfume at 120° F.

Formula No. 6
Quick Melting, Absorption

Beeswax	7%
Spermaceti	8%
Lanolin Absorption Base	20%
Water	50%
Mineral Oil	15%

Melt the waxes and add the mineral oil. Warm the absorption base to 40° C. and the water likewise; then slowly add the water with steady but not violent agitation. Then add the melted waxes which should be of the same temperature.

Formula No. 7
Quick Melting, Absorption

Beeswax	7%
Spermaceti	8%
Mineral Oil	50%

Lanolin Absorption Base	20%
Water	15%

Procedure same as above.

Formula No. 8
Quick Melting, Oily Type

Petrolatum	40%
Anhydrous Lanolin	10%
Mineral Oil	50%

Melt the petrolatum, add the lanolin and mineral oil. Mix and perfume.

Liquid Cleansing Cream

Stearic Acid Triple Pressed	12 parts
White Mineral Oil	30 parts
Triethanolamine	4 parts
Glycerin	4 parts
Perfume	½ part
Water	80 parts

Cold Cream
Formula No. 1

White Mineral Oil	73.5%
White Beeswax U.S.P.	16.5%
Paraffin Wax	5.0%
Ozokerite	5.0%

The above should be heated to a temperature slightly above its melting point. The proper type of perfume is then added.

In a separate container, 1½ pound borax is dissolved in water. The amount of water can vary from 26 to 36 pounds per 100 pounds of finished cold cream.

The borax dissolved in the water and heated almost to the boiling point (approximately 200° F.) is then added all at one time to the above mixture. Stir during the addition of the borax water and continue stirring until a complete emulsion has been formed. The cream should be filled into the jars while still warm; not hot but just warm enough to pour easily so that it will have a perfectly smooth and even appearance at normal temperature.

Caution: If the water is not hot enough when added to the base, the cream will separate drops of water upon standing. If the stirring is not properly done or not continued long enough, the cream will likewise separate water.

Formula No. 2

White Mineral Oil	45.0%
White Beeswax U.S.P.	13.0%
Spermaceti	6.0%

Heat all together until liquid; then add under stirring

Powdered Borax	1.0%
Water	35.0%

which has been heated previously to about

200° F. In the summer time or warm weather, it is advisable to increase percentage of Beeswax and decrease percentage of Spermaceti, reducing the amount of water slightly.

Formula No. 3

Stearic Acid	120 parts
Lanolin (Anhydrous)	24 parts
Beeswax (White)	20 parts
Mineral Oil (White)	42 parts
Olive Oil	90 parts
Paraffin	12 parts
Triethanolamine	15.2 parts
Diethylene Glycol Ethyl Ether	64 parts
Water	380 parts

This formula is stated in parts by weight.

Melt the first six constituents and heat to about 70° C. Heat the triethanolamine and water to boiling, and stir vigorously into the hot solution of waxes. Add perfume and stir into the above cream after it has cooled somewhat. Stir the mixture until it is homogeneous and pour into jars while still warm.

Formula No. 4

Stearic Acid	30 parts
Lanolin	6 parts
Paraffin Wax	16 parts
Paraffin Oil (White)	50 parts
Triethanolamine	7 parts
Carbitol	15 parts
Water (Distilled)	00 parts
Perfume Oil	Sufficient

Melt together the stearic acid, lanolin, paraffin, and paraffin oil. Add this mixture to the triethanolamine and water, which have been mixed and heated to the boiling point, stirring constantly. As the cream begins to thicken, add the carbitol and perfume, continuing the stirring until a smooth cream is obtained.

Formula No. 5

Olive Oil (Benzoated)	10 parts
Beeswax	35 parts
White Petroleum Jelly	35 parts
Paraffin Wax	10 parts
Liquid Paraffin	80 parts
Water	60 parts
Borax	1 part

Melt the fats and wax. Boil the water and add the borax, cool a little and mix.

Tissue Cream

White Mineral Oil	20 lb.
Boracic Acid	6 lb.
Glycerin	14 lb.
Lanolin	20 lb.
Water	120 lb.

Heat the white mineral oil and lanolin with half water to about 190° F.; cool gradually, stirring continuously until mass is smooth and homogeneous. Then add balance of water together with boracic acid and glycerin. Stir cold and add perfume. Use distilled water which has been heated to a temperature of 15 to 20 degrees higher than melted waxes.

Liquefying Creams

	1	2	3	4	5
Paraffin Wax	18	5	0	0	15
Ceresin	0	0	0	3	0
Beeswax, White	0	0	5	0	10
Ozokerite	0	0	0	4	0
Petroleum Jelly	18	2	85	0	0
Mineral Oil	64	16	10	48	75
Borax	0	0	0	0	1
Water	0	0	0	0	12

Melt the ingredients at the lowest possible temperature, using a water bath to prevent burning. Add any color or tint desired. Stir well, and allow the mixture to cool slowly. Stir at 5 or 10 r.p.m. to avoid the formation of a crust at the side of the container. After the cream has cooled considerably, add the perfume and stir again. By now the cream is quite cool, and ready to be poured. If the jars are previously warmed to about 37° C. or about 99° F., the cream will form a regular and smooth surface on top. If the cream is poured too hot, a hole will remain down through the center of the jar. If this happens, remelt the cream. Too much wax will make the cream crack away from the side of the jar. Regarding sweating, it can be said that practically all liquefying creams sweat. Formulas 1 and 2 obviate the condition somewhat. This is due to the contraction of the physical jell structure of the set cream, causing the oil to ooze out. If the mass is cooled previous to pouring, the condition can be obviated to a great extent. Use of low melting waxes along with petroleum jelly will help too.

Liquefying cleansing cream can do nothing more than cleanse the skin. Any additional claims are unfounded. A cream of this type can scarcely have multiple purposes. It is not a skin food, nor a powder base. To state on the label that the cream should be removed with soap and water after the usual removal with tissue is another good point. Refrain from using aromatic oils that will burn the skin. Be careful in your manipulation and along with the type formulas suggested, little trouble will be experienced.

Liquid Turtle Oil Cream

Trihydroxyethylamine	
Stearate	20 parts
Turtle oil	50 parts
Paraffin Oil (White)	40 parts
Diglycol Stearate	5 parts
Carbitol	20 parts
Water (distilled)	300 parts
Preservative	5 parts
Perfume Oil	To Suit

Melt together the first four of these ingredients, and slowly add the water, which has been heated to 50° C., with constant stirring. Continue stirring until cool, and then add carbitol, containing the perfume and preservative.

Vanishing Cream (Greaseless)

Stearic Acid (Triple	
Pressed)	25 parts
Potassium Hydroxide, U.S.P.	1 part
Water	75 parts
Glycerin, U.S.P.	8 parts
Perfume	To Suit

Melt stearic acid and heat to 80° C., dissolve the potassium hydroxide in half of the water and heat to the same temperature and add to the acid while stirring. Mix the glycerin with the remainder of the water and heat to 80° C. and add to the mixture. If the process is carried on slowly and in small quantities, water will have to be added because of evaporation. While warm ordinary hydrogen peroxide may be added and a peroxide cream produced. This is a beautiful pearly cream. The pearliness is due to stearic acid. The glycerine prevents drying out of the skin that is so common for vanishing creams. Without perfume this makes an excellent brushless shaving cream by increasing the potassium hydroxide to 2 parts.

Peanut Oil Massage Cream

This product is excellent for undernourished parts, cold feet, stiff joints, and as an aid to any flabby parts of the body.

White Beeswax	20 parts
Spermaceti	20 parts
Peanut Oil	45 parts
Lanolin	12 parts
Water	45 parts
Borax	3 parts
Camphor	1 part

Melt the wax over a water bath, add the lanolin and oil, and heat to 80° C. The camphor is dissolved in the oil with gentle heat, the borax is dissolved in the water and heated to 80° C. At this temperature the water is added to the wax oil mixture at the same temperature with vigorous stirring; best done with a motor stirrer.

Creme Kaloderma

Gelatin	25 parts
Distilled Water	275 parts
Honey	100 parts
Glycerin	600 parts
Preservative	15 parts

Astringent Cream

Formula No. 1

Diglycol Stearate	10 parts
Lanolin Anhydrous	5 parts
Mineral Oil	3 parts

Heat to 170° F.

Tannic Acid	2 parts
Water	70 parts

Heat to 170° F. and add to above at that temperature with agitation.

Perfume as desired.

Formula No. 2

Paraffin	15 parts
Lanolin	10 parts
Mineral Oil	60 parts
Cetyl Alcohol	15 parts

Heat to 150°. Then add a heated solution of:

Alum	15 parts
Tannic Acid	15 parts
Water	280 parts

Agitate until the cream congeals.

Quinto Cream

Quince Seed	90	gr.
Boric Acid	30	gr.
Salicylic Acid	20	gr.
Glycerin	1.5	oz.
Cologne Water	4	oz.
Boiling Water	4	oz.
Spirit of Lemon	Sufficient	

Triturate the quince seed with the boiling water, add the acids, and strain through muslin.

Emollient Cream

Very soft emollient creams, containing glycerin, can be made by the aid of vanishing cream. The method of production is very simple. All that is necessary is thoroughly to mix the vanishing cream with glycerin of starch with which a little zinc oxide has been previously incorporated. If zinc oxide is objected to because of the possibility of its reacting with the free stearic acid present, it may be advantageously replaced by titanium dioxide; and the following new formula for a cream of the type in question is suggested, which is based on a vanishing cream:

Glycerate of Starch	40 parts
Titanium Dioxide (Pure)	2 parts
Vanishing Cream	58 parts

To prepare this cream, first triturate the titanium dioxide with the glycerin of starch thoroughly in a mortar. Then work in the vanishing cream, a little at a time. The product is a beautifully white very soft emollient cream, excellent for the hands in cold weather. As the glycerin of starch exercises a softening effect on the vanishing cream, this should not be itself too soft in texture. The method may, indeed, be employed for using up vanishing cream which has become somewhat hardened by evaporation. The vanishing cream should itself be perfumed before use, and a larger proportion than that indicated may be incorporated if desired.

Benzoin and Almond Cream

Stearic Acid	20 parts
Potassium Hydroxide	1 part
Curd Soap	3 parts
Water	150 parts
Almond Oil	2 parts
Tincture of Benzoin	2 parts

Dissolve the soap in 40 parts of water (hot); form a vanishing cream with the remainder of the formula, using 80 parts water; mix the two and heat again, adding the remainder of the water when cold. A better sheen would no doubt result if the almond oil were omitted. Two or three days will elapse before the sheen appears.

Almond Meal

Borax	2 lb.
Bicarbonate of Soda	2 lb.
Powdered Castile Soap	2 lb.
Fine Yellow Corn Meal	33 lb.
Wheat Flour	61 lb.
Perfume	5 oz.

Mix all the dry ingredients together and spray the perfume oils into the mixture with an atomizer.

Sulphur, Castor Oil and Tragacanth in a Lotion

Tincture Quillaia	3	parts
Salicylic Acid	1	part
Precipitated Sulphur	2	parts
Alcohol	20	parts
Castor Oil	10	parts
Tragacanth	0.5	part
Water to make	100	parts

Triturate the salicylic acid, precipitated sulphur and tragacanth with the castor oil, and the tincture of quillaia, and gradually add most of the water, triturating constantly. Add the spirit, shake vigorously, and make up to volume with water.

Larkspur Lotion

Larkspur, Coarse Powder	100 parts
Potassium Carbonate	10 parts
Alcohol	500 parts
Water	500 parts

Boil the larkspur for 15 minutes in an aqueous solution of the potassium carbonate. Filter and while cooling pass the alcohol through the drug while in the filter. Add enough water through the filter to bring up to 1000 cc.

Mud Beauty Mask
French Patent 738,808

Tannic Acid	60 g.
Acacia	60 g.
Egg	1
Storax	15 g.
Eau de Cologne	15 g.
Barley Meal	120 g.
Starch	90 g.
Borax	20 g.

This powder is colored with carmine. For use a paste is made by mixing one part with half a part of olive oil and one part of water. The face is first cleansed with cold cream, and the mud pack is applied fairly thickly. The face is fanned to aid the drying of the mask, which is removed with a damp swab of cotton wool or a trowel.

Facial Mask

Mix to a paste about 15% bentonite with witch hazel or other aromatic water. A balsamic perfume may be added, or bay rum may be used instead. The preparation should be tinted as the mass is of a dull gray-green color. Powdered calamine may be added to both color the preparation and increase its value as a mask.

Facial Clay

Colloidal Clay	50%
Citric Acid	5%
Hydrogen Peroxide	5%
Magnesium sulphate	5%
Glycerin	10%
Water	25%

Bleach Cream
Formula No. 1

Hydrogen Peroxide	10 parts
Lanolin (Anhydrous)	30 parts

Formula No. 2

Citric Acid	2 parts
Bismuth Oxychloride	2 parts
Rose Ointment	30 parts

Skin Bleach

A mild skin bleaching lotion that is absolutely harmless may be prepared from lemon juice with addition of 15 per cent alcohol to preserve it and 1 per cent glycerin.

A very good bleach for removing summer tan may be made by mixing 5 per cent of zinc peroxide or sodium perborate with 95 per cent magnesium carbonate or calcium carbonate. This should be mixed to a paste with water and then applied.

Freckle Preventive Treatment

Sulphur lotions should not be used as these tend to increase the pigmentation, neither should tar preparations, ichthyol, or resorcin be ingredients of the lotions. Before the hot weather comes the following ointment should be used:

Quinine Bisulphate	1.5 parts
Aesculin	1.0 part
Simple Ointment	27.5 parts

The following ointment, to be applied twice a day, left on for thirty minutes, and then wiped off with a cleansing tissue, is recommended by Continental dermatologists:

Zinc Peroxide	20 parts
White Soft Paraffin	70 parts
Anhydrous Lanolin	10 parts

The affected parts are then powdered with a powder composed of magnesium peroxide 30 parts, talc 50 parts, zinc oxide 20 parts.

Wrinkle Remover

Wrinkling is caused by loss of elasticity of the skin through degeneration of its fibers. No treatment can cure it or restore to the skin its youthful elasticity and smoothness. Slight inflammation causes swelling and temporarily smooths out the wrinkles, but with its subsidence they return. Such an inflammation, followed by exfoliation, can be caused by ultra-violet radiation, by freezing with carbon dioxide snow or by chemical irritants. A formula of this type is the following:

Betanaphthol	10 g.
Precipitated Sulphur	40 g.
Soft Soap, U.S.P.	25 g.
Petrolatum	25 g.

If this turns dark it does not indicate any loss of activity. It must not be used in cases in which the kidneys are impaired, and during its use the urine must be watched for signs of kidney irritation. It must be used on small areas only and not given to the patient to apply; but she must come daily to the doctor for the application.

The area should be washed with ether or benzine and the paste should be spread thickly and allowed to remain from twenty to thirty minutes, when it is removed thoroughly. Soon after application a slight burning sensation is felt, but this ceases in a few minutes. After the removal of the paste, the skin is red for a few hours. The treatment should be repeated each day until a tightening sensation or the onset of exfoliation shows that treatment has been sufficient. Five days is usually enough. During treatment no soap or water is to be used on the areas treated, but they may be cleansed by 0.5 per cent salicylic acid in alcohol.

Wrinkle Lotion

Alum	20 grains
Zinc Sulphate	5 grains
Glycerin	2 fl. dr.
Tincture of Benzoin	2 fl. dr.

Perfume to suit.
Distilled water sufficient to make 1 quart.

Sun-Burn Preventive Preparations
Formula No. 1 (Cream)

Quinine Sulphate	3 %
White Ceresin Wax	5.5%
White Petrolatum	20.5%
Mineral Oil	19.5%
Lanolin, Anhydrous	15.0%
Water	35.5%
Oil of Cassia	1.0%

Heat the water to 70° C. and dissolve the quinine sulphate in it. Melt the ceresin, petrolatum, and lanolin together, stir in the mineral oil, bring the mixture to 65° C. and stir in the quinine solution. Continue stirring until the temperature drops to 45° C. and then add the oil of cassia.

Formula No. 2 (Cream)

Quinine Bisulphate	2.5%
Cholesterin Absorption Base	25.5%
Mineral Oil	12.5%

Alcohol	10.0%
Water	49.5%
Perfume	.5%

Dissolve the quinine bisulphate in the alcohol. Melt the absorption base and the mineral oil together. Heat the water to the same temperature and stir it into the melted fat. Continue stirring until the temperature drops to about 45° C. and add the quinine solution and the perfume.

Formula No. 3 (Cream)

Sodium Naphthol 6:8 disul-	
phonate	3.0%
Borax	3.0%
Beeswax, White	20.0%
Mineral Oil	20.0%
Water	37.5%
Petrolatum	16.0%
Perfume	.5%

Heat the water and dissolve the borax in it. Add the sodium naphthol 6:8 disulphonate and stir until dissolved. Melt the beeswax, petrolatum, and mineral oil together. Add the liquids to the fats with constant stirring and when the temperature drops to about 45° C. add the perfume.

Formula No. 4 (Lotion)

Quinine Bisulphate	3.00%
Glycerin	5.50%
Gum Tragacanth Powder	2.50%
Alcohol	15.50%
Citric Acid	.75%
Water	72.25%
Perfume	.50%

Mix the gum tragacanth powder with one-half of the alcohol and add it to one-half of the water. Dissolve the quinine bisulphate in the remainder of the alcohol and add the perfume. Dissolve the citric acid in the remainder of the water. Add the glycerin to the citric acid solution then add the quinine solution and finally the mucilage. Mix well.

Formula No. 5 (Lotion)

Aesculin	3.5%
Glycerin	3.5%
Borax	4.0%
Alcohol	15.0%
Water	73.5%
Perfume	.5%

Heat the water, dissolve the borax in it and add the aesculin. Stir until dissolved. Mix the perfume with the alcohol, add the glycerin and stir this into the aesculin solution. Filter. If it is desired to color the solution add a sufficient quantity of spirits soluble brown.

Formula No. 6 (Oil)

Quinine Oleate	4.5%
Olive Oil	30.0%
Peanut Oil	64.5%
Oil of Cassia	.5%
Perfume Oil	.5%

Dissolve the quinine oleate in the peanut oil; add the other oils and perfume. If it is desired a sufficient quantity of pigment brown may be added to this preparation.

Formula No. 7 (Powder)

Quinine Sulphate	3%
Zinc Stearate	10%
Titanium Dioxide	7%
Talc	58%
Colloidal Clay	13%
Precipitated Chalk	5%
Suntan Color Base	4%

Dissolve the quinine sulphate in a small quantity of alcohol and rub up this solution with the colloidal clay. Allow the alcohol to evaporate; add the other ingredients and proceed as in the manufacture of ordinary face powder. This powder differs from the suntan formulas given in the chapter on face powders in that it is much more opaque and possesses a substance which will prevent sunburn.

Suntan Cream
Formula No. 1

Petrolatum	87.5%
Lanolin	6.0%
Pigment Brown	6.0%
Perfume	.5%

Grind the color with a small quantity of petrolatum. Melt the petrolatum together with the lanolin, strain in the color base, mix, strain and perfume.

Formula No. 2

Ceresin Wax	15.0%
Beeswax	8.0%
Lanolin	10.0%
Petrolatum	10.0%
Mineral Oil	56.5%
Perfume	.5%
Color	Sufficient

Melt the waxes; add the lanolin and petrolatum. Mix, strain and perfume. Then add a sufficient quantity of brown.

Suntan Oil
Formula No. 1

Olive Oil	50%
Peanut Oil, Refined	49%
Oil of Bergamot	1%
Color	Sufficient

Mix the oils and add a sufficient

quantity of pigment brown for a suitable shade.

Formula No. 2

Olive Oil	50%
Peanut Oil, Refined	43%
Sesame Oil	5%
Oil of Thuja	1%
Oil of Bergamot	1%

Mix the oils, filter and add a very small quantity of oil soluble brown.

Suntan Lotion

Alcohol	15.0%
Glycerin	5.0%
Water	79.5%
Perfume	.5%
Color	Sufficient

Add a sufficient quantity of spirit soluble brown to this lotion to produce the required shade. Then mix and filter the preparation. Owing to the difficulty of applying such lotions uniformly they are not particularly satisfactory.

Some manufacturers prefer to use a lotion with a sulfonated oil or sulfonated fatty alcohol base together with a small quantity of alcohol and water. This is colored with a spirit soluble color. Such a preparation is easy to apply and it does not give the skin an oily appearance because the sulfonated materials being water soluble are dissolved as soon as the user goes into the water leaving a brown precipitate on the skin.

Sunburn Healing Preparations

Formula No. 1 (Healing Oil)

Tincture of Benzoin	4%
Borax	4%
Olive Oil	40%
Lime Water	52%

Dissolve the borax in the lime water; add the olive oil with rapid agitation. When an emulsion is formed add the tincture of benzoin.

Formula No. 2 (Sunburn Lotion)

Zinc Sulphocarbolate	1%
Alcohol	8%
Glycerin	5%
Spirit of Camphor	1%
Rose Water	85%

Dissolve the zinc sulphocarbolate in the alcohol. Add the glycerin and the spirits of camphor, mix with the rose water.

Formula No. 3

Boric Acid	4%
Acetic Acid	2%
Citric Acid	1%
Alcohol	10%
Glycerin	5%
Water	78%

Heat part of the water and dissolve the boric acid in it. Dissolve the citric acid in the alcohol and the acetic acid in the remainder of the water. Add the glycerin, boric acid solution, the citric acid solution, mix and filter.

Formula No. 4 (Sunburn Salve)

Benzocaine	4%
Boric Acid	5%
Lanolin	15%
Petrolatum	76%

Mix the benzocaine and part of the petrolatum. Add the boric acid and run through an ointment mill until a smooth and impalable mass is formed. Melt the remainder of the petrolatum and lanolin, strain and add the milled medicament base.

Formula No. 5 (Sunburn Powder)

Boric Acid	10%
Bismuth Subnitrate	5%
Magnesium Stearate	10%
Isobutyl ester of Para-amino- benzoic Acid	5%
Talc	70%

Make the same as talcum powder.

Formula No. 6 (Sunburn Lotion)

Picric Acid Solution (20%)	1.5%
Alcohol (90%)	10 %
Water	88 %
Perfume	.5%

Dissolve the perfume in alcohol, add to the water and then dissolve the picric acid in the solution. This is excellent for sunburn but as it is will stain the skin yellow; it may be found objectionable for facial sunburn. As dry picric acid is explosive when heated quickly or subjected to percussion, it is safer to purchase the 20%.

Sun Tan Liquid

Agar-agar	4 parts
Glycerin	40 parts
Rose-water	60 parts
Æsculin	5 parts
Extract of Tormentilla	40 parts

Sun-Tan Oil

Quinine Hydrochloride	3.0%
Peanut Oil	96.5%
Color Oil (Soluble)	As desired
Perfume	.5%

Sun Proof Cream

Zinc Oxide	26%
Lanolin	46%
Water	28%

Rub together the zinc oxide and lanolin and then work in the water gradually.

Depilatory Cream

Titanium Dioxide	15 g.
Barium Sulphide	37 g.
Starch	50 g.
Phenol	1 g.
Lanolin	26 g.
Stearic Acid	5 g.
Triethanolamine	1.6 g.
Water	137.4 g.

Add the triethanolamine and the stearic acid to half the water. Heat the mixture until the stearic acid melts and then stir until a creamy soap forms. Allow the mixture to become lukewarm and then stir in starch and continue to stir until all lumps have disappeared. Dissolve the barium sulphide in the rest of the water and bring to a boil. Then stir in the soap-starch solution and continue stirring until the mixture thickens. Add the melted lanolin and stir. Then slowly sift in the titanium dioxide and mix until smooth. Finally add the phenol and perfume.

Depilatory Cream
Hungarian Patent 107,485

The cream consists of barium sulphide 40, American petrolatum 300, spermaceti 100, stearin 70, tincture of iodine 15, potassium carbonate 15 and water 400 grams, made up to creamy consistency.

Depilatory
British Patent 400,980

42 per cent strontium sulphide, 17 per cent chalk, 20 per cent starch, 10 per cent albumen, 10 per cent potassium sulphocyanide, and 1 per cent perfume.

Depilatory

Calcium Sulphide	4 lb.
Wheat Starch	12 lb.
Powdered Acacia	1 lb.

FACE POWDER

Poudre De Riz, Light

Osmo-Kaolin	47 lb.
Titanium Dioxide (Finely Pulverized)	8 lb.
Italian Talcum	15 lb.
Rice Starch	20 lb.
Magnesium Stearate	5 lb.
Magnesium Carbonate	5 lb.
Perfume Oil	Sufficient
Powdered Color	Sufficient

Add to perfume oil equal amount of alcohol. Then place in a mortar the magnesium stearate and magnesium carbonate, rubbing the perfume into these two ingredients. When thoroughly mixed, if it is found that the powder mixture is still damp, add enough osmo-kaolin until you have a dry mixture.

Place all other ingredients in the mixer and start the mixing machine running. Then add the balance of the ingredients containing perfume. Run mixer for about two hours and after the machine has run for about one hour, add powdered color. Take powder from mixer, putting it into tightly covered galvanized cans and allow to age for one month. Then place powder back in mixer and run for thirty minutes.

Put powder in bolting machine and continue bolting until thoroughly blended. The powder then should be immediately boxed and sealed.

Poudre De Riz, Medium

Best Italian Talcum	600 lb.
Zinc Stearate	130 lb.
Rice Starch	175 lb.
Osmo-Kaolin	50 lb.
China Clay (Pulverized and Bolted)	20 lb.
Magnesium Carbonate	10 lb.
Titanium Oxide (Finely Pulverized)	25 lb.
Perfume Oil	Sufficient
Powdered Color	Sufficient

Add to perfume oil equal amount of alcohol. Then place in a mortar the magnesium stearate and magnesium carbonate, rubbing the perfume into these two ingredients. When thoroughly mixed, if it is found that the powder mixture is still damp, add enough Osmo-Kaolin until you have a dry mixture.

Place all other ingredients in the mixer and start the machine running. Then add the balance of the ingredients containing perfume. Run mixer for about two hours and after the machine has run one hour, add powdered color. Take powder from mixer, putting it into tightly covered galvanized cans and allow to age for one month. Then place powder back in mixer and run for thirty minutes.

Put powder in bolting machine and continue bolting until thoroughly blended. The powder then should be immediately boxed and sealed.

Liquid Face Powder

Osmo-Kaolin	3½ oz.
Precipitated Chalk	2½ oz.
Glycerin	2½ oz.
Spirit	1 oz.
Water	20 oz. or 25 oz.

If suspension is deemed necessary mucilage of tragacanth 1 oz. may be used, but a portion of the powders must first be rubbed down with a mucilage, otherwise, if the tragacanth is added last, the deposit is much more noticeable. The following are suitable colors: White, no addition; pearl white, 10 minims of erythrosin solution (1–80); pink, 40 minims of the same solution; naturelle, 20 minims of the above solution and 1 dram of yellow ochre; rachel, 1½ drams of yellow ochre.

Poudre Azyade

Rice Starch	10 g.
Talc	40 g.
Basic Bismuth Nitrate	70 g.
Perfume	2 gr.

"Cooling" Talcum Powder

Menthol	1–4 parts
Alcohol	10 parts

Dissolve the above and add to:

Talcum Powder	100 parts

Mix until uniform.

Talcum Powder

Talc	71 g.
Precipitated Chalk	20 g.
Zinc Stearate	3 g.
Boric Acid	5 g.
Perfume	1 g.

Amandine

Honey	16 c.c.
Soft Soap	8 g.
Balsam of Peru	1 c.c.
Oil of Bergamot	1.5 c.c.
Oil of Bitter Almonds	1.5 c.c.
Oil of Cloves	1 c.c.
Oil of Sweet Almonds	56 c.c.
Preservative	1.5 g.

Lait D'Iris (Milk of Iris)

Spermaceti	3 g.
White Wax	3 g.
Marseilles Soap	3 g.
Sweet Almonds	40 g.
Distilled Water	100 c.c.
Alcohol (90%)	50 c.c.
Salicylic Acid	.1 g.
Perfume	.5 g.
Rose Water	100 c.c.

Lait Virginal (Virgin Milk)

Rose Water	900 c.c.
Tincture of Myrrh	10 c.c.
Essence of Opoponax	10 c.c.
Tincture of Benzoin	10 c.c.
Tincture of Quillaia	Sufficient to emulsify

Mix and strain.

Acqui Di Lubin (Lubin Water)

Alcohol	2000 c.c.
Tincture of Orange Peel	350 g.
Tincture of Musk Seed	300 g.
Tincture of Tonka Bean	100 g.
Tincture of Tuberose	50 g.
Tincture of Styrax	50 g.
Tincture of Benzoin	50 g.
Tincture of Vanilla	30 g.
Oil of Lemon	40 g.
Oil of Bergamot	4 g.
Oil of Neroli	1 g.
Tincture of Musk	4 g.
Tincture of Civet	3 g.
Orange Flower Water	250 g.

Eau De Botot (French Type)
(Botot Water)

Anise	10	oz.
Cochineal	.75	oz.
Mace	150	grains
Cloves	150	grains
Cinnamon	2.75	oz.
Alcohol	6	pints
Oil of Peppermint	.75	oz.

Eau De Botot (English Type)
(Botot Water)

Tincture of Red Cedar Wood	8	pints
Tincture of Myrrh	2	pints
Tincture of Rhatany	2	pints
Oil of Lavender	.75	oz.
Oil of Peppermint	1	oz.
Oil of Rose	150	grains

Pot-Pourri
Formula No. 1

Rose Leaves	16 oz.
Lavender Flowers	16 oz.
Orris Root (Coarse Powder)	8 oz.
Cloves (Coarse Powder)	2 oz.
Cinnamon (Coarse Powder)	2 oz.
Allspice (Coarse Powder)	2 oz.
Table Salt	16 oz.

Formula No. 2

Sandal Wood	16 oz.
Gum Benzoin	2 oz.
Orris Root	12 oz.
Cloves	2 oz.

Mace	1 oz.
Tonka Beans	2 grains
Musk	40 grains
Oil of Rose	40 drops
Oil of Lavender	1 dram
Oil of Bergamot	2 drams
Oil of Lemon	2 drams

Formula No. 3

Powdered Cloves	2	oz.
Powdered Pimento	2	oz.
Powdered Benzoin	2	oz.
Essence of Musk	1	oz.
Essence of Bergamot	4	drams
Oil of Lavender	4	drams
Oil of Cloves	2.5	drams
Oil of Cassia	2.5	drams
Oil of Rose	80	drops
Rose Leaves	4	oz.
Powdered Jamaica Pepper, to make	48	oz.

Antiseptic Dusting Powder
Formula No. 1

Potato Starch	700	parts
Light Magnesium Carbonate	1750	parts
Precipitated Chalk	350	parts
Talc	350	parts
Quinosol	5.25	parts
Tincture Benzoin	17.5	parts

Formula No. 2

Thymol	2	parts
Boric Acid	50	parts
Precipitated Chalk	100	parts
Zinc Oxide	300	parts
Talc	450	parts
Almond Oil	20	parts

Formula No. 3
Petrolatum Dusting Powder

Yellow Petrolatum	176	parts
Lanolin	44	parts
Rice Starch	1000	parts
Talc	800	parts
Zinc Oxide	400	parts

Rice starch, talc and zinc oxide are first very thoroughly mixed. Petrolatum and lanolin are melted in kettle at moderate temperature, being well stirred with wooden paddle. Powder mixture is gradually introduced into the still hot fat melt with vigorous agitation. Addition of more powder gradually converts mass into pasty condition, and then coarsely grained mass is obtained. Remainder of powder is added and mixture ground in roller mill to convert it into fine granular state. Uniform distribution of fatty particles, which is an important prerequisite, is thus obtained.

Antiseptic Soap
U. S. Patent 1,946,079

282 parts oleic acid are caused to react by heat with 119 parts of methyl di-ethanolamine.

The soap obtained in this way has the property, when in alcoholic solution, of dissolving mercuric oxide and giving a relatively stable product soluble in water.

Cosmetic "Oxygen" Soap

Coconut Oil	15	g.
Castor Oil	7	g.
Tallow	77	g.
Cimol-Neutral	2	g.
Sodium Cholate	.5	g.
Sodium Perborate	5	g.
Preservative	.3	g.
Magnesium stearate	1	g.

Jones' Perfume for Liquid Soap

Syringeol Synthetic	5	c.c.
Artificial Oil of Rose	.5	c.c.
Artificial Oil of Jasmine	.5	c.c.
Oil of Rose Geranium	.5	c.c.
Oil of Clove	.5	c.c.
Terpineol	7.5	c.c.
Artificial Musk	.5	g.
Alcohol, to make	20	c.c.

Bath Powder

Use a mixture of equal parts of sodium bicarbonate and borax.

Pine Needle Bath Oil

Mixing equal parts of pine needle oil with sodium sulfo-ricinoleate, gives a fine concentrate to be added to the bath. The product so made is completely dispersed in the bath water. To thin it out, add either water or similar diluent.

Methyl Salicylate in Bath Preparations

The problem in using methyl salicylate in making bath essences and the like is to obtain the latter in such form that it is readily soluble in water and the methyl salicylate itself in a very fine dispersed emulsion. Turkey red oil, that is sulphonated castor oil, is found to be suitable for this purpose. It is used with the addition of glycerin and potassium carbonate and is obtained thereby in clear solution in water. This solution emulsifies with water without any difficulty. Eucalyptol and menthol may also be used in the place of methyl salicylate o mixed with it in various proportions. An example of a composition is as follows:

Sulphonated Castor Oil	150 parts
Methyl Salicylate	150 parts
Eucalyptol	45 parts
Menthol	5 parts
Potassium Carbonate	50 parts
Glycerin	100 parts
Water	500 parts

Aromatic Bath Salts

Potassium Bromide	1 g.
Calcium Carbonate	1 g.
Sodium Sulphate	5 g.
Sodium Phosphate	8 g.
Sodium Carbonate (exsicc.)	300 g.
Oil of Lavender	1 g.
Oil of Rosemary	1 g.
Oil of Thyme	1 g.

Use about 300 grams in each bath.

La Rouce Bath Cream

Tannic Acid	4 g.
Expressed Oil of Almonds	160 g.
Hydrous Lanolin	240 g.

Melt, mix and beat until smooth.

The preparation made from this French recipe is much used to close the pores, constrict the skin and make the flesh firm after the hot or Turkish bath. It is also used as a wrinkle cream.

Sulphur Toilet Waters

Sulphur in proper physical state in toilet waters is claimed to be highly useful for keeping skin in good condition and also for treatment of acne and other common skin troubles. To obtain sulphur in proper condition in such preparations, mix:

Borax	5 parts
Water	850 parts

add:

Sodium Thiosulphate	50 parts
Glycerin	50 parts

then add:

Eau d'Cologne	50 parts

Sulphur is said to be present in nascent state. When face lotion is used, the sulphur is precipitated on skin and its action is most effective under such circumstances.

Magoffin's Perspirine

Powdered Talc	5 lb.
Corn Starch	5 lb.
Boric Acid	10 oz.
Oil of Rose	1 dram

Mix the first three ingredients. Triturate the oil with two ounces of the mixture and then mix all together. Run through a No. 60 sieve at least five times.

Deodorant Powder

Zinc Oleate, Special	8 lb.
Sodium Perborate	3 lb.
Boric Acid, Powder	5 lb.
Sodium Bicarbonate	2 lb.
Magnesium Carbonate	4 lb.
Italian Talc	18 lb.

Mix powders thoroughly and perfume to suit.

Sanitary Napkin Deodorant
U. S. Patent 1,950,286

One part or layer of the inlay or pad is impregnated with sodium acetate and another part with sodium bisulphate, the pad then being placed in a wrapper, the two impregnated layers being separated by an intermediate layer. When the pad becomes moistened during the discharging period acetic acid is released through reaction, which as is well known has strong deodorizing and disinfectant properties. Ethereal oils are especially adapted as an aromatic addition, which may be used and which will evaporate slowly along with the acetic acid.

It is easily seen that the above described process has many advantages over the use of any poisonous disinfectant and both of the compounds used are non-poisonous and non-volatile which also makes the preparation of the pads very much easier and much safer for the workers.

The easy solubility and reactive power of the salts used in this process makes possible during the discharge of the secretions a continuous development of the deodorizing and disinfectant substance which in consequence of its volatility exercises its effect not only at the place of reaction but throughout the whole napkin and in this way acts as a complete deodorizer and disinfectant.

Powdered Shampoo Mixtures

Powdered shampoos can be prepared by drying soap stock as usual and mixing to a fine powder with other ingredients, according to the formulas.

Formula No. 1

Soap Powder	60%
Coconut Oil Soap Powder	10%

Disodium Phosphate	10%
Sodium Bicarbonate	10%
Borax	10%

Formula No. 2

Soap Powder	50%
Sodium Lauryl Sulphate	15%
Sodium Bicarbonate	5%
Borax	30%

Formula No. 3

Soap Powder	65%
Sodium Cetyl Sulphonate	5%
Borax	10%
Sodium Bicarbonate	15%
Trisodium Phosphate (powd.)	5%

A second method of preparation is to add the ingredients to the soap mixture before it is sent to the cooling press, as with the following:

Formula No. 4

Soap from the Kettle	75%
Protein or Gall Soap	3%
Borax	12%
Sodium Bicarbonate	10%

Formula No. 5

Soap	60%
Sodium Lauryl Sulphate	10%
Sodium Bicarbonate	20%
Borax	10%

Shampoo Powder

Sodium sesquicarbonate, a compound intermediate in character between sodium carbonate and sodium bicarbonate, is the alkali par excellence for incorporation in shampoo powders. It may be used in quite large proportions. Forty per cent of soap, 50 per cent of sodium sesquicarbonate, and 10 per cent of borax form a good combination.

Henna Shampoo Powder

Castile Soap (in powder)	33	parts
Coconut Oil Soap (in powder)	7	parts
Sulphated Fatty Alcohol	20	parts
Sodium Sesquicarbonate	33	parts
Egyptian Henna Powder	7	parts
Perfume	0.5	parts

As the sulphated alcohol is most effective, it is also possible to make good shampoo powders containing about 20 or 30 per cent of this constituent plus a filler. Alternatively sulphated alcohol may be used alone, save for perfume, and, if desired, chamomile or henna powder. For example, the following simple formula yields a chamomile shampoo powder of superlative quality:

Chamomile Shampoo Powder

German Chamomile, powdered	20 parts

Sulphated Fatty Alcohol	80 parts
Perfume	1 part

In using this powder, the quantity should be reduced to about one-third of that generally employed with a shampoo powder of the ordinary type.

"Soapless" Shampoo
U. S. Patent 1,946,272

Secondary Sodium Phosphate	16 oz.
Sulphonated Coconut Oil	2 oz.
Saponin	2 g.

First make the secondary sodium phosphate fluid by subjecting it to as approximate temperature of about 100 degrees Centigrade. To the fused secondary sodium phosphate add the other ingredients, either in their natural state or dissolved in a minimum quantity of hot water. The mixture is then thoroughly stirred while cooling, to prevent separation of the ingredients. After careful commingling of the ingredients, the mixture is cooled and crystallized. As so made up, the cleansing composition is in suitable condition to pack and sell.

The method of use comprises initially the dissolution of the crystallized mixture in water. While the relative proportions of crystallized composition and water may vary within reasonable limits, a solution of one ounce of composition to one-half pint of water is used. The water to which the composition is added may be at, or only slightly above, room temperature.

After the solution has been formed it is further prepared for use by being violently beaten or whipped. This violent beating or whipping of the dissolved mixture results in the formation of a relatively great mass of intermingled film or suds. This mass of film is produced primarily from the sulphonated oil, and carries the dissolved cleansing salt. In use a mass of the suds may be gathered in the hands and worked into the scalp in the same manner that a soap lather is applied. In its proximate use, also, as distinguished from its ultimate effects, the suds of this cleansing composition differ in a marked manner from soap suds. This marked difference is in the fact that they do not tend to adhere to the hair in the form of a greasy or oily coating, but may be washed from the hair by a single rinsing of short duration without leaving any deposit whatever.

As to the ultimate effect of the cleansing composition, since both the sulphonated oil and the cleansing salt are non-alkaline, the composition does not remove from the hair, sulphur, or the finer gloss of the hair itself.

As to other uses of the composition, its properties render it suitable for washing the finest laces, cleansing tinted walls and mural decorations, and even for the cleansing of oil paintings.

It will be noted that throughout the preceding discussion but little mention has been made of the saponin, which is included in a small quantity in the formula. That is for the reason that its inclusion in the composition is not necessary to the composition. When present in a small quantity, it does, however, serve to make the carrying suds or film more stable, thus improving the vehicle for the specifically detergent salts. It is itself substantially neutral chemically.

Soapless Shampoos
Formula No. 1

Sapamine Acetate	20	%
Boric Acid	0.5	%
Perfume	0.5	%
Water	79	%

Enzymes are supposed to have a certain emulsifying and purifying action. Shampoos containing enzymes are as follows:

Formula No. 2

Soap Powder	80%
Sodium Bicarbonate	10%
Borax	8%
Pancreatin	2%

(Maximal enzyme action at 40° C. and pH 7.8–8.0)

Formula No. 3

Sodium Lauryl Sulphate	63%
Sodium Phosphate	5%
Sodium Bicarbonate	20%
Powdered Borax	10%
Pancreatic Lipase	2%

These preparations are quite stable in dry form, but enzyme activity causes deterioration in solution. Dry shampoos may be rubbed into the hair and then brushed out, without the use of water. They consist largely of starch, as in the following:

Formula No. 4

Rice Starch (defatted)	94%
Sodium Bicarbonate	1%
Powdered Borax	5%

Formula No. 5

Powdered Silica Gel	10%
Defatted Starch	85%
Powdered Trisodium Phosphate	5%

Soapless Liquid Shampoo

Saponin (Pure)	2 parts
Rose Water (Diluted)	80 parts
Perfume	1 part
Alcohol	15 parts
Water (Distilled) to make 100 parts	

To prepare this lotion, the saponin should be dissolved in the rose water, the alcohol containing the perfume added, and the mixture made up to the requisite volume with distilled water. It may be tinted yellow with a trace of tartrazine. Distilled water can be substituted for rose water, and a little phenyl ethyl alcohol added.

Pine Tar Shampoo

Pine Tar Shampoo can be made by adding a small proportion of solution of coal tar and pine oil to a good shampoo liquid. The following formula would serve as a base for the shampoo liquid:

Potassium Carbonate	12 oz.
Water	320 oz.
Alcohol	480 oz.
Dry Extract of Quassia	1 oz.
Saponin	2 dr.

Pine Tar Shampoo

Pine Tar	25 parts
Yellow Soft Soap	300 parts
Industrial Spirit	200 parts
Water	475 parts

Bay Rum Shampoo

White Castile Soap	½ oz.
Rose Water	1 oz.
Solution of Ammonia	1 oz.
Bay Rum	2 oz.
Distilled Water to make	2 pt.

Dissolve the soap in 30 oz. of water by heating; cool to about 100° F., and add the rest of the ingredients.

Hair Treatments
Formula No. 1

Resorcinol	.8%
Beta Naphthol	.8%
Chloral Hydrate	1.5%
Tincture of Capsicum	4.0%
Castor Oil	2.0%
Alcohol	90.0%
Oil of Nutmeg	.9%

Dissolve the chloral hydrate in the castor oil. Dissolve the resorcinal in part of the alcohol, the beta naphthol in the remainder of the alcohol. Add the oil of nutmeg to this solution, then the tincture of capsicum, the castor oil mixture and finally the resorcinol. Mix well and filter.

Formula No. 2

Quinine Sulphate	.25%
Pilocarpine Hydrochloride	.05%
Chloral Hydrate	1.00%
Castor Oil	4.00%
Chloroform	7.00%
Spirits of Formic Acid	12.50%
Alcohol	74.95%
Perfume	.25%

Dissolve the chloral hydrate in the castor oil. Dissolve the pilocarpine in the formic spirits. Mix the chloroform with about one-half its weight of alcohol and dissolve the quinine sulphate in it. Add the formic spirit solution, mix and add the formic spirit solution. Mix again. Finally add the perfume. Mix well and filter.

Formula No. 3

Resorcinol	.28%
Soft Soap	.46%
Pine Tar Oil Rectified	2.70%
Potassium Sulphide	3.00%
Water	93.31%
Perfume	.25%

Dissolve the potassium sulphide in half of the water. Dissolve the resorcinol in the remainder of the water. Mix the perfume and pine tar oil with the soft soap and then rapidly mix this with the resorcinol solution. Finally add the potassium sulphide solution.

Formula No. 4

Resorcinal Monoacetate	3.00%
Castor Oil	7.00%
Spirits of Formic Acid	20.00%
Alcohol	69.75%
Perfume	.25%

Dissolve the resorcinol monoacetate in the formic spirits, the castor oil in the alcohol. Mix the two solutions and add the perfume. Filter.

Formula No. 5

Colloidal Sulphur	5.00%
Pulverized Camphor	.40%
Tincture of Cantharides	5.00%
Resorcinol Monoacetate	1.00%
Oil of Nutmeg	.50%
Alcohol	72.85%
Glycerin	15.00%
Perfume	.25%

Rub off the sulphur with the glycerin until a smooth mixture is obtained. Dissolve the resorcinol in the alcohol. Mix the pulverized camphor with the tincture of cantharides. Add this to the alcohol mixture and finally add the sulphur mixture. Mix thoroughly and add the perfume. Do not filter.

Hair Fixatives

Fixative jellies for the hair may be prepared by mixing tragacanth (5 parts), with alcohol (5 to 10 parts) and glycerin (5 to 10 parts), coloring and perfuming the mixture and then adding, all at once, 75 to 85 parts of water. Alternatively, the jellies may be made without the use of alcohol, but the process is more troublesome, and the use of a mechanical mortar is recommended, in order to get a perfect mixture of the glycerin and the gum. The addition of a preservative is desirable, especially if the jelly does not contain alcohol. Sodium benzoate is often employed, but sodium fluoride and esters of para-hydroxy-benzoic acid are more efficient. The use of perfumes containing clove oil or eugenol are also recommended. In a series of experiments with a jelly which, after one month, showed growth of Asperigillus, 1 per cent of sodium benzoate was not found to be an effective preservative; 0.2 per cent of "nipagin," however, was sufficient to keep the jelly, as was also 0.1 per cent of sodium fluoride.

Nursery Hair Oil

Benzoin	½ oz.
Alkannin	½ oz.
Oil of Stavesacre	1 oz.
Almond Oil	1 pint

Macerate for a week, shaking daily, filter, and add

Perfume	½ dram
Mix.	

Hair Oils
Jasmin

Benzyl Acetate	10.0 parts
Tolyl Acetate	5.0 parts
Benzyl Propionate	3.0 parts
Phenyl Ethyl Alcohol	8.0 parts
Rhodinol	2.0 parts
Nerol	1.0 part
Geraniol	2 parts
Alpha-amyl Cinnamic Aldehyde (pure)	2.5 parts
Para Methyl Quinoline	0.3 part
Musk Ketone	0.5 part
Aldehyde C13 (1–10)	0.3 part
Linalyl Acetate	1.0 part
Linalol	1.0 part
Hydroxycitronellol	3.0 parts
Methyl Ionone (Alpha)	5.0 parts
Methyl Anthranilate	2.0 parts
Phenyl Acetic Acid	0.1 part

Rose

Rhodinol	20.0 parts
Nerol	5.0 parts

Geraniol	5.0 parts	Rose, Artificial	1.0 part
Phenyl Ethyl Alcohol	8.0 parts	Jasmine, Artificial	1.0 part
Citronellol	3.0 parts	Heliotropin	1.0 part
Geranyl Acetate	2.0 parts	Citronella Hydrate	2.0 parts
Phenyl Ethyl, Methyl Ethyl		Musk Ambrette	0.2 part
Carbinol	3.0 parts	Methyl Heptine Carbonate	0.1 part
Terpeneless Rose Geranium	2.0 parts	Bergamot	3.0 parts
Guaiacum Wood Oil	1.5 parts	Aldehyde C13 (1–10)	1.5 parts
Terpeneless Petitgrain	0.5 part	Cassia Absolute	1.5 parts
Nonyl Aldehyde (1–10)	0.5 part		
Decyl Aldehyde (1–10)	0.2 part		

Build the formula in the order shown, special attention being given to the effect of each addition. Rhodinol, particularly if a natural isolate, has a decided rose odor, but would lack power and body as a scent by itself. Other constituents of a rose-like odour, or that have a decided value in the synthesis, must be added. Nerol is next in importance; note the effect of this addition; repeat with the addition of each item until the terpeneless rose geranium has been included. Up to this stage the compound will not have shown any definite character; the items that follow are included to act as toning agents. Guaiacum wood oil has a slight orris note; this is the reason for the inclusion, but at the same time it can be classed as a fixative. Any ingredient with a slight odor of orris can be used in this formula as a modifier, so the question arises: Why not orris itself? Oil of orris (orris liquid) can be used, of course, but the other serves the purpose well at a greatly reduced cost; and orris concretes need a deal of preparation. However, liquid orris appears in the violet formula, so those who wish to try it must use only 1 gm. Terpeneless petitgrain is also used as a modifier; it greatly helps to impart the somewhat bland note necessary in "rose." The aliphatics are best manipulated in a 10 per cent solution. Benzyl benzoate is the fixing agent. Rhodinol (ex Bourbon geraniol), geraniol (ex palmarosa) and citronellol are the isolates to use for the best results. Finally, the addition of 5 per cent pure virgin otto is left to the compounder's discretion.

Violet

An entirely synthetic violet leaves a lot to be desired, and the formula given below is not of outstanding merit, so the inclusion of cassia absolute is advised:

Ionone (100 per cent)	4.0 parts
Ionone Alpha	2.0 parts
Ionone Beta	2.0 parts
Methyl Ionone Alpha	2.0 parts
Orris Liquid	1.5 parts

Manufacturers have a wide divergence of opinion as to what constitutes ionone. Samples vary so much that the first experiments should be made with the smallest possible quantities. Many experiments can be made by the alterations of the ionones, but alpha-methyl-ionone should never be exceeded beyond the given amount, and the total to be about 50 per cent of the formula. Aldehyde C13 and methyl heptine carbonate can be slightly increased or decreased, at will of the compounder.

Hair Cream

Formula No. 1
(Without Oil)

Gum Tragacanth (pulv.)	1.5 parts
Castor Oil (1st Grade)	8.0 parts
Glucose	8.0 parts
Alcohol	14.0 parts
Sodium Benzoate	0.2 part
Formaldehyde	0.4 part
Water to	160.0 parts
Perfume	Sufficient

Formula No. 2
(With Oil)

Gum Tragacanth (pulv.)	1.5 parts
Castor Oil	4.0 parts
Glycerin	8.0 parts
Alcohol	14.0 parts
Sodium Benzoate	0.2 part
Formaldehyde	0.4 part
Water to	160.0 parts
Perfume	Sufficient
Mineral Oil	48.0 parts

The best method of compounding is to place together the tragacanth, castor oil, spirit, and perfume, stir well and briskly, add the glycerin, again stir, and slowly add 16 parts of water (the cream will not thicken at this stage). Then continually and briskly stirring, run in the remainder of the water at a fairly fast rate. Should the cream show any sign of not being homogeneous (in other words, being lumpy) at any time during the operation, cease the flow of water until the cream rectifies itself. Finally complete the formula.

Tincture of benzoin as an aid to an opaque cream has not been overlooked, but in this formula it is not required. It has been remarked that most men prefer a fugitive odor, and this can only be obtained by the use of essential oils. Lavender and eau de Cologne are the most popular, but offer nothing to distinguish one cream from another. At the same time, there is little to choose when using essential oils; from the following a choice can be made for experiment:

Suggested Perfume Combinations

	A	B	C	D	E
Lavender	4	2	1	—	—
Bergamot	4	4	6	6	10
Geranium	2	2	1	2	1
Petitgrain	—	—	2	1	3
Bois de rose	1	—	1	—	—
Lemon	—	—	1	1	1
Orange	—	—	½	½	½

Hair Cream

White Wax	10 parts
Liquid Paraffin	125 parts
Borax	1 part
Distilled Water	14 parts

Allow the wax to dissolve in 60 parts of liquid paraffin, then stir in the remainder of the liquid paraffin. Dissolve the borax in the distilled water, add to the previous solution, and stir the cream so formed very thoroughly.

Hair Grooming Cream

Flaxseed (Whole)	1 lb.
Boric Acid	2 oz.
Glycerin	12 oz.
Water	1 gal.

Color and perfume to suit.
Boil the flaxseed with the water until syrupy; filter by squeezing through a linen bag. Discard the residue. Add the boric acid and glycerin (above) to the liquid.

Brilliantine (Separating)

Almond Oil	49%
Alcohol	50%
Perfume	1%

Brilliantine (Non-Separating)

Castor Oil	49%
Alcohol	50%
Perfume Oil	1%

Brilliantine (Non-Greasy)

Glycerin	59%
Alcohol	50%
Perfume	1%

Greaseless Brilliantines for Men

There is a slight difference between the composition of such preparations used by men and those used by women. Those applied to the man's hair are merely required to keep it in place without giving is a greasy appearance. A very satisfactory product is obtained by using the following formula:

Gum Tragacanth (in ribbons)	2 lb.
Sodium Benzoate Powder	1 lb.
Water	55 gal.
Perfume Oil	6 oz.

Yellow, water-soluble dye sufficient to obtain an intensive yellow color.
Take a small vessel (10 gallon capacity); place into it the gum tragacanth in ribbons and 5 gallons of the water required; soak the gum overnight; next morning place the ''jelly'' in a kettle with a high speed mixer and let it run for five minutes; then add the balance of the water (cold—50 gallons) slowly while the mixer is still in motion.
After the water is all mixed, add the sodium benzoate, the color, and the perfume. Strain before using through a wide-meshed cheesecloth.

Brilliantine
Formula No. 1 (Liquid)

Mineral Oil (Heavy)	3.5 gal.
Peanut Oil	1.5 gal.
Perfume Oil	5 oz.

Color to suit.
This formula is best used with coloring material, as it looks slightly yellowish when uncolored.

Formula No. 2 (Solid)

Petrolatum Yellow	20 lb.
Ceresine Wax	3 lb.
Mineral Oil (Heavy)	4 pt.

Color and perfume to suit.
It is stated that the product made from this formula has a smooth texture and a fine luster and will never turn rancid.

Formula No. 3 (Solid for Hot Countries)

Ceresine Wax	2 lb.
White Wax	0.1 lb.
Mineral Oil (Light)	20.0 lb.
Oil of Lemon	0.32 oz.
Oil of Geranium	0.16 oz.
Oil of Bergamot	0.64 oz.
Color	Sufficient

Quinine Hair Dressing

Quinine, Sulphate	20 g.
Castor Oil	1 oz.
Tinct. Cantharides	½ oz.
Ext. Jasmine	3 drams

Eau de Cologne	3	oz.
Oil Bitter Almonds	5	drops
Oil Bergamot	½	dram
Alcohol	8	oz.

Mix and color with tincture of alkanet if desired.

Cholesterol Hair Tonic
Formula No. 1

| Cholesterol | 10 parts |
| Mineral Oil | 50 parts |

Heat to 130°.

| Sodium Choleate | 5 parts |
| Glycerin | 50 parts |

Heat to 130° and add to above slowly with agitation.
Then add slowly with agitation a solution of

| Borax | 5 parts |
| Water | 1000 parts |

That has been heated to 130° C. Stir until cold.

Formula No. 2

A much more stable lotion results by replacing the mineral oil in Formula No. 1 with

Lanolin	10 parts
Peanut Oil	10 parts
Cocoa Butter	30 parts
Preservative	1 part

Hair Tonic, Foaming

Carbonate of Potash	½	oz.
Borax	½	oz.
Dissolve in water	2	gal.

Then add:

Glycerin	1	pint
Ethyl Alcohol	1⅞	gal.
Hair Tonic Perfume Oil	1	oz.

Tint to suit.

Hair Tonic

| Deodorized Castor Oil | 10 parts |
| Alcohol (Specially Denatured) | 90 parts |

Perfume to suit (flower type).
Dissolve the oil in the alcohol and add the perfume.

Hair Tonic
British Patent 399,007

A preparation for stimulating hair growth comprises castor oil 8, peach kernel oil 2, verbena oil 4, and surgical spirits (industrial spirit containing small amounts of castor oil and boric acid) 2 parts by volume.

Hair Tonic for Dandruff

Alcohol (Specially Denatured for Hair Tonics)	90 parts
Glycerin	10 parts
Capsicum, Tincture	15 parts
Oil of Red Rose	2 parts

Add in the above order.
Any perfume may be used. Color green with chlorophyll.

Hair Luster Powder
Formula No. 1

Alum	60 parts
Tartaric Acid Powdered	30 parts
Adipic Acid	10 parts

Perfume as desired.

Formula No. 2

| Hexamethylene Tetramine | 9 parts |
| Adipic Acid | 40 parts |

Warm until ester forms. Then add:

| Tartaric Acid | 50 parts |
| Perfume | 1 part |

Then grind.

"Falling Hair" Ointment

Salicylic Acid	1.0 g.
Resorcinol Monoacetate	1.5 g.
Precipitated Sulphur	1.5 g.
Ointment of Rose Water to make	30.0 g.

Rub into the scalp vigorously once a day.

Hair Pomade and Straightener

Yellow Petrolatum	6 parts
Yellow Beeswax	1 part
Peanut Oil (Refined)	2 parts
Perfume Oil	Sufficient

Melt the petrolatum in double boiler and add the other ingredients. The perfume should not be added until the mixture begins to solidify around the edges. This is an amber product. If a water white product is desired use white petrolatum and white beeswax and substitute equal parts of deodorized castor oil and mineral oil for the peanut oil. By adding more wax and less of the oils a product can be obtained that will straighten the most unruly hair.

Hair Curling Fluid
U. S. Patent 1,933,021

Keratin is used in combination with a solvent or vehicle such as an aqueous solution of ammonia and alcohol in which the keratin is dissolved or suspended or otherwise held in suitable manner so that when applied to the hair and heated the keratin will enter the pores of the hair, the solvent or vehicle being such that it will not damage the hair when so applied.

Keratin	1 oz.
Ammonia Water (fort. B.P.)	8 oz.
Alcohol	8 oz.

Then take of this solution 6 drams and make it up with ammonia water (B.P.) to 4 ounces liquid measure. This solution is the strength that is used upon the hair.

Waving Fluid

Many small manufacturers of hair waving fluid, made from quince seed, do not wash the seed in cold water previous to extracting the mucilage. This procedure cleans the seed of clay, dirt or sand, as well as wormy seeds. Then, to get a clean fluid of light color, the extraction of the mucilage should be made with cold water too. About 3% of seed gives a thick slime, of sufficient viscosity to be used along with 2% borax and an aromatic agent of some type. Dissolve the borax in a little water, then add to the quince slime. Esters of para-oxybenzoic acid are good preservatives for these lotions. Hot water produces a turbid mixture, use cold water throughout.

Wave Setting Lotion

Mix:

Madagascar Quince Seed	10 lb.
Water	35 gal.

Keep at 180° F. for 24 hours. Then strain.

Add:

Borax	1 lb.
Alcohol	5 gal.
Perfume	1 pt.

Color as desired.

Hair Wave Lotion

Formula No. 1

Gum Tragacanth	1 part
Alcohol	100 parts
Rose Water	300 parts
Potassium Carbonate	4 parts
Borax	1 part
Perfume	2 parts

Formula No. 2

Gum Acacia Powdered	3 parts
Borax	20 parts
Alcohol	100 parts
Water	900 parts

Hair Waving Solution

Formula No. 1

Sodium Sulphite Heptahydrate	25	grains
Borax	12½	grains
Sodium Carbonate Decahydrate	2	grains
Ammonium Carbonate	1	grain
Water to	1	fl. oz.

Formula No. 2

Sodium Carbonate Anhydrous	4.52%
Potassium Bicarbonate Anhydrous	3.32%
Sodium Chloride Anhydrous	0.2 %
Borax	0.34%
Glycerin about	2.0 %
Alcohol, perfume, and coloring matter	traces

Finger Waving Solution

High Grade Quince Seed	3.00%
Borax	3.00%
Formaldehyde	.15%
Water	93.85%

Boil half the water, add the quince seed, soak 10 to 12 hours and strain. Dissolve the borax in the rest of the water, filter and add to the mucilage. Mix well, add formaldehyde and perfume.

Permanent Waving Solution

Formula No. 1

Crystalline Sodium Carbonate	11.7 parts
Commercial Potassium Carbonate	1.9 parts
Ammonium Ammonium Carbonate	3.8 parts
Dilute Solution of Ammonia	9.5 parts
Water to	100.0 parts

This formula is expressed in parts by weight.

Formula No. 2

The following is a suitable formula for a permanent waving solution of the non-ammonia type:

Potassium Carbonate	40 g.
Borax	10 g.
Tragacanth Mucilage	100 cc.
Alcohol	100. cc.
Rose Water to make	1000 cc.

Milky Permanent Wave Solution

Sodium Sulfite	6.5 parts
Ammonium Carbonate	2.0 parts
Ammonium Hydroxide 26%	5.0 parts
Water	86.5 parts
Carnauba Wax Emulsion	1-2 parts

Permanent Waving Solution
British Patent 391,355

A permanent waving solution is a mixture of a solution of borax 40–70, potash 10–20, ammonium chloride 5–15 parts in 700–1000 parts distilled water with 70–130 parts alcohol containing phenolphthalein 1, glycerin 10–15, lavender oil 2–5, and nitrobenzene 1 part.

Non-Poisonous Hair Dye
U. S. Patent 1,937,365
Formula No. 1

Satisfactory hair-dyeing solutions may be prepared by dissolving 80 grams of bismuth nitrate in a liter of solution containing 80% glycerin and 20% water by volume, and by similarly dissolving 170 grams of sodium thiosulphate in a liter of solution containing 80% glycerin and 20% water by volume. The two solutions are stored in separate containers and are mixed just before the hair is to be treated. The mixed solutions are distributed evenly over the hair, and after a brief time the reaction occurs with the formation of a colored pigment which darkens the color of the hair.

Formula No. 2

As another example prepare a solution containing the following:

Glycerin	66.6 cc.
Water	33.3 cc.
Bismuth Chloride	7.2 g.
Hydrochloric Acid	1.7 cc.

and another solution:

Glycerin	66.6 cc.
Water	33.3 cc.
Sodium Thiosulphate	15.3 g.

The two solutions are stored in separate containers and mixed just before the hair is to be treated. The mixture is applied as in the preceding embodiment of the invention.

As before indicated, the proportions of the ingredients employed may be varied. With increasing concentrations of thiosulphate up to between 10% and 12%, the speed of the reaction increases. After 12%, the speed of the reaction decreases. With thiosulphate up to about 12%, the color produced on the hair is reddish brown. Above 12%, the red begins to disappear and the color is a neutral brown. A darker shade is obtained by adding to the solution of bismuth compound, such as the nitrate, ammonium acetate in the proportion of about 12½ grams per liter, or equivalent amounts of ammonium hydroxide and acetic acid.

The advantages of using bismuth compounds, and particularly bismuth nitrate, are that bismuth produces none of the irritating or poisonous effects of lead.

Hair Dye

Good results have been obtained by a combined silver-nickel-cobalt dye prepared according to the following formula:

Silver Nitrate	3.5 g.
Cobalt Nitrate	1.5 g.
Nickel Nitrate	3.0 g.
Ammonia (0.880 sp. gr.)	Sufficient
Water, sufficient to make	100 cc.

After application of the solution of metallic nitrates the color in the hair is developed by means of a 3 to 4 per cent of pyrogallol dissolved in water.

Bismuth Hair Dyes

First, 50 parts of bismuth nitrate are dissolved in 250 parts water and 700 parts alcohol and sufficient ammonia is added to form solution number 1; second solution contains 100 parts sodium thiosulfate in one liter of water. In another preparation 50 parts bismuth citrate are dissolved in 33 parts alcohol, 200 parts rose water, 300 parts water and sufficient ammonia to give solution number 1, while second solution contains 120 grams sodium thiosulfate dissolved in 400 grams water. In another preparation 12 parts bismuth subnitrate are mixed with 58 parts water and mixture poured into porcelain mortar and 18 parts nitric acid are gradually added until solution takes place. This solution is then poured into solution of 9 parts tartaric acid, 8.5 parts sodium bicarbonate dissolved in 900 parts water. Precipitate is collected on filter cloth, washed well with water and dissolved in sufficiently strong ammonia solution. Then 6 parts sodium thiosulfate, 30 parts glycerin are added as well as sufficient water to obtain total of 240 parts.

White "Henna"

White henna is usually a paste of magnesium carbonate, ammonia water and 17-volume hydrogen peroxide. This formula varies according to user. Some use just ordinary peroxide. Others leave out the peroxide or/and ammonia. Still others add some magnesium peroxide to the carbonate, this mixture comprising the white henna. Still others add some henna to the bleaching mixture, to the extent of about 10%.

White Henna Powder

Sodium Perborate (powd.)	18 lb.
Henna Leaves (powd.)	2 lb.

Walnut Hair Dye

Green Walnut Shells	2 oz.
Alum	2 drams
Olive Oil	4 oz.

Heat all in water bath until all water has been expelled. Express, filter and perfume.

Eye Lash Grower (Darkener)

Yellow petroleum jelly is supposed to have the property of stimulating the growth of lashes and brows, as well as darkening them at the same time. Two formulas will indicate the type:

Formula No. 1

Yellow Petrolatum Jelly	50%
Turtle Oil	50%

Melt and Perfume.

Formula No. 2

Yellow Petrolatum Jelly	50%
Castor Oil	49%
Paraffin Wax	1%

Either No. 1 or No. 2 can be modified to give a darker-looking preparation by the addition of burnt sienna and/or umber. In this instance the usual precaution will have to be observed regarding packaging products containing suspended materials. Use a preservative where there is a chance for the oils to decompose. Para-oxy-benzoic acid esters are suitable for this. Perfume with lavender or bergamot, or both.

Eye Cream

The purpose of such a cream is to soften wrinkles, commonly called crowsfeet, around the eyes. The preparation must be rich in the so-called nutritives. The following are type formulas:

Formula No. 1

Lanolin	50%
Olive Oil	20%
Castor Oil	10%
Turtle Oil	20%
Preservative	Sufficient

Perfume as desired.

Formula No. 2

Lecithin	5%
Cholesterol	1%
Lanolin	10%

Expressed Almond Oil	59%
Beeswax	25%
Preservative	Sufficient

Perfume as desired.

One preparation of this type now on the market is perfumed with violet, the net result being that it is much different from the usual cream.

Eye Oil (Balsamic)

Some women prefer to use an oil preparation on the lashes and brows. For this purpose the following is a type:

Castor Oil	24.5%
Sweet Almond Oil	75.0%
Perfume	.5%

The perfume should consist of some mild stimulant, such as camphor oil, along with balsams. Oil storax, 10; oil camphor, 45; rose oil synthetic, 10. The preparation should be allowed to stand for some time before bottling. It should be crystal clear and just slightly aromatic. The addition of .5 per cent caritol is a wise one, especially so because vitamin A is very much concerned with the health of the eyes. Caritol is a .3 per cent solution of carotene in oil.

Plucking Cream

It is useful to apply an anæsthetic cream on those parts of the eyebrow to be plucked. The instructions are to rub the preparation in well and leave on for half an hour previous to plucking. For manufacturing such a preparation use a cold cream base and 1 per cent ethyl amino benzoate. The resulting cream will also facilitate the plucking of undesirable hairs.

For Brittle Finger Nails

The following preparation is recommended for brittle finger nails:

Almond Oil	25 parts
Soft Paraffin	20 parts
Water	35 parts
Glycerin	5 parts
Stearic Acid	5 parts
Triethanolamine	4 parts

This is applied at night and allowed to dry. In the morning the nails are rubbed (polished) with the following powder: Tin oxide 7, talc 2, zinc oxide 1.

Nail Bleach

Formula No. 1

Glycerin	420 c.c.
Hydrogen Peroxide (20 vol.)	2000 c.c.
Rose Water	1500 c.c.

Formula No. 2

Hydrogen Peroxide (20 vol.)	130	parts
Water	68	parts
Ammonium Hydroxide	0.1	part
Alcohol	.1	part
Perfume Compound	0.5	part

Amor Nail Polishing Paste

Ceresin	6 parts
Olein	44 parts
Precipitated Chalk	50 parts
Color and perfume to suit.	

Nail Polish and Powder

Finest Powdered Silica	800 g.
Talcum (Extra Fine)	180 g.
Starch Rice	70 g.

The powder is tinted with a solution of eosin and perfumed suitably rose or muguet.

Nail Polish Powder

Formula No. 1

Stannic Oxide	1600 g.
Talcum	420 g.
Zinc Oxide	210 g.
Bengal Red	Trace
Perfume	Sufficient

Formula No. 2

Pumice Powder	40%
Talc	15%
Stannic Oxide	45%

Formula No. 3

Titanium Dioxide	65%
Talc	10%
Pumice Powder	25%

Nail Polish Liquid

Nail polish liquids are essentially of the same composition, plus water and glycerin. The abrasive is kept uniformly suspended in the liquid by a colloidal agent such as china clay. The following is a typical formula:

Stannic Oxide	450 g.
Talc	450 g.
Glycerin	75 c.c.
Colloidal China Clay	150 g.
Gum Tragacanth or Gum Arabic	3 g.
Water	1000 c.c.
Perfume	Sufficient

Nail Polish Stick

The stick or pencil type polish makes up in convenience what it lacks in efficiency. It is prepared by adding to a gum mucilage (such as tragacanth) an abrasive mixture of 90 per cent stannic oxide, 5 per cent pumice powder, 5 parts zinc oxide, dyed carmine. The mixture is moulded into pencils of the required shape and size, which are then wrapped in tinfoil to prevent excessive drying and cracking. A proportion of magnesium carbonate can also be incorporated in the formula to absorb the gum. If this method is not suitable a little zinc oleate or stearate might be used, or, alternatively, a mixture of fat and wax. Glycerin is another possible binding agent, but the tendency of this product to sweat should always be carefully watched.

Finger Nail Polish

Part No. 1 (Solids)

Cellulose Nitrate (½ sec.) on Dry Basis (by weight)	65%
Dibutyl Phthalate (by weight)	15%
Ester Gum (Low Acid Number and Pale) (by weight)	20%

Part No. 2 (Solvents)

Acetone (by volume)	15%
Methanol	15%
Benzol	25%
Ethyl Acetate	30%
Butanol	5%
Butyl Acetate	10%

The finished polish contains 15% of Part No. 1 to 85% of Part No. 2.

The solvents are mixed and the plasticizer added. If the cellulose nitrate is purchased submerged in water, it is carefully dried at low temperatures and added to the above mixture. The ester gum is powdered by beating or in mortar with pestle, or in a coffee grinder or similar mill. This will hasten its passing into solution. If the cellulose nitrate is purchased already in solution, the solvents it contains, will have to be deducted from the above formula.

Finger Nail Polish Remover

A low priced solvent mixture which is more effective than acetone alone, and useful on a wider variety of polishes is the following, which is stated in parts by volume:

Acetone	30	%
Methanol	24	%
Benzol	40	%

Ethyl Lactate	5.5%
Methyl Salicylate	.5%

If another odor is desired, it may be substituted for the methyl salicylate. It may also be given a very faint color by the use of either oil or spirit (alcohol) soluble dyes.

Nail Enamels

A typical nail enamel may be made from the following formula:

Stock Solution

½ Second Cellulose Nitrate (Damped with Butyl Alcohol)	20 lb.
Amyl Acetate	40 lb.
Ethyl Acetate	40 lb.

Take 10 pounds of the above and add to the following:

Safranine (0.5% Solution in Alcohol)	1 lb.
Ethyl Acetate	10 lb.
Castor Oil	1 lb.

The nitrocellulose is dissolved in the solvents by shaking, and allowed to stand until bright, when the other ingredients are added. The tinting is very important, and dyes are specially prepared by various makers for this purpose—safranine and carmoisine, for example. Castor oil serves as a plasticizing agent and prevents too rapid drying (which gives a streaky finish) and improves the adhesion of the enamel. Butyl alcohol may be added if the enamel is poor in adhesion and flakes off quickly.

Nail Enamel Removers

The remover is a simply formulated article, consisting solely of a mixture of good nitrocellulose solvents:

Amyl Acetate	20 lb.
Acetone	60 lb.
Ethyl Acetate	20 lb.

This is perfumed as desired.

Pencil for Whitening Surface Under the Nails

British Patent 383,965

A manicuring implement for cleaning and whitening or tinting the underside of finger nails comprises a stick of a compressed composition containing little or no waxy component and consisting of two parts lithopone, zinc sulphide, zinc oxide, to one of filler, such as starch, chalk, kaolin, talcum, and not less than 8% of water-absorbed binder such as gum tragacanth or arabic, Irish moss, casein, or dextrin.

Cuticle Skin Cream

Petroleum Jelly (Pale Yellow)	1½ oz.
Deodorized Coconut Oil	1 oz.
Hard Paraffin	1 dram
Stearic Acid	2 drams
Lanolin, Hydrous	1 dram
Water	2 drams
Borax	5 grains

Cuticle Remover

Caustic Potash	1 part
Glycerin	10 parts
Water	40 parts
Perfume	Sufficient

Cuticle Remover

British Patent 394,949

Preparations for removing or loosening cuticle, which contain a tissue-disintegrating substance such as potassium hydroxide, are made with soap. Using glycerin 74.56 grams, stearic acid 8.38 grams, caustic soda 1.18 grams, potassium hydroxide 3 grams, and water 12.88 grams, the acid is heated with the soda in the presence of the glycerin to which the water has been added; the heating is carried up to, say, 93° C., with stirring, and until effervescence occurs. The potassium hydroxide is added to the mass after it has been cooled.

Cuticle Remover

Distilled Water	178 lb.
Caustic Potash C.P.	4 lb.
Glycerin	10 lb.
Perfume Compound	Sufficient

White Cosmetic Which Turns Pink on the Skin

This is a white, greasy cream or ointment, which in contact with the skin turns pink.

Almond Oil	180 parts
Spermaceti	30 parts
White Wax	30 parts
Water	50 parts
Alloxan	5 parts

This is prepared in a similar manner to cold cream; the alloxan is dissolved in alcohol and added last. The product should be stored in well-closed containers.

"Bear Grease" Cosmetic
Formula No. 1

Beef Marrow	4	parts
Veal Suet	2	parts
Preservative	.1	part

Color and perfume as desired.

Formula No. 2 (Stick Form)

Beef Marrow	4	parts
Veal Suet	2	parts
Ceresine	2	parts
Preservative	.1	part

Shaving Cream
U. S. Patent 1,940,026

Formula No. 1

Soap (or Potash Soap)	60 parts
Alginate of Soda (7½% Solution)	20 parts
Talc (Soapstone)	20 parts
Water	20 parts

Formula No. 2

Soap (or Potash Soap)	60 parts
Alginate of Soda (7½% Solution)	15 parts
Talc Soapstone	30 parts
Water	26 parts

Shaving Cream

An excellent shaving cream can be made by the following formula: 15 parts by weight of stearin, 5 parts arachis oil, 7 parts coconut oil, 16 parts 28° Bé. caustic potash and 16 parts water. To use fatty acids instead, take 15 parts stearic acid, 8 parts lauric acid, 16 parts 50° Bé. potash solution, 50 parts water, and 1 part perfume. The potash is dissolved in hot water and the fatty acid mixture poured in, with stirring. A creamy product is formed which, after a short time, exhibits the silvery sheen of stearin crystals.

Shaving Creams

The following formulas illustrate creams of various consistency employing different ingredients. Consistency can be adjusted to suit requirements by increasing or decreasing the percentage of water.

Formula No. 1

Stearic Acid T. P.	33.6%
Coconut Oil (cochin)	6.4%
Glycerin	4. %
Sodium Lauryl Sulphate	3. %
Boric Acid	1.4%
Potassium Hydroxide (42°Bé.)	18.4%
Sodium Hydroxide (42°Bé.)	2.8%
Water	29.9%
Perfume	.5%

Melt one-half of the stearic acid and the coconut oil in a kettle. Put the water, boric acid and sulphate glycerin into another kettle and heat until the boric acid dissolves. Add to this solution the potassium hydroxide and sodium hydroxide and mix. Run the melted stearic acid into a steam jacketed mixer, and then add the hot alkali solution with the agitator in motion. Keep the steam on for a half hour while mixing. Meanwhile, in a separate kettle, have the remainder of the stearic acid melted and now run it slowly into the soap. Continue the stirring, not too rapidly, until the soap is smooth and homogeneous. Then add the perfume and mix it in well. Dump the batch and age it for three days.

If the cream appears spongy this is due to the incorporation of air owing to excessive rapid mixing or the use of improperly shaped paddles. Flat paddle mixers, whether of the epicyclic; the pony or interacting type, are best. If air has been incorporated it can be removed by running the cream through an ointment mill. In large production the cream is sometimes run over chilling drums.

Formula No. 2

Stearic Acid T.P.	28.0%
Cetyl Alcohol	2. %
Coconut Oil	6. %
Potassium Hydroxide 50°Bé.)	18.8%
Sodium Hydroxide (20° Tw.)	1.6%
Glycerin	3. %
Water	38.5%
Boric Acid	1.6%
Perfume	.5%

Melt one-half of the stearic acid and coconut oil in one kettle. In another kettle melt the remainder of the stearic acid and the cetyl alcohol. In a third kettle heat the water; dissolve the boric acid in it, and add the glycerin. Then to this solution add the potassium and sodium hydroxides. Strain the melted stearic acid and coconut oil into a steam packeted mixer. Start the agitator and run in the alkali solution. Mix for a half hour with the heat on and then add the stearic acid, cetyl alcohol mixture very slowly. Continue the mixing until the soap is smooth; add the perfume and mix it in well. Dump the batch in portable tanks and allow to age for three days.

Formula No. 3

Palm-Kernel Oil	5.72%
Stearic Acid T.P.	25.50%
Coconut Oil	6.36%
Potassium Hydroxide (85%)	8.57%
Sodium Hydroxide (85%)	.63%
Glycerin	6.54%
Boric Acid	1.59%
Water	44.00%
Bay Oil	.87%
Menthol Crystals	.22%

Put one-half of the stearic acid, all of the coconut and palm-kernel oils into one kettle. Melt the remainder of the stearic acid in another. Put the water, boric acid and glycerin into another kettle and heat until the boric acid dissolves. Then add the alkalies and stir with a wooden paddle until completely dissolved. Run the mixture of stearic acid, palm-kernel and coconut oils into a hot steam jacketed mixer. Start the agitator and add the hot alkali solution. Mix for a half hour with the steam on. Shut the steam off and slowly add the remainder of the melted stearic acid from the other kettle. Mix until a smooth, homogeneous cream is obtained. Then add the menthol dissolved in the bay oil. Allow the cream to stand for three days and mill if desired.

Formula No. 4

Olive Oil	2.0%
Stearic Acid T.P.	25.0%
Coconut Oil (Cochin)	8.0%
Potassium Hydroxide (42° Bé.)	18.0%
Sodium Hydroxide (42° Bé.)	2.5%
Water	39.0%
Boric Acid	1.5%
Glycerin	3.5%
Perfume	.5%

Melt one-half of the stearic acid and coconut oil in one kettle. Melt the remainder of the stearic acid and olive oil in another. Heat the water; add the boric acid and glycerin and when dissolved add the potassium and sodium hydroxides. Strain the stearic acid, coconut oil mixture into a mixing kettle; start the agitator and run in the alkali solution. Keep the heat on for one-half hour and then add the stearic acid, olive oil mixture. Mix in the perfume; age for three days.

Formula No. 5

Stearic Acid T.P.	25.00%
Coconut Oil (Cochin)	5.13%
Sodium Cholate	.12%
Lecithin	1.00%
Potassium Hydroxide (50° Bé.)	21.15%
Water	42.50%
Glycerin	4.60%
Perfume	.50%

Melt one-half of the stearic acid and coconut oil in one kettle and the remainder of the stearic acid and lecithin in another. Dissolve the sodium cholate in the water and glycerin; then proceed with the mixing of the soap as before.

Formula No. 6

Stearic Acid T.P.	28.50%
Cetyl Alcohol	4.00%
Sodium Cholate	.20%
Coconut Oil (Cochin)	12.80%
Potassium Hydroxide (85%)	9.57%
Sodium Hydroxide (85%)	.63%
Boric Acid	1.50%
Glycerin	5.00%
Water	37.30%
Perfume	.50%

Melt one-half of the stearic acid and the coconut oil in one kettle, and the remainder of the stearic acid and cetyl alcohol in another. Heat the water; dissolve in it the boric acid; and add the glycerin and sodium cholate. Then add the alkalies and proceed as before.

Brushless Shaving Cream
Formula No. 1

Stearic Acid	20.00%
Cetyl Alcohol	1.50%
Mineral Oil	2.00%
Ethylene Glycol	1.50%
Triethanolamine	1.65%
Borax	1.85%
Water	71.00%
Perfume	0.50%

Melt the stearic acid, add the cetyl alcohol and mineral oil, bringing the temperature to about 70°C. Put the borax and triethanolamine into the water and bring to a boil. Then add the melted fats with rapid agitation. When the temperature drops to about 40°C., add the perfume mixed with the ethylene glycol.

Formula No. 2

Stearic Acid (Triple Pressed)	14.0%
Cetyl Alcohol	2.0%
Potassium Hydroxide (Sticks)	1.0%
Glycerin	5.0%
Water	77.5%
Perfume	0.5%

Melt the stearic acid, cetyl alcohol and mineral oil. Dissolve the potassium hydroxide in the water; add the glycerin and stir in the melted fats. Mix until a smooth white cream is formed and when the temperature drops to about 40°C., add the perfume. It will not evaporate at that heat.

Brushless Shaving Cream

Formula No. 1

Stearic Acid Triple	21.00 lb.
Lanolin, Anhydrous	3.50 lb.
White Mineral Oil	.80 lb.
Alcohol	1.50 lb.
Triethanolamine	1.60 lb.
Borax	1.51 lb.
Distilled Water	70.00 lb.

Melt the stearic acid together with the oil, add the lanolin and heat to 70°C. Dissolve borax and triethanolamine in water and heat to boiling point. Add the fat solution while agitating rapidly. When smooth add perfume dissolved in alcohol. Stir at slow speed until cold. Consistency of cream can be regulated by changing quantity of water.

Formula No. 2

Stearic Acid	15.00%
Spermaceti	2.00%
Sulphonated Castor Oil	4.00%
Ammonia (26°)	6.75%
Sodium Hydroxide (Sticks)	.25%
Ethyl Ether of Diethylene Glycol	7.00%
Water	64.50%
Perfume	.50%

Mix the ammonia with the water; add the sodium hydroxide, and ethylene glycol and when it dissolves bring the temperature of the solution to about 70° C. Melt the spermaceti and stearic acid and add the sulphonated castor oil. Run the melted fats into the alkali solution and mix rapidly until emulsified. When the temperature drops to 40° C. add the perfume.

Formula No. 3

Stearic Acid T.P.	19.00%
Cocoa Butter	2.00%
Potassium Hydroxide (Sticks)	1.25%
Glycerin	7.00%
Alcohol	5.00%
Gum Tragacanth Powder	1.50%
Water	63.75%
Perfume	.50%

Mix the gum tragacanth with the alcohol, add a part of the water. Dissolve the potassium hydroxide in the remainder of the water, add the glycerin. Melt the stearic acid and coca butter. Heat the alkali solution to the temperature of the melted fats acid and add them very rapidly. When emulsified add the gum tragacanth. Mix thoroughly and when the temperature drops to about 40° C. add the perfume.

Formula No. 4

Glyceryl Monostearate	6.5%
Stearic Acid T.P.	6.5%
Mineral Oil	2.0%
Sulphonated Olive Oil	6.0%
Glycerin	10.0%
Potassium Hydroxide (Sticks)	.2%
Water	68.3%
Perfume	.5%

Dissolve the potassium hydroxide in water, then add all the ingredients with the exception of the perfume, and with continuous mixing heat until the mass becomes homogeneous. Continue the mixing until the temperature drops to 40° C., then add the perfume. Adjust consistency with water.

Formula No. 5

Stearic Acid	20.00%
Lanolin	1.50%
Mineral Oil	2.00%
Ethylene Glycol	1.50%
Triethanolamine	1.65%
Borax	1.85%
Water	71.00%
Perfume	.50%

Melt the stearic acid; add the lanolin and mineral oil, bringing the temperature to about 70° C. Put the borax and triethanolamine into the water and bring it to a boil; then add the melted fats with rapid agitation. When the temperature drops to about 40° C. add the perfume, mixing with the ethylene glycol.

Formula No. 6

Stearic Acid T.P.	20.0%
Anhydrous Lanolin	1.8%
Peanut Oil	3.2%
Triethanolamine	1.7%
Borax	1.9%
Water	70.9%
Perfume	.5%

Melt the stearic acid; add the lanolin and the peanut oil, bringing the temperature to 70° C. Put the triethanolamine and borax into the water; bring to the boiling point and add the melted fats. Mix rapidly until smooth, white cream results. Add the perfume at 40° C.

Formula No. 7

Stearic Acid T.P.	20.0%
Hydrogenated Cotton Seed Oil	4.5%
Ethylene Glycol	1.5%
Triethanolamine	1.5%

Alcohol	1.0%
Water	71.0%
Perfume	.5%

Melt the stearic acid; add the hydrogenated cotton seed oil and the ethylene glycol. Put the borax and triethanolamine into the water; bring to the boiling point and add the melted fats with vigorous stirring. Continue stirring until the temperature drops to 40° C. and add the perfume.

Formula No. 8

Stearic Acid	10.00%
Sulphonated Olive Oil	15.00%
Sodium Lauryl Sulfonate	1.00%
Lecithin	2.00%
Triethanolamine	1.25%
Water	65.25%
Glycerin	5.00%
Perfume	.50%

Melt the stearic acid, add the sodium lauryl sulfonate and the lecithin and then the sulphonated oil. Dissolve the triethanolamine in the water. Add the glycerin and bring to a temperature of 60 or 70° C. Add the hot oil and stir rapidly until cool. Perfume.

After Shave Lotions
Formula No. 1

Glycerol	50 cc.
Alcohol	650 cc.
Water	300 cc.
Menthol	5 g.
Salicylic Acid	1 g.
Ferric Chloride (Dilute Solution)	1 drop

Formula No. 2

Glycerol	100 cc.
Alcohol	500 cc.
Water	400 cc.
Menthol	2 g.
Salicylic Acid	1 g.
Ferric Chloride (Dilute Solution)	1 drop

Formula No. 3

Glycerol	200 cc.
Alcohol	400 cc.
Water	400 cc.
Menthol	1 g.
Salicylic Acid	1 g.
Ferric Chloride (Dilute Solution)	1 drop

For most skins, Formula No. 1 is desirable; it dries quickly and has a pleasant after glow. Formula No. 3 gives an oily product which should be wiped off with a towel. The intensity of the color can be varied by the amount of ferric chloride used.

Dental Cream
U. S. Patent 1,943,856
(Finest Grade)

Bentonite	50 parts
Magnesium Oxide	50 parts
Sodium Fluoride	1 part
Sodium Phosphate	5 parts
Soap Powder	5 parts
Oil of Eucalyptus	1 part
Gum Camphor	1 part
Tincture of Benzoin	5 parts

Water in sufficient quantity to give a paste of the right consistency.

The sodium fluoride, sodium phosphate, and soap powder may be eliminated, if desired. In making this preparation, the bentonite, magnesium oxide and soap powder are thoroughly mixed together. The eucalyptol, camphor and tincture of benzoin are thoroughly dissolved in each other. The sodium fluoride is dissolved in the water, and the three mixtures are then combined into one and worked into a paste of suitable consistency. Water may be omitted, using practically the same formula to produce a dry dental preparation.

Dental Cream

Precipitated Chalk	100 g.
Sugar	10 g.
Magnesium Carbonate	50 g.
Glycerin	10 g.
White Soap (Powdered)	50 g.
Oil of Peppermint	5 g.
Carmine	0.3 g.

Dental Cream

Powdered Soap	2000 parts
Precipitated Chalk	800 parts
Glycerin	Sufficient
Oil of Peppermint	10 parts
Alcohol	100 parts

The soap and chalk are mixed with a sufficient quantity of glycerin to make a paste of the desired consistency and then the oil of peppermint, dissolved in the alcohol, is added.

Dental Cream

Magnesium Hydroxide	3.700 parts
Glycerin	31.300 parts
Corn Starch	4.700 parts
Saccharine (Soluble)	0.015 part
Calcium Carbonate (Precipitated)	46.900 parts
Soap	2.300 parts
Flavoring	0.700 part
Water	10.440 parts

Dental Cream

Sodium Carbonate	3.7 parts
Sodium Bicarbonate	45.8 parts
Glycerin	32.1 parts
Soap	4.3 parts
Water	12.7 parts
Flavoring	1.4 parts

Milk of Magnesia Dental Cream

Milk of Magnesia	23 parts
Precipitated Chalk	42 parts
Castile Soap (Powd.)	1 part
Glycerin	21 parts
Gum Tragacanth	0.5 part
Water	12 parts

Ribbon Dental Cream

Glycerin	28.0 parts
Soap, Powdered	5.0 parts
Calcium Carbonate	35.7 parts
Precipitated Chalk	7.6 parts
Sodium Benzoate	2.1 parts
Flavoring Oils	0.9 part
Corn Starch	6.5 parts
Water	14.2 parts

Tooth Paste

Calcium Carbonate, Precipitated	32.00 parts
Magnesium Carbonate Precipitated	3.50 parts
Soap, Powdered	5.60 parts
Glycerin	30.00 parts
Water	26.60 parts
Saccharine	0.06 part
Flavoring	1.80 parts
Gum Karaya	0.30 part
Irish Moss	0.30 part
Color	Sufficient

Tooth Paste

Glycerin	25.5 parts
Calcium Carbonate Precipitated	30.1 parts
Neutral Soap Powd.	1.2 parts
Magnesium Hydroxide	4.9 parts
Gum Tragacanth Powd.	0.9 part
Flavoring	1.7 parts
Saccharine	0.07 part
Water, to make	100 parts

Tooth Paste

Precipitated Calcium Carbonate	55.00 parts
Soap, Powdered	6.00 parts
Glycerin	34.00 parts
Petrolatum	1.25 parts
Saccharine	0.25 part
Oil of Peppermint	0.10 part
Water	2.26 parts
Potassium Iodide	0.24 part

Tooth Powders

Dentifrice powders consist of an abrasive, a soap and an odor. Abrasives may be chalk (either precipitated or prepared); activated carbon; calcium tri-phosphate; calcium di-phosphate; calcium oxide; magnesium carbonate; magnesium oxide; sodium bicarbonate; borax; sodium perborate. Some makers have used calcium suphate but this material delivers such a coarse gritty product that its use is not advisable. All powders should be of the finest possible grain.

Characteristic formula:

Calcium Tri-Phosphate	90 lb.
Magnesium Carbonate	3 to 5 lb.
Soap (Fine Powder)	2 to 4 lb.
Saponine	1 lb.
Odor—flavor	1 to 1½ lb.

Tooth Powder

Calcium Carbonate Precipitated	60.0 parts
Magnesium Carbonate	1.0 part
Magnesium Oxide	2.0 parts
Sodium Bicarbonate	30.0 parts
Castile Soap (Powd.)	6.0 parts
Sodium Chloride (Powd.)	5.0 parts
Saccharine	0.2 part
Oil Wintergreen	1.0 part
Oil Peppermint	0.4 part

Tooth Powder

Salt	59.5 parts
Sodium Bicarbonate	19.8 parts
Magnesium Carbonate	4.9 parts
Sodium Perborate	14.9 parts
Oil of Cloves	0.3 part
Methyl Salicylate	0.6 part

Denture Powder

Cuttlefish Bone	1 oz.
Pumice (Fine Powder)	2 oz.
Borax	2 oz.
Precipitated Chalk	16 oz.
Thymol	10 grains
Oil of Peppermint	10 drops
Oil of Aniseed	10 drops

Salt Dentifrice

Salt	25%
Calcium Carbonate	33%
Orris Powder	6%
Soap	3%
Glycerin	33%

Dentrifice

Glycerin	24.50 parts
Dicalcium Phosphate (400 mesh)	54.00 parts

Galactonic Acid Lactone	3.00 parts
Water	10.00 parts
Alcohol	4.00 parts
Petrolatum	1.00 part
Gum Tragacanth	0.75 part
Gum Karaya	0.75 part
Malic Acid	0.25 part
Salt	0.60 part
Saccharine	0.07 part
Menthol	0.05 part
Oil of Peppermint	0.75 part

Tooth Paste

Soap	33.00 g.
Precipitated Chalk	25.00 g.
Absolute Alcohol	20.00 g.
Glycerin	15.00 g.
Benzoic Acid	3.00 g.
Oil of Eucalyptus	2.00 g.
Oil of Peppermint	2.00 g.
Saccharin	0.50 g.
Thymol	0.25 g.

Tooth Paste

Sodium Bicarbonate	11.0 g.
Borax	4.0 g.
Powdered Soap	14.0 g.
Precipitated Chalk	4.0 g.
Magnesium Carbonate	10.0 g.
Powdered Cuttlefish Bone	10.0 g.
Glycerin	50.0 g.
Menthol	0.4 g.
Oil of Anise	2.0 g.
Oil of Peppermint	3.0 g.
Oil of Cinnamon	2.0 g.
Carmine	0.4 g.

Dentifrice Massing Fluid

Merely add a massing fluid to tooth powder.

Dentifrice Massing Fluid
Formula No. 1

Water	4
Gelatin	1 to 2
Glycerin	7

Dissolve gelatin in water by heat and add glycerin.

Formula No. 2

| Glycerin | 1 part |
| Mucilage of Acacia | 1 part |

Dental Paste

| Tooth Powder | 600 parts |
| Massing Fluid | 300 to 400 parts |

Cleaning Agents for Dentifrices

These consist of sulfocyanides soluble in water provided with a protective coating to make them inactive as far as hygroscopicity is concerned. Coatings used are gum tragacanth, gelatin, albumen, dextrin and the like. Sometimes they are hardened by treatment with tannic acid or formaldehyde. Chalk and other suitable bases are used in making the tooth paste. The addition of carbonates is desirable since carbon dioxide is evolved on reaction with the sulfocyanic acid, producing rapid distribution of the dentifrice in the case of mouth washes and gargles. An example of a dentifrices made with sulfocyanides involves the use of 450 grams of saponin, 225 grams of calcium sulfocyanide and 111.3 grams of saccharin dissolved in one liter of water. The solution is agitated in a mixing machine. Then 12.9 kilograms of anhydrous gypsum are added gradually in small proportions and the mixing is continued until powder is obtained, which is only slightly moist and non-dusting. The mixture is then dried at a temperature of 70 to 80° C. cium sulfocyanide is perfectly prote. in the powder by the gypsum and the saponine from coming in contact moisture. Another powder is made from 525 parts of tartaric acid intimately mixed with 900 parts of gum tragacanth. About 100 to 200 parts of ethyl alcohol are added and worked into the mixture until a crumbly, slightly moist mass is obtained, which is then dried at a moderate temperature. The two powders are then mixed together to give the finished dentifrice.

EMULSIONS

Causes of Separation

Many manufacturers are completely at sea when a preparation separates. The causes of separation are not so numerous, and some of them of course are due to physical reactions about which little is known. Among the causes encountered are:

Decomposition: This usually occurs when mucilages, albumen and animal products are used as emulsifying agents, and is due to the lack or an insufficiency of preservative.

Incompatibility: This occurs when the emulsion contains an ingredient which is incompatible with the emulsifying agent and causes it to precipitate. Borax in an emulsion containing gum acacia; mineral acids in emulsions containing albumen, etc. Another very common instance of incompatibility occurs in connection with an attempt to make an acid emulsion with a soap emulsifying agent. Alcohols and solutions containing electrolytes are likewise detrimental to emulsions. Similarly metallic compounds in solution react with soap emulsifiers and cause separations.

Insufficient dispersion: This is due to failure to reduce the oil globules to a very minute state. It is a deficiency caused by lack of adequate equipment. There are any number of emulsions that will separate on standing when mixed in the ordinary way but will stand up indefinitely when mixed rapidly, as for instance in a colloid mill or homogenizer.

Excessive dispersion: This is the reverse of the foregoing and is due to such fine dispersion that the properties of the emulsifying agent are destroyed. This condition frequently occurs in connection with preparations containing vegetable mucilages.

Electrolytes: The presence of salts especially those of calcium, magnesium, aluminum, copper, iron, zinc, bismuth mercury, etc., are likely to destroy emulsions, but in some instances very small quanties can be added *after* the emulsion has first been formed.

Temperature: Numerous emulsions will break if exposed to low temperatures; others will break if exposed to high temperatures. Emulsions containing soap are likely to be sensitive to cold; those containing animal products are likely to be sensitive to heat. Sometimes a temperature variation of twenty degrees one way or another will cause separation, as for instance, when emulsions are heated by attrition in going through a colloid mill, or by pressure in a homogenizer.

Insufficient viscosity: Many emulsions will separate if not sufficiently viscous. This can be remedied by the addition of an emulsifying agent that will bring the viscosity up.

Improper procedure: Some emulsions must be handled slowly; others must be handled rapidly. Some will require the making of a *primary* emulsion first which can be diluted after emulsification has been effected. Chemicals like phosphorus, sulphur, camphor, creosote, etc., cannot be added to the batch directly but if first dissolved in a fixed oil can be emulsified very readily.

Failure to distinguish types: Attempts to add water to water-in-oil type emulsions or oil to oil-in-water types will cause separation. Many manufacturers fail to distinguish between these two emulsion types. They assume that because the emulsion contains more oil than water they can add more oil to it, whereas an emulsion can consist of ninety per cent oil and ten per cent water and still be an oil-in-water emulsion.

Improper emulsifying agent: No one emulsifying agent is suitable for all emulsions. The manufacturer should know what emulsifying agents are best for his particular emulsion. Often a stable emulsion can be secured by changing its type, that is, from an oil-in-water to a water-in-oil, an effect which can be secured by changing the emulsifying agent. Some materials are better emulsifying agents than others.

Emulsion Formulas
U. S. Patent 1,930,853
Formula No. 1 (Montan Wax Emulsion)

Bleached Montan wax is fused together with an equal amount of a solid, wax-like polymerization product of ethylene oxide (softening point about 50° C.) obtained by allowing ethylene oxide to stand at about 10° below zero C. with about 1 per cent its weight of sodium oxide or by heating ethylene oxide with about 0.1 per cent of caustic soda solution to from 50° to 60° C., the melt then being heated in ten times its weight of water while stirring. An emulsion is obtained which after cooling becomes very viscous and which, if desired, may be further diluted with water while heating, without deflocculation of the wax, and may be employed as a polish. Instead of Montan wax, stearic acid, montanic acid, tallow, wool fat and similar substances may be emulsified in an analogous manner.

Formula No. 2 (Gum Mastic Emulsion)

Gum mastic is fused together with an equal weight of the ethylene oxide polymerization product specified in Formula No. 1 and the melt is then warmed with five times its weight of water while stirring. A highly dispersed stable emulsion is formed which may be employed, for example, for sizing paper.

The gum mastic may also be replaced by other resinous products as for example colophony and the like.

Formula No. 3 (Pyroxylin Emulsion)

100 parts of a lacquer obtained by dissolving in 60 parts of the acetate of ethylene glycol monoethylether with the addition of 10 parts of trinormal-butyl phosphate, 20 parts of a pyroxylin which is readily soluble in alcohol to give solutions of low viscosity and which has been moistened with 10 parts of butanol, are treated for about 48 hours at room temperature in a suitable emulsifying apparatus, as for example a ball mill, together with a solution of 15 parts of the ethylene oxide polymerization product specified in Formula No. 1 in 35 parts of water. A highly dispersed, stable emulsion is thus obtained, which may be employed for coating purposes.

Formula No. 4 (Castor Oil Emulsion)

1 part of the ethylene oxide polymerization product specified in Formula No. 1 is dissolved while heating in 10 parts of castor oil. The solution solidifies to a stiff paste when cooled and this paste yields a milky emulsion when treated with hot water without deflocculation of the oil. The preparation may be used as a softening agent, for example for the product prepared according to Formula No. 3.

Formula No. 5 (Tricresyl Phosphate Emulsion)

A solution of 1 part of the ethylene oxide polymerization product specified in Formula No. 1 in 5 parts of tricresyl phosphate solidifies to a pulp when cooled and this pulp yields a stable emulsion by treatment with water while heating. The preparation may be used as a softening agent, for example for the product prepared according to Formula No. 3.

Formula No. 6 (Lacquer Emulsion)

100 parts of a very viscous nitrocellulose lacquer, consisting for example of 20 parts of a collodion cotton the solutions of which show a low viscosity, 7.5 parts of benzyl butyl phthalate, 2.5 parts of resin obtained by condensation of methyl cyclohexanone with the aid of alkali, 2.5 parts of Lithol Fast scarlet and 68.5 parts of ethylene glycol monobutyl ether, is mixed with 10 parts of dimethyl cellulose and 10 parts of a polymerization product from propylene oxide which is soluble to a slight extent in water and has been obtained by allowing propylene oxide to stand at about 0° C. with about 1 per cent its weight of sodium oxide. The mixture is introduced into 200 parts of water while stirring and a very stable emulsion is obtained which may be employed for coating purposes.

Formula No. 7 (Bleached Montan Wax Emulsion)

1 part of bleached Montan wax is melted down together with 0.5 part of a wax-like polymerization product of ethylene oxide melting at 52° C. After cooling the melt to 60° C., about 10 times its weight of ethyl alcohol are stirred into it, whereby a milky dispersion is obtained which, on cooling, solidifies to a pulpy paste.

Formula No. 8 (Mixed Wax Emulsion)

5 parts of a solid wax-like polymerization product of ethylene oxide having a softening point of about 54° C. are fused with 4 parts of ozocerite, 6 parts of bleached Montan wax and 3 parts of a preparation consisting of 10 parts of Nigrosine Base (Schultz, Farbstofftabellen, 1923, No. 698), 10 parts of stearine and 10 parts of crude Montan wax. While stirring, 35 parts of water are introduced into the melt. A homogeneous paste free from granular matter is obtained which may find useful application as a boot polish.

Formula No. 9 (Wax Oil Emulsion)

5 parts of the polymerization product referred to in the foregoing example are fused together with 5 parts of Montan wax, 5 parts of carnauba wax and 2 parts of castor oil until a homogeneous mass is obtained which is then incorporated with

50 parts of water, while stirring. A semi-solid mass is obtained which may find useful application as a floor polish. Wood or linoleum polished with the preparation show an excellent gloss and a good resistance to water.

Formula No. 10 (Lanolin Wax Emulsion)

60 parts of a white, wax-like polymerization product of ethylene oxide, prepared by dissolving a crude polymerization product of ethylene oxide in benzene and precipitation with the aid of ethyl ether, are fused with 60 parts of purified wool grease. A solution of 20 parts of boric acid in 60 parts of glycerol and 20 parts of water is then slowly introduced into the melt. A stiff paste is obtained which may find useful application as a base for the production of salves or ointments owing to the physiological inactivity of the polymerization product.

Formula No. 11 (Emulsion Paint)

100 parts of green earth are made into a paste with from 50 to 100 parts of water and 1 part of a polymerization product of ethylene oxide having a melting point of 54° C. dissolved in water in the ratio of 1:10, and then 2.5 parts of a basic dyestuff, as for example New Fuchsine (Schultz, Farbstofftabellen 1923, vol. 1, No. 513), dissolved in 250 parts of water are added. A color lake is obtained which contrasted with the green earth lake prepared without the employment of the said polymerization product is distinguished by greater clarity and brightness of shade and also an improved fastness to light and can be employed for cheap paints such as lime distempers.

Formula No. 12 (Lake Color Emulsion)

The following substances are mixed while stirring in the order given: 20 parts of heavy spar, 10 parts of aluminium sulphate (10 per cent solution in water), 5 parts of calcined soda (10 per cent aqueous solution), 12.5 parts of barium chloride (10 per cent aqueous solution), and 5 parts of powdered Hansa green G (Color Index 1924, page 354). 0.5 part of a polymerization product of ethylene oxide having a melting point of 50° C. (10 per cent aqueous solution) is then added and the whole is precipitated with 6 parts of barium chloride and 0.5 part of aluminium sulphate (10 per cent aqueous solution). A color lake is obtained which contrasted with the color lake prepared without the employment of the said polymerization product gives purer and more brilliant shades when brushed on. The fastness to light is very good and the preparation may be employed for wall-paper paints.

Formula No. 13 (Lake Color Emulsion)

10 parts of aluminium sulphate (containing 18 per cent of Al_2O_3 and in the form of a 10 per cent aqueous solution), 5 parts of calcined soda (10 per cent aqueous solution), and 50 parts of water are mixed at 50° C., boiled for an hour, allowed to cool to 50° C. and then 3.9 parts of disodium phosphate (10 per cent aqueous solution) and 3 parts of calcined soda (10 per cent aqueous solution) are added at 50° C. Then 18 parts of a 20 per cent aqueous paste of Alizarine red (Schultz, Farbstofftabellen 1923, vol. 1, No. 778) which have been intimately mixed with 1.44 parts of a polymerization product of ethylene oxide having a melting point of 48° C. (10 per cent aqueous solution) and to which have been added 3.9 parts of Turkey red oil are added and the whole stirred for 1 hour. The whole is allowed to stand for about 12 hours and 550 parts of water are added, the whole is slowly heated up to the boiling point during the course of 6 hours and kept at the same temperature for 3 hours. Towards the end 0.4 part of aluminium sulphate (10 per cent aqueous solution), 0.1 part of calcined soda (10 per cent aqueous solution), and 20 parts of heavy spar are added to the boiling mixture and boiling continued for some time. A color lake is obtained the shade of which is equally as strong as that of color lakes obtained without the employment of the said polymerization product but containing about 15 to 30 per cent more of coloring matter.

The product may find useful application for light-resistant paints for wall-papers.

Formula No. 14 (Pigment Emulsion)

13.8 parts of paranitraniline are made into a paste with 10 parts of water and 7.5 parts of sodium nitrate and mixed with 50 parts of water. The whole is diluted with a further 250 parts of cold water and 40 parts of hydrochloric acid of 20° Baumé strength are allowed to flow in. The resulting diazo solution is allowed to flow slowly while stirring well into a solution which has been prepared from 15 parts of beta-naphthol dissolved in 50 parts of water, 13 parts of caustic soda of 40° Baumé strength and 10 parts of calcined soda diluted with 100 parts of water and which also contains 3.4 parts of a polymerization product of an alkylene oxide, as for example of ethylene oxide, having a melting point of 57° C. and is dissolved in from 1 to 10 times its weight of water. When the coupling is completed the whole is worked up in the usual manner. In order to prepare adulterated preparations, 400 parts of heavy spar for example may be added. The preparation of the pigment may also be carried out in the presence of one of the usual emulsifying agents, as for example Turkey red oil, or the sulphonic acids referred to above, such as propyl or butyl naphthalene sulphonic salts, resin soaps and the like, about 3 parts being usually employed. The resulting pigment which may be employed for oil paints has great strength of color and the shades of the adulterated products have an improved fastness to light.

Formula No. 15 (Pigment Emulsion)

50 parts of heavy spar, 100 parts of a 4% aqueous paste of aluminum hydrate and 10 parts of 25% aqueous paste of Autol red RLP (Schultz, Farbstofftabellen 1923, vol. 1, No. 106) are stirred together and a concentrated aqueous solution of from 0.25 to 1.25 parts of a polymerization product of ethylene oxide having a melting point of 54° C. are then added and the whole intimately mixed. A pigment suitable for printing wallpapers is obtained which yields prints having excellent covering power and beauty of shade.

Formula No. 16 (Color Lake Emulsion)

20 parts of heavy spar, 10 parts of aluminum sulphate (10% aqueous solution), 5 parts of calcined soda (10% aqueous solution) and 12 parts of barium chloride (10% aqueous solution) are mixed together. The precipitate is washed three times with water and then 4.5 parts of the tungsto-molybdic compound of Brilliant Wool blue FFR extra (Schultz, Farbstofftabellen 1923, vol. 2, page 24) in the form of its 1% aqueous solution, 0.45 to 1.8 part of an ethylene oxide polymerization product having a melting point of 52° C. (10% aqueous solution), 7 parts of barium chloride (10% aqueous solution) and 0.75 part of aluminum sulphate (10% aqueous solution) are added one after another. After separation by filtration the dye paste obtained is washed and worked up in the usual manner. A color lake is obtained the shade of which is as equally strong as that of color lakes obtained without the employment of the said polymerization product but containing about 15 to 30% more of coloring matter. Moreover, the brilliancy is in-

creased at the same time and the fastness to light is improved.

Formula No. 17 (Color Lake Emulsion)

12 parts of heavy spar are made into a paste with 100 parts of an aqueous 4% suspension of alumina hydrate and 15 parts of lead acetate dissolved in 150 parts of water, from 2 to 3 parts of a polymerization product of ethylene oxide having a melting point of about 54° C. dissolved in five times their quantity of water are added and a solution of 6.2 parts of sodium bichromate in 62 parts of water is then introduced. The precipitate is filtered off, washed and dried, the coloring material having a purer and more greenish yellow shade than products obtained without the aid of the polymerization product of ethylene oxide. The coloring material may be employed for oil or glue paints.

Formula No. 18 (Lake Color Emulsion)

100 parts of Lithol Rubin B in the form of powder (Schultz, Farbstofftabellen 1923, vol. 1, No. 152) are intimately mixed with from 1 to 5 parts of a polymerization product of ethylene oxide having a melting point of 54° C. and made into a paste with from 50 to 100 parts of water. The polymerization product may also be added, in solution, as for example as a 10% aqueous solution, and the water may be replaced by organic solvents, as for example wholly or partially by ethyl alcohol. Contrasted with pastes prepared without the addition of the said polymerization product, the said addition produces an increased wetting action so that the further working up is considerably facilitated. The paste may be employed for paints or in the manufacture of colored writing chalk.

Formula No. 19 (Lake Color Emulsion)

100 parts of heavy spar, 4 parts of Helio Fast red RL powder (Schultz, Farbstofftabellen 1923, vol. 1, No. 73), 5 parts of a polymerization product from ethylene oxide having a melting point of 57° C. are made into a paste with 100 parts of water. Contrasted with a mixture free from the said polymerization product, the addition of the said polymerization product renders it possible to wet the mixture about four times as quickly. Even in cases when mixtures are used which are brightened with from say 1 to 2 parts of mineral oil, petroleum jelly and the like, (as for example a mixture of 100 parts of heavy spar), 4 parts of Helio Fast red RL powder (Schultz, Farbstofftabellen 1923, vol. 1, No. 73), 1 to 2 parts of petroleum jelly, 0.2 to 0.5

part of a polymerization product from ethylene oxide having a melting point of 57° C. and from 2 to 5 parts of water prepared by intimately stirring, a more ready and intimate wetting is obtained by the said additions. Other substrata, such as mixtures of heavy spar and precipitated calcium sulphate or mixtures of gypsum and whiting and the like may be employed instead of heavy spar. The preparation may be employed in the manufacture of colored writing chalk.

Formula No. 20 (Bronze Powder Emulsion)

From 5 to 10 parts of a polymerization product of ethylene oxide having a melting point of 48° C. are added to 100 parts of a metal bronze, which has or has not been mordanted with alum for the removal of paraffin oil, and stirred with 200 parts of water.

Alternatively, 100 parts of a metal bronze, 66 parts of a 15% aqueous paste of the phosphotungsto-molybdic compound of methyl violet (Schultz, Farbstofftabellen 1923, vol. 1, No. 515) and from 200 to 300 parts of water are intimately stirred together.

In each case an increased wetting action combined with a good dispersion of the dyestuff is obtained.

Dehydration of Petroleum Emulsions

Petroleum emulsions are normally of the water-in-oil type, and consist of fine droplets of naturally occurring waters or brine dispersed in a more or less permanent state throughout the oil which constitutes the continuous phase of the emulsion. They are obtained from producing wells and from the bottom of oil storage tanks and are commonly referred to as "cut oil," "roily oil," "emulsified oil," and "bottom settlings or B. S."

Little difficulty is encountered with emulsions in the Pennsylvania oil fields, but in certain Mid-Continent, Gulf coast, and California fields a considerable portion of the crude oil production requires treatment to free the oil of emulsified water to meet pipe line and refinery specifications.

A condition favoring emulsification exists in any field that produces water with the crude oil, and the stability of the emulsion and degree of emulsification depends upon the nature of the crude oil, emulsifying substances present, and the mechanical agitation in the process of production, whether wells are flowing, produced by the air or gas-lift method, by swabbing, or by pumping.

Emulsified oil may be treated by any of several methods developed for the purpose. These methods include (1) mechanical separation; (2) dehydration by heat; (3) chemical dehydration; (4) electrical dehydration; and (5) a combination of two or more methods. The selection of the correct treating process for a particular emulsion can best be determined by its laboratory examination.

Many emulsions contain the water in a coarsely divided state and will settle out on standing, or on being slightly heated and allowed to stand. Others require special treatment due to the presence of emulsifying materials in the crude oil which tend to concentrate in the interface between the oil and the water, and which have the property of preventing the drops of water from coalescing. Chemical separation of emulsions is based upon the action of the chemical upon the emulsifying agent in the interface to modify the properties of the emulsifier and allow demulsification, and in many instances on the fact that emulsifiers of contrary types counteract each other. Thus, by adding an emulsifier of the contrary type, one can destroy or diminish the protective action of the emulsifier already in the interface, by means of which the emulsion can be made to coagulate. In other words, one might use an emulsifier which forms an oil-in-water type of emulsion to demulsify a water-in-oil emulsion. Since sodium oleate usually emulsifies oil-in-water and calcium oleate emulsifies water-in-oil, a mixture of the two oleates will behave differently, depending on the relative amounts present. Usually we get oil-in-water emulsions if the oriented molecule or particle of the emulsifying agent at the interface is chiefly in the water phase, and water-in-oil if the oriented molecule is chiefly in the oil phase.

To be effective the demulsifier must be able to reach the oil-water interface surrounding the dispersed droplets. This is usually accomplished by stirring, by adding the chemical at the well, or while the emulsion is being flowed through the pipes to the treating tank, and also by selecting materials which are miscible in both oil and water, or are rendered miscible by the use of proper solvents.

By the chemical treatment of emulsions the factors obstructing coagulation are largely eliminated, however, viscosity, difference in specific gravity of the water and oil, size of the dispersed droplets and the temperature of treating are important factors in the time required for demulsification. A high viscosity and a fine dispersion may require a very long period for demulsification. Sometimes coagulation can be accelerated by stirring, but the possibilities are limited, because such stirring may have the contrary effect upon emulsions of great viscosity, and the technique of stirring is sometimes very important. Centrifuging is usually beneficial in the case of emulsions of low viscosity and having an appreciable difference in the specific gravities of the oil and water.

Certain important factors that affect the recovery of oil from emulsified crude are (1) base of crude, (2) gravity, (3) viscosity, (4) asphaltic content or carbon residue, and (5) size of dispersed water droplets.

When a treating problem is encountered, much preliminary work can be accomplished by a study of these factors. The base of the crude should be determined. A paraffin or mixed base crude should be classified further as to gravity and viscosity. Usually if the specific gravity is below 0.85, and the viscosity below 55 seconds Saybolt Universal at 100° F., the cut oil may be treated by steaming plants run below 125° F., by settling plants, by centrifuges operated at normal temperatures, or by chemical treatment at ordinary temperatures. If the specific gravity is above 0.85 and the viscosity above 55, similar methods of treatment may be used, but the treatment must be carried out at a much higher temperature. For example, the temperature in the steaming plant may need to be 150° F. or higher. For either oil the efficiency of treating will be directly affected by the degree to which the water particles have been dispersed. If the oil has a naphthene base, it should be examined with reference to its asphaltic content, which, if high, will usually necessitate high temperature dehydration or electrical dehydration. If the asphaltic content is low, as in Gulf coast crudes, dehydration may be effected by chemical demulsification, high temperature treatment, or electrical dehydration.

As in other crudes the size of the dispersed water particles will affect the efficiency of treating.

The chemical dehydration of petroleum emulsions has been the subject of numerous patents covering the use of a wide variety of chemicals. The Tret-O-Lite process and the various Tret-O-Lite chemicals are probably the most widely used in the field to-day. Modified fatty acids; sulfonated or sulfated derivatives of such materials as castor oil, oleic acid, Turkey red oils, rosin, sperm oil, etc.; sulfates of alcohols of high molecular weight, e.g., Gardinol; certain petroleum sulfonates; alkyl naphthalene sulfonates; and naphthenic acids are types of materials mentioned in patents. The material used has to be prepared according to certain specifications to possess the necessary properties required of effective demulsifying agents. The alkali or ammonium salts are usually employed, although triethanolamine salts may be used in some cases.

The following compositions are examples of materials now being used in the Mid-Continent oil fields:

	Compounds (Patented)				
	A	B	C	D	E
Modified fatty acids partially or totally saponified with ammonia	39.0	61.0	35.0	53.0	48.0
Oil-soluble sodium petroleum sulfonate	31.0	16.0	23.0	13.0	15.0
Water-soluble solvent, water, dilute alcool (10–40%), etc.	20.0	17.0	18.0	17.5	10.0
Oil-soluble solvent, kerosene, crude oil, solvent naphtha, cresol, etc.	8.5	4.0	22.5	15.5	25.0
Inorganic sulfate, sodium sulfate, sodium sulfite	1.5	2.0	1.5	1.0	2.0

The modified fatty acids are obtained by the hydrolysis of fatty acid sulfates derived from castor oil, oleic acid, or sperm oil, or mixtures of these materials, in the manner conventionally employed to produce Turkey red oils. The hydrolysis is carried out at about 60° C. The petroleum sulfonates are obtained in the manufacture of white oil, medicinal oil, certain lubricating fractions, etc., and have oil soluble properties.

The sodium salt of the petroleum sulfonate is mixed with the modified fatty acid material with cautious heating. To this mixture, while warm, is added the kerosene and the 10% alcohol solution containing the ammonium or sodium sulfate, with stirring until mixing is complete. After cooling to 30° C. or less strong ammonia or alkali is added to partially neutralize the modified fatty acids present and to give the desired properties of oil and water solubility. As the ammonia is added the mixture is frequently tested until water solubility is obtained. The material must still be soluble or miscible with oil.

After treatment, the emulsion is allowed to stand in a quiescent state, usually in a settling tank, and at a temperature varying from atmospheric temperature to about 200° F., so as to permit the water or brine to separate from the oil, it being preferable to keep the temperature low enough to prevent the volatilization of valuable constituents of the oil. If desired, the treated emulsion may be acted upon by one or more of the various kinds of apparatus now used in the operation of breaking petroleum emulsions, such as homogenizers, centrifuges, or electrical dehydrators. The amount of treating agent that may be required to break the emulsion may vary from approximately 1 part of treating agent to 500 parts of emulsion, up to 1 part of treating agent to 20,000 parts of emulsion. The approximate amount of treating agent is determined by laboratory tests.

Acamulsia (Emulsifier)

Powdered Acacia	5 parts
Powdered Tragacanth	5 parts
Sugar	5 parts
Starch	5 parts
Boric Acid	1 part

Mix intimately. Use 1 part of above powder to every 32 parts of emulsion to be made. Eight ounces of the oil to be emulsified is put into a dry 32-ounce bottle, and shaken with ½ ounce of acamulsia. When the powder is evenly suspended, 8 ounces of water are added at once, and the mixture is well shaken until a perfect emulsion is formed.

Emulsifier and Detergent
German Patent 547,895

10 kilograms of casein and 10 kilograms of citric acid are stirred up and heated with 100 liters of water, and the solution is then diluted with a further 200 liters of water, the undissolved parts being removed. Potash lye is then added hot until the concentration is pH 7.4, and the solution is evaporated over hydrogen. A homogeneous residue is thus obtained which is completely soluble in water, and is admirably adapted either as an emulsifying agent or detergent, either by itself or with fat solvents or neutral soap, especially for wool scouring.

Easily Dilutable Emulsions
U. S. Patent 1,873,580

Formula No. 1

2.5 parts of good bone glue are dissolved in 22 parts of water together with 1¼ parts of the product obtained by the alkylation and sulphonation of mineral oil and 5½ parts of urea. 100 parts of neutral tar oil of b.p. 280°–360° C. are emulsified in this by good stirring in a suitable machine or other emulsifying device.

Formula No. 2

21 parts of alkylated sulphonated mineral oil as used in Formula No. 1 are dissolved in a solution of 60 parts gelatin in 444 parts of water. To the resulting solution are added 60 parts of paraffin wax, and the mixture heated until the wax has melted when it is stirred, the wax rapidly becoming emulsified. 90 parts of urea are now added, and the product allowed to cool, stirring being continued.

The product is a thin cream which may be readily poured, and which dilutes easily with cold water, in contradistinction to the product obtained when the preparation is carried out without the addition of urea, etc. The viscosity of the product may be varied, within limits, by the amount of agent employed.

Bentonite ("Volclay") Oil Emulsions

In any emulsion, it is important that the oil be reduced to globules of the smallest possible size. Therefore, the more thorough and rapid the agitation, the more stable the emulsion will be. Simple emulsions like kerosene, water and bentonite can be made by shaking in a jar by hand, but more rapid agitation is recommended. An electric soda-fountain stirrer is excellent for laboratory experiments. A colloid mill produces the best results, but is not usually necessary. It is advantageous to have the ingredients warm, and necessary to do so in the case of heavy oils. The usual method is to make a dispersion of bentonite and water, warm it to 180° F. and then pour in the warm oil. Grease, paraffin, etc., must of course be heated to fluidity.

Disperse one part bentonite to fifteen parts water (by weight), mix thoroughly until a smooth and uniform gel results, and allow to stand six hours or longer. Warm this gel and pour the oil into it gradually while agitating or stirring, using 5 to 10 parts of the bentonite gel to 100 parts oil. Then add 100 parts of water during continued stirring or agitation.

The proportions may be varied greatly. Stable emulsions with light oils have been made between these ranges:

Bentonite	4.5%	Bentonite	1.34%
Water	62.2	Water	18.66
Oil	33.3	Oil	80.00

The emulsion may be thickened by the addition of various ingredients, which can be determined by empirical experiments. If it is desired to do so, try 3 to 5% calcium hydroxide, or smaller percentages of boric acid, liquid chlorine, sodium chloride, magnesium oxide, caustic calcined magnesite, calcium sulphate, sodium carbonate, sodium or ammonium hydroxide.

In some cases, 50 grams of soap per gallon of mixture is added, but it is seldom necessary. With grease, paraffin, stearate, etc., the procedure is practically the same except that slightly larger proportions of bentonite are required and the ingredients should be hot. A stable buttery emulsion can be obtained with 5 cubic centimeters of the bentonite dispersion mixed with 60 grams hot Montan wax and 40 cubic centimeters hot light oil, creosote for instance, and then with 100 cubic centimeters hot water. A similar emulsion is obtained substituting 40 grams of wax and 60 cubic centimeters of oil, or 15 grams of wax and 85 cubic centimeters of oil.

The stability of emulsion is effected by the pH, and also by chemical activities of other ingredients. Such activities are influenced by the degree of dilution or concentration. In some cases, the emulsion will recede evenly after standing, leaving a layer of clear water at the top which can be siphoned off, and the balance will be stable, due to a different degree of concentration.

Bentonite ("Volclay") Wax Emulsions

Make a 5% suspension of bentonite, that is, 95% water and 5% bentonite. The easiest way to do this is to put all the water to be used in a vessel and sprinkle a little of the bentonite on the surface. Allow to stand about three minutes or until the bentonite sinks to the bottom. Repeat. Don't stir or agitate. Allow to stand for at least four hours, then stir or agitate.

Heat the wax to fluidity. Heat the bentonite dispersion to the same temperature or slightly higher. Pour the wax into the water and stir. Stirring should be particularly vigorous at the point where the wax begins to solidify. A mechanical agitator is the best. In the laboratory use an electric soda-fountain stirrer.

The final mixture may be between 20% to 50% wax. However, it is best not to use over 40% wax, otherwise the emulsion will be too viscous.

The pH of bentonite in water is 7.5–7.8. In most emulsions a slightly higher pH is desirable, for instance, about 8.5. It would be well to adjust the water dispersion to the higher pH by very slight additions of alkali. We recommend calcium oxide, calcium hydroxide, caustic calcined magnesite, magnesium oxide.

In some cases it will be found that a thicker dispersion of bentonite and water is desirable. This can be accomplished by using 6% bentonite or by adding very small percentages of other ingredients to the 5% dispersion. These other ingredients should be added *after* the dispersion is made. They will combine the advantages of thickening the dispersion and at the same time raising the pH. Add minute quantities and stir, using just enough to thicken the gel.

In relation to the quantity of bentonite in the dispersion, proper proportions would be as follows:

Calcium Oxide	3%
Calcium Hydroxide	3%
Chloride of Lime	2%
Caustic Calcined Magnesite	½ to 1%
Magnesium Oxide	½ to 1%

For instance, with 95 cubic centimeters of water and 5 grams of bentonite, you would use 15/100 of a gram of calcium oxide, that is 150 milligrams. The magnesia compounds react more slowly. The other mentioned, reacts practically immediately.

Emulsifying Wax
U. S. Patent 1,932,643

Melt together 65 parts of spermaceti with 25 parts of cetyl alcohol and 10 parts of stearic acid. This product has a fusing point of about 50° C. and in the molten state is capable of combining with as much as 600% of its own weight of a 2% sodium carbonate solution.

Wax Emulsion Spray

Montan Wax	160 parts
Ammonium Linoleate	40 parts
Water	600 parts

Dissolve the ammonium linoleate in the water and heat to 95–100° C.; pour the wax into this slowly while stirring vigorously. While still hot run through a colloid mill. A stable emulsion that sprays easily results.

Defoaming and Defrothing Agents
U. S. Patent 1,947,725
Formula No. 1

5 parts of aluminium stearate is dissolved in 95 parts of pine oil, to give a substantially clear fluid solution.

Formula No. 2

5 parts of aluminium stearate is dissolved in a mixture of 78.75 parts of pine or eucalyptus oil and 26.25 parts of a mixture of higher alcohols obtainable as a by-product during the manufacture of methanol.

The addition of one per cent or less of either of the above to a solution of 10 parts of glue and 10 parts of dextrine in 100 parts of water completely prevents the development of foam or froth when the mixture is shaken.

Formula No. 3

5 parts by weight of aluminium stearate is dissolved in 95 parts of the fraction boiling from 195° C. and up of the liquid oxygen-containing organic compounds obtainable as a by-product in the known synthesis of methanol under pressure from hydrogen and oxide of carbon.

The addition of 1 per cent or less of the above solution to a solution consisting of 20% of animal glue in water completely prevents the development of foam or froth when the mixture is shaken.

Formula No. 4

5 parts of zinc stearate is dissolved in 95 parts of pine oil at 95–100° C.

Formula No. 5

3 parts of calcium oleate is dissolved in 97 parts of pine oil at 95–100° C.

The addition of 0.25% of either of the above two agents to a solution of 10 parts of glue and 10 parts of dextrine in 100 parts of water considerably reduces the amount of foam formed when the mixture is shaken as compared with the amount normally formed.

When 0.25% of pine oil alone is added, the amount of froth formed is also sensibly reduced but still remains about double that formed according to the above process.

Formula No. 6

5 parts of barium stearate is dissolved in 95 parts of oleic acid at 95–100° C.

The addition of $\frac{1}{4}$% of this agent to glue-dextrine solutions reduces the froth formed on shaking to half the amount formed when $\frac{1}{4}$% of oleic acid alone has been added.

Emulsifying Agent
German Patent 547,895

While most gelatinous substances, even those difficultly soluble in water, easily dissolve in alkaline media, yet this is not always the case in using neutral and acid solutions. In any case there are several compounds which, together with difficultly soluble gelatinous substances, can form neutral or acid solutions. Such, for example, are the salts of organic sulphoacids, together with a number of organic and inorganic compounds, which have a solvent action on gelatinous bodies. Such are organic acids, tartaric and citric acids, and the salts of organic acids, potassium butyrate, potassium acetate, sodium oxalate and others. Phosphites and hypophosphite may also be used. By way of example, 10 kilograms of casein and 10 kilograms of citric acid are stirred up and heated with 100 liters of water, and the solution is then diluted with a further 200 liters of water, the undissolved parts being removed. Potash lye is then added hot until the concentration is pH 7.4, and the solution is evaporated over hydrogen. A homogeneous residue is thus obtained which is completely soluble in water, and is admirably adapted either as an emulsifying agent or detergent agent, either by itself or with fat solvents or neutral soap, especially for wool scouring.

Emulsifying and "Wetting" Agent
German Patent 581,955

55 parts of lauryl amide are mixed with 50 parts of formaldehyde (30 per cent), 0.5 part of potassium hydroxide solution (50 per cent concentration) and 300 parts of water. The mixture is agitated for half an hour at a temperature of 60° C. It is agitated for another hour after the addition of 50 parts of phenol and 10 parts of concentrated solution of hydrochloric acid. The viscous oil, obtained from this treatment, is an excellent emulsifier. It is mixed with 200 parts of sulfuric acid (monohydrate) at a temperature of 60° to 70° C., and after the sulfonated mixture has been agitated for several hours, it is mixed with 200 parts of ice and slowly heated with agitation to a temperature of approximately 70° C. After agitation is stopped, the sulfonate formed quickly separates in the state of a viscous oil over an aqueous solution of sul-

furic acid. The latter is removed and the oily sulfonic acid is dissolved in water, neutralized and evaporated. A product is finally obtained which possesses high detergent powers both in hard and soft water and is resistant to the action of hot acids and alkalies.

Wetting and Dispersing Agents
U. S. Patent 1,951,469
Formula No. 1

325 kilograms of octadecenylamine are sulphonated at 25° C. with 400 kilograms of concentrated sulphuric acid. After the product has become water-soluble, it is pressed on to ice, washed with saturated Glauber's salt solution in order to remove the excess of sulphuric acid and finally neutralized with caustic soda solution.

Formula No. 2

380 kilograms of octadecanolamine hydrochloride are introduced into 200 kilograms of trichlorethylene and sulphonated at 10° C. with 120 kilograms of chlorsulphonic acid. The product is pressed on to ice, washed with saturated Glauber's salt solution in order to remove the excess of sulphuric acid and acetic acid and neutralized with caustic soda solution.

"Wetting-Out" Agents
U. S. Patent 1,947,951
Formula No. 1

100 parts by weight of olive oil and 100 parts by weight of diethylene-triamine are heated to about 180–200° C. until a test portion of the reaction mixture is soluble in dilute hydrochloric acid. After distilling off the excess of diethylene-triamine, advantageously under reduced pressure, there remains a strongly viscous mass, the hydrochloric acid solution of which has great foam forming properties.

A similar product is obtainable by heating free oleic acid with a large excess of diethylene-triamine under the same conditions.

Formula No. 2

310 parts by weight of the ethylester of oleic acid are heated with 286 parts by weight of triethylenetetramine at about 160° C. for about 12 hours when a homogeneous solution is formed; the alcohol formed and the excess of triethyl-enetetramine are distilled off. The residual reaction product forms a brown oil, which is difficultly soluble in water and readily soluble in alcohol, benezene and dilute hydrochloric acid.

A quite similar product is obtainable by heating free oleic acid with an excess of triethylenetetramine to about 180–200° C. and distilling off the excess of triethylenetetramine under reduced pressure.

Formula No. 3

200 parts by weight of olive oil are heated at 180–200° C. with 300 parts by weight of a mixture of bases, which is obtained by the action of ammonia on ethylene chloride at 80–120° C. under pressure of 10 atm. and after distilling off the ethylene diamine said mixture of bases boiling at about 15 mm mercury between 90 and 300° C. When a test portion of the reaction product is smoothly soluble in dilute hydrochloric acid the water formed and the excess bases are distilled off under reduced pressure and a yellowish brown oil is obtained, a solution of which in dilute hydrochloric acid can be used as a washing or wetting agent.

Formula No. 4

Equal parts by weight of stearic acid and triethylene-tetramine are heated at 200–210° C. (oil bath temperature) until a test of the reaction product is soluble in dilute hydrochloric acid. After distilling off the excess triethylenetetramine there remains a nearly colorless mass which solidifies after cooling, being soluble in alcohol and dilute hydrochloric acid. The hydrochloric acid solution shows pronounced foam forming properties.

Instead of stearic acid other higher fatty acids, for example linseed oil, ricinoleic acid, acids of wool fat and the like, may be used.

Formula No. 5

Equal parts of linseed oil and the mixture of bases (described in Formula No. 3) are heated for about 5 hours to 200–205° (oil bath temperature) and the excess free bases and the glycerine formed during the reaction are distilled off under reduced pressure. The remaining oil is easily soluble in dilute hydrochloric acid and shows good foaming and emulsifying properties.

Instead of linseed oil also other fats, for example, train oils, wool fat, beef fat and the like may be used.

Solubilizing Water-Soluble Materials
U. S. Patent 1,956,415
Formula No. 1

70 parts of a resin obtained by treating 63 parts of phenol with 17.2 parts of chloracetone at a raised temperature, are slowly introduced in small portions, while well stirring into the sulfonating mixture consisting of 106 parts of tetrahydronaphthalene and 144 parts of sulfuric acid of 98 per cent strength.

The product dissolves clearly in water not only in an acid state, but also when adding a caustic soda solution.

Formula No. 2

138 parts of commercial cresol are condensed to a resin with 51 parts of asymmetric dichlorether at a temperature between 90° C. and 100° C. This resin is dispersed in small portions at 100° C., while well stirring, in the sulfonating mixture prepared from 222 parts of tetrahydronaphthalene and 301 parts of sulfuric acid of 98 per cent strength.

With reference to its solubility in water, the product behaves in the same manner as that obtained in Formula No. 1.

Formula No. 3

50 parts of crude benzene are run, while stirring, at a temperature of 40° C. to 50° C., into 140 parts of sulfuric acid of 100 per cent strength and the whole is finally heated at 100° C. until the sulfonation is complete. After cooling the mass to 60° C. to 70° C., 26.4 parts of tetrahydronaphthalene are run into the sulfonation mixture while again stirring, whereupon the mass is again heated to 100° C. until the tetrahydronaphthalene has also become water-soluble.

93 parts of a liquid resin are run into the sulfonation mixture at about 90° C.; the resin being obtained from 2 mol. of phenol and 1 mol. of formaldehyde in the presence of an alkaline condensing agent by slowly heating to about 100° C. until the formaldehyde odor has disappeared and subsequently separating the excess of water by the addition of sodium chloride.

When the product which is clearly soluble in water is to be used as a tanning material, it is neutralized with 181 parts of caustic soda solution of 20 per cent strength. The leather tanned therewith becomes light colored, soft and full-bodied. In using the product for tanning purposes it shows a remarkable power of dissolving phlobaphene when mixed with a vegetable tanning substance as for instance "quebracho."

Formula No. 4

78 parts of a-chloronaphthalene are sulfonated at about 130° to 150° C. with 122 parts of fuming sulfuric acid of 5 per cent strength.

57 parts of the liquid phenolic resin referred to in Formula No. 3 are then run into the mixture thus obtained at a temperature between 95° C. and 97° C. while well stirring.

In case the product is to be used as a tanning agent, it is so much neutralized with caustic soda solution of 20 per cent strength that about 10 cc. of N caustic soda solution are required for completely neutralizing 10 g. of the final product.

Formula No. 5

70 parts of a liquid resin obtained by an alkaline condensation of 2 mol. of phenol and 1.2 mol. of formaldehyde and subsequent separation of the water present by the addition of sodium chloride are introduced, at a temperature of about 95° C., while well stirring, into 200 parts of the sulfonating mixture obtained from 81.5 parts of α-methylnaphthalene and 118.5 parts of sulfuric acid of 98 per cent strength.

When to be used as a tanning material, the product is so much neutralized with caustic soda soultion of 20 per cent strength, that 10 cc. of N caustic soda solution are required for completely neutralizing 10 g. of the final product. The leather tanned in this manner feels fine and soft and is of light color.

Formula No. 6

108 parts of commercial cresol are subjected, while gradually raising the temperature to 100° C., to a reaction with 50 parts of formaldehyde of 30 per cent strength in the presence of a solution of 30 parts of phosphoric acid in 200 parts of water which solution serves as a condensing agent.

105 parts of the fluid resin so obtained, after having separated therefrom the aqueous layer, are introduced, drop by drop, at 90° C. to 100° C., while well stirring, into 250 parts of the sulfonating mixture obtained as indicated in Formula No. 1 and prepared from tetrahydronaphthalene and sulfuric acid.

The resulting product is easily soluble in water in any dilution and can be employed as a resist in dyeing.

Formula No. 7

128 parts of naphthalene are sulfonated at 165° C. with 205 parts of sulfuric acid of 94 per cent strength. A mixture of 148 parts of n-butyl-alcohol

and 200 parts of sulfuric acid of 98 per cent strength is introduced, at 120° C., into the sulfonating product.

After separation of the waste acid, 54 parts of the liquid resin obtainable as indicated in Formula No. 3 from phenol and formaldehyde, are slowly run, at a temperature between 90° C. and 100° C., into 217 parts of the butylated naphthalene-sulfonic acid thus prepared. The resulting product is clearly soluble in water and can, if necessary, after it has been partly or completely neutralized, be used in dyeing.

Formula No. 8

150 parts of tetrahydronaphthalenesulfonic acid, obtained by sulfonating 140 parts of tetrahydronaphthalene with 180 parts of concentrated sulfuric acid, are heated to 70° C. and into the fused mass 70 parts of methylenediphenylether are slowly introduced, while well stirring. As is known, methylenediphenylether becomes very quickly resinous in the presence of an acid. The temperature raises, while the methylenediphenylether is introduced, to 90° C. and is kept at 95° C. to 100° C. until a test gives a clear solution in water.

Formula No. 9

A solution of 14 parts of paraformaldehyde in a mixture of 40 parts of commercial cresol and 26 parts of phenol (crystallized carbolic acid) which corresponds to the composition of the so-called washed carbolic acid, is introduced, while well stirring, at 90° to 95° C., into 150 parts of the crude melt of naphthalene sulfonic acid. The reaction is rather vigorous and accompanied by a rise of the temperature. Therefore, the product is introduced so slowly that the temperature does not rise above 100° C. and that no separation of resin occurs. After both components have mixed to a homogeneous syrup, the mass is still stirred for about one hour, i.e., as long as a test clearly dissolves in water.

Formula No. 10

Into 250 parts of a tetrahydronaphthalene-sulfonation mixture obtained as indicated in Formula No. 1 are introduced, while stirring, at a temperature of 90° C. to 100° C., 100 parts of the liquid condensation product, prepared by slowly heating to 75° C., while stirring, 25 parts of crystallized phenol and 75 parts of cresol with 65 parts of formaldehyde solution of 30 per cent strength to which 0.2 parts of sodium carbonate have been added; this mixture is then stirred until the smell of formaldehyde

has disappeared therefrom. After having introduced 20 parts of sodium chloride, the aqueous layer is separated and the liquid resin is used for being dispersed by means of tetrahydronaphthalenesulfonic acid. The reaction mass is partly neutralized with 230 parts of caustic soda solution of 20 per cent strength, and the so obtained product, which forms a greenish-brown, partly crystallized mass, can be used as a material for tanning animal skins.

Formula No. 11

170 parts of benzylchloride are condensed at 90° C. to 130° C. with 185 parts of naphthalene and then sulfonated at a temperature gradually rising from 60° C. to 110° C. with 250 parts of sulfuric acid containing 20 per cent of SO_3 which can be distilled off. 160 parts of a phenol-formaldehyde condensation product prepared as described in Formula No. 3 are run, at a temperature between 90° C. and 100° C., while stirring, into the aforementioned sulfonation mixture. The so obtained reaction mass is partly neutralized at about 50° C. with 410 parts of caustic soda solution of 20 per cent strength. A tanning material by means of which a very light-colored plump leather is obtained is produced.

Asphaltic Emulsions
British Patent 400,409

A part of the molten asphalt is added to a hot, dilute, aqueous solution of caustic soda, an emulsifying agent, e.g., oleic acid, is added, and then the remainder of the asphalt, all with agitation; at an appropriate time of the operation, e.g., before adding the asphalt, a small proportion of colloidal clay is added as a stabilizer. A suitable emulsion may contain asphalt 48–52%, water 46–48%, oleic acid 1%, sodium hydroxide less than 1%, bentonite 1%.

Asphalt Emulsion
U. S. Patent 1,931,072

This consists in dissolving substantially 9 parts, by weight, of soap in substantially 78 parts by weight of warm water. To the soapy water thus formed is slowly added, with vigorous agitation, about 20 parts, by weight, of a low grade fuel oil or a crude oil having an asphaltic base, to produce a dispersion of the oil, in the warm soapy water.

The oil that is preferred for the practice of this invention is low grade fuel oil, by which is meant an oil that is low in cost and comparatively low in heat units, but having a substantial asphaltic content (preferably about 10%), because such oils contribute to the reduction in the cost of the material produced; because they soften and make workable the asphalt; and because they enable one to use a relatively tough asphalt for the production of an emulsion. Unless specifically stated, however, the invention is not to be construed as limited to the use of a low grade fuel oil, because other oils may be used in lieu of fuel oil with considerable success.

The low grade fuel oil which is preferably used should have an asphaltic content of substantially 10% or more as measured by precipitation with 86° naphtha; should contain a rather high percentage of unsaturated hydrocarbons, showing an iodine value of 40 or more; and should have a specific .gravity of from .95 to .98. For convenience herein, the term "emulsive" oil is used to include fuel oil, crude oil, or any of the relatively involatile heavy oils usually used as liquefiers or fluxes for the asphalts in paving mixtures, and "fuel oil" is herein used to define the common oils on the market and used for burning in plants or residences for heating purposes; among which are those oils known in the trade as "Bunker C Gulf oil" and "Sun Bunker oil."

After the emulsion or emulsive oil has been made in the soapy water, a relatively small quantity of a metallic salt of a fatty acid is added to the emulsion and distributed or dispersed therein with thorough agitation or stirring. The salt preferably used in the process is aluminum oleate, but aluminum stearate or calcium oleate or stearate, or the stearates or oleates of zinc, iron and similar metals similarly increase the ductility of the asphalt and lower the penetration index or number thereof, and hence any of these metallic salts of the fatty acids may be used successfully in the practice of this process, and are fully within the scope and purview of this invention.

The amount of oleate or stearate so added to the emulsion is small, preferably from 1 to 2 parts by weight.

The salt of the fatty acid having been uniformly distributed throughout the emulsion, asphalt, preferably heated, is now slowly added thereto, with agitation, until substantially 296 parts, by weight, have been added thereto and distributed uniformly throughout the mixture. During the

mixing of the asphalt with the emulsion and the dispersing of the asphalt therein, the materials are thoroughly stirred or otherwise agitated and the best results will be obtained if the mixture is kept heated. The temperature at which the mixture is maintained is not very important so long as it is below the boiling point of water, for it is undesirable to substantially reduce the water which constitutes the menstruum of the emulsion prior to the application of the emulsion to the aggregate.

Asphalt Emulsion
U. S. Patent 1,932,648

Asphalt	64.05%
Water	35.00%
Caustic Soda	.08%
Corn Gluten (or Soy Bean) Meal	.64%
Green Acid Soap	0.23%

The asphalt for use in this formula may be produced from Mid-Continent petroleum; it may have a melting point of about 110° F., and a penetration of about 130 at 77° F. My formula may be readily adapted to other asphalts, however, the proportions of emulsifying agents being slightly increased for asphalts of increased melting points.

Green acid soap is a composition of preferentially water soluble sulfonic salts. This soap may be prepared by the neutralization of green acids, which are well known in the petroleum industry. Green acid soap is used in refineries for breaking emulsions and a "spent G. A. soap" which results from this use may also be employed. If the green acid soap contains appreciable amounts of oil, a harder asphalt should be emulsified, to produce a residue of given penetration, than is necessary when the green acid soap is free from oil. The term "green acid soap" as used herein includes concentrated soap, spent soap, soap containing oil and other forms in which this substance is available. In the formula the percent of green acid soap is on the dry soap basis. Caustic soda is included to increase the alkalinity of the composition and may obviously be substituted by equivalent alkaline salts. In some cases it may be omitted altogether.

The farinaceous protein emulsifier is preferably corn gluten meal or soybean meal, both of which contain about 40% protein and about 40% carbohydrates with 1–4% of fat aid 4–7% fiber. Starch has some effect as an emulsifier but protein is decidedly superior thereto and is the active principle of the farinaceous emulsifier.

In preparing the emulsion the procedure is as follows: The caustic, farinaceous emulsifier, and green acid soap are mixed with the water and heated to a temperature of about 200° F. This hot solution or mixture is placed in a suitable stirrer, agitator or mixer and is beaten by paddles, circulated by centrifugal pumps or dispersed between suitable rotors moving at high velocities.

The melted asphalt at about the same temperature is slowly stirred into the solution and further agitated until complete emulsification has taken place. When high melting point asphalts are used it may be necessary to increase the temperature at which the asphalt is added, but it is desirable to keep the temperature of the emulsion below the boiling point of water so that the foaming due to the production of steam may be prevented.

Copal-Bitumen Emulsion

Bitumen	10 parts
Copal (Powder)	10 parts
Potassium Hydroxide	1 part
Crude cresols	0.1 part

Melt the above together and stir until uniform; while stirring add 75 parts boiling water.

Cresol Emulsion

Mix the cresol (500 cubic centimeters) with the linseed oil (180 grams) and heat to about 80° C. Dissolve the potassium hydroxide (42 grams) in distilled water (40 cubic centimeters). Mix the two solutions and heat the mixture at about 90° C., with frequent stirring, until a small portion dissolves in water without the separation of oily drops. Cool and add sufficient distilled water to produce 1000 cubic centimeters of the liquor.

Cresylic Acid Emulsion

Cresylic Acid	50	parts
Diglycol Oleate	10½	parts
Pine Oil	2¼	parts
Potassium Hydroxide (25% solution)	16½	parts
Oleic Acid	20	parts
Alcohol	4½	parts
Water	7¼	parts

Mix thoroughly to obtain a clear "soluble" oil. This emulsifies on stirring in water.

"Cumar" Emulsion

Naphtha	20 parts
Cumar	20 parts

Heat above until dissolved; while at 90–100° C. and following solution (at 90–100° C.) slowly with vigorous stirring.

Ammonium Linoleate	
Paste	10 parts
Water	40 parts

Lecithin Emulsions
U. S. Patent 1,934,005
Formula No. 1

30 parts of vegetable lecithin, obtained by extraction of soya bean, with an oil content of about 30% are mixed with 10 parts of benzyl alcohol in the cold, whereupon 60 parts of linseed oil are admixed. 400 parts of water are added thereto and strongly agitated therewith, whereby a stable emulsion is obtained which is permanent for several days.

Formula No. 2

A mixture of 30 parts of soya lecithin with an oil content of about 30%, 8 parts of benzyl alcohol, 4 parts of triethanolamine and 58 parts of train oil is produced and the mixture is dispersed in 900 parts of water, whereby immediately a stable emulsion is obtained.

Formula No. 3

For the production of margarine or for the improvement of dough products a mixture of 90 parts of soya lecithin with an oil content of about 30% and 10 parts of benzyl alcohol is produced; this is dispersed in the desired proportions in the water fat emulsion of the margarine or in the dough worked up from meal or flour and water.

Oleic Acid Emulsion

Oleic Acid	45	parts
Turkey Red Oil	5	parts
Glue	1.5	parts
Water	48.5	parts

Free Rosin Emulsion
U. S. Patent 1,958,470

Caustic Soda	¼ lb.
Water	100 lb.
Rosin	2½ lb.

Boil together and stir until uniform.

Olive Oil Emulsion
U. S. Patent 1,930,845

100 parts of olive oil are mixed, while stirring, with 20 parts of a 50 per cent

aqueous solution of the salt of butylated naphthalene sulphonic acid and octodecyl ethanolamine. 180 parts of water are then gradually added, whereby a stable emulsion is obtained.

"Opal Wax" Emulsion

Add desired weight of wax to weighed portion of water (emulsions having 40% wax, 60% water form a thick cream. Higher proportions of wax will be solid on cooling). Heat the water to boiling and continue boiling until wax has completely melted and formed a homogeneous oil film on the surface of the water. Remove from hot plate and after boiling has ceased add NH_4OH while stirring until the melt has a slight but definite ammoniacal odor. Place in a mechanical agitator which will give rapid and thorough stirring (e.g., for small lots the type of mixer used at soda fountains is entirely satisfactory) and continue this stirring until emulsion has cooled below 60° C. Slow cooling in the range from 85° to 70° C. is advantageous.

Any froth or foam which forms will break upon standing except in the case of the higher wax contents (30% and above), when it may be necessary to apply vacuum to break the foam completely.

In the case of the higher wax contents, only very slow cooling while stirring will prevent the formation of a few small granules of solid wax.

Paraffin Rosin Emulsions
U. S. Patent 1,940,432

In producing paraffin dispersions, various proportions of paraffin and rosin may be used, for instance, from as high as 85% paraffin and as low as 15% rosin to as low as 15% paraffin and as high as 85% rosin. Assuming that a mixture of about 50% paraffin and about 50% rosin is employed, the mixture is melted and heated to about 220° F., which temperature is materially above the melting point of the paraffin (130° F.) and above the melting point of the rosin (about 180° F.). The heat-liquefied thermoplastic mixture is then commingled as a regulated stream flowing, say, at the rate of 15 pounds per minute with a regulated stream of 5% caustic soda solution at 110° F., flowing, say, at the rate of 20 pounds per minute. Under these conditions, an aqueous dispersion of a solids content of about 45% and at a temperature of about 155° F. is produced, which when suddenly chilled to a temperature below 130° F. to prevent coalescence of dispersed particles, is of a creamy consistency. In producing asphalt dispersions, at least about 35% rosin, based on the total weight of thermoplastic material, should be used in order to produce a stable dispersion of fine particle size. Assuming that a mixture of about 50% asphalt having a melting point of about 150° F. and about 50% rosin is employed, the mixture should be heated to about 300° F., whereupon it is brought in contact with a caustic soda solution under practically the same conditions as in the case of the paraffin dispersion, except that the solution should be at 150° to 160° F. The resulting dispersion has a temperature of about 200° F., and when permitted to cool, is of a smooth, soft, paste-like consistency.

Resin Emulsion

Rosin, Ester Gum, Cumar, or Other Resin, (Melted)	3 lb.
Oleic Acid	1 oz.
Triethanolamine	2 oz.
Casein Solution	2 lb.
Water to make	1 gal.

Tallow Emulsion

Tallow	190 parts
Ammonium Linoleate	8 parts
Water	200 parts

Heat to 160–170° F. for ½ hour while stirring.

Kerosene Emulsion

Diglycol Oleate	7	parts
Alcohol	3	parts
Caustic Potash (45%)	½	part
Pine Oil	1½	parts
Water	4½	parts
Kerosene	66	parts

Mix the above until clear then while mixing vigorously with a high speed stirrer run in slowly

Water	99	parts

Benzol Emulsion

Diglycol Oleate	7	parts
Alcohol	3	parts
Caustic Potash (45%)	½	part
Pine Oil	1½	parts
Water	4½	parts
Benzol	66	parts

Procedure as in preceding formula.

Water	99	parts

Orthodichlor Benzol Emulsion

Diglycol Oleate	9.8 parts
Alcohol	4.2 parts
Caustic Potash (45%)	0.7 part
Pine Oil	2.1 parts
Water	6.3 parts
Orthodichlor Benzol	66 parts

Procedure as in preceding example.

Water	99 parts

Carbon Tetrachloride Emulsion

Diglycol Oleate	14 parts
Alcohol	6 parts
Caustic Potash (45%)	1 part
Pine Oil	3 parts
Water	9 parts
Carbon Tetrachloride	66 parts

Proceed as above.

Water	99 parts

Linseed Oil Emulsion

Same as Carbon Tetrachloride Emulsion except that linseed oil is used.

Chinawood Oil Emulsion

Same as Carbon Tetrachloride Emulsion except that chinawood oil is used.

FOOD PRODUCTS, BEVERAGES, AND FLAVORS

Preservation of Yeast
British Patent 406,398

Attack of pressed yeast by chromo-genic organisms is prevented by mixing permanently with the yeast small amounts of aliphatic alcohols, e.g., 0.05–0.25% of butyl alcohol, 0.20–4% of ethyl alcohol or propyl alcohol. To prepare the mixture addition of a 2% ethyl alcohol soslution to the washed, pressed cake is effective; or the yeast is allowed to remain in 1% alcohol for 1 hour, or may be treated with (the alcohols) before pressing.

Bleaching Molasses
U. S. Patent 1,933,830

The process of bleaching molasses for subsequent use in the production of yeast, consists in diluting the molasses until it contains about 15% of sugars, adding ½ to 1% of a non-poisonous mineral acid, subjecting the mixture to the action of direct electric current until the solution is bleached to a degree of about 60%, filtering the molasses to remove impurities, adding a reagent to neutralize the acid content of the molasses, and decanting the solution from the precipitate if any.

Bean Flour
U. S. Patent 1,956,913

The process of preparing a bread improver consists in treating fresh beans by wetting them with a green malt infusion, maintaining the wet beans for eight to twelve hours at approximately 60° F., then grinding and bolting the product.

Bread Improver
U. S. Patent 1,953,332

1¼ pounds of a 28% solution of hydrochloric acid is sprayed into 1,000 pounds of dry starch. The amount of hydrochloric acid may be varied. The lactic acid is preferably mixed with the hydrochloric acid and added simultaneously. The amount of lactic acid which is employed may also be varied. Use 12,250 cubic centimeters of a 50% solution of lactic acid for a thousand pounds of starch. Acetic acid or citric acid is added in about the same amounts as lactic acid. After the acids have been added and thoroughly mixed with the starch, the starch is then heated in a steam jacketed converter for 35 to 45 minutes with its temperature gradually rising up to a maximum temperature of 240° F. By this time the conversion of the starch will have been carried sufficiently far and there will be present in the finished dextrine between three and, perhaps, fifteen per cent of dextrose together with other partially converted starch derivatives such as dextrin. The steam should then be shut off and while the dextrine is still hot it is preferable to add the ammonia gas while agitating the dextrine to thoroughly diffuse the ammonia throughout the dextrine. Ammonia is added until the acid condition of the dextrine is substantially neutralized. The dextrine is at this time a white dry non-hygroscopic powder which may be stored for long periods of time, without deterioration, without discoloration or without absorbing moisture and becoming lumpy.

It is found that by using this product the amount of yeast which would otherwise have been employed in the absence of this product may be materially diminished when this product is employed. This is a considerable saving in itself. Also the amount of sugar normally required for a dough batch is partially supplied by the sugars in the dextrine, hence the sugar supplied in other manners may be diminished. Due to the stimulating action of the ammonia which is released during the fermenting period the activity of the yeast is greatly increased and the fermenting period may be greatly shortened. In commercial operations where the time factor is very important this reduction of the fermentating period is highly desirable.

Bread Improver
Austrian Patent 132,388

Gelatin is added to dough for making bread, etc., in an amount corresponding to about 20 grams per liter of water used in making the dough.

Baking Powder
U. S. Patent 1,936,636

A composition of matter adapted for use with yeast and approximately 100 parts of flour in making leavened bread comprising the following substances in the proportions specified:

Calcium Acid Phosphate	.2 part
Ammonium Sulphate	.028 part
Potassium Bromate	.0005 part
Potassium Iodate	.0004 part

Sodium Aluminum Sulphate Type Baking Powder

Sodium Aluminum Sulphate	485 lb.
Sodium Bicarbonate	509 lb.
Starch, Corn (5% moisture)	166 lb.

Powder, sift, and mix.

Easter-Bun Spices
Formula No. 1 (London)

Nutmeg	6	oz.
Mace	1	oz.
Red Pepper	2	oz.
Cinnamon	4	oz.
Ginger	8	oz.

Formula No. 2 (Provincial)

Mace	2	oz.
Ginger	1	oz.
Oil of Cloves	6	drops

Formula No. 3 (Scotch)

Ginger	5	oz.
Coriander	5	oz.
Caraway	3½	oz.
Cloves	1	oz.
Pimento	6	drams
Cassia	6	drams
Nutmeg	4	drams

The spices are to be powdered, mixed and sifted, and one ounce of the mixture is to be used with 7 pounds of flour in baking the buns.

Curry Powder
Formula No. 1

Coriander	5	lb.
Turmeric	1.5	lb.
Fenugreek	12	oz.
Black Pepper	8	oz.
Cumin	8	oz.
Mustard	8	oz.
Dill	4	oz.
Pimento	4	oz.
African Ginger	4	oz.

Table Salt	1.5	oz.
Capsicum	1.25	oz.

Grind all the ingredients together to a fine powder.

Formula No. 2

Turmeric	2	oz.
Coriander	1	oz.
Ginger	2	oz.
Cardamom	½	oz.
Capsicum	½	oz.
Cumin	½	oz.
White Pepper	1	oz.
Lemon Peel	1	oz.

Formula No. 3

Coriander	13	oz.
Black Pepper	5	oz.
Capsicum	1	oz.
Cumin	6	oz.
Fenugreek	6	oz.
Turmeric	6	oz.

Grind all together and sift.

Shortening (Crisco Type)

Cottonseed Oil (Hydrogenated to an Iodine Value of 70)	88%
Cottonseed Oil (Fully Hydrogenated)	2%
Palm Oil (Hydrogenated to an Iodine Value of 24)	10%

After hydrogenation, these oils are melted together, neutralized with dilute caustic soda, washed and deodorized. The product is then run over a lard roll, well beaten to incorporate air; and pumped into tierces.

Cottage Cheese Cake

Cottage Cheese	1½	cupfuls
Butter	1	tablespoonful
Salt	½	teaspoonful
Vanilla	¼	teaspoonful
Milk	⅔	cupful
Cornstarch	1	tablespoonful
Sugar	⅔	cupful
Eggs	2	

Blend the cheese, butter, salt, and vanilla until smooth. Mix the milk and cornstarch and bring to a boil, stirring to prevent lumping or sticking. Add the sugar to the cooked milk and cornstarch and heat. Pour the hot mixture over the slightly beaten eggs, stirring thoroughly, and heat to a custard consistency. Add the cottage mixture to the custard and pour into a deep baking dish which has previously been lined with short-cake dough. Bake until crust is done and slightly browned.

Cottage Cheese Pie

Cottage Cheese	1	cupful
Sugar	⅔	cupful
Milk	⅔	cupful
Egg Yolks (Beaten)	2	
Melted Fat	1	tablespoonful
Salt	½	teaspoonful
Vanilla	¼	teaspoonful

Mix the ingredients in the order given. Bake the pie in one crust, prepared in the usual way. Cool it slightly and cover it with meringue made by adding 2 tablespoonfuls of sugar and ½ teaspoonful of vanilla to the beaten whites of 2 eggs and brown it in a slow oven.

Almond Fig Paste

Corn Syrup	32 lb.
Sugar	16 lb.
Water	2 gal.

The above is cooked to 240° F., and then add 80 pounds ground figs. Cook to 242° F. Now add 16 pounds chopped roasted almonds, cook for one minute, and then remove from kettle.

Fig Paste

Sugar	45	lb.
Corn Syrup	20	lb.
Ground Figs	7½	lb.
Water	2½	gal.

Bring to the boiling point. Then strain into it 4½ pounds cornstarch that has been dissolved in 4 gallons cold water. Cook up to 222° F.

Fig Filling

Fig Paste	60 lb.
Figs	60 lb.
Sugar	110 lb.
Corn Syrup	140 lb.
Salt	1 lb.
Sodium Benzoate	6 oz.

Cover fig paste and figs with water and soak overnight. In the morning grind fine. Place in the kettle, cook up a little, then add the sugar, corn syrup, salt and sodium benzoate. Cook to approximately 224° F.

Dried Apricot Paste

Apricot Pulp	85 lb.
Sugar	15 lb.

The apricots are first pitted, and then exposed to burning sulphur fumes for 2 hours. Then pass the fruit through the pulping machine. The pulped fruit and sugar is spread on wooden salt trays, covered with paraffine paper. Spread about 2 pounds over one square foot. Heavy layers retard drying. The trays when exposed to the sun rays, brings drying about in 24–30 hours. At 150° F., in a dehydrator, drying is effected in 9–12 hours. Pack in moisture proof containers when dried.

Dried Apple Paste

The following procedure is recommended:

Cut fruit in half or quarters.
Steam for 8 minutes.
After partial cooling expose to sulphur dioxide for 2 hours.
Run through pulping machine.
Add 10% sugar.
Spread on trays and dry.

Apple Pie Filling

Evaporated Apples	50 lb.
Cane Sugar	50 lb.
Corn Syrup (43° Baumé)	150 lb.
Corn Starch	6 lb.
Tapioca Starch	4 lb.
Mixed Spices	9 oz.
Salt	1 lb.
Sodium Benzoate	8 oz.

Cover apples with water and soak overnight.

Make up cold starch solutions with 3 gallons cold water.

Add the apples and the starch solutions to a kettle, then add the cane sugar. Cook up until very heavy. Then shut off the steam and add the corn syrup and the spices and salt. Stir vigorously. Now the sodium benzoate, which should be dissolved in a quart hot water is stirred in.

Blueberry or Huckleberry Pie Filling

Berries	110	lb.
Cane Sugar	85	lb.
Corn Syrup	85	lb.
Citric Acid	9	oz.
Sodium Benzoate	4	oz.
No. 80 Powdered Pectin	1½	lb.

Dissolve the pectin in 8 gallons of water. Place in the kettle along with the fruit and cane sugar, and bring to 218° F. Shut off steam, and add the corn syrup, acid and sodium benzoate, the latter first dissolved in 1 pint hot water. Agitate the contents of the kettle, and then run out into containers.

Lemon Pie Filling

Rice Flour	2	oz.
Wheat Flour	1	oz.
Powd. India Gum (Karaya)	½	oz.
Sugar	4	oz.
Salt	30	grains.
Citric Acid	30	grains.
Oil Lemon U. S. P.	4	drops.
Imitation Butter Flavor	1	drop.
Certified Lemon Yellow Color	2	grains.

To use mix above quantity with 1 quart of water, add 2 tablespoonfuls of butter or shortening, bring to a boil stirring constantly to avoid burning.

Dried Pear Paste

The addition of sugar to the pulped paste prevents cracking and splitting. By adding 0.50% of citric acid, more sugar can be added to the paste. The directions for making are same as for apricot paste.

Orange-Pineapple Cake Filling

Pineapple (No. 10 Tins)	1	doz.
Cane Sugar	100	lb.
Sodium Benzoate	2	oz.
Agar (Dissolved in 2 Gallons Water)	½	lb.
Pure Orange Extract	40	oz.
Orange Color Solution	4	oz.
Galagum C.	½	oz.

Put pineapple pulp into kettle; add cane sugar and agar. Heat to 190° F. Then add flavor, color, galagum C., and sodium benzoate, with mixing. Heat up to 210° F. Shut off heat, and fill into containers.

Prune Paste

Corn Syrup	75	lb.
Water	15	lb.
Ground Prunes	20	lb.
Number 80 Pectin	1	lb.

Bring the above corn syrup, water and prunes to a boil. Mix the pectin with 5 pounds of cane sugar. Add to the kettle while cooking, with constant stirring. Heat to 220° F. Remove from kettle.

Peach Pie Filling

Evaporated Peaches	75	lb.
Cane Sugar	75	lb.
Corn Syrup	75	lb.
Sodium Benzoate	5	oz.

Cover peaches with water and soak overnight. In the morning add to kettle fruit, and cane sugar. Cook to 222° F. Shut off heat then add the corn syrup and sodium benzoate with mixing.

Raisin Pie Filling

Seeded Muscatel Raisins	100	lb.
Cane Sugar	60	lb.
Corn Syrup	100	lb.
Corn Starch	10	lb.
Tapioca Starch	2	lb.
Salt	1	lb.
Sodium Benzoate	6	oz.
Amaranth Color Solution	2	oz.
Citric Acid	8	oz.
Pure Lemon Extract	20	oz.

Cook raisins until soft with about 15 gallons of water: Then add salt. Place in about 5 gallons of this, the starches, citric acid and sodium benzoate and mix into a smooth paste. Add this to the raisins in the kettle. Now also add the cane sugar and heat the kettle to about 222° F., with mixing. Shut off heat and add the corn syrup and color, mixing very thoroughly.

WHITE ICINGS

	Simple Type				Cream and "Fudge" Type					"Butter Cream Type"		
	1	2	3	4	5	6	7	8	9	10	11	12
Fat	...	...	1.85	17.1	15.0	15.3	19.3	7.6	6.6	31.0	50.0	36.0
Powdered Sugar	75.0	75.0	68.0	68.5	60.0	62.0	34.0	71.0	77.0	61.5	25.0	56.0
Invert Syrup	10.0	15.0	...	...	10.0	10.2	...	...	...	...	...	...
Fondant Sugar	...	...	14.2	...	...	...	14.6	13.0	...	...	...	...
Corn Syrup	...	...	...	...	...	...	...	...	3.15	...	...	...
Dry Skim Milk	...	...	4.2	4.3	5.0	5.1	8.7	...	.25	...	...	...
Gelatin	.5	.5	...	...	...	.64	...	...	...	...	...	...
Gum	...	...	.85	...	...	...	.88	.78	.8	...	...	...
Egg White	7.0	...	...	...	...	...	...	...	...	...	...	...
Whole Eggs	...	...	...	...	...	...	13.1	...	...	7.5	25.0	8.0
Water	7.0	10.0	11.0	10.1	10.0	7.7	8.75	8.8	11.8	...	...	...

COCOA AND CHOCOLATE ICINGS

	Simple Type		Cream and "Fudge" Type								
	1	2	3	4	5	6	7	8	9	10	11
Fat	3.5	3.5	5.0	6.9	7.8	8.7	9.4	10.0	12.0	15.0	15.6
Cocoa Powder	2.0	...	...	8.7	...	...	12.4	14.0	...	8.75	6.25
Sweet Chocolate	...	7.1	...	...	...	11.2	...	...	12.0	...	...
Chocolate Liquor	...	...	20.0	...	14.4	...	...	...	...	...	...
Powdered Sugar	73.0	71.0	50.0	62.0	57.5	65.0	58.0	56.0	58.0	55.0	62.5
Invert Syrup	3.5	3.5	10.0	...	10.4	5.0	...	10.0	8.0	10.0	...
Corn Syrup	...	...	...	8.3	...	...	3.2	...	...	...	...
Dry Skim Milk	7.1	7.1	...	...	...	...	...	...	...	3.7	4.0
Gelatin	...	...	...	...	...	...	...	...	...	...	...
Gum	...	...	...	...	...	...	.8	...	...	.6	...
Agar-Agar	...	...	...	...	...	...	...	.1	.1	...	...
Water	11.2	7.1	18.0	15.2	10.4	10.0	16.1	9.9	10.0	7.5	12.5

Cream Walnuts

Sugar Syrup (67° Brix)   ½ gal.
Walnut Pieces   4½ lb.

Put nuts in jar. Cover with ½ gallon syrup. If a maple walnut is wanted add 1 ounce of maple flavor to each gallon of syrup.

Jelly Preparations
British Patent 396,749

The rate of setting of confectioners' pectin jelly products can be retarded by adding the salt of a strong base and weak acid e.g., 0.35–0.48% of sodium acetate, to the batch.

"Kosher" Vegetable Jelly
U. S. Patent 1,946,649
Formula No. 1

15 parts of granulated or other suitable sugar are mixed with 1 part of agar agar, 1 part of gum karaya, ³⁄₄₀ part of certified strawberry flavor, ³⁄₈₀ part of dry certified strawberry color and ⁴⁄₁₀ part of tartaric acid. In making the product, a portion of the sugar has added to it the flavor in which the color is dissolved and the whole mixed together. To this mixture the remainder of the sugar is added. After thoroughly mixing until fine, powdered agar agar in ground form and gum karaya are added to the mixture. Finally, powdered tartaric acid is added and the whole is thoroughly mixed. These formulas are stated in parts by weight.

Formula No. 2

Sugar	15 parts
Agar Agar	1 part
Gum Karaya	1 part
Raspberry Flavor	³⁄₄₀ part
Dry Color	³⁄₈₀ part
Tartaric Acid	⁴⁄₂₅ part

Formula No. 3

Sugar	15 parts
Agar Agar	1 part
Gum Karaya	¾ part
Cherry Flavor	¹⁄₄₀ part
Dry Certified Color	³⁄₈₀ part
Tartaric Acid	⁴⁄₅ part

Formula No. 4

Agar Agar	1 part
Gum Karaya	½ part
Cocoa	3½ parts
Sugar	15 parts

Formula No. 5

Sugar	15 parts
Agar Agar	1 part
Pectin	¹⁄₁₀ part
Raspberry Flavor	³⁄₈₀ part
Color	³⁄₈₀ part
Citric Acid	⁴⁄₂₅ part

The mixtures described in the foregoing examples are dissolved in about 100 parts of hot water and are completely dissolved therein. After the mixture is dissolved, the aqueous solution is allowed to stand and to set. In practice this solution will set at ordinary room temperatures to a stiff and elastic jelly-like mass in about one hour's time. A low temperature, such as can be obtained in an ice box, however, will effect stiffening in a shorter time, although it is not necessary.

Grapefruit Jelly

Peeled Grapefruit	1	lb.
Water	2	pints
Sugar	¾	lb.

Cut up grapefruit and cook with water until thoroughly soft. Drain juice through bag, bring to a boil, add the sugar, and boil to 220° F. Seal in glasses covered with melted paraffin wax.

Apricot Jam

Evaporated Apricots	50 lb.
Cane Sugar	85 lb.
Sodium Benzoate	2 oz.

Add enough water to cover the fruit. Let soak overnight. Following day, grind it, and place the ground fruit and sugar in a steam jacketed kettle. Cook to 224° F. Then dissolve the 2 ounces of Sodium Benzoate in 6 ounces hot water, and add to the cooked jam with stirring.

Sour Orange Jelly

Peeled Sour Oranges	1 lb.
Water	2 pints
Sugar	1 lb.

Prepare the orange juice as described under sour orange marmalade. Then cook the juice, water and sugar, to 220° F. Seal in glasses and cover with melted paraffine wax.

Orange Pectin Juice

White Portion Orange		
Peel	½	lb.
Lemon Juice	2	tablespoonfuls
Water	1	pint

Cut or grate the yellow from the peel of the orange, pass the remaining white portion through a food chopper. For each ¼ lb. of white portion add ½ pint water, add the lemon juice. Mix and allow to stand 5 hours. Now add 1¼ pints water and let stand overnight. In the morning boil 10 minutes. Let cool and then squeeze through a flannel jelly bag. This juice may be used as a foundation in making jellies from fruits which are deficient in pectin.

Orange Pectin Jelly

Powdered Pectin (No. 80)	1½	lb.
Cane Sugar	70	lb.
Corn Syrup	35	lb.
Egg Color Solution	2	oz.
Pure Orange Extract	40	oz.
Water.	8	gal.

Heat water in kettle, then add the pectin, cook to full boil. Then add the cane sugar, and cook to 216° F. Now shut off steam and add corn syrup, color and flavor. Mix. Now run jelly into a 30 pound pail to which has been added 5 ounces of citric acid solution. (This citric acid solution has been prepared by dissolving 4 pounds of citric acid in 6 pints of water.) The jelly will set shortly.

Mint and Orange Pectin Jelly

Orange Pectin Juice	1 part
Lemon Juice	2 tablespoonfuls
Sugar	1 lb.
Oil of Mint	2 drops
Green Vegetable Color	2 drops

Heat the orange pectin juice and lemon juice and sugar to the boiling point. Now add the color. The boiling is now continued to 220° F. Then add the mint oil. Seal in glasses and cover with melted paraffin.

Glucose Jelly

Water	15	gal.
Cane Sugar	10	lb.
Powdered Pectin (No. 80)	3½	lb.
Corn Syrup (43° Baumé)	300	lb.
True Fruit Flavor (Raspberry)	10	oz.
Red Certified Color Sufficient.		

Mix the cane sugar with the pectin. Heat the water in a kettle, then add the mixed pectin powder. Bring to a boil with mixing. Now add the corn syrup, color and flavor. Heat to 220° F. Then run the hot jelly into 30 pound wooden pails, each pail having in it 5 ounces of acid solution prepared as follows:

Phosphoric Acid (85%)	4 lb.
Water	6 lb.

The jelly sets shortly.

Whole Kumquat Preserve

Whole Kumquats	2 lb.
Sugar	2 lb.
Water	1 quart

Clean the kumquats, then scald with boiling soda water (about 1 tablespoon soda for 1 quart fruit). Then cover with boiling water. Let stand for 10 minutes. Then rinse the fruit with cold water three or four times. Slit each fruit, to facilitate penetration of syrup later on. Now cook kumquats until tender. Drain. Boil sugar and water together for 10 minutes. Add the drained kumquats, and cook until the fruit is shining, clear and transparent. Then cover tightly and allow to plump 24 hours.

Florida Conserve

Grapefruit Pulp	2	cups
Orange Pulp	2	cups
Nut Meats	¾	cup
Orange Peel (Chopped) from 1 orange		
Seeded Raisins		½ cup
Grated Pineapple (Canned)		½ cup
Sugar	2	cups

Add one cup water to orange peel, boil gently for 10 minutes, cover, set aside to cool. Mix grapefruit, orange pulp and orange peel, boil gently for 20 minutes, add the sugar. When sugar has dissolved add the pineapple. Cook until the mass thickens and will give the jelly test. Add nuts and raisins, boil for 2 minutes, pour into glasses and seal.

Sour Orange Marmalade

Peeled Sour Oranges	1	lb.
Sugar	1½	lb.
Water	2	pints
Peel Removed from Oranges	⅓	

Wash fruit. Remove peel. Save ⅓ of peel and cook until tender. Now cut fruit into small pieces and boil with water until it thoroughly disintegrates. Strain juice through bag. Now take the juice and peel and bring to a boil. Add 1½ lbs. of sugar for each pound of fruit. Continue the boiling until the jellying point has been reached which is indicated by the flaking or sheeting from spoon.

Sweet Orange Marmalade

Peeled Sweet Oranges	1	lb.
Sugar	⅞	cup
Water	2	pints
Peel Removed from Oranges	¼	

Wash fruit, remove ¼ of the peel. Cut this peel into thin slices and then boil for about 10 minutes until tender. Now cut fruit into small pieces, add to it the white of the Orange Peel, and boil for about 20 minutes. Now strain the juice, then place in a kettle and bring to a boil. For each pound of fruit taken add ⅞ cup of sugar. When this comes to a boil, add the peel which has been cooked tender, and cook to the jellying point.

Treatment of Edible Nuts
U. S. Patent 1,887,256

Removal of the skins of Brazil and other nuts, with or without pretreatment with fatty material, is effected by treatment with an alkaline solution (6 ounces per gallon) at 71–93° for 1–7 minutes, washing with water under pressure (70–80 pounds) treating with hydrochloric acid or edible acid to remove excess alkali, washing with water, and air-drying. Glycerin may be added to the lye bath as wetting agent.

Prevention of Mold Decay in Oranges

Immerse fruit for 4 minutes at 110° F. in a solution of 8 pounds of borax per 100 pounds of water. The treated should not be washed to get the best protection.

Butter Coloring

A solution of annatto in oil constitutes probably the best and most satisfactory butter coloring. The following formula may be used:

Annatto, (Powder)	3 oz.
Cottonseed Oil	16 oz.

Mix, heat to 212 deg. F. for some time, set aside for twenty-four hours, strain and filter.

Purified annatto, or annattoin, yields a still finer preparation, somewhat less being required in proportion, although "strength" is a matter for individual judgment.

The following compound annatto powder is also largely used:

Annattoin	5 oz.
Turmeric (Powder)	6 oz.
Saffron (True)	1 oz.
Lard Oil (Odorless)	16 oz.
Alcohol	4 oz.

Rub the annattoin and turmeric with the oil, which may be deodorized by filtration through charcoal and macerate for several days. Prepare a tincture with the alcohol and saffron. After sufficient maceration separate the solids from the oil by filtration, adding more oil through the filter to keep the quantity, and mix with it the tincture of saffron, afterwards driving off the alcohol with gentle heat.

Microscopic Examination of Butter

The following procedure is a suggested method for the microscopic examination of butter:

Melt carefully a representative sample of butter by heating to 45° C. (113° F.).

Centrifuge 10 cubic centimeters of the melted butter in a separatory funnel until the serum is separated from the fat.

Draw off the serum.

Spread 0.01 cubic centimeter of the thoroughly mixed serum measured with a Breed pipette over a definite area (from 1 to 8 or more square centimeters) on a microscope slide and allow to dry.

Stain as in the microscopic count for milk.

Examine under the microscope for the general character of the flora.

If an estimation of the number of organisms per cubic centimeter of butter is desired, determine the number per microscopic field of the serum and then calculate the number per cubic centimeter of serum and finally the number per cubic centimeter of butter.

Preparation of Edible Emulsions of Solid in Fat

U. S. Patent 1,894,677

From 0.3 to 3.5% of finely-powdered gelatine, together with less than 1% of water, is added to chocolate in the melangeur. It is claimed that less fat can be used, milling time is reduced, covering power is increased, and that, in the finished chocolate, both sugar bloom and fat bloom are retarded.

Cottage Cheese Salad Dressing

Cottage Cheese	½ cupful
Worcestershire Sauce	1 tablespoonful
Vinegar	3 tablespoonfuls
Salt	⅛ teaspoonful
Catsup	2 tablespoonfuls
Olive Oil	3 tablespoonfuls
Pepper	⅛ teaspoonful
Paprika	⅛ teaspoonful

Cream cheese with catsup, gradually adding other ingredients. This is especially good with vegetable salads.

Low Acid Rennet Cottage Cheese

Pasteurize the skim milk at 145° F. and hold 30 minutes. Maintain the setting temperature, 72° to 80° F., throughout the fermentation period of 5 to 10 hours. Add from 1 to 5 per cent of good starter. Add 1 cc. of rennet per 1,000 pounds of skim milk. Acidity of the whey at time of cutting the curd should be 0.50 to 0.55 per cent. The curd should be firm, but not hard or brittle. Fill the jacket of the vat with water at 105° to 115° F. Cut the curd into ½-inch cubes. Run 2 inches of water at 105° to 115° F. over the cut curd and repeat in one-half hour with water at 120°. Stir the curd carefully until it will stand stirring without breaking. Keep the water in the jacket 25° to 35° warmer than the curd. Cook the curd slowly to a temperature of 118° to 130° F. in 1 to 3 hours. Small batches should be cooked to a slightly higher final temperature. When the curd can be gently squeezed in the hand and will hold its original shape, draw off the whey. Wash the curd two or three times in cold water, depending upon the acidity of the curd.

After the curd has been washed, ditch it, allow it to drain for one hour, and put it on trays, or in cans or in tubs with holes in the bottom, and place in a cooler at 30° to 40° F. for about 12 hours. Add 1 pound of salt per 100 pounds of dry curd at time of mixing. At the end of about 12 hours the curd will be firm and dry. To each 100 pounds of curd add 60 to 75 pounds of a mixture of milk and cream containing 10 per cent butterfat. Then stir the mixture vigorously by hand, or for five minutes in a mechanical mixer operated in low gear. Then put the cottage cheese in a cooler for a few hours at from 30° to 40° F. It is then ready to package. The curd should be firm, and even after it is mixed the curd particles should be distinct; however, the curd should not be tough or hard a few hours after mixing. Careful cooking is necessary to get the proper degree of firmness.

Identification of Cold-Storage Eggs

The eggs for cold-storage are sprayed with a 1% solution of phenolphthalein in alcohol, which has no effect on their appearance. They are then readily identified later by the coloration given by a sodium carbonate solution.

Butterscotch Pudding Powder

Corn Starch	2 lb.
Sugar (Powdered)	8 lb.
Old Fashioned Butter Scotch Flavor	1 oz.
Salt	¼ oz.
Certified Food Color (Yellow or Brown)	24 grains

Mix well. Pack in 5 ounce portions. To use, mix 1 package with 1 pint of milk, rub smooth and bring to a boil.

Custard Type Dessert Powder

Rice Flour	2 lb.
Arrow Root (Powdered)	1 lb.
Certified Egg Color	30 grains
Egg Custard Flavor	30 minums
Salt	1 oz.

Mix well. Put up in 1 ounce packages.

To use, mix contents of package with ¼ pint of milk. Rub smooth and add to ¾ pint of boiling milk. Continue to heat until it begins to boil. Stir to prevent burning. Can then be baked in oven to form crust.

Chocolate Dessert

Cane Sugar	24 parts
Corn Sugar	40 parts
Corn Starch	24 parts
Cocoa Powder	12 parts

Caramel Dessert

Cane Sugar	30 parts
Corn Sugar	52 parts
Corn Starch	18 parts
Caramel flavor.	

Vanilla Dessert

Cane Sugar	31 parts
Corn Sugar	52 parts
Corn Starch	18 parts
Vanillin flavor.	

Gelatine Dessert, Lemon Flavor

Cane Sugar	88 parts
Edible Food Gelatine	8.6 parts
Tartaric Acid	2 parts

Pure fruit lemon flavor. Vegetable color archel.

Gelatine Dessert, Cherry Flavor

Cane Sugar	75 parts
Corn Sugar	14 parts
Edible Food Gelatine	8 parts
Tartaric Acid	1½ parts

Pure fruit cherry flavor. Certified color amaranth.

Chocolate Peanut Bars

Sugar	3 lb.
Corn Syrup	3 lb.
Water	1 pint

Cook to 240° F., then add 8 pounds of roasted peanuts, and cook for 10 minutes. Remove from fire, and roll out on slab. Cut into small pieces, and then dip in chocolate.

Coffee Caramels

Sugar	4 lb.
Corn Syrup	3 lb.
Cream	½ gal.
Good Java Coffee	½ gal.

Cook the above to 238° F. Then add one quart sweet cream and cook to about 242° F. Remove from fire.

Butterscotch Squares

Sugar	10 lb.
Corn Syrup	4 lb.
Water	3 pints

Cook to 285° F.; add 4 ounces of butter, ½ teaspoonful salt. Stir and cook to 295° F. When cooled down flavor with oil of lemon. Pour on greased slabs.

Everton Toffee

Sugar	4 lb.
Water	3 cupfuls
Cream of Tartar	½ teaspoonful
Butter	2 oz.

Put the water, sugar and cream of tartar into a pan and stir until they boil, but no longer. Add the butter after the other comes off the fire, but do not stir it in.

Licorice Drops

Sugar	12.5	lb.
Water	3	pints
Cream of Tartar	1	dram
Powdered Extract of Licorice	1	oz.
Oil of Wintergreen	30	minims
Oil of Anise	20	minims
Powdered Charcoal	1	dram

Cook the sugar, water and cream of tartar to 340° F.; pour out on an oiled slab; add the other ingredients, and fold and knead until well mixed; then stamp into drops.

Fruit Jelly Candies

Fruit Juice	2 qt.
Sugar	3 lb.
Corn Syrup or Invert Syrup	3 lb.
Commercial Pectin Syrup	1 qt.

Mix fruit juice and pectin syrup. Add the sugar and corn or invert syrup. Boil to 222–223° F. Remove from fire. Pour into starch molds or oiled pans, about ½ inch deep. Allow to harden 24 hours. Remove from molds and dust with powdered sugar or dip in chocolate.

Fruit Caramels

Sugar	2 lb.
Chopped Walnuts	½ pt.
Corn Syrup	1¼ lb.
Ground Dried Fruit	1 lb.
Butter	2 oz.
Cream (Whipping)	2 qt.

Cook sugar, corn syrup, and ½ of cream to 238° F., then add ½ of remaining cream and cook to 242° F., add remaining cream, dried fruit, butter and chopped nuts and cook to 248° F. Pour into an oiled pan or onto a slab to cool and harden.

Fruit Marshmallows

Powdered Edible Gelatine	4 oz.
Hot Water	1 pt.
Corn Syrup	2½ lb.
Powdered Sugar	2½ lb.
Dried Apricots, Figs, Prunes, Pears, Chopped or Whole	
Raisins	3 lb.

Dissolve the gelatine in hot water. Cook corn syrup to 250° F. Beat into it the dissolved gelatine and powdered sugar until the mixture is light. Flavor with vanilla. Then add the dried fruit. Beat a short time and pour on oiled paper or a slab to harden. Fruit syrup may be used instead of dried fruit— ¾ pint.

Molasses Wafers

Sugar	5 lb.
Corn Syrup	1 lb.
Water	1½ pt.
New Orleans Molasses	1 pt.
Butter	¼ lb.

Stir and cook to 290° F. Roll out or a greased slab.

Maraschino Cherries

Place in a paraffin lined barrel, or a crock, or a bottle, all depending upon amount, firm ripe Royal Ann Cherries. Do not remove the stems. Cover the cherries with the solution made as follows:

Water	1	gal.
Bisulphite of Soda	1	gal.
Citric Acid	½	oz.
Salt	6	oz.

Dissolve the above ingredients. Let stand until cherries are bleached, which may take 2 weeks or longer. Then discard the solution. Stem and pit the fruit. Boil the fruit in 5 or 6 changes of water until free from sulphur taste and until tender. Prepare a 40% sugar solution, and for each 10 gallons of cherries dissolve ⅛ ounce of certified red color, Ponceaux 3R. Also add ½ teaspoonful citric acid. Now cover cherries with this sugar colored solution, and cook for 10 minutes. After 24 hours, cook cherries and syrup again, after adding 10 pounds sugar. Repeat this again on third day. Then allow to remain in crock for 2 weeks. Now add ¼ ounce of bitter almond oil dissolved in a little alcohol, for flavor. Instead of the red, a green food color may be used and instead of wild cherry, mint flavor may be added.

Wild Cherry Drops

Sugar (Crystal A)	4 lb.
Water	1 pint
Cream of Tartar	24 grains
Extract of Bitter Almonds	3 drams
Powdered Orris	3 drams
Red Color	Sufficient

Cook the sugar, water and cream of tartar to 335° F. Pour out on an oiled slab; let it cool a little; add the extract, the orris, and the color. When cool enough, work thoroughly and stamp into drops.

Aniseed Drops

Sugar (Crystal A)	5 lb.
Water	22 oz.
Cream of Tartar	40 grains
Oil of Anise	40 minims
Red Color	Sufficient

Put the sugar in a suitable container; add the water and the cream of tartar. Cook to 335° F.; pour out on an oiled slab; as it cools add the oil of anise and the color; fold over until cool enough to handle, then work thoroughly and stamp into drops.

Nut Glaces

Sugar	10 lb.
Corn Syrup	2½ lb.
Water	3 pints

Cook sugar and water to 290° F. Shut off heat and drop in nuts to be sugar coated or glaced. Remove with fork.

Revivifying Dry Popcorn

Popcorn which pops poorly because of having become too dry may be restored to good popping condition by the following method:

Put 40 pounds of the corn into a 10 gallon can.

Add 1 to 3 pounds of water, according to the dryness of the corn as indicated by the way it pops. If its popping yield is less than one-third the normal yield of the variety, add 3 pounds of water; if it is two thirds normal, add only 1 pound. For intermediate degrees of popping add intermediate quantities of water. For different varieties of popcorn the normal popping yield varies from 15 to about 30 volumes.

Put on the cover, using a rubber, and clamp it down tightly.

Shake thoroughly.

Leave stand two days or longer before popping.

The poor popping quality of some lots or corn is due to other causes than lack of moisture. In such cases the popping cannot be improved by adding water. The above method of restoring poppability is applicable only to small quantities of corn. When it becomes necessary to increase the moisture content of a large quantity of popcorn, it should be done by storing the corn in a cool, damp place, as, for example in a shed outside during winter, or in a basement room during summer.

Popcorn Fruit Crisp

Sugar	3	lb.
Corn Syrup	1½	lb.
Water	1	pt.
Popcorn (popped)	4½	qt.
Butter	3	tbsp.
Dried Fruit Chopped	¾	lb.
Salt	Sufficient to flavor	

Cook sugar, corn syrup and water to 285° F. Add the butter, salt, and fruit and stir. Now stir in the popcorn. More or less popcorn can be added to suit.

Salted Peanuts

In a kettle on a gas flame, add coconut oil and heat to about 300° F. In a basket add shelled Spanish peanuts. Immerse this basket in the hot oil, and roast until brown. Now pour peanuts over a fine sieve, and when the oil has drained off add the salt.

Candied and Glacé Fruits

Use firm ripe fruit. Peel peaches, pit, and cut in half. Peel pears, cut in half and core. Stem and pit cherries. Do not pit apricots, plums, and prunes, but puncture to the pit in several places with a silver fork. Figs require no treatment. Cut oranges, lemons and grapefruit in half and scoop out the pulp, retain the peels only for candying. Citron cut in half, scoop out pulp store peel three weeks in brine of 1 lb. salt to gallon water.

Canned fruit is good for use in candied fruit. Drain off syrup; to each two cups of syrup add one cup of corn syrup, return it to the fruit and boil the fruit and syrup for 3 minutes. Let it stand for 24 hours. The orange, lemon, grapefruit, and citron peels are boiled in water, first to modify texture of peels, so that they will absorb syrup, without shriveling. Secondly to get rid of bitterness, with several changes of boiling water. When the peels are thoroughly cooked and tender, place them in the first syrup (35%) sugar solution; boil for 5 minutes and set aside 24 hours.

Second Boiling: (This is for canned fruit also) At end of 24 hours after first boiling, drain off the syrup. Prepare a mixture of equal parts of cane sugar and corn syrup. Add enough of this mixture, to the drained syrup to increase the sugar to 40%. Now add the fruit to this 40% sugar solution, boil the mixture from 2–3 minutes. Return it to storage vessel and allow to stand 24 hours.

Subsequent Boilings: At 24-hour intervals drain the syrup from the fruit and add enough of the above cane syrup mixture to increase sugar percentages as follows:—50%, 60%, 70% and finally 74% on successive days. Boil the syrup and fruit together each day for 2–3 minutes and return it to the storage vessel.

Store the fruit in the final syrup of 74%, for at least two weeks to permit the fruit to become plump.

After two weeks storage or longer in final syrup, remove the fruit. Dip it momentarily in hot water and place it free from adhering syrup. Place it on screen trap, and allow to dry; or dry the candied fruit in a dehydrator at 120°– 130° F., 4 to 6 hours. Pack in boxes after drying.

Chewing Gum Base
U. S. Patent 1,930,436

100 parts of rubber latex containing 38% of rubber is treated with 100 parts of water containing 10 parts of caustic soda in solution (this may be at room temperature). Agitate well (which gives a liquid containing about 5% of sodium hydroxide or reaction products thereof) and let stand until separation occurs, into two layers, a creamy layer which floats on an amber colored lower layer. Draw off the bottom layer. To the creamy layer, which carries the solid, add about 10 parts of solid caustic soda (or dissolved caustic soda may be used), settle again and draw off the aqueous liquid. This latter step may be repeated one or more times. The creamed product may now consist of 115 parts, and 50 parts of the water may now be added. 100 to 150 parts of coumarone resin are now added, as a powder, and mixed at room temperature. This forms a pasty mass. Then heat with agitation until the water is driven off and the resin melts, forming a homogeneous mass. Wash this with water to remove most of the remaining alkali, the last traces of which can be neutralized with phosphoric acid, tartaric acid or a fruit acid.

The product is then mixed with hard

hydrogenated oil, say 125 to 185 parts, and up to 15–25 parts of powdered cocoa bean, with or without fillers, and/or softening oils, say hydrogenated coconut oil (1 to 15% each).

Unwrapped Licorice Caramels

Cane Sugar	60 lb.
Corn Syrup (43°)	50 lb.
Cream (40%)	4 gal.
Fresh Unsweetened Whole Condensed Milk (10%)	5 gal.
Sweetened Whole Condensed Milk (8%)	3 gal.
Table Salt	6 oz.
Fat (140 m.p.) (Hydrogenated Peanut—or Cottonseed Stearine)	2 lb.
Licorice Syrup	8 lb.

Put all ingredients except color and licorice in kettle. Cook to a soft ball. Add licorice syrup and black color. Cook batch to hard ball in cold water. Pour on slab and let stand overnight. Cut to size desired. No wrapper is required for this excellent "stand-up" caramel. Put in waxed paper cups.

Licorice Taffy

Cane Sugar	15 lb.
Licorice Syrup	8 lb.
Cocoa Butter	1 lb.
Corn Syrup (43°)	13 lb.

Place in kettle; cook to soft ball in cold water and add licorice syrup; cook to hard ball, add black color and spread out on marble to desired thickness. Let cool and cut to suit. Wrapper should be about 20 pound white waxed.

This is a splendid number for wax-wrapped five- and ten-cent bars. May also be used to increase variety and give novelty to chocolate and bonbon centers. Makes an excellent center for pan goods.

Licorice Jelly Gum Drops

Gum Arabic	50 lb.
Cane Sugar	80 lb.
Licorice Syrup	12 lb.
Corn Syrup (43°)	5 lb.
Water	4 gal.

Soak gum in hot water, using 1 pound of gum to 1 quart of water. Let stand in hot room overnight. Skim off and strain through very fine sieve.

Cook sugar, corn syrup and water to 250° F. Add licorice syrup and black color and continue cook to 255°. Add gum and cast in starch. Put in hot room for 48 hours. Take out and let stand until cool. When cool, remove from starch and place in crystal pans. Use crystal cold cook to 232°. Leave in crystal overnight.

The smaller sizes of licorice jelly gum drops make excellent centers for pan work. For this work they are, of course, not crystallized. Put them in the pan after removing them from starch and handle them the same as you would jelly beans or cordial drops. Do not color the syrup but finish white.

Chocolate or bonbon-coated, this luscious jelly gum drop makes an excellent addition to the conventional cordial box or summer package.

Hard Licorice Drops (Pastilles)

Gum Arabic	60 lb.
Cane Sugar	36 lb.
Glycerin	3 oz.
Licorice Syrup	12 lb.
Corn Syrup	6 lb.

Melt gum in the proportion of 1 pound to 1 quart of hot water, and let set overnight in hot room. Bring to a boil next day and strain through a very fine sieve.

Cook sugar and corn syrup to 280° F., add gum, licorice syrup, and black color, and put in hot room over night. Next morning, skim off and cast in starch.

Place in hot room again and allow to set for three days (temperature of hot room to be maintained between 130° and 140° F.). After removing goods from the hot room, let stand for about 24 hours to cool. Clean off starch and return to hot room until goods are hot, rub in refined petrolatum (white vaseline) until the desired shine is obtained.

Wrapped Licorice Caramels

Corn Syrup (43°)	10 lb.
Cane Sugar	30 lb.
Cocoa Butter	3 lb.
Cream (40%)	4 gal.
Licorice Syrup	6 lb.

Put all ingredients except black color and licorice syrup in kettle. Cook to a soft ball. Add licorice syrup and black color. Cook to a hard ball in cold water. Pour on slab and allow to stand over night. Cut and wrap.

Licorice Marshmallow Caramels

Cane Sugar	60 lb.
Corn Syrup (43°)	50 lb.
Cream (40%)	4 gal.
Fresh Unsweetened Whole Condensed Milk (10%)	5 gal.
Sweetened Whole Condensed Milk (8%)	3 gal.

Table Salt	6 oz.
Licorice Syrup	8 lb.

Put all ingredients except color and licorice in kettle. Cook to a soft ball. Add licorice syrup and black color. Cook batch to a hard ball in cold water.

Turn out ⅛ inch thick. On this lay a sheet of sheet marshmallow. Put top piece of licorice caramel on when batch is cool enough to handle.

Cut and pack in waxed paper cups, straight-packed or in assortments, alternating rows with plain licorice caramels.

Licorice Caramel-Nougat Center for Chocolates or Bonbons

Corn Syrup (43°)	10 lb.
Cane Sugar	30 lb.
Cocoa Butter	3 lb.
Cream (40%)	4 gal.
Licorice Syrup	4 lb.

Mix half licorice caramel and half vanilla nougat. Do not add any additional color. The resulting batch will be lighter than licorice caramel but of good color.

The licorice-caramel-nougat mixture makes a particularly fine eating piece of candy and is confidently recommended to those who seek something distinctive in the way of chocolate or bonbon centers.

Licorice Turkish Paste

Cane Sugar	75 lb.
Corn Syrup (43°)	15 lb.
Cooking Starch	9 lb.
Cream of Tartar	6 oz.
Licorice Syrup	20 lb.
Water	4 gal.

Put sugar, corn syrup and water in kettle and bring to boil. Add cooking starch which has previously been soaked in 1 gal. of cold water.

After batch has boiled 10 minutes, add cream of tartar (previously melted in 6 ounces of water). Next add licorice syrup and continue boiling until batch is finished (cook until it drops easily from a knife).

Two Oriental delicacies in one. A licorice confection of mild character and velvety texture.

Licorice Strings, Opera Drops or A. B. Gum-Work

Cane Sugar	50	lb.
Corn Syrup (43° Bé.)	135	lb.
Cream of Tartar	2½	oz.
Tartaric Acid	1½	oz.
Special Thick-Boiling Starch	20	lb.

Licorice Syrup	12	lb.
Water	8	gal.

Put sugar and corn syrup in kettle with water. Bring to a boil. Add starch, which has been previously melted in 3 gallons of cold water, and cream of tartar. Cook 1 hour. Add licorice syrup; cook for another half hour, making a total cooking time of 1½ hours. Turn off steam, add vegetable black color, and finally add acid which has been melted in a little hot water.

Cast in blood-warm starch (98°–100° F.); leave in hot room (at 140° F.) for three days; take out of hot room and let stand until cold.

When cold, sugar up and let stand in hot room or a good dry room for 24 hours.

Cook crystal to 233° F., apply on goods hot; leave goods in crystal for 10 hours; drain off crystal and empty crystal pans into boards. Be sure goods are thoroughly dry before packing.

This formula makes a tender and delicious-eating string or drop. It can also be used for A.B. gums, if desired.

Licorice Paste for Bag Work

Powdered XXXX Sugar	10 lb.
Licorice Syrup	1 lb.
Whites of eggs, well beaten	8 oz.

Beat egg whites to a stiff fluff. Add sugar and Licorice Syrup, and sufficient black vegetable color to get the desired shade.

Keep this paste in a covered container and use as needed.

Sets up brittle and "crunchy" like a biscuit. A deliciously flavored paste for making "dominoes," "clubs and spades," Hallowe'en witches and black cats, and all sorts of holiday bag work and bag work novelties.

Licorice Bonbon Fondant (Natural Color)

Cane Sugar	80	lb.
Egg Whites	1	pint
Water	2½	gal.
Licorice Syrup	8	lb.

Cook sugar and water to 242° F., pour on ball beater and let cool. Beat egg whites to a stiff fluff, add to batch and start beater. When thoroughly mixed, add licorice syrup. Use no color. When fondant is finished, put it in a cream tank. Do not use a wet cloth to cover fondant; use a metal or wood cover instead.

A mild flavored bonbon coating of natural light brown color. Adds a new piquant note to the character of bonbons; advances but does not dominate the flavor of the center.

Cocoanut Center for Licorice Bonbons

Fresh Grated Cocoanut	55	lb.
Corn Syrup (43°)	35	lb.
Cane Sugar	15	lb.
Table Salt	¼	oz.

Put all of the above ingredients in steam kettle and cook to a soft ball. Add 10 pounds of nougat cream and 4 ounces vanilla extract. Pour on marble, let cool; cut or roll to size desired. Fork-dip in licorice bonbon coating.

A delicious fresh cocoanut bonbon for the high-class retail trade. A less perishable center can be made using desiccated instead of fresh cocoanut.

Low-Priced Wrapped Licorice Caramels (Natural Color or Colored)

Corn Syrup (43°)	60	lb.
Cane Sugar	40	lb.
Cocoanut Butter (92°)	8	lb.
Full Cream Caramel Paste	25	lb.
Whole Sweetened Condensed Milk	25	lb.
Vanilla Extract	4	oz.
Licorice Syrup	8	lb.

Put all the ingredients except licorice syrup and vanilla extract in kettle. Cook to a soft ball. Add licorice syrup and vanilla extract. Add black color if desired. Cook to a hard ball in cold water. Pour on slab and allow to stand over night. Cut and wrap.

An inexpensive caramel of excellent stand-up and eating qualities.

Licorice Wrapped Kisses (Semi-Nougat Type)

Corn Syrup (43°)	20	lb.
Cane Sugar	10	lb.
Cocoa Butter	¾	lb.
Whole Sweetened Condensed Milk	1¼	lb.
Licorice Syrup	1½	lb.
Color	Sufficient	

Put all materials together in kettle. Cook to 255° F., pour on slab and let cool. Pull for five minutes, spin out and cut to suit.

A pleasing departure from the ordinary run of wrapped kiss formula, this licorice kiss is inexpensive to make, will stay soft and is easy to eat.

Almond Licorice Nougat

Cane Sugar	80	lb.
Corn Syrup (43°)	70	lb.
Almonds	15	lb.
Egg Albumen (Good Beating Quality)	1¼	lb.
Licorice Syrup	10	lb.
Water	8	gal.

Soak egg albumen in water overnight. Then put in nougat kettle and beat 5 minutes. Cook sugar, corn syrup and water to 240° F. Take 3 quarts of this mixture and beat into egg fluff very slowly. Cook balance of sugar and corn syrup batch to 300° F. Add slowly to batch in nougat kettle until all are thoroughly mixed. Now add licorice syrup, almonds and sufficient vegetable color to get a good black color.

If a short nougat is desired, the batch should be beaten about one-half hour. If a chewey nougat is preferred, fifteen minutes beating will be sufficient.

Turn out nougat into steel or wood trays and cut the following day.

This makes a high-class cut nougat which is without equal for adding color contrast and flavor variety to pallid vacation numbers, summer packages and straight nougat assortments.

Genuine Licorice "Anthracite"

Licorice Syrup	12	lb.
Granulated Cane Sugar	50	lb.
Corn Syrup (43°)	60	lb.
Hard Fat (hydrogenated peanut or cottonseed stearine)	2	lb.
Whole 8% Sweetened Condensed Milk	5	gal.
Salt	½	lb.
Cocoanut Butter (92° m.p.)	½	lb.

Put all raw materials in kettle, with the exception of licorice syrup and cook to a crack. Add licorice syrup and black vegetable color to suit. Cook to a hard crack. Pour on marble slab and allow to cool. Cut into irregular chunks resembling coal.

A rich, delicious-eating piece of the butter-toffee variety. Has the novel appearance of anthracite and a savory background of licorice.

Genuine Licorice "Black Birds"

Cane Sugar	12	lb.
Refined Dextrose (Corn Products "Cerelose" or equivalent)	20	lb.
Corn Syrup (43°)	30	lb.
Thin Boiling Starch	9	lb.
Gelatine	3¾	lb.
Licorice Syrup	5	lb.

Dissolve cane sugar, dextrose and corn syrup in 4½ gallons of water. Cook as for a regular high cooked gum drop. Suspend 9 lbs. of thin boiling starch in 2 gallons of water and add gradually to the boiling batch. The cooking is continued until a high cooked gum drop will result when finished.

Dissolve 3¾ pounds of gelatine in 5½ quarts of water and warm up. Mix this in with the remainder of the batch toward the end of the boil. Finally, add licorice syrup and black vegetable color to suit.

It is customary to give this piece a high gloss after taking it out of starch.

The popular chewey black licorice drop in the appealing all-licorice flavor. Those desiring a still chewier drop will find that they can vary the above formula by adding to the gelatine solution an equal amount of gum arabic dissolved in water.

Sugar Coating for Candy

Cocoa Butter Substitute	35	lb.
Powdered Skim Milk	15	lb.
Powdered Sugar	50	lb.
Salt	2	oz.
Vanillin	½	oz.

This may be colored and flavored with oil soluble colors. After melting the cocoa butter substitute add the other materials gradually in a dough mixer and finally run over finishing rolls at a temperature of 110° F. In using, run on enrober at 104° F.

Chocolate Coating

Cocoa Butter Substitute	28	lb.
Cocoa Powder	10	lb.
Skim Milk Powder	10	lb.
XXXX Powdered Sugar	52	lb.
Vanillin	½	oz.
Salt	2	oz.

Melt fat and mix thoroughly. May be used at a temperature of about 110° F. For summer use, add 2% cottonseed stearine (140° m.p.).

Catsup

Tomatoes (After Removing Skins, Seeds, and Green Spots)	30 lb.
Salt	1¼ to 1½ cupfuls
Redistilled Vinegar*	2¼ to 3⅓ cupfuls
Sugar	5 to 8 cupfuls

* If redistilled vinegar cannot be obtained, use 4½ to 6 1/3 cups of white vinegar 5 per cent strength, or 6 to 8 cups of cider vinegar, 4 per cent strength.

Spice Formula No. 1 (Mild-Spiced Catsup)

For a mild-spiced product, add to the above catsup mixture the following:

Celery Seed	3 tsps.
Mustard	3 tsps.
Cinnamon	2 tsps.
Paprika	3 tsps.
Cayenne Pepper	2 tsps.
Onions	4

Spice Formula No. 2 (Heavily-Spiced Catsup)

For a heavily-spiced product, add to the catsup mixture the following:

Mace	2 tsps.
Cassia	2 tsps.
Chillies	2 tsps.
Cayenne Pepper	2 tsps.
Paprika	4 tsps.
White Pepper	4 tsps.
Celery Seed	2 tsps.
Grated Garlic	¼ tsps.
Mustard	2 tsps.
Onions	8

The spicing of catsup is very largely a matter of individual taste and can therefore be varied at will, but at least the minimum quantities of vinegar, sugar, and salt, given in the recipes must be used. The higher quantities mentioned give a heavier catsup which of course will not require as much boiling to concentrate to a desired consistency. In some of the popular commercial catsups as much as 8 cups of sugar, 1⅔ cups of salt, and 4½ cups of vinegar is added to each 30 pounds of tomatoes.

To produce a catsup of proper consistency, 30 pounds of tomatoes should be boiled down to give 16 pounds of catsup. However, depending upon whether the low or high quantities of sugar are used and upon the amount of water in the tomatoes at the start, boiling may have to be continued until the material is reduced to 13 pounds. The consistency of catsup, like the flavor, is largely a matter of personal preference.

To determine the point where boiling should be discontinued, the kettle and contents should be weighed before and after boiling, as for example

Before boiling	
Weight of kettle	2 lbs. 8 ozs.
Weight of tomato juice	30 lbs.
Total weight	32 lbs. 8 ozs.
After boiling	
Weight of catsup	16 lbs.
Weight of kettle	2 lbs. 8 ozs.
Total weight	18 lbs. 8 ozs.

Chop Relish

Black Pepper	1 oz.
Allspice	4 drams
Salt	1 oz.
Horseradish	4 drams
Shallots	4 drams
Walnut Ketchup	20 oz.

Steep for fourteen days, strain, and put into small bottles.

"Soaking" Dried Fish

Dried fish suitably trimmed, are soaked in 10-fold the amount of water whereby there is a 2.2–2.5 increase in weight, with loss of protein and ash constituents. This process takes 3–7 days according to the size of the fish and the temperature. The water is renewed every other day. The fish are then cut into 4–5 centimeter pieces, carefully washed and treated with lye. The lye is made from the ash of birch or from pure sodium or potassium hydroxide. The lye strength is 0.06 normal or greater; the amount used is 10–15 times the original weight of the fish. After remaining in the lye for 4–7 days, the fish are then carefully washed.

Preservation of Meat
British Patent 404,871

Sides of beef are cooled in about 20 hours to 1.7° C. by cold air, quartered, and dipped in a bath of rectified cotton-seed oil at 100° for 1½–2½ minutes. Thereafter the meat is stored at 1.7–0° C. until required for consumption, when it is removed from cold-storage and the protective coating, which then becomes fluid, is wiped off.

Cervelat Wurst Flavor with Pork
Flavoring for 50 Kilograms.

Salt	1700 g.
Saltpetre	20 g.
Pepper (Powdered)	120 g.
Whole Pepper	75 g.

Cervelat Wurst Flavor Extra
Flavoring for 50 Kilograms.

Salt	1700 g.
Saltpetre	20 g.
Sugar	100 g.
Pepper (Powdered)	120 g.
Whole Pepper	80 g.

Cervelat Wurst Flavor Thuringen
Flavoring for 50 Kilograms.

Salt	1800 g.
Saltpetre	20 g.
Sugar	100 g.

Pepper (Powdered)	120 g.
Whole Pepper	80 g.

Cervelat Wurst Flavor Braunschweiger
Formula No. 1
Flavoring for 50 Kilograms.

Salt	1600 g.
Saltpetre	20 g.
Sugar	50 g.
Pepper (Powdered)	150 g.
Oil Cardamom	0.25 g.

Formula No. 2
Flavoring for 50 Kilograms.

Salt	1600 g.
Saltpetre	20 g.
Sugar	50 g.
Pepper (Powdered)	100 g.
Whole Pepper	40 g.
Oil Cardamom	0.25 g.

Balsenwurst Flavor
Flavoring for 50 Kilograms.

Salt	1800 g.
Saltpetre	20 g.
Sugar	75 g.
Coarse Ground Pepper	200 g.

Mettwurst Flavor Braunschweiger
Formula No. 1
Flavoring for 50 Kilograms.

Salt	1300 g.
Saltpetre	20 g.
Sugar	100 g.
Pepper (Powdered)	100 g.
Paprika (Powdered)	30 g.
Oil Cardamom	0.25 g.

Formula No. 2
Flavoring for 50 Kilograms.

Salt	1700 g.
Saltpetre	20 g.
Sugar	100 g.
Pepper (Powdered)	150 g.
Oil Caraway	1.2 g.

Mettwurst Westphalen
Flavoring for 50 Kilograms.

Salt	1500 g.
Saltpetre	15 g.
Sugar	50 g.
Black Pepper (Powdered)	150 g.
Oil Allspice	2.625 g.

Salami & Plock Flavor Wurst
Flavoring for 50 Kilograms.

Salt	1800 g.
Saltpetre	20 g.
Sugar	100 g.
Pepper (Powdered)	200 g.

Knackwurst Flavor Thuringen
Flavoring for 50 Kilograms.

Salt	1500	g.
Saltpetre	20	g.
Sugar	100	g.
Pepper (Powdered)	200	g.
Oil Sweet Marjoram	0.10	g.
Oil Caraway	5.00	g.

Thee Wurst
Flavoring for 50 Kilograms.

Salt	1300	g.
Saltpetre	15	g.
Sugar	80	g.
Pepper (Powdered)	75	g.
Oil Allspice	1.75	g.
Oil Mace	5	g.
Oil Cardamom	1	g.

Liverwurst
Formula No. 1
Flavoring for 50 Kilograms.

Salt	1250	g.
Sugar	100	g.
Pepper (Powdered)	150	g.
Oil Allspice	1.75	g.
Oil Mace	3.75	g.
Oil Sweet Marjoram	0.25	g.
Oleoresin Ginger	2	g.
Oil Cardamom	1	g.

Formula No. 2
Flavoring for 50 Kilograms.

Salt	1250	g.
Sugar	50	g.
Pepper (Powdered)	150	g.
Oil Allspice	0.875	g.
Oil Thyme	1.200	g.

Liverwurst Housmacher
Flavoring for 50 Kilograms.

Salt	1400	g.
Sugar	50	g.
Pepper (Powdered)	150	g.
Oil Allspice	1.75	g.
Oil Mace	5	g.
Oil Sweet Marjoram	0.25	g.
Oil Thyme	1	g.

Liverwurst Hildeshimer
Flavoring for 50 Kilograms.

Salt	1250	g.
Pepper (Powdered)	150	g.
Oil Mace	6.25	g.

Liverwurst Thuringen
Formula No. 1
Flavoring for 50 Kilograms.

Salt	1250	g.
Pepper (Powdered)	150	g.

Oil Allspice	1.75	g.
Oil Mace	5	g.
Oil Sweet Marjoram	0.375	g.

Formula No. 2
Flavoring for 50 Kilograms.

Salt	1200	g.
Pepper (Powdered)	150	g.
Oil Allspice	1.75	g.
Oil Sweet Marjoram	0.375	g.
Oil Thyme	0.400	g.
Oleoresin Ginger	0.800	g.
Oil Cardamom	0.750	g.
Oil Cassia	0.250	g.

Blood & Tongue Wurst Thuringen
Flavoring for 50 Kilograms.

Salt	1800	g.
Oil Allspice	1.75	g.
Black Pepper (Powdered)	200	g.
Oil Sweet Marjoram	0.75	g.
Oleoresin Ginger	2.40	g.
Oil Cloves	8.50	g.
Oil Caraway	2	g.

Paprika Wurst
Flavoring for 50 Kilograms.

Salt	1300	g.
Saltpetre	16	g.
Paprika (Powdered)	100	g.

Mortadella
Flavoring for 50 Kilograms.

Salt	1300	g.
Saltpetre	25	g.
Sugar	100	g.
Pepper (Powdered)	150	g.
Oil Mace	3.75	g.
Oleoresin Ginger	2.40	g.

Liver Pastaten
Flavoring for 50 Kilograms.

Salt	1300	g.
Sugar	50	g.
Pepper (Powdered)	150	g.
Oil Allspice	1.75	g.
Oil Mace	6.25	g.
Oil Sweet Marjoram	0.25	g.
Oleoresin Ginger	2	g.
Oil Cardamom	1	g.
Oil Cassia	0.20	g.

Frankfurters Wiener
Flavoring for 50 Kilograms.

Salt	1300	g.
Saltpetre	15	g.
Sugar	50	g.
Pepper (Powdered)	150	g.
Oil Mace	1.25	g.
Oil Cardamom	1	g.

Sausage American Pork

Flavoring for 50 Kilograms.

Salt	1300 g.
Pepper (Powdered)	100 g.
Oil Sage	4 g.
Oleoresin Capsicum	10 g.

Sausage German

Flavoring for 50 Kilograms.

Salt	1000 g.
Saltpetre	15 g.
Sugar	100 g.
Pepper (Powdered)	150 g.
Oleoresin Ginger	4 g.
Oil Cardamom	1 g.
Oil Cloves	5 g.

Mince-Meat

The following formula will make 1,685 pounds of mince-meat.

Fresh Apples (Peeled and Cored)	320	lb.
Seeded Raisins	75	lb.
Powdered Mustard	3	lb.
Kidney Suet	75	lb.
Currants	96	lb.
Seedless Raisins	96	lb.
Candied Citron Peel	45	lb.
Candied Lemon Peel	25	lb.
Candied Orange Peel	15	lb.
Salt	13	lb.
Ground Cinnamon	9½	lb.
Ground Allspice	1½	lb.
Ground Nutmeg	1	lb.
Ground Cloves	⅜	lb.
Ground Ginger	¼	lb.
Ground Black Pepper	¼	lb.
100-grain Distilled Vinegar	11	gal.
Cane Sugar Syrup	26½	gal.
Cane Sugar	380	lb.
Corn Beef	150	lb.

Peel and core a good grade of cooking apples, a variety used for making apple sauce. Then remove by hand all of the peel not removed by the machine. The apples should be free from rot and worms. Do not use the variety of apples usually selected for canning, such as Newtowns. This variety is not suitable for mince-meat as the apples are too solid to become sufficiently cooked when pies are made.

The raisins and currants should be free from worms and in perfect condition. The suet should be fresh and without any trace of rancidity.

Broken candied peel will fulfil the purpose as well as the fancy candied peel and costs much less. A good grade of powdered mustard and spices should be used as they give a more satisfactory flavor and act as a preservative to prevent fermentation.

The 26½ gallon quantity of cane sugar syrup is made by placing 182 pounds of cane sugar in a steam jacket kettle with 14½ gallons of water. Stir and heat to 212 degrees Fahrenheit and let stand until cold before using. Run the apples, meat, and candied peel through a meat chopper and then place in a mixing machine. Then add the powdered mustard, spices, raisins, currants, salt, vinegar, sugar, and cold syrup.

The suet should be thoroughly chilled by allowing it to stand in ice water. Run the chilled suet through a machine having a plate with ⅛ of an inch perforations. Such a machine is generally used by large meat dealers. Then place the prepared suet in the mixer with the other ingredients and let remain twenty-five minutes.

Mince-meat prepared according to the above formula has been made and shipped for the last several years in glass and barrels without any loss from spoilage.

Gravy Aid Kitchen Bouquet Type

Caramel	½	gal.
Water	3¾	pints
Oil Spice Composite	¼	oz.
Gum Arabic (Powdered)	¼	oz.
Water	4	oz.

Mix the Caramel with the 3¾ pints of water. Emulsify the oil spice composite with the powdered gum arabic and 4 ounces of water. Mix this emulsion with the caramel solution.

Sauer Kraut

Cabbage	40	lb.
Salt	1	lb.

Remove leaves which are dirty. Cut cabbage in quarters and then into shreds. Mix the salt and shredded cabbage. Pack, not too tightly into a 10 gallon crock until nearly full. Cover with a cloth and place heavy weight on same. If temperature is kept around 86° F. fermentation may be complete in 6 to 8 days. The scum which forms on top of kraut should be skimmed off from time to time.

Essential Oils

The flavoring of food products with essential oils replacing ground spices is steadily gaining favor.

The products in which oils are used

as flavors are brighter in appearance and more uniformly flavored. Their application to the materials in work present no difficulties and the cost of flavoring is lower.

From the following table of equivalents one can readily calculate the quantity of essential oils represented by the known weight of spice.

Quantity	Spice	Ess. Oil
100 lb.	Allspice	3½ lb.
100 lb.	Bitter Almonds	½ lb.
100 lb.	Angelica Root	¾ lb.
100 lb.	Angelica Seed	1 lb.
100 lb.	Anise Seed (Russian)	2½ lb.
100 lb.	Anise Star (Chinese)	3 lb.
100 lb.	Calamus Root	2½ lb.
100 lb.	Caraway Seed	5 lb.
100 lb.	Cardamom Seed	5 lb.
100 lb.	Cassia Cinnamon	1 lb.
100 lb.	Celery Seed	3 lb.
100 lb.	Cinnamon Ceylon	1 lb.
100 lb.	Cloves	17 lb.
100 lb.	Coriander Seed	½ lb.
100 lb.	Cumin Seed	3 lb.
100 lb.	Dill Seed	3½ lb.
100 lb.	Estragon	½ lb.
100 lb.	Fennel Seed	5 lb.
100 lb.	Horse Radish	1 oz.
100 lb.	Laurel Leaves (Bay)	2 lb.
100 lb.	Lovage Root	½ lb.
100 lb.	Mace	12½ lb.
100 lb.	Mustard Seed	¾ lb.
100 lb.	Nutmeg	12½ lb.
100 lb.	Parsley Seed	3 lb.
100 lb.	Sage	2 lb.
100 lb.	Sweet Basil	3/20 lb.
100 lb.	Sweet Marjoram	½ lb.
100 lb.	Wild Marjoram	2 lb.
100 lb.	Thyme	2 lb.
100 lb.	Valerian Root	1 lb.
100 lb.	Blackpepper	Oleoresin 6 lb.
100 lb.	Capsicum	Oleoresin 8 lb.
100 lb.	Ginger	Oleoresin 10 lb.
100 lb.	Garlic	Imit'n flavor 4 oz.
100 lb.	Onion	Imit'n flavor 4 oz.

The oil can be added directly to the products and distributed during the mixing operation, or if the quantity of oil is small it can be mixed with a quantity of a fixed oil such as corn oil to increase bulk to a desired quantity.

Thus, a calculated amount, say 3 drams 49 minims can be dissolved in sufficient corn oil to make the entire bulk one fluid ounce, which would be the quantity to add and which would contain just the 3 drams 49 minims of the flavoring oil.

Other methods are also applicable. An emulsion can be made in calculated proportions to make the addition of any desired quantity to a batch. Or, the oils could be absorbed on salt, sugar, flour, starch or any other dry material used in the process.

For pickle makers a solution of the oil in 75% acetic acid forms a terpeneless solution of the flavor bodies that is readily distributed in the various liquors and products.

True Maple Flavor
U. S. Patent 1,961,714

Maple Syrup	100 parts
Alcohol	400–100 parts

Add the alcohol slowly with stirring. The sugar precipitates and the flavor is dissolved by the alcohol which is filtered off from the sugar.

Imitation Lemon Flavor

Citral	2½	drams
Rhodinol	6	drops
Glycerin	1	oz.
Alcohol	58	oz.
Water to make	1	gal.

Non-Alcoholic Lemon Flavor

Citral	3 oz.
Glycopon XS	40 pints
Dissolve and add	
Water to make	8 gal.

Non-Alcoholic Imitation Lemon Flavor

Citral	⅜ oz.
Glycopon XS	5 pints
Water to make	1 gal.

Household Root Beer Flavor

Caramel	½ gal.
Water	3¾ pints
Oil Root Beer C.	¼ oz.
Gum Arabic (Powdered)	¼ oz.
Water	4 oz.

Mix the caramel with the 3¾ pints of water. Emulsify the Oil Root Beer C. with the gum arabic and 4 ounces water. Mix this emulsion with the caramel solution. Put up in 4 ounce bottles.

To use mix 4 ozs. of the above flavor with 3¾ gallons of water; add 3½ pounds of sugar and one cake of yeast. Bottle and allow to ferment.

Temperature Readings for Syrups

Light String	226° F.
Heavy String	230° F.
Soft Ball	238° F.
Medium Ball	240° F.

Stiff Ball	244° F.
Hard Ball	250° F.
Light Crack	254° F.
Brittle Crack	275° F.
Hard Crack	290° F.

Caramel Syrup for Sundaes

Sugar Coarse	13	lb.
Maple Sugar	6½	lb.
Granulated Sugar	6½	lb.
Corn Syrup	3	lb.
Sweet Cream	1¼	gal.
or		
Butter	1½	lb.

Cook until of thick consistency. Then add 1¼ gallons sweet cream and cook to 226° F. While cooling off flavor with vanilla. Color with burnt sugar.

Chocolate Syrup Concentrate

Chocolate Liquor	34	lb.
Granulated Sugar	30	lb.
Nulomoline	6¾	gal.
Water	2¾	gal.

The above formula yields approximately 135 pounds of syrup, or approximately 12 gallons of Chocolate Syrup Concentrate. The method of manufacture is as follows.

Place 2¾ gallons of water in a 25 gallon steam jacketed copper kettle and bring water to a boil. Add 30 pounds of granulated sugar and stir until dissolved. Add 6¾ gallons of Nulomoline and stir until dissolved. Add 34 pounds of melted chocolate liquor, a little at a time with constant stirring, until thoroughly dispersed. Bring the mixture to a boil and boil for 5 minutes. Run syrup through a strainer, and cool to 160° F. Put syrup through a homogenizer at approximately 3000 pounds pressure. The chocolate syrup is run into a holding kettle and then run into cans. When the syrup is cooled down to 90° F. put the covers on cans and store cans in a cool place preferably not over 75° F. It is necessary to keep everything spotlessly clean to avoid mold contamination. This is accomplished by thoroughly washing machinery, floors, etc.

The above syrup can be used as follows:

It serves as a cake ingredient, filling between layers, and as an icing. For thin chocolate syrup for fountain use, dilute one gallon can with one gallon simple sugar syrup. Use as is or dilute as above for Malted Milk and Milk Cocoas. It could be used as a confection by mixing with chopped nuts and powdered sugar and cutting into squares. For

Ice Cream Frappes, warm syrup and pour over ice cream.

Chocolate Syrup
Formula No. 1

Nulomoline Syrup	80	lb.
Chocolate Liquor	20	lb.
Water	15	lb.

Heat Nulomoline to 200° F. Shut steam, and add the chocolate liquor. Stir until melted. Add the water. Heat for 3 minutes.

Formula No. 2

Nulomoline Syrup	1330	lb.
Cocoa Powder	170	lb.
Chocolate Liquor	70	lb.
Salt	8	lb.
Water	145	lb.
Vanillin Crystals	1	lb.

Formula No. 3

Sugar	210	lb.
Nulomoline	350	lb.
Water	145	lb.
Cocoa Powder	110	lb.
Salt	1½	lb.

Heat water and sugar to 180–190° F. Cool to 150° F. Then add the Nulomoline. Sift in cocoa and salt. Heat to 180° F.

Ginger Ale Syrup

Simple Syrup	1	gal.
Ginger Ale Extract	1	oz.
Citric Acid	½	oz.
Color with Caramel.		

Ginger Fizz Syrup

Simple Syrup	1	gal.
Ginger Ale Extract	2	oz.
Lemon Extract	1	oz.
Citric Acid	½	oz.

Grape Syrup

Simple Syrup	5	pints
Grape Juice	3	pints
Tartaric Acid	¼	oz.
Benzoate of Soda	⅛	oz.

Synthetic Honey Syrup

Invert Syrup	75	parts
Malt Extract Syrup	10	parts
Water	15	parts

Lemon Syrup

Simple Syrup	1	gal.
Lemon Extract	2	oz.
Citric Acid	¾	oz.
Color Yellow with Tartrazine.		

Lemon Syrup Fresh

Grate the peels and squeeze the juice of 10 lemons into 1 gallon of warm Simple Syrup. Let stand overnight. Then strain.

Maple Syrup

Dissolve 7 pounds of Maple Sugar in ½ gallon water. Bring to a boil and strain.

Mint Syrup

Simple Syrup	1 gal.
Essence Peppermint	½ oz.
Color	Green

Orange Syrup

Simple Syrup	1 gal.
Orange Extract	2 oz.
Citric Acid	¼ oz.
Color	Orange

Pear Syrup

Simple Syrup	1 gal.
Pear Extract	2 oz.
Citric Acid	¼ oz.
Color	Light Yellow

Pure Pineapple Syrup

Chopped Pineapple	1 quart
Citric Acid	1 oz.
Simple Syrup	3 quarts

Root Beer Syrup

Simple Syrup	1 gal.
Root Beer Extract	2 oz.

Color with Caramel.

Preparation of Simple Syrup

Granulated Sugar	7 lb.
Water	5 pints

Bring water to a boil, then add sugar, stir until dissolved.

Invert Sugar Syrup

Sugar	100	lb.
Water	44	lb.
Tartaric Acid	1⅞	oz.

Heat the above ingredients for 30 minutes. The batch should not be boiled violently, as too much evaporation results.

Table Syrup
Formula No. 1

Simple Syrup	75 parts
Maple Sugar	25 parts

Formula No. 2

Simple Syrup	75 parts
Corn Syrup	25 parts

Formula No. 3

Simple Syrup	75 parts
Invert Syrup	25 parts

Wintergreen Syrup

Simple Syrup	1 gal.
Essence Wintergreen	½ oz.
Color	Light Pink

Burnt Sugar Coloring

Place sugar in an old sauce pan, and set on fire. The sugar starts to melt, then stir, and when sufficiently brown, remove from heat. Stir in a little water to dissolve the burnt sugar. By adding corn syrup instead of water a caramel paste will form.

Preserving Citrus Oils

Dehydration with sodium sulphate, followed by a rapid filtration through a Chardin filter, gives a product which keeps for 12 to 15 months in glass vessels with ground stoppers in a cool place. Such purified oils keep for at least two years if mixed with 0.02 per cent of quinol or p-diphenol or indefinitely if dissolved in 20 to 25 per cent weight of ethyl alcohol; deterpination is of less advantage; sweet orange oil is less changeable than lemon oil, and its keeping qualities are improved by the same treatments. Bitter orange oil keeps almost indefinitely in a cold place if previously dehydrated. Mandarin oil should be kept in the dark to prevent coloration.

Vanilla Extract

Vanilla beans (Comminuted and Crushed)	10 g.
Sugar	200 g.

Mix

Ethyl Alcohol (96%)	650 cc.
Water	350 cc.

Moisten vanilla beans with 100 cubic centimeters of the dilute alcohol and let stand in tightly closed container for 12 hours. In mortar mix mass with the sugar, grind to a pulp; add 750 cubic centimeters of the dilute alcohol, allow to macerate for 7 days in bottle. Filter, wash filter with dilute alcohol to make total volume 100 cubic centimeters.

Imitation Extract Pineapple

Add ½ pint non-alcoholic imitation oil

pineapple to 5 pints alcohol (95%). Mix and add distilled water to make 1 gallon.

Imitation Extract Strawberry

Add ½ pint non-alcoholic imitation oil strawberry to 4½ pints alcohol (95%). Mix and add distilled water to make 1 gallon.

Imitation Extract Raspberry

Add ½ pint non-alcoholic imitation oil raspberry to ½ gallon alcohol (95%). Mix and add distilled water to make 1 gallon.

Imitation Extract Banana

Add ½ pint non-alcoholic oil banana to 4¼ pints alcohol (95%). Mix and add distilled water to make 1 gallon.

Imitation Extract Peach

Add ¾ pint non-alcoholic imitation oil peach to 4½ pints alcohol (95%). Mix and add distilled water to make 1 gallon.

Imitation Extract Root Beer

Add ½ pint non-alcoholic imitation oil root beer to 5½ pints alcohol (95%). Mix and add distilled water to make 1 gallon.

Imitation Extract Grape

Add 1 pint non-alcoholic imitation oil grape to 4¼ pints alcohol (95%). Mix and add distilled water to make 1 gallon.

Imitation Extract Cherry

Add ½ pint non-alcoholic imitation oil cherry to ½ gallon of alcohol (95%). Mix and add distilled water to make 1 gallon.

Imitation Extract Apricot

Add ¾ pint non-alcoholic imitation oil apricot to 4¼ pints alcohol (95%). Mix and add distilled water to make 1 gallon.

Imitation Extract Blackberry

Add ½ pint non-alcoholic imitation oil blackberry to 4½ pints alcohol (95%). Mix and add distilled water to make 1 gallon.

Imitation Extract Red Currant

Add ½ pint non-alcoholic oil red currant to 4 pints, alcohol (95%). Mix and add distilled water to make 1 gallon.

Imitation Extract Black Currant

Add ½ pint non-alcoholic oil black currant to ½ gallon alcohol (95%). Mix and add distilled water to make 1 gallon.

Apricot Basic Ether

Jasmine Oil, True	⅙	dram
Anethol	4	drams
Eugenol	6	drams
Pettigrain	14	drams
Amyl Alcohol	1½	oz.
Vanillin	3	oz.
Peach Flavor Oil	10	oz.
Ethyl Valerate	1	lb.
Ethyl Acetate	1¼	lb.
Amyl Acetate	1½	lb.
Ethyl Butyrate	1½	lb.
Butyl Butyrate	1¾	lb.

Apple Basic Ether

Cognac Oil White	¾	oz.
Benzaldehyde	1	oz.
Clove Oil	3	oz.
Ethyl Oenanthate	11¼	oz.
Ethyl Butyrate	1	lb.
Amyl Acetate	1½	lb.
Amyl Valerate	1½	lb.
Ethyl Valerate	2	lb.
Ethyl Acetate	3	lb.

Cherry Basic Ether

Ceylon Cinnamon Oil	½	oz.
Cognac Oil	1½	oz.
Clove Oil	2	oz.
Benzyl Benzoate	2½	oz.
Vanillin	2½	oz.
Benzaldehyde	7	oz.
Amyl Alcohol	8	oz.
Ethyl Oenanthate	12	oz.
Amyl Formate	2¼	lb.
Ethyl Acetate	3½	lb.

Cherry Brandy Basic Ether

Neroli Oil	⅛	dram
Vanillin	12	drams
Clove Oil	12	drams
Cinnamon Oil	1	oz.
Cognac Oil	1½	oz.
Amyl Alcohol	1½	oz.
Benzaldehyde	4½	oz.
Ethyl Pelargonate	6	oz.
Benzyl Benzoate	½	lb.
Ethyl Oenanthate	1½	lb.
Ethyl Acetate	4	lb.

Grape Basic Ether

Methyl Salicylate	1¼	drams
Cognac Oil	4¾	drams
Cardamom Oil	6	drams
Mace Oil	12	drams

Ethyl Pelargonate	3½	oz.
Amyl Butyrate	11	oz.
Ethyl Oenanthate	2½	lb.
Ethyl Acetate	4½	lb.

Grenadine Basic Ether

Clove Oil	6	drams
Honey Flavor Oil	12	drams
Ceylon Cinnamon Oil	14	drams
Neroli	1½	oz.
Pettigrain Oil	2½	oz.
Benzoyl Benzoate	16	oz.
Peach Flavor Oil	¾	lb.
Amyl Valerate	1¼	lb.
Amyl Acetate	1½	lb.
Ethyl Acetate	3½	lb.

Peach Basic Ether

Pettigrain Oil	5	drams
Pimento Oil	6	drams
Cardamom Oil	13	drams
Sweet Orange Oil	2	oz.
Vanillin	2	oz.
Benzaldehyde	2½	oz.
Lemon Oil	4	oz.
Amyl Butyrate	4	oz.
Peach Flavor Oil	1	lb.
Ethyl Butyrate	1	lb.
Amyl Acetate	2½	lb.
Ethyl Acetate	2½	lb.

Pineapple Basic Ether

Vanillin	1¼	oz.
Lemon Oil	1½	oz.
Butyric Acid 100%	1½	oz.
Ethyl Sebacate	1¾	oz.
Ethyl Pellargonate	4	oz.
Amyl Acetate	6	oz.
Ethyl Acetate	1¼	lb.
Ethyl Butyrate	3	lb.
Amyl Butyrate	4¾	lb.

Raspberry Basic Ether

Rose Oil	⅛	dram
Jasmine Oil 10%	1⅛	drams
Ceylon Cinnamon Oil	6¾	drams
Clive Oil	12	drams
Pettigrain Oil	12	drams
Vanillin	3	oz.
Orris Oil Solution 10%	5	oz.
Benzyl Benzoate	6	oz.
Ethyl Formate	½	lb.
Ethyl Acetate	1½	lb.
Amyl Acetate	2	lb.
Iso Butyl Acetate	3	lb.

Strawberry Basic Ether

Coumarin	12	drams
Ceylon Cinnamon Oil	12	drams
Amyl Formate	4	oz.

Neroli	4½	oz.
Ethyl Benzoate	6	oz.
Amyl Butyrate	½	lb.
Methyl Salicylate	½	lb.
Ethyl Butyrate	1	lb.
Ethyl Acetate	1	lb.
Benzyl Acetate	1½	lb.
Amyl Butyrate	1½	lb.

Tangerine Basic Ether

Lemon Oil	1	lb.
Bitter Orange Oil	1	lb.
Amyl Acetate	1	lb.
Ethyl Acetate	1½	lb.
Mandarin Oil	2	lb.
Amyl Butyrate	3½	lb.

Creme de Cocoa

Cocoa Tincture	¼	oz.
Vanilla Tincture	3½	oz.
Liqueur Body	11	gal.
Brown color	Sufficient	

Creme de Menthe

Oil Peppermint	32	drops
Alcohol	52	fl. oz.
Sugar	56	oz.
Water to make	1	gal.

Dissolve the oil in alcohol, the sugar in ater, mix and color green with chloro-phyll or synthetic coloring.

Absinthe Essence

Wormwood	11	lb.
Anise Italian	11	lb.
Star Anise	2¾	lb.
Fennel Seed	¾	lb.
Coriander	¾	lb.
Mace	½	lb.
Cinnamon	2½	oz.
Alcohol (95%)	17½	gal.
Water	8¾	gal.

Soak for 8 days and distill over.

Barbado Essence

Lemon Peel	14	oz.
Orange Peel	14	oz.
Cinnamon	4	oz.
Cloves	3	oz.
Mace	1	oz.
95% Alcohol	22	gal.
Brown color	Sufficient	

Cacao Essence

Cacao Nibs	67	lb.
Cinnamon	14	oz.
Cloves	10	oz.
Vanilla	1½	oz.
95% Alcohol	40	gal.

Chartreuse (Essence)

Angelica Root	10	oz.
Calamus	3	oz.
Coriander	12	oz.
Cassia	5	oz.
Anise Seed	5	oz.
Cardamom	5	oz.
Cinnamon	6	oz.
Cinchona	2	oz.
Lemon Peel	12	oz.
Orange Peel	12	oz.
Hyssop	8	oz.
Melissa	18	oz.
Peppermint	12	oz.
Poplar Buds	8	oz.
Ginger Jam	10	oz.
Thyme	6	oz.
Tonka Beans	4	oz.
Mace	4	oz.
Cloves	2	oz.
Wormwood	16	oz.
Alcohol 95%	7	gal.
Water	3½	gal.

Macerate 14 days, press, and filter or distill.

Lemon Essence

Alcohol (95%)	2½	quarts
Lemon Juice	2½	quarts

Mix until incorporated, and filter.

Spearmint Essence

Spearmint	70	lb.
Peppermint	8½	lb.
Melissa	4	lb.
Alcohol (95%)	22	gal.

Waldmeister Essence

Woodruff	20	lb.
Tonka Beans	3½	oz.
Alcohol (95%)	3	gal.
Green color	Sufficient	

"Angostura" Bitters

Formula No. 1

Angostura Bark	70	oz.
Angelica Root	4	oz.
Ceylon Cinnamon	8	oz.
Gentian	6	oz.
Galangal	21	oz.
Hops	6	oz.
Ginger	1½	oz.
Cardamoms	8	oz.
Cloves	1½	oz.
Pimento	10	oz.
Orange Peel	35	oz.
Raisins	19	lb.
95% Alcohol	18	gal.

Color to suit. Macerate for 2 weeks and filter.

Formula No. 2

Angostura Bark	6	lb.
Gentian	2½	lb.
Galangal	5	lb.
Cardamoms	6	lb.
Catechu	2½	lb.
Coriander	2½	lB.
Caraway	2½	lb.
Curcuma	35	lb.
Dandelion Root	2½	lb.
Mace	1	lb.
Nutmeg	2½	lb.
Pimento	5	lb.
Orange Peel	10	lb.
Snake Root Con.	2½	lb.
Licorice	2½	lb.
Wormwood	2½	lb.
Cinnamon	2½	lb.
Red Sandal	7	lb.
Alcohol (65%)	29	gal.

Formula No. 3

Angostura Bark	12	oz.
Gentian	8	oz.
Galangal	8	oz.
Ginger	1½	oz.
Cardamom	11	oz.
Cinnamon	11	oz.
Clover	1	oz.
Bitter Orange Peel	14	oz.
Sandal Red	30	oz.
Tonka Beans	20	oz.
Zedoary Root	8	oz.
60% Alcohol	22	gal.
Caramel Color	7½	oz.

Macerate 2 weeks and filter.

Formula No. 4 (Very Bitter)

Angostura Bark	17½	lb.
Cardamoms	3½	lb.
Cloves	14	oz.
Cassia Buds	8¾	lb.
Water	20	gal.
Alcohol (95%)	20	gal.

Formula No. 5

Bitter Orange Peel	1	oz.
Angostura Bark	1	oz.
Calamus	1	oz.
Cardamom	160	g.
Alcohol	10	fl. oz.
Water	6	fl. oz.

After this mixture has stood one week it is diluted to two to four gallons with 80 proof alcohol and filtered. It may be colored red with cochineal or cudbear.

Formula No. 6

Galangal Root	28	oz.
Sweet Flag Root	28	oz.
St. John's Bread	70	oz.
Cardamom	70	oz.
Ginger Root	70	oz.
Quillac Root	70	oz.

Lemon Peel	70 oz.
Cloves	98 oz.
Tonka Beans	140 oz.
Ceylon Cinnamon	140 oz.
Bitter Orange Peel	140 oz.
Angostura Bark	350 oz.
Sugar Syrup	11 gal.

Add alcohol and water to give 38% concentration of alcohol.
Color—Red Brown.

Aromatic Bitters

Angostura Bark	4	oz.
Chamomile Flowers	1	oz.
Cardamom Seeds	2	drams
Cinnamon	2	drams
Orange Peel	1	oz.
Raisins		
Dilute Alcohol (50%)	2½	gal.

Baker's Bitters

Bitter Orange Peel	1½	oz.
Quassia	1	oz.
Calamus	1	oz.
Cardamom	160	g.
Catechu	½	oz.
Alcohol (95%)	8	oz.

Allow to stand one week, then add water 8 oz. and filter.

Boker's Bitters

By substituting quassia bark 1 oz. for the angostura this popular bitters may be prepared using the formula given above.

Spanish Bitters

Herbs

Marjoram	8 oz.
Cloves	16 oz.
Camellia	16 oz.
Pimento	24 oz.
Sweet Flag	24 oz.
Coriander	24 oz.
Peppermint	32 oz.
Cascarilla Bark	32 oz.
Gentian Bark	48 oz.
Vermouth	48 oz.
Angelica Root	48 oz.
Galangal Root	52 oz.
Calamus Root	80 oz.
Bitter Orange Peel	80 oz.

Oils

Roman Camellia	1 oz.
Cassia	4 oz.
Angelica Root	4 oz.
Vermouth Green	4 oz.
Cloves	8 oz.

Juniper	4 oz.
Calamus	20 oz.
Bitter Orange	20 oz.
Alcohol (38%).	

Stomach Bitters

Formula No. 1

Elecampane	7	oz.
Anise	10	oz.
Calamus	8	oz.
Coriander	4	oz.
Dill	2½	oz.
Fennel	8	oz.
Galangal	4	oz.
Caraway	5½	oz.
Mace	4	oz.
Nutmeg	4	oz.
Cloves	5½	oz.
Pimpinella	2½	oz.
Orris	4	oz.
Cinnamon	12	oz.
Zedoary	4	oz.
Alcohol (95%)	22	gal.
Color to suit.		

Formula No. 2

Angelica Root	7	oz.
Cardamus	7	oz.
Calamus	10	oz.
Lemon Peel	25	oz.
Coriander	7	oz.
Cardamoms	1	oz.
Galangal	7	oz.
Ginger	7	oz.
Marjoram	7	oz.
Orange Peel	25	oz.
Rosemary	7	oz.
Thyme	7	oz.
Tonka Beans	17½	oz
Orris Root	7	oz.
Juniper Berries	7	oz.
Alcohol (95%)	22	pints
Color to suit	25	gal.

Formula No. 3

Cloves	12 oz.
Angelica Root	12 oz.
Vermouth	16 oz.
Gentian Root	16 oz.
Galangal	16 oz.
Coriander Seed	16 oz.
Calamus Root	24 oz.
Aloe Root	24 oz.
Quillac Root	32 oz.
Bitter Orange Peel	40 oz.
Rhubarb	40 oz.
Star Anise	56 oz.
Licorice Root	96 oz.
St. John's Bread	80 oz.

Color—Dark Brown. Sugar Content—no added sugar. Alcohol (42%).

Swedish Bitters

Aloes	6 g.
Rhubarb	1 g.
Gentian	1 g.
Zedoary	1 g.
Saffron	1 g.
Diluted Alcohol	200 g.

Turbidity in Vermouth Bitters

Turbidity is avoided if the catechu and licorice are extracted with 10% alcohol instead of water; allowed to stand until a precipitate forms; and then filtered with a suitable filter aid.

In preparing these liqueurs and brandies the oils are dissolved in alcohol and the sugar in water. The two solutions are then mixed and allowed to stand in a warm place to blend the flavors. If not clear they should be filtered. The use of talc and chalk as clarifying agents is common.

Absinthe

Formula No. 1

Oil Wormwood	16 drops
Oil Bitter Almonds	2 drops
Oil Anise	1 drop
Spirit Nitrous Ether	2 drams
Sugar	3 oz.
Alcohol	56 fl. oz.
Water to make	1 gal.

Dissolve the oils in the alcohol, the sugar in the water, mix, and color green with chlorophyll.

Formula No. 2

Oil Wormwood	1	oz.
Oil Juniper Berries	½	oz.
Oil Cinnamon	60	grains
Oil Coriander	60	grains
Oil Ginger	60	grains
Oil Nutmeg	30	grains
Bitter Orange Peel	30	grains
Alcohol (95%)	3	pints

Allow to stand one week. Then add water to make one gallon and filter.

Formula No. 3

This is a stronger drink.

Oil Wormwood	64 drops
Oil Anise	48 drops
Oil Coriander	32 drops
Oil Fennel	32 drops
Oil Thyme	16 drops
Alcohol (95%)	1 gal.
Water	1 gal.

Absinthe (Turin)

Anise Seed	44	oz.
Bitter Almonds	20	oz.
Fennel	28	oz.
Calamis	5½	oz.
Coriander	14	oz.
Peppermint	3	oz.
Sassafras	12	oz.
Wormwood	5½	oz.
Sugar	12	lb.
95% Alcohol	22	gal.

Anisette Liqueur

Concentrated Russian Oil of Anise	40	g.
Alcohol (90%)	45½	l.
Normal Syrup*	4	l.
Water	51½	l.

Lemon Liqueur

Oil of Lemon	75	g.
Alcohol (90%)	44.5	l.
Normal Syrup*	10	l.
Water	45.5	l.
Tincture Spanish Saffron	Sufficient	

* Normal Syrup (For above two formulas)

Sugar	50 kg.
Water	25 l.

The mixture of the above is heated slowly until boiled down to about 50 liters. The scum forming at the top is removed from time to time.

Cacao Liqueur (Chocolate)

Formula No. 1

Cocoa powder	4 oz.
Cinnamon	1 oz.
Cassia buds	1 oz.
Cardamom	90 g.
Cloves	90 g.
Milk	6 fl.oz.
Alcohol	52 fl.oz.
Sugar solution to make	1 gal.

Mix the cocoa with the milk and alcohol. After 24 hours add the powdered spices. After standing a further 24 hours it is filtered and 14 to 28 ounces of sugar in 42 ounces of water is added. A creme requires 56 ounces of sugar.

Formula No. 2

Cocoa (Powdered)	4 oz.
Oil Cinnamon	1 oz.
Oil Cassia	1 oz.
Oil Cardamom	90 g.
Oil Cloves	90 g.
Milk	6 oz.
Alcohol	42 oz.

Allow to stand 48 hours agitating occasionally. Add 14–24 ounces sugar in 36 to 42 ounces water and filter.

Chartreuse

Oil Angelica	3 drams
Oil Mellisa	32 drops
Oil Hyssop	32 drops
Oil Cinnamon	10 drops
Oil Mace	42 drops
Oil Coriander	20 drops
Alcohol	72 fl.oz.
Sugar	14–28 oz.
Water to make	1 gal.

White chartreuse is made with 14 ounces of sugar, yellow requires 28 ounces of sugar and is colored with saffron. Green chartreuse requires 21 ounces sugar and is colored with indigocarmine.

Grand Chartreuse (Green)

Lemon Mint (Dried)	1	gram
Dried Tops of Flowering Hyssop	250	grams
Peppermint (Dried)	250	grams
Alpine Wormwood (Swiss Tea)	250	grams
Costmary	125	grams
Thyme	50	grams
Angelica Seeds	150	grams
Angelica Root	150	grams
Arnica Flowers	20	grams
Buds of Balsam Poplar (North American)	20	grams
Chinese Cinnamon	15	grams
Mace	15	grams
Spirits of Wine, (80 o.p.) American	58.5	litres

Macerate for 24 hours: distill; rectify; and collect 60 litres. Prepare a syrup by melting under heat 25 kilograms of refined sugar in 24 litres of water. Mix the syrup with the distillate, and reduce the product with 100 litres of water: *trancher* (as previously described); and color with blue and saffron (or caramel) according to the shade wanted. Fine; and filter.

Curacao (Triple Sec)
(Extra Finest Quality)

Prepare 24 ounces of extra-thin genuine fresh curacao peels and 12 ounces of extra-thin fresh orange peels. Macerate this mixture for two days with 2½ gallons of alcohol (190 proof). Collect and save for a later stage of the process 2 gallons of this extract. To the remaining macerate add 20 pounds of extra-thin fresh curacao peels, 15 pounds of extra-thin fresh orange peels, 10 ounces of mace, 2 ounces of cloves, 38 gallons of alcohol (190 proof), and 40 gallons of distilled water.

Digest for six hours with very little heat, place in still, add another 5 gallons water, distil slowly for two hours with return flow, until all the alcohol is driven over. Rectify the raw distillate to 35% alcohol by volume and filter clear over Kieselguhr; clean still and then rectify the raw distillate to 58 gallons, 60% alcohol by volume.

Prepare the finished curacao by the following formula:

Rectified Distillate (60%) (as prepared above)	58 gal.
Extract (Taken out from above process)	2 gal.
Genuine Jamaica Rum (74%)	1 gal.
Grape Distillate (60%)	4 gal.
Port Wine	2 gal.
Glucose (42° Bé.)	5 gal.
Syrup*	18 gal.
Distilled Water	13 gal.

* This syrup is made from 250 pounds of the best grade sugar, 25 pounds of milk sugar and 1 pound of citric acid, C. P.

Curacao (Good Quality)

Prepare 1½ pounds of extra-thin curacao peels, macerate for two days with 1½ gallons of alcohol (190 proof). Collect and save for a later stage of the process 1 gallon of this extract. To the remaining macerate add 20 pounds of extra-thin curacao peels, 10 pounds of dried expulped curacao peels, 10 ounces of mace, 2½ ounces of cinnamon, 2½ ounces of cloves, 13 gallons of alcohol, and 13 gallons of distilled water.

Digest for six hours with very little heat, place in still, add another five gallons of water, distil slowly for two hours with return flow, until all the alcohol is driven over. Rectify the raw distillate to 35% alcohol by volume and filter clear over Kieselguhr; clean still and then rectify the raw distillate to 20 gallons, 60% alcohol by volume.

Prepare the finished curacao by the following formula:

Rectified Distillate (60%) (as prepared above)	20	gal.
Extract (Taken out from above process)	1	gal.
Genuine Jamaica Rum 74%)	½	gal.
Grape Distillate (60%)	2½	gal.
Port Wine	2	gal.
Alcohol (190 Proof)	24½	gal.
Syrup*	23	gal.
Distilled Water	29½	gal.

Color with Caramel to Light Brown.

* This syrup is made from 300 pounds

of sugar, 20 pounds of milk sugar and 1 pound of citric acid C. P.

The curacao liqueurs when newly manufactured contain a taste similar to the peel, and are very raw. However, by the addition of a small quantity of milk sugar to the syrup the rawness is not so pronounced. By heating the liqueur to approximately 130° F. in a vacuum the age of this really fine product will be hastened.

Gin
Formula No. 1

Oil Juniper	64 drops
Oil Cassia	8 drops
Alcohol	56 fl.oz.
Sugar	16 oz.
Water to make	1 gal.

If a dry gin is desired the sugar may be omitted.

Formula No. 2

Oil of Juniper	10	oz.
Oil of Coriander	1½	oz.
Oil of Angelica	½	oz.
Oil of Bitter Orange	¾	oz.
Oil of Lemon	1	dr.

To this add sufficient Alcohol to make 1 pound.

The resulting compound when added 1 ounce to 10 gallons of 90 proof alcohol gives a specific type of gin.

The blender can round out the flavor of the product so that it will approach the quality of a distilled product by aging for several months or one can hasten this aging process by keeping the finished product at the warm temperature of 150° F. for a period of 48 hours. Agitation during this aging process is helpful. The heat combined with agitation assists in blending the oils into their proper bouquet and aroma.

Sloe Gin

To the above add six to twelve crushed sloes and allow to macerate for 1–3 weeks. Filter.

Gold Water

Spirit of Lemons	10	l.
Spirit of Oranges	9	l.
Spirit of Coriander	2	l.
Spirit of Daucus	2	l.
Spirit of Fennel	2	l.
Orange Flower Water	2	l.
Spirits of Wine (80 o.p. American)	11	l.
Refined Sugar	50	kg.
Water	25.5	l.

Genuine gold leaf is added to the liqueur in the bottles, after being broken by stirring with a glass rod or simply shaking in a small quantity of liqueur. Distilled Goldwaters are sold uncolored; when prepared synthetically they carry a sherry tint. The sugar density must be such as will keep the bits of gold leaf floating in the finished liqueur.

Danziger Goldwasser
Herbs

Angelica Root	⅓ oz.
Cloves	½ oz.
Peppermint	⅓ oz.
Caraway	½ oz.
Fenchel	1 oz.
Star Anise	3 oz.

Oils

Rose	1/20 oz.
Coriander	⅕ oz.
Lemon	⅕ oz.
Calamus	½ oz.
Cloves	½ oz.
Cinnamon	½ oz.
Caraway	1 oz.
Bitter Orange	1 oz.
Anise	1⅓ oz.

Cherry Juice	10 quarts
Sugar Syrup	40 quarts
Alcohol	32%

Kümmel
Formula No. 1

Oil Caraway	16 drops
Alcohol	3 pints
Sugar	10 oz.
Water to make	1 gal.

Formula No. 2

A kümmel of somewhat fuller body may be made as follows.

Oil Caraway	12 drops
Oil Anise	1 drop
Oil Celery	2 drops
Vanilla Extract	12 drops
Spirit Nitrous Ether	2 drams
Alcohol	60 fl.oz.
Sugar	10 oz.
Water to make	1 gal.

Formula No. 3

Seeds of Dutch Kümmel (carvi)	4	kg.
Aniseed	1.5	kg.
Coriander Seeds	.6	kg.
Fennel	1	kg.
Peppermint	.8	kg.
Spirits, 80% by volume (60 o.p. American)	30	l.

To the distillate is added several drops of rose essence; 50 litres of sugar syrup:

36°; 21 litres of spirits, 95 per cent (90 o.p. American); and water to complete 100 litres. The strength of the liqueur should be 40 per cent by volume, (20 o.p. American).

o.p. = overproof.

Rock and Rye

Rye Whiskey	3 pints
Sugar Syrup	1 pint
Tincture Tolu	1 fl.oz.

Mix the tincture and whiskey, filter and add the syrup.

Swedish Elixer

Elecampane	2½	oz.
Gentian	2½	oz.
Zedoary	3¼	oz.
Cinnamon	1¾	oz.
Saffron	1	oz.
Alcohol (95%)	22	gal.

Trojanka
Formula No. 1

Orange Peel	1	oz.
Cut Gentian	1	oz.
Cardamom	½	oz.
Galangal	½	oz.
Star Anise	½	oz.
Caraway Seed	½	oz.
Centaury	1	oz.
Red Clover Blossoms	½	oz.
Blood Root	½	oz.
Cinchona	1	oz.
Cinnamon	1	oz.
Cloves	½	oz.
Senna Pods	½	oz.
Orange Flowers	½	oz.
Nutmeg	1	oz.

Formula No. 2

Gentian	50	oz.
Galangal	50	oz.
Sarsaparilla	50	oz.
Red Clover Blossoms	50	oz.
Centaury	50	oz.
Orange Peel	15	oz.
Star Anise	8	oz.
Cinnamon	8	oz.
Nutmeg	4	oz.
Nux Vomica	2	oz.

Vermouth

Oil Wormwood	6	drops
Oil Angelica	2	drops
Oil Bitter Almonds	2	drops
Spirit Nitrous Ether	100	drops
Alcohol	60	fl.oz.
Sugar	32	oz.

Vermouth Italian

Angelica Root	8½	oz.
Valerian Root	4	oz.

Carduus Bened.	8	oz.
Cardamom	3	oz.
Guaiac	8	oz.
Orange Peel	20	oz.
Peppermint German	25	oz.
Centaury Minor	25	oz.
Wormwood Pont.	35	oz.
Alcohol (95%)	22	gal.

Brown color to suit.

Vermouth Formula of the French Academy

Wormwood	4	oz.
Gentian	2	oz.
Angelica Root	2	oz.
Carduus Benedictus	4	oz.
Calamus Aromatic	4	oz.
Elecampane Root	4	oz.
Centaury Leaves	4	oz.
Germander	4	oz.
Nutmegs	2	oz.
Orange Peel	2	oz.
Rectified Spirit	9	pints
Sweet White Wine	20	gal.

Macerate for 15 days and filter.

Cardinal Cordial

Orange Peel	17⅜	lb.
Orange Fruit Buds	10½	lb.
Lemon Peel	14	oz.
Alcohol (95%)	22	gal.

Benedictine

Anise Italian	15	oz.
Cardamoms	12	oz.
Calamus	10	oz.
Coriander	12	oz.
Galangal	10	oz.
Gentian	10	oz.
Ginger	8	oz.
Cinchona	10	oz.
Sweet Orange Peel	16	oz.
Bitter Orange Peel	10	oz.
Wormwood	12	oz.
Angelica Root	10	oz.
Celery Seed	10	oz.
Tonka Beans	8	oz.
Mace	4	oz.
Myrrh	3	oz.
Alcohol (95%)	25	gal.

Benedictine (de Fécamp)

Cloves	200	grams
Nutmegs	200	grams
Cinnamon	300	grams
Balm Mint	250	grams
Peppermint	250	grams
Fresh Angelica Root	250	grams
Alpine Wormwood	250	grams
Calamus Aromaticus	150	grams
Cardamoms, Minor	500	grams

Arnica Flowers	80 grams
Spirits of Wine	
(80 o.p., American)	40 litres

Cut, bruise, or grind the hard materials as necessary, and macerate along with the herbs for two days in the spirits of wine. Add 20 litres of water; distill, and draw off 40 litres of fine distillate. Prepare a syrup with 30 kilograms of refined sugar and ten kilograms of honey in sufficient water to produce 100 litres of liquid. Mix the syrup with the distillate. Cook until yellow; and filter.

Bishops Miter

Curacao Peel	17 oz.
Cloves	4 oz.
Orange Peel	4 oz.
Cinnamon	3 oz.
Alcohol (95%)	11 gal.
Cherry Juice	11 gal.
Sugar	40 lb.

Yellow color to suit.

Simple Tinctures for Flavor

Angelica Root	52 oz.
Angostura Bark	35 oz.
Aniseed	56 oz.
Benzoin Tears	21 oz.
Calamus	85 oz.
Curacao	100 oz.
Cardamom	56 oz.
Cassia	35 oz.
Celery	7 oz.
Cinnamon	35 oz.
Cloves	52 oz.
Cocoa	70 oz.
Fennel	56 oz.
Juniper	90 oz.
Lavender Flowers	35 oz.
Mace	28 oz.
Marjoram	35 oz.
Melissa	87 oz.
Nutmegs	35 oz.
Orange Peel	140 oz.
Orris Flowers	35 oz.
Peppermint German	35 oz.
Rose Flowers	52 oz.
Sage	88 oz.
Star Anise	56 oz.
Spearmint	35 oz.
Thyme	87 oz.
Valerian	85 oz.
Vanilla	3 oz.
Woodruff	70 oz.
Wormwood	35 oz.

Macerate for 2 weeks in 2¾ gallons 95% Alcohol and filter.

Artificial Brandy

Tincture of Galls (1–5)	10 parts
Aromatic Tincture	5 parts
Acetic Acid	5 parts
Sweet Spirit of Nitre	10 parts
Acetic Ether	1 part
Alcohol (68%)	570 parts
Distilled Water	400 parts

To be set aside for a few days and filtered.

Apricot Brandy

Oenanthic Ether	1 fl.dr.
Chloroform	1 fl.dr.
Amyl Butyrate	1 fl.dr.
Saturated Alcoholic Solution	
Tartaric Acid	1 fl.dr.
Glycerin	4 fl.dr.
Amyl Alcohol	2 fl.dr.
Valerianic Ether	5 fl.dr.
Butyric Ether	10 fl.dr.
Alcohol (50%) to make	10 gal.

Apricot Brandy and Liqueur

White Wine	44 fl.oz.
Apricots, sliced	6 oz.
Cinnamon	½ oz.
Alcohol	36 fl.oz.
Water	36 fl.oz.

Macerate the wine, fruit and cinnamon with 18 ounces of water for 7 days Add the rest of the water and filter. I₁ a liqueur is desired dissolve 26 ounces of sugar in the latter portion of water

Cherry Brandy

Ancient practice recommended a casl half-filled with cherries, filled up witl proof spirit. For improving the flavo of grain spirit, half an ounce of cinna mon, one ounce of cloves and about thre pounds of sugar is added to each 2₁ gallons used in steeping. The choices cherry brandy is drawn from the casl after a steeping of 18 or 20 days. ₁ quantity of spirit, roughly equivalent t two-thirds of that used in the first place is then added to the cask, and after ₁ month's maceration a second extract i drawn off, the cherries being exhauste of absorbed spirit by pressure.

Genuine cherry brandy is still produce by old methods. For liqueurs of th finest quality the cherries are carefull picked and all bruised and blemishe fruits rejected. Stalks are cut to two thirds of their length. The cherries a₁ washed in cold water to harden then To expedite the steeping each cherry ₁ pricked all round with a needle. Th fruit is filled into wide-mouthed bottl₁ or carboys, and spirits of wine (70 o.₁ American) sufficient to cover the fruit added along with 100 grammes of sug₁ to the kilograms of cherries. A perfumed spirit composed of:

Spirit of Coriander	2 l.
Spirit of Chinese Cinnamon	1 l.
Spirit of Cloves	½ l.
Tincture of Mace	½ l.

is added for each hundred litres of alcohol poured upon the fruit.

"Orangeade Punch"

Citric Acid Solution (Containing 200 grams per liter of citric acid)	300 cc.
Sugar Syrup (Containing 500 grams per liter of sugar)	5,000 cc.
Water	10,000 cc.
Orange Extract (Containing 5 cubic centimeters sweet orange oil per 100 cubic centimeters of alcohol)	10 cc.
Certified Food Color—Orange (Containing 1 gram color per 20 cubic centimeters)	20 c.c.

Squeeze and slice in 2 oranges, juice and peel. Dilute to taste.

Water Ice

Cane Sugar	25.0 lb.
Corn Sugar	7.0 lb.
Agar	0.2 lb.
Gum Tragacanth or High Grade India Gum	0.4 lb.
Water, Fruit, Fruit Acid, Flavor and Color	67.4 lb.

Grape Ice

Unpasteurized Grape Juice	40.00 lb.
Sugar	23.46 lb.
Agar (99.80 Grams)	.22 lb.
Gelatin (99.80 Grams)	.22 lb.
Water	36.00 lb.
Dry Citric Acid (Dehydrated)	.10 lb.

In making a 50 pound batch which is ample for a 10 gallon freezer, 50 grams of powdered agar is added to 4 or 5 pounds of the cold water called for in the formula. The water and the agar are boiled to put the agar into solution. Powdered agar can be mixed with a portion of the sugar in the formula and the mixture slowly added to boiling water while it was being stirred. Either method will dissolve the agar. Fifty grams of gelatin is added to 1.5 pounds of cold water and the mixture is warmed to 145° F. to dissolve the gelatin. The sugar, and the water of the formula, which was not used for dissolving the agar and gelatin are mixed and then the warm gelatin solution and hot agar solution is stirred into this mixture. The dry citric acid crystals are then added but they may also be conveniently added in a solution. This mixture is then put into the freezer at a temperature of 100° F. and frozen for about 4 or 5 minutes or until it is partially stiff. Then the cold grape juice is added to the mixture in the freezer. The ice can be drawn in approximately four or five minutes when a 0° F. (−17.78° C.) brine is used for freezing.

Roman Punch Ice

Formula No. 1 (Plain)

To 1 gallon of plain lemon ice mix add ½ to 1 pint of Jamaica rum. Freeze as usual.

Formula No. 2 (Sherbet Type)

To one gallon of lemon ice mix, in which ice cream or milk has been incorporated, add Jamaica rum as in No. 1 and freeze.

Formula No. 3 (With Champagne)

In addition to rum as in No. 1 and 2, add 1 pint to 1 quart of champagne. Note, when champagne or effervescent wine is used, it is best to work it, a little at a time, into the freshly frozen punch just before delivery.

Formula No. 4

Roman punch with cooked or plain meringue or beaten egg whites. Meringue or egg whites beaten added in proportion to ¼ or ⅓ of the volume may be worked in just before delivery.

(a) Plain Meringue—Twelve to 15 egg whites to 1 pound of powdered sugar. Whip stiff, then add sugar, a little at a time, using wooden spoon or wooden spatula for mixing.

(b) Cooked Meringue—Six to 7 egg whites to 1 pound of sugar. Add to sugar just enough water to prevent scorching and cook to soft ball or 238° F. Pour hot sugar into whipped egg whites in a small stream stirring constantly. Cook 2–3 minutes over low heat and stir all the time.

Lalla Rookh Ice

Formula No. 1

To 1 gallon of custard ice cream add ¼–½ pint of Jamaica rum or brandy. While ice cream is still soft work in ⅓–½ of its volume of sweetened whipped cream.

Formula No. 2

Same as No. 1 except that cooked meringue ¼–⅓ by volume instead of whipped cream is worked in.

Punch Richelieu Ice

To one gallon of raspberry ice mix add ½ pint (more or less) brandy, 1 pint claret, 3 or 4 ounces candied cherries chopped fine, and previously soaked in brandy several hours.

A dash of mixed spices is often added. If egg whites or meringue is used, the amount should be less than used in the Roman Punch.

Claret Punch Ice

To one gallon water ice of orange or lemon flavor add 1 pint or 1 quart of claret. Usually a little red color is used.

Generaly speaking the use of sufficient liquor to give ice cream a "kick" will spoil it from a flavor angle. Rum, brandy, and sherry are the acoholic beverages most commonly used for flavoring fancy ice cream, and they should be used for flavoring only. Where Tutti Frutti or Nesselrode ice creams are flavored, the canned fruits are usually soaked several hours in brandy or rum.

Sherbet Using Ice Cream Mix

Cane Sugar	25.0 lb.
Corn Sugar	7.0 lb.
Agar	0.2 lb.
Gum Tragacanth or High Grade India Gum	0.2 lb.
Ice Cream Mix, Without Sugar or Gelatin	10.0 lb.
Water, Fruit, Fruit Acid, Flavor and Color	57.6 lb.

Sherbet Using Milk

Cane Sugar	25.0 lb.
Corn Sugar	7.0 lb.
Agar	0.2 lb.
Gum Tragacanth or High Grade India Gum	0.2 lb.
Whole Milk	50.0 lb.
Water, Fruit, Fruit Acid, Flavor and Color	17.6 lb.

Apricot Milk Sherbet

Dried Apricots	¼ lb.
Water	½ lb.
Sugar	¼ lb.
Salt	Few grains
Lemon Juice	100 cc.
Evaporated Milk	½ lb.

Wash the apricots to remove all grit. Soak in the water several hours. Cook 5 minutes in the water in which they have been soaked. Add sugar and salt and cook 5 minutes longer. Cool and rub through a sieve. There should be ½ pound of pulp and syrup. If not, add water. Cool. Add lemon juice and combine with milk. Freeze with 1:8 salt-ice mixture.

Banana Milk Sherbet

Large Bananas	6
Sugar	135 oz.
Evaporated Milk	200 cc.
Orange Juice	600 cc.
Salt	Few grains

Select full ripe bananas, well flecked with brown spots. Remove peelings and outside fibrous portion. Mash bananas or press through a coarse sieve. There should be 240 cubic centimeters pulp. Combine ingredients in order given. Let stand 20 minutes in a cold place, stirring occasionally to dissolve sugar. Freeze with 1:8 salt-ice mixture. If 20 cubic centimeters lemon juice is added to ingredients given above, and the orange juice is increased to 120 cubic centimeters and the sugar to 160 grams, another very good sherbet results.

Cocoa Milk Sherbet

Sugar	¼ lb.
Cocoa	30 g.
Flour	5 g.
Salt	½ g.
Boiling Water	⅜ pint
Egg	1
Vanilla	5 cc.
Evaporated Milk	½ pint

Mix sugar, cocoa, flour and salt. Add water and boil 3 minutes, stirring constantly. Pour over well beaten egg slowly. Cook over boiling water 2 minutes, continuing stirring. Cool, then add vanilla and milk, freeze with 1:8 salt-ice mixture.

Excellent served in combination with orange milk sherbet.

Orange Milk Sherbet

Evaporated Milk	180 cc.
Sugar	240 g.
Salt	Few grains
Water	120 cc.
Orange Juice	240 cc.
Lemon Juice	40 cc.

Chill milk in ice cream can. Boil sugar, salt and water about 5 minutes. There should be 240 cubic centimeters of syrup. Add orange and lemon juice and chill. Pour orange mixture slowly into cold milk. Freeze with 1 : 8 salt-ice mixture.

If a mild flavored sherbet is desired increase evaporated milk to 240 cubic centimeters.

Peach Milk Sherbet

Mashed Peaches	⅜	lb
Sugar	¼	lb.
Salt		Pinch
Water	¼	lb.
Evaporated Milk	¼	lb.
Vanilla Extract	1	cc.

Ripe full flavored peaches are used; add sugar, salt and water to peaches; then the milk and vanilla. Stir well. Put in cold place for 20 minutes, stirring from time to time to dissolve sugar. Freeze with 1 : 8 salt-ice mixture.

Pineapple Milk Sherbet

Sugar	120	g.
Salt		Pinch
Water	120	g.
1 No. 2 Can Crushed		
Pineapple	275	g.
Lemon Juice	40	cc.
Evaporated Milk	180	cc.

Add sugar, salt, water and lemon juice to pineapple. Let stand in refrigerator 20 minutes, stirring occasionally to dissolve sugar. Chill milk in ice cream can. Add pineapple mixture slowly. Freeze with 1 : 8 salt-ice mixture.

To make Pineapple Mint Sherbet: add 6 drops oil of peppermint and a little green coloring to mixture before freezing.

Prune Milk Sherbet

Prunes	180	g.
Cold Water	240	cc.
Sugar	120	g.
Salt	½	g.
Evaporated Milk	240	cc.
Orange Juice	60	cc.
Lemon Juice	40	cc.

Wash Prunes to remove all grit. Soak in the 2 cups cold water several hours. Cook prunes in water in which they have been soaked 2 minutes. Add sugar and cook 2 minutes longer. Cool. Pit and rub through a sieve. There should be 2 cups of pulp and syrup. If not add water. If prunes are very acid, more sugar may be needed. Combine all ingredients and cool before freezing. Freeze with 1 : 8 salt-ice mixture.

Strawberry Milk Sherbet

Strawberries	1	quart
Sugar	¼	lb.
Salt		Pinch
Lemon Juice	40	cc.
Evaporated Milk	½	lb.

Select well ripened, perfect berries. Pick and wash thoroughly. Crush berries with wire potato masher. Press juice and pulp through a coarse sieve. There should be ¼ pound of juice and pulp. Add sugar, salt and lemon juice and set in refrigerator to cool, stirring occasionally to dissolve the sugar. Pour milk into ice cream can. Chill thoroughly, then add. the cold strawberry mixture. Freeze with 1 : 8 salt-ice mixture.

Mint Stick Sherbet

Peppermint Stick Candy	½	lb.
Boiling Water	¼	lb.
Evaporated Milk	1	lb.
Salt		Pinch
Eggs	2	

Crush candy fine. Add water, milk and salt, and heat in double boiler until candy is dissolved. Pour over slightly beaten eggs, beating constantly. Cook 5 minutes longer over boiling water, stirring well. Cool. Freeze with 1 : 8 salt-ice mixture.

Three-in-One Sherbert

No. 2½ Can Apricots	1	
Oranges 4 = Orange		
Juice	2½	cupfuls
Lemons 2 = Lemon		
Juice	1½	cupfuls
Water	1	pint
Sugar	½	lb.
Evaporated Milk	½	lb.
Egg Whites	2	

Press apricots through a coarse sieve. There should be ⅔ pound of pulp and syrup. Add orange and lemon juice. Boil sugar and water together 2 minutes. Add to fruit juice. Cool. Freeze with 1 : 8 salt-ice mixture to a mush, then add the stiffly beaten egg whites and milk and finish freezing.

Raspberry Sherbet

Cane Sugar	22	lb.
Gum Tragacanth	¼–½	lb.
Powdered Tartaric Acid	4	oz.
Raspberry Puree, No. 12		
Can	1	
Cerelose	5	lb.

Mix cane sugar, cerelose and gum tragacanth together, then pour in milk can containing six gallons of water. When sugar has dissolved, add one No. 12 can of Raspberry Puree and four ounces of powdered tartaric acid. Fill can to the top with water, making about ten gallons. Add sufficient amount of raspberry color to suit the desired shade. Stir well, place in a freezer and freeze. It should be very stiff when drawn out of the freezer. Do not shut brine or ammonia off until water ice is ready to pull.

Pistachio Parfait

Sugar	¼ lb.
Water	⅛ lb.
Egg Whites	3
Evaporated Milk	½ lb.
Vanilla	5 cc.
Ground Pistachios	¼ lb.
Salt	Pinch

Boil sugar and water until it spins a thread. Pour slowly into stiffly beaten egg whites, beating constantly. Cool. Chill milk thoroughly, then whip until stiff. Fold in the cold egg white mixture, vanilla, nuts and salt. Turn into a cold mold quickly. Seal and pack in a 1:3 salt-ice mixture 2 to 3 hours.

Small Experimental Batches of Ice Cream and Sherbet

The following formulae are sufficient for one gallon frozen ice cream:

Formula No. 1 (Vanilla)

Cream (20%)	4 lb.
Sugar	⅞ lb.
Gelatin	2 tablespoonfuls
Vanilla Extract	1 teaspoonful

For fruit or nut creams, add as directed. Bisque may be made by adding broken cakes, macaroons, grapenuts, etc.

Formula No. 2 (Vanilla)

Cream (20%)	3 lb.
Condensed Milk	1 lb.
Sugar	⅞ lb.
Gelatin	2 tablespoonfuls
Vanilla Extract	1 teaspoonful

For tutti fruitti, add a combination of fruits with shredded cocoanut.

Formula No. 3 (Sherbet)

Water	4½ lb.
Sugar	2 lb.
Lemon Juice	½ lb.

If other flavors are desired, use ⅕ pound lemon juice and ⅘ pound of juice of the flavor desired. For sherbets, pineapple and grape flavors are in demand.

Frozen Cream for Ice Cream

A sweet cream frozen for ice cream purposes is improved by adding one part of sugar (sucrose) or 1⅓ parts of invert sugar to 10 parts of cream. It should be frozen at -15° F.

Frozen Egg Yolk for Ice Cream

Obtain yolks from fresh eggs. Add 5 per cent of corn sugar and 0.5 per cent of gelatin or one per cent of glycerine. Freeze and store at 0° F. Thaw out for use at room temperatures. Use one pound of yolk per 100 pounds of mix.

Classification and Method of Making Ice Creams

I. Plain Ice Creams.
II. Nut Ice Creams.
III. Fruit Ice Creams.
IV. Bisque Ice Creams.
V. Parfaits.
VI. Mousses.
VII. Puddings.
VIII. Aufaits.
IX. Lactos.
X. Ices.
  1. Sherbets.
  2. Milk Sherbets.
  3. Frappes.
  4. Punches.
  5. Souffles.

I. Plain ice cream is a frozen product made from cream and sugar with or without a natural flavoring.

Formulas are given for making ten gallons of finished ice cream. Fillers may be added in definite proportions.

Vanilla Ice Cream

Cream	5 gal.
Sugar	8 lb.
Vanilla Extract	4 oz.

Chocolate Ice Cream

Cream	5 gal.
Sugar	10 lb.
Bitter Chocolate	1½ lb.
or Cocoa	1 lb.
Vanilla Extract	4 oz.

Maple Ice Cream

Cream	5 gal.
Cane Sugar	4 lb.
Maple Sugar	4 lb.

Caramel Ice Cream

Cream	5 gal.
Sugar	8 lb.
Caramel	12 oz.

Coffee Ice Cream

Cream	5 gal.
Sugar	8 lb.
Extract from 1 lb. coffee	

Mint Ice Cream

Cream	5 gal.
Sugar	8 lb.
Concentrated Creme de Menthe Syrup	1 pt.

Few drops green coloring.

II. Nut ice cream is a frozen product made from cream and sugar and sound non-rancid nuts.

Walnut Ice Cream

Cream	5 gal.
Sugar	8 lb.
Vanilla Extract	4 oz.
Walnut Meats	4 lb.

According to this general formula the following nut ice creams may be prepared by substituting different kinds of nut meats.

Chestnut Ice Cream.
Filbert Ice Cream.
Hazelnut Ice Cream.
Pecan Ice Cream.
Peanut Ice Cream.
Almond Ice Cream.
Pistachio Ice Cream.

At times pistachio ice cream is made from oil of pistachio instead of from the nuts. If thus prepared, it will come under the head of plain ice creams.

III. Fruit ice cream is a frozen product made from cream, sugar, and sound, clean, mature fruits.

Strawberry Ice Cream

Cream	5 gal.
Sugar	8 lb.
Crushed Strawberries	1 gal.

Employing the same formula the following creams may be made by merely substituting fruits and berries for the strawberries. The amount of sugar may be varied according to the acidity of the fruit.

Pineapple Ice Cream.
Raspberry Ice Cream.
Cherry Ice Cream.
Peach Ice Cream.
Apricot Ice Cream.
Currant Ice Cream.
Grape Ice Cream.
Cranberry Ice Cream.

Preparation of lemon and orange ice creams cannot be included under this general rule. These creams may be prepared as follows:

Lemon Ice Cream

Cream	5 gal.
Sugar	10 lb.
Lemon Juice	2 pt.

Orange Ice Cream

Cream	5 gal.
Sugar	10 lb.
Orange Juice	2 qt.
Lemon Juice	½ pt.

IV. Bisque ice cream is a frozen product made from cream, sugar, and bread products, marshmallows or other confections, with or without other natural flavoring.

Macaroon Ice Cream

Cream	5 gal.
Sugar	8 lb.
Vanilla Extract	4 oz.
Ground Macaroons	5 lb.

From this formula we can make:

Grape Nut Ice Cream.
Nabisco Ice Cream.
Sponge Cake Ice Cream.
Marshmallow Ice Cream.

V. Parfait is a frozen product made from cream, sugar, and egg yolks, with or without nuts or fruits and other natural flavoring.

Walnut Parfait

Cream (30%)	4 gal.
Egg Yolks	120
Sugar	14 lb.
Vanilla Extract	4 oz.
Ground Walnut Meats	4 lb.

From this formula by substituting the nut meats we can make:

Filbert Parfait.
Almond Parfait.
Peanut Parfait.
Hazelnut Parfait, etc.

By substituting the same proportion of fruits as are used for fruit ice creams, for the vanilla extract and nut meats, fruit parfaits such as strawberry, raspberry, and cherry parfaits, and others may be prepared.

Coffee Parfait

Cream (30%)	4 gal.
Egg Yolks	120
Sugar	14 lb.
Extract from 1 lb. Coffee.	

Maple Parfait

Cream (30%)	4 gal.
Egg Yolks	120
Maple Sugar	6 lb.
Cane Sugar	8 lb.

Tutti Frutti

Cream (30%)	4 gal.
Egg Yolks	120
Sugar	14 lb.
Vanilla Extract	4 oz.
Candied Extract	3 lb.
Candied Assorted Fruit	3 lb.
Pineapple	3 lb.

VI. Mousse is a frozen whipped cream to which sugar and natural flavoring has been added.

Cranberry Mousse

Cream (30%)	2 gal.
Sugar	4 lb.
Cranberry Juice	1 qt.
Lemon Juice	¼ pt.

From the same formula combinations may be made with various other fruit juices and natural flavors, such as coffee, vanilla, maple, caramel, pistachio, etc.

Sultana roll, as indicated by the name, is made in a round mould. The center of the mould is filled with tutti frutti and the outside with pistachio mousse.

VII. Pudding is a product made from cream or milk, with sugar, eggs, nuts, and fruits, highly flavored.

Nesselrode Pudding

Cream (30%)	3 gal.
Eggs	120
Cane Sugar	10 lb.
Vanilla Extract	4 oz.
Maraschino Cherries	4 lb.
Maroons	4 lb.
Almonds	2 lb.

Manhattan Pudding

Cream (30%)	3 gal.
Eggs	120
Sugar	12 lb.
Orange Juice	2 qt.
Lemon Juice	1 pt.
Walnut Meats	4 lb.
Cherries and Assorted Fruits	4 lb.

Plum Pudding

Cream (30%)	3 gal.
Eggs	120
Sugar	10 lb.
Chocolate	2½ lb.
Cherries and Assorted Fruits	4 lb.
Raisins	2 lb.
Figs	2 lb.
Walnut Meats	1 lb.
Ground Cinnamon	3 tablespoonfuls
Ground Cloves	½ teaspoonful

VIII. Aufait is a brick cream consisting of layers of one or more kinds of cream with solid layers of frozen fruits.

Fig aufait may be made from three layers of cream of various flavors with two layers of whole or sliced preserved figs. It is most satisfactory to slice the figs lengthwise in halves. The following illustration shows how a very delicious brick of fig aufait may be made.

Other aufaits may be made from a variety of preserved fruits and berries and combined with different creams.

IX. Lacto is a product manufactured from skimmed or whole sour milk, eggs, and sugar, with or without natural flavoring.

X. Ices are frozen products made from water or sweet skimmed or whole milk, and sugar, with or without eggs, fruit juices, or other natural flavoring. Ices may for convenience be divided into water sherbets, milk sherbets, frappes, punches, and souffles.

A. A water sherbet is an ice made from water, sugar, egg albumen, and natural flavoring, and frozen to the consistency of ice cream.

Lemon Sherbet

Water	6 gal.
Egg Whites	24
Sugar	20 lb.
Lemon Juice	6 pt.
Orange Juice	1½ pt.

Water, eggs, and sugar will be the same for other sherbets; the flavoring material only will change.

Orange Sherbet

Orange Juice	6 qt.
Lemon Juice	1 pt.

Pineapple Sherbet

Crushed Pineapple	1 gal.
Lemon Juice	1 qt.

Peach Sherbet

Peach Pulp	6 qt.

Cherry Sherbet

Cherry Juice	6 qt.
Lemon Juice	1 pt.

Ginger Sherbet

Ginger Syrup	4 pt.
Chopped Preserved Ginger	4 lb.
Lemon Juice	4 pt.

Strawberry Sherbet

Strawberry Juice	6 qt.
Lemon Juice	1 pt.

B. Milk sherbet is an ice made fro

sweet skimmed or whole milk with egg albumen, sugar, and natural flavoring, frozen to the consistency of ice cream.

Pineapple Milk Sherbet

Milk	6 gal.
Sugar	16 lb.
Egg Whites	24
Pineapple Pulp	1 gal.
Lemon Juice	1 qt.

Milk sherbets of various flavors may be prepared according to above formula by using the same amount of flavoring material as is used for plain sherbets.

C. Frappe is an ice consisting of water, sugar, and natural flavoring and frozen to a soft semi-frozen consistency. Same formulas as are given for sherbets will answer for frappe by omitting the egg albumen.

D. Punch is a sherbet flavored with liquors, or highly flavored with fruit juices and spice.

E. Souffle is an ice made from water, eggs, sugar, and flavoring material. It differs from sherbets mainly in that it contains the whole egg.

Pineapple Souffle

Water	2½ gal.
Eggs	48
Sugar	12 lb.
Grated Pineapple	1 gal.
Lemon Juice	1 qt.

Following the same formula, souffles or other flavors can be made.

The formulas presented have been given mainly for the purpose of making clear the difference between the various groups. Numerous other formulas may be prepared based on the same general outline.

Caramel Ice Cream Flavor

For 100 pounds of mix use 15 pounds of sugar which has been heated with one quart of "coffee" cream in a one-piece utensil until the mixture turns a deep brown. Cool and add to 100 pounds of mix.

Chocolate Ice Cream Flavor

Sugar	19 lb.
Cocoa	13 lb.
Water	4½ gal.

Mix and heat to 175° F. for 15 minutes. Use 7 pounds of syrup to 40 pounds of mix.

If chocolate liquor is used, add 2.25 pounds of chocolate instead of 1.5 pounds of cocoa.

Honey Ice Cream Flavor

Seventeen to 18 pounds of honey will provide sufficient sugar as well as flavor to the 100 pounds of mix, but since this mix is difficult to whip, it may be desirable to replace only 50 to 75 per cent of the cane sugar. In this case 1.3 pounds of honey should be used to each pound of sugar replaced.

Fruit Ice Creams

In presenting these directions for making fruit ice creams, no endeavor has been made to give complete directions but rather to stress the points of variation from standard practice so that the ice cream manufacturer can improve present methods.

Fruit ice cream formula containing 14 per cent fat, 11 per cent solids not fat, 14 per cent sugar prepared with 30 per cent cream

Cream	46.3 lb.
Skim Milk Powder	4.9 lb.
Skim Milk	34.5 lb.
Gelatin	.50 lb.
Sugar	14 lb.

Formula No. 1—Strawberry Ice Cream

A strawberry variety that has been found to give good results in freezing tests should be used. Strawberry varieties that retain their color, firmness, fragrance, and flavor in frozen desserts are: Big Joe, Klondike, Brandywine, Blakemore and Redheart. The berries should be sliced. A ratio of berries to sugar of 2 to 1 or the use of a 75% sugar solution (2.5 pounds of berries to 1.6 pounds of syrup) will give good results. It is desirable to freeze the fruit quickly to 0° F. and to hold it at 0° or below to retain the maximum flavor. Small containers, such as the 30-pound tin single-service container, can be more readily frozen and handled than barrels and proper proportions of berries to syrup can be secured more easily. Before using the berries they should be thawed at a temperature not exceeding 40° F. and soaked in their syrup for 12 to 24 hours to soften them.

It is desirable to make a special mix for fruit ice creams in which the fat content is about 2 per cent higher and the serum solids 1.5 per cent higher than in regular vanilla ice cream to make allowance for the diluting effect of the syrup. The sugar content of the mix should be about 2 per cent below that of vanilla ice cream, as 5 to 6 per cent sugar is added in the fruit juice. The use of 20 per cent of fruit gives a very desirable, evident flavor and plenty of visible fruit. The syrup should be drained off the berries and the berries alone placed in the hardening room for about a half hour to chill well. The syrup and color should be added to the mix in the freezer. The berries can be readily mixed by hand with the frozen ice cream. A hopper is desirable, but a large can, such as a 10-gallon tin container, may be used for this purpose. It is essential to pre-chill the utensils and the berries to avoid increasing the temperature of the ice cream, thereby lessening the degree of smoothness and creaminess of the finished product.

Formula No. 2—Raspberry Ice Cream

The general directions for packing and freezing raspberries should be the same as for strawberries, except that the berries ought not to be sliced.

Formula No. 3—Peach Ice Cream

Fresh, ripe peaches of a standard variety with a fairly pronounced flavor should be dipped in boiling water, skinned, and ground or very finely sliced. The pulp should be mixed with sugar at the rate of 5 to 1 and immediately frozen to and held at 0° F. or below.

A day before using the frozen peaches should be held at 40° F. to thaw. The color and the peaches may be added to the ice cream mix in the freezer as there is little advantage in endeavoring to retain pieces of peach in the ice cream due to the blending of color with the ice cream and to lack of a pronounced flavor. About 25 per cent of peaches gives a recognizable, mild peach flavor. Peaches do not increase the moisture and sugar content of the ice cream as much as strawberries or raspberries, so they may be advantageously added to the regular vanilla mix, but the flavor will be best if the mix is rather rich.

As previously mentioned, approximately 20 to 30 per cent of the fresh, ripe peaches may be skinned like an apple and, after grinding the skins finely and adding 20 per cent of sugar, cooked for 5 minutes just below the boiling point. The cooked skins with pulp may then be frozen and used to increase the intensity of the peach flavor by replacing 5 to 10 per cent of the peaches in the ice cream. The skins must be fine enough to avoid detection in the ice cream.

One per cent of apricot added to peach ice cream increases the intensity of the flavor noticeably.

Honey Ice Cream

The most satisfactory recipe thus far employed is the following (the quantities given being those required for 10 gallons of finished ice cream):

Cream (17%)	40 lb.
Extracted Honey	11 lb.
Gelatin	4 oz.
Color, lemon yellow.	

French Nougat Ice Cream

Formula No. 1

To 5 gallons of regular mix, when practically frozen and almost ready to draw, add

Drained Red Pineapple Prisms	1 pt.
Drained Green Pineapple Prisms	1 pt.
Miniature Marshmallows (400 to pound)	2 lb.

Flavor with 3 ounces pure lemon extract, or pure orange extract and color a light "peach."

Formula No. 2

Marshmallow Creme	½ gal.
Milk (to thin Marshmallow)	⅛ gal.
Pure Lemon or Orange Extract	3 oz.
Color light "peach."	

Add late, or stir in, one pint red pineapple prisms, and one pint green pineapple prisms "drained."

Chocolate Marshmallow Ice Cream

Use regular chocolate ice cream mix, adding three pounds Miniature Marshmallows to each ten gallons finished ice cream.

Cherry Marshmallow Ice Cream

Use regular vanilla ice cream mix, adding to it two and one-half pounds Miniature Marshmallows and one quart chopped maraschino cherries.

Nut Marshmallow Ice Cream

Use regular vanilla ice cream mix, adding to it two and one-half pounds Miniature Marshmallows and one-half can English walnut pieces in maple flavored syrup.

Best results can be obtained when about one-half pound of the Miniature Marshmallows are added to the freezer just before drawing the ice cream from the freezer, and the rest of the Miniature Marshmallows sprinkled into the ice cream as it is run from the freezer into the can before hardening.

Preventing Greenish-Black Discoloration of Ice Cream

Use cocoa low in tannins to prevent the discoloration. Also use cans that are well tinned or lined with paper liners.

Malting Process

U. S. Patent 1,914,244

The grain is sprinkled with an acid solution such as acetic acid or lactic acid of about 10% concentration by weight, corresponding to an acid concentration of about 1.11 grammolecular weights per liter, at the rate of about 6 gallons of acid solution per 448 pounds. After the drain is germinated and modified it is treated in an organic acid solution, such as a solution of lactic acid for about 24 hours at a temperature of about 15° C. This treatment facilitates malting and increases the yields.

Cocoa Malt Powder

Fruit Sugar	55	lb.
Malt Powder (Mild Flavor)	19½	lb.
Skim Milk	12½	lb.
Cocoa	13	lb.
Vanillin	2	oz.
Salt	6	oz.

The above formulas are cocoa, chocolate, and malt powders met with commercially. The powders are prepared by mixing the ingredients thoroughly in a suitable device and then running the mixed ingredients through a coarse sieve. These mixtures are usually put up in 1 to 1¼ ounce glassine envelopes. One envelope is sufficient for a 6 or 8 ounce glass or cup.

Sweet Cocoa Powder

Cocoa	175	lb.
Powdered Sugar	325	lb.
Vanillin	5	oz.

Sweet Milk Cocoa Powder

Cocoa	24	lb.
Powdered Sugar	41	lb.
Milk Powder	35	lb.
Vanillin	1	oz.
Salt	1	oz.
Bicarbonate of Soda	1	oz.

Sweet Cocoa Powder and Skim Milk

Cocoa	33	lb.
Fruit Sugar	51	lb.
Skim Milk Powder	16	lb.
Vanillin	1	oz.

Sweet Chocolate Powder and Skim Milk

Chocolate Liquor	35	lb.
Powdered Sugar	50	lb.
Skim Milk	15	lb.
Vanillin	1	oz.
Salt	2	oz.

Chocolated Milk

For commercial manufacture, make a syrup by bringing to a boil five parts by weight of cocoa and 18 parts by weight of sugar with 11 parts by weight of water. This is cooled, a very small amount of salt added, flavored to taste with vanilla extract, and then added to milk at the rate of one part of syrup to eight parts of milk.

Cocoa-Milk Beverages

Formula No. 1

Cocoa Syrup	1	gal.
Skim Milk	10	gal.
Tapioca Flour	10	oz.

Formula No. 2

Cocoa Syrup	1	gal.
Water	5	gal.
Skim Milk	5	gal.
Tapioca Flour	14	oz.

Formula No. 3

Cocoa Syrup	1	gal
Dry Skim Milk	8	lb.
Water	9	gal.
Tapioca Flour	14	oz.

Fluid skim milk or skim milk powder and water can be used. The tapioca flour may be incorporated into the cocoa syrup in the dry milk cocoa powder mixture or it may be added to the mix separately. The process involves sterilization of the product by heating to 240° F. for a period of 20 minutes. The product is heated in bottles. The rotary type of sterilizer is superior to the stationary type because the tapioca flour must be agitated during the heating process.

Honey Cream

Honey cream is a mixture of high-test cream and strained honey. This product is usable as a spread on bread, biscuits, waffles and the like.

It is important that the cream test

about 75 per cent fat, for otherwise the mixture will not solidify properly. The high-test cream for this study was secured by using a De Laval cream separator equipped with a special cream cover to handle the heavy cream. It is advisable to separate the milk at the pasteurizing temperature with the screw adjustment set for a high-testing cream, and with the rate of inflow reduced to about one-third normal.

Various kinds of honey were tried, and the milder flavored ones, such as sweet clover, cotton, tupelo, white thistle, white orange and alfalfa, proved most popular with the judges. Additional flavors may be added, such as maple, coffee, chocolate, orange, sorghum, and raspberry. It is necessary to reduce the amount of honey when using flavoring in order to obtain a product which will spread properly.

If raw honey is used it should be heated momentarily to at least 155° F. to destroy enzymes that may be present and which might cause the honey cream to become rancid. It is also advisable to separate the cream from pasteurized milk for the same reason.

While the honey and cream are still warm, mix in the proportion of 42 parts of honey and 58 parts of the high-test cream. Pour into small glass or paper containers and store in a refrigerated room (40°–50° F.). If more than 42 per cent of honey is used the mixture will not be firm enough for spreading. If much less than 42 per cent of honey is used the spread will lack honey flavor when used on biscuits, waffles and the like.

Honey cream is perishable and should be kept refrigerated until used. The rather high fat content (about 43.5 per cent) makes the product likely to develop rancid or tallowy flavors if kept too long. Tallowiness is best controlled by avoiding copper and iron contamination as much as possible. The higher the proportion of honey used the less likely will tallowy and rancid flavors develop.

Honey Milk Drinks

Formula No. 1—Honey Blossom Drink

To each pint of milk, add three ounces of honey and 15 drops of lemon or orange extract. Shake until dissolved. Color a lemon shade.

Formula No. 2—Egg Drink

A number of egg drinks may be made using honey with milk. For one of these, beat an egg thoroughly with three ounces honey and add to a pint of milk. Mix thoroughly and serve cold.

Formula No. 3—Egg Drink

For another, beat two eggs with an egg beater and add to a shaker. If this is not available, a common fruit jar may be used. Add a pint of milk and three ounces of honey and shake well. The product is then ready to serve. A pleasing variation to this milk shake is made by adding a tablespoonful of malted milk.

Egg Nog and Juleps

Egg nog is a milk favorite of long standing. It is made by beating the mixture of an egg, one teaspoon sugar, a sprinkle of cinnamon and nutmeg, a few drops of vanilla, and mixing thoroughly with a glass of milk. This drink is especially desirable and nutritious, being unusually rich in proteins and minerals because of its egg and milk content. Other flavors that may be had in egg nog are obtained by using lemon, orange, or prune juice.

Spiced milk is made by adding a mixture of a teaspoon of sugar and a sprinkle each of powdered cloves, cinnamon, and nutmeg to a glass of hot milk. The mixture is beaten smooth and is served hot.

If a portion of a carbonated bottled drink such as ginger ale or root beer is added to a glass of cold milk a pleasing flavored product may be obtained. These give the mixture a snap because of their carbon dioxide content.

Buttermilk lemonade is made by adding sugar and lemon juice to fermented milk. A quantity slightly in excess of the amount used in lemonade is necessary to be predominant over the flavor of the soured milk. This drink is best when served very cold.

A convenient method for the quick preparation of certain milk drinks is to make a syrup of medium thickness by boiling a mixture of sugar and water. This is divided into portions and flavored heavily with extracts of the variety desired. These are held as stock materials to be used when desired. Flavors recommended are strawberry, pineapple, vanilla, orange, lemon, cherry, maple, root beer, and grape. Simple milk drinks are made by adding one and one-half ounces, three tablespoonfuls, of any syrups to three-fourths of a glass of milk. Juleps are made in the same manner except than an egg is beaten into the mixture. In any of the drinks, a small amount of charged water will add zest and character.

Tomato-Milk Drink

Milk	1½ cupfuls
Tomato Juice	2¼ cupfuls
Salt	½ teaspoonful

Combine milk and tomato juice gradually, while stirring. Add salt. Celery salt or onion juice may be added.

Butter Scotch Shake

Mix together one quart coffee cream, one and one-half pounds sugar, one pound Karo corn syrup, and one-half pound butter. Heat to 240° F. and cook until it cracks. Cool and add a dash of lemon and orange extracts. Thin with water until a syrup results. Add four ounces of this mixture to each pint of milk. Color with burnt sugar.

Stain for the Direct Microscopic Examination of Milk

The following stain is used to give red color to the bacteria and leucocytes, and the background in green:

Ethyl Alcohol (95%)	54	cc.
Tetrachlorethane	40	cc.
Acetic Acid (Glacial)	6	cc.
Neutral Red (Powder)	1	g.
Brilliant Green (Powder)	0.5	g.

The alcohol is warmed to 70° C. (158° F.) and the tetrachlorethane is added. Then the dyes are added, and when cold the acetic acid is added. The mixture is then filtered.

The milk is fixed on the slide in the usual way, then dipped in the stain twice. After it is dry it is washed twice by giving two whirls in a glass of water. The slide is drained and dried.

The method gives results comparable to those obtained with the Breed-Brew method.

It is preferable to use 2 square centimeters rather than 1 square centimeter for the milk spreads better. This necessitates reading twice as many fields as with either area only 0.01 cubic centimeter of milk is used.

Lemon Juice Evaporated Milk

Formula No. 1

Evaporated Milk	14 oz.
Water	16 oz.

Corn Syrup	2 oz.
Lemon Juice	5 teaspoonfuls

Formula No. 2

Evaporated Milk	14 oz.
Water	20 oz.
Corn Syrup	1.5 oz.
Lemon Juice	5 teaspoonfuls

The water for both formulas should be boiled and cooled before using. Both formulas are fed to under-nourished infants. Formula No. 2 is used in connection with breast milk.

Citric Acid Milk for Infant Feeding

By adding 4 grams of dehydrated citric acid to 1 quart of milk it is enhanced in feeding qualities for infants. The citric acid can be added to hot or cold milk. It has, in this respect, an advantage over lactic acid which is only successfully used to modify milk cold.

Corn Sugar as a Sweetener for Condensed Skim Milk

Corn sugar can be used as the preserving agent of sweetened condensed skim milk with sucrose, in fact it is slightly more efficient than cane sugar for this purpose, although there is a tendency to produce thickening. The thickening is overcome largely by fore-warming the milk and sugar separately.

One-half of the cane sugar may be substituted with corn sugar. The sugar is made into a syrup by adding enough water to make a 60 per cent solution. The syrup is then heated to 175° F. The water should be free from O H-ions (not any free alkalinity) and should not be contaminated with milk or milk stone.

This syrup is drawn into the pan when the milk has been condensed to approximately the correct concentration.

To figure the amount of sugar needed for 5,000 pounds of skim milk use the following formula:

$$\frac{5,000 \times 0.09}{28} \times 42 = 676.4 \text{ pounds}$$

This is for a finished product testing 28 per cent milk solids not fat and 42 per cent sugar.

The Baumé reading for the correct concentration would be 34.2 at 120° F.

The concentrated product should be drawn from the pan and cooled rapidly to 93° F. Then 4 to 5 ounces of milk sugar added to each 1,000 pounds of condensed milk and mixed thoroughly.

Then it should be cooled to 75° F. in 45 to 60 minutes with constant agitation. Then it is cooled to the storage temperature.

"Acidophilus" Milk

The manufacture of pure acidophilus milk requires the greatest care, both in the preparation of the milk before inoculation and throughout the entire process of manufacture. It must be remembered that L. acidophilus is not a natural inhabitant of milk and must be acclimated to this medium. Even though it be acclimated to milk, its growth is nevertheless such that it is unable to overcome a slight contamination. Laboratory facilities, close supervision, and adequate equipment are essential to the successful production of acidophilus milk.

The first step in the manufacture of acidophilus milk is to obtain from a reliable source a pure culture of the bacterium Lactobacillus acidophilus of proved therapeutic value. After a pure culture is obtained it must be kept pure as a starter. All chances of contamination must be avoided. Florence or Erlenmeyer flasks are the most suitable containers in which to carry the starter. The flasks should be about half filled with fresh skim milk and plugged with cotton. They are then sterilized in an autoclave at 15 pounds pressure for 20 minutes. The milk after sterilization has a slightly caramelized appearance. After the milk has cooled, a flame from a gas burner should be passed over and around the mouth of the flask and the culture introduced into the flask after the mouth of the culture tube is passed through the flame. The cotton plug from the flask should be held in the hand during this operation and replaced in the mouth of the flask immediately after inoculation. The inoculated milk is incubated at 98° to 100° F. until the milk has curdled. A starter should show a small quantity of whey on top of the curd which should appear firm with no evidence of gas. The curd is broken up by rotating the flask vigorously. The starter should be of a creamy consistency and possess a clean acid flavor and a characteristic aroma. This starter, or mother starter, is carried on by daily transfers from flask to flask. The same procedure as described above is followed except that usually the first bit poured over the lip of the flask is discarded as it serves to wash off the lip. The care necessary to maintain a pure starter of L. acidophilus cannot be overemphasized. The starter should be examined frequently for purity either microscopically or by plates. Should it develop a bad flavor or odor, it must be discarded and a new culture obtained. In any case it is advisable to get a new culture at regular intervals. Kopeloff advises the use of a new fecal culture every two to four months to assure the therapeutic value of the product.

The bulk starter, which is the starter to be used to inoculate the batch, is prepared in the same way, as the mother starter; and the same precautions against contamination must be observed. Florence or Erlenmeyer flasks of 2-liter capacity or larger are most satisfactory in which to prepare this bulk starter. These flasks permit the autoclaving of the milk and are most easily handled under the conditions necessary to prevent contamination. The starter should be 18 to 24 hours old when used to inoculate the batch and should possess all the qualities indicating purity previously mentioned for the mother starter.

Equally as great care must be exercised in the treatment of the milk as in the preparation of the starter. Only fresh milk of the best quality should be used in the manufacture of acidophilus milk. Whole milk, milk which is partly skimmed, or skim milk may be used. The pasteurizer employed in the preparation of the product must be so fitted that the milk may be heated and cooled without removing it from the vat, which should be equipped with an efficient mechanical stirrer. The vat must be used not only as a pasteurizer but also as an incubator for the milk. The characteristics of the bacterium Lactobacillus acidophilus make it necessary that the bacteria in the milk be killed before the starter is added. The milk is heated to about 205° F. for an hour and a half or longer and then cooled to 98°. It should have a slightly caramelized appearance after heating; however, this in itself is not an indication that the necessary destruction of the bacteria has been accomplished. Bass recommends heating the milk to 190° to 195° for one hour; then cooling it to 98°, at which temperature it is held for three or four hours; again heating it to 190° to 195° for one hour; then cooling it to 98°; and inoculating. After the milk has been cooled to 98° to 100° it is inoculated with about 2 per cent of the bulk starter. The greatest care must be observed in the inoculating process to prevent the contamination of the batch. The mouths of the flasks should be passed through a flame before pouring the starter. The cover of the vat is lifted just sufficiently to permit the pouring of the starter and is replaced immediately. After inoculation the milk is stirred for a few minutes to distribute the inoculum, and the batch is then allowed to incubate. The temperature of incubation, 98° to 100°, should be maintained throughout the entire incubation period. The batch is incubated until a firm curd has been formed, which usually requires about 18 to 24 hours. The length of the incubation period is dependent upon the size of the inoculation, the temperature of incubation, and the activity of the starter. After the milk has curdled it is broken up by stirring, cooled to room temperature, and bottled.

Kefir

The yeast is prepared by adding a half teaspoonful of sugar to a 6-ounce or 8-ounce bottle of boiled and cooled water. Half a yeast cake is added to this sugar solution and set in a warm place overnight. This will give an active culture of the yeast and obviate the necessity for adding the yeast cake directly to the milk. This yeast culture should be ready at the time the buttermilk is received or, if made at home, at the time it is curdled.

One to one and a half per cent of sugar is added to the buttermilk. On the quantity of sugar added to the buttermilk will depend the extent of the alcoholic fermentation. Theoretically about one-half of fermented sugar may be converted into alcohol; that is, milk to which 1 per cent of cane sugar has been added may contain after the fermentation 0.5 per cent of alcohol. The quantity of sugar added should be governed by the amount of carbon dioxide it is desired to have in the finished product. This should be sufficient to make the kefir distinctly effervescent and impart to it the peculiar, sharp taste of charged water, but should not be developed enough to blow the fluid out of the bottles when the stoppers are removed. Experience shows that 1 to 1.5 per cent of sugar will give the proper amount of gas. This may be approximated by adding sugar in the proportion of 2 even teaspoonfuls of sugar to each pint of milk. When the buttermilk and the yeast cul-

ture are ready, the sugar is dissolved in the buttermilk.

The yeast culture is added to the buttermilk in the proportion of 1 teaspoonful to 1 quart of buttermilk.

The buttermilk is mixed thoroughly and bottled. The bottles should be very strong, as sufficient gas pressure is sometimes generated to break ordinary bottles. The heavy bottles used for ginger ale or other carbonated drinks answer this purpose very well. They should be carefully cleaned and boiled or steamed before being filled and then stoppered tightly. The stoppers should be wired or tied securely in place.

The product is put in a cool place to ferment. If the fermentation is too active the kefir will have a yeasty taste, and the curd is likely to become lumpy and filled with large gas bubbles. A temperature of 18° to 21° C. (65° to 70° F.) will be found satisfactory for kefir which is to be used on the third or fourth day. The floor of a cool cellar is a convenient place to ferment kefir made in the home. The bottles should be shaken as often as may be necessary to keep the curd in a finely divided condition. The finished product should be smooth and creamy, effervesce rapidly when poured from the bottle, and have the pleasant, acid taste of buttermilk, with the added sharpness caused by the gas and the trace of alcohol. Kefir 2 or 3 days old may have a yeasty taste, but if it has been properly made this will disappear as the fermentation of the sugar nears completion; made under these conditions, it should be used when 3 to 5 days old, but if put on ice it may be held for a week or even longer.

Determining the Vitality of Lactic Cultures

Cultures used for cheese making or cultured milk produce acid at different rates. The following formula is used to determine the vitality of lactic cultures. A pint of milk is placed in a sterile ground stoppered bottle and heated to 100° F. Then 5 cubic centimeters of the lactic culture is added and 30 minutes later 1 cubic centimeter of rennet is added. One hour after adding the rennet the curd is cut and 2 hours after cutting the curd the whey is drained off. At periods of 2 and 3 hours after draining the whey acid tests are made on the whey surrounding the curd. A "slow" culture will produce whey titrating 0.16 and 0.23 while an active culture will produce whey titrating 0.42 and 0.63 per cent of acid at the two consecutive titrations. The titrations are made with tenth normal sodium hydroxide using phenolphthalein as an indicator. The result is expressed as lactic acid.

Quickly Soluble Lactose

Dissolve ordinary milk sugar in water so as to have a 10 to 20 per cent solution. Dry on a roller drum drier (the type used to produce dry skimmilk). A steam pressure of 75 pounds and 5 revolutions per minute of the drum will produce a sugar containing 99 per cent of beta lactose. This is an easy way to convert insoluble alpha hydrate (ordinary milk sugar) to a soluble form beta lactose.

Increasing Viscosity of Cream

Cream is cooled to 5–6° for 2–3 hours, warmed to 25–35°, and again cooled. Heating and cooling may be repeated and the cream may be agitated during warming.

Sterilized Sweet Cream

Heat cream to 176° F. and homogenize at 3000 pounds pressure. Then bottle in ordinary soda or pop bottles. Sterilize at 244° F. for 12 to 14 minutes. This product has a slightly cooked flavor. It will keep indefinitely without refrigeration. It is good for coffee and cereal purposes. It will not whip.

Fast Frozen Sweet Cream

"A high grade milk or cream produced in strictly sanitary dairies in earthen containers was pasteurized at 145° F. for 30 minutes, after which the milk or cream was put through a homogenizer, then filled into the receiving reservoir of a de-aerating freezing unit, where a vacuum of 29.80 inches was maintained for 20 minutes, after which the milk or cream was passed into the freezer where it was frozen to a slush in 4½ minutes still under a high vacuum. From the freezer the milk or cream slush was filled into one-gallon stone jugs of thermos type and sealed under 15 inches of vacuum. The jugs were placed in a freezing unit, which freezes at a temperature of minus 45° F., and frozen hard in 1 hour and 55 minutes. The product was then placed in cold storage and held at 0° F.

This product was examined at intervals of 30 days and the quality of the product was found to be excellent and comparison with fresh milk at each interval indicated very little, if any, change in the product."

In addition to earthen and stone vacuum jugs, sulphite paper containers, cellophane lined containers, and lacquered tin cans have been used successfully.

Improving the Whipping Properties of Cream and Milk
U. S. Patent 1,939,326

The material is heated with addition of about 0.3% of sodium citrate.

Whipping Cream

Select clean raw or pasteurized cream containing 32 to 36 per cent fat. The cream should be aged at least 4 hours at 40° F. before whipping. The cream and utensils should be at 40° F. when whipping. Use a turbine type whipper, avoid over-whipping, and when using a small or large unit avoid over-loading it. Cream properly whipped can be stored at 40° F. for 48 hours or more.

Whipping Cream

Points for successful cream whipping are as follows:

Use a turbine whipper.

Suit the size of each batch to the whipper.

Stop whipping at the right time.

Use cream containing 30 to 35 per cent of fat, if possible.

Age the cream at 40° F. for four hours or more and whip it at that temperature. Ordinary whipping cream, as sold by milk companies, usually has sufficient fat and has been properly aged before sale.

Store in a cold refrigerator as near 40° F. as possible.

Whipped cream when stored should be held in a refrigerator.

Caffein-Free Coffee
U. S. Patent 1,957,358

1000 kilograms of air-dried coffee-beans are filled into a vertically arranged preferably cylindrical extraction vessel, whereupon the solvent (for instance trichlor ethylene, chloroform, benzol or the like) is added. Thereupon 400 kilograms of water are added and the vessel is closed. The contents of the vessel is heated and both fluids, i.e., solvent and water, are caused to circulate and wallow rapidly in order to form the emulsion.

The fluids may for instance be inter-mixed by means of a propeller or another stirring device or by means of a pump of sufficient capacity. If a pump is used the fluids may for instance be sucked off at great velocity from the bottom of the container and be reintroduced into the same at the upper level of the charge. This circulating movement may of course also be reversed. The vertical flow caused by this circulation will effect a thorough intermixing of the substances in the vessel and the emulsion is rapidly formed. The specific weight of the solvent used is without importance. The extraction may therefore if desired be carried out with solvents which are heavier than water or with solvents which are lighter than water.

When the fluids have been intermixed for about 30 minutes the coffee beans will have absorbed the water completely. The extraction may now be continued in the usual manner. After the extraction has been completed the coffee beans are heated in order to remove the last remains of the solvent from the same. The caffein may be separated from the solvent by slowly evaporating the latter.

Lecithin from Soy Beans

A 37% lecithin preparation can be extracted directly from soy beans by boiling alcohol in the proportion 1:5; the lecithin content can be increased by recrystallizing from alcohol. The lecithin content of soy beans is 1.5 to 1.6%. It is advisable to use meal from which the oil has already been extracted. The alcohol method has several advantages for extracting lecithin.

Deodorizing Soy Beans

For products of the type of roasted coffee, deodorization is accomplished by roasting the beans at 150° until completely dry. For products for which a stable emulsion is not required the moist beans are heated to 110–120°. (In both cases the beans are shelled, soaked, and ground.) If the ability to emulsify must not be destroyed, as in the preparation of soy-bean milk, deodorization is accomplished by treating the ground beans with saturated steam for ½ hour and drying at 60–65° F. The milk can be deodorized by prolonged boiling in an open vessel. Hydrogen peroxide is not a suitable deodorizing agent.

FUELS

Solidified Alcohol Fuel
U. S. Patent 1,934,860
Formula No. 1

Dissolve 2.5 parts of a nitrocellulose, preferably one having a nitrogen content of about 12.1%, in 49 parts of absolute ethyl alcohol at a sub-zero temperature of about —30 degrees C. The formation of this solution may be facilitated by stirring, and the time required will be about two hours. There is then added 51 parts of ethyl alcohol containing about 6 parts of water. The aqueous alcohol is chilled to about —30 degrees C. previous to the addition. The stirring is continued and when a clear sol is obtained, it is filled into suitable containers, as for example, metal cans of desired size. Gellation is then caused by warming to ordinary atmospheric temperature, i.e. about + 20 degrees C.

The nitrocellulose dissolves quite readily in the cold alcohol and after the addition of the aqueous alcohol it solidifies quickly upon becoming warm. There results a firm jelly which does not contract or pull away from the sides of the container and which exhibits very little, if any, syneresis, i.e., exudation of liquid. The fuel so produced burns with a high and intense flame and with little soot deposition.

It has been found that the amount of water which may be employed varies with the variety of nitrocellulose used. The higher the nitration, the less the amount of water; for example, with a nitrocellulose having a nitrogen content of about 12.1%, satisfactory use may be made of 5% water, whereas, with a nitrocellulose having about 11% nitrogen content, use may be made of 8% water.

Formula No. 2

Dissolve 2.8 parts of a nitrocellulose, having a nitrogen content of about 11.5%, in 47 parts of absolute ethyl alcohol. Then cool the solution to about —30 degrees C. and admix with 46 parts of 90% aqueous ethyl alcohol, also chilled to —30 degrees C. The admixture is stirred until a clear sol is obtained, whereupon it is filled into suitable containers and gelled in accordance with the procedure set forth in Example I.

Formula No. 3

Dissolve 2.8 parts of a nitrocellulose, having a nitrogen content of about 12%, in a mixture containing 79 parts of absolute ethyl alcohol and 20 parts synthetic methanol at a temperature of about —30 degrees C. The mixture is stirred until a clear sol is obtained and then the cold solution is filled into suitable containers, wherein it is solidified by warming to ordinary atmospheric temperature.

As a preferred range of nitrogen content of the nitrocellulose to be utilized, mention may be made of a nitrogen content approximately within the range of 10.5% to 12.5%, it being understood that the range not only includes the figures mentioned but figures approximating the same on either side thereof.

Solidified Alcohol

Alcohol	4 gal.
Stearic Acid	300 grams
Caustic Soda	40 grams
Water	4 oz.

Heat alcohol on steam bath to about 160° F. Add the molten stearic acid in a thin stream to the hot alcohol and stir properly. Now add the caustic soda, dissolved in the water, slowly while stirring. Fill in cans and let cool.

Solid Alcohol Fuel
U. S. Patent 1,934,725

About 2.5-5.0% of a nitrocellulose, insoluble in the alcohols used at ordinary temperatures, is dissolved in anhydrous methyl or ethyl alcohol at a low temperature of about —30° to —60° C. and the temperature is then permitted to rise to normal temperature without evaporation to produce a solid gel suitable for use as a fuel.

Alcohol Motor Fuel

Alcohol	15%
Benzol	20%
Gasoline	65%

This mixture will not separate and has high anti-knock properties.

Solidified Gasoline

Coconut Oil	50 g.
Caustic Soda Solution (10%)	46 cc.

Water	120 cc.
Benzol	60 cc.
Gasoline	5 gal.

Mix the coconut oil, caustic solution, alcohol, and 20 cubic centimeters of the water, and heat over a Bunsen burner until completely saponified. Then add the remaining 100 cubic centimeters of the water and stir out. Add to this the benzol and stir until completely emulsified. Then add the gasoline in small portions while stirring vigorously with a mechanical stirrer. The resulting jelly-like mass forms a very stable emulsion and may be kept in glass jars or tubes for various purposes.

Composition for Treating Fuels Such As Gasoline for Minimizing Carbon Accumulations in Engines
U. S. Patent 1,925,048

The composition consists of benzene 30, trinitrotoluene 2, o-nitrochlorobenzene 8, a high heat-resisting lubricating oil 30, castor oil 25, α-naphthylamine 16, acetone 11 and an acetate such as butyl acetate 6 parts.

Gaseous Fuel for Engines
U. S. Patent 1,936,155

A fuel suitable for internal-combustion engines of lighter-than-air craft contains hydrogen about 52% and butane about 48%.

Gaseous Fuel
U. S. Patent 1,936,155

A gaseous fuel for internal combustion engines of lighter-than-air craft consists of about 52% hydrogen and about 48% butane.

Gaseous Fuel
U. S. Patent 1,936,156

A gaseous fuel for internal combustion engines of lighter-than-air craft consists of about 27.5% hydrogen and about 72.5% propane.

Fuels for Internal Combustion Engines
British Patent 396,427

To a light Diesel fuel oil is added not more than 2% (preferably less than 1%) of an organic acid of high molecular weight (e.g. oleic acid or a mixture thereof with lubricating oil).

Non-Detonating Fuel
U. S. Patent 1,893,021

A mineral hydrocarbon oil is mixed with 0.1–1.0% of iron carbonyl.

Anti-Knock Fuel
U. S. Patent 1,903,255

Para-cymene (more than 2% by volume) is added to gasoline.

Diesel Fuel
British Patent 396,427

Fuel for Diesel type engines consists of a light Diesel oil with up to 2% of an organic acid containing at least 14 carbon atoms added to obtain improved lubrication of the pistons and cylinders. Naphthenic acids and higher fatty acids are specified, e.g., 3 kilograms of a 30:70 mixture of oleic acid and lubricating oil or 1 kilogram of oleic acid to 500 kilograms of 0.8 petroleum Diesel oil.

Fuel
British Patent 395,282

A composition for treating coal or preparing fuel briquets comprises sodium chloride, sodium chlorate and potassium chlorate and/or potassium permanganate. In an example, 1 pound of a mixture comprising crushed rock salt passing a 7–mesh sieve 27%, dried medium salt passing a 12–mesh screen 40%, sodium chlorate 15% and potassium permanganate 3%, is dissolved in 8 gallons of water and distributed over 2 tons of coal.

Removal of Hydrogen Sulfide from High-Sulphur Gases

The concentration of hydrogen sulphide in gas varies from a few grains per 100 cubic feet to 10–20% by volume. A low-sulphur gas is one containing up to 500 grains per 100 cubic feet. Such a gas can be cleaned by a usual Seaboard or ferric oxide processes, but stronger gases require a different treatment. Of the weaker bases used for recovery, triethanolamine gives the most satisfactory results and is considerably more economical. A 40% solution is used. Ammonia (30 grams per liter), lime in salt solution (2.4 grams of calcium hydroxide per liter of 10% sodium chloride solution) or sodium carbonate (130 grams per liter) can be used but the cost of power and steam is 4–20 times as high as for triethanolamine.

Composition for Treating Coal
British Patent 395,282

Mixture of potassium permanganate 3%, potassium chlorate 15%, sodium chlorate 15%, and sodium chloride 67%, is dissolved in water (1 pound in 8 gallons) and sprayed onto the coal (2 tons) to improve its burning powers.

Briquetting of Coal
British Patent 398,007

The coal dust is mixed with 1½ to 2% of sorghum meal and the mixture is treated with steam at 120° to 150° and briquetted.

Briquetting Molds
German Patent 588,403

Permanent briquetting molds are prepared by molding and sintering a mixture containing tungsten carbide 50 to 60%, chromic carbide 20 to 30%, silicon carbide 10 to 15%, beryllium 5 to 10% and aluminum 0.5 to 2%.

Coating of Coal
U. S. Patent 1,902,642

For spraying on coal an aqueous suspension of colored cellulose pulp distributed in a water repellent substance, e.g., 10–75% of paraffin wax dissolved in benzol, is used.

Coal Binder
U. S. Patent 1,887,183

Briquettes are made from a mixture of coal 10 parts and hydraulic cement, less than 1 part by volume, moistened with a hygroscopic salt solution.

Coloring Coal
U. S. Patent 1,952,180
Formula No. 1

To 100 gallons of water add 2.5 pounds of ferric chloride crystals and 3 pounds of potassium ferricyanide. It is better to dissolve these salts separately in a little water which can readily be done because they are freely soluble and then add these concentrated solutions to the larger bulk of water. Either the dry or wet coal is then immersed in this bath for a short time and if desired it can then be washed. The color display appears on the coal while it is still in the bath but shows up more brilliantly after the coal is dry. The immersion time may be as short as thirty seconds if the temperature of the bath is not too low, thus permitting the passage of a large bulk of coal through a standing bath in a continuous operation. A solution like the above can also be sprayed on the coal or be applied to it by other means. The proportions of chemicals used can be varied over wide limits, thus obviating the necessity of rigid control.

The deposit on the coal resulting from the reduction of the ferric ferricyanide or potassium ferric ferricyanide solution prepared as above, is an insoluble film of Prussian blue or related chemical compound, so thin as to produce the rainbow effect as described. The relationship of the different colors can be altered by varying the conditions, for instance orange and yellow can be made predominant, or blue and violet. The film is very adherent.

Formula No. 2

Another method of producing the rainbow effect by chemical reduction is as follows: Mix together twenty volumes of a one per cent solution of potassium permanganate, ten volumes of a one per cent solution of ferric chloride and two volumes of a one per cent solution of sulphuric acid. Immerse the coal in this at normal temperature for a minute or more.. Remove and allow to dry. A thin brown film of oxides or hydrated oxides of manganese and iron forms on the surface of the coal which by reflected light appears as a rainbow effect. The above proportions of chemicals, and the time and temperature can be widely varied, and the ferric chloride can be omitted or other chemicals added such as aluminum sulphate or sodium bichromate and a rainbow effect can still be obtained.

The more detailed procedure in regard to the formation of deposits by oxidation reactions is as follows:

(1) Five-tenths of a pound of the dye Ciba blue 2B is vatted with three-tenths of a pound of caustic soda, and one pound of sodium hydrosulphite in 3 gallons of water at 165° F. in much the ordinary way of preparing a solution of the leuco-compound of this dye for coloring textile fibers. The above "stock vat" is then diluted with 190 gallons of water which in textile parlance has been "sharpened" with a very little caustic soda and hydrosulphite of soda. The diluting water can be around 80° to 90° F. and need not be at the usual dyeing temperature for Ciba blue 2B. The coal is immersed in the above dye solution for a minute or more until a very thin blue deposit is formed thereon, which if sufficiently thin exhibits the rainbow effect.

Froth-Flotation of Coal

The froth-flotation process has been applied within recent years to the separation of coal from the ash forming materials. Briefly described, the process involves the air agitation of raw coal, all sizes up to $1/10$ inch and special coals up to $1/4$ inch, with three to ten times its weight of water and a small quantity of suitable reagents. The amount of reagent may vary from 0.5 to 2.0 pounds per ton of dry raw coal, depending on the method of operation and impurities present. The reagents ordinarily used are coal-tar derivatives such as creosote oil, cresylic acid, tar-oil distillates, and naphthalene or anthracene oils. Gas oil, petroleum oil, wood-tar distillates, and Flotation Aldol are also used. Iron sulfates when used in slightly acidic solution have been recommended for depressing pyrite flotation. The reagent with air and water forms bubbles that selectively attach themselves to the coal particles. The coal particles are buoyed up and float at the surface, where they may be removed, while most of the refuse material is wetted and remains behind.

For example, the coal is crushed to size, usually about 0.1 inch, and fed into the first box of the flotation machine. Water amounting to three and one-half times the weight of the coal is added to the mixing box, and the reagents, comprising a mixture of refined cresylic acid (0.77 pound per ton of coal), and petroleum-gas oil (0.37 pound per ton of coal), are added as required. The quality of the washed product is controlled by the amount of cresylic acid and gas oil that was added as reagents.

Removal of Soot from Furnaces and Flues

Throw $1/2$ a cup of common salt on the fire after it has been banked at night —of especial benefit where the flues soot readily.

GLASS, CERAMICS, ENAMELS, ETC.

This formula for glass fuses at a fairly low temperature and is easy to handle.

Phosphoric Acid

(60 deg. Bé.)	76.1	59.4 parts
Boric Acid	....	18.7 parts
Dicalcium Phosphate	45.7	.... parts
Magnesium Phosphate Hydrated	11.4	45.5 parts
Aluminum Phosphate Hydrated	....	30.1 parts
Titanium Oxide	2.0	1.6 parts

Add a little sugar to reduce the melt. These formulae will produce a very good workable glass.

"Dope" for Cutting and Filing Glass

Make a saturated ether solution of gum camphor and add a volume of spirits of turpentine equal to the volume of ether used.

Keep the surface of glass moistened with this solution by frequently dipping file in solution.

Pencils for Cutting Glass

These pencils are used by first starting a crack with a file or diamond cutter. Then apply the pencil which has previously been made red hot, in the direction of the cut.

Wood Charcoal	90 lb.
Sodium Nitrate	3 lb.
Powdered Gum Benzoin	1 lb.
Ammonium Nitrate	2 lb.
Tragacanth	4 lb.

Make into a paste, roll into pencils and dry.

Glass Etching Solution

Hydrochloric Acid	6 lb.
Sulphuric Acid	1 lb.
Water	3 lb.

Protective Solution for Coating Glass with Above Solution

Asphalt	50 lb.
Stearin	15 lb.
Beeswax	30 lb.
Lampblack	4 oz.
Turpentine	200 lb.

Glass Etching or Frosting Compound

Formula No. 1

Hot Water	19.0%
Ammonium Bifluoride	69.5%
Sodium Fluoride	2.5%
Hydrofluoric Acid (30%)	9.0%

Formula No. 2

Hot Water	18%
Ammonium Bifluoride	40%
Sodium Fluoride	10%
Molasses	20%
Hydrofluoric Acid (60%)	12%

These formulas frost well as slightly above room temperature, and weigh approximately 12 pounds per gallon. Mix in order given in a lead lined vessel. The mixture should be kept somewhat warmed until the ingredients are almost all dissolved, and must be thoroughly stirred before use. A copper wire basket may be used to dip glassware. Bottles should be fitted with rubber stoppers, and must not be frosted on the inside, as this makes them liable to break easily. Immerse glass into solution for one minute. Remove, drain ten seconds, and wash off at once with hot water. Re-immerse one minute, remove and wash off with hot water as before. Dry. If the frost is not sufficiently opaque, make the first immersion for two minutes, instead of one minute. Use goggles when mixing solution. Adequate ventilation must be maintained as the vapors of Hydrofluoric Acid are exceedingly dangerous. The excess Ammonium Bifluoride provides a reserve etching capacity. After being in use for some time, the solution may be further fortified by the addition of Hydrofluoric Acid.

Glass Etching Ink

Hot Water	12%
Ammonium Bifluoride	15%
Oxalic Acid	8%
Ammonium Sulfate	10%
Glycerine	40%
Barium Sulfate	15%

This formula is stated in percentages by weight.

Molasses may be substituted for the glycerine, and talc for the barium sulfate. If the ink does not readily adhere to the glass, add an additional very slight amount of water to cut the viscosity. It is not advisable to add free hydrofluoric acid, as this causes the ink to run and to blur. The addition of about two per cent of sodium fluoride sometimes improves the quality of the ink. The glass should be slightly warmed before writing upon it. Allow the ink to act for about two minutes, then wash off thoroughly with hot water and dry. Good legible writing should be obtained easily in not more than 30 seconds when the glass is warm. Use an ordinary steel pen. Wash ink from pen when through. Keep in hard rubber or lead bottles.

Glass Frosting Liquid

Hydrofluoric Acid (60%)	40%
Ammonium Bifluoride (32% Hydrogen Fluoride)	30%
Ammonium Carbonate (28% Ammonia)	10%
Sodium Carbonate (Anhydrous)	5%
Water	15%

This formula is stated in percentage by weight. To use the product, immerse the glass from 1 to 3 minutes. It will develop a particular type of matt on a glass of a certain, definite composition. Any variation from the formula will result in a different etch. If the matt etch is to be kept of constant type, careful attention must be given to the purity of the materials and to the accurate weighing and mixing of the batch.

Ultra Violet Transparent Glass
German Patent 585,816

The glass contains 15–25 parts of boric acid, 1–2 parts of beryllium carbonate, and an amount of alkali carbonate equivalent to 3–6 parts of lithium carbonate. The glass is also useful for lenses, prisms and mercury vapor lamps.

Glass Transparent to Ultra Violet Rays
German Patent 583,001

Glass of high transparency to ultraviolet rays contains barium oxide at least 10%, and boron oxide not more than 10%, and not more than 3½ gram-molecular weights of acid (total silicon dioxide and boron oxide) per 1 gram-molecular weight of base (total barium oxide, sodium oxide, and/or potassium oxide, with or without zinc oxide). The glass contains no arsenic trioxide and practically no ferric oxide, and preferably no calcium oxide. A specified composition

is silica 59%, boron oxide 4%, sodium oxide 7%, potassium oxide 7%, barium oxide 19%, and zinc oxide 4%.

Multicellular Glass
British Patent 406,179

A spongy mass (density 0.80) is produced by allowing a fused mass (e.g., sand 100 parts, sodium carbonate 32 parts, boric acid 76 parts, aluminum hydroxide 5.5 parts, titanium oxide 5.0 parts, ammonium chloride 1.5 parts, powdered charcoal 0.7 part) to swell while being maintained near its softening point (e.g., 700°). Alternatively, the fused mass is rapidly cooled and gradually reheated.

Opaque Glass
U. S. Patent 1,956,176

A cream colored opaque glass is formed by fusing together a white glass batch which includes a fluorine compound and an aluminum compound and coloring ingredients comprising rouge, sodium uranate and selenium in the following proportions, batch about 200 pounds, rouge 4 to 10 pounds, sodium uranate 2 to 7 pounds, and selenium 2 to 10 ounces.

Prevention of Blurring or Dulling of Sheet Glass
British Patent 394,635

The dulling or corrosion of sheets or plates of glass, due to the action of condensed moisture, is prevented by treating their surfaces with a medium, such as cupric sulphate, ferrous sulphate, citric acid, boric acid, that neutralizes the alkalies extracted by the moisture. The medium may be incorporated with a carrier adapted to be spread thinly on the surface, such as asbestos, sawdust, etc. Paper, saturated with a solution of 1 gram of ferrous sulphate per liter of water, may be used for separating glass sheets when packed or stacked.

Electric Lamp Coloring Solution

Dissolve 25 parts of bleached shellac, 8 parts of powdered rosin, and 1 part of gum benzoin in 75–100 parts of denatured alcohol, and add an alcohol-soluble aniline dye (light-fast) of the color desired.

Removing Carbon Residue from Glassware

Instead of the more usual procedure of using scouring or abrasive material and soap, or by dipping in solutions of hydrofluoric acid or strong soda lye, a quicker and safer method is found by use of the following solution:

Tri-Sodium Phosphate	2 tbsp.
Sodium Oleate	1 tbsp.
Soft Water	1 quart

Set the vessel to be cleaned in this solution several minutes, remove and brush off the incrustation with a moderately stiff bristle brush, rinse with water. This solution may be used to much advantage to cleaning apparatus used in testing oils, greases, paints, etc.

Mirror Silvering

Prepare the following three solutions and stock separately, number two preferably in an amber colored bottle.

Solution No. 1

Dissolve 25 grams of sugar in about 250 cubic centimeters distilled water and add 3 grams of tartaric acid. Boil the solution for 10 minutes, cool, add 50 cubic centimeters of ethyl alcohol and dilute with water to 500 cubic centimeters.

Solution No. 2

Dissolve 20 grams of silver nitrate and 30 grams ammonium nitrate in water and make up to 500 cubic centimeters.

Solution No. 3

Dissolve 50 grams sodium hydroxide in water and make up to 500 cubic centimeters.

Thoroughly clean the glass free from all grease and finger marks. Rapidly mix equal volumes of each of the above solutions and flow the mixture onto the glass. In a few moments a well adhering coat of metallic silver will be deposited on the glass when same may be rinsed well with distilled water and allowed to dry.

Silvering Automobile Headlight Bulbs

A. Preparation of Lamps

The lamp rack is placed on the drain board, bottom side up and filled with lamps. A new stiff scrubbing brush is dipped into the cleaning solution, sprinkled with powdered ammonium bicarbonate and the lamps then thoroughly scrubbed. A new brush appears to be necessary for about every five batches, as due to the softening action of the cleanser the brush becomes so soft and flexible that pin-holes otherwise develop. Although pumice may be used on ordinary glass plates the use of this or of any other abrasive should not be applied to bulbs, due to the resulting mirror showing every scratch, however slight, thereby producing a rather unsightly mirror having no sale value. The rack is then stood on end and by means of a hose connected to the faucet is thoroughly washed with warm water, after which the excess water is shaken off and the rack placed over a tray containing (1 plus 2) nitric acid for five minutes. The acid is of such depth that when the lamps are placed over it the acid comes up over the glass to a greater depth than the silver mirror to be made. (It must not be neglected to take into consideration that many of the glazes on enamel ware such as that which is found on ordinary trays contain tin oxides, in which trays nitric acid should not be placed, because nitrates of tin may be formed resulting in the ruining of the acid bath for the purpose for which it is used.) The rack is then removed from the acid tray, the acid thoroughly washed off and again scrubbed as above, placed over the acid tray and again thoroughly washed with warm water. The rack still on end is then bathed with the stannous chloride solution by means of the siphon and immediately washed off with the warm water spray. This washing is repeated about four or five times in order to remove the last traces of stannous chloride. Much better results are obtained by bathing the lamps with the stannous chloride using the siphon rather than by dipping them into a tray containing the stannous chloride, in which a skin quickly forms on the surface of the solution. The bulbs, in being lowered down below the surface of such a solution appear to receive a minute film which results in every poor mirrors being produced.

A final washing with distilled water from a siphon insures the removal of the last traces of the chlorides present in the hydrant water. The rack is then ready for the plating operation which follows under "Tray Plating Solution."

B. Precautions

Too much stress cannot be laid on the caution that rubber mounted goggles should at all times be worn by those inexperienced in the preparation of alkaline silver nitrate solutions. Those solutions containing either sodium or potassium hydroxides are especially dangerous. It is prudent to consider that any solution of alkaline silver nitrate which shows a slight mirror on the bottle in which it is kept, or in which dark sediment is noted, regardless of how little, is a potential danger and that judgment should be used in handling. It must be very carefully decanted behind a reinforced glass plate taking care that the sediment is not touched with the end of the siphon.

Solution No. 1 (Potassium Hydroxide)

Dissolve 200 grams C.P. pellets in distilled water and dilute to 200 cubic centimeters.

Solution No. 2 (Iodine Solution)

Dissolve five grams iodine in pure 95% alcohol and dilute to 100 cubic centimeters.

Solution No. 3 (Double Cyanide of Mercury and Potassium)

Potassium cyanide	0.5 g.
Mercuric cyanide	0.5 g.

Dissolve in distilled water and dilute to 100 cubic centimeters.

Solution No. 4 (Nitric Acid for Tray)

Concentrated nitric acid (sp. gr. 1.42)	1 volume
Distilled Water	2 volumes

Solution No. 5 (Stannous Chloride)

Into a clean five pint bottle arranged with siphon and pinchcock place five grams C.P. Crystallized Stannous Chloride and almost fill the bottle with distilled water. Shake several times during the first five minutes after adding the water, until no crystals remain on the bottom of the bottle.

Solution No. 6 (Glass Cleaning Solution)

Dissolve 100 grams cleaner in distilled water and dilute to 1500 cubic centimeters. Pour out into a tray or other shallow vessel as needed. This cleaner is a mixture of detergent alkaline salts and may be obtained from electroplating supply houses.

Ammonium Bicarbonate

Use the fine white crystals or powder.

Lacquer

A dilute colorless lacquer solution is made up from a heat resisting resin. The solvent is preferably toluol, acetone, alcohol, or a mixture of all three. Their prices are reasonable and they have a desirable rate of evaporation in the drying oven.

Solution No. 7 (Stock Silver Solution)

Into a five pint glass stoppered bottle place 150 cubic centimeters concentrated C.P. ammonium hydroxide, Sp. Gr. 0.90. Weigh out 90 grams C.P. crystallized silver nitrate and add to the ammonium hydroxide in the bottle This procedure

is important, and the use of goggles is again suggested at this point. It is advisable to add the nitrate in amounts of about thirty grams, shaking vigorously with a circular motion immediately after each addition until a clear solution results, before making further additions. When all the silver nitrate is completely dissolved, the solution will be found to be quite warm or even hot. About 200 cubic centimeters distilled water are then added followed by 450 cubic centimeters potassium hydroxide solution. The entire solution is then diluted to 1800 cubic centimeters with distilled water. A scratch previously made on the bottle at the 1800 cubic centimeter mark will facilitate the dilution. Place the stopper in the bottle and shake thoroughly, after which the solution is ready for use. It will be noted that this silver solution is crystal clear, in distinction to silvering solutions in general, due to the excess ammonium hydroxide present, and it will not be necessary to filter it. When made according to this method this solution has been kept for periods as long as five days without any apparent deleterious effects on its silvering capacity. It does, however, deposit a very slight mirror on the sides of the bottle during this time. All bottles or containers in which solutions are stored or into which they come into contact must have no trace of grease, oil or paraffin present.

Solution No. 8 (Stock Reducing Solution)

Into a large glass or porcelain vessel place 20 cubic centimeters concentrated nitric acid, Sp. Gr. 1.42 and 5000 cubic centimeters distilled water. Add 450 grams cane sugar and stir with a glass rod until dissolved. Heat to boiling and allow to boil for five minutes. Cool, stir in 15 cubic centimeters iodine solution and place in five pint glass stoppered bottles. This formula with the exception of the iodine is that commonly known as Brashear's reducing solution. In place of the iodine solution it has been found that 15 cubic centimeters of a one per cent solution of the double cyanide of mercury and potassium may be used with equally successful results. In so doing, however, it will be noted that the deposit in the bottom of the plating tray is quite different than that produced when the iodine is used and that each is different from that produced from the unmodified Brashear formula.

Solution No. 9 (Tray Plating Solution)

Into a 1000 cubic centimeter graduate place 330 cubic centimeters stock silver solution. Dilute to 1000 cubic centimeters with distilled water and pour into a wide mouth one gallon glass bottle or jug. Measure out another 1000 cubic centimeter distilled water in the graduate and add to the solution in the gallon jug making a total volume of 2000 cubic centimeters. This procedure will insure not only getting practically all of the silver into the jug but will prevent the gradual deposition of dark silver deposits on the inside of the graduate. Place this jug next to the plating tray. Fill a 100 cubic centimeter graduate to 80 cubic centimeters with the stock reducing solution and likewise place this graduate next to the tray beside the jug. When the lamp rack is ready for plating add the 80 cubic centimeters reducing solution to the 2000 cubic centimeters silver solution in the jug, shake the mixture with a circular motion for a moment to insure thorough mixing and pour out into the tray. At once, place the lamp rack over the solution very carefully so as to avoid ripple marks at the border of the mirror, and leave for fifteen minutes. When iodine or the double cyanide is used, the solution when first mixed preparatory to placing in the plating tray does not turn black almost instantly as in the unmodified Brashear solution, but remains clear for about ten to fifteen seconds, at which time it turns to an orange, purple and finally black color. Experience has shown that if the tray to be used for silvering, if first given an initial coating of silver by running a bath without bulbs, it will produce more satisfactory results in subsequent silvering.

Recovery of Silver Residues

When the rack is removed from the bath at the end of this period the bottom of the tray will be found to be covered with a dark spongy deposit which is washed out into a large porcelain lined container or tub. A quantity of common salt is added to the tub at the beginning of each day's work to insure the precipitation of all silver. On the following morning all silver values will be found in the bottom of the tub. The clear supernatant liquor is decanted off and discarded, and the residue washed into a five gallon bottle into which mossy zinc has been placed, for the reduction of the residues to metallic silver, which are then refined, converted to the nitrate and used over again.

Lacquering Bulbs

The rack of bulbs, immediately after removal from the plating bath is washed off with warm hydrant water several times and followed by a final treatment with distilled water. As much water as possible is shaken from the rack which is then placed in a drying oven. The oven is heated by means of electric hot plates which are on continuously. After a fifteen minute period in the oven the rack is removed, allowed to cool and then dipped into the lacquering tray. The lacquer in this tray is of such depth that it comes up about one millimeter above the edge of the mirror when the bulbs are immersed. A cover is kept over the lacquer tray at all times when not in use to prevent the evaporation of solvent into the laboratory. After dipping in the lacquer the rack is carefully shaken to eliminate the excess lacquer and is then placed in the lacquer drying oven which is identical in construction to the first mentioned oven. On removal from this oven after a twenty minute period, the lamps are allowed to cool, removed from the rack, and each individual lamp tested for imperfections in the mirror by lighting up the filament on six volts. Imperfections in the mirror which are indistinguishable at three to four volts are quite apparent at six volts.

Temporary Protective Coating for Porcelain, Etc.

U. S. Patent 1,936,152

In some cases it is desirable to use a paste which has not a great degree of adhesive power but which is harmless to enameled surfaces. For example, enameled fixtures, such as the bath tub, are usually installed in buildings before plastering, painting and the finishing operations are conducted. It is desirable to protect the enameled surface from being scratched or damaged during these later operations, and one favorite method is to paste paper over all the enameled surfaces. The paste for this purpose need not have great adhesive power, in fact it will be found advantageous to have a paste with a rather low adhesive power to facilitate the removal of the paper when desired. It is desirable, therefore, to use a paste which does not dry hard, as do starch pastes, thereby rendering the removal of the paper and adhering paste an easier task.

A formula found suitable for the purpose of attaching a covering for the protection of enameled surfaces as described above is twenty-seven parts of clay, six and one-half parts of zinc oxide, and six and one-half parts of vegetable starch preferably corn or wheat flour.

For some purposes substitute all or a portion of the clay in the above formula with the substance known as bentonite.

It will be found that this paste is absolutely harmless to enameled surfaces, and that the covering attached by means of this paste, as well as the residuum of paste, is easily removed, as it dries soft, is somewhat flocculent in texture, and does not harden excessively as does a pure starch paste.

It will be understood that in case the prevention of fermentation is not important, as, where the paste is to be used as soon as mixed with water and is to be used upon surfaces not affected by organic acids, the zinc oxide may be omitted and a mixture of paste and clay used.

Batches for Pottery Bodies

Formula No. 1—Sanitary Ware Firing at 1285° C.*

* Cone 11.

English China Clay	16 parts
Florida Kaolin	4 parts
Tennessee Ball Clay	12 parts
Kentucky Ball Clay	5 parts
American Flint	27 parts
Feldspar	36 parts

Formula No. 2—Sanitary Ware Firing at 1250° C.*

* Cone 9.

English China Clay	26 parts
Florida Kaolin	8 parts
English Ball Clay	10 parts
American Flint	31 parts
Feldspar	25 parts

Formula No. 3—Vitreous China Firing at 1260° C.*

* Cone 10.

English China Clay	24 parts
Florida Kaolin	8 parts
North Carolina Kaolin	8 parts
English Ball Clay	7 parts
American Flint	35 parts
Feldspar	15 parts
Whiting	3 parts

Formula No. 4—Vitreous China Firing at 1260° C.*

* Cone 10.

English China Clay	20 parts
Florida Kaolin	15 parts
Kentucky Ball Clay	15 parts
American Flint	32 parts
Feldspar	18 parts

Formula No. 5—Semi-Vitreous China Firing at 1225–1250° C.*

* Cones 8–9.

English Ball Clay	5 parts
English China Clay	21 parts
Florida Kaolin	12 parts
Tennessee Ball Clay	5 parts
American Flint	30 parts
Feldspar	26 parts
Whiting	1 part

Formula No. 6—Semi-Vitreous China Firing at 1225–1250° C.*

* Cones 8–9.

Florida Kaolin	9 parts
Georgia Kaolin	16 parts
Missouri Ball Clay	14 parts
Kentucky Ball Clay	14 parts
American Flint	31 parts
Feldspar	16 parts

Formula No. 7—Hotel China Firing at 1285° C.*

* Cone 11.

English China Clay	16 parts
Florida Kaolin	4 parts
Tennessee Ball Clay	12 parts
Kentucky Ball Clay	5 parts
American Flint	27 parts
Feldspar	36 parts

Formula No. 8—Hotel China Firing at 1260–1285° C.*

* Cones 10–11.

North Carolina Kaolin	35 parts
Florida Kaolin	23 parts
Tennessee Ball Clay	9 parts
American Flint	20 parts
Feldspar	12 parts
Whiting	1 part

Formula No. 9—High Tension Electrical Porcelain Firing at 1260–1285° C.*

* Cones 10–11.

English China Clay	20 parts
English Ball Clay	30 parts
American Flint	20 parts
Feldspar	20 parts

Formula No. 10—High Tension Electrical Porcelain Firing at 1410° C.*

* Cone 15.

English China Clay	30 parts
Kentucky Ball Clay	25 parts
American Flint	25 parts
Feldspar	15 parts

Formula No. 11—Floor Tile Firing at 1310° C.*

* Cone 12.

English China Clay	2 parts
North Carolina Kaolin	8 parts
Tennessee Ball Clay	14 parts
American Flint	39 parts
Feldspar	36 parts
Magnesium Carbonate	1 part

Formula No. 12—Wall Tile Firing at 1260° C.*

*** Cone 10.**

English China Clay	34 parts
English Ball Clay	25 parts
Pebble Flint	28 parts
Cornwall Stone	13 parts

Formula No. 13—Crucible Body Firing at 1310° C.*

*** Cone 12.**

New Jersey Fire Clay	45 parts
Crushed firebrick grog passing a 60 mesh screen	55 parts

Note: Clays and feldspars are of somewhat variable composition. Even those coming from the same mine may have decidedly different properties.

Preparation of Bodies for Forming Plastic Body

Crush lumps of clays, stir with water (blunging) until a thin slip is formed and all the clays are completely in suspension. Add the flint and feldspar with continued blunging until everything is in suspension and a smooth slip of the consistency of light cream is produced. Screen through a 100 mesh screen. Stir frequently while slip is in storage to prevent settling. Filter press at 80–90 pounds pressure. If small batches are being made, concave plaster bats may be used to absorb the excess water as a substitute for filter pressing.

Casting Body for Casting in Plaster Molds

Using dry body which has been processed by above method, add gradually to water while blunging. The water should contain the deflocculents in solution. The following makes a satisfactory consistency:

Dry Body	100
Water	35
Soda Ash	0.1
Sodium Silicate	0.15

The casting slip is poured into the plaster of Paris mold and allowed to remain until a sufficiently thick layer of clay has been absorbed on the walls of the mold. The excess slip is then poured out and the mold permitted to drain. The cast piece is left in the mold until it breaks free from the plaster by shrinkage, then removed for trimming and drying.

Glaze Batches

Formula No. 1—Raw Lead Glaze for Whiteware

Florida Kaolin	4.5 parts
White Lead	24.8 parts
Feldspar	44.4 parts
Whiting	20.3 parts
Zinc Oxide	6.0 parts

Fire at Cone 6 (1180° C.).

Formula No. 2—Raw Porcelain Glaze

Feldspar	51.5 parts
Cornwall Stone	20.6 parts
Flint	6.2 parts
Whiting	15.5 parts
Ball Clay	6.2 parts

Fire at Cone 8 (1225° C.).

Formula No. 3—Raw Porcelain Glaze

Whiting	14.8 parts
Feldspar	35.2 parts
Ball Clay	8.2 parts
Flint	41.8 parts

Fire at Cones 10–11 (1260–1285° C.).

Formula No. 4—Fritted Glaze, Leadless

Frit:

Flint	42.0 parts
Feldspar	13.7 parts
China Clay	6.2 parts
Whiting	12.8 parts
Magnesium Carbonate	2.1 parts
Borax	23.2 parts

Glaze:

Frit	100 parts
Ball Clay	7 parts

Fire at Cone 5 (1180° C.).

Formula No. 5—Raw Bristol Glaze (Opaque, Glassy)

Feldspar	58.5 parts
Whiting	10.5 parts
Zinc Oxide	11.2 parts
Ball Clay	9.3 parts
Flint	10.5 parts

Fire at Cone 8 (1225° C.).

Formula No. 6—Raw Opaque Enamel

White Lead	34.3 parts
Feldspar	29.6 parts
Whiting	8.0 parts
Florida Kaolin	3.3 parts
Tin Oxide	12.8 parts
Flint	12.0 parts

Fire at Cone 05 (1030° C.).

Preparation of Glazes

Raw glazes are prepared by grinding the various materials together in a ball mill with 40–45% of water added. The glaze may be applied by dipping the fired ware in the glaze slip or by spraying the glaze slip on the ware.

In fritted glazes the soluble materials are fused with other portions of the glaze to produce an insoluble glass. The glass is then pulverized and used as any other raw material would be in the above preparation.

German Chemical Porcelain

Feldspar	20%
Kaolin	50%
Calcined Kaolin (Calcined at Cone 15)	20%
Flint	10%

Fire this mixture to Cone 16.

Glaze for Chemical Porcelain

Potassium Oxide Content	.2	molecular equivalents by weight
Calcium Oxide Content	.7	molecular equivalents by weight
Magnesium Oxide Content	.1	molecular equivalents by weight
Aluminum Oxide Content	.75	molecular equivalents by weight
Boron Oxide Content	.1	molecular equivalents by weight
Silicon Dioxide Content	7.5	molecular equivalents by weight

American Chemical Porcelain

Feldspar	25%
Ball Clay	15%
Kaolin	25%
Calcined Kaolin	25%
Flint	10%

Fire this mixture to Cone 14. The glaze matures with the body. All chemical porcelain is biscuited at about 950° C. before glazing.

Calcined Kaolin	12%
Flint	22%

Fire this mixture to Cone 15.

American Electrical Porcelain

Feldspar	32%
Ball Clay	35%
China Clay	17%
Flint	16%

Fire this mixture to Cone 12 in 36 to 40 hours.

American Electrical Porcelain Glaze

This product contains the following proportions of the various elements, which are expressed below in terms of their oxides:

Potassium Oxide Content	.3	molecular equivalents by weight
Calcium Oxide Content	.5	molecular equivalents by weight
Magnesium Oxide Content	.05	molecular equivalents by weight
Barium Oxide Content	.05	molecular equivalents by weight
Zinc Oxide Content	.10	molecular equivalents by weight
Aluminum Oxide Content	.70	molecular equivalents by weight
Boron Oxide Content	.10	molecular equivalents by weight
Silicon Dioxide Content	6.	molecular equivalents by weight

German Electrical Porcelain

Feldspar	22%
Kaolin	44%

Mature at Cone 12 with the body.

Pyrometer Tube Body

Kaolin	54%
Calcined Alumina	21%
Artificial Silliminate	25%

Fire this mixture to Cone 18.

American Hotel China Body

Feldspar	17 %
Calcium Carbonate	0.5%
Magnesium Carbonate	0.5%
Ball Clay	10 %
Florida Kaolin	10 %
English China Clay	12.5%
American China Clay	12.5%

This is biscuitfied at Cones 10–11, and glazed at cones 4 to 6.

German Dinnerware Body (True Porcelain

Feldspar	22%
Whiting	1%
Kaolin	55%
Flint	22%

Fire this mixture to Cone 15.

Jasperware Body

Cornwall Stone	25%
Gypsum	3%
Clay	27%
Barite	45%

This is biscuited at Cone 8, and glazed

with a soft glaze of the lead borosilicate type, maturing at Cone 3.

Ceramic Insulation Binder

When the silicate clays which are usually used are mixed with water prior to molding they tend to lose water and crumble. If Diglycol Stearate in the proportion of 6% by weight of the clay mixture is taken up with boiling water and thoroughly incorporated the clays act as a binder and help to retain sufficient moisture to prevent crumbling. When the forms are baked the Diglycol Stearate volatilizes and disappears, thus not affecting the resistance of the finished product.

Transparent Siliceous Materials
French Patent 749,436

A material suitable for insulation in spark plugs, etc., is made by fusing together crystallized quartz and beryl ($Si_6O_{18}Al_2Be_3$). The best proportion is 3-5% of beryl. Aluminum oxide may be added if desired.

Acid Resisting Frit

Frit	100 lb.
Clay	6 lb.
Tin Oxide	4 lb.
Magnesia	4 oz.
Sodium Aluminate	2 oz.
Water	40 lb.

This frit ground to a fineness of 5 to 6 grams dry on a 200 mesh sieve from a 100 gram sample of wet enamel as taken from the mill may be allowed to stand over a long period of time without loss of set.

Coating Incandescent Globes Green
U. S. Patent 1,941,990

Kaolin or china clay 50 grams; hydrated chromic oxide or "Guignet Green" 200 grams; cadmium sulphide 50 grams; boric acid 160 grams and sodium silicate 1000 cubic centimeters of 1.015 to 1.035 specific gravity.

The cadmium sulphide is employed to tone the color of the green pigment and may be dispensed with as desired.

Glazes Without Lead for Terra Cotta

A successful leadless glaze for Seger cone 0.5 is one corresponding to the Seger formula: 1.12 silica; 0.5 alumina; RO; 0.15 F. Its composition is: 39.40 feldspar, 29.42 quartz, 11.48 cryolite, 8.69 zinc oxide, 6.96 sodium carbonate, 4.05

chalk. A preliminary fritting of the batch is recommended.

Glazed Pottery, Tiles, Etc.
British Patent 391,646

Glazed tiles, pottery, etc., are formed from clay etc., with the addition of 1 or more fusible materials prior to forming. A glaze containing gum tragacanth is applied to the surface and the whole is submitted to a single burning. For example, the body is composed of a ground mixture of china pitchers, earthenware pitchers, tiles, silica bricks or sand 72, clay, e.g., ball, china or fire clay 52, fusible material, e.g., boric acid, soft or fritted glaze, 1.25 and sodium silicate of d. 140° Tw. 8 parts, and the formed articles are dipped in a glaze of gum tragacanth 1 and normal glaze 36 parts.

Ceramic Glaze Maturing Below 900° C.
(Cherry Red)

Ground Flint	16 parts
English China Clay	9 parts
White Lead	25 parts

Refractory Articles
U. S. Patent 1,897,183

A mixture of crystalline x-alumina 50-60, finely-divided mullite (artificial or calcined sillimanite, cyanite, or andalusite) 30-40, and highly aluminous bond clay 10% is fired at 1650° for 90-150 hours. The ingredients should be as free as possible from glassy material.

Refractories
British Patent 382,295

Refractory compositions that can be shaped are made from silicon or silicides, e.g., of carbon, copper, chromium, or iron, and a ceramic binding agent that sinters or melts below 1000°, e.g., low-melting glass, water glass, sulfates or chlorides. The composition is sintered at less than 1000°, e.g., 900°, until hard enough to be worked, machined to the required shape and glowed at 1200-1250° to sinter the silicon or silicides. Other binding agents melting above 1000°, e.g., clay, quartz, feldspar, kaolin, zirconium oxide, may be present. In an example 60 parts of 90% silicon are mixed with clay 20, sodium-potassium silicate 15 and a mixture of potassium and sodium chlorides 5 parts.

Modeling Clay

This is a mixture of clay, sulfur and suitable greases or waxes plus grease.

The clay mix consists of 100 mesh Florida Kaolin and finely ground sulfur (Rubber Makers' Grade). Clay and sulfur are thoroughly mixed. To this is added the greases or mixture of greases in the proportions listed below. For certain colors, lithopone is added to the clay-sulfur mix. Then the colorant is added and mixed in the mixer until evenly distributed throughout. The mix is wedged to exclude all the air; allowed to age several days and then cut into shape. The colorants are the oxides alone finely ground or oxides ground in mineral oil.

Formula No. 1

Clay	67 parts
Rubber Makers Sulphur	33 parts

Formula No. 2

Clay	55 parts
Sulphur	25 parts
Lithopone	20 parts

The amount of vehicle is that which will have the proper feel when mixed with the clay. It is approximately one part of vehicle to three parts of clay-sulfur or clay-sulfur-lithopone mix.

Vehicles

1. No. 2 Petrolatum
2. No. 3 Cup Grease — 50%
   National Refining Special T & R grease — 50%
3. Palm Oil — 80%
   Japan Wax — 20%
4. Lanolin — 60%
   Glycerine — 40%
5. Lanolin — 60%
   Palm Oil — 20%
   Glycerine — 20%
6. Palm Oil — 80%
   Lanolin — 20%

For the different colors use the following:

Green—Chrome Oxide in Formula No. 1.
Orange—Oil Ground Chrome Yellow Orange in Formula No. 1.
Blue—Oil Ground Ultramarine Blue in Formula No. 1.
Yellow—Oil Ground Chrome Yellow in Formula No. 1.
Brown—Oil Ground Burnt Sienna in Formula No. 1.
Black—Oil Ground Drop Black in Formula No. 1.
Gray—Oil Ground Drop Black in Formula No. 2.
Pink—Ruby Red Dry Powder in Formula No. 2.
White—Formula No. 2 alone.

Refractory Paint for Firebrick

To prevent oxidation of metal under heat for painting inside of firebrick walls, etc., use:

Flint	50 parts
Clay	45 parts
Sodium Silicate	5 parts

Dyes for Coloring Clay Slips or Glazes Prior to Firing

Formula No. 1

To 700 pounds of slip, add 30 grams of fuchsine.

Formula No. 2

To 700 pounds of slip, add 8 grams of aniline blue.

Black Paint for Marking Refractories Subjected to Heat

Mixture No. 1

Cobalt Oxide	160 parts
Flint	280 parts
China Clay	60 parts

Mix with water to 20 ounces per pint. To $\frac{2}{3}$ of Mixture No. 1 in slip form add $\frac{1}{3}$ Albany Slip in slip form.

Eighty parts of the cobalt oxide shown above may be replaced with 80 parts iron oxide.

Enameling Composition

U. S. Patent 1,944,938

An enameling or glazing composition for direct application on metal initially comprising in the raw batch the following ingredients substantially in parts by weight viz: sodium zirconium silicate 10 to 43; aluminum hydrate 0 to 3; potash feldspar 0 to 48; quartz 0 to 24; sodium carbonate 0 to 11; sodium nitrate 3 to 5; calcium carbonate 0 to 8; cryolite 0 to 7; zinc oxide 4 to 14; fluorspar 0 to 11; and borax 23 to 47.

Enameling or Glazing Metals

U. S. Patent 1,944,938

A coating suitable for use on sheet iron and steel is formed from a batch comprising sodium-zirconium silicate 10–43, aluminium hydroxide 0–3, potassium feldspar 0–48, quartz 0–24, sodium carbonate 0–11, sodium nitrate 3–5, calcium carbonate 0–8, cryolite 0–7, zinc oxide 4–14, fluorspar 0–11 and borax 23–47 parts.

Production of Enameled Iron Articles

British Patent 401,021

A satisfactory white opaque ground coat contains antimony oxide, zinc oxide, and zirconium oxide in the form in antimonates made, e.g., by calcining at 700–800° a mixture of antimony oxide 70, zirconium oxide 30, ammonium nitrate 25, clay 10 parts.

Enameled Gold-Alloy or Platinum Tooth Crowns
Hungarian Patent 106,893

A mixture of silica 6.5, borax 2.0, crystallized soda 1.65, sodium nitrate 0.30, cryolite 1.20 and tin oxide 0.50% is heated for 2 hours to 1000°, powdered and 10 parts of the powder mixed with 0.5 part white clay and an opacifier, the mixture colored with the usual metal oxides and after addition of about 0.75 liters water per kilograms ground in a mill for 24 hours. An aqueous or turpentine pulp made from the dust is applied to the metal crowns and fired.

Enamel for Paper

French Zinc White	13½	lb.
Pigment of Desired Color	1	oz.
Ground in Nitrocellulose	1½	lb.
Camphor	1½	lb.
Alcohol	3	gal.
Castor Oil	7	gills

Vitreous Enamels
U. S. Patent 1,949,479

For producing an opaque frit for vitreous enamels resistant to acids mix a raw batch of enameling materials substantially in the following proportions in parts by weight; sodium zirconium silicate 7.08; quartz 40.58; borax 18.94; soda ash 19.65; sodium nitrate 3.50; titanium oxide 13.86; red lead 3.04; antimony oxide 6.93; and fluorspar 5.94, and then heat the raw batch so mixed to fusion and formation of opaque enamel frit.

Vitreous Enamel
U. S. Patent 1,933,437

An enamel consisting of flint approximately 29.236 per cent, boron approximately 13.127 per cent, nitrate soda approximately 5.727 per cent, soda ash approximately 10.740 per cent, red lead 14.920 per cent, barium carbonate approximately 7.757 per cent, fluor spar approximately 6.563 per cent, antimony oxide approximately 7.160 per cent.

Acid Resistant Vitreous Enamels
U. S. Patent 1,949,479

An opaque frit for vitreous enamels resistant to acids is formed by fusion of a raw batch comprising sodium zirconium silicate 7.08, quartz 40–58, borax 18.94, sodium carbonate 19.65, sodium nitrate 3.5, titanium oxide 13.86, lead oxide 3.04, antimony oxide 6.93 and fluorspar 5.94 parts.

Acid-Proof Enameled Apparatus

The enamel contains 9.75 sodium oxide, 0.25 calcium oxide, 0.075 aluminium oxide, 2.8 silica, 0.08 boron oxide, 0.2 fluorine, corresponding with sand 527, kaolin 65, borax 57, calcium carbonate 85, sodium carbonate 230, sodium silico fluoride 42.

White Bobbin Enamel

Lithopone	3 pints
Castor Oil (No. 3)	6 gal.
White Shellac (4 lb.)	3 pints
Denatured Alcohol	3 pints
Toluol	24 lb.

Pink Bobbin Enamel

Ceres Red	4 oz.
Ceresine Orange	9 oz.

Aniline dye previously dissolved in the coal tar solvent (Toluol).

Purple Bobbin Enamel

Similarly, the White is shaded with Oil Violet dye.

Green Bobbin Enamel

Milori Green (Chrome) C.P.	11¼ lb.
Castor Oil (No. 3)	3¾ pints
Orange Shellac (4 lb.)	15 gal.
Toluol	3¾ pints

Brown Bobbin Enamel

Iron Oxide (Rouge) (95–98%)	2 lb.
Castor Oil (No. 3)	3 pints
Orange Shellac (4 lb.)	6 gal.
Toluol	3 pints

and finally shaded with 2½ ounces nigrosine dye previously cut in the toluol.

Coloring Fluid for Wire Enamel

This method covers the process to be followed in preparing coloring fluid for air-drying wire enamel.

Wood Alcohol (97%)	1	gal.
Castor Oil	16½	fl. oz.

Venice Turpentine   16½ fl. oz.
Emerald Green Crystals,
    No. 233   3   oz.

Thoroughly mix the castor oil and venice turpentine; add half of the wood alcohol gradually until the castor oil and venice turpentine are completely dissolved.

Dissolve the emerald green in the remaining wood alcohol. Allow to stand for at least one hour and then filter thoroughly to remove any sediment or insoluble material. Add the filtered solution gradually to the mixture of castor oil, venice turpentine and wood alcohol, stirring thoroughly.

INK, CARBON PAPER, CRAYONS

Hectograph Ink

Cane Sugar	100 g.
Gelatin	150 g.
Glycerin	600 cc.
Distilled Water	500 cc.

Mix the ingredients in a vessel and heat over a steam bath, with stirring, until homogeneous. Filter through cloth into a flat pan. Allow to cool. Wash the surface with clean water before using.

Hectograph Paste

Water	10 parts
Glycerin	20 parts
Clay, Soft White	70 parts

Mix water and glycerin and knead with clay to consistency of putty. Pack in waterproof paper.

Safety Ink
U. S. Patent 1,951,076

To fifty gallons of water add fifty grains of tannic acid, in which water the acid quickly dissolves. To this add 640 fluid ounces of decolorized tincture of iodine and 2,637 grains of glacial acetic acid. The result is a clear, slightly opalescent, liquid. The stains may be made lighter by increasing the proportion of water and they may be intensified by decreasing the water proportion. Conversely the stain or discoloration may be darkened by increasing the proportion of tincture of iodine used. The shade of the penetrating stain may be darkened and the sensitivity of the paper to chemicals increased by increasing the proportion of tannic acid. The exact proportions of water, tannic acid, iodine and glacial acetic acid to be used are governed by the paper to be treated, the kind of safety paper desired, the degree of stain desired, et cetera.

Inasmuch as it is often desirable not only to have attempts to eradicate writing show, as by discoloration of the paper, but also to know what letter or figure it was attempted to eradicate, consequently a paper having an ink set feature is desirable. By ink set we mean that any attempts to eradicate writing on a paper having this ink set feature results in the writing becoming set in the paper and clearly visible. This ink set feature in safety paper is produced with the same ingredients heretofore described in producing a safety paper, and is determined by the proportion of tannic acid to glacial acetic acid. In a gallon of water, for each grain of tannic acid used there should be not less than 40 grains or more than 90 grains of glacial acetic acid, in order to endow the paper with this ink set feature.

In a 24-pound buff colored paper for each gallon of water and grain of tannic acid 57 grains of glacial acetic acid should be used to produce the ink set feature, and that for canary, pink, gray and goldenrod colored papers the proportion of glacial acetic acid should be from 48 to 57 grains. A solution for blue paper may contain as much as 90 grains of glacial acetic acid, while for a green paper not more than 48 grains should be used. For a 28-pound paper an extra 8 grains of glacial acetic acid should be used to provide the paper with the ink set feature. These results are obtained when the proportion of decolorized tincture of iodine to water is one to ten, although the proportion of said iodine to water may be varied as stated above, as the ink set feature depends more particularly on the ratio of tannic acid to glacial acetic acid.

Inasmuch as the cost of decolorized tincture of iodine is relatively high the proportion of iodine may be materially reduced by using a cheap alkali such as borax and doubling the quantity of tannic acid and glacial acetic acid used. In other words the borax provides sufficient alkali to overcome the acid effect resulting from doubling the quantities of tannic acid and glacial acetic acid. Such a mixture to obtain the stain effect in the paper would consist of fifty gallons of water, one hundred grains of tannic acid, 128 fluid ounces of decolorized tincture of iodine, 5,275 grains of glacial acetic acid and 4,070 grains of borax. To obtain the ink set feature it is merely necessary to use tannic acid and glacial acetic acid in the same proportions previously outlined.

Green Ink

The following gives a green which has not only a lively green color but is very durable:

Bichromate of Potash	10 parts
Hydrochloric Acid	10 parts
Alcohol	10 parts
Gum	10 parts
Water	30 parts

Mix the bichromate, finely powdered, with the acid, and let it stand for an hour. Into the red solution thus obtained the alcohol is slowly poured, with constant stirring. The reaction is very vigorous, and the liquid froths and gets very hot, and gradually turns to a dark green. If the action gets too violent, a little cold water is put in. To avoid boiling over, it is best to add the alcohol in portions, waiting till the frothing after each addition is over before adding the next. The next step is to add carbonate of soda till all effervescence has ceased and a greenish precipitate just begins to form. The liquid is then left covered up for a week, filtered from the salts which have crystallized out, and diluted to the desired color. Finally the gum is dissolved in it. This ink penetrates the paper deeply, and gives green writing which is absolutely permanent and is very difficult to efface.

Finger-Printing Ink

Glycerin	112 parts
Ferric Chloride	10 parts
Colloidal Carbon Black	1 part
Acetone	90 parts

Shading Ink

Add gum arabic to ordinary fountain pen ink until fluid is suitably heavy.

Red Ink (Eosin Type)

Eosin (Water-Soluble)	2	g.
Thymol	0.5	g.
Gum Arabic	1	g.
Water	100	cc.

Soak the gum arabic over night in the water, and then heat to dissolve thoroughly. Add the thymol and eosin; when dissolved filter. A few drops of dilute caustic soda may be added to insure materials remaining in solution.

"Emergency" Red Ink

Mercurochrome	2 parts
Water	100 parts

This gives a clear red ink, excellent in writing qualities, and permanent.

Blue Ink, Acid-Proof

Dissolve in 2 quarts of water, 2½ ounces of Tiemann's blue and 1 ounce of oxalic acid. Dissolve in two quarts of warm water 1½ ounces of gum arabic. Mix the two solutions and strain.

Blue-Black Writing Ink

Ferrous sulfate, 6.5 grams; gum arabic, 6.5 grams; glycerin, 7 cubic centimeters; carbolic acid, 1.5 grams; dil. sulfuric acid, 15 grams. Dissolve all the above in 8 ounces of water; tannic acid, 4 grams; gallic acid, 4 grams; dissolve these two in 5 ounces water. Mix, add enough water to make 20 ounces. Dissolve 1.5 grams Tiemann's Blue (soluble) in the mixture. Let stand for a week; strain quickly and bottle.

Magoffin's Black Ink

Extract of Logwood	8	oz.
Potassium Dichromate	0.5	oz.
Potassium Ferrocyanide	0.25	oz.
Alcohol	8	oz.
Oil of Cloves	1	oz.
Boiling Soft Water	5	gal.
Cold Soft Water	1	gal.

Dissolve the extract of logwood in the boiling water, and add the cold water, and the alcohol in which the oil has been dissolved. When cold, strain through flannel and keep in tightly corked bottles.

Scoville's Black Ink

Tannic Acid	3	oz.
Gallic Acid	1	oz.
Ferrous Sulphate	2	oz.
Ferric Chlor. Sol., U.S.P.	11	fl.oz.
Indigotin	1½	oz.
Acacia	60	grains
Phenol	60	grains
Water	1	gal.

Dissolve the tannic and gallic acids and indigotin in six pints of warm water. Dissolve the iron salts, acacia and phenol in the remaining two pints of water, and mix with the first solution. Shake frequently during several days. Allow the ink to stand at least two weeks and then filter.

Quick-Drying (Writing) Ink
U. S. Patent 1,897,071

To 3 parts of a known ink 1 part of a mixture of diacetone alcohol 10, ethylacetate 30, acetone 30, and ethylene glycol mono-ethylester 30 is added.

Permanent Quick-Drying Writing Ink
U. S. Patent 1,932,248

This ink is water resistant and ''non-feathering.''

Direct Pure Blue 6B Ex. Con. (Color index No. 518, Schultz No. 424)	1.6	parts
Flake Caustic Soda	1.8	parts
Ammonium Meta-Vanadate	0.35	part
Amyl Xanthate	0.02	part
Corn Starch	0.05	part
Bentonite (Wilkinite)	0.20	part
Water	100.0	parts

Printing Composition
U. S. Patent 1,931,485

A fluid printing ink suitable for printing cellulosic plastics contains 100 parts low viscosity cellulose nitrate, 200 parts dimethyl phthallate and 75 to 200 parts pigment.

Printing Ink Varnishes

Varnish used in printing inks is generally a heat bodied linseed oil. In the following table are average chemical and physical properties of printing ink varnishes.

No. of Varnish	Viscosity 25° C.	Specific Gravity 25° C.	Refractive Index	Acid Number
000	2.20	.9460	1.4845	8.2
00	4.40	.9560	1.4875	9.3
0	7.50	.9587	1.4878	9.8
1	12.0	.9596	1.4883	10.8
2	19.0	.9636	1.4890	12.5
3	61.0	.9694	1.4910	15.8
4	101.0	.9724	1.4918	18.0
5	180.0	.9732	1.4923	19.0

Marking Porcelain and Glassware

Porcelain and silica ware may be marked with platinum, when properly done producing very neat and legible characters. About a decigram of scrap platinum may be dissolved in aqua regia in a small porcelain crucible and the solution evaporated nearly to dryness. About one ml. lavendar oil is added and the mixture stirred until permanently mixed. The ware is slightly warmed and this mixture applied with a rubber stamp or thinly painted on with a fine brush, sharpened match or something of the kind. The ware is gently heated to burn off the oil and finally strongly ignited to fuse the metal into the surface, producing a distinct black mark. If applied too thickly the characters may be blurred. If the mixture becomes gummy after long standing, it may be thinned with a little more of the oil.

Ceramic Ink for Marking Glass and Porcelain

A very good ceramic ink can be made by using the following common reagents:

Potassium Carbonate	1 part
Sodium Tetraborate	1 part
Litharge	2 parts
Cobalt Nitrate	2 parts

Grind these together in a mortar to a fine powder, dry if necessary, mix intimately with copaiba balsam mixture composed of four parts copaiba balsam, one part clove oil and one part lavendar oil. If such a mixture is not available use raw linseed oil or any other drying oil. Mix the powder glass color with just enough oil so it will run slowly from a pen. The desired markings are made on the clean surface with a pen or a small brush. The marked article is then warmed evenly over a flame to dry the oil and also prevent cracking in the final heating. The place where the mark has been made has been heated by holding against the side of a flame of a burner; the flame touching the marking on a tangent, the article being rotated part of a circle. The marking will first turn black, then it will begin to glow a dull red. At this point the article is removed, allowed to cool a little, and reheated till the characters, not the glass, begin to glow. Markings thus obtained present a smooth and shiny surface which cannot be removed by mechanical or usual chemical means.

Ceramic Inks for Porcelain

A solution which can be made with the use of materials commonly found in general analytical or research laboratories and which produces a permanent marking on porcelain vessels is made up as follows:

Sodium Carbonate	1 part
Sodium Tetraborate (Borax)	2 parts
Potassium Chromate	3 parts
Water	40 parts

In the marking process the borate and carbonate function as fluxes and chromate as the source of color. The viscosity of this solution does not differ materially from that of ordinary writing ink and is therefore easily applied by means of a pen.

The marking is made on the clean porcelain surface and held near a bunsen flame to evaporate the water, after which it is strongly heated in an oxidizing flame to develop the green color of the chromium compound. In this manner a vessel may be marked in about one minute, since ordinarily sufficient solution runs from the pen to impart a suitable depth of color to the characters. When a blast lamp is employed, the markings are greenish black, probably due to the formation of a chromium silicate.

Ceramic Ink
Canadian Patent 337,714

A ceramic ink comprises Cobalt oxide 3, borax 1 part, lead oxide and a sufficient amount of linseed oil varnish to cause the ink to have the consistency of printer's ink. The composition of the ink may be widely varied in accordance with the color desired and the nature of the surface to which it is to be applied. Thus vitreous enamel fluxes other than borax may be used, such as glass or lead borate, and any of the following substances, as well as mixtures, thereof, are suitable as pigments, manganese oxide, nickel, iron, chromium, copper and selenium red.

Ink for Writing on Glass, Tin, etc.

Shellac (Bleached)	2 oz.
Venice Turpentine	1 oz.
Gum Sandarac	¼ oz.
Gum Mastic	¼ oz.
Oil of Turpentine	3 oz.

Dissolve by warming. Then add one of the following for different colors according to strength of color required.

Lamp Black
Ultramarine
Brunswick Green
Vermilion

Quick-Setting Printing Ink
U. S. Patent 1,954,627

A mixture is made up, employing grinding rolls for instance, comprising 30 pounds dry Royal blue pigment, 15 of heat-treated linseed oil, commonly known as thin plate oil, 6 of double-boiled linseed oil, 8 of iron blue dry pigment, 8 of petrolatum, 8½ of litho-varnish, 7 of dry purple toner pigment, ½ of kauri, 16 of alkali blue red shade in paste form, 32 of dry aluminum hydrate, 16 of rosin oil, second run, 16 of paraffin wax, 32 of cobalt drier, and 32 of kerosene. A mixture of 324 of heat-treated linseed oil, commonly known as heavy plate oil, and 128 of dry carbon black, is separately ground, and this mixture and the fore-going are then ground together, and 152 of copal varnish (9½ per cent of copal, with 35½ per cent vegetable drier and 55 per cent mineral spirits) are incorporated, and 76 of benzyl alcohol.

In the use of such compositions, immediately after the impression is made, heat should be applied, and most advantageously this may be accomplished by a suitable heater, electric, gas, etc., arranged on or adjacent the press, so that the delivered printed impression is subjected to a substantial degree of heat to complete the setting action.

To Print on Wax or Glassine Paper Bags With Ordinary Printing Ink

Mix from 5% to 30% of any good quick drying liquid glue (not mucilage). with any good bond ink. The mixing must be thorough using a piece of glass and a spatula or ink knife. Do not allow to dry on the press.

Textile Printing Ink
British Patent 406,324

Curd soap (25%), lithographic varnish (75%), and an oil-soluble water-insoluble dye.

Ink for Marking Fabrics

Copper Sulphate	200 g.
Dextrin	110 g.
Aniline Chloride	300 g.
Glycerin	60 g.
Distilled Water	1000 g.

The copper sulphate and the dextrin are first triturated together and are then mixed with the aniline chloride. Then the glycerin is added and finally the water stirring the mixture continuously.

Kayser's Marking Paste for Fabrics

Copper Sulphate	20 g.
Aniline Chloride	30 g.
Dextrin	10 g.
Glycerin	5 g.
Water	Sufficient

The dry ingredients are first mixed together thoroughly and are then mixed with the glycerin and sufficient quantity of water to make a paste that may be applied with the aid of a fine brush.

Silk Screen-Printing

Printing by the silk screen is very practical in the decoration of expensive materials such as silks and satins; especially so in the ornamentation of scarfs and similar novelties, where it is essential to produce multi-colored and often complicated designs with great accuracy in rather small quantities.

Due to the fact that until recently these screens were made by hand, working with complicated designs required great skill and plenty of time, as these designs had to be painted on the screen by means of paint and brush.

With the aid of photography which enables us to transfer designs from paper onto the screen it is possible to achieve much better results in accuracy and brilliance of design, in less time and with less effort and skill. A brief discussion of the photographic method, the manipulations involved and the equipment required, follows.

A piece of fine bolting cloth (of about 150 meshes per linear inch) is mounted on a frame of hard wood. The cloth, which should be well stretched, is fastened to the wood by carpet tacks. The bolting cloth is permeable to color paste and fluids of high consistency. It is the objective to render it impermeable by the application of certain chemicals. If some parts of the cloth are protected against the effects of such chemical treatment and left permeable, the screen becomes a negative. The permeable points form a given design, which are now able to be transferred to any medium by simply spreading color paste over the entire screen.

The first step is the sensitizing of the screen.

For the preparation of the light sensitive emulsion the following formula will lead to desirable results:

Fish Glue	85 g.
Le Page's Belting Cement	45 g.
Ammonium Bichromate	10 g.
Albumen (or the Whites of Three Eggs)	85 g.
Distilled Water	½ l.

Dissolve each of these substances separately in warm water, then mix in the order given above, add one to three drops of strong ammonia and stir well. After this emulsion has been allowed to "ripen" in a dark place, where it has been kept for about 24 hours, it is ready to be spread on the screen.

If the sensitizer is applied to the screen by means of a brush, great care should be taken to avoid streaks, uneven coating or pin holes. A sure and efficient way of coating is this:

Raise one end of the screen about 5 inches and pour the emulsion over it, catching it at the lower end; repeat this once or twice. The screen is then placed on the table, and a scraper squeegee (or the edge of a ruler), the length of which is equal to the width of the screen, is moved up and down along the screen until the coating becomes even and the emulsion sets. After the inside of the screen has been coated in the same manner, the screen is dried by a fan. The whole procedure should be repeated after the screen has been dried, for the purpose of assuring the uniformity and strength of the coat and also the absence of pin holes.

The screen is now allowed to dry for about 24 hours, for the drier it gets, the more sensitive it becomes. It is understood, of course, that the entire operation is performed in the dark room.

The design which is to be produced on the screen, is first drawn on any drawing board and then traced in opaque ink on transparent paper. Great care must be taken that the lines on the tracing paper be clear and that there be no holes in the ink-covered area. Since for each set of lines of the same color one screen is needed, it is evident that the number of screens required for one design depends upon the number of colors the design contains.

———

Talcum powder is spread over the inside of the screen and velvet or soft flannel placed on top of it, and the latter is covered with a cardboard, 1 inch shorter than the screen in length as well as in width, in such a way as to make the velvet exert a slight pressure on the bolting cloth; then heavy logs are placed on top of the cardboard in order to keep it in place.

There are two methods of exposure at our disposal: the design (inked side upward) can be pasted on a glass plate resting on a pair of horses or on a wooden frame and the screen put on top of it. The source of light should be underneath the glass plate.

This method has two shortcomings: (a) contact between screen and design is not perfect, (b) the source of light must be placed underneath the screen. This is impractical when exposing to arclight and impossible when making use of sunlight. The following method is therefore employed:

A very thin film of colorless axle grease is spread on the screen and the design put on top of it with the inked side of the paper touching the silk. Then, by means of a scraper squeegee, all excess grease and air bubbles are removed, and the tracing paper smoothly pasted on the screen. Now the screen may be stood up against a wall and exposed to sun-arc or electric light.

The time of exposure will vary with the light used. From 30 to 60 minutes' exposure to arclight will be sufficient, whereas shorter and longer periods of exposure, respectively, will be needed with sun and electric lights. Overexposure will make developing very difficult. The correctly exposed screen should be gold-brown.

Card board, velvet, and tracing paper are now removed from the screen, and the grease is washed off by means of a piece of cotton dipped in acetone.

The action of the light has rendered the exposed ammonium-bichromate-gum coating insoluble in water, while the unexposed coating can be dissolved by a mild stream of tepid water. Any parts of the unexposed coat which remain undissolved (usually a result of over-exposure or of letting too much time elapse between sensitizing and exposing), can be rubbed away by a fine brush which has been dipped in stronger ammonia. While developing one must never rub the screen, because that would remove the entire coat.

In order to harden the coat it is immersed in the following bath for about 5 minutes:

Water	4 l.
Methyl Alcohol	½ l.
Bichromate of Ammonia	14 g.
Chromic Acid	1 g.

The coat will not be affected by paints containing turpentine, benzine, oils, or alcohol. If it is desired to print with pastes of aniline dyes the emulsion must be protected against the action of the water and of the small amounts of acids present, which would gradually dissolve the whole coat. This protection is easily achieved by coating the screen with an enamel paint. The latter is spread over the screen in the same manner as the sensitive film and removed from the uncoated parts of the screen by means of a piece of cotton which is soaked in benzine or in any other suitable dissolvent of the enamel and rubbed against the reverse side of the screen. The enamel covering the emulsion will, of course, remain undisturbed. The enamel coat is applied on both sides of the screen.

Finally it is advisable to tape the inner and outer edges of the frame so as to cover up the spaces between silk and wood, where paint might accumulate and make the cleaning of the screen very difficult.

Imitation Engraved Printing
U. S. Patent 1,928,668

A good quality light orange shellac ground to the desired fineness; for a medium grade, the powder should pass through 160 copper wire mesh. With each pound of this powder, there is thoroughly mixed approximately 1/64 lb. of best carbon black. To approximately each pound of the mixed shellac and color there is quickly sprayed or even poured on about ½ gill of wood alcohol or other suitable solvent and then quickly the materials are incorporated and worked together as by being lightly rubbed by hand or mechanical means, so as to cause the fine color particles to more or less completely coat the grains of shellac.

The quantity of color used is so small in proportion to the amount of shellac that the tenacity and resiliency of the shellac are not injuriously affected. Furthermore, the application of the coloring simply as a coating, instead of in solution with the shellac, apparently makes this coloring keep its position as an outside coating on the shellac when it subsequently fuses on the printing, giving even more definite printing effects than are attained with the bleached shellac.

In the second method of bonding or binding the color to the shellac, heat is used as the dissolving agent with portions of graded orange shellac and color mixed as above. Only sufficient heat is applied to partially fuse the shellac.

Among practical methods, a vibrating machine, such as is used for grading powders is employed, with a closed bottom tray substituted in place of the vibrating screen and subjected to gas or electric heat distributed uniformly. As an instance, in a tray of approximately 25″ x 25″ exposed heating surface, a ten pound mixture of orange shellac and color is vibrated for about one minute to insure substantially uniform admixture of shellac and color. Then a fusing heat is turned on for approximately 10 or 15 minutes or until proper amalgamation has been accomplished. Care is exercised to actually effect adhesion of the color particles to the shellac granules, without causing fusion of the mass. When the heat is turned off and the vibration stopped, the tray should be quickly removed to overcome tendency of accumulated heat to fuse the mass.

The color-coated powder, as in the first instance above, should be graded to eliminate over-size particles, which may have been produced by fusion of granules and then be passed through a fine sieve of about 250 mesh to eliminate any color that may not have become properly amalgamated.

For the so-called embossing compounds, the same procedures may be followed, the grading being heavier in accordance with understood practice in the use of embossing compounds.

Lithographic Ink
U. S. Patent 1,935,629

The ingredients of which the ink is composed are as follows: Preferably use a mixture of plate oil and wool fat which are triturated in the proper proportions with a dry color pigment ground therein. To this mixture is added an emulsion of gum-arabic which is obtained by dissolving the gum-arabic in water and charging with chromic acid for inks of blacks and yellows, gallic or other suitable acids being used for light colored inks. The gum-arabic solution or dispersion with the chromic or gallic acid is triturated with a heavy lithographic varnish and a small amount of glycerine is added thereto.

The plate oil is made by burning linseed oil which is the base of lithographic varnish with hot irons which are immersed therein until the oil has lost some of the greasy characteristics thereof which are troublesome in lithographic printing.

The wool fat is used because it has a low setting point working well at low temperatures without crumbling and supplies the necessary greasy nature to the ink and also keeps the design on the lithographic plate from wearing while by incorporating the gum-arabic and chromic or gallic acid in an emulsion with the heavy lithographic varnish such as No. 3 or heavier and grinding with the color pigment scumming and false tinting of the planograph or analogous printing plates are eliminated.

The relative proportion of the constituents of the composition to be used in the manufacture of the improved ink for use with cold lithographic or analogous printing plates may vary within rather wide limits. The following examples illustrate proportions which have been found to be effective:

For blue ink: Preferably use 35 pounds of blue pigment (milori blue), 18 pounds of plate oil, 1 pound of wool fat and 3 pounds of the gum-arabic emulsion.

For red ink: 50 pounds of red pigment (litho red), 45 pounds of a light form of lithographic varnish such as is known in the art as No. 1 lithographic varnish, 1 pound of wool fat and 3 pounds of gum-arabic emulsion.

For yellow ink: 57 pounds of yellow pigment (lemon yellow), 21 pounds of No. 1 lithographic varnish, 1 pound of wool fat and 4 pounds of gum-arabic emulsion.

Gum-arabic emulsion is. obtained by dissolving 5 pounds of gum-arabic in 1 gallon of water and mixing with equal parts by weight of No. 3 lithographic varnish. The gum-arabic emulsion used in making the yellow ink requires the addition of one ounce of chromic acid just before the mixture is triturated for grinding.

Ink for Marking Tin Cans
Use a solution of copper sulphate.

Soap Proof Laundry Black Ink

Phenolic Resin	10.7%
Dimethyl Phthalate	10.7%
Sudan Black Dye	3.8%
Methyl Cellosolve	39.1%
Furfural	35.7%

Ink for Branding Meat Products
U. S. Patent 1,895,641

An ink for use on greasy or moist surfaces comprises alcohol-soluble nigrosine 7 parts, glycerin 19 parts, glacial acetic acid 6 parts, and ethyl alcohol 68 parts.

Ink for Printing on Wet Lumber

Starch	90–100
Glycerin	8– 16
Vinegar	60– 70
Aniline Color	To Suit

Intaglio Printing Ink
U. S. Patent 1,962,823

Rosin	2	parts
Caustic Potash (10% solution)	1.6	parts
Casein	0.1	part
Ammonium Hydroxide	0.24	part
Water	4	parts
Turpentine	0.2	part

Copper-Plate Ink Varnish

Gilsonite	40 parts
Toluol	60 parts

Rotogravure Ink Varnish

Gilsonite	33 parts
Toluol	15 parts
Light Benzine	17 parts

Pale-Colored Intaglio Ink
Formula No. 1 (For Copper-Plate)

Dammar	50 parts
Toluol	50 parts

Formula No. 2 (For Rotogravure)

Dammar	50 parts
Toluol	25 parts
Light Benzine	25 parts

Formula No. 3

Another formula of modern type, as far as the resins are concerned, which is recommended for pale-colored rotogravure inks comprises:

Dammar	26 parts
Beckacite K	24 parts
Toluol	25 parts
Light Benzine	25 parts

After breaking up the asphalt or dammar in a gum breaker the fragments are dried upon trays to eliminate water. The resin or asphalt is then charged into a filling machine equipped with a sprinkler device. Large pieces of dammar or asphalt dissolve very slowly, while any attempt to achieve solution of smaller fragments en masse results in agglomeration and no advantage is gained. With the aid of the sprinkler, however, the resin is conveniently introduced into the solvent in small instalments which readily go into solution. The sprinkling apparatus is arranged over a gum mixer with a capacity of about 1,000 kilograms. The measured quantity of toluol and other solvent is filled into mixer, and the stirrer is started up prior to commencing gradual addition through the sprinkler of the resin which has been reduced to coffee-bean size. About eight to ten hours is required for complete solution, when the varnish is put through a 120- or 14-mesh silk sieve prior to storage.

In addition to the above so-called standard varnishes, the manufacturer requires to have on hand a series of stock (base) inks which are generally referred to as intermediate products.

In conjunction with the above formula for "Copper Plate Ink Varnish," the following comprises a suitable base ink:

Black Varnish 1	15 parts
Blanc Fixe	1 part
Alumina	8 parts

Some indication of German practice in the manufacture of colored rotogravure inks is given by the following formula for a brown ink suitable for illustrated periodicals:

Base Ink

Varnish 1	30 parts
Carbon Black	8 parts
Toluol	9 parts

Brown Ink From Above Base Ink

Varnish 1	145 parts
Base Ink	20 parts
Sudan Brown RRW (Dissolved in Toluol)	1 part

As an example of the grinding procedure which can be adapted in general to all intaglio colored inks, the varnish and base ink are first thoroughly incorporated before adding the dyestuff, preferably in solution form in toluol. Finally, incorporation is most efficiently achieved with the aid of two superimposed cone mills. Good service has been rendered in this connection with the totally enclosed water cooled cone mills with porcelain discs. The mix is given a second run through the two mills before issuing ready for the press, although, as a final precaution, it should be passed through a 140-sieve.

Water Marking Varnish

Magnesium Oxide	18.75 parts
Blanc Fixe	12.50 parts
Castor Oil	62.50 parts
Linseed Oil	6.25 parts

Indelible Transfer Ink
Formula No. 1

Asphalt Emulsion	200 parts
*Para Cumarone Resin	20 parts
Ozokerite	20 parts
Carbon Black	20 parts

* Varnish Grade.

Heat together till free from water.

Formula No. 2

Cellulose Acetate	50 parts
Butyl Tartrate	50 parts
Triphenyl Phosphate	50 parts
Mineral Oil	5 parts

Formula No. 3

Amberol	150 parts
Carnauba Wax	84 parts
Ozokerite	32 parts
No. 6 "Litho" Varnish	35 parts
Blown Castor Oil	60 parts
Cobalt Drier	1 part
Butyl Carbitol	10 parts
Cadmium Red	215 parts

Cellulose Ether	15 parts
Ethyl Abietate	10 parts
*Para Cumarone Resin	10 parts
Blown Castor Oil	10 parts
Butyl Carbitol	15 parts
Pigment	15 parts

* Varnish grade.

Formula No. 5

Cellulose Acetate	170 parts
Triacetin	200 parts
Phenolic Resin	200 parts
Pigment	250 parts

Transfer Composition
U. S. Patent 1,954,451

Two parts of stearic acid, twenty-four parts of white Ozokerite, ten parts of carnauba wax, and forty parts of paraffin wax resin are melted in a container until the above mentioned constituents are wholly in liquid form.

Ninety-five parts of turpentine or any other suitable volatile solvent are then added to the mixture of molten substances previously specified. The solution thus prepared may be designated as solution "A."

A second solution is then made from three parts of gum-tragacanth, well mixed with twenty-five parts of turpentine to which one hundred and fifty parts of water are added. Likewise, two parts of triethanolamine are added. This second solution, which may be designated as solution "B," is heated to approximately the same temperature as solution "A."

Then three parts of the dye to be utilized are dissolved in one hundred and fifty parts of water, the water being heated to the same temperature as solutions "A" and "B." Solution "A" is mixed with solution "B." The mixture is vigorously made so as to emulsify the wax. Likewise, triethanolamine stearate is formed in the combined solutions, and since this is a soap, it acts as a protective colloid to keep the wax in dispersed or emulsified condition. The solution of the dye in water is then added and the mixture is stirred until it is cool. The wax thus remains suspended in finely divided form in an aqueous solution of gum-tragacanth and dye. This emulsion is a thick creamy mixture which can be printed by any suitable means upon the thin paper base which is ordinarily used for making transfers. The mixture dries upon the paper, and it can be transferred from the paper to the fabric by pressing with a hot iron in the ordinary manner. Since the dye-stuff is present in finely divided form, it gives the same appearance as though the fabric had been directly dyed and the marking is then fixed upon the fabric by means of a steaming operation. The dry mixture may be designated as being insoluble in water, although said mixture contains some tragacanth which forms a colloidal suspension in water.

Solutions "A" and "B" can be heated to 65° C. which is also the temperature of the aqueous solution of the dye when the final mixtures are made.

The method and composition above set forth are very useful in forming dye patterns on fabrics, since it is possible thereby to produce a non-aqueous transfer marking on a paper base, and the marking composition can include a dye or dyes. Likewise, the steaming operation can be utilized to form a marking of any desired appearance. The wax is soft and flexible so that the marking on the fabric does not have a harsh appearance.

Transfer Composition
U. S. Patent 1,954,450
Formula No. 1

"Amberol"	10 parts
Ozokerite	4 parts
Blown Rapeseed Oil	2 parts

All ingredients are fused and intermixed until the melted composition is then thoroughly homogeneous. Any suitable coloring material is then added to the composition with thorough intermixing, and the composition is then ready for use on the machines.

Formula No. 2

Another useful composition can be made by melting together the following:

Cumar Varnish (Grade)	100 parts
Ozokerite	12 parts
Blown Rapeseed Oil	20 parts
Lithographic Varnish (No. 6)	12 parts
Butyl Carbitol	5 parts
Cobalt Drier	2 parts

When the ingredients of this composition have been thoroughly intermixed while the fusible ingredients thereof are in the molten condition, color is stirred in and the composition is then panned.

In place of the butyl carbitol in the above formulae, the following may be used:

Pine Oil.
Diethyl Oxalate.
Mono-Butyl Tartrate.
Nitrobenzol.
Carbitol (Mono-Ethyl Ether of Diethylene Glycol).
Butyl Oxalate.

Mimeograph Ink

Red Oil	¼ oz.
Balsam Copaiba	9 oz.
Carbon Black	3 oz.
Indigo	5 drams
Prussian Blue	5 drams
Indian Red	6 drams
Yellow Soap (Dried and Powdered)	2 to 3 oz.

Rub well together until a smooth, creamy paste results.

Engraver's Black Stenciling Ink

Paraffin Wax	¾ lb.
Pigment Black	10 oz.
Kerosene	½ gal.
Nitrobenzol	½ oz.

Melt the wax. Stir in the pigment black until well dispersed and then add the kerosene and nitrobenzol to the cooling wax.

Stencil Paper
U. S. Patent 1,916,203

A mixture of aluminium stearate 2 parts, 45% phenol-formaldehyde resin solution 16 parts, chlorinated naphthalene (Halowax) 14 parts, and maize oil 13 parts, is applied to a fabric backing and the sheet is heated to harden the resin.

Stencil Sheet
U. S. Patent 1,935,875

The following process produces a stencil sheet which is not broken or torn easily. It does not require wetting before use. It may be twisted, crushed or rolled without damage and is unaffected by inks and neutral solvents.

Manufacture an acid anhydride by reacting on 280 parts of stearic acid and 3 parts of concentrated sulphuric acid, and 90 parts of acetic anhydride, parts being by weight. The stearic acid is introduced warm, into a closed, acid proof container and while in agitation the mixture of sulphuric acid and acetic anhydride is run in, heating and agitation being continued for about one hour. The reaction takes place practically at atmospheric pressure and the temperature of the mix is maintained at from 120 to 140 degrees F. Double relief valves for the purpose of preventing material increase of pressure and formation of a partial vacuum are desirable for the container. At the end of one hour, a cold, saturated solution of brine is introduced, preferably heated to water bath temperature and agitation is continued for the purpose of washing the anhydride free of acids and impurities, this being repeated two or three times, each time the mixture being allowed to stand about fifteen minutes to allow the liquids to separate, the lower salt solution being then drawn off. At the third washing it is preferable to allow the mixture to stand about twelve hours as a more complete separation will be obtained.

After complete washing the warm, finished stearic anhydride is then drawn off, filtered and thus being freed from slight amounts of adhering salt by being allowed to stand in a warm condition, is ready for use.

Then prepare a mixture as follows, parts being by weight: Water 20 parts, denatured alcohol 60 parts, glacial acetic acid 10 parts, glycerin 40 parts, No. 1 ground gelatin 10 parts. This mixture is allowed to stand for about one hour to allow the gelatin to swell or may be agitated by suitable means to save time. The mixture is then warmed to about 100 degrees F. with agitation until the gelatin is dissolved. The solution is then filtered, a suitable coloring matter being added if desired. There is then added to the above solution 40 parts of the anhydride of stearic acid warm, and dissolved in about 20 parts of benzine and 4 to 5 parts of formaldehyde, this latter substance serving to harden the gelatin somewhat. The whole is then well agitated and again warmed if necessary to bring it to about 90 degrees F

In the next step the above solution is entered into a suitable pan or tray equipped with a scraping rod or straight-edge, the solution being maintained at around 80 to 90 degrees F. The sheets of Yoshino paper are then drawn over the surface of the solution, the surplus being removed by the scraper, and suspended to dry. On evaporation of the solvents the film sets to a homogeneous compound and in two or three hours the stencil sheets are ready for use.

Sheets are typed in the well-understood manner, being superimposed on heavy paper, a carrier sheet, and are then used in duplicating machines operated by inking pads or rollers for the purpose of making impressions.

Manifold Sheet
U. S. Patent 1,950,982
Formula No. 1

Glue or Gelatine	2 parts
Water	2 parts
Magnesium Chloride	3 parts
Glycerin	4 parts

Formula No. 2

Ferric Chloride	4 parts or
Ferric Sulphate	4 parts or
Ferric Potassium Sulphate	4 parts
Magnesium Chloride	4 parts
Glycerin	6 parts

Formula No. 3

Gallic Acid	1 part or
Tannic Acid	1 part
Magnesium Chloride	1 part
Glycerin	8 parts

Formula No. 4

Sodium Ferrocyanide	1 part
Magnesium Chloride	1 part
Glycerin	4 parts
Water	2 parts

Formula No. 5

Casein	1 part
Water	12 parts
Ammonia	1 part

Formula No. 6

Lanolin Anhydrous, 33 to 50 per cent of the weight of the mixtures prepared by Formulas No. 2, No. 3, or No. 4.

Duplicator Sheet Base
U. S. Patent 1,938,927

A durable non-sticky duplicator base is made of 10 pounds of commercial glue in 25 pounds of water, mixing in approximately 65 pounds of glycerin, maintaining the mixture at a temperature somewhat above its melting point, dissolving a quantity of formaldehyde in a mixture of water and glycerin, adding formaldehyde and caustic soda to the mixture of glue and glycerin, the amount of formaldehyde used being approximately .045 by weight of the entire mixture, the amount of caustic

soda used being sufficient to give a pH of approximately 8.5 to the mixture, and chilling the resulting mixture within less than one hour of the making of the mixture to a temperature below the melting point.

Novelty Burning Ink

A 20% solution of potassium nitrate is used for writing on thin paper. When dry a flame is applied to one end of the writing (which must be continuous). The paper burns and chars only where it has been written on, leaving the writing exposed as a black char.

Wax Crayons

Color used: Lithopone, Red Oxide, Chrome Yellow, Milori Blue, Scarlet 618, Violet 2101, Pink 685, Blue Lake 2520, Green Lake 486, Ultramarine Blue, Cobalt Blue, Violet 1000, Light Brown, Chrome Orange 51, Purple Lake, Magenta 1000, Medium Green 601, Violet Red 1080, English Vermilion Extra Pale, Red Toner 99, Red Quinone 5018, Red 338, Flamingo Red 18%, Lamp Black.

Typical Red Material for Wax Crayons

China Clay	48.6%
English Vermilion	42.4%
Geranium Lake	2.0%
Red 5018	5.0%
Red Toner 99	2.0%

Typical Formulae for Crayons
Formula No. 1 (Red)

China Clay	71.3%
Red 5018	21.4%
Red Toner	7.3%
*Gum (added percentage)	

Formula No. 2 (Red)

China Clay	42.0%
English Vermilion	45.5%
Pink	12.5%
Gum (added percentage)	2.21%

Formula No. 3 (Red)

China Clay	58.8%
Red 5018	29.4%
Red Toner 99	11.8%
Gum (added percentage)	2.2%

Formula No. 4 (White)

| White Material | 98.54% |
| Gum | 1.46% |

Formula No. 5 (Light Blue)

| Light Blue Material | 98.34% |
| Gum | 1.66% |

Color Mixing

Color Desired	Parts	Parts	Parts	Parts
Blue Gray	100 White	3 Prussian Blue	1 Black	
Bright Blue	20 White	1 Cobalt Blue		
Blue Black	9 Black	4 Prussian Blue		
Bronze Green Light	3 Raw Umber	1 Chrome Yellow	1 Dark Orange	
Bronze Green Dark	20 Black	2 Chrome Yellow		
Bottle Green	5 Chrome Green	1 Black		
Cherry Red	50 Vermilion	50 Carmine		
French Blue	5 Cobalt Blue	2 White		
	5 Lemon Yellow and	1 each of Chrome, Green and Black		
Fern Green	100 White	5 Chrome Yellow	3 Ultramarine Blue	
Green Blue	5 Chrome Yellow	1 Light Venetian Red		
Gold Russet	50 Light Indian Red	50 French Ochre		
Indian Brown	3 Dark Ochre	1 Light Venetian Red		
Mahogany	5 Venetian Red	1 Black		
Maroon Light	9 Dark Indian Red	1 Black		
Maroon Dark	7 Light Ochre	50 Orange Chrome Yellow		
Olive Green	50 Burnt Sienna	1 Ultramarine Blue		
Orange Brown	2 Rose Pink	1 Rose Pink		
Purple	3 Black	1 each of Ultramarine Blue and Black		
Purple Black	5 Dark Indian Red		50 Black	
Purple Brown	5 Dark Indian Red			
Royal Blue Dark	18 Ultramarine Blue	2 Prussian Blue (to lighten, use white)		
Royal Purple	2 Ultramarine Blue	1 Carmine or Lake		1 Rose Lake
Russet	14 Chrome Yellow	1 Medium Chrome Green		1 Venetian Red
Royal Blue	34 White	19 Ultramarine Blue		
Snuff Brown	50 Burnt Umber	50 Ochre		
Terra Cotta	2 White	1 Burnt Sienna	2 Ochre	
Tuscan Red	9 Indian Red	1 Rose Pink		
Turquoise Blue	20 White	3 Ultramarine Blue	2 Prussian Blue	
Yellow, Amber	20 Medium Chrome Yellow	7 Burnt Umber	1 Chrome Yellow	
Yellow, Canary	15 White	2 Lemon Yellow	3 Burnt Sienna	
Yellow, Golden	10 Light Chrome Yellow	3 Deep Orange Chrome	1 Chrome Yellow	
Apple Green	4 Orange	9 Chrome Green	5 White	
Amber	4 Burnt Sienna	4 Burnt Umber	1 Orange	
Apricot	30 Chrome Yellow	2 Vermilion	1 Crimson	
Blood Red	Vermilion (tone with Blue and Yellow)			
Photo Brown	4 Bluish Red	1 Brown Black		
Maroon Brown	8 Bluish Red	1 Blue		
Chocolate Brown	12 Deep Red	2 Deep Blue (tone with Black)		
Chestnut Brown	3 Venetian Red	3 Black (tone with Orange)		
Citron	2 Yellow	1 Red	1 Blue	

Green Tints

Color Desired	Parts (White)	Parts / Ingredient	Parts / Ingredient	Parts / Ingredient
Apple Green	50 White	1 Medium Chrome Green		
Citron Green	100 White	3 Medium Chrome Yellow		
Emerald Green	10 White	1 Paris (Emerald) Green		
Nile Green	50 White	6 Medium Chrome Green		
Olive Green	50 White	? Medium Chrome Yellow	8 Raw Umber	1 Black
Pea Green	50 White	1 Light Chrome Green		
Sage Green	100 White	3 Medium Chrome Green	1 Prussian Blue, 3 Raw Umber	1 Black
Sea Green	50 White	1 Dark Chrome Green	5 Medium Chrome Green	1 Black
Water Green	15 White	10 French Ochre	1 Raw Umber, 1 Dark Chrome Green	

Brown Tints

Color Desired	Parts (White)	Parts / Ingredient	Parts / Ingredient
Chocolate	25 White	3 Burnt Umber	1 Black
Cinnamon	10 White	2 Burnt Sienna	1 Ochre
Cocoanut	50 White	50 Burnt Umber	
Dark Drab	40 White	1 Burnt Umber	
Fawn	25 White	3 Burnt Umber	
Golden Brown	20 White	4 Ochre	2 Ochre, 1 Burnt Sienna, 1 Medium Chrome Yellow
Hazelnut Brown	50 White	5 Burnt Umber	
Purple Brown	50 White	6 Indian Red	
Seal Brown	30 White	5 Burnt Umber	
Snuff Brown	25 White	1 Burnt Umber	

Red Tints

Color Desired	Parts (White)	Parts / Ingredient	Parts / Ingredient
Cardinal Red	50 White	50 Scarlet Lake	
Carnation Red	15 White	15 Scarlet Lake	
Deep Rose	10 White	10 Red Lake	2 Ultramarine Blue
Deep Purple	5 White	1 Ultramarine Blue, 2 Red Lake	1 Medium Chrome Yellow
Deep Scarlet	15 Bright Vermillion		1 Ochre
Flesh Pink	100 White	1 Orange Chrome Yellow	1 Rose Pink
Lavender	50 White	2 Ultramarine Blue	5 White, 1 Red Lake
Light Pink	50 White	1 Bright Vermilion	1 Red Lake
Lilac	5 White	1 Rose Pink	
Purple	10 White	2 Ultramarine Blue	1 Red Madder Lake, 1 Yellow Ochre
Red Brick	2 White	3 Light Venetian Red	
Reddish Terra Cotta	50 White	5 Burnt Sienna	3 Rose Lake
Salmon	50 White	4 Deep Orange Chrome	1 Black
Violet	15 White	4 Ultramarine Blue	

Gray Tints

Color Desired	Parts (White)	Parts / Ingredient	Parts / Ingredient	Parts / Ingredient
Ash Gray	30 White	2 Ultramarine Blue	1 Burnt Sienna	
French Gray	150 White	2 Black	1 Orange Chrome Yellow	
Lead Color	50 White	1 Black	1 Medium Chrome Green	
Lustrous Gray	10 White	1 Graphite (Plumbago)		
Olive Gray	200 White	2 Black		1 Chrome Red
Pure Gray	100 White	1 Black		
Pearl Gray	100 White	1 Ultramarine Blue	1 Black	
Silver Gray	150 White	2 Black	3 Ochre	
Warm Gray	100 White	3 Black	2 Ochre	1 Light Venetian Red

Greens, by Two-Color Mixtures

Below is a list of greens by two-color mixtures. The yellow to be used may be a chrome yellow for darker shades of greens and lemon yellow for lighter shades of greens. Any of these colors may be deepened by adding a touch of black, or lightened by adding white.

Color Desired	Parts	Parts
Bronze Green	2 parts Bronze Blue	3 parts Yellow
Bright Green	1 part Deep Blue	5 parts Yellow
Bluish Green	2 parts Blue	4 parts Green
Dark Green	2 parts Blue	1 part Yellow
Emerald Green	1 part Green	3 parts White
Green Black	3 parts Blue Green	2 parts Black
Green Tint	2 parts Bright Green	30 parts White
Grass Green	1 part Bronze Green	1 part Bright Green
Light Green	1 part Green	1 part Yellow
Olive Green	1 part Blue	4 parts Orange
Olive Green	4 parts Yellow	1½ parts Black
Sea Green	1 part Blue	3 parts Green
Yellowish Green	1 part Deep Blue	100 parts Yellow

Purples, by Two-Color Mixtures

Bright Violet	1 part Purple	1 part Blue
Light Purple	1 part Purple	1 part White
Regular Purple	10 parts Rose Lake	1 part Blue
Typewriter Purple	1 part Purple	4 parts White
Violet	4 parts Rose Lake	1 part Ultramarine Blue

Browns, by Two-Color Mixtures

Following is a list of Browns by two-color mixtures. To mix browns, lemon yellow, Persian orange, bright scarlet, flag red, and scarlet lake may be used to best advantage.

Brown Tint	1 part Reddish Brown	40 parts White
Bronze Brown	2 parts Bronze Red	1 part Bronze Blue
Chocolate Brown	12 parts Red	2 parts Deep Blue
Maroon Brown	15 parts Red	2 parts Black
Photo Brown	1 part Vermilion	1 part Black
Sepia Brown	20 parts Orange	1 part Black
Snuff Brown	8 parts Vermilion	1 part Deep Blue
Tan Brown	2 parts Yellow	1 part Purple
Yellow Brown	4 parts Yellow	3 parts Light Brown
Yellow Brown Tint	1 part Yellow Brown	30 parts White

Grays, by Two-Color Mixtures

Blue Gray	2 parts Regular Gray	1 part Blue
Dark Gray	12 parts White	2 parts Black
Light Gray	20 parts White	1 part Black
Regular Gray	12 parts White	1 part Black
Green Gray	6 parts Gray	1 part Green
Lead Gray	15 parts Gray	1 part Deep Blue
Purple Gray	8 parts Gray	1 part Purple
Pink Gray	12 parts Gray	1 part Red
Red Gray	6 parts Gray	1 part Red
Warm Gray	10 parts Gray	1 part Vermilion
Yellow Gray	8 parts Gray	1 part Yellow

Miscellaneous

All colors in the left-hand column are used as base colors; by adding either white or black, the following shades of color in the right-hand columns will be obtained:

Base Color	Result of Adding White	Result of Adding Black
Vermilion	Yellowish Pink	Russet
Scarlet	Salmon Buff	Brown
Orange Red	Salmon Buff	Yellowish Brown
Orange	Salmon	Terra Cotta
Emerald Green	Pea Green	Sage
Bluish Green	Sea Green	Myrtle
Ultramarine	Azure	Blue Slate
Any Lake Red	Pink	Maroon

Formula No. 6 (Violet)

China Clay	83.2%
Violet	14.8%
Gum	2.0%

Formula No. 7 (Yellow)

China Clay	65.00%
Primrose Yellow	35.00%
Gum (added percentage)	5.38%

Formula No. 8 (Yellow)

China Clay	29.5%
Yellow	70.5%
Gum (added percentage)	2.0%

Formula No. 9 (Green)

Green Material	98.4%
Gum	1.6%

Formula No. 10 (Blue)

Blue Material	97.95%
Gum	2.05%

Formula No. 11 (Pink)

Pink Material	98.22%
Gum	1.78%

Formula No. 12 (Black)

Black Material	98.84%
Gum	1.16%

*The gum in each of the above cases is Gum Tragacanth which is given as the dry percentage weight in the formulae, but is used in the form of a plastic made by soaking the dry flakes in hot water and then ripened in a steam heated closet for 24-28 hours or until a smooth, stiff paste is formed (10% solution).

The color materials are milled together wet in a pebble mill, the sludge is dried and ground for use. The color material and gum are mixed in a dough mixer and the resulting plastic milled in rollers, or collenders; rough pressed through a fine holed die in hydraulic presses and extruded hydraulically as a continuous crayon. This is cut in lengths, straightened, air dried and heat dried. The dried lengths are cut to size and pointed then soaked in hot wax made of mutton tallow, stearic acid, Japan wax and hard paraffin.

Wax Crayons

Formula No. 1 (White)

Titanium Pigment	85–80%
IG Wax O	15–20%

Formula No 2 (Yellow)

Red Lead or Litharge	85–80%
IG Wax O	15–20%

Formula No. 3 (Yellow)

Litharge	35%
Titanox	50%
IG Wax O	15%

Formula No. 4 (Red)

Red 394	90–85%
Carnauba Wax	10–15%

Formula No. 5 (Red)

Red 394	85%
IG Wax O	15%

Formula No. 6 (Maroon)

Maroon	75%
Quicklime	2%
Zinc Oxide	4%
Stearic Acid	4%
Paraffin Wax	15%

Green Crayon Leaving Orange Mark on Hot Metal

Chrome Green	75%
Quicklime	2%
Zinc Oxide	4%
Stearic Acid	4%
Paraffin Wax	15%

Hot Metal Crayon
Formula No. 1 (White)

Quick Lime	2%
Zinc Oxide	4%
Titanium Dioxide	70%
Japan Wax	5%
Stearic Acid	4%
Paraffin	15%

Formula No. 2 (Yellow)

Quick Lime	2%
Zinc Oxide	4%
Titanium Dioxide	60%
Lead Chromate	12½%
Lead Dichromate	2½%
Japan Wax	5%
Stearic Acid	4%
Paraffin	10%

Formula No. 3 (Yellow)

Quick Lime	2%
Zinc Oxide	4%
Japan Wax	5%
Stearic Acid	4%
Paraffin	10%
Red Lead	75%

The waxes are melted together and the lime and zinc oxide stirred in; when the foaming ceases, stir in the pigments. The mixture is milled and extruded hot.

Molded Drawing Crayon

Paraffin Wax	50%
Mutton Tallow	40%
Carnauba Wax	3%
Pigment	7%

Pressed Drawing Crayon

Clay	70%
Glue Solution	10%
Pigment	20%

Knead and force through hydraulic press into desired shapes.

Shoe Repair Crayons (Wax Base)

Stearic Acid	60 lb.
Candelilla Wax	15 lb.
White Ceresin Wax	7½ lb.
Ozokerite Wax	7½ lb.

Mix and mould 1½ oz. of above wax base with ¼ oz. of lamp black oil.

Bottom and Shank Blacking

No. 1 Carnauba Wax (Cut with 46 oz. Castile Soap in 11½ Gallons Water)	23 lb.
Garnet Shellac (Cut with 100 oz. Borax in 13 Gallons Water)	26 lb.
Nigrosine (In 10½ Gallons Water)	21 lb.

Mix the three solutions and add water to make 50 gallons.

Burnishing Wax

No. 3 North Country Carnauba Wax	9 oz.
Shellac Wax	4 oz.
Ozokerite Wax	1½ oz.
Montan Wax	1 oz.
Rosin	½ oz.
Paraffin Wax	½ oz.
Liquid Oil Black	3 oz.

Melt waxes and add liquid oil black and mould.

Edge Blacking for Shoes

Dissolve 1½ pounds of green soap in 2 gallons of water, and add 11¼ pounds of No. 3 carnauba wax. Melt wax and cook at slow boil for 15 minutes after wax has melted, add rapidly with stirring 4 gallons cold water. Cook till smooth, then cool. Dissolve 30 ounces of Nigrosine in 2 gallons of water and stir in 2½ ounces of Venetian Red.
Cool and add to wax solution.
To this mixture add

Liquid Glue	30 fl. oz.
Glucose	50 fl. oz.
Oil of Hemlock	5 fl. oz.
Ammonia	20 fl. oz.

Strain and make up to 10 gallons with water.

Indelible Copying Leads (Violet)
Formula No. 1

Methyl Violet	33%
Ceylon Graphite	48%
C & M Graphite	17%
Gum Tragacanth	2%

Formula No. 2

Methyl Violet	33%
Ceylon Graphite	48%
C & M Graphite	16%
Gum Tragacanth	3%

Formula No. 3

Methyl Violet	33%
Ceylon Graphite	25%
C & M Graphite	22%
Gum Tragacanth	3%
China Clay	17%

Formula No. 4

Methyl Violet	30%
Ceylon Graphite	28%
C & M Graphite	25%
Gum Tragacanth	1.5%
China Clay	15.5%

Formula No. 5

Methyl Violet	33%
Ceylon Graphite	40%
C & M Graphite	20%
Gum Tragacanth	1%
China Clay	6%

Indelible Copying Leads (Red)
Formula No. 1

Magenta	30%
Ceylon Graphite	20%
C & M Graphite	25%
China Clay	23%
Gum Tragacanth	2%

Formula No. 2

Magenta	30%
Ceylon Graphite	30%
C & M Graphite	20%
China Clay	18%
Gum Tragacanth	2%

Indelible Copying Leads (Yellow)

Auramine DO Conc.	30%
Ceylon Graphite	30%
C & M Graphite	20%
China Clay	18%
Gum Tragacanth	2%

Indelible Copying Leads (Black)
Formula No. 1

Copying Black SK	30%
Ceylon Graphite	20%
C & M Graphite	25%
China Clay	23%
Gum Tragacanth	2%

Formula No. 2

Ceylon Graphite	20%
C & M Graphite	25%
China Clay	23%
Gum Tragacanth	2%
Copying Black STK	30%

Self-Hardening Pencil "Lead"
U. S. Patent 1,937,105

In carrying out the process, stock is prepared by grinding a mass composed of a mixture of water, graphite and clay, to which the carbonaceous substance such as lamp black or pyrographitic acid may be added if a very black lead is required. The mass so formed is then divided into smaller portions or lumps to facilitate drying, the drying being performed by exposing the lumps to the action of air, or by any of the other drying processes common to this art. When dried, the lumps are heated or calcined, which greatly adds to their porosity. These calcined lumps are then immersed in molten wax for a period of time sufficient to cause the wax to penetrate the pores of the mixture and surround and adhere to the graphite particles therein and to the particles of the carbonaceous substance, if used. Six or seven hours is usually sufficient to cause a thorough penetration of the wax into the pores of the mixture and a coating of the graphite and carbonaceous substance particles by the wax. The lumps now composed of a mixture of calcined clay, graphite and possibly the carbonaceous material and wax are then reduced to a powder by grinding. The ground or powdered mass is then mixed with a water-soluble organic compound which serves as a binder, or a bonding element. When employed as herein set forth three to twenty per cent by volume is used in the mixture. The more alkyl cellulose binder employed the harder will be the mixture. Instead of alkyl cellulose binder, gum tragacanth may be employed in the same proportions, although its use is not particularly recommended due to its tendency to ferment readily.

In addition to alkyl cellulose binder, an inorganic material such as bentonite may be mixed with the calcined clay, graphite and carbonaceous material (if used) and wax. Bentonite is a plasticizing agent and serves to render the mixture plastic. Two to five per cent by volume of bentonite, and about eighteen to twenty per cent water is used. The resultant plastic mass consists of graphite, a carbonaceous material (if used), calcined clay, wax, alkyl cellulose binder, bentonite and water. This mass is dried in any suitable way, such as by passing it through a screen which acts to break the mass up and facilitate drying it. By making suitable tests during the drying operation it can be easily ascertained when the mixture is in its proper condition for extrusion to form the lead. Tests indicate that the mass is satisfactory for extrusion when it contains approximately twelve to twenty per cent water. When the drying has progressed to the required extent, the extrusion is performed which converts the mass into a lengthy, flexible, rod-like section of lead. Since it is soft, the lead section can be formed into a roll or coil and can be kept from hardening for a reasonable period of time, such as from two to three weeks by storing it in an airtight container.

Another satisfactory method of making the improved self-hardening graphite lead is substantially as follows:

Graphite is ground to the proper degree of fineness in water. To this graphite and water mixture is added chlorinated naphthalene and a wax. Since the chlorinated naphthalene is later removed from the mixture by distillation the amount employed is relatively unimportant. The amount of wax employed is from five to ten per cent by volume of graphite.

The mixture, consisting of graphite,

chlorinated naphthalene and wax is agitated, the chlorinated naphthalene cleaning and purifying the graphite so as to cause the wax to intimately coat or adhere to the graphite particles.

Instead of using chlorinated naphthalene in the mixture, it is entirely feasible to use a volatile oil mixing it with wax and a benzol compound, such as xylol, with satisfactory results.

When the mixture has been prepared the water therein, containing the impurities cleansed from the graphite by the action of the chlorinated naphthalene, is removed by decantation and then the chlorinated naphthalene (or oil and xylol, if used) is removed by distillation. This leaves a powdered mixture of wax-coated graphite to which may be added, if desired, a carbonaceous material, such as lamp black or pyrographitic acid, in such quantity as necessary to produce a lead of required blackness. This mixture is then mixed with alkyl cellulose binder, bentonite and water, as previously described in connection with the first method of carrying out this invention. The resultant mass is dried to the required extent and is extruded to form it into a soft, self-hardening lead in the form of a continuous flexible rod of indefinite length which will retain its flexibility, when kept in an airtight container, for several weeks. Such lead, since sheathed while soft, does not crack or break and consequently losses from waste are small. This self-hardening lead can be handled in very lengthy form since it can be conveniently rolled or coiled, and it enables pencils, and particularly those which have a sheath composed of a plastic material, to be made by an extrusion method not possible when hard, short-length leads are used.

White Chalk

Precipitated Chalk	85%
Lithopone	10%
Glue Solution	5%

Knead and extrude through hydraulic press.

INSECTICIDES, EXTERMINATORS, DISINFECTANTS

Fly Spray
Formula No. 1
Powdered Pyrethrum Flowers	8 oz.
Paraffin Oil	1 gal.

Macerate for 48 hours, strain and add 1 ounce of safrol.

Formula No. 2
Pyrethrum Oleoresin	5 g.
Paradichlorbenzene	5 g.
Citronella Oil	10 g.
Paraffin Oil	1 l.

Odorless Fly Spray
Odorless Pyrethrum Extract (20 lbs. Flowers per gal.)	1 part
Deodorized Light Petroleum Oil	19 parts

Scented Fly Spray
Odorless Pyrethrum Extract (20 lbs. Flowers per gal.)	1 part
Deodorized Light Petroleum Oil	19 parts
Perfume Oil (per 100 gal. Finished Spray)	1½ to 3 oz.

Odorless Insecticide for Vaporizing Machines
Odorless Pyrethrum Extract (20 lbs. Flowers per gal.)	1 part
Deodorized Light Petroleum Oil	3 parts

Fly Poison
Arsenic Trioxide	2 parts
Water	100 parts
Sugar	5 parts

Add the arsenic trioxide to the water and warm and stir until complete solution results. Then add the sugar and stir until all is dissolved. May be used as a poison water or may be used to saturate paper and used as poison paper.

"Injun Jo's" Fly Dope
Oil of Sassafras	4 drams
Oil of Tar	1 oz.
Castor Oil	1½ oz.

An application will protect from the attacks of sand flies, midges, and black flies.

Carpet Beetle Control
By using a mixture of 90 parts of propylene dichloride with 10 parts of carbon tetrachloride in a pressure spray gun on stuffed furniture, bedding or rugs, the clothes moths, Tinea pelionella and T. biselliela, and the cigaret beetle, Lasioderma serricorne, are destroyed.

Roach Spray
Kerosene	50 gal.
Methyl Salicylate	51 oz.
Rosin Oil	2 lb.
Propylene Dichloride	1 gal.
Carbon Tetrachloride	1 gal.

Rodent Poison
Formula No. 1
Barium Carbonate Powder	5 oz.
Bread or Rolled Oats	14 oz.

Formula No. 2
Red Squill Powder	1 oz.
Bread or Rolled Oats	1 oz.

Stir in just enough water to allow proper mixing to form a crumbly mixture.

Rat Poison
Barium Carbonate	75%
Red Squill	10%
Dried Cereal	15%

Rub and mix well, spread on buttered bread or hard crackers.

Rat Extermination
Food or poison should always be put out at the same hour in the evening,

preferably at eight o'clock. Only new paper pie plates should be used for the food or poison. The first night, put out a dozen or more, depending on size of the premises, of plates of fresh hamburger. If they do not eat this the first night, leave it a second night. The next night put out about the same number or a few more plates of fresh ground liver. The next night, as many or a few more plates of cheap pink salmon. The next night, some more hamburger and the next night, liver or liver and salmon. By this time the rats will be waiting and squealing for the food at eight o'clock. The fifth or sixth night, all three foods should be mixed separately with 1½ ounces of "Red" Red Squill to the pound of food and put on the paper plates separately, and three or four times as many plates of the poisoned food set out than before of the unpoisoned food.

Care should be taken that the poisoned and unpoisoned food is not touched by human hands or anything that hands have touched. A clean stick or other utensil should be used for mixing it with the poison.

This method will exterminate all the rats in or near any particular building.

Insecticide

Ethylene Dichloride	3 oz.
Carbon Tetrachloride	1 oz.

"Tuma's" Insecticide

Carbon Disulphide	200 cc.
Oil of Turpentine	100 cc.
Completely Denatured Alcohol	200 cc.
Oil of Cloves	10 cc.

Bed Bug Spray

Deodorized Light Petroleum Oil	99 %
Cresylic Acid	1 %

Bed Bug Fluid
Formula No. 1

Paraffin Oil	50 gal.
Ortho-Dichlorbenzene	2 gal.
Methyl Salicylate	1 gal.

Formula No. 2

Paraffin Oil (Boiling Point 170° C. to 240° C.)	1000 parts
Oil of Mirbane (Nitrobenzene)	2 parts

Cresol	2 parts
Pyrethrum Flowers (Ground)	10 parts

Mothproofing Composition
British Patent 389,860

Chlorohydroxy-m-Xylene	3-5%
Trinitroisobutyl-m-Xylene	3-5%
Magnesium Carbonate	94-90%

Mothproofing Composition
Canadian Patent 338,896

A mothproofing composition comprises a chlorinated hydrocarbon ext. of cubé incorporation in a mixture of light hydrocarbon oil 9, and chlorinated hydrocarbon 1 part.

Mothproofing Fluid
U. S. Patent 1,901,960

A composition is made of a volatile solvent (water), a soluble fluoride (0.5% of sodium fluoride), 0.2% of sodium tauro- and glyco-cholate and carbon dioxide dissolved under pressure sufficient to cause the spray to penetrate the goods.

Fireproof Moth Spray

Paradichlor Benzene	1.0 lb.
Alcohol	3.5 lb.
Carbon Tetrachloride	5.0 lb.

Tineol Moth Preventive

Naphthalene	80 g.
Chloroform	150 g.
Oil of Bergamot	5 g.
Oil of Cloves	10 g.
Oil of Lavender	15 g.
Benzine	1 kg.

Treating Textile Materials To Proof them Against Moths and Mildew
U. S. Patent 1,921,926

A composition for treating cotton, wool or rayon is prepared by mixing egg albumin 6 pounds dissolved in water 125 gallons with a rare earth acetate such as cerium acetate 66 pounds dissolved in 250 gallons of water.

Compound for Repelling Moths from Garments
U. S. Patent 1,924,507

Use pellets consisting of a mixture of parachlor-nitrobenzene and para-dichlorbenzene in the ratio of 4-1: 1-4 (4: 1).

The preferred mixture does not stain fabrics at 69°.

Moth Briquette

Fine Cedar Wood Shavings	75 %
Vetivert Oil	2 %
Camphor	3 %
Stearic Acid	5 %
Paraffin Wax	15 %

Melt the stearic acid and paraffin. Add the oil to the camphor and warm until dissolved. Then add this to the cooling wax mixture, stir and pour into molds.

Fumigation with Propylene Dichloride Mixture

Commercial propylene dichloride containing 10% carbon tetrachloride killed larvae of the European corn borer in corn stalks when used in the proportion of 2 pounds per 100 cubic feet of space at 15.6° to 25.6°, the exposure time being 24 hours. The method appears applicable to the treatment of truck-crop produce prior to movement from infested to noninfested areas.

Mosquito Larvae Killer

Two-tenths of one per cent of a mixed potassium oleate and coconut oil soap kills mosquito larvae and pupae.

"Nash's" Mosquito Repellant

Oil of Citronella	1 oz.
Spirit of Camphor	1 oz.
Oil of Cedar	½ oz.

Fumigant
British Patent 396,004

A composition, which may be applied to match-heads, contains gum benzoin 2, balsam of tolu 2, gum olibanum 2, powdered sandalwood 2.5, potassium nitrate 2.25, gum tragacanth 1.3 and water 16 pts. Carbon black may be added as coloring matter.

Fumigating Cones

Charcoal	46%
Cascarilla	15%
Gum Benzoin (Siam)	13%
Cardamoms	4%
Cubebs	3%
Myrrh	1%
Saltpetre	5%
Bergamot Oil	3%
Peru Balsam	3%
Cassia Oil	2%
Sandalwood Oil	2%

Patchouli Oil	1%
Phenylethyl Alcohol	1%
Ionone (100 per cent)	1%

Sometimes the cones are colored, when, of course, they must be made without charcoal, or with only very slight amounts.

Rotenone Emulsion Insecticide
Formula No. 1

1 gram of pure rotenone is dissolved in 100 cubic centimeters of pyridine. 1 cubic centimeter of this solution is added to 100 cubic centimeters of distilled water and shaken gently. A pale opalescent colloidal solution results which shows no evidence of separation over an extended period of time.

Formula No. 2

5 grams of pure rotenone is dissolved in 100 cubic centimeters of pyridine. 1 cubic centimeter of this solution is added to 100 cubic centimeters of distilled water and shaken gently. A deeply opalescent colloidal solution results which shows no evidence of separation over an extended period of time.

Formula No. 3

1.2 grams of pure rotenone is dissolved in 100 cubic centimeters a-picoline. 1 cubic centimeter of this solution is added to 100 cubic centimeters distilled water and shaken gently. A colloidal solution of medium opalescence results which shows no evidence of separation during a considerable period of time.

Extracting Rotenone
U. S. Patent 1,942,104
Method No. 1

One gram of rotenone is dissolved in 50 cubic centimeters of carbon tetrachloride at a temperature of 50° C. This solution is allowed to cool to 20° C., when crystallization will occur. The crystalline material which separates is filtered off and dried in the air. The product so obtained is the pure addition compound of rotenone and carbon tetrachloride.

Method No. 2

Fifty grams of the roots of Derris (Deguelia) sp. (tuba root) is completely extracted in a continuous extractor with carbon tetrachloride. The extract is evaporated to a volume of 25 cubic centimeters and cooled in a refrigerator. When crystallization is complete the separated material is filtered, excess solvent removed by suction and the needle-like crystalline product dried in the air.

Method No. 3

Five kilograms of the roots of Lonchocarpus nicou (cube root) is percolated with 30 to 40 liters of carbon tetrachloride at a temperature of 50° C. The extract so obtained is evaporated to a volume of 1 liter. This evaporated extract is cooled until crystallization occurs. The separated material is filtered, excess solvent removed by suction and the crystalline mass dried in air.

The product obtained by the method outlined in Methods No. 2 and 3 is substantially the addition compound of rotenone and carbon tetrachloride in an impure state, but possesses insecticidal properties, and may be purified by the method outlined above without losing its insecticidal value.

Soluble Pine Oil Disinfectant

Raw Pine Oil	60%
Sulfonated Castor Oil (50%)	30%
Red Oil or Oleic Acid	9%
Potassium Hydroxide (Solid)	1%

Mix the pine oil and the sulfonated castor together. Then add the red oil and dissolve the potassium hydroxide in the mixed oils. These figures are by weight not by volume. This product will give a milky emulsion in water which will not separate out on standing.

"Soluble" Cresylic Acid Disinfectant

Cresylic Acid or Cresol	50%
Sulfonated Castor Oil (50%)	24%
Red Oil or Oleic Acid	7%
Sodium or Potassium Hydroxide	1%
Water	18%

Mix the red oil and cresol warm. Dissolve the hydroxide in the water warming if necessary. Mix these two solutions and add the sulfonated castor. This product will give a stable emulsion.

"Milky Disinfectant" for Cleaning Glassware

Light Coal Tar Oil (Sp. Gr. about 1.02 at 60° F.)	69.4%
Rosin (Grade F)	18.4%
Caustic Soda Solution (Sp. Gr. 1.3)	9.6%
Water	2.6%

This formula is stated in percentages by weight.

Melt rosin and add oil, mix well. Add soda and mix. Add water.

This solution is especially useful for cleaning glassware that has become dirty from tar and its by-products.

Disinfecting Solution
U. S. Patent 1,903,614

Iodine	1000 parts
Dissolved in Aqueous Solution of	
Sodium Iodide	1104 parts
Potassium Iodide	48 parts
Calcium Iodide	32 parts

The iodides being in approximately the same proportion as in human blood.

Disinfectants

Pine Oil	57.00%
Rosin	25.00%
Caustic Potash (25% Solution)	8.50%
Glucose	1.00%
Water	8.50%

The caustic potash and water are mixed and heated. When boiling the rosin is added slowly and the heating is continued for one hour taking care to stir mixture occasionally, then the pine oil is added and the heating is continued until a sample that has been withdrawn and poured into water does not show separation of oil. When the test is satisfactory the heat is removed and the glucose is added.

For use add one ounce to gallon of water. This disinfectant can also be used for a cattle dip in the above proportions.

Coal Tar Disinfectant

Tar Acid Oil	66.00%
Rosin	20.00%
Caustic Soda (25% Solution)	10.30%
Water	3.70%

This disinfectant is made same as above and used in same manner.

Cresylic Disinfectant (B.P.)

Cresol	50	cc.
Linseed Oil	17	g.
Oleic Acid	1	g.
Potassium Hydroxide	4.2	g.
Distilled Water to	100	cc.

The oil and acid are heated to the maximum temperature of the water-bath and a solution of the potassium hydroxide in 25 mils. of water, heated nearly to boiling, is added.

The mixture is well stirred, and heated with frequent stirring until saponification is complete. If too much evaporation occurs the soap may aggregate to a mass and float on the surface of the oil. This is remedied by the addition of water. The dish is finally removed from the water-bath and allowed to cool for five minutes. The cresol is then dissolved in the soap without heat and the liquid adjusted to volume.

Cesspool Deodorant and Disinfectant

Formula No. 1

Calcium Oxide (Quick-lime pwd.)	10 lb.
Chlorinated Lime (Bleaching Powder)	2 lb.
Potassium Carbonate	1¼ lb.

Mix and sift thoroughly the three ingredients. Put up in air-tight metal or glass containers. Sprinkle liberally over the excrement which will liquefy and deodorize. The colloidal matter will be completely destroyed. Be careful not to get the compound in contact with the skin.

Formula No. 2

Calcium Hydroxide	9 lb.
Ammonium Chloride	6 lb.
Water	1 gal.

Mix the calcium hydroxide with the water. This is best done by placing a pail, containing the gallon water in cold water bath before the calcium hydroxide is added as a great amount of heat is evolved when the calcium is added to the water. When cold, and the ammonium chloride and stir until completely dissolved. Pour off the clear solution after settling and keep in tightly closed containers. Pour a liberal amount over the excrement which will completely liquefy and deodorize. Be careful not to get the compound in contact with the skin.

INSULATING AND ELECTRICAL
SPECIALTIES

Electrolytic Condensers, Semi-Dry

In the preparation of these, the compounds are mixed together and then heated to a boil after which the condensers are impregated.

Formula No. 1

Glycerin	3 parts
Sodium Borate	1 part

Formula No. 2

Ethylene Glycol	100 g.
Sodium Borate	50 g.
Boric Acid	50 g.

Electrolytic Condensers
British Patent 399,762

The electrolyte of a condenser has an acid, e.g., citric, added thereto to render it acid within defined limits. An example gives glycerol 1 liter, potassium hydrogen phosphate 40 grams, citric acid 40 grams.

Electron Emitting Body
U. S. Patent 1,948,445

The process of making an electron emissive cathode which has a more constant and more lasting electron emission than "thoriated tungsten," comprises forming an intimate mixture at least containing thorium and osmium of which there is between 100 and 50 times more osmium than there is thorium, heating mixture in a non-oxidizing environment to a temperature above the melting point of thorium to form an intimate and uniform alloy between the thorium and osmium, and then raising the temperature close to the melting point of osmium whereby the mixture is consolidated and becomes uniform to give a resulting body which is pliable in the cold state.

Resistance Rods

C & M Graphite	37%
Washed Crucible Clay	63%

Resistance about 900 ohms.

Electrical Resistor Rod
U. S. Patent 1,918,317

The surface of silicon carbide resistor rods is coated with a glaze consisting of titanium oxide 30% and aluminimum oxide 70%, or titanium oxide 20% and calcium oxide 80%, which prevents rapid burning of the rod when used at temperatures above 1100° C.

Electrical Resistances
U. S. Patent 1,947,692

A satisfactory mixture for electric resistances developing temperatures up to 2000° C. may be made up as follows:

7 parts by volume of silicon carbide containing about 30–35% of free carbon.

3 parts by volume of calcined zirconium oxide.

3½ parts by volume of protected carbon.

The foregoing materials are mixed dry, and to the resulting mixture is added such an amount of the neutralized potassium silicate as to produce an earth-wet mass. This mass is then compacted, as by extruding, into the desired shape or form for the electric resistance and subjected to drying at 150–250° F.

When silicon carbide is used as a refractory material in conjunction with silicate as the bonding agent, the latter breaks down and becomes an ineffective binder when the mixture is heated to about 1000° C. This difficulty may be overcome by adding a small amount, not exceeding 0.5% by volume, of a saturated aqueous solution of potassium permanganate, or other similar oxidizing agent, to the mixture prior to compacting.

For temperatures up to about 2000° C., preferably protect the carbon particles with an oxide of manganese, while for higher temperatures preferably use an oxide of tungsten. For the lower temperatures (up to 2000° C.) use refractory materials like silicon carbide, zirconium oxide, etc., since these produce a tough, crystalline-like structure. For temperatures above 2000° C., very satisfactory results are gotten with corundum, oxides of titanium or tungsten, and the like, as the refractory high electrical resistance material.

The following mixture has given excellent results in practice at temperatures approaching 3000° C.:

9 parts by volume of corundum.
1 part by volume of calcined fluorspar.
5 parts by volume of carbon protected by oxide of tungsten.

The dry mixture is made earth-wet with a solution of gum arabic or dextrin. Hollow resistance rods are formed by extrusion. The extruded rods are heated in a vacuum at a temperature gradually rising from room temperature to 180° F. After exposure to this heat treatment for several hours, the temperature is raised to 260° F., or somewhat higher, for approximately one hour. The thus dried rods are then immersed in an alcoholic solution of tungsten nitrate, preferably in a vacuum. After this treatment the rods are placed in a muffle furnace and raised to a temperature of about 2000° C. as promptly as possible.

Another mixture giving very satisfactory results at high temperatures is as follows:

30 parts by volume of corundum or corundum carbide.
25 parts by volume of cerium fluoride or calcium fluoride.
5 parts by volume of calcined fluorspar.
18 parts by volume of calcined tungstic acid.
22 parts by volume of protected carbon.

Make earth-wet with neutralized potassium silicate.

Electric resistance or heating elements made in accordance with the invention are mechanically strong and have a remarkably long useful life in constant use, or under conditions of repeated heating and cooling. Such resistances heat up very promptly upon the passage of an electric current therethrough to unusually high temperatures. At such high temperatures and exposed to atmospheric conditions, substantially no oxidation or deterioration of the resistance takes place. These resistances are of special advantage in electric resistance furnaces where high temperatures are desired. The resistance may

have any appropriate physical form, such as a hollow rod, a flat strip or ribbon, a wall of a heating chamber, or a vessel such as a crucible.

(Electric) Arc Carbon
U. S. Patent 1,920,921

A core for a positive electrode producing a light similar to that from an incandescence lamp comprises calcium fluoride 40, strontium fluoride 10, sodium (or potassium) silicate 5, and carbon flour 45 pts., with tar as binder.

Heat-Transferring Media
British Patent 399,762

The media comprises mixtures of molten metallic chlorides containing zinc chloride, for example, zinc chloride 75%, potassium chloride 15% and sodium chloride 10%, or zinc chloride 70%, potassium chloride 12%, sodium chloride 8%, and lithium chloride 5%, ferric chloride 5%, having melting points of 180° and 140°, respectively.

Heat-Energy Transfer Medium
U. S. Patent 1,893,051

The medium comprises diphenyl oxide with 25% naphthalene, pyrene, or p-hydroxydiphenyl.

Insulation
U. S. Patent 1,888,437

A thick, tacky insulating composition, particularly for refrigerators. Reclaimed rubber 40 pounds (vulcanized rubber may be used if sulphur is removed by heating). Bentonite 60 pounds. Water 1 gallon (only enough water is used to make the composition tractable). Shellac or rosin gum, 1 pound. (First dissolved in a non-drying solvent.)

Insulators
British Patent 386,652

The parts of composite electrical insulators are joined by a cement comprising cellulose acetate and a plasticizer with or without 1 or more fillers and without a volatile solvent, the cement being packed in position and the article heated to 120°. Pressure is not employed. A suitable composition is powdered porcelain 87, acetyl cellulose 11 and triphenyl phosphate 2 parts.

Alternative fillers are steatite, stoneware, kaolin, kieselguhr, etc.

Inorganic Transformer Insulation
U. S. Patent 1,924,311
An inorganic coating for transformer laminations which will withstand annealing temperature comprises a refractory oxide and a binder, e.g., aluminium oxide 40–60, sodium silicate 30–2, and water 20–80%.

Oil Resistant Insulation
A copper wire with tin coating was coated with a layer of ordinary rubber and an outer layer of a mixture of rubber 40, sulphur 7, magnesium carbonate 40. The wire showed no more increase in volume after soaking 200 hours in neutral mineral oil or 60 hours in gasoline; in the latter case, there was a slight decrease in volume after this time, which may have been the result of a partial solution of the rubber. The wire showed therefore excellent resistance to oil, though it softened slightly, not, however, so that it cracked when bent.

Insulating Compound
British Patent 382,147
5 kilograms of crude crepe rubber is melted at 230°–270° C. and mixed with 65 kilograms of oxidized bitumen of Kramer and Sarnow softening point 90°–100° C. and 30 kilograms of fine-fibred asbestos, the mixture being heated to about 200° C. The product has a softening point above 130° C.

Underground Cable Insulating Compound
U. S. Patent 1,935,323
An intimately intermingled mixture including 10% to 40% vegetable pitch, 10% to 15% blown asphalt, 30% to 45% ground vulcanized rubber, about 10% of ground silica and 10% to 30% of material selected from the group, gilsonite and asbestos fibres, said material being substantially free from hygroscopic materials and cellulose fibres.

Sheathed Underground Cables
U. S. Patent 1,935,323
Metal-sheathed cables are provided with an outer covering of a mixture comprising "vegetable pitch" 10–40, blown asphalt 10–15, ground vulcanized

rubber 30–45, ground silica about 10 and gilsonite or asbestos fibers 10–30%.

Insulating Material
U. S. Patent 1,949,087
The method of making insulating material, comprises mixing 1500 pounds of soft water and 100 pounds of mineral wool, stirring into this mixture 10 pounds of asphalt emulsion and then molding to the shape desired.

Insulating Material
French Patent 756,220
A porous insulating construction and refractory material is made by mixing organic fibrous material 3, asbestos flour, aluminium oxide, etc., 5–7.5, and sodium silicate, of 36–8° Baumé 5–7.5 parts, pressing and drying.

Insulating Materials
British Patent 389,816
A composition, particularly for coating insulating tapes, comprises a wax having a melting point such as that of montan or carnauba wax or ozokerite 40–50, colophony or a derivative thereof, e.g., an ester gum, 32–40 and a nondrying oil, e.g., castor oil, 10–28 parts.

Ignition Insulation Composition
U. S. Patent 1,892,105
A composition that may be applied to damp surfaces and is therefore suitable for telephone cables etc. contains boiled linseed oil 52, japan dryer 26, olive oil 20, pumice stone 1, and pigment 1%.

Insulated Copper Conductors
British Patent 393,070
Copper electrical conductors are covered with an insulating layer of basic phosphates, acetates, borates, carbonates or mixtures thereof, produced by making the copper the anode in an electrolytic bath of a salt of the acid desired, the conductivity being improved by the addition of other salts, e.g., ammonium chloride, sodium chloride. A suitable solution is 5% ammonium chloride and 5% ammonium dihydrogen phosphate in 90% water. The coated conductor is heated to 300–400° to vitrify the coating. Fillers, e.g., asbestos or quartz or mica powder, are applied to the coating previous to heating and incorporated thereby.

Magnetic Insulation
U. S. Patent 1,943,115

With every 100 grams of magnetic dust particles, 0.13 gram of colloidal clay, dry or in a water suspension, is first mixed; then 0.70 gram of the milk of magnesia in dilute water suspension and 0.21 gram of sodium silicate in dilute water solution are added to the mixture. The resulting mixture is then evaporated to a condition of complete dryness with constant stirring to apply the first insulating coating to the magnetic dust particles. The coated magnetic particles are then mixed with an additional 0.13 gram of colloidal clay, an additional 0.70 gram of milk of magnesia in a dilute water suspension and an additional 0.21 gram of sodium silicate in a dilute water solution are added thereto and the whole evaporated again to dryness with constant stirring to form the second layer of insulation on the magnetic dust particles. Then the resulting coated magnetic dust particles are mixed with 0.13 gram of colloidal clay, the remaining 0.70 gram of milk of magnesia in a dilute water suspension and the remaining 0.21 gram of sodium silicate in dilute water solution are added to the mixture and the whole evaporated to dryness to form the third layer of insulation on the magnetic particles.

The magnetic dust particles insulated in the above described manner are then mixed with 0.38 gram of colloidal clay, placed in a mould and compressed into core parts under a pressure of approximately 200,000 pounds per square inch. The core parts are then transferred to an annealing furnace where they are annealed at a high temperature preferably in hydrogen or in an inert atmosphere to relieve the internal stresses set up in the material by the pressing operation, thereby producing a core having low hysteresis loss. Where the annealing heat treatment is carried out in air, the core parts are preferably subjected to a temperature of approximately 500° C. for about 45 minutes. When the annealing heat treatment is carried out in hydrogen which enables higher annealing temperatures to be used, the core parts are preferably subjected to a temperature of approximately 650° C. for about 60 minutes. The usual loading coil toroidal windings are wound on the single core parts thus produced or on a plurality of said core parts stacked coaxially. The number of core parts used in a given core will depend upon the existing electrical characteristics of the telephone circuit with which the loading coils are to be associated.

A large number of core parts made by the above described methods and in which the magnetic material is an alloy containing approximately 81 per cent nickel, 17 per cent iron and 2 per cent molybdenum were tested by well known methods and found to have permeabilities ranging from 120 to 160 and sufficiently low hysteresis and eddy current losses as to be satisfactory for use as cores in coils for loading voice frequency telephone circuits.

Wall Coverings
French Patent 752,912

An antiphonetic, insulating and incombustible covering for walls is made by causing ammonium carbonate to react on gypsum and adding a silicate, preferably potassium silicate, to give sufficient solidity after setting, and naphtha oil in amount sufficient to harden the product and prevent hydration. An example contains ammonium carbonate 3.5–4.5, potassium silicate 10–15, naphtha oil 15–20, gypsum 450–420 and water 500–1000 parts.

Wood Impregnation

Wood intended for impregnation should be carefully selected sap wood free of knots, twists and well seasoned for from 20 to 30 months. The moisture content of not more than 10 to 12% in order to obtain good impregnation and avoid checking.

The wood used mostly is maple and copper-beech but a number of others may be used to advantage particularly if mechanical strength is not of prime importance.

Process No. 1
(Mineral Oil Impregnation)

The roughly finished pieces are placed in a drying oven and kept at a temperature of about 40° C. for six to ten days.

The dried pieces are then carefully brushed off and placed in a suitable impregating tank filled with mineral oil of about 35° C. The temperature of the oil is then brought up to about 100° C. and this temperature maintained for from 50 to 200 hours depending on the sizes of pieces involved.

The time may be considerably shortened if one of the following methods is used:

1. The dried pieces are placed in the

empty tank and a vacuum of 20 to 30 inches at a temperature of 90° C. maintained for 15 to 25 hours. The oil is heated to 90° C. before drawn into the tank. The vacuum and temperature is maintained from 25 to 100 hours.

2. Same as 1 but after the oil has been drawn in a pressure of 75 to 100 lbs. per square inch is substituted for the vacuum oil stage and maintained from 15 to 40 hours.

The gain in weight is due to the impregnation should be at least 35% and an impregnated piece of wood should withstand 100,000 volts for 5 minutes between two electrodes spaced 10 inches apart without sign of stress or local heating.

After the impregnating stage the oil bath is brought down to room temperature, the pieces removed, finished to size and coated with a suitable insulating varnish. The finished pieces should be stored in a warm dust-free place at as near even temperature as possible to prevent moisture absorption by breathing.

Method No. 2 (Linseed Oil Impregnation)

The dried and cleaned pieces are placed in a tank and further dried under vacuum for 20 to 60 hours at a temperature of about 90° C. Clean moisture free boiled Oil at a temperature of 90° C. is then drawn into the tank and boiled under vacuum from 50 to 200 hours with a temperature of about 100° C.

The increase in weight should not be less than 35%.

After the oil bath has been cooled down to room temperature, remove wood pieces, allow to drip off and place in a warm dust-free room for storage. The impregnated wood should be left in the storage until the oil is thoroughly dried before finishing.

In laying out an impregnating plant great care must be taken to avoid condensation or moisture absorption in the pipe lines or tanks.

In placing wood in a vacuum tank it should be kept in mind that much less heat is conducted to a piece of wood placed in the middle of the tank unless special provision such as metal shelves are provided for the direct conduction of heat to the pieces.

Inorganic Electrical Insulation
U. S. Patent 1,946,146

The method of insulating metal sheets adapted for use in electrical apparatus consists of the steps of passing such sheets through a solution consisting of about 2 parts of sodium silicate, about 1 part of water and about 1.8 parts of chromic acid, passing them through a set of rolls to wipe off excess solution, drying and baking at a temperature high enough to drive off excess water, to form a precipitate and to bake the precipitate when formed onto the sheets.

Insulating Electric Cables
British Patent 382,667

Electric cables are insulated by vulcanizing thereon a rubber composition containing so much ultra-accelerator that vulcanization is very rapid and so much softening agent that extrusion onto the conductor may be effected rapidly at a low temperature. Such a composition is crude rubber 22, reclaimed rubber 20, mineral rubber 5, whiting 44.7, zinc oxide 2.5, antioxidant 1.5, sulphur 1, pine tar oil 3 and tetramethylthiuram mono or di-sulfide 0.3%.

Insulation for Electric Cables
U. S. Patent 1,952,923

An extrudable ozone-resistant insulation adapted for high tension electrical cables contains fillers together with crude rubber about 5–10, reclaim rubber (about 25% rubber content) about 15–25 and ozone-resistant vulcanized pitch about 25–45% (the vulcanized pitch comprising the products of vulcanization of palm-oil pitch, rape-seed oil and sulphur).

Heat and Sound-Insulating Material
U. S. Patent 1,892,138

A mixture is made of "tree sand" (waste material from redwood bark) 3 parts and exfoliated zonolite (to prevent excessive packing) 1 part.

Heat-Insulating Material

The conditions for a concrete heat insulation are: It must have a small bulk and heat transfer coefficient, sufficient mechanical strength and non-permeability to water and it should have a satisfactory resistance against temperature. The following formula was developed and proved to be satisfactory for preparing a cement foam of sufficient stability, namely: animal glue 143 grams, light rosin 46 grams, solution of calcined soda of 18° Bé. 125

cubic centimeters and water 860 cubic centimeters. The above ingredients are charged into a kettle and heated to boiling for 10–20 minutes, with constant agitation until the entire mixture is converted into a colloidal solution of a specific gravity of 1.4–1.5 grams per cubic centimeter. A cement mortar is then added to the emulsion in water (1: 70 by weight) with constant agitation and the mixture is ready for molding, etc.

Heat Insulating Cement
U. S. Patent 1,899,473

A composition adherent to metal comprises a mixture of soot and a hydraulic cement, e.g., more than 50% of flue soot and less than 50% of asbestos-magnesia cement.

Heat Insulating Cement
U. S. Patent 1,933,271

Slag wool 73, asbestos fibre 13, bentonite 6, a modifying clay such as ''medium fat'' clay 7 and sodium carbonate 1% are used together.

Insulating Cement Blocks

This method covers the manufacture of blocks from cement and sand, to be used for electrical purposes to replace soapstones, slate and in some cases marble.

The mixture shall consist of 50% of Portland cement and 50% of clean sharp sand thoroughly mixed together with a sufficient quantity of water to make the mixture of such consistency that it will pour readily into the mold. This usually requires about 20% of water when the sand and cement are dry. If either one or both are damp, the percentage of water will be proportionately less. The mixing should be very thoroughly done to make sure that the sand and cement are uniformly mixed and of even texture throughout.

The molds should be made from cast iron or malleable iron with the inside surface machined. The thickness of metal in small molds should be about ¼ inch and for large molds should not exceed ½ inch. The parts of the molds should be so arranged that they can be taken apart and reassembled readily, so that after the cement has set the mold can be taken from the block.

All holes that are necessary to put through the block should be molded in, and where the cement is intended to be used around porcelain, the porcelain may be put in place and the cement molded around it.

The mold should be given a coat of mineral oil thickened with vaseline, all over the inside surface to keep the cement from sticking to the mold.

Pour the concrete, of proper consistency, slowly into the mold, and allow to set in place for at least 18 hours, keeping the surrounding air moist. Remove the block from the mold, and place the block under water for three days. Dry the block in air for three days, then put it into a gas oven and raise the temperature gradually to about 200° C. and hold at this temperature for 14 hours, after which boil the block in paraffin at 120° for 14 hours.

Inductance Coils
British Patent 400,277

An inductance coil having a negligible external magnetic field is made by filling a wound coil with a mixture of latex and dust of magnetic material, e.g., iron dust 80 and latex 20 parts, and then solidifying the latex, e.g., by vulcanization.

Mercury Switch

Where mercury cups are used for making connections in electrical circuits in laboratory apparatus, the contacts can often be improved by dipping the ends of the copper wires into a solution of mercurous nitrate and wiping to leave the copper coated with a deposit of clean mercury. A mercurous nitrate solution made slightly acidic with nitric acid and containing a little free mercury is suitable.

To Prevent Sulphation of Storage Batteries

When replacing evaporation from storage batteries, if phosphoric acid be added to the distilled water in the proportion of 1 teaspoonful per quart, sulphation of the cells will be greatly retarded and the life of the battery considerably prolonged.

Storage Battery Separator
U. S. Patent 1,937,205
Formula No. 1

Air-Dried Jute Fibre	250 g.
Sodium Silicate (42.5° Bé. factor 1 Na_2O to 3.25 SiO_2)	1000 g.

Benzol (in which is dis-
solved 200 grams Hard
Asphalt, Ball and Ring
melting point 240° F.,
penetration, 150° F., 100
grams, 5 seconds)   200 g.
Hydrochloric Acid (1 part
concentrated acid to 1
part water)   450 cc.
Filler   500 g.

The 42.5° Bé. silicate solution is poured into a paper mill beater, then the jute fibre is furnished together with enough water to circulate the stock. Immediately after the fibre furnishing is completed the filler is added. Up to this point and until the fibre is properly refined and mixed in with the filler, the beater roll has been manipulated. From now on it is necessary to run with the beater roll up, or with a very minimum of manipulation, so that the addition of the asphalt in benzol solution at this point will result in a mere mixing action.

The process now enters a different stage, wherein enough hydrochloric acid is added directly to the contents of the beater to precipitate almost all the sodium silicate as a silicon hydroxide gel, leaving but enough to maintain a slight alkalinity, which protects the beater and other paper-making machinery from corrosion. The weight of acid given in the formula is but approximate and needs control by a chemical indicator such as litmus paper. The precipitated gel surrounds and permeates the fibres and after dehydration in later stages of the process deposits finely divided silica in the pores of the cells.

After the reaction with acid, the pulp is treated like ordinary paper making pulp and can be formed into webs on the paper machine; it can be molded into sheets on vacuum molding apparatus; it can be pressed into shape with molds having porous plates and embossing plates. In other procedures the wet sheet on the first driers of the paper machine can be embossed by fluted rolls so that after passage over the remaining driers, a dried sheet comprising elongated separators side by side is produced and can be cut into dimensions. Another procedure is to form the pulp into a web on a paper machine, dry it, then feed the rolls or individual sheets of the paper through an embossing machine operating somewhat on the principle of floor covering printing machines, whereby, after moistening the sheet with water, the embossing dies come down and form the sheet into the shape of separator desired. Beside producing a silica gel the reaction between hydrochloric acid, sodium silicate and various impurities in the pulp, produces water soluble compounds such as sodium and iron chloride, etc. These substances must be removed from the finished separator to prevent them from adversely affecting the accumulator. Part of this is removed when the pulp is diluted and formed into a sheet on the paper making screens, the remainder of the substances is removed by washing the finished separator in water before inserting it in the accumulator.

Formula No. 2

As a variant of the process outlined above use the same formula, essentially, as that given above, but introduce the asphalt otherwise than in benzol solution. To operate with this formula, which avoids the fire hazard of benzol, it suffices to keep the temperature of the beater water down to 70 or 80° F. and add the asphalt in molten form at 400–450° F. to the cooled contents of the beater. The rapid chilling of the asphalt crackles it, and the action of the beater roll reduces it to powdery fragments of irregular sizes and shapes, which fragments, after the formation of the pulp into sheets, are incorporated into the melted mass by fusion and penetration into those fibres immediately around each fragment. Where the formula calls for 200 parts asphalt to the finished product, it is necessary to use an excess during the mixing to compensate for those large particles that drop to the bottom of the beater and remain there.

Formula No. 3

A third procedure involves introducing the bitumen in association with a preformed pulp. This may be done in several ways, as by pulverizing bitumen and mixing it with a pulp, or by emulsifying it and mixing it with a pulp, with or without a flocculating agent. A satisfactory formula for use in this third exemplary process is:

Air-dried Jute   250 g.
Sodium Silicate Solution
(42.5° Bé. gravity (1:
3.25 factor))   2000 g.
Asphalt (Testing: Pene-
tration 77/100/5 = 1,
Ball and Ring Melting
Point of 184° F.)   125 g.
Newspapers   125 g.

Filler   500 g.
Hydrochloric Acid (1 part
  Concentrated Acid to 1
  part Water)   600 cc.

The asphalt and newspapers are mixed to a watery bituminous pulp in accordance with the teachings of either of the copending applications referred to, and after the beating off of the jute and filler in the silicate soultion, the bituminous pulp is mixed in, preferably without beating. Finally, the acid is added, and the separator pulp is ready for fabrication into final shape as here described.

Solution for (Lead) Storage Battery
U. S. Patent 1,906,784

To the ordinary sulfuric acid is added 4% of a solution of acetic acid 1.6% and uric acid 0.4%.

Condensers for Ignition Apparatus

Split the mica with a knife, using a good quality of mica which splits without great loss. Each piece must be free from breaks, wrinkles and other imperfections. All splittings must gauge not less than .0015 inch and not over .0025 inch in thickness.

Build the mica and tinfoil together to form the condenser. Use a suitable mold to hold the mica and foil while building. Apply a thin coat of shellac between each layer when building, taking care to keep dust and dirt out.

Place a number of condensers in a suitable jig, inserting a piece of mica between condensers. Apply as much pressure as can be exerted by screwing down on the jig using a 12-inch wrench. Wipe off the excess shellac that squeezes out in pressing. Bake in an oven at a temperature of 150° C. for 8 to 12 hours.

Remove from the jig and clean with a knife. Trim the tinfoil to the dimensions shown on the drawing, making sure to leave no chance for breakdown around the edges and corners, due to overlapping foil. Clean the jig by boiling in hot potash.

Igniter Charge for Blasting Caps
U. S. Patent 1,890,112

The charge comprises potassium ferricyanide 30, potassium chlorate 30, and nitrocellulose 40%.

Contact Carbon
British Patent 396,250

Contact pieces for electric switches, etc., and carbon brushes are made of carbon with which is incorporated a powdered base metal coated with an unoxidizable metal, e.g., copper coated with silver. A suitable powder may be made by dropping powdered copper through a 300-mesh sieve into a silver nitrate solution containing 2½ grams of silver per liter.

Carbon Remover
U. S. Patent 1,786,860

Aniline	1 part
Ethyl Alcohol	1 part
Benzol	1 part
Naphthalene	1 part

20 to 40 cubic centimeters of the mixture is injected into the hot cylinder of an automobile engine and the engine is left idle for several hours. Later when the engine is started the loosened carbon is blown out.

Canvas "V" Rings for Commutators

This method covers the process to be followed in the building of commutator "V" rings from cloth or canvas.

In building up the "V" rings, only enough shellac to produce a firm stiff wall in the product should be used. The duck used shall be ten ounce duck and shall be of a uniform thickness and free from flaws and dirt. The electric linen shall be approximately .012 inch thick and shall be of a uniform thickness and free from flaws and dirt. No. 1 White Mica splittings shall be used. The shellac shall consist of D.C. shellac of .90 specific gravity.

The steps in this process shall be followed in the order as given.

The duck and the electric linen shall be shellacked with a brush. In shellacking, the cloth should be laid on a clean table and a uniform coat of shellac applied on each side. The material shall be hung up to dry, in some place where it is not affected by any heating apparatus. This shellacking and drying shall then be repeated until the cloth has received four coats on each side.

The shellac cloths and the mica shall be cut out round to the proper diameter corresponding to the size of the "V" ring to be made. In cutting, cloth showing flaws or broken threads shall not be used.

The sizes for Type "R" work are as follows:

Type "R" No. 1—2½ inches diameter
  "   "   "   2—2¾   "   "
  "   "   "   3—3   "   "
  "   "   "   4—3½   "   "

The electric line when used on the inside of the "V" ring shall slit uniformly around the circumference in at least eight places.

The mica, after being cut to the required diameter, shall have eight radial slots cut uniformly about the circumference.

The cloth and mica sheets shall be laid together in such a manner as to make the finished "V" rings as follows:

Rings No. 1 and No. 2 shall have shellacked duck on the outside and shellacked electric linen on the inside of the "V" rings with two layers of mica in the middle.

Rings No. 3 and No. 4 shall have shellacked electric linen on the outside, three layers of mica in the middle, and shellacked duck on the inside of the "V" rings.

The assembled sheets shall be laid on a steam table, heating first on one side and then on the other, until the shellac has slightly softened. They shall then be accurately placed in the die and pressure applied. The whole operation shall be performed in as short a space of time as possible. The sheets shall not be left on the steam table more than half a minute; and more than three sets should not be on the steam table at one time. The pressure shall be applied as rapidly as possible, and shall be left on at least 10 seconds. No shellac shall be used throughout the process, except as specified under "Preparation of Cloth."

LEATHER SKINS, FURS, ETC.

Preventing Deterioration of Leather

Leather intended for book-binding or upholstery is treated with 5–8% precipitated chalk and 0.25% calcium chloride. These are added to the exhausted dye bath and drumming is continued until absorbed by leather. Mineral acid should not be used in preparing these leathers for dyeing.

Prevention of Deterioration of Vegetable-Tanned Leathers

The leather is impregnated with 5% solutions of sodium salts of organic acids, e.g., sodium formate, sodium acetate, to increase the content of nontans and buffer the sulphuric acid absorbed by the leather from the atmosphere.

Artificial Leather
U. S. Patent 1,958,821

A piece of cotton batten weighing about 220 grammes per square metre is placed on a screen and another screen is placed on the piece. The piece with the screens is slowly moved through a reservoir containing the following mixture:

Latex (40%)	100 kg.
Water	40 l.
"Wetting Agent"	250 g.
Paraffin-Oil Emulsion	11 kg.
Cumaron-Resin Emulsion	6 kg.

The time for which the piece or body moves through this mixture is so determined that the batten will be thoroughly impregnated. The impregnating period is comparatively short, frequently under one minute.

The body and the two screens are now taken through a pair of rollers where the excess binder is removed and the fibres are compressed. The screens are then removed and the body is brought into a precipitating bath of about 45 grammes of acetate of alumina per litre, and is about 80° C. warm. After the precipitation, the body is again squeezed and then dried, whereupon it may be finished in any suitable manner.

Preparing and Bleaching Reptile Leather

Soak in five parts of water at 15 to 20° C. for 20 hours. Lime with water 500%,* of the wet leather, calcium oxide (50%) 10 grams per liter, sodium sulphide (62%) 2 grams per liter, for four days at 15° C. for lizard skin and for three days for snake skin. Delime with water 300%, sodium bisulphite 2%, hydrogen chloride 0.25% of the weight of the raw skin at 35° C. for one hour. Soften with water 400%, manganese dioxide 0.5% of the raw skin at 35° C. For vegetable tanning treat with sumac solution of 0.3° Bé. in the first and 3° Bé. in the last vat (4° Bé. for lizard skin in the last vat) for 4–5 days. Bleach with acetic or oxalic acid; fat liquor with an emulsion of alizarin oil (2%), castor oil (2%), preserved egg yolk (3% with 10% fat), and water 100% of the tanned skins, at 35° C. for 20 to 30 minutes. For formalin-chrome tanning give a preliminary treatment with water 200%, formalin 3%; add to the same solution after 1 hour, sodium carbonate 4% and water 50% of the second weight of the skin and let stand 2 hours. Then treat with water 200% and chromic oxide 2% of the second weight of the skin having a viscosity according to Schorlemmer of 35%. Neutralize fat liquor and finish. For S-chamois tanning, the skins are pickled for 1 hour after softening in a solution of 3% hydrogen chloride, 150% water, 10% sodium chloride. The thiosulphate bath is prepared from 150% water, 5% sodium thiosulphate, and 6% sodium chloride, of the second weight of the raw skins; duration 1 to 1½ hours. For chamois tanning use water 40%, alum 7%, flour 10%, sodium chloride 3%, egg yolk 20% (with 10% fat) dry and finish. The lizard leather prepared by one of the above methods has a breaking strength of 2.18 kilograms per square millimeter when dry, and 1.79 kilograms per square millimeter when wet. Potassium permanganate is the best bleaching agent. First put the

* Percentages are based on weights of wet raw leather or skins.

kip into a paddle vat containing a solution of sodium sulphite. This solution should contain 25 pounds of the sulphide to each hundred gallons of water. Although this solution will dispose of the hair in less than an hour, the kips should be allowed to remain in it for a few hours, but not over 4 or 5 hours, depending on the weight of stock. The kips are then transferred to another vat for liming. The lime solution should be well saturated with lime, even containing some lime that has not yet been dissolved. Allow the kips to remain in this solution for at least 24 hours.

Blood Albumen Leather Finishes

There are three kinds of blood albumen:—black, brown, and pale. Their quality is easily tested by their solubility in cold water. Good blood albumen should be completely soluble when vigorously shaking it with cold water in which it has been soaking for 1 to 2 hours. The finest qualities are used in leather finishing. The free tannin in vegetable tanned leather causes the finish to be firmly anchored to the leather and the blood albumen is rendered quite insoluble. The finish must be rendered insoluble on other tannages by stoving at high temperature or by subsequent formaldehyde treatment. Blood albumen finishes are very suitable for colored leathers in which the defects have been accentuated by the dyeing since they tend to hide some of the defects. The following are two recipes—colored leather: 150 grams pale blood albumen, 150 grams gelatine, 1,500 cubic centimeters milk; 25 liters water. Black leather: 150 grams nigrosine, 1.8 liters milk, 30 grams glycerine, 60 grams dark blood albumen, 25 grams phenol, 14 liters water.

Tanning Liquors for Light Leathers

Bichromate of Soda	50 lb.
Sulphuric Acid (66°)	40 lb.
Glucose Syrup (42°)	25 lb.

Tanning Liquors for Heavy Leathers

Bichromate of Soda	50 lb.
Sulphuric Acid (66°)	40 lb.
Glucose Syrup (42°)	25 lb.

In the above formulas, in place of glucose syrup, granulated sugar may be used, using only 80% of the amount of glucose syrup indicated.

Tanning Liquor

A liquor of convenient strength for tanning is made as follows:

First dissolve the bichromate of soda in the ratio of 2 pounds of bichromate to 1 gallon of water, using a tank or receptacle of sufficient size to hold at least twice the volume of the liquor after reduction. Then add the sulphuric acid, with stirring, and then the glucose syrup, very slowly, adding only about one-tenth of the amount at the start and until violent ebullition of the liquor takes place. As the reaction goes on, the remaining syrup should be added in small amounts. Allow the liquor to stand over night and the next morning add whatever water is necessary to have the finished liquor represent a strength of 1 pound of bichromate of soda in every gallon of liquor. This gives a liquor containing .35 pound of actual chromium (Cr) per gallon.

Care should be exercised when making the liquor that not too much syrup is added at the start as the reaction is a very violent one, although taking a little while to get started, and if too much syrup is added at the beginning the solution is likely to boil over. However, lively boiling of the liquor during reduction is desirable.

Processing Suede Gloving Leathers

Suede gloving leathers are manufactured from alum sheep, the raw skins for which are imported from several different countries. The skins are soaked in water overnight and this usually suffices to soak them back effectively. To complete satisfaction, it is advisable to use some chemical assistant in the soaks. Only acids or salts may be used on this class of goods. Alkalies and sulfides must be barred because of their deleterious effect on the wool. Of course, if the wool is of no consequence, then an alkali or sulfide soak may be used.

The acids for the acid soak may be chosen from the following:—Hydrochloric, liquid bisulfite or formic acid. Usually tanners prefer the hydrochloric acid because of cheapness. Two and one-half parts of commercial hydrochloric acid per 1,000 parts of water (i. e., 2½ pounds per 100 gallons) is usually found to be sufficient.

The skins are soaked for twenty-four hours in clean water, then transferred to a weak acid soak in the case of wool skins, or a weak sulfide soak in the case of hair sheep. At the end of the soaking period, the skins are transferred to a drum and drummed for 30 to 60 minutes in running water. The skins should be examined as they are removed from the drum and the workman emptying the drum should have instructions to throw out any skins which are not completely softened, particularly in the neck and shanks. Such throw-outs should be broken over on the beam in order to complete the softening process.

The soft soaked skins are allowed to drain and are then painted on the flesh side with a paint made up from:
100 pounds burnt lime and 11–13 pounds red arsenic.

The painted skins are folded down the middle of the back, with the flesh side in and left overnight. Next morning the wool is loose and is pulled. The pelts are washed up in cold water then paddled in lime liquors made from freshly slaked lime with or without the addition of some red arsenic, according to the nature of the pelts. The paddles should not be kept in motion continuously but allowed to turn for ten minutes every hour during the working hours, and after two to four days of this treatment the liming should be complete.

The pelts are scudded lightly on the grain side to remove any remaining hair or wool, particularly the edges and shanks, and are then turned over and fleshed. After having been washed in soft tepid water for one-half hour, the fleshed pelts are puered or bated. According to most authorities, puering can be entirely dispensed with in favor of using artificial bating materials. The following are the particulars as supplied by a well-known firm of artificial bate manufacturers:—

The washed lime pelts are well drained and weighed. The bating is done in a paddle, which is run up with water at 95° F. 10–14 ounces of artificial bate is weighed out for every 100 pounds of drained pelt, 12–15 ounces for hard natured skins, e.g., Arabians. The bating material is sprinkled into the paddle and, after five minutes' running, the skins are entered and the paddle kept running for one hour, allowed to stand for a quarter of an hour, then paddled for two or three minutes every half-hour. Bating will be complete in about four hours.

An alternative method is described, in which the bating is carried out overnight. The bating temperature is lowered to about 85° F. The bating

liquors are not thrown away after use but kept for the next pack. It is warmed up to 90° F., the skins paddled in it for twenty or thirty minutes and then thrown out to drain. Meanwhile, the old liquor is run off with the exception of about 10 per cent and the paddle filled up with fresh hot water. The necessary amount of artificial bate is thrown into the water, the skins entered and the main bating process carried through.

The bating process is complete when the skins are flaccid, the scud is loose and grain silky. Only experience in handling skins in this stage will enable the operator to be certain that the process is complete. Pelts for gloving leathers should be very well bated. The elastin fibre should be completely destroyed in order to produce the stretch necessary in gloving leathers. Nearly five hours' bating is required to destroy the elastin fibres, so that one should not expect to hurry the bating unduly.

The bated skins are well scudded and bran drenched, and are then ready for the actual tanning process. The usual alum, salt, flour and egg yolk tawing mixture can be applied in the form of a thin paste:
5 per cent alum; 4 per cent flour; 3 per cent salt; 2 per cent egg yolk.

The pelts are drummed in this paste for several hours in a polygonal or box drum. Some dressers add a second lot of paste of the same composition and continue the drumming for another three or four hours. Eight to ten hours' drumming with a suitable taw mixture usually suffices to convert the pelt into leather. The tawed skins are horsed up overnight to drain and then hung up to dry.

The dry leather is usually very hard and crusty. It should be conditioned in clean damp sawdust, well and carefully staked, allowed to dry out and staked again. The skins at this stage should be extremely soft, perfectly white and free from any discoloration. The skins should be carefully wheeled on the flesh side to produce a suitable fine nap.

The speed of running of the wheels must be carefully adjusted to meet the manufacturer's own requirements. A quick running wheel will have a drastic cutting action on the leather. A slow running wheel will raise the nap, but not cut away much leather. The fineness of the emery powder on the wheel influences the cutting action. A much greater action is afforded by a coarse powder than that given by a fine powder. Some glove leather dyers have two different types of gloving wheel, a coarse wheel covered with a coarse emery running slightly faster, about 120 revolutions per minute, and a fine wheel, covered with fine emery powder running at not more than 80 revolutions per minute.

Other manufacturers use carborundum wheels for producing the nap. The coarse flesh layer is removed by means of a fluffing wheel and the skins are then finished off on a carborundum wheel.

Either at this stage or before the wheeling process, the skins should be put on one side for several weeks to allow the leather to "age." This fixes the alum in the leather to a certain extent.

After the aging process the skins are selected for dressing. They cannot be mordanted and dyed like vegetable tanned sheep skins because a certain amount of the tannage is removed every time the leather is soaked in water. Special measures have to be adopted to counteract this loss of the tannage.

Alum leathers do not readily wet back. Although a portion of the tannage is easily washed out, other parts do not wet back and a mere soaking in water leaves white spots, which will not dye and finish satisfactorily if left in that state.

The aged selected skins should be drummed in an extremely weak ammonia solution, 1 per cent of weak ammonia (20 per cent) should be added to some warm water at 25–30° C. and the skins drummed with this in a slowly revolving drum (14–16 revolutions per minute). After thirty minutes, the skins will generally be found to be completely wet back. They are now ready for re-egging with 20–25 grams of egg yolk for every small skin.

The re-egged skins are struck out flesh side up on a slate or marble table, and brushed first with an ammonium carbonate solution (one per cent). This is followed by a brush application of fustic, fiset wood, bresiline or rhamnetine crystal solution (five per cent) according to the shade required. It is then possible to produce certain brown shades by applying a "striker," e.g., copper sulfate, titanium potassium oxalate, ferrous sulfate alum or potassium bichromate (one-half to one per cent solution). This will "strike" a color with the previously applied natural dyestuff and experience will indicate what striker or mixture of strikers is required to produce the desired shade.

Another mode of procedure is to apply a weak solution of potassium bichromate first, e.g., a one per cent solution, and then without allowing the skin to dry, apply five coats of a solution of an acid or basic dye. Some acid dyestuff solutions are readily and evenly fixed by the leather. A one per cent solution or weaker will prove quite strong enough for the purpose. The solution is applied with a soft bristle brush and dabbed into the leather. The solution is repeatedly dabbed into the skin, going over different parts several times. This dabbing forms miniature pools of dye on the surface of the leather and it will be found that 10–15 minutes of such treatment will have dyed the flesh.

Should any difficulty be experienced in obtaining a level dyeing, then the bichromate application should be succeeded by a weak solution of fustic paste, myrobalans extract or suitable natural dyestuff. This will provide a mordant base for the coal tar dye solution which can subsequently be brushed on by the usual method of leather staining.

There is another method of preparing alum gloving leather for dyeing, in which the skins are first soaked overnight in a weak solution of salt and lactic acid as follows:

Lactic Acid (30%)	3 oz.
Salt	5 oz.
Water	6 gal.

Next morning, the skins are drummed for one hour in the same solution, drained and immediately retanned for one hour with two ounces of gambier per skin, which is subsequently fixed with one-twentieth of one ounce of potassium bichromate per skin. This is added to the liquor in which the gambier treatment has been given and allowed twenty minutes drumming.

The skins are now rinsed and drummed in a fresh lot of water to which the dye solution is added at intervals. They are allowed to drum in this for one hour, after which a solution of suitable metallic salts is added to strike and fix the color. The leather is drummed with the striker for twenty minutes, rinsed and re-egged in a fresh bath with

Egg Yolk	1 oz.
Salt	1/3 oz.

per skin (medium size). Half an hour suffices for this and then the skins are struck out, air dried, conditioned, staked and finished. The following recipes give some idea of the type of dyestuffs and strikers used.

Formula No. 1

Dye:

Euchrysin RR. (per dozen skins)	1 oz.

Striker:

Copperas	1/16 oz.
Bluestone	1/16 oz.
Bichromate	1/16 oz.

Formula No. 2

Dye:

Bismarck Brown	1 oz.

Striker:

Copperas	1/16 oz.
Bluestone	1/16 oz.
Bichromate	1/16 oz.

Formula No. 3

Dye:

Methylene Blue	3/4 oz.
Bismarck Brown	1/4 oz.

Striker:

Copperas	1/16 oz.
Bluestone	1/16 oz.
Bichromate	1/16 oz.

Formula No. 4

Dye:

Gambier	5 oz.
Fustic Crystals	1/3 oz.
Hematine Crystals	1/4 oz.
Nigrosine	1/4 oz.
Acid Violet 4BN	1/40 oz.

Striker:

Copperas	1/8 oz.
Bluestone	1/10 oz.
Bichromate	1/200 oz.

Most glove leather manufacturers have experienced difficulty in producing a satisfactorily dyed gloving leather without chrome retanning, unless the leather is dyed with natural dyestuffs. The natural dyestuffs contain filling matter, and vegetable tanning which plump the skins, whereas the ordinary alum leathers finish out very much thinner if they are dyed with coal tar dyestuffs only. The new chrome dyestuffs offer advantages in this respect, because if they are applied on a table by means of a brush the leather is not unduly "distressed," neither is it excessively plumped, as happens if natural dyestuffs are used.

Preparing Glove Splits

Colored Splits for Gloves

After placing 1,000 pounds of shaved and trimmed splits in the blue state into the drum, add sufficient water to float and then add 10 pounds of bicarbonate of soda. Mill one-half hour, wash the stock one-half hour with cold water and one-half hour with water at 120° F. Prepare color and fat-liquor as follows:

Paraffin Oil	60 lb.
Sulphonated Cod Oil	40 lb.
Borax	2 lb.
Water (at 140° F.)	20 gal.

Mix well with a motor stirrer, add 50 to 100 pounds of the desired pigment and 20 gallons of water; stir the whole mass well, place into the drum and mill for one hour.

Remove stock from drum and pack into a centrifuge where it is whirled until dry. Place into a heated dry mill and mill until dry and soft. After removing from the mill, sort stock and stake the harder splits.

Amount of pigment to use varies with the color desired, although for glove stock the amount of fat liquor remains the same. The addition of 2 per cent soft soap to the fat liquor will result in a softer split, and in order to firm the stock it is advisable to add from 3 to 5 per cent raw cod oil.

Crusted Glove Splits

When market conditions do not permit ready movement of splits, it is advisable to "crust" the stock. This will be a little more expensive but better merchandise is obtained by crusting.

After splitting, shaving, trimming and sorting in the blue, make the stock up into lots of 1,000 pounds. Place into the tan drum and float with water at 70° F., to which 50 pounds of salt has been added. Mill for 15 minutes. Add the equivalent of 2 per cent bichromate of soda in 35 per cent basic glucose reduced chrome liquor and mill for 2½ hours. Dissolve 7½ pounds bicarbonate of soda in 20 gallons of water and add to drum; continue milling for one hour, dump the stock and pile up to drain for 25 hours.

After the stock has drained, place in the fat liquor drum and float with water at 70° F., add 5 pounds bicarbonate of soda dissolved in 20 gallons of water, mill for one-half hour, drain the drum, and wash stock the first half hour at 70° F., gradually raising the temperature to 130° F. so that the stock is washed at the higher temperature the last ten minutes of the second half hour. In the meantime, prepare the fat liquor.

Fat Liquor for Crusted Stock

Fig Soap	20 lb.
Water	30 gal.
Waterless Moellon	30 lb.
Sulphonated Cod Oil	20 lb.
Paraffin Oil	30 lb.
Water to	50 gal.

Heat the soap and water (30 gal.) to boiling with steam until dissolved. Add the moellon and stir with a motor mixer, then add the cod oil. Continue mixing, but turn off the steam. Finally, add the paraffin oil, stir for one-half hour, and add sufficient water to make 50 gallons.

Do not let the temperature go below 140° F. Add 50 gallons of water at 140° F. to drum and then add emulsion. Mill stock in this fat liquor for 45 minutes, dump the stock and pile up to drain for 24 hours. Set on serial table machine; tack the stock on boards and allow it to dry slowly.

After removing from the boards, the stock is piled down to crust and should be allowed to crust at least ten days before it is worked.

Stock is sorted out of crust for heavy, medium, light medium and light weights and then buffed. Make up each weight into lots of 500 pounds. Place into drum, float with water at 140° F., and mill for 45 minutes. Drain the drum and float stock again in water at 140° F. Add pigment to the drum and mill for 30 minutes; add 5 pounds of sulfonated cod oil in ten gallons of water and mill 15 minutes longer. Remove from the drum, centrifuge, and place into a heated dry drum, milling until the stock is dry and soft. Remove from the dry drum and stake.

Limed Splits

The stock as received is limed for 24 hours, washed, bated, pickled and tanned.

Pickled Splits

Place the stock into the tan drum with water at 70° F. Add 8 per cent salt and ¼ of 1 per cent sulfuric acid and mill for 30 minutes. Tan.

After the splits are tanned, limed or pickled, drain for 24 hours and set out on the serial table machine. Split and shave.

Common practice is to trim limes or pickled splits when they are received at the plant, the trimming to be scant and final trimming to be done after the stock is split and shaved. Coloring and fat liquoring is done as outlined for splits in the blue state. Buffing is usually practiced on stock in crust only.

Practically all splits are colored with pigments with the exception of the brighter shades of blue and red, which are, however, principally used for slipper stock. Large users of pigments for coloring splits make up the various shades as needed. Equipment for this purpose consists of a dry powder mixer such as is used in flour mills.

Leather Dye Solution

The following composition is made for dyeing leathers such as are used for the uppers of shoes, gloves, and leather goods of that type. Dyestuffs soluble in alcohol and benzol can be used, as well as colors other than black. This formula is stated in percentages by weight:

Methanol	65%
Benzol	20%
Black Dye	5%
Carbitol	5%
Castor Oil	5%

The outstanding advantage of this solution is that it dries quickly and leaves no after odor of nitro benzene which characterizes most other leather dyes. It has good penetration and the solvent mixture is sufficiently active to cut under some glazes. It carries a leather softening oil and does not leave the leather in a hard condition.

Deglazing Fluid

This solution is used for removing the glaze from leather prior to dyeing. It also cleans the leather and makes for uniform dyeing and freedom from mottling and spots. This formula is stated in percentages by weight:

Benzol	35%
Methanol	20%
Acetone	15%
Turpentine	10%
Neatsfoot Oil	10%
Silica 200-mesh	10%

This composition must be shaken in order to get a portion of the silica which is the mild abrasive on the rag used to rub over the leather.

Black Leather Dye

Deodorized Kerosene	20 parts
Benzol or Toluol	80 parts
Pylam 77 Black Dye	4 parts
Pylam 512 Black Dye	3 parts

Brown Leather Dye

Deodorized Kerosene	20 parts
Benzol or Toluol	80 parts
Pylam 123Y Dye	4 parts
Sudan 5Ba Dye	2 parts

Paste Shoe Polish

Carnauba Wax	25 g.
Ceresin	10 g.
Triethanolamine	1 c.c.

Melt together and stir; add following slowly:

Turpentine	30 c.c.
Deodorized Kerosene	5 c.c.
Gasoline	30 c.c.

For black polish add 1% oil black dye.

For brown polish add brown iron oxide sufficient to color.

The above mix is poured into cans at 35° C. and allowed to set.

Calf Leather Cleaner and Polish

Water	20	gal.
Potassium Oleate	7.5	lb.
Trisodium Phosphate	.5	lb.
Beeswax, Yellow	6	lb.
Carnauba Wax	6	lb.
Turpentine	4.5	gal.
Pine Oil	.5	gal.
Terpineol	.25	gal.

Greasy Leather Cleaner

Water	10	gal.
Castile Soap	.75	lb.
Trichlorethylene Soap	3.5	lb.
Methyl Acetone	.5	gal.
Lemon Grass Oil	.15	lb.

Dissolve soap in water by heating and stirring. Cool and stir in other ingredients.

Leather Finish Remover

Ethyl Acetate	60%
Butyl ''Cellosolve''	20%
Butyl Acetate	20%
or Ammonium Hydroxide	20%

Removing Mildew from Leather

Make a thick paste of bicarbonate of soda and rub into the leather, and stand in the sun for a day. This will kill the mildew, though the leather will doubtless require painting with a new leather finish.

Shoe-Stiffener Material (Fire-Resistant)
U. S. Patent 1,893,924

Such material (e.g., cotton fabric impregnated with nitro-cellulose) can be fireproofed, without impairing its capacity for being softened when treated with nitro-cellulose solvents, by impregnating it at 100° C. with an aqueous solution containing 20 parts sodium silicate and 30 parts trisodium phosphate.

Leather Softening Emulsion

Water	15 gal.
Castor Oil	5 qt.
Casein Solution	4 qt.
Methylated Spirits	1 qt.
Benzol	1 qt.
Lactic Acid	½ pint

Carnauba Leather Finish Emulsion

Carnauba Wax	300 lb.
Dry Castile Soap	40 lb.
Water	50 gal.

Boil soap in water for 30 minutes: add wax and boil for 30 minutes. Add water to make ˙150 gallons. This is mixed with casein shellac solutions for leather finishes.

Leather Stain Remover

Make a solution with clear water containing from ½% to 2% oxalic acid. If the stain is simply a surface one, then ½% of the acid will be sufficient; however, if the stain has penetrated through the fibre of the leather, then 1% or even 2% will be necessary. First swab the stained areas with the solution and then after drying, give each side or piece of leather a bath in the solution for a few moments in order to impart a uniform appearance.

White Leather Polish and Cleaner
U. S. Patent 1,932,262

The following formula gives a mixture which must be shaken before use. It is applied with a cloth and allowed to dry. The excess powder is rubbed off with a cloth. If a gloss is desired, rub briskly.

Formula No. 1

Lithopone	25	oz.
Trisodium Phosphate	1.8	oz.
Soap	.13	oz.
Calcium Carbonate	.25	oz.
Sodium Benzoate	.1	oz.
Phenolphthalein	.1	oz.
Water	100	oz.

This gives a pink product which dries white on normal acid leather. If the leather is not acid, it may be rendered so by daubing with weak acetic acid. Variation of the above can be made as follows:

Formula No. 2

Lithopone	20	to 30	%
Trisodium Phosphate	1.4	to 2	%
Diglycol Stearate	.1	to	.2%
Calcium Carbonate	.2	to	.5%
Sodium Benzoate	.1	to	.3%
Phenolphthalein	.05	to	.2%
Water	1	qt.	

White Shoe Dressing

Trihydroxethylamine Stearate	100 g.
Titanium Dioxide	5 g.
Latex (60%)	5 g.
Ammonium Alginate Solution (5%)	5 g.
Ammonia (28%)	2 g.

The titanium dioxide is triturated in a mortar with the ammonium alginate solution until in the form of a paste and then stirred into the emulsion. The ammonia is then added and the latex is stirred in last.

White Shoe Dressing

Casein	5 parts
Titanium Dioxide	20 parts
Stearic Acid	½ part
Trisodium Phosphate	1 part
Water	75 parts

Dissolve the trisodium phosphate and the casein in the water. Heat, and add the stearic acid, which has been heated to the melting point. Agitate until saponification is complete. Allow to cool, and add the titanium dioxide.

Cleaner, for White Leather Shoes

The following formula is similar to that of the better grades of dressings for white leather and duck shoes and other leather goods. It consists of:

Lithopone (or Other White Pigment)	40 %
Shellac (Bleached, Dry)	3 %
Borax	.8%
Water	56.2%

About a fourth of the water is heated to about 60° C., and the borax added. When solution is complete all of the shellac is macerated with a portion of this solution and this paste stirred into all of the borax solution, and continued heating and stirring until the shellac is all dissolved. The white pigment which may also be zinc oxide, zinc sulfide or titanium dioxide, is thoroughly wetted with the above solution and finally all the water added. Mix thoroughly. It is then passed through a fine sieve to remove the small lumps.

Calf Leather Cleaner and Polish

Water	20 gal.
Potassium Oleate	7½ lb.
Trisodium Phosphate	½ lb.
Yellow Beeswax	6 lb.
Yellow Carnauba Wax	6 lb.
White Spirits, Turpentine, or Benzol	4½ gal.
Pine Oil	½ gal.
Terpineol	1 qt.

Dissolve first three materials and bring to a boil; other materials are melted together and run into first solution while stirring vigorously.

Greasy Leather Cleaner

Water	10 gal.
Castile Soap	¾ lb.
Trichlorethylene Soap	3 pints
Methyl Acetone	½ gal.
Lemongrass Oil	2½ oz.

Dissolve the castile soap in the water, heating until complete solution is obtained, cool and stir in the acetone, trichlorethylene soap and lemongrass oil.

Neutral Shoe Cream

Carnauba Wax	6 parts
Ceresin Wax	3 parts
Candellila Wax	3 parts
Turpentine	5 parts
Lemongrass Oil	½ part
Light Mineral Oil	1 part
Diglycol Stearate	5 parts

Heat together below 90° C. and stir till clear. Pour slowly while stirring with high speed stirrer into

Water (Boiling)	100 parts

Stir until temperature falls to 70° C. The finished product is snow-white, applies easily and rubs to a high gloss.

Tanning Bear and Lion Skins

This class of skins is first soaked and washed with soap-water. Then they are partially dried. Next treat the flesh-side with a heavy oil, such as cod. Be sure that this oil enters the fibre of the skin and is not left on the flesh side. A mixture of bran and salt water is then applied to cause swelling of the skin and loosening of the flesh. The skin is then thoroughly cleaned of all flesh by using a beam knife and beam table. Next prepare a paste by mixing French chalk in a dilute solution of caustic soda. Apply this paste to the flesh-side, fold the skins and allow them to lie for 10 to 12 hours. Then put skins into a dilute solution of calcium chloride and leave overnight. Next morning give the skins a thorough washing in a drum or paddle, first with clear water and then in water containing from ¾% to 1% of 22% lactic acid.

The skins are next rubbed on the flesh side with seal oil. They are then folded and either put into a drum fitted with "breaking" fins or into a tramping machine. The drum or machine is run about three hours in order to work the seal oil into the fibre. The skins are next hung up in a warm room to dry. Then apply another coat of oil to the flesh side and give the skins a second drumming for three hours. A third application of oil is sometimes necessary for extra heavy skins. After the skins are sufficiently oil-tanned they are rinsed in a soda solution to remove any excess oil. Then wash in clear warm water and hang them up to dry.

There are a number of variations to this process as no two tanners of skins use exactly the same materials, the same length of time or similar apparatus. All kinds of oils are used beside cod and seal, such as paraffin, neatsfoot, olive, castor and cottonseed. These oils in sulphonated form are also used. Vaseline, glycerine or egg-yolk are sometimes added. These oils and fatty substances may be used singly or in any number of mixtures. An emulsion of any oil with soft soap is also favored by many tanners.

Skins can also be tanned with alum, chrome, iron salts, formaldehyde and vegetable tans, but the above described oil process is the best as it imparts softness, strength and elasticity to the fibre.

Preparation of Reptile Skins for Leather

Recommendations as to methods of preparation of reptile skins for the market are as follows: The skins must be cut open down the belly, with the tails cut down the back, the heads and legs being opened out so as to obtain maximum spread of skin. The only exceptions to this are Iguanas and Chameleons, where there is a fringe down the center of the back. These species must be cut down the back.

After the skin has been removed it should be carefully washed, great care being taken to remove all flesh from the skin, also from the head, tail, etc.

Skins should preferably be shipped in dry condition and should be carefully stretched and staked out to their full natural width, great care being taken not to over-stretch the skin, which must then be dried in the shade. Dry skins can be packed in cases and bales, and the use of naphthalene is recommended as a precaution against insect damage.

Before drying, it is preferable, if means are available, to put the skins through an arsenic wash, to minimize the possibility of insect damage.

Alternatively to being shipped in dry condition they may be shipped in wet-salted state, when a plentiful use must be made of finely crushed salt. In preparing the salted skins it would be preferable in the first place to sprinkle the skins with coarse salt and stack them and allow them to drain thoroughly for 12 hours; they should then be packed in barrels, flesh to grain, with a plentiful application of finely crushed salt on each skin.

Lizard skins are required 8 inches and over in width, 10 inches and over being preferred.

Snake skins should invariably be cut open down the belly and prepared in the same way as lizards. They should be carefully staked out to their full width without undue stretching, and shipped in dried condition. Pythons and large snakes should be rolled up singly. The smaller and shorter species of snakes should be shipped flat in bale or cases.

Large snakes such as pythons, boas and anacondas are required in widths of skin from 8 to 12 inches.

Only the belly part of the crocodile or alligator skin is usable for leather. The hard bony scutes on the back are of no value. The belly part of the skin, including the head, whole of the tail and legs, must, however, be taken off at the fullest possible width, leaving about two rows of scutes on each side of the skin. In taking off skins, cut as near to the middle of the back as possible over the legs so as to allow maximum width of skins at the legs, i.e., the skins when properly taken off should be as wide at the legs as in the center of the body. The leg portion should also be left as long as possible and cut exactly in the middle of the large scales on the top of the leg. In taking off skins great care must be taken to avoid butcher cuts and spear-holes, as a skin damaged in this way is classed as a ''second'' and valued approximately at 15 to 20 per cent less. All flesh must be carefully removed and the skins properly washed, drained and thoroughly salted.

They should be shipped in bags, bales, or barrels, preferably the latter. The belly part of the skin must never be folded down the middle. Skins should have the soft part of the sides and legs folded in, keeping the main portion of the skin (via the belly) flat. When shipping in bags or barrels, the skins after folding must be rolled with the scales outwards, and tightly packed. Only heavy bags should be used, which must be well roped. If put in bales, the soft part of the sides must be folded in with the flesh outwards and packed so that the body part will lie flat. The bales should be covered with bagging so as to prevent damage to the skins. When shipping in bales skins must never be folded with the scale side outwards as this side would be exposed to possible damage during transit.

Skins are sold per skin, either by the length from tip to tail or belly width in inches measured between the hard bony scutes. Skins are de-classified as seconds owing to damage, holes or cuts, badly cut open, viz., not full width and proper shape, insufficient salting, or for loose scales, i.e., top skin peeling off.

Tanning Reptile Skins

Pickle all skins well to remove any free lime or sodium sulfide. A good ''pickle'' is:

Salt	6–8%
Formic Acid	0.5–1%
Water	100–150%

based on the weight of the skins.

Then tan in drum or paddle with 20% of a synthetic tanning agent after pretanning as follows:

Skins are put into drum with an equal weight of water and a mixture of equal parts of the following two solutions added in five portions during the course of an hour:

Solution No. 1

Water	30%
Formaldehyde	3%

Solution No. 2

Water	30%
Soda Ash	1%

these are figured on bate weight.

Run for five hours and horse up overnight. Place in drum and run for one hour with three portions of following solution:

Water	100%
Synthetic Tan	15–20%

Add oxalic acid to bring pH value to 3.5. Wash skins well and horse-up overnight. Fat liquor with an emulsion of a neutral soap and oil or 4% Turkey Red Oil of very light color. Work for half hour, then treat with aluminum acetate solution. The aluminum soap which is thus precipitated helps to water-proof the leather.

Home Tanning of Woodchuck and Other Skins

Add 8 ounces of quicklime to a gallon of water. Immerse hide 10 to 20 days. As soon as hair starts to slip readily, remove and scrape off hair and flesh. Then immerse in a solution containing 8 ounces of salt and 4 ounces of alum per gallon of water. Leave 10 to 20 days. Remove hide, place in box and cover with hardwood sawdust. Tramp with feet until hide becomes soft and pliable. To color hide white add two gallons of flour or white clay to sawdust before tubbing. To color yellow like buckskin smoke over a willow smudge before tubbing. Remove hide from sawdust, place over a tightly stretched rope and see-saw hide flesh side down till soft dry and pliable. Rubbing the hide with castor oil will increase its pliability.

Quebracho Tannage for Sheepskins

There are several variations of a que- bracho tannage for sheepskins, depending on what purpose the leather is to be used; that is, for gloves, mittens, shoe linings, hatbands, skivers, roller stock, etc., also whether the finish is to be dull black, colored, embossed morocco, suede, kid, etc.

General Method

When the skins are ready for coloring, apply a weak quebracho liquor to them in the drum. Start the drum and add the liquor through the gudgeon, running the drum for at least 45 minutes. This weak liquor will be sufficient to just color the skins uniformly. In preparing this liquor use one pound of 35% clarified liquid quebracho extract for each dozen skins, diluting to 12 to 15 pailfuls of water heated to 100 degrees Fahrenheit or slightly over. It is possible to use 2 or 3 times this quantity of extract without drawing the grain of the skins, but more than one pound per dozen will make them much darker in color which would prevent the skins from being used for extremely light colored stock.

After the skins have been colored, the tanning process can be entirely done in the drum, but most tanners prefer a combination of paddle and drum. When the drum is used alone it is quite necessary that a very highly treated quebracho extract and plenty of salt be applied to the stock. The first liquor must be weak and mellow and as the tannage proceeds the liquors are stronger and are warmed. To make the first liquor, add enough extract to water to obtain a 2 or 3 degree barkometer solution. Drum the skins in this liquor until they have a uniform color and are well struck through with tannin. Then either strengthen this first liquor or change the skins into a fresh liquor showing 4 to 6 degrees barkometer strength. After several hours drumming, this second liquor is strengthened to 6 to 8 degrees. Again after several hours the liquor is strengthened to 8 to 10 degrees. Continue applying liquors until the stock is completely tanned. The salt is added when the skins begin to absorb the quebracho. At least one pound of salt is used for each dozen skins and is mixed in with the tan liquor. Skins are apt to be much firmer when the drum is used alone for tanning instead of the paddle and drum. To overcome this firmness it is best to retain the skins with sumac, using 1 to 2 lbs. of sumac for every dozen.

To make a soft, light-colored leather that can be easily colored and finished in nearly every desired manner, the following is a good paddle and drum process:

After the skins have been colored, take five dozen and weigh them. Then calculate the weight per dozen and the quantity of clarified liquid quebracho extract by multiplying the number of dozen of skins put in the paddle by the weight per dozen and divide by 35. The answer is the total quantity of liquid extract to be applied to the skins during six portions. Put the skins in a salt solution of about 25 degree barkometer strength. Add one sixth of the calculated quantity of liquid extract dissolved in hot water and first cooled with some of the salt solution taken from the paddle wheel pit.

The skins remain in this extract-salt liquor for 12 hours, stirring 5 minutes every hour.

The skins are then taken out of the paddle and drained and fleshed. They are then returned to the paddle and sufficient water and salt are added to fill the paddle pit and keep the liquor up to a strength of at least 25 degrees barkometer. Then another one-sixth of the calculated quantity of extract is dissolved in hot water and added to the pit solution, first being cooled with some of the pit solution. At intervals of 12 hours apart, three more quantities of one-sixth of the calculated amount of liquid extract are added as before. The skins should continue to be stirred by the paddle every hour for five minutes.

Next the skins are taken out of the paddle and folded and piled so as to allow the excess liquor to drain for 24 hours back into the paddle pit.

The skins are then ready for the final one-sixth of the calculated quantity of liquid extract. This application is made in the drum, using one pail of warm water to each 2 dozen skins. The extract is dissolved in this water which should be at least 100 degrees Fahrenheit. Mill the skins 45 minutes, applying the tanning solution through the gudgeon. Then remove the skins from the drum, fold and pile, ready to be set out and hung up to dry.

After the skins are dried and sorted for colors, those that are to go into colors are drummed in two-thirds of a pound of sumac for every dozen. The sumac can be fixed by a trace of tartar emetic.

Degreasing Sheepskin

The following give best results.

Soda Ash   2% or
Lactic Acid   15% or
Lactic Acid   15% and Acetone or
Carbon Tetrachloride or
Carbon Tetrachloride
and Lactic Acid 15%

New Mineral Tannings
U. S. Patent 1,940,610
Method No. 1

A solution of zirconium sulphate containing the equivalent of 20% ZrO_2 is used to tan pickled sheepskin as follows: 100 pounds pickled sheepskin showing pH less than 2 is drummed in 300 pounds of 5% solution of sodium chloride, and 35 pounds of tanning solution added in one feed. After drumming for 6 hours the skins feel tanned. The liquor is drained, and the skins washed with water until a pH of 4½ is reached, then dried out. By re-wetting and staking good white leather is obtained.

Method No. 2

A solution of zironium chloride is prepared by dissolving basic zirconium carbonate in an amount of concentrated hydrochloric acid calculated to give a composition corresponding to the formula $ZrOCl_2$, and diluting to a solution containing the equivalent of 20% ZrO_2.

100 pounds pickled sheepskin showing pH 2¼ is drummed in 250 pounds 10% solution of sodium chloride and 25 pounds of tanning solution added in 3 feeds at half hour intervals. Drum 2 hours, lay overnight in liquor, then a solution of sodium bicarbonate added in successive portions till a pH of over 4 is reached. Skins are washed well, and dried out.

Method No. 3

A solution of zirconium nitrate is prepared in a manner similar to the chloride by dissolving basic zirconium carbonate in nitric acid and diluting to a solution containing the equivalent of 20% ZrO_2.

100 pounds pickled sheepskin showing pH 2 is drummed in 300 pounds 10% solution of sodium chloride and 30 pounds of tanning solution added in 5 feeds at ½ hour intervals. Drum four hours after the last addition. The liquor is drained, and the skins drummed in 300 pounds of 5% salt solution to which is added 3 pounds gluconic acid, then a solution of borax in successive feeds until a pH of 5 is reached. The skins are washed, then fatliquored with 5 pounds sulphonated cocoanut oil and dried out.

Method No. 4a

100 pounds of calfskins are taken in the pickled condition and given 6 pounds of a one third basic chromium sulphate, then 5 pounds of the tanning solution described in Method No. 1. After drumming for 5 hours, the skins are neutralized to pH 4, washed well, fatliquored with 5 pounds sulphonated neat's-foot oil, and dried out.

Method No. 4b

100 pounds of calfskins are taken in the pickled condition and given 10 pounds of the tanning solution described in Method No. 1. After drumming 3 hours, sodium carbonate is added until a pH of 3½ is reached, then 3 pounds of a one third basic chromium sulphate added. After drumming 2 hours sodium carbonate is added very slowly until a pH of 4 is reached, and the skins allowed to lie

in the liquor overnight. In the morning they are processed as is customary for chrome tanned calfskin and give good leather of bluish white color.

Method No. 5

100 pounds of pickled sheepskins are drummed in 300 pounds of 5% solution of sodium chloride, and given 20 pounds of the tanning solution described in Method No. 1. Drum for 4 hours, drain, and in a fresh liquor of 5% salt given 8 pounds of aluminum sulphate. Drum 4 hours, allow to lie in liquor overnight, and next day sodium bicarbonate added in successive feeds until a pH of 4½ is reached. Skins are rinsed, fatliquored with a mixture of 4 pounds neat's-foot oil and 3 pounds Ivory soap, and dried out.

Method No. 6

100 pounds of bated kidskins are given a light pickle with sudphuric acid and salt, then treated with 10 pounds of a 37% solution of formaldehyde. Drum for 3 hours, then neutralize to a pH of 7 and allow to lie in the liquor overnight. Next day give 5 pounds of the tanning solution described in Method No. 2, drum for 3 hours, then wash until a pH of 4½ is reached, fatliquor as in Method No. 5, and dry out.

Method No. 7

100 pounds of pickled calfskins showing pH of 2½ are drummed with 300 pounds of a 5% solution of sodium chloride and 20 pounds of the tanning solution described in Formula No. 1 added. Drum for 4 hours, then neutralize with sodium bicarbonate to a pH of 3½. Liquor is drained and the skins rinsed lightly, and divided into 4 parts. In fresh salt solution 5 pounds of solutions of various synthetic tanning agents are added:

(a) Consisting of the product of condensation of phenol sulphonic acid and formaldehyde;

(b) Consisting of the product of sulphonation of the viscous resinous liquid produced by condensation of phenol with acetaldehyde;

(c) Consisting of the product of condensation of naphthalene sulphonic acid and formaldehyde;

(d) Consisting of the product of sulphonation of 4.4′–dihydroxydiphenlydimethylmethane.

After drumming with the synthetic tanning agent for 3 hours, the skins are washed well, fatliquored, and dried out.

Method No. 8

100 pounds of side leather which had been tanned by slow addition of a blend

of quebracho and chestnut extracts is treated towards the conclusion of tannage with 5 pounds of the tanning solution described in Formula No. 1. After washing well, the skins are fatliquored, and dried out in the usual way.

Preserving Lizard Skins

Zinc Chloride	2 parts
Salt	10–20 parts
Water	100 parts

Treating Rabbit Skins

After the coarse hairs have been cut off, the skins are treated with an 8–12% solution of caustic soda, dried, soaked in water at 18° for 20–30 hours, fleshed, immersed in a solution of 2.0% caustic soda at 30°, washed for 20 minutes, immersed in a second bath at 25°, containing 0.4% sodium bicarbonate, for 1–2 hours, washed for 15–20 minutes, immersed in a third bath containing 0.4% sodium bicarbonate, 0.4% boric acid, 3% ''sozhal'' (prepared by fermentation with Aspergillus orizale) at 35°. This method yields a satisfactory skin as well as good down. Use of pancreatic extract produces a good soft leather, but the down is of a lower quality.

Softening Squirrel Skins

After the skins have been fleshed and before they are tanned and dyed apply to the flesh side a paste made by mixing a very dilute solution of caustic soda with French chalk. Then fold the skins flesh side in and allow them to lie for several hours or even over night. The next day they are put into a dilute solution of calcium chloride and again left over night. Then wash the skins in a paddle or drum, first with fresh water and then in a water solution containing ½% to 1% lactic acid. This solution will remove all traces of lime or other substances which have the tendency to make the skins stiff and tinny.

Ferment for Hides

Raw hides or skins are treated at 37° C. with a catheptic ferment 0.1 part of powdered liver, spleen, kidneys or the mucous membrane of the stomach is extracted with 100 parts of water. The development of bacteria is retarded by an addition of 0.5 per cent boric acid. The catheptic ferment loosens the hair in forty-eight hours. After unhairing the skins they are treated with a neutral solution of salt before tanning. A softer leather is obtained if the skins are treated with a 0.2 per cent solution of sodium hydrosulfide or a sodium salt of thioglycollic acid.

Enzyme for Unhairing (Hides)

Hides are treated, without pretreatment with acids or alkalis, with a 1% solution of papain in a sulphurous acid solution of pH 5 at 30° for 24–40 hours.

Unhairing of Hides for Chrome Leather With Recovery of the Hair

The complete cycle of operations requires 6 days. On the 1st day the hides are soaked in a 1 gram per liter of lime bath at 18°; on the 2nd they are fleshed, resoaked in a bath of the same composition, and finally spread on screens; on the 3rd they are unhaired in a bath at 25° consisting of: water 400–500%, calcium oxide 6% (both on the weight of the hides), sodium sulphide 1 gram per liter, 100% bisulphite 0.67 gram per liter (so as to neutralize only ½ of the sodium hydroxide formed on solution of the sodium sulphide). The total time of this treatment is 72 hours. On the 6th day removal proper of the hair is easily carried out, without damage, the sodium bisulphite probably acting as passivating agent to protect the hair from hydrolysis. The concentration given corresponds to the most favorable hydrosulphide to hydroxide ratio for intensive unhairing.

Carroting of Fur
U. S. Patent 1,919,141

Yellowing of the fur is avoided by carroting with an aqueous liquor containing mercuric nitrate (1.6%), nitric acid (1.0), ammonium fluoride (0.03), and hydrogen peroxide (0.75).

Fur Carroting Solutions
U. S. Patent 1,955,678
Solution No. 1

To 20 parts, by volume, of 100 volume hydrogen peroxide, add 80 parts, by volume, of water (thus reducing the hydrogen peroxide to 20 volume strength) and to the solution so obtained add 100 parts, by volume, of 3° Baumé solution of mercury nitrate.

Solution No. 2

To 1 part, by volume, of 100 volume hydrogen peroxide, add 1 part, by volume, of 20° Baumé solution of mercury nitrate.

Solution No. 3

To 1 part, by volume, of 200 volume hydrogen peroxide, add 1 part, by volume, of 30° Baumé solution of mercury nitrate.

Glazing Furs
U. S. Patent 1,952,137

Add 10 parts of ammonium linoleate to 500 parts of water and to this mixture is added 50 parts of melted carnauba wax.

The liquid used for treating furs is made by adding 4 ounces of the above wax emulsion to a gallon of water and in some cases add to the emulsion so employed 2 ounces of a mordant, such as aluminum acetate for each gallon of water used. The complete mixture will therefore consist of 1 gallon of water, 2 ounces of aluminum acetate and 4 ounces of wax emulsion prepared as above described.

The furs are immersed in this luke warm emulsion for approximately 15 to 20 minutes, after which time the superfluous liquid is drained off by centrifuging or otherwise. The furs are then dried, drummed with sawdust and finished in the usual way by which operations a slight protective film of waterproof wax is left upon the surface of the fur and skin.

It should be noted that instead of immersing the furs in this luke warm emulsion, the hair of the furs can be brushed or sprayed with the same liquid above described, the subsquent processes of drying and drumming with sand or sawdust or similar material and finishing being the same as now generally employed.

Fur Mordant

Chrome Acetate Mordant (20° Baumé)	250 cc.
Caustic Soda Solution (38° Baumé) (32.5%)	320 cc.
Glycerine (30° Baumé) (95%)	10 cc.

The solution of these substances is brought up to a volume of 1 liter by the addition of 420 cubic centimeters of water.

Tanning Material for Furs
Canadian Patent 335,058

A tanning solution consists of water 200, sodium thiosulphate 5, U. S. P. fornaldehyde 5, and soap 5 parts by weight. This solution is used to tan fur skins which have been previously softened, fleshed and pickled.

Dressing of Fur Skins

The skins are well soaked with water and then immersed in the following solution for 36 hours:

Water	10 gal.
Potash Alum	5 lb.
Chrome Alum	10 oz.
Salt	5 lb.

The skins are taken out, drained well and immersed in a solution of 10 pounds of stainless sumac in 8 gallons of water, and left for 24 hours. They are drained again and oiled on the flesh with one part water and one part sulphonated neatsfoot oil and hung up to dry. When dry the skins are tumbled with hardwood sawdust for 3 to 4 hours and then whipped with a beater.

Bleaching Furs
U. S. Patent 1,894,277

Furs and textile materials are bleached with a 1½% aqueous alkaline solution of hydrogen peroxide made more reactive by the addition of sodium pyrophosphate (1¼%) and a persulphate (1¼%).

Bleaching Dark Colored Fur Skins
U. S. Patent 1,564,378

Treat for 1 hour with:

Water	1	gal.
Salt	0.25	lb.
Soda Ash	20	g.
Sal Ammoniac	16	g.

Wring, wash, wring.

Mordant overnight with:

Water	1	gal.
Salt	0.50	lb.
Copperas	40	g.
Sal Ammoniac	20	g.
Tartar Emetic	4	g.

Wring, wash, wring.

Bleach for 4 hours with:

Water	0.50	gal.
Hydrogen Peroxide	0.50	gal.
Sodium Pyrophosphate	20.0	g.
Potassium Carbonate	8.0	g.

Wring, wash, wring.

Treat for 1 hour with:

Water	1	gal.
Salt	0.50	lb.
Ammonium Bifluoride	15	g.
Oxalic Acid	20	g.

Seal Black Topping Solution
Solution No. 1

Water	1	gal.
Aniline Salt	0.50	lb.
Sal Ammoniac	0.25	lb.

Solution No. 2

Water	1	gal.
Bluestone	100	g.
Potassium Bichromate	40	g.
Potassium Chlorate	100	g.

Mix equal parts of solutions No. 1 and No. 2 just before using.

Fur Skin Dressing
U. S. Patent 1,845,341

Soak the skins overnight in:

Water (on weight of skins)	95%
Gluconic Acid (on weight of skins)	5%

Wring and flesh.

Pickle for 4 hours in:

Water	195%
Sulphuric Acid	0.50%
Salt	16%

Tan for 3 hours in:

Water	195%
Formaldehyde	5%
Sodium Thiosulphate	5%
Soap	5%

Logwood Black on Fur Skins

Treat for 2 hours with:

Water	10	gal.
Salt	2	lb.
Sal Soda	280	g.

Wring, wash, wring.

Dip overnight at 95° F. in:

Water	10	gal.
Logwood Extract	5	lb.
Turmeric	1	lb.

Pull skins. Hang to oxidize 8 hours.

Add to bath:

Sumac Extract	280	g.
Roasted Nutgalls	280	g.
Iron Liquor	150	cc.
Copperas	75	g.
Bluestone	300	g.

Dip overnight at 95° F.
Pull and oxidize 8 hours.
Dip overnight at 95° F.
Pull and oxidize 8 hours.
Dip overnight at 95° F.
Pull and oxidize 8 hours. Wash thoroughly and finish.

Bluefox Shade on Fur Skins

Treat for 1 hour at 80° F. with:

Water	1	gal.
Salt	0.25	lb.
Ammonia (28%)	15	cc.
Soda Ash	20	g.

Wring.

Mordant overnight at 85° F. with:

Water	1	gal.
Salt	0.5	lb.
Copperas	18	g.
Tartar Emetic	4	g.
Sal Ammoniac	12	g.

Wring, wash, wring.

Dye for 2 hours at 85° F. with:

Water	1	gal.
Salt	0.25	lb.
Para Aminophenol	0.25	g.
Para Phenylene Diamine	1.5	g.
Pyrogallol	0.33	g.
Ammonia (28%)	1.0	cc.
Hydrogen Peroxide	30	cc.

Grey Shade on Fur Skins

Treat for 1 hour at 80° F. with:

Water	1	gal.
Salt	0.25	lb.
Ammonia (28%)	10	cc.

Wring, wash, wring.

Mordant overnight at 85° F. with:

Water	1	gal.
Salt	0.50	lb.
Copperas	40	g.
Tartar Emetic	10	g.
Sal Ammoniac	10	g.

Wring, wash, wring.

Dye for 2 hours at 85° F. with:

Water	1	gal.
Copperas	10	g.
Tannic Acid	2.0	g.
Pyrogallol	2.0	g.
Sumac Extract	5	cc.

Bark Tanned Calfskins (With Hair)

The skins are first soaked, washed, and fleshed.

Pickle for 4 hours with:

Water	12.5	gal.
Salt	18	lb.
Sulphuric acid	6	oz.

25 lbs. of skins.

After pickling, horse up to drain for 24 hours.

Tan with the following quantities (per 100 lbs. of pickled skins).

Water	100 lb.
Salt	4 lb.
Sumac Extract	80 lb.
Quebracho Extract	34 lb.
Sellatan A (Geigy)	5 lb.

Add the above materials to the tan mill and mill until tanned through. Pull and drain for 24 hours. Wash for 20 minutes.

Drain and fat liquor by milling for 30 minutes with:

Water	10 gal.
Sulphonated Cod Oil	25 lb.
Fig Soap	10 lb.

Furrier's Moth Spray

Carbon Tetrachloride	60 lb.
Toluol	11 lb.
Paradichlor Benzol	30 lb.
Oil Colonial Bouquet	1 oz.

Weight about 11⅛ pounds to a gallon used by furriers to spray garments before storage. Leaves a fine deposit of para crystals where spray is applied.

LUBRICANTS, OILS, ETC.

Lubricating Composition
U. S. Patent 1,937,462

A mineral oil is used with the addition of a small proportion (suitably about 5%) of an amide of a fatty acid such as stearanilide serving as a stiffening agent, and of a metallic soap (suitably about 0.75% of sodium soap and 2.75% of zinc soap) to raise the melting point.

Lubricating Compound
U. S. Patent 1,944,273

A wire drawing compound comprises substantially one part of soluble alginate, four parts of tallow, two parts of soap, and one hundred and ninety-five parts of water.

Lubricants
British Patent 395,867

Fluid lubricants of increased thermal conductivity and especially suited for the cylinders of internal-combustion engines are prepared by dissolving a metal soap and an ammonium soap in 1 or more vegetable oils and combining the solution with 1 or more mineral oils. Thus ammonium oleate 1, ammonium stearate 1, copper oleate 0.25, a vegetable oil mixture, e.g., of equal weights of olive, peanut and soy-bean oils, 7.75 and mineral oil 90 pounds are used.

Lubricant for Removal of Asphaltic Molded Products

By incorporating from 3% to 18% of aluminum stearate in the asphaltic binder easy removal from the mould follows. Butyl stearate, dibutyl phthallate and tricresyl phosphate can also be used by melting up the asphaltic binder and stirring in the lubricant. These, however, lower the melting point of the asphalt somewhat.

Lubricant for Removal of Plaster of Paris Art Objects

Dissolve stearic acid technical in either kerosene or gasoline, the amount of acid depending entirely on circumstances.

Any good grade of white soap that does not contain sodium silicate as a filler can be dissolved in hot water and used as a lubricant.

Ordinary beef tallow dissolved in kerosene or gasoline gives quite satisfactory results as a lubricant for both plaster of paris and concrete products.

Dibutyl phthallate or butyl stearate used either as spray or brushed on also gives very good results as a lubricant. A thinner coat can be had on the molds by dissolving the lubricant in a volatile carrier such as gasoline or naphtha. These two lubricants have the added advantage of not reacting with the plaster chemically. Plaster of paris products thus made can be lacquered or painted without fear of scaling due to lubricant being present on the plaster.

Die Lubricant
U. S. Patent 1,946,121

Powdered graphite is suspended in carbon tetrachloride and sprayed on dies used for magnesium alloys.

The wrought articles produced from dies lubricated with the graphite suspension have a highly polished surface and the dies even after extended service are not appreciably scored or disfigured in any way. The metal flows readily and more intricate wrought articles may be made with a considerable reduction of the practical difficulties heretofore encountered.

Ammunition Lubricant
U. S. Patent 1,953,904

Loaded cartridges are assembled into place and then dipped momentarily into the lubricating bath. Regulation of the depth to which the cartridges are immersed in the lubricant, is maintained so that only the bullet of each cartridge is coated with lubricant. Regulation of the depth to which the cartridges are immersed in the lubricant may also be maintained so as to coat the bullet and also the mouth end of the shell. The coating of the mouth end of the shell will thus act as a seal to moisture and at the same time facilitate the extraction of the fired case. The cartridges are then passed over a felt mat in order to drain the excess lubricant adhering to the nose of the bullet. They are then dried in a current of air.

The application of the lubricant may also be accomplished by spraying, brushing or barrel tumbling.

The bath of lubricant comprises a high melting-point wax dissolved in a suitable solvent. Any wax melting above 145° F. may be used. Use carnauba wax, candelilla wax, ceresin wax, Montan wax or the synthetic waxes of the chlorinated naphthalene type. Volatile solvents such as naphtha, benzene or carbon tetrachloride may be used to reduce the viscosity of the waxes. A water soluble fatty acid soap such as aluminum stearate may be added to the lubricant. The addition of small amounts of aluminum stearate will hold the lubricant in a solid state and not allow the wax to flow, even though the lubricant be exposed to temperatures in excess of the melting-point of the wax.

A lubricant which has proven successful is one prepared according to the following formula:

Carbon Tetrachloride	90 parts
Ceresin Wax (Melting-Point 176° F.)	9 parts
Aluminum Stearate	1 part

In proceeding to make the mixture the aluminum stearate is first dissolved in the wax at a temperature of approximately 260° F. with thorough stirring. The wax is then cooled and dissolved in the solvent at somewhat elevated temperatures, approximately 130° F.

In applying the lubricant the lubricating bath is maintained at a temperature of approximately 120° F.

Low-Pour-Point Lubricating Oils

U. S. Patent 1,896,342

A lubricating distillate which contains 2–10% of animal or vegetable oil is heated to 149–176° and admixed with 0.05–0.3% of a soap, e.g., aluminium stearate.

High Speed Lubricant

U. S. Patent 1,937,463

A typical example of a lubricant suit-able for machines having high speeds and low bearing pressures is:

Zero mineral oil, 91.5%, viscosity 500° Saybolt at 100° F. sodium soap .75%, zinc soap 2.75%, stearanilid 5%.

In producing solidified oils the oil, preferably a mineral oil, is thoroughly mixed with a wax-like solidifying agent and with a small percentage of metallic soap, if such soap is used at a temperature above the melting point of the solidifying agent, and sufficiently high to enable all of the ingredients to mix in liquid form. The liquid mixture is then delivered into a suitable cooling device, which may be of any suitable form capable of rapidly and uniformly cooling the mixture. The liquid mixture is subjected to the cooling operation until the temperature of the mixture has been reduced to a point at which the mixture is in semi-solid condition. The semi-solid mixture may then be delivered into suitable receptacles or containers and allowed to cool slowly to atmospheric temperature. The rapid and uniform cooling of the mixture is essential to the production of a homogeneous compound.

In mixing the ingredients, the stearanilid is placed in the bottom of the mixing tank and the tank is heated to melt the solidifying agent. The metallic soap is first added to the solidifying agent and thoroughly incorporated therein by means of a rotating mixer blade within the tank. The mixture in the bottom of the tank is maintained at a predetermined temperature by a steam jacket and the lubricating oil is slowly added and stirred into the mixture in the tank at a rate such that the temperature of the mixture is not materially reduced. When stearanilid is used as a solidifying agent, together with a small quantity of metallic soap, the mixture in the mixing tank will be maintained at a temperature of from 250° to 350° F., depending upon the proportions of the ingredients used. After the desired amount of oil has been added in the mixing tank, the discharge outlet is opened and the liquid mixture is permitted to flow to the cooling cylinder. In passing over the cooling cylinder, the temperature of the mixture is lowered to a temperature of approximately 120° and in passing through the discharge trough the temperature of the mixture is lowered to approximately 100°, at which temperature the composition may be delivered into suitable containers, which when filled, are allowed to stand for twenty-four hours or longer before the containers are sealed for shipment or before the material in the containers is transferred to other containers for shipment.

Lubricating Block

Madagascar Graphite or	
Ceylon Superfine	98%
Colloresine Dry	2%

Make colloresine to a 10% paste in cold water.

Metal Drawing Lubricants
U. S. Patent 1,948,194

By using a beeswax emulsion (e.g., one given in the chapter on emulsions) a very efficient wire drawing lubricant is made. This reduces the power, required for drawing, as much as 40% as compared with lubricants of tallow and soap emulsion type.

Greaseless Lubricating Pencil

For lubricating hinges, automobile doors, etc., the following is useful as it will not run off and produce stains.

Beeswax	100 parts
Graphite Powder	50–100 parts

Melt together and stir until a little above room temperature. Pour into molds and allow to cool.

Spring Lubricant
British Patent 396,195

Porous flexible elements for insertion between the leaves of a laminated spring are made by mixing copper 10, sponge iron 88.5 and graphite 1.5 parts, together with a small amount of volatile lubricant, e.g., petroleum, stearic acid, compressing under high pressure, sintering at about 2100° F. in a non-oxidizing or reducing atmosphere and then quenching with oil before cooling or immersing in oil after cooling. The porous elements may also be similarly made from a mixture of powdered copper 85, powdered tin 13 and graphite 2 parts, with a small amount of stearic acid.

Hub and Gear-Box Lubricant
British Patent 398,936

Lubricants, suitable for hubs and gearboxes of automobiles, consist substantially of a cellulose ester or ether, vegetable and/or fish oils and a heavy mineral lubricating oil. For example 6 pounds of triethyl cellulose is soaked 15–25 hours in 64 pounds of castor oil, dissolved by heating to 210°, cooled and masticated and then masticated with 30 pounds of crank-case oil.

Lubricant for Ball and Roller Bearings
Swedish Patent 75,225

A mixture of mineral oil and one or more solid fatty acids or their glycerides is saponified with a mixture of 1–2 parts of caustic soda and 1 part of caustic potash, water is removed by heating and more mineral oil added to a total content of about 85% the mass is finally cooled rapidly.

Upper Cylinder Lubricant
British Patent 405,145

1–20 per cent by volume of colloidal graphite, e.g., Oildag, is mixed with 99–80 per cent by volume of mineral oil, 1 per cent by volume of an oil-soluble color, and 1–2 per cent by volume of xylene. 1 part of this mixture is then mixed with 9 parts of naphthalene and pressed into tablets under constant pressure at raised temperature to form a solid top lubricating oil, preparation suitable for adding to motor fuel.

Water Soluble "Drawing" Lubricant
U. S. Patent 1,952,973

A composition found to be particularly suitable for drawing deep stampings is composed of 70% of 43° Bé. corn syrup, 15% of 000 multicel infusorial earth, and 15% of water.

Viscous Emulsion Lubricant

Heavy Mineral Oil	10 parts
Diglycol Stearate	10 parts
Water	40 parts

Heat together to 60° C. and stir until cold. A heavy cream results. The color is dependent on the color of the oil used.

Lubricating Stick

Petrolatum	10 parts
Paraffin Wax	10 parts

Melt together, mix well and pour into cardboard moulds. The sticks are used for the lubrication of door latches of automobiles, etc.

Steering Gear Lubricant

Blown Rapeseed Oil	2 parts
Red Mineral Oil	
(750 Viscosity)	8 parts

The mineral oil should be a Gulf coast or other asphaltic base type. Paraffin base oils will generally separate. This oil will have a low cold test and will be suitable for year round use.

Cutting Lubricants
(For general lathe and drilling work)

Sal Soda	¼ lb.
Lard Oil	½ pint
Soft Soap	½ pint

Water enough to make 10 quarts

Boil the above ingredients for one-half hour and place in suitable containers for use as needed.

Cutting Lubricants
(For cast iron)

For hard cast iron use strong sal soda water. For hand reaming use a mixture of tallow and graphite. (For brass and babbitt), kerosene and turpentine give good results in hand reaming with somewhat dull reamers. (In boring babbitt-rod boxes) in lathe or boring mill, kerosene and lard oil lessen the chips forming into hard balls and tearing of the metal.

Lubrication of Ground Glass Joints

The following method is used to lubricate ground glass joints used at elevated temperatures, since under those conditions stop-cock greases are not ordinarily satisfactory.

This method consists of applying a little phosphorus pentoxide to the joint and allowing the latter to remain exposed to the air for a few moments. This causes the phosphorus pentoxide to become moist, and if the joint is now put together satisfactory lubrication is obtained. This type of lubrication has been used with good results on ground glass joints which were exposed to boiling glacial acetic acid for seventy-two hours.

Chassis Lubricating Oil
U. S. Patent 1,944,164

A small proportion (suitably about 6–15%) of blown rape-seed oil is admixed with a base of petroleum oil made from asphaltic crude oils and having a viscosity of about 3000 S. U. V.

Cylinder Oil
U. S. Patent 1,924,211

A heavy steam-refined lubricating oil is mixed with about 0.5–4.0% of asphaltic material and about 0.2–1.5% of aluminum stearate, in order to form a steam cylinder oil.

Soluble Oils
U. S. Patent 1,938,804

Formula No. 1

Potash Soap	30 parts
Olein	7 parts
Cyclohexanol	2 parts
Paraffin Oil	32 parts

Formula No. 2

Potash Soap	32 parts
Turpentine	60 parts
Olein	6 parts
Methylhexalin	2 parts

Artificial Drying Oil
U. S. Patent 1,918,599

A mixture of rosin 25, animal fat (beef tallow) 12, lubricating oil (density 1.28) 3, cobalt acetate ¼–3, lead oxide 2–10, water 2 parts, is heated to 150–290° (optimum 250° approx.), quickly cooled, and 65 parts of stove oil (density 1.26–1.38 are added. A small proportion of melted rubber may be added to increase the elasticity of the film.

Rust-Proofing Oil
U. S. Patent 1,943,808

About 2% of naphthalene is mixed with mineral white oil.

Penetrating Oil

Fuel Oil (No. 2 Furnace)	3 parts
200 Viscosity Pale Oil	1 part

Graphite may be added in proportion of from 2 to 10% as desired. Nitrobenzine should be added in proportion of from .05% to .1% to give suitable odor.

Extreme Pressure Oil

Lard oil is treated with 20 parts of sulphur monochloride by adding the sulphur chloride and heating the mixture to 200° F. After this mixture has stood 24 hours it is blended in proportion of 1 part to 10 parts of steam refined cylinder stock. The resultant compound is a suitable lubricant for worm and hypoid type automotive rear axles.

Sulphurized Cutting Oil
U. S. Patent 1,604,068

7% of sulphur monochloride is added to mineral spindle oil and the temperature is raised to 170° F. The mixture is agitated by air until all gases are removed. The resultant compound is superior to lard oil for cutting and threading iron and steel.

Norwood's Machine Oil

Sperm Oil	3 parts
Kerosene	1 part

Bleaching Palm Oil

Palm oil can be bleached by adsorption of the coloring matter by fuller's earth, by chemical reaction such as oxidation or reduction, and by air with or without the use of a catalyst. Best method involves use of fuller's earth at a temperature of 100° C. under a vacuum. Avoidance of the presence of air removes the danger of destruction of the vitamin A content of the oil. However, it is much simpler to bleach the oil by means of air. The easiest method is to allow the air to bubble through the heated oil, live steam being used as the heating medium. Bleaching is completed within a few minutes at a temperature of 240° C., within three hours at 150° C. and within seven hours at 110° C. The speed of the air current is at the rate of one-tenth of total volume used per second. Addition of catalysts, such as borates or resinates of cobalt, nickel, or manganese, allows the reaction temperature to be decreased below 100° C. The proportion of catalyst used is 0.01 per cent of the weight of the oil.

Antioxidant for Oils and Fats

Gum guaiac appears to be non-toxic and is an effective antioxidant for lard (e.g., at 0.05% concentration), the preservative effect persisting in the presence of water, and in bakery goods made from the lard. It is a suitable antioxidant for cracklings, fatty stock foods, etc., but not sufficiently active to protect cottonseed oil from autoxidation.

Dewaxing Oils
Canadian Patent 338,604

Methylene chloride 1–4 volumes, is mixed with 1 volume of a wax-oil mixture and the resulting mixture chilled below 0° F. The precipitated wax is filtered from the mixture and thereafter the Methylene chloride is separated from the oil.

Dewaxing Mineral Oil
U. S. Patent 1,956,780

A process of removing mineral wax from mineral oil comprises dissolving approximately 100 barrels of wax-bearing oil in approximately 95 barrels of acetone and 80 barrels of benzol, chilling the solution thus formed to approximately +40° F., filter pressing the chilled solution to remove a relatively high melting point mineral wax, adding approximately 75 barrels of benzol to the filtrate, chilling the modified filtrate to approximately —10° F., and filter pressing to remove a relatively low melting point mineral wax.

Dewaxing Petroleum Oils
U. S. Patent 1,947,359

The oil is dissolved in a mixture of isopropyl ether 60% and acetone 40% and the solution is cooled to separate the wax by precipitation.

Removing Wax from Petroleum Oils
U. S. Patent 1,960,617

The process comprises diluting a wax-containing oil with a quantity of naphtha amounting to approximately 60% of the volume of resultant mixture, chilling the diluted oil to approximately 5° F., adding ground rice hulls of approximately 200 mesh to the chilled mixture at the rate of one pound of hulls to approximately one gallon of chilled oil, stirring the resultant mixture, filtering the mixture under a vacuum to separate the oil and diluent from the congealed wax and rice hulls, distilling the oil and diluent to separate the diluent from the oil, adding naphtha of approximately 120° F. to 160° F. to the mixture of wax and rice hulls, agitating the resultant mixture until the wax is dissolved and filtering the last mentioned mixture to separate the rice hulls from the naphtha and wax.

Treatment of Crude Oil
U. S. Patent 1,891,987

A composition suitable for removing water and bottom settlings comprises castor oil 25, nitric acid 15, salt 10, diluent (petrol) 50%.

Thickening Lubricating Oil
U. S. Patent 1,918,403

Lubricating oil is heated with 8% of nitronaphthalene at 150° for 30 minutes and 0–1% of a compound obtained by treating a 2–5% solution of caoutchouc in toluol with 5% of xylidine. The mixture is again heated at 150° under pressure and cooled to obtain a lubricating jelly.

Increasing Lubrication of Oils
U. S. Patent 1,945,614

The addition of 35–80 grams of dihydroxy stearic acids per gallon of lubricating oil increases its lubricating value and even causes it to penetrate the micropores of metals.

Green Bloom Coloring for Lubricating Oil
U. S. Patent 1,944,851

A gas oil distillate of naphthenic type, ranging in gravity between 25° and 30° Bé., is charged to a still and subjected to distillation in the presence of from 2.5 to 3 per cent of anhydrous aluminum chloride, by weight on the charge of oil used. The distillation is continued until the oil bottoms, exclusive of coke and solid aluminum chloride hydrocarbon compounds, amount to about 10 per cent by volume of the charge to the still. During this distillation the green bloom sludge formed originally is decomposed to coke by heating, the active green bloom matter being extracted by the residual oil present.

The bottoms from the above process are cooled down by a pump circulation through a cooling coil and delivered to a storage tank. The time for this particular run amounts to about 60 hours. All suspended aluminum chloride-coke particles are allowed to settle out of the oil which has been delivered to the storage tank, and the so clarified oil is then removed for further use. Alternatively the aluminum chloride-coke might have been completely removed by filtration, by centrifuging or by washing with a little dilute sulfuric acid followed by neutralization. The clarified oil thus produced is a dark brownish-green substance capable of imparting brilliant green bloom to lubricating oils when added thereto in small amounts, usually less than 1 per cent.

Lubricating Oil Base
U. S. Patent 1,939,170

Compound with 5% glycerine, 12% sperm oil and 8% castor oil, and to this compound 75% of metallic soap such as aluminum stearate introduced while thoroughly stirring the product, and the resulting base produced from this compounding will be of a granular or powdered character having slight greasy or oily characteristics.

In the production of a predetermined amount of heavy oil or grease from the powdered base, approximately 3% or more of the base is incorporated with the computed amount of oil or grease to be produced in the following manner. The desired percentage of base is first compounded with about one-third of the mineral oil, subjecting the same to heat at about 220° Fahrenheit, thoroughly and continuously stirring the mass in this operation until the base is dissolved in the oil and the resulting product is clear, after which the balance of the two thirds of the oil is slowly added, while continuing to stir the mass until all of the oil has been thoroughly incorporated and the resulting product is clear. Depending upon the amount of mineral oil and the percentage of the base incorporated therein, the resulting product will resemble a very heavy oil or a grease, both of which products will be of a stringy consistency.

A product produced in this way will possess substantially constant viscosity and will be found to be of exceeding value as a lubricant for apparatus or machinery of various kinds and classes, wherein the rotating or reciprocating parts subjected to the lubricant will draw or carry the lubricant with the rotating or reciprocating element, by virtue of the viscous or stringy characteristics of the oil, thus insuring better lubrication.

Lubricating Grease
British Patent 398,402

A saponifying agent in the shape of crystalline barium hydrate is specified. For example, 56 kilograms of the barium compound are reacted with 100 kilograms colza oil in an autoclave under three atmospheres pressure, and the resulting soap is dissolved or emulsified in a suitable mineral oil at about 100° C. According to the consistency of the grease in view, 150 kilograms of the soap are treated with 400 to 1,000 kilograms mineral oil.

Manufacture of Lubricating Grease
British Patent 406,399

Mineral oil is mixed at 71–127° with anhydrous aluminium palmitate or stearate (15–45%; 5–50% of the whole) and the mixture is rapidly chilled.

Lubricant Grease
U. S. Patent 1,936,632

A composition of matter for lubricants comprises a high viscosity steam-refined paraffin base residual mineral oil approximately 150 Saybolt viscosity at 210° F. and approximately 10% aluminum oleate and aluminum stearate which is solid at 212° F.

Stopcock Lubricant (Sticky, Clinging Type)

| Petrolatum | 100 g. |
| Raw Crepe Rubber | 5 g. |

Shred the rubber as finely as possible, and stir into the melted petrolatum. Keep at a moderate heat (125 to 150° C.) for several hours, or until dissolved. If the temperature be allowed to go too high the rubber will decompose.

Stopcock Grease

Vaseline	16 parts
Pure Gum Rubber	8 parts
Paraffin	1 part

This formula is stated in parts by weight.

Stir while warming gently until all are dissolved.

Stopcock Grease (High Vacuum)

| Petrolatum | 1 lb. |
| Coagulated Rubber Latex | 1½ lb. |

Heat gently for several days, stirring at convenient intervals, until all of the rubber is in solution.

Stopcock Grease (General Purpose)

| Petrolatum | 1 lb. |
| Coagulated Rubber Latex | 1 lb. |

Prepare as above.

Air Seal Grease for Glass Joint

Melt together beeswax and castor oil in proportions to suit use—should be tacky but not too stiff.

Soaps for Grease Bases

Soaps used in the manufacture of solid greases are, according to a British chemical patent, prepared by saponifying fatty substances by means of alkaline earths, for example, solutions of baryta. In this specification the use of crystallized barium hydrate is advocated. Soaps thus prepared may be kept for long periods without deterioration. The amount of barium used depends on the weight and saponification number of the fatty substance. Thus, for 100 kilograms of tallow 56 kilograms of barium hydrate are used, while for 100 kilograms of colza oil 50 kilograms of hydrate are sufficient, the actual saponification taking place in an autoclave at three atmospheres pressure. The soap is then dissolved in mineral oil at 1,000 deg. C. The properties required of the grease determine the soap to oil proportions. A thick grease, for example, would be made by dissolving 150 kilograms of soap in 400 kilograms of oil.

Removing Metallic Soap from Lubricating Emulsions
U. S. Patent 1,955,522

A process for removing metallic soap from a lubricant emulsion containing particles of such metallic soap, comprises superimposing upon such lubricant emulsion and in contact therewith, a layer of molten paraffin wax or paraffin wax stocks, so that when the metallic soap particles of said lubricant emulsion float upwardly therein and then into the molten paraffin wax or paraffin wax stocks they will be dissolved thereby.

Corrosive Protection Grease

Neutral Petroleum Grease	100 lb.
Nitrogenous Base	1 lb.
Zinc Chromate	2½ lb.

The nitrogenous base is a derivative solvent from coal tar, composed of various quinolines. It is a very powerful reactant against corrosive acids.

MATERIALS OF CONSTRUCTION

Artificial Stone

Cement or any other suitable binding agent and any one of a group of relatively coarse aggregates such as sand, marble chips and the like are used and, if preferred, a suitable pigment for the purpose of lending color to the composite mass. This last named group of aggregates, including the Portland cement and the color, if employed, will be suitably mixed so as to produce a uniform mixture and then reduced to a plastic state by the addition of any one or more of the solutions specified for the purpose of rendering plastic the first named group of materials set forth for forming the body of the slab or like article.

This last named group of materials will also be molded or pressed into the desired form and permitted to harden sufficiently so that the finishing coat hereinafter more fully explained may be applied.

With the body of the slab or like article formed in either of the manners hereinbefore described, the next step in the manufacture of the artificial stone product is the application of a suitable finish or finish coat to the article thus formed so that the same may truly resemble the physical appearance of natural marble and also possess its desirable physical characteristics such as nonporosity and the like.

A finish closely resembling natural marble may be produced on the hereinbefore described previously prepared slabs or like articles by polishing the same with a suitable quantity of oxalic acid placed on the surface of the articles during the time the polishing operation is carried on. The oxalic acid will, during the process of reducing imperfections in the surface of the article, produce a resultant finish having the physical appearance which most closely resembles natural marble. Instead of applying the oxalic acid either in solution or crystalline form to the surface of the slabs or articles being polished during the polishing operation, a somewhat more desirable finish may be attained in the following manner:

A quantity of chalk may be mixed with solution of oxalic acid to produce a plastic mass which is spread onto the slab or like article formed in the manner previously described in order to form a thin layer or surface coating of such material on the body portion. After this layer has set sufficiently to be worked, the same may be polished by a coating either of oxalic acid and fluoride and/or a solution of sodium silicate.

A particular function of the oxalic acid employed in the manner specified in addition to the provision of the smooth finish is the formation of calcium oxalate crystals which produce in the finished surface of the product a multi-faceted appearance which has light reflecting properties identical with those of natural marble so that the finished product most closely resembles the natural stone. These oxalate crystals produce the desired finish even though the actual smooth finish is provided by either the fluoride or the waterglass solutions, or both in combination.

Slag Cement
U. S. Patent 1,912,815

The cement is composed of a finely ground mixture of blast-furnace slag 100, calcium hydroxide 30, sodium hydroxide 3, magnesium silicofluoride 2, and calcium fluoride 2 parts.

Cement Waterproofing Compound
U. S. Patent 1,913,430

Bitumen-water emulsions for admixture with hydraulic cement are stabilized by naphthenic soap and colloidal clay, e.g., oil 22.5, bentonite 25, sulphur 2.5, water 50%.

Sulpho-Aluminous Cement

A slag cement can be obtained by grinding together granulated blast-furnace slag 80, calcium sulphate (anhydrite) 12–15, and calcium oxide 1–2%. Hardening depends on the formation of sulpho-aluminates. The cement is resistant to sulphate solutions and does not cause corrosion in reinforced concrete.

Arc-Shield Cement

This cement has high heat resisting qualities and is used for repairing arc shields. It is stated below in percentages by weight:

Barytes	52.2%
Silica (Powdered)	17.4%
Asbestos No. 205-B	2.2%
Potassium Silicate 2 parts } Sidium Silicate 1 part	28.5%

Mix the barytes, silica and asbestos well, then add the sodium, potassium silicate and sodium thoroughly.

To apply scrape away burnt or glazed surfaces before using the cement. Apply the cement with a knife blade, like putty, and press it well into the recessed parts of the shield. Air dry for 24 hours. The cement will continue to harden for two or three days.

Waterproofing Concrete, Cement, Etc.

Dissolve 1–2 pounds of ammonium stearate (25% dry soap basis) in every 7½ gallons gaging water.

Protecting Concrete Surfaces During Curing
U. S. Patent 1,899,576

Green concrete is treated with sodium silicate containing an amount of weak non-caustic electrolyte insufficient to cause actual precipitation of silica, the addition being (a) acetic acid or sodium acetate, and (b) ammonium chloride.

Hardening Concrete

Use "S" Brand or "A-Syrup" silicate of soda. Dilute by adding four pails of water to one of the silicate. On extra dense concrete use still more water. Put on successive applications a day apart until the concrete will absorb no more. It will usually take from two to four applications. The action is the formation of a hard mass within the concrete which greatly increases its resistance to wear and reduces its absorption of water or oil.

Concrete Hardener

To harden concrete surfaces, clean and flush surface thoroughly, allow to dry and finally flush with magnesium fluosilicate (20° Bé.). Allow to dry for several days, and repeat flushing with the hardener once or twice more, or until the desired hardness is obtained.

Integral Concrete Hardener and Waterproofer

This consists of a solution of calcium chloride (30° Bé.). It is mixed as follows:

(a) For topping (for 100 square feet of surface 1 inch thick)

Cement	1 barrel
Sand	2 barrels
Concrete Hardener	1 gal.

(b) For mass concrete (1–2–4)—¾ gallon of concrete hardener to 1 barrel of cement.

(c) For mass concrete (1–2–5)—1 gallon of concrete hardener to 1 barrel of cement.

Hardening and Waterproofing Concrete
U. S. Patent 1,951,186

There are first prepared the following batches:

Batch No. 1

A solution of calcium chloride in water constituting a 26% solution of 77–80 flake calcium chloride in water. It is preferred that this solution have a specific gravity of 1.250, taken at the time when the batch is first prepared, the temperature being from about 150 to 170° F., due to the heat of solution. It will be observed that though this concentration is preferred, particularly in connection with the preferred proportions hereinafter set forth, for certain purposes the concentration of the calcium chloride solution may be varied so as to obtain a specific gravity varying from 1.5 to 1.1

Batch No. 2

A solution of technical tannin and water is made comprising a volume of tannin to a volume of water. To this is added lampblack and venetian red in proportions varying from 1/20 to 1/8 volume of lampblack add 1/10 to 1/4 volume of venetian red. Where a distinctive color is desired, these ingredients are varied within the range noted to obtain a bluish purple or plum color.

Batch No. 3

A dilute solution of sodium silicate is formed and to this is added amorphous silica, preferably in the form of infusorial earth. The sodium silicate solution and the infusorial earth are mixed volume for volume. There are then additionally added twelve volumes of water. Although it is preferred that the sodium silicate and amorphous silica be added volume for volume, a variation of 25% above or below the equal quantities spec fied may be made so that there is a 50–50 mixture by volume of these ingredients or approximately 65 parts of one to 35 parts of the other ingredient, or vice versa. The water quantity, however, for these combined ingredients should be within close limits of the amount specified. The concentration of sodium silicate is that obtained in the form of N brand of the Philadelphia Quartz Co. and this constitutes a 41° Baumé solution of sodium silicate containing approximately 8.9% Na_2O and 29.0% $(SiO)_2$.

Twelve ounces each of the mixtures prepared in Nos. 2 and 3 are added to a sufficient quantity of the calcium chloride solution, as prepared in Batch No. 1, to constitute five gallons, this amounting to approximately 2% each of mixtures 2 and 3, the order of addition being preferably first to add No. 3 and then to add No. 2.

As an alternative of the above method of mixing the ingredients, the twelve ounces each of Batches Nos. 2 and 3 are first mixed together before addition to the calcium chloride solution is made constituting Batch No. 1. This will make a mixture of approximately 96% of No. 1 and 4% combined Nos. 2 and 3 mixtures.

By the order of addition as specified, the solution may be immediately packaged in suitable containers and sealed, without the formation of objectionable bubbles and when so added, particularly combined by vigorous agitation of a mixing machine, the ingredients which do not go into solution, such as lampblack, venetian red and infusorial earth, will be suspended and maintain a uniform composition throughout prolonged periods of storage, requiring no objectionable amount of stirring or mixing when ready for use.

It will be observed further that it is preferred that the calcium chloride solution as utilized above be clarified and this may be obtained by allowing the calcium chloride solution as above prepared to stand for about two hours in a settling tub, which includes a draining faucet about an inch or so from the bottom, with suitable provision to clear out the sediment at periods of time to assure the desirable clear solution of calcium chloride.

The product so prepared constitutes the preferred form of indurating composition and for purposes of use in concrete mixtures, it may be added in proportions of one quart of this mixture to seventeen quarts of the water used in the cement or concrete batch. It is preferred to calculate the indurating composition, for best results, in the proportions of one quart of the mixture to each bag of the cement used in making up the concrete or cement batch, calculated upon 94 pounds of cement to the bag. This quantity may be increased where acceleration in set is desired and an addition of twice the quantity of indurating composition, to wit, two quarts of the mixture per bag of cement, will give approximately 50% greater acceleration in set, that is, whereas the addition of one quart of the mixture per bag of cement will cause the cement mixture to set in approximately three hours, two quarts of the mixture will cause this batch to set in approximately two hours. It will be observed that smaller amounts of the mixture may be added and there will be obtained proportionately decreased speeds of setting.

The quantity of the calcium chloride may be increased, for purposes of accelerating the speed of setting of the cement, up to a limit based upon absolute quantities of 3.3% of calcium chloride to the absolute quantity of cement. Increase of the quantity of the calcium chloride beyond this percentage will cause a general tapering off of the compressive strength of the resultant cement matrix.

The indurating composition as above provided, when added to cement mixtures, including concrete, will be strengthened to the extent of producing a compressive strength averaging 200 pounds per square inch in excess of untreated concrete. This will be obtained in addition to the acceleration of the speed of setting and of hardening of the concrete, for instance, final hardening will be accelerated to be obtained in two days as compared with seven days in ordinary untreated concrete. When mixed with the water necessary for preparing the concrete, the concrete mixtures may be worked at subnormal temperatures and as low as 15° above zero F. without danger of freezing.

Though the calcium chloride accelerates the time of set and hardening, the ingredients used with this agent will not undesirably affect the workability of the batch and, in fact, render the entire batch more flowable and prevents separation of the aggregate used in the concrete mixture. The material will flow into the mold much more readily.

The product also has the desirable properties, in addition to the strength above specified, of being exceedingly dense, not only of the physical characteristics of the cement product wherein the pores of the cement are filled, without any non-uniform separation of the impalpable fillers but reactions will occur tending to render insoluble the normally soluble constituents present. By the addition of the indurating composition, cement mortar may be troweled to a highly polished, glasslike surface, impervious to the disintegrating effect of oils or dilute acids, of a character unattainable by untreated mortars.

Hardening of Lime Mortar
U. S. Patent 1,898,358

Plaster of Paris and minute quantities of magnesium chloride and magnesium sulphate should be added to the mortar.

To Accelerate Setting of Plaster of Paris

To the mixing water dissolve from ¼ ounce to 3 ounces of ordinary potassium alum, or crystallized potassium sulphate. If it is desired to further accelerate the setting heat the mixing water to 110° to 135° F.

To Retard Setting of Plaster of Paris

To the mixing water dissolve about one tablespoonful of ground glue to the gallon. The amount of glue used will depend on just how much it is desired to retard the setting.

Another method is to add one tablespoonful of ordinary whiting or precipitated chalk to the gallon of water used in mixing the plaster.

Preventive of Efflorescence on Brick
Formula No. 1

Cement	1 part
Lime	¾ part
Sand	4 parts

Formula No. 2 (for Stronger Cement)

Cement	1 part
Lime	¾ part
Sand	3 parts

Impregnation of Mine Pilings

Wood piles are treated with 0.25–0.37% solution of mercuric chloride under 60–96 pounds pressure. The piles are vacuumized before and after impregnation.

Artificial Marble
U. S. Patent 1,935,985

One mixture used may comprise in parts by weight, chalk, 25, cement, 10, and pigment, ½. These relative quantities, however, can be varied considerably without materially changing the resulting product.

The above named ingredients in the proportions specified or in any other suitable proportions found desirable, are ground to a fineness so that the separate constituents cannot be readily recognized in the mass. The mass is then reduced to a plastic state by the addition of a solution which may be water, a fluoride such as magnesium fluoride, a solution of sodium silicate or a solution of oxalic acid. After the materials have been so reduced to a plastic mass, it will be understood that by mixing therein suitable coloring materials so as to produce veins or local areas of contrasting colors, the mass is molded and permitted to set to form a slab or like article of desired shape.

After the material has sufficiently hardened, the same will be polished or provided with a finished surface in the manner hereinafter more fully explained.

The hereinbefore outlined group of ingredients combined in the manner specified may be employed for the production of an artificial stone article having certain characteristics which resemble and closely approach the physical characteristics of natural marble. Another group of ingredients which may be employed for the purpose of producing the body of the slab or like article to be formed may include Portland.

Artificial Marble

Dissolve 3½ parts of ammonium aluminum sulphate in one part of water by means of heat. Continue to heat until the boiling point is reached and pour into moulds. The material may be colored as desired by adding any water soluble aniline dye. By making two or more separate batches of different color and mixing the two together before pouring in the mould, a graining will result that closely imitates marble. A low price filler can also be used such as plaster of paris or talc.

Clay Surfacing Composition
British Patent 401,304

A composition for surfacing artificial cricket pitches, etc., consists of fibrous material, plastic material of the nature of clay and very finely divided non plastic, preferably siliceous, material. A typical composition contains asbestos

16–44, sawdust 0–24, clay 6–12%, with SiO_2 and fine sand as remainder.

Light Bricks with a High Sound-Insulating Power
Norwegian Patent 52,994

One part (by volume) of coke slag, 1 part of grout and 1 part of cork are mixed and added to a mixture containing 0.1 part of kieselguhr and 0.1 part of dry plaster of Paris stirred out in a suitable amount of thin grout, after which all is mixed intimately, molded, dried and burned. The resulting bricks or blocks have an apparent specific gravity of 0.75–1.4 kilograms per liter.

Artificial Stone
British Patent 381,694

A mixture of fiber, e.g., asbestos, cement, e.g., portland cement, and a thermoplastic material, e.g., resin, copal, pitch, bitumen, asphalt, sulphur, is molded with water, e.g., as a sheet, and allowed to set, heated to diffuse the thermoplastic material, impregnated with a solution of, or the constituents of, an artificial resin and again heated to harden the resin. For example the composition consists of portland cement 200, asbestos 100 and powdered pitch 60 pounds, the resin being a phenol-formaldehyde resin with, if desired, 5% aniline oil.

Artificial Stone
British Patent 381,035

Fibers, e.g., textile waste, wool, asbestos, are soaked in water 12–24 hours, dried and ground with powdered resin in the proportion of 1–1.5 parts resin to 1 of fibers. The mixture is molded under heat and pressure to form building slabs, etc. A small addition of Glauber's salt gives an improved color with asbestos fibers.

Interior Stucco
Formula No. 1

Marble Flour (20 mesh)	50 parts
Mica	20 parts
Keenes Cement	10 parts
Good Bone Gelatin or Hide Glue	8 parts

Mix with hot water to form a thin paste, let stand one hour and apply to walls, texturing as desired.

Formula No. 2

The following is a smoother product now on the market as "Craftex":

China Clay	90 parts
Casein	10 parts

Waterproofing Stucco
U. S. Patent 1,942,601

A 5% solution of sodium stearate soap, in water, to which is added 2% of melted suet. In order to incorporate the suet in the solution, the latter is heated to about 50° C. and the melted suet is then stirred in. There is then added to the soap and suet mixture, the "Lysol" or soft soap solution of cresols, in the proportion of approximately one tablespoon (½ fluid ounce) of the "Lysol" to each two gallons of solution.

When this preparation is ready for use one part thereof is mixed with three parts of hot water and while it is still hot or warm it is applied to the wall surface to be treated, by spraying, pouring or brushing it on.

The application of this solution to stuccoed or plastered exterior or interior walls, or any walls that have a stucco wash coat on them, will render the walls absolutely waterproof, by the action of the stucco upon the solution which forms an insoluble compound which will preserve the stucco by keeping the moisture out of the same. Since stucco is porous and absorbs moisture readily, the solution, which prevents this absorption, not only prevents the breaking down of the stucco through the action of the water thereon but also prevents the same from becoming dirty from dirty water which may come into contact therewith.

The preferred method of applying the solution is by spraying until the stucco is thoroughly saturated. The solution has no covering qualities and no coloring qualities. It is only necessary to saturate the stucco with it to obtain the desired results.

Premoulded Expansion Joint
U. S. Patent 1,911,139

Compositions for joints in masonry comprise: (a) gilsonite 5, blown bituminous matter 70, rubber (with solvent) 10, sulphur 5, and cottonseed pitch 10%, together with fillers if desired.

Waterproofing Surfaces of Building Walls
U. S. Patent 1,925,214

Asphalt of "semi-mastic type" is mixed with a solvent and with over 40 pounds asbestos fiber per 50 gallons of the asphalt; the mixture is heated and pressure is applied, and the mixture is sprayed at high velocity onto the surface to be waterproofed.

Plaster Composition
U. S. Patent 1,897,956

A finishing coat is composed of finely-ground plaster of Paris 20–40 (30), calcium hydroxide 10–30 (20), finely ground limestone 35–55 (47), ground asbestos 2–5 (3%) and a retarder.

Plaster
U. S. Patent 1,894,628

A plaster suitable for building construction and for wallboards consists of a mixture of Portland cement 675, ground stone 1090, and vegetable fiber (of the sugar-cane group) 235 pounds.

Sound Insulating Plaster
British Patent 402,810

A mixture of shredded vulcanized rubber (10–33%) and gypsum plaster or Portland cement (90–67%) is used as a sound-insulating plaster.

Porous Plaster
U. S. Patent 1,951,691

Anhydrite	650 parts
Aluminium Sulphate	28 parts
Portland Cement	9 parts
Potassium Sulphate	6 parts
Chalk	3 parts

The mix is gauged with 235 pounds of water and mixed for about two minutes, by which time it becomes substantially homogeneous. The mixture is then poured into a suitable mould, which comprises a number of compartments corresponding to the particular size of block required. If a finished block of about 12 inches height is desired the initial wet mixture should reach about 7 inches in height in the mould. Expansion sets in immediately and at the end of about 5 minutes, when the evolution of gas has practically ceased, the material in the mould has risen to a height of about 13 inches. After 1 hour the plaster is set and the block removed from the mould. The surfaces of the block can be trimmed or finished in any required manner. If the Portland cement is omitted the expanded mixture falls back about 2 or 3 inches when gas evolution ceases so that the mould is not filled and blocks of the required size are not obtained. The method used for plastering in situ is as follows: The dry solids in the quantities given above are carefully mixed together beforehand in the factory. On the job where the plaster is to be used it is put into a suitable mixer and mixed with 250 pounds of water. The mixing is carried on for 2 minutes until the mass is homogeneous, when it is poured into the space or mould which it is required to fill. It will be understood that this mould may take different forms according to the work in progress. For example, in the construction of a composite floor the mould would take the form of shuttering erected specially for the purpose. In the case of filling a cavity wall, the mould would, of course, be the space between the two halves of the wall. The process, however, is similar in all cases, the fluid mixture described above being poured out and allowed to expand. The volume after expansion will be approximately twice that before expansion.

Manufacture of Porous, Fireproof Insulating
British Patent 405,166

A mixture of organic fibrous material (3), powdered mineral filler (5–7.5), and sodium silicate solution, density 1.325–1.35 (5–7.5 parts), is pressed between perforated sheet-metal plates and dried.

Fibrous Compositions
British Patent 396,652

A composition for making building elements, e.g., bricks, slabs, etc., comprises at least 3 parts of a hydraulic binder and 1 part of sugar-free bagasse. In an example the binder comprises cement 18.75, lime 917 and trass 6 kilograms. This is added to 6 kilograms begasse and made into mortar by mixing with water.

Dust-Preventing Composition for Tennis Courts
Austrian Patent 135,143

The composition, which is heavier than water, comprises crude or waste oil mixed with about 2% of powdered asbestos and about 8% of a solution of bitumen in solvent naphtha.

Bituminous Road-Making Compositions
British Patent 395,288

The interstices between paving blocks are filled by a composition prepared by mixing sand with bitumen in a hot state. In an example the filling material comprises fine sand 100, bitumen 3–5 and calcium hydroxide 0.5–0.75 parts. The sand is perfectly dry and heated to 150–200° and mixed with liquid bitumen at the same temperature, the calcium hydroxide being added when the mixture is cold. The blocks preferably comprise calcium oxide sandstone or a mixture of calcium oxide and silica or silicic acid, molded and impregnated with bitumen, the proportion of bitumen being substantially higher than in the filling composition. The molded blocks are treated with steam under pressure for a short time to produce a thin skin of calcium hydrosilicate.

Road-Building Compositions
British Patent 387,184

A road pavement, etc., is formed by applying to a layer of rolled dry road metal, or of rolled tar macadam, a binding material in the form of mortar consisting of a stabilized bituminous emulsion and a fine filling material, excluding cement, lime, plaster of Paris or the like, such mortar being applied in such a fluid state as to penetrate between the road material and completely fill the voids. The mortar may consist of calcareous sand 650, calcareous filler (impalpable stone powder) 60, sodium oxalate (to stabilize the emulsion) 0.2 and common bituminous emulsion 100 parts. The road may be surfaced with a mortar consisting of calcareous sand 500, stone powder 40, sodium oxalate 0.2, gravel 370, water 60 and common bituminous emulsion 100 parts.

Inflexible Waterproof Composition
U. S. Patent 1,949,229

Mix 15 to 45 parts bituminous material, 15 to 40 parts clay, 15 to 50 parts sand, and fibre with water, shape the plastic mass so produced, dry the shaped mass and thereafter heat the material at a temperature between that of the melting point of the bituminous material and 800° F.

Glass Substitute
U. S. Patent 1,931,518

Wire or cloth netting is covered with a transparent film formed from:

Cellulose Acetate	100	parts
Tricresyl Phosphate	10–20	parts
Dibutyl Tartrate	10–20	parts

This is permeable to ultra-violet light.

Preservative for Wood
U. S. Patent 1,957,872

A preservative for wood comprises 40% of crystalline sulphate of iron, 40%

of sodium fluoride, 10% of arsenious acid and 10% of sodium bisulphite.

Wood Preservative
U. S. Patent 1,957,873

A water soluble preservative for wood fibre consists essentially of a water soluble salt of hydrofluoric acid to which a water soluble chromium salt is added as a fixative, the two being in about equal quantities so that more than 30% of the preservative will be fixed on the wood fibres and will not wash out.

Wood Preservative
Canadian Patent 335,328

A wood preservative consists of calcium fluoride 70, phenol 9, Scheele's green or arsenious acid 6, oxalic acid 6, oils (oil of butter, oil of tar, linseed oil, olive oil or creolin) 6, and sulphuric acid 3%. Twenty pounds of this composition is mixed with 50 gallons of hot water with or without 1% sodium sulphate heated to at least 70° F. and not more than 120° F. Wood treated with this solution will not crack, warp or dry rot, the life is increased, the wood strengthened and fungus growth stopped.

Wood Preservative
Belgian Patent 393,291

Five per cent crude phenol is mixed with 95% carbon tetrachloride.

Wood Preservative
U. S. Patent 1,890,650

Rosin	96 parts
Wax	64 parts
Creosote	32 parts
Quicklime	10 parts
Castor Oil	1 part

Melt together and impregnate wood hot. This will set hard and not sweat out.

Preservation of Wood
U. S. Patent 1,905,327

The wood is impregnated with a solution containing 15% of sodium silicate and 1% of sodium arsenate and then treated with acid (e.g., 2% sulphuric acid).

Impregnating Wood for Interiors of Buildings
U. S. Patent 1,939,186

Wood such as that for doors or flooring, etc., is impregnated with a mixture comprising fish oil 22, turpentine 22, mineral lubricating oil 22, mineral turpentine substitute 22, oleic acid 4, calcium chloride 2 and scenting material such as oil of citronella about 2 parts, and may then be coated with a composition which dries and hardens to a transparent coating, such as varnish or lacquer.

Wood Figure Casting Mold

Tar Oil	3 oz.
Orange Shellac	6 lb.
Soapstone	4 lb.
Emery Flour	4 lb.
French Chalk	4 oz.

Melt the shellac and add the tar oil. Add the soapstone and mix thoroughly. Mix separately the dry emery powder and chalk; then pour this into the previous melt, stirring thoroughly and vigorously. Place the master carving or pattern flat side down in a box, then pour the casting mixture over same. When cooled, the result will be a mold into which can be cast the material for molding the composition ornaments. A typical composition is glue, rosin, linseed oil, glycerin, and whiting to form a putty mass.

Wood Veneers
British Patent 388,593

In the preparation of wood veneers for the manufacture of wallpapers the veneer is divided into widths and steeped in a bath consisting substantially of cellulose acetate 15, 14% chrome alum solution 10, formaldehyde 5 and water 70 parts, dried and steeped in a bath of 25% glycerol 30, gelatin 35 and water 45 parts, dried and seasoned. They may then be dyed and further dried and seasoned.

Fireproofing Wood Fibre Board
U. S. Patent 1,942,977

The first stage in the treatment consists in immersing the "boards" in a bath containing an aqueous solution containing about 30 to 35% by weight of mono-ammonium phosphate, a small proportion of di-ammonium phosphate and a trace of boric acid. The boards are generally immersed in this solution for a period of from ten to twenty minutes. The time of immersion depends upon the thickness of the board, the porosity of the board and the absorbent properties of the cell structure of the board.

In the second stage of the process the boards are removed from the above-mentioned bath and transferred to a bath containing 15 to 20% of the soluble acid phosphate of calcium. The acid phosphate of calcium reacts with the di-ammonium phosphate to precipitate the insoluble calcium phosphate within the cell structure of the material.

The precipitated calcium phosphate confers a white color on the material, which remains impregnated with the mono-ammonium phosphate and is thereby rendered fireproof.

As a final operation the boards are immersed in a bath containing a solution of sodium aluminate and/or sodium silicate with or without addition of colloids such as carragheen or viscose.

Alternatively the second stage of the process may consist in immersing the boards in a solution containing acid phosphate of zinc together with acid phosphate of magnesium.

This invention when applied to the treatment of jute, for example in the form of sacking, may be carried out by the process described above for the impregnation of "boards."

A solution of acid phosphate of zinc suitable for use in the present process may be prepared for example by mixing 8 parts by weight of zinc oxide with 26 parts by volume (39 parts by weight) of phosphoric acid, H_3PO_4 (s. g. 1.500) and adding sufficient water to bring the volume up to 100.

Fire-Resisting Fibre Board
U. S. Patent 1,910,469

Wood pulp is saturated with a 10% sodium hydroxide solution (1) at 49–65° and mixed in the beater with mineral wool to the extent of 50–90% of the total, maintaining the same concentration of (1). Boards or blocks may be made from this stock.

Extruded Wood
Formula No. 1

Wood Flour	45 parts
Casein	20 parts
Copper Hydroxy Cellulose	10 parts
Calcium Stearate	3 parts
Infusorial Earth (Fine)	2 parts
Kaolin or Asbestos Powder	10 parts

Formula No. 2

Wood Flour	50 parts
Casein	18 parts
Copper Hydroxy Cellulose	20 parts
Jute Fibres	2 parts

Formula No. 3

Wood Flour	60 parts
Casein	20 parts
Copper Hydroxy Cellulose	20 parts

Formula No. 4

Wood Flour	60 parts
Zinc Chloride Cellulose	20 parts
Casein	20 parts

Solutions for Dissolving Cellulose

Formula No. 1

| Zinc Chloride Solution | 40–50% |

Use this solution at a temperature of 80 to 100° C.

Formula No. 2

| Zinc Chloride | 1 part |
| Hydrochloric Acid (35%) | 2 parts |

Formula No. 3

| Copper (in the form $CuO \cdot nH_2O$) | 2½–3½% |
| Ammonia (in the form of Ammonium Hydroxide) | 15% |

Fire-Extinguishing Apparatus
British Patent 397,380

A container for compressed carbon dioxide is sealed by a wire for a low-melting alloy which is soldered into a bore of a copper ferrule screwed into an opening in the steel body of the container. Alloys used are (1) cadmium 3 parts, tin 4 parts, lead 8 parts, bismuth (melting point 68°) 5 parts; (2) tin 3 parts, lead 8 parts, bismuth 9 parts (melting point 79°); (3) tin 3 parts, lead 5 parts, bismuth (melting point 94.5°) 8 parts; (4) cadmium 3 parts, tin 8 parts, lead 8 parts, bismuth (melting point 70°) 15 parts.

Fire-Extinguisher Composition
U. S. Patent 1,910,653

An aqueous solution of potassium carbonate is mixed with less than 10% of a polyhydric alcohol, for example, ethylene glycol (I), to depress the freezing point, and, if desired, with less than 3% of sodium dichromate as anticorrosive. For transport (I) may be absorbed into the solid materials to form a dry powder.

Foam Fire-Fighting Methods

For chemical foam the first solution is water containing about 13% aluminum sulphate, the second solution contains about 8% sodium bicarbonate and a stabilizer like saponin, licorice or Turkey-red oil to strengthen the bubble walls. It is held that in this system most of the chemicals are wasted in creating carbon dioxide and are themselves without fire-extinguishing value; 100 pounds of powder with 100 gallons of foam, and a saving made of 90% of the cost; the product is more stable and less dangerous to the fabric of the hose.

METALS AND ALLOYS

Removal of Beer Stains from Aluminum

A hot 5% solution of a mixture of 95 parts tartaric acid and 5 parts sodium fluoride is recommended for severe cases of deposit. Several gallons of this hot solution poured into a container and agitated for 5–10 minutes will loosen all deposited matter and remove all stains. The container should then be thoroughly flushed with hot water.

Coating Aluminum
Canadian Patent 337,120

Aluminum to be coated is immersed, without the application of externally applied electrical energy, in a solution containing sodium silicate 2.5 to 12.0%, sodium hydroxide 2.0 to 6.0% and ammonia 1.5 to 2.5% by weight. An adherent insulating coating is produced.

Aluminum Foundry Flux
U. S. Patent 1,950,967

A flux for use in the preparation and founding of aluminum and alloys thereof, consists of potassium chloride from 10 to 65%, sodium chloride from 15 to 75%, and cryolite from 5 to 65%.

Welding Rod for Aluminum Alloys
U. S. Patent 1,913,394

The welding rod is composed of aluminum 92.5, copper 4–5, silicon 1.20, zinc 0.25, iron 1.20, magnesium 0.25%, manganese, trace.

Woods Metal Heating Bath

Bismuth	4 parts
Lead	2 parts
Tin	1 part
Cadmium	1 part

Melting point 141° F.

When using a Woods metal bath in heating glass distilling apparatus for distilling organic liquids in the laboratory, the Woods metal may be prevented from sticking to the glass by passing a luminous flame over the bottom and sides of the flask before immersing it into the molten metal.

Electric Arc-Welding Process for Welding Aluminum or Aluminum Alloys
British Patent 401,487

Three-phase current is used, the work being connected to one terminal and two graphite electrodes to the others; during welding the work is covered with a flux composed of potassium fluoride 30, sodium-aluminum fluoride 30, and Al_2O_3, $2SiO_2$, $2H_2O$, 40%.

Aluminum Cutting Oil

For use on hand or automatic screw machines.

Kerosene or ''Mineral Seal'' Oil	50–75%
No. 1 Lard Oil	50–25%

Protection of Aluminum Against Corrosion

The best protection in moist atmosphere is obtained by anodic treatment in an aqueous bath containing sodium carbonate decahydrate 125, sodium chromate 8, concentrated ammonium hydroxide 25 cubic centimeters per liter at 50–60° for 10 minutes.

Corrosion Resistant Aluminum Alloy
U. S. Patent 1,932,795

Copper	.05– .5 %
Chromium	.75–1.5 %
Cobalt	.25– .5 %
Silicon	.75–3 %
Magnesium	.5–1.25%

Balance commercial aluminum.

Treatment of Non-Ferrous (Aluminum) Alloys to Protect them Against Corrosion
British Patent 396,746

Aluminum alloys with less than 80% aluminum are immersed in a hot (e.g. 130°) 6% solution of phosphoric acid in ethylene glycol, trimethylene glycol, or glycerol and the coated alloy is heated at 100–200°.

Hardening of Aluminum
U. S. Patent 1,930,463

The method of increasing the hardness of an aluminum article at or near the surface, comprises packing the aluminum article in close and intimate contact with comminuted magnesium metal and heating the packed article at temperatures above about 250° C. and below about 460° C.

Heat Treatment of Aluminum-Magnesium Alloys
U. S. Patent 1,945,737

The method of treating an aluminum base alloy containing from about 5 to 20% magnesium to improve the physical properties thereof, comprises heating the alloy for about 5 to 20 hours at a temperature from about 250° to 450° C., and cooling the alloy rapidly enough to preserve the benefits of the heating.

Working Aluminum-Magnesium Alloys
U. S. Patent 1,926,057

Aluminum base alloys containing magnesium about 5–15% are preheated to above about 288° but below the temperature of incipient fusion, cooled to a working range which is below about 315° and is also below the preheating temperature but not lower than about 245°, and working is effected within such temperature range.

Aluminum-Base Alloy
Canadian Patent 337,970

An aluminum alloy contains manganese 0.75–3.0, magnesium 0.2–0.5%. The addition of the magnesium inhibits grain growth.

Soldering Aluminum
French Patent 757,901

A product for soldering aluminum contains borax 0.1–0.5, boric acid 0.1–0.4, resin 0.3–0.8, tin 15–22, pure aluminum 20–28 and fused aluminum 50–65%.

Solder for Aluminum
French Patent 758,394

The solder contains zinc 8, tin 1, copper 0.5, aluminum 1.0, and bismuth 0.025 kilogram.

Aluminum Solder

Zinc	24 parts
Tin	12 parts
Mercury	4 parts
Aluminum	1 part

Aluminum Solder
U. S. Patent 1,927,052

A solder is formed of zinc 24 parts, tin 12 parts, mercury 4 parts and aluminum 1 part.

Solder for Aluminum and Its Alloys
U. S. Patent 1,926,853

Tin 66–69, zinc 27.5–28.5, and aluminum 2.5–6.5% are used together.

U. S. Patent 1,926,854

A solder for a similar use comprises tin 47.5%, zinc 47.5–49% and aluminum 2.5–5%.

U. S. Patent 1,926,855

This solder comprises tin 37–45%, lead 37–45%, zinc 9–21%, and aluminum 1–5%.

Swedish Iron Finish on Aluminum

Probably the best way to produce an imitation Swedish iron finish on aluminum is to plate the work in a black nickel solution and then relieve the highlights and lacquer the work.

Formula for black nickel solution:

Double Nickel Salts	8 oz.
Sodium Sulphocyanide	2 oz.
Zinc Sulphate	1 oz.
Water	1 gal.

Use the solution at room temperature with ¾ to 1 volt.

Surface Treatment for Aluminum and Its Alloys

A high resistance to corrosion of pure aluminum and to copper-free alloys can be attained by a treatment described below.

The process consists in placing the aluminum parts to be protected in an aqueous solution of anhydrous sodium carbonate and anhydrous sodium chromate at a temperature of from 90° to 110° C.; typical proportions of the salts are 5 per cent and 1.5 per cent respectively. The solution is maintained at boiling point for from three to five minutes; the aluminum is then washed in hot water and dried. Tests have shown that about 17 square yards of surface can be protected per gallon of solution. The plant may also be operated at lower temperatures with a somewhat different composition of the solution. It is claimed that the M.B.V. film cannot be detached by rolling or bending, and allows the use of these metals for a large range of chemical material.

The subject is still being studied, but the indications are that it will open up a new field for the use of aluminum and its alloys.

Aluminum Base Alloys
U. S. Patent 1,947,121

An aluminum base alloy having a low thermal coefficient of expansion comprises a predominant amount of aluminum 7% to 20% silicon, .2% to 6% magnesium, and from .1% to 5% manganese, said alloy being comparatively free from other elements having a higher specific gravity than aluminum.

Aluminum Alloys
French Patent 750,703

Alloys for pistons, etc., contain aluminum 90–2%, copper 4–5%, magnesium 1–2%, nickel 0.5–0.7%, manganese 0.06–0.1%, titanium 0.05–0.1%, chromium 0.1–0.3%, and tungsten 0.5–1%.

Aluminum Alloys
U. S. Patent 1,921,089

In tempering age-hardenable aluminum alloys containing magnesium in a proportion not exceeding 2% and which may contain copper 4.2%, in order to increase the elongation while preserving a high yield point the alloy is annealed for 20–40 hours at about 100–125° after the alloy has been heated to the high temperature, quenched, completely aged at room temperature and cold-worked.

Aluminum Alloys

These alloys are characterized by high physical and tensile strength at elevated temperatures.

U. S. Patent 1,932,837

About 4.0 per cent by weight of magnesium, about 0.5 per cent by weight of manganese, and about 0.2 per cent by weight of calcium, the balance being aluminum.

U. S. Patent 1,932,838

1.0 to 15.0 per cent by weight of magnesium and 0.2 to 3.5 per cent by weight of cobalt, the balance being aluminum.

U. S. Patent 1,932,839

3.0 to 7.5 per cent by weight of magnesium, and 0.2 to 2.0 per cent by weight of nickel, the balance being aluminum.

U. S. Patent 1,932,840

3.0 to 8.0 per cent by weight of magnesium, 0.5 to 4.0 per cent by weight of manganese and 0.5 to 4.0 per cent by weight of nickel, the balance being aluminum.

U. S. Patent 1,932,841

3.0 to 8.0 per cent by weight of magnesium, 0.5 to 4.0 per cent by weight of manganese, 0.5 to 4.0 per cent by weight of nickel, and 0.1 to 2.0 per cent by weight of cobalt, the balance being aluminum.

U. S. Patent 1,932,842

3.0 to 8.0 per cent by weight of magnesium, 0.5 to 4.0 per cent by weight of nickel, 0.5 to 4.0 per cent by weight of manganese, 0.1 to 2.0 per cent by weight of chromium, the balance being aluminum.

U. S. Patent 1,932,843

3.0 to 8.0 per cent by weight of magnesium, 0.5 to 4.0 per cent by weight of nickel, 0.5 to 4.0 per cent by weight of manganese, and 0.1 to 0.75 per cent by weight of at least one of the class of metals titanium, xirconium, vanadium, molybdenum and tungsten, the balance being aluminum.

U. S. Patent 1,932,844

3.0 to 8.0 per cent by weight of magnesium, 0.5 to 4.0 per cent by weight of manganese, 0.5 to 4.0 per cent by weight of nickel, and 0.5 to 4.0 per cent by weight of copper, the balance being substantially aluminum.

U. S. Patent 1,932,845

3.0 to 7.5 per cent by weight of magnesium, 0.2 to 2.0 per cent by weight of nickel, and 0.05 to 0.4 per cent by weight of at least one of the class of elements antimony and bismuth, the total amount of the antimony and/or bismuth being not greater than 0.4 per cent by weight, the balance of the alloy being aluminum.

U. S. Patent 1,932,852

3.0 to 8.0 per cent by weight of magnesium, 0.5 to 4.0 by weight of nickel, 0.1 to 3.0 per cent by weight of cobalt, and 0.05 to 0.4 per cent by weight, of at least one of the class of metals antimony and bismuth, the total amount of the antimony and/bismuth being not greater than 0.4 per cent by weight, the balance being aluminum.

U. S. Patent 1,932,854

2.0 to 10.0 per cent by weight of magnesium, 0.2 to 5.0 per cent by weight of nickel, and 0.05 to 2.0 per cent by weight of calcium, the balance being aluminum.

U. S. Patent 1,932,855

3.0 to 7.5 per cent by weight of magnesium, 0.2 to 2.0 per cent by weight of nickel, 0.05 to 0.4 per cent by weight of at least one of a class of elements composed of antimony and bismuth, and 0.05 to 2.0 per cent by weight of calcium, the balance being aluminum.

U. S. Patent 1,932,856

3.0 to 8.0 per cent by weight of magnesium, 0.5 to 4.0 per cent by weight of nickel, 0.5 to 4.0 per cent by weight of manganese, and 0.05 to 2.0 per cent by weight of calcium, the balance being aluminum.

U. S. Patent 1,932,857

3.0 to 8.0 per cent by weight of magnesium, 0.5 to 4.0 per cent by weight of nickel, 0.5 to 4.0 per cent by weight of manganese, 0.05 to 2.0 per cent by weight of calcium, and 0.05 to 0.4 per cent by weight of at least one of a class of elements composed of antimony and bismuth, the balance being aluminum.

U. S. Patent 1,932,858

2.0 to 15.0 per cent by weight of magnesium, 0.05 to 0.4 per cent by weight of a class of elements composed of antimony and bismuth, and 0.05 to 2.0 per cent by weight of calcium, the balance being aluminum.

U. S. Patent 1,932,859

2.0 to 10.0 per cent by weight of magnesium, 0.2 to 5.0 per cent by weight of nickel, 0.5 to 3.5 per cent by weight of chromium, and 0.05 to 2.0 per cent by weight of calcium, the balance being aluminum.

U. S. Patent 1,932,860

3.0 to 8.0 per cent by weight of magnesium, 0.5 to 3.5 per cent by weight of nickel, 0.5 to 3.5 per cent by weight of chromium, 0.05 to 0.4 per cent by weight of at least one of a class of elements composed of antimony and bismuth, and 0.05 to 2.0 per cent by weight of calcium, the balance being aluminum.

U. S. Patent 1,932,861

2.0 to 10.0 per cent by weight of magnesium, 0.2 to 5.0 per cent by weight of nickel, 1.0 to 6.0 per cent by weight of copper, and 0.05 to 2.0 per cent by weight of calcium, the balance being aluminum.

U. S. Patent 1,932,862

3.0 to 8.0 per cent by weight of magnesium, 0.5 to 5.0 per cent by weight of nickel, 1.0 to 6.0 per cent by weight of copper, 0.05 to 0.4 per cent by weight of at least one of a class of elements composed of antimony and bismuth, and 0.05 to 2.0 per cent by weight of calcium, the balance being aluminum.

U. S. Patent 1,932,866

3.0 to 8.0 per cent by weight of magnesium, 0.5 to 3.5 per cent by weight of chromium, 1.0 to 6.0 per cent by weight of copper and 0.05 to 2.0 per cent by weight of calcium, the balance being aluminum.

U. S. Patent 1,932,872

2.0 to 9.0 per cent by weight of magnesium, about 0.05 to 2.0 per cent by weight of calcium, about 0.1 to 4.0 per cent by weight of manganese, and about 0.05 to 0.4 per cent by weight of at least one of the class of metals composed of antimony and bismuth, the balance being aluminum.

U. S. Patent 1,932,873

1.0 to 15.0 per cent by weight of magnesium, 0.2 to 3.5 per cent by weight of cobalt, and 0.05 to 0.4 per cent by weight of at least one of the class of metals antimony and bismuth, the total of the antimony and/or bismuth being not greater than 0.4 per cent by weight, the balance being aluminum.

Aluminum-Barium Alloys

Aluminum is melted with an electric current in a Pythagoras crucible surrounded with magnesium oxide, and when the temperature has reached 900°, small pieces of barium are immersed in the melt with an iron needle. The whole is then stirred with a rod of magnesium oxide until the barium has dissolved. The alloy obtained in this matter is highly contaminated with oxides and nitrides. By using a surface layer of barium chloride and potassium chloride it was possible to prepare an alloy containing 7% barium without contamination. As a better method barium oxide was added to molten aluminum at 1000°–1100°, the reduced barium being taken up by the excess aluminum to form an alloy. Barium chloride or a mixture of 83% barium chloride and 17% barium fluoride was used as a flux to dissolve out aluminum oxide and undecomposed barium oxide. Alloys containing as much as 35% barium were obtained by this method.

Aluminum-Beryllium Alloy
U. S. Patent 1,952,048

An aluminum base alloy contains from about 0.025% to 1.0% beryllium, about 0.1% to 1.0% silicon, about 0.1% to 0.5% magnesium, and characterized by high hardness and beneficial age hardening properties.

Aluminum-Beryllium Alloy
U. S. Patent 1,952,049

An aluminum base alloy containing from about 0.025% to 1.0% beryllium, about 0.025% to 1.0% beryllium, about 1.0% to 10.0% magnesium, and characterized by high hardness and beneficial age hardening properties.

Aluminum Silicon Alloy
Canadian Patent 337,971

A comminuted mixture of aluminum 10–80% and silicon 20–90% is subjected to pressure to form a dense, homogeneous mass which is heat-treated above 300° and below the temperature at which the most fusible constituent of the alloy will melt. The addition of magnesium and beryllium decreases the specific gravity without impairing the low thermal expansivity of the high compressive strength of the original mixture. Additions of iron, manganese and the like modify the properties of the compressed product for a variety of uses.

Hardening Copper-Zinc Alloys
British Patent 399,177

Copper-zinc alloys, containing zinc up to 37, nickel at least 2 and aluminum at least 0.5%, are heated to about 900°, rapidly cooled and reheated to about 300–500°. Cold-working may be interposed between the cooling and the reheating. The alloys preferably contain 1–3% aluminum and at least 3 times as much nickel.

Aluminum-Silicon Alloy
U. S. Patent 1,934,281

An aluminum base casting alloy consisting of silicon between about 3 and 15 per cent, titanium between about 0.1 and 0.5 per cent, and the balance aluminum, characterized, in the cast condition, by substantial freedom from surface shrinks or draws.

Aluminum-Silicon Alloy
U. S. Patent 1,921,195

An alloy which has good casting properties comprises aluminum together with silicon 4–25, and zirconium 0.1–2.0%.

Bright Dipping of Brass
Solution No. 1

Oil of Vitriol, 66°	1 gal.
Muriatic Acid, 42°	2 gal.

Solution No. 2

Oil of Vitriol, 66°	2 gal.
Muriatic Acid, 42°	1 gal.
Muriatic Acid 24°	1 pint

Cover with soot and on the next day add carefully ½ gallon of water, then it is ready for use.

After removing all grease in alkaline cleaners, the brass articles are first dipped in Solution No. 1 for a few seconds, rinsed in cold water and then dipped in Solution No. 2 which gives the required brightness. Upon removal from Solution No. 2, rinse immediately in cold water.

Bright Dip for Brass Castings

Nitric Acid (concentrated) 80%
Sulphuric Acid (concentrated) 20%

These two acids mixed together and used as a dip followed immediately by a water rinse proves very satisfactory for a bright dip on brass castings.

Black Finish on Brass Plated Ware

A black finish may be produced on brass or brass plated articles, providing the brass deposit is heavy enough, in the following solution:

White Arsenic	8 oz.
Yellow Antimony Sulphide	1/8 oz.
Sodium Cyanide	8 oz.
Water	1 gal.

Use at boiling temperature.

Brass Cutting Oil

For use on hand or automatic screw machines. An uncompounded pale mineral oil having an approximate viscosity of 100 seconds at 100° F. (Saybolt).

Oxidized Brass and Bronze Finish

The casting is bright dipped, then polished and buffed. It is cleaned in an alkaline cleaning solution, and then oxidized in the following solution:

White Arsenic	1 lb.
Caustic Soda	1 lb.
Sodium Cyanide	1 oz.
Water	1 gal.

Operate solution at 80° F., with 1½ to 2 volts. Use steel anodes.

After the work is oxidized, it is relieved on a rag wheel using pumice and water, then dried and lacquered.

Coloring Bronze Brown

To produce a brown color on the alloy, prepare a solution of 1 ounce of liquid sulfur and 1 ounce of yellow barium sulphide to 1 gallon of water. Use at 100° to 120° F., and immerse the work for 15 to 30 seconds. If the color is not dark enough, increase the proportions of chemicals, the temperature, or the length of time the work is left in the solution.

After the desired shade is produced, scratch brush dry, using a fine crimped brass wire wheel operated at 800 to 1000 r.p.m.

In cleaning the work after polishing, if a smooth finish is desired clean in an alkaline cleaning solution before using the bronzing solution. If a dull finish is desired, sandblast the work and then clean in the alkaline solution before using the bronzing solution.

Green Finish for Bronze

A dark brown or black finish can be produced by heating with a torch and applying a sulphur solution while the work is hot.

The sulphur solution is made by using one or two ounces of liquid sulphur to one gallon of water. After the bronze or black color is produced, the work can be painted with the following formula for a verde finish:

Copper Nitrate	8 oz.
Ammonium Chloride	4 oz.
Acetic Acid	4 oz.
Chromic Acid	1 oz.
Water	1 gal.

Allow the work to dry thoroughly, and if the verde is uneven, paint lightly a second time. Waxing will help to preserve the finish.

"Bright Dip" for Phosphor Bronze or Beryllium Copper

Sulphuric Acid	2 gal.
Nitric Acid	1 gal.
Muriatic Acid	1 oz.
Water	1 quart

Black Finishing Chromuim-Plated Metal

U. S. Patent 1,937,629

Chromium plated articles are treated in a molten salt bath containing an alkali metal or alkaline earth metal cyanide, e.g., in a bath formed from sodium cyanide 45, sodium carbonate 35, and sodium chloride 20%.

(Chromium Iron-Nickel Wire) Filament

U. S. Patent 1,924,543

A radio-tube filament consisting of a chromium 15, iron 20, nickel 65%, alloy coated with barium oxide is made.

Coating Copper

British Patent 393,039

Copper (alloy) is coated with green patina of basic copper sulfate by anodic treatment in an electrolyte containing a soluble sulfate and an oxidizing agent. A suitable electrolyte contains magnesium sulphate, magnesium hydroxide 2 and potassium bromate 2%. A temperature of 95° and current density of 5 volts are suitable. Cathodes of stainless steel or carbon may be used.

Royal Copper Finish

Royal copper finish can only be produced upon copper or copper plated work. Brass work to receive this finish should be heavily copper plated, preferably in an acid copper solution, and then copper colored. Then it is cleaned and plated in a lead solution for a few minutes to obtain a thin uniform deposit of lead on the surface. Dry this and place in a molten sodium nitrate bath for a few minutes to produce an oxide of copper. When taken from the sodium nitrate bath, let the work cool, then wash in hot water and dry in hardwood sawdust; and then color lightly on a muslin buff, using a lime composition. Formula for lead solution:

Lead Carbonate	2 oz.
Caustic Soda	6 oz.
Water	1 gal.

Use lead anodes; temperature, 175° F.; plate with 2 to 4 volts.

Black Finish on Copper

Dissolve 1 ounce of copper nitrate in 4 ounces of water and apply an even coat to the surface to be blackened. Then heat over a gas flame to produce the black color, and if color is not uniform repeat the oxidizing and heating operation. A light scratch brush operation will help produce an even finish.

Coloring Copper or Brass
Canadian Patent 334,996

An adherent green patina on copper or its alloys is produced by wetting with a 2½–15% water solution of ammonium sulphate. Dry with a hot air or oxygen blower. After a suitable depth of color is produced, blow steam on it or immerse in hot water to develop the desired shade of green.

Coloring Copper Alloys Gray

The rate of coloring increases with temperature but 45° C. should not be exceeded for a steel-gray color. Chemically pure hydrogen chloride can be used. Coloring is accelerated by repeated rinsing and drying of the ware; the gray tone is obtained more easily on mat surfaces than on polished. The pickling rate can be increased 50% by using 100 grams of hydrogen chloride,

21 grams of ferrous sulphate heptahydrate, 14 grams of arsenic trioxide at room temperature. A solution of 11.0 grams of ferrous sulphate heptahydrate and 10.3 grams of arsenic trioxide in 100 grams of hydrogen chloride reduces coloring time from 25 minutes to 10 seconds; it produces a gray tone on zinc sheet, brass, tombac, red brass, cast bronze, rolled bronze and copper sheet.

Tarnish-Proofing Copper
U. S. Patent 1,942,923

Copper is dipped in a 2% solution of sodium azide for 25 minutes. After washing it is dipped in a 1% solution of potassium ferrocyanide containing 0.1% of acetic acid for 30 seconds.

Copper Solder
German Patent 581,748

A solder for sintered or fused carbide, nitride, silicide and boride alloys consists of copper with 0.1–10% aluminum or manganese.

Antique Finish on Copper

Suspend thoroughly cleaned article in closed container having strong ammonia water at bottom (but not touching article) for several hours. Remove and polish.

Imitation Gold Alloy

The following alloy, with minor modifications sometimes, is the color of pure gold and has satisfactory physical properties for working:

Copper	90–93%
Aluminum	7–10%

The minor modifications sometimes consist of the addition of a very small (less than 1%) amount of one or more of the following: nickel, silver, iron.

Hardening Copper
U. S. Patent 1,943,738

The copper to be treated is preferably reduced to small pieces to facilitate melting and is then placed in a crucible and heated to its melting temperature of 1083° C. As soon as the copper begins to melt a mixture of pulverized copper sulphate (blue stone), sodium chloride and borax is added. While these latter ingredients could be readily mixed with the comminuted copper prior to placing the same in the crucible, it is preferable to add such ingredients after the copper begins to melt thereby avoiding any deleterious effect on the ingredients prior to their entrance into the molten mass.

A typical formula is as follows:

Copper	1 lb.
Copper Sulphate	2 oz.
Salt	2 oz.
Borax	2 oz.

As previously stated, the copper is reduced to small pieces and brought to a melting temperature. The copper sulphate is previously reduced to a comminuted form and mixed with the proper proportions of salt and borax. This mixture is added to the copper as soon as the latter begins to melt, and when thoroughly incorporated, the material is poured into molds of the proper design for casting bearings or other articles desired to be manufactured of hardened copper, and allowed to cool.

Copper treated in this manner is extremely hard and durable and is useful in the manufacture of many articles heretofore made of steel.

Brass or Copper Tinning

Brass or copper may be tinned by boiling with tin filings in a solution of caustic alkali or cream of tartar.

Copper Alloy
British Patent 397,697

Containers for liquid oxygen capable of withstanding 100 pounds per square inch and moderate shock and vibration at less than 180° are made of an alloy containing copper 92.5–96.5 (94.5), silicon 3–6.5 (4.5), and manganese 0.5–1.5.

Copper Alloy
Japanese Patent 98,986

A hard copper alloy contains zinc 0.1–20, nickel 5–7 and tin 6–10%. It is quickly cooled below solidification temperature and then heated at 250–600°.

Copper Alloys
U. S. Patent 1,938,172

Alloys containing no nickel or at most substantially less than 1% nickel are formed of copper 75–95, aluminum 0.5–4.0, tin 0.25–2.0%, the balance being mainly zinc. These alloys are resistant to corrosion by saline waters.

Copper Alloys
French Patent 753,562

Hard alloys of copper are prepared by heating together iron, aluminum and copper to a temperature at which an exothermic action is produced, whereby the temperature of the molten metal is notably increased. An example contains copper 80, iron 1, aluminum 2, zinc 13 and tin 4%.

Copper-Zinc Alloys
French Patent 750,947

A white alloy of high resistance contains approximately copper 47, zinc 38, nickel 12.5, manganese 1.5, lead 0.6, and aluminum 0.4%.

Copper Alloys Containing Iron, Silicon and Zinc
U. S. Patent 1,924,581

Alloys which may be drawn, rolled or extruded contain copper 88–93 together with iron 0.1–0.6 and silicon 0.4–5%, the balance being approximately all zinc, sufficient silicon being used to cause substantially all the iron to be combined with silicon to form iron silicide, and the zinc in no case exceeding about 10% (the iron silicide being dissolved in a base of copper, zinc, and silicon).

Alloy of Copper and Lithium
U. S. Patent 1,923,955

An alloy which has improved electrical conductivity is formed of copper together with lithium 0.002–0103%.

Thermal Treatment of Copper Alloys
German Patent 580,304

Copper alloys containing chromium 0.2–3, with or without aluminum 1–10%, are heated to above 700°, chilled, and then kept for some time at 400–700°. The hardness of the alloys is increased.

Copper Alloy for Chill and Die Casting
U. S. Patent 1,954,003

A chill or die casting of an alloy consisting of about 65% and up to 94% copper, from 2% to 6% silicon, from 3% to 28% zinc, and an appreciable amount of aluminum not more than 2%.

Dental Alloy
French Patent 755,827

An alloy for metal bases of dental plates contains tin 24, antimony 4, copper 1 and lead 20 parts. Its melting point is 330°.

Dental Amalgam Alloy

Silver	67 parts
Tin	30 parts
Zinc	2 parts

Alloy for Surgical Needles
U. S. Patent 1,942,150

Surgical sewing needles, dental instruments and the like which require considerable stiffness and elastic hardness, are made from an alloy containing iron up to 25%, beryllium 0.3 to 3%, chromium 10 to 20%, molybdenum 2 to 5%, tungsten 3 to 8%, and cobalt 1 to 5%, the remainder being nickel.

Reclaiming Dissolved Gold

To reclaim the gold from the solution add ¼ ounce of No. 80 mesh metallic zinc to each gallon of solution and let it stand for a day or two, giving it an occasional stirring. The zinc will precipitate the gold from the solution.

Syphon out or decant the solution, and wash the zinc with water. Then dissolve the zinc with dilute nitric acid (1 part of acid, 1 part of water). A little heat will hasten the action. After the zinc is all dissolved, wash thoroughly with water, and dry the fine gold powder. The gold can then be dissolved in aqua regia, precipitated with ammonia, washed, and used to make the new gold solution.

Alloys of Gold, Palladium, Silver, Copper and Zinc
U. S. Patent 1,924,097

Alloys which are suitable for dental work are formed of gold 30–40, palladium 35–50, silver 10–23, copper 4–20 and zinc 2–6 parts.

Gold Alloys
German Patent 584,549

An alloy consists of 5–25% of metals of the platinum group, 0.05–5% of metals of the iron group and the rest gold. Alternatively, the gold may be replaced by a mixture of gold and silver containing at least 60% gold. Also copper up to 5% may be added. The alloy is heated just below its metling point (900°), chilled, and tempered at 500–550°.

Coating Iron with Aluminum
U. S. Patent 1,941,750

For this purpose the objects are immersed into a melted flux bath containing calcium chloride, zinc chloride, an alkali chloride and an alkali aluminum fluoride before they are transmitted to the metal bath after having first been mechanically cleaned to remove fatty substances, oxides and slag, or having been subjected to mordanting in diluted acid and, if necessary, having also been subjected to another treatment for volatilizing superficial impurities. The proportions of the substances forming the melted flux bath may lie within the following limits (stated in parts by weight):

Anhydrous Calcium Chloride	2–50 parts
Anhydrous Potassium Chloride	1–30 parts
Anhydrous Sodium Chloride	½–20 parts
Anhydrous Sodium Aluminum Fluoride	¹⁄₁₀–5 parts
Anhydrous Zinc Chloride	½–30 parts

While the objects are immersed in the melted flux bath they may be kept in rotation or stirred in another manner or they may be subjected to the action of brushes or other mechanical means for facilitating the removing of the dust particles.

The melted flux bath may have such a temperature that the objects contained therein will be heated so much that they can be transmitted to the molten bath of aluminum or aluminum alloy without their temperature sinking beneath the melting-point of the said bath.

The treatment for volatilizing impurities which may precede the immersing into the melted flux bath consists preferably in that the objects are dipped into a melt of ammonium chloride and zinc chloride or they are immersed into an aqueous solution of the double salt, and then they are dried by heating. By the heating the chloride is transformed, with oxidic impurities on the surfaces of the objects, into impurities which volatilize, and the objects will be coated by an anhydrous layer of salt which preserves against oxidation.

Welding Rod
Canadian Patent 340,263

A welding rod contains cobalt 40–65, chromium 25–35, tungsten 17, carbon

9-2.75, manganese not more than 0.8 and silicon 0.3-2.0%. The proportion of silicon to manganese is at least 1 to 1.25.

Flux (For Iron Welding Rods)
U. S. Patent 1,922,692

Iron rods for arc-welding are coated with a 1: 1: 1: 1 mixture of magnesium sulphate, potassium nitrate, calcium carbonate, and magnesium carbonate.

Short Cycle Annealing for Malleable Iron
U. S. Patent 1,902,475

White cast iron is heated at 900-1000°/12 hours to cause cementite to go into solution, cooled at 720° during 6-11 hours, and reheated to and held at 730° for a few hours. A number of quick coolings and heatings between 730° and 720° may be effected.

Malleabilizing White-Iron Castings
U. S. Patent 1,925,855

White-iron castings having a combined carbon and silicon content of less than 3.2% are heated to above 750° to effect solution of the cementite, and the castings are then gradually cooled to between 675° and 750° and maintained at such temperature for several hours, then further cooled, reheated to above 750° but not above 1020°, held at such temperature for a few hours, again cooled to 675°-750° and maintained in such temperature range for several hours.

Surface Hardening Malleable Iron

The process consists of heating a bath of cyanide consisting of 53% sodium cyanide and 47% potassium cyanide to a temperature of 1300° F. The castings or parts are then placed in this bath and held at this temperature two to three hours then removed and quenched in cold water. The resultant surface is file hard and is quite resistant to corrosion. The core will not exceed 140 Brinell hardness.

Elimination of Embrittlement of Malleable Iron

Elimination of blue heat embrittlement of malleable iron can be readily obtained by heating the parts to a temperature of 1300° F, and quenching in water or air. After this treatment galvanizing or Japanning can be done without embrittlement or loss of ductility.

High Strength Semi-Malleable Iron

High strength semi-malleable can be made from regular malleable iron by a secondary heat treatment. This heat treatment consists of reheating the parts to a temperature of 1450° F. holding at this temperature 2 to 3 hours followed by cooling in the furnace. The resultant product will have a tensile strength of 75000-85000 pounds per square inch and an elongation of 8 to 10%.

Molds Suitable for Iron-Foundry Use
U. S. Patent 1,935,362

Permanent molds are formed of chrome ore 3, fused silica 1 and bentonite 1 part.

Mold for Casting Iron Articles
British Patent 392,390

Unhardened iron articles are cast in a centrifugal metal mold lined with a substance confaining aluminum. A mixture comprising graphite, 3-10% aluminum and 40-80% powdered iron or steel may be used. The iron and aluminum may be mixed and heated to 400-600° away from air to make them adhere.

Casting Mold
British Patent 399,562

A mold for casting steel ingots is formed of pig iron containing 0.1-0.5% molybdenum. The iron may contain 3-4% carbon and the usual manganese, silicon, phosphorus, and sulphur. The molybdenum is added as ferromolybdenum to the fused iron in the ladle.

"Corrosionless" Cast Iron
German Patent 590,058

Cast iron which does not deteriorate or corrode contains 0.1-3% titanium, as in German Patent 564,681, and also nickel up to 3, chromium 1.5, aluminum 0.5, molybdenum 0.5, and titanium 0.2%, singly or severally.

Improving Iron Castings
British Patent 404,180

Iron castings are converted into malleable iron by heating them at 800-1150° for 3-5 hours while embedded in a mixture of alumina, aluminum, and silicon carbide or calcium carbide containing 2–5% of chromic chloride and 1–2% of ammonium chloride, and then cooling slowly during 24–48 hours out of contact with air.

Cast Iron Solder

Tin	44%
Lead	44%
Zinc	12%

The tin and the lead are melted together and the zinc incorporated until a good alloy is obtained. This material works very good on cast iron in places where the temperature does not exceed 300° F.

Refining Iron and Steel
British Patent 400,593

A composition for refining iron or steel comprises a mixture of alkali metal chlorate, preferably potassium chlorate, manganese dioxide and a concentrated scavenging agent which comprises alkali metal fluoride, borax or sodium carbonate or a mixture of any two of these. In an example are potassium chlorate 12–20 parts, borax and (or) boric acid 2–16 parts, sodium fluoride and (or) soda ash 2–8 parts and manganese dioxide 1–4 parts.

Rustless Iron or Steel
British Patent 394,806

To a bath of 500 kilograms of molten iron or steel with the desired carbon content is added chromite (48% chromic oxide) 300, nickelous oxide 30, calcium oxide 150, fluorspar 40 and bauxite 30 kilograms and to this mass, when molten to form a slag is added a reduction mixture of metal oxides and reducing agents comprising chromite 820, nickelous oxide 110, aluminum 316 and iron-silicon 82 kilograms.

(A) Rust-Proofing of (B) Production of Protective Coatings on Iron and Steel Articles
U. S. Patent 1,895,568

A bath producing a strongly adherent olive-green coating, insoluble in 10% acetic acid or concentrated sodium acetate, comprises: (A) oxalic acid, 1–10%, and an accelerating agent consisting of sulphuric acid 0.5–1.0% and a ferric salt 1%, the treatment being applied at 100° and for 5 minutes; and (B) a substantial amount of ferric

oxalate and acetic acid (or equivalent acid) to prevent hydrolysis but insufficient to act as a pickle.

Stainless Iron, Nickel, and Chromium Alloy
U. S. Patent 1,922,038

Iron 74, nickel 8 and chromium 18%, in the form of clean-surfaced particles of 200-mesh size or smaller are intimately mixed, subjected to a pressure of 20,000 pounds per square inch or higher and heated in a nonoxidizing atmosphere to 900–1200° to form a substantially homogeneous product.

Removing Enamel from Cast Iron

Articles are placed in bath of molten caustic soda kept at 450–550° C. The bath is rejuvenated with live steam.

Treatment of Gray Cast Iron and Malleable Iron
U. S. Patent 1,894,752

The iron is hardened by heating at 843–871° in a bath of sodium chloride (16 parts) and potassium ferrocyanide (1 part) with subsequent cooling to 712°, quenching in water, and tempering at 204–343°.

"Pickling" Iron

A very convenient way of pickling small iron pieces which have some grease on their surfaces consists of adding 1 to 2 per cent Gardinol W A to hydrochloric acid containing an inhibitor. The Gardinol enables the acid to penetrate the grease and remove oxide, without previous degreasing, and without any but the simplest equipment, where an elaborate set-up is not desired.

Pickling Wrought Iron and Steel
(Patented I.G. Farbenindustrie)

According to the nature of the iron surface, one or two pickle baths will suffice. Slightly rusted iron or iron with little mill scale is left in a bath of phosphoric acid of 6–10% for from 5 to 30 minutes, the bath preferably having a temperature of 70–90° C. After this, WITHOUT BEING RINSED, THE IRON IS DRIED QUICKLY. Iron with heavy mill scale may be dipped to better advantage in two pickle baths, that is to say, first in a phosphoric acid bath of 10–15% having a temperature of 80–90° C., when the mill scale usually disappears completely within half an hour, and then, after being rinsed in hot water, in a bath of dilute phosphoric acid of 2–3% for 8 to 10 minutes, also at a temperature of 80–90° C. For the latter bath one may use the exhausted liquid of the first pickling process. When the tubes are removed from this bath (WITHOUT BEING RINSED), they usually dry very rapidly due to their own heat. The actual anti-corrosive layer is formed on the tubes while they are drying. The tubes have a bluish-gray color, and before being painted are only rubbed with a dry cloth. Provided they do not come into contact with liquid water, such tubes are resistant to corrosion for appreciable periods of time. Moreover, the extremely fine film of ferro-ferric phosphate which has been formed on them is an ideal foundation for subsequently applied coats of paint, as it adheres firmly to the iron and does not split off readily. This process is slightly more expensive than the sulphuric acid process, but has advantages in simplicity of operation, etc.

Iron Alloys
German Patent 575,000

A hard iron alloy for casting consists of carbon 2–3, silicon 0.5, manganese 1.5–2.5, phopshorus 0.1, sulphur 0.03, and at least 20% of spongy iron.

Alloys of Iron
French Patent 748,323

Austenitic alloys of iron which may be easily worked while preserving their excellent mechanical properties are made by adding to the other usual constituents of the alloys 0.10–0.75% of arsenic. Examples containing chromium 12–20, nickel 6–15, carbon 0.05–0.25, arsenic 0.10–0.75 and up to 2% of the following: molybdenum, tungsten, cobalt, titanium, zirconium.

Iron or Steel Alloys Containing Chromium
British Patent 402,430

A mixture of chromite (48% chromium) 65, calcium 23, ferrosilicon (80% silicon) 2, aluminum 4, calcium fluoride 4, and calcium oxide 2% is smelted in an electric furnace to obtain a chromium alloy containing chromium 60, iron 39, carbon 0.25, and silicon etc., 0.75%,

which is added to molten iron under a layer of chromite 30, calcium oxide 25, bauxite 20, calcium fluoride 15, and iron ore 10%.

Corrosion-Resistant Iron Alloy
Swedish Patent 74,798

An austenitic or austenite-martensitic iron-nickel alloy with 22–35% nickel and 0.05–1.5% carbon contains 0.5–7.0% silicon and 0.2–25% copper and, if desired, an amount of chromium not exceeding 3%.

Removal of Lead (Coatings) from Tubes and Other Articles
U. S. Patent 1,918,817

The articles are immersed at 80–85° in saturated sodium chloride solution containing 0.5–10% sulphuric acid and 1–10% ferric chloride or 2–10% sodium nitrate.

Melting and Refining of Magnesium Alloys
British Patent 404,563

A flux for use in melting and refining magnesium and its alloys consists of magnesium chloride 21–38, potassium chloride 6–9, sodium fluoride 0.5–3, calcium fluoride 0.5–3, and manganese dioxide 0.2–10 parts.

Magnesium Alloy
Japanese Patent 98,956

Copper less than	2%
Aluminum	0.1–0.8%
Calcium	0.1–0.8%
Manganese	0.1–0.8%
Magnesium	Balance

This alloy has a high thermal conductivity.

Magnesium Base Die Casting Alloys
U. S. Patent 1,946,069

A magnesium base alloy consists of aluminum in amount from 8 to 15%, manganese from 0.5 to 0.1%, silicon from 0.35 to 1.0%, the balance being magnesium.

Improving Properties of Magnesium-Manganese Alloys
U. S. Patent 1,936,550

Such alloys are heat treated at 250–500° C.

Magnesium Flux for Refining
French Patent 759,481

A flux for the fusion and refining of magnesium and alloys having a basis of magnesium contains a magnesium halide and a large proportion of magnesium compounds, rich in oxygen, particularly manganesic oxide or manganese dioxide. An example contains magnesium chloride 21–38, potassium chloride 6–9, sodium fluoride 0.5–3, calcium fluoride 0.5–3, and manganesic oxide or manganese dioxide, 0.2–10 parts.

Chilled Roll
U. S. Patent 1,948,244-7
Formula No. 1

A high carbon chilled iron alloy roll contains a small amount of silicon, chromium and manganese to ensure a definite chill, said amounts running from about .15 to about .65 silicon and from about .15 to about .27 manganese and from about .20 to about .50 chromium, said roll also containing a substantial quantity of nickel to secure toughness of said chill, said nickel running from about 2.50 to about 5.0.

Formula No. 2

The method of making chilled iron alloy rolls consists in ensuring a chill by using small quantities of silicon, manganese and chromium running from about .25 to about .35 silicon and from about .25 to about .35 chromium and from about .18 to about .22 manganese and securing hardness of said chill by using carbon and nickel as desired from about 3.15 to about 3.20 carbon and from about 3.50 to about 4.0 nickel.

Formula No. 3

The method of making chilled iron alloy rolls consists in ensuring a chill by using small quantities of silicon, manganese and chromium and securing hardness of said chill by employing carbon and nickel from about 3.25 to about 3.35 carbon and from about 4.75 to about 5.0 nickel.

Formula No. 4

The method of making chilled iron alloy rolls consists in ensuring a chill by using small quantities of silicon, manganese and chromium running from about .15 to about .25 silicon and from about .25 to about .35 chromium and from about .15 to about .20 manganese and securing hardness of said chill by using carbon and nickel as desired from about 3.25 to about 3.35 carbon and from about 4.75 to about 5.0 nickel.

Fine Lead Coatings

Gelatin	10	g.
Water	100	l.
Gelatin	10	g.
Water	100	l.

Current Density 1 ampere per square decimeter.
Bath voltage 0.15–0.2 volts.

Lead of High Resistance Against the Action of Hot Sulphuric Acid
U. S. Patent 1,939,799

Soft lead alloy of high resistance against the action of hot concentrated sulphuric acid containing nickel in an amount of about 0.01 to about 0.05% and selenium in an amount from 0.01 to 0.1%, the balance consisting of commercial soft lead.

Nickel Alloys
French Patent 42,520

An example of an alloy suitable for watch springs, particularly for thermo-compensated oscillating systems, contains nickel 60, iron 16, chromium 15, manganese 2, molybdenum 6.5 and beryllium 0.5%.

Magnetic Alloys
British Patent 393,287

In the manufacture of magnetic cores the alloy used is composed of nickel 60–80 and cobalt 20–40%.

(Nickel Copper Silver Alloy) Contact Material
U. S. Patent 1,883,650

An alloy of silver 65, copper 30, and nickel 5% is used.

Platinum Alloys
German Patent 578,676

The mechanical properties of platinum alloys containing 9 up to 10% rhodium, nickel, iron, chromium, tantalum or tungsten, or of gold alloys containing 5–25% platinum and 10% rhodium, are improved by heating to over 700°, chilling, and hardening at 400°.

Recovering Silver from Scrap

Silver scrap is dissolved in (1 + 2) nitric acid and diluted with 10 volumes hot water, then decanted and filtered. Either hydrochloric acid or a filtered solution of common salt (not the household grade) is added in excess until no further precipitation takes place. Allow the precipitate to settle and test the supernatant liquor for the presence of dissolved silver by adding more hydrochloric acid or salt solution. If no further precipitate is formed all the silver is in the form of the insoluble chloride. Wash the precipitate with hot water, allow to settle and decant several times on successive days, to remove the soluble substances.

For each 10 ounces of silver scrap originally taken, add to the precipitate 3 ounces of mossy zinc and one-fourth ounce concentrated sulfuric acid, then partly fill the vessel with water. Stir occasionally and after two days the silver chloride will be found to be reduced to finely divided metallic silver having a gray appearance. Wash with hot water and decant several times as before. Dry and mix the precipitate with a small amount of soda ash as a flux and melt down in a furnace. The product will be pure silver. Do not attempt to reduce silver chloride in a furnace as it is not only somewhat volatile but easily comes through the pores of the crucible and becomes reduced on the outside in the form of globules.

Silver Solder

Tin	66⅔%
Silver	33⅓%

This metal has a melting point of around 875° F., and can be used satisfactorily for filling defects in brass and cast iron castings, where a high temperature melting alloy is necessary.

Stainless Silver Alloy
British Patent 401,527

The alloy consists of silver 60–92, cadmium 20–4, and tin 20–4%.

Hardening Silver Alloys
U. S. Patent 1,928,429

Alloys containing 50–90% silver with beryllium 0.1–2.5% and the remainder of copper are heated for at least 2 hours at 745–760° C., quenched and reheated for not less than 2 hours at 300–355° C.

Steel
British Patent 396,438

Cylinders or bottles for storing compressed gases are made of steel containing at least 98% iron with carbon about 0.42, nickel about 0.62, silicon at least 0.2 and sulphur and phosphorus together about 0.06%, the remainder consisting of varying amounts of manganese, silicon, chromium and molybdenum. The steels may be made by melting steel containing carbon and silicon with ferro-alloys containing manganese, chromium, nickel, and molybdenum prepared in an electric furnace or aluminothermically.

Corrosion Resistant Steel
U. S. Patent 1,943,782

An oil cracking tube composed of a ferrous alloy containing about 16% to 22% of chromium, about 6% to 16% of manganese, and carbon, the carbon content being not more than about 0.3%, and the balance being substantially all iron.

High Speed Steel
U. S. Patent 1,955,529

A stabilized high speed tool steel alloy of fine grain and having high elastic limits and high resistance to frictional wear under high temperature working contains iron alloyed with 15 to 18 parts tungsten, 2.5 to 5 parts chromium, 1 to 2.5 parts titanium and 0.1 to 0.8 parts carbon in 100 parts of the alloy, the remainder being principally iron.

Pickling Stainless Steel
U. S. Patent 1,939,241

The method of pickling stainless steel comprises subjecting it to the action of an acid mixture containing substantially the following ratio of acetic, nitric and hydrochloric acids:

Acetic Acid (100%, CH_3COOH)	40–50%
Nitric Acid (100%, HNO_3)	10–15%
Hydrochloric Acid (100% HCl)	4–8%
Water	Balance

(Inhibitor For) Acid Pickling Baths
(For Steel)
British Patent 397,553

A long chain alkyl quaternary ammonium compound, e.g., cetyl-pyridinium bromide, is used.

Spring Steel
U. S. Patent 1,943,347

A steel comprising from about 0.30% to about 5% beryllium, about 0.35%

but not more than 1.5% molybdenum; about 1% but not more than 1.5% chromium; about 0.35% but not more than 0.50% carbon, about 1% but not more than 1.5% manganese; and the balance iron.

Steel for Nitriding
U. S. Patent 1,943,348

A steel for nitriding comprising carbon approximately from 0.15% to 0.5%, chromium from 1.5% to 3.0%, manganese about 0.5%, molybdenum from about 0.5% to 5%, beryllium from about 0.35% to about 5% and the balance iron with its common contaminants.

Identification of Obliterated Numbers on Iron or Steel

Smooth the surface with emery removing as little of metal as possible. Dilute concentrated hydrochloric acid to 50% with pure water. Moisten felt or cotton swab with the hydrochloric acid, then rub first on copper-ammonium-chloride and then on metal surface. It may be necessary to repeat this for several hours. If this treatment fails, heat the metal gradually up to a cherry red.

Steel Cleaning Composition
U. S. Patent 1,897,913

The surface is sprayed or brushed with a mixture of 85% phosphoric acid (1), ethyl alcohol (1) and a water-soluble oil solvent (¼-2 parts) preferably an ether of a polyglycol.

Tempering Bath (For High Speed Steel
U. S. Patent 1,919,846

A fused salt bath for the heat-treatment of tungsten steel at 1350° consists of the reaction product of 8 parts of barium carbonate and 5 parts of boric acid with or without a proportion of barium chloride.

To Remove Tempering Colors from Steel

Immerse the finished ground steels tempered at a temperature from 400 to 700° F. in a solution of 50% acetic acid at room temperature to which is added 2% boric acid by weight. The color on the steel ranges from light yellow to blue, depending upon the tempering temperature. As soon as the color on the steel disappears it should be removed from the solution and washed in hot water. The solution restores the original luster and finish without harming the steel.

To Impart Extreme Hard Case to Steel

For this purpose a low temperature cyanide bath consisting of 53% sodium cyanide and 47% potassium cyanide and melting at 935° F. is employed.

Method No. 1

Plain carbon and alloy steels, after being finished are immersed in this bath for one to two hours at 1050° F. Then they are removed, cooled in air and washed in hot water. The piece will be free of distortion on account of low temperature and cooling in air and the surface would be file hard.

Method No. 2

Tools made of high speed steel, after usual hardening from 2350° F., are tempered in this bath at 1050° F. from thirty minutes to two hours, depending upon the depth of case required. With this temper the surface will be considerably harder than the one obtained in the usual way. A Rockwell hardness of 70C. has been obtained by this method.

Thick-Coated Electrodes for Steel Welding

A coating containing aluminum oxide 1-1.4, ferric oxide 22-33.6, pyrolusite (manganese 60, silicon dioxide 20, calcium oxide 4%) 50.4-64, and calcium oxide 13-15% is recommended. The finely powdered materials are made into a paste with sodium silicate solution (density 1.16), and the paste is applied 0.5 millimeter thick for a 3-millimeter rod, or 1 millimeter thick for a 4-6 millimeter rod. The electrodes, which are dried and heated for 1-2 hours at 150-200°, can be used for welding steel containing not more than 0.3% C.

Bluing Steel

The quality of blue obtained by any method depends primarily upon the polish given the object previously. A good cleaning solution contains 4 ounces of lye per gallon water. Boil object in lye solution 5 minutes, and wash with water.

Emergency Blue. (Too Hot for Springs)

Heat object to desired blue in even mixture of equal parts fine clean sand and powdered charcoal.

High Temperature Blues. (Do Not Rust; But Too Hot for Springs.)

Formula No. 1

Potassium Permanganate	1 part
Sodium Nitrate	2 parts
Potassium Nitrate	2 parts

Formula No. 2

Manganese Dioxide	1 part
Sodium Nitrate	9 parts

With either of these formulae, fuse and hold just above fusing point while object is immersed for five minutes. Remove article, wash in water and rub with oil.

Warm Blue

To each pint of distilled water, add and dissolve one ounce of bichloride of mercury, one ounce of potassium chlorate, and one ounce of potassium nitrate. Raise to 100° F., and add 1½ ounces of spirits of nitre. Dip cleaned object in boiling water 5 minutes, then in solution (kept at 110° F. or above) five minutes, then polish with steel wool or wire brush. Repeat several times, then rub with oil.

Cold Blue

Crystallized Iron Chloride	2 parts
Solid Antimony Chloride	2 parts
Gallic Acid	1 part
Water (Distilled)	4 parts

Apply with sponge, allow to dry, polish, and repeat process until desired blue is obtained. Then rub with oil.

Rubbing oil for blued steel may be either pure boiled linseed oil, or a mixture of equal parts light paraffin oil and lard oil.

Blacking Steel

Dip thoroughly cleaned and polished object into a physical solution of sulphur in turpentine, then remove and hold in non-oxidizing flame until dry. Repeat until article is an even black, then rub with oil.

Treating Metal (Steel)

U. S. Patent 1,890,485

Nitrogenous impurities are removed from steel by means of a flux comprising bauxite 0.75, feldspar 1, calcium oxide 1%; or aluminum oxide 0.525, silicon dioxide 0.80, calcium oxide 1.05%.

Bronze Finish on Steel

The steel is brass plated quite heavily, and then oxidized by using the following solutions:

Solution No. 1

Ammonium Polysulfide	1 oz.
Water	1 gal.

Solution No. 2

Copper Sulphate	2 oz.
Water	1 gal.

Dip the work in No. 1 solution, and then without rinsing dip in No. 2 solution. Then rinse in clean cold water and repeat the dipping operations until the bronze color is produced.

The work is then dried and relieved by using a greaseless composition on a loose leaf buff wheel. The work must be lacquered to prevent tarnishing.

Bluing Steel (Gun Barrels)

Mercuric Chloride	4 parts
Potassium Chlorate	3 parts
Alcohol	8 parts
Water	85 parts

Mix in glass, chinaware or enameled vessel.

The steel to be blued must be free from rust, brightly polished and chemically clean. The lye solution for cleaning and water used for heating article, must be boiled to remove dissolved air *before* article to be blued is placed in them. Remove rust and old finish with fine grades of steel wool or sandpaper. Polish with rouge paper if desired. The higher the polish the shinier the final finish. To clean article boil for five minutes or longer in a lye solution made by dissolving one tablespoonful of lye to a quart of cold, clean, rain or distilled water. Rinse well. Article is clean if water spreads uniformly over surface. Handle article with tongs or wire hooks. Do not touch with hands. Article is now ready for bluing.

Use enamel ware, iron or glass vessel and sufficient clean rain or distilled water to completely immerse article to be blued. Keep the water boiling vigorously at all times and the bluing solution hot. Place article in the boiling water and allow to remain long enough to reach the boiling temperature of the water. Then remove article, dry quickly with clean cloth and immerse in bluing solution. Return at once to the bluing solution in boiling water, boiling one minute between coats. If immersion in bluing solution is impracticable article may be coated with bluing solution by means of a clean cotton cloth swab tied on a stick. Polish lightly between coats for purposes of inspection and to remove any imperfections. Repeat the process until desired shade is obtained, usually one to four coats being sufficient. After the final application boil five minutes, remove, wipe dry and while still hot apply a coat of boiled linseed oil on a rag. The final blue-black finish will vary somewhat with the type of steel used.

Prevention of Sticking of Steel Sheets
U. S. Patent 1,933,519

The method of preventing sticking in box annealing sheets and plates which consists in chemically treating the rolled and pickled sheets or plates by immersion in an aqueous solution of zinc sulphate of a concentration of 0.001 and 0.20 per cent, piling the treated materials and then box annealing the piles.

Silicon Steel Sheets for Electrical Work
British Patent 403,950

The hot-rolled sheets are pickled in dilute sulphuric acid containing an inhibitor, washed, sprayed with milk-of-lime, scoured, slightly cold-rolled, coated with calcium oxide, and slowly heated in an electric furnace to 1050° at which temperature they are kept for 3 hours. After very slowly cooling to 530–540° in the furnace, the charge is removed and allowed to cool under a cover.

Preventing Formation of Roll Scale
U. S. Patent 1,916,677

In hot-rolling steel rails the rolls are sprayed with cupric sulphate solution or potassium dichromate solution to prevent adhesion of the scale to the rolls.

Steel Containing Chromium, Manganese and Silicon
U. S. Patent 1,929,554

A steel which has high ductility contains chromium 0.2–75, manganese 0.6–2.5, silicon 0.2–2.5 and carbon less than 0.1%.

Alloy Steel
U. S. Patent 1,934,520

A ferrous base alloy having a toolproof essentially austenitic structure which is not softened sufficiently by tempering to substantially decrease its tool resistance, contains about 1.25 to 1.75 per cent of carbon, about 1 to 3 per cent of chromium, about 1 to 3 per cent of molybdenum, about 3.5 to 7.0 per cent of nickel, the remainder being iron except for elements present normally and in amounts not adversely affecting the characteristics of the alloy.

Steel Alloy for Dies
U. S. Patent 1,938,221

A die for use in the die-casting of aluminum base alloys is composed of a steel alloy containing:

Carbon	.20% to .50%
Silicon	.10% to 1.50%
Manganese	.10% to 1.00%
Chromium	3.00% to 8.00%
Molybdenum	1.50% to 2.75%
Tungsten	.50% to 2.00%
Nickel	.50% to 2.00%

and the remainder mainly iron.

Softening Cobalt Alloy Steel
U. S. Patent 1,936,406

The method of softening cobalt alloy steel to render it easily machinable comprises heating the steel to a temperature of about 800° C. to about 850° C. and quickly cooling.

Hardening and Tempering Combined

Straight carbon steels, instead of being quenched in brine and tempered afterwards can be quenched from the desired heat in an oil bath at about 120° F. This quenching will produce the desired hardness and toughness and eliminate the necessity of additional operation of tempering. This is recommended where the tool is not to be used for cutting purposes. In the latter case a brine quenching followed by a tempering at 350° F. is recommended.

Atmosphere for High Speed Steel Furnace

In cases where the high speed tool is to be ground after hardening and the surface, cleanliness and appearance are of more importance, a furnace atmosphere consisting of 9 to 10% carbon monoxide and 4 to 5% carbon dioxide is recommended. From this atmosphere the work comes out smooth and clean. However, in cases where the high speed tool is not ground afterwards and is used as it comes out of the fire and no surface decarburization can be tolerated a furnace atmosphere consisting of 3% carbon monoxide and 10% carbon dioxide is recommended. Work comes out from this atmosphere without any soft skin.

Alloys of Tin and Nickel
U. S. Patent 1,924,244

For producing alloys of good hardness, an alloy is formed containing nickel together with tin 5–25% and the alloy is heated to between 900° and its melting point then rapidly cooled and subsequently annealed at 400–800°.

Casting Metals
French Patent 753,698

An intense heat is produced on the surface of cast metal by means of a chemical composition containing e.g., aluminum powder 15–40, aluminum oxide powder 15–42, copper oxide powder 1–6, iron oxide powder 0.5–4.5, silicon dioxide 4–16% and a small amount of zinc oxide, calcium oxide, carbon and moisture.

Coatings for Metal-Casting Molds
British Patent 406,582

A suspension of cork dust (1) in an alkali solution, e.g., ammonium carbonate, calcium hydroxide or ammonium hydroxide, is sprayed on to the mold and heated until (1) is carbonized and a uniform, bright black coating is produced. Suitable mixtures are obtained by dispersing in 160 grams of water, 60 grams of casein, 20 grams of disodium hydrogen phosphate, 10 grams of sodium sulphite, 30 grams of calcium oxide and 100 cubic centimeters of cork dust finer than 80-mesh.

Cleaning and Rustproofing Metal
U. S. Patent 1,949,921

In preparing an efficient metal cleaning and rust preventing bath, mix 1 volume 85% phosphoric acid with 1 volume denatured alcohol or similar solvent and 1½ volumes water and from 1 to 5% of iso-propyl-ether. Rusted and oily sheet iron, when dipped in such solutions, is readily wetted, the rust dissolves and the metal coats with a thin and even film of phosphate which imparts to the whole surface of the article an excellent rust resistance.

Heat Treating Ferrous Metals
U. S. Patent 1,942,937

A process of heat treating ferrous metals, consisting in first preparing a mixture of finely ground slow-burning carbon-producing material, charcoal, manganese, chromium and borax, then packing the metal articles to be treated in such mixture, then raising the temperature of the entire body of material to from 1850° F. to 1950° F. effecting first, the combustion of the charcoal followed by the burning of the slow-burning carbon material to liberate gases which will reduce the silicon content, and open the grain and maintaining such temperature for a period to effect the desired penetration of the carbon, manganese and chromium into the grain of the metal.

Ornamenting Metal Surfaces
U. S. Patent 1,937,146

In carrying on this method, a silhouette photographic film negative is made by means of an exposure thereof in an ordinary camera. To make said silhouette negative, a powerful source of light is placed behind a semi-transparent white screen and the subject placed between this screen and the camera in such a way that a silhouette image is obtained on the negative. The film negative thus obtained is then developed in the usual way and dried.

The metal surface to be ornamented is then covered by a coating of light-sensitive material. A smooth grained oval copper plate has its face thoroughly cleaned by polishing with charcoal or powdered pumice which is then covered with a coating compound. For practical purposes, this coating may consist of the following emulsion:

Water	30 oz.
Photo-Engravers Glue	8 oz.
White of Twelve Eggs	
Bichromate of Ammonia	25 grains
Aqua Ammonia Concentrated	18 drops

After the face of the plate is thoroughly covered with this coating, it is revolved over a gentle heat to cause even and rapid drying. This procedure should be carried out in a dark room or in a very weak light and until the emulsion coating is thoroughly dry. The heating is then discontinued and the plate allowed to cool.

After plate has cooled, the photographic negative prepared as stated above is placed in contact with the coated side of the plate and the negative then exposed to a source of bright light. The exposure to the light will cause the emulsion under the transparent portions of the negative to become insoluble, the other portions, that is, the portions around the silhouette image remaining soluble.

After this exposure, the plate is washed in water to remove the soluble portions of the coating, thereby leaving the surface of the plate coated only to correspond with the silhouette image. The resultant plate is then slowly dried by baking under heat and baking is continued until the image becomes uniformly black. The plate is then placed on a slab and allowed to cool.

To finish the plate, it is dipped in the following solution:

Water.	1 gal.
Sodium Bichromate	5.6 oz.
Commercial Sulphuric Acid	5.6 oz.

It is permitted to remain in this solution for a few minutes and until a satin finish is obtained on the uncoated portions or silhouette background of the plate. The plate is then dried and sprayed with lacquer to prevent injury or corrosion and is then completed for use.

Type Metal
German Patent 590,686

An alloy suitable for type metal consists of 20–30% antimony, 5–12% tin, 0.5–2% copper, 0.5–3% cadmium, 0.1–0.5% arsenic, 0.01–0.02% aluminum, 0.01–08% alkali and (or) alkaline earth metal, including magnesium, 0–2.99% zinc, and the rest lead.

Refining Type Metals
British Patent 400,258

A powder for refining lead, tin or lead-tin alloys, which may contain antimony, comprises finely powdered ammonium chloride 37, sodium chloride 14, calcium chloride 11, carbon 25 and naphthalene 14 parts by volume. The powder is mixed with the metal at 300–750° F., being used particularly for alloys used in type-casting machines and stereotyping foundries and comprising approximately tin 4–6, antimony 12–13 and lead 80–85%.

Type Metals

	Tin	Antimony
Linotype	4.25%	12.00%
Special Line Casting	6.00%	12.00%
Newspaper Stereotype	6.00%	13.50%
Autoplate	7.00%	13.50%
Standard Monotype	7.25%	16.50%

Except for small amounts of impurities, the balance is lead.

Protecting Light Metals Against Corrosion
U. S. Patent 1,957,354

Process for the formation of coatings for preventing the oxidation on articles made of light metals or alloys such as aluminum, magnesium, consists in immersing the articles in a solution comprising about 2 parts of double fluoride of potassium and titanium, 1 part of chromium fluoride and ½ part of fluoride of sodium for 1000 parts of water.

Casket Trimming Metal

Antimony, Approximately 13%
Lead, Balance.

Graphitized Metals and Alloys

Antifriction bronze containing graphite is produced by mixing the graphite with the powdered alloy, briquetting under a pressure of 1–3000 kilograms per square centimeter, and heating to a temperature between the solidus and the liquidus in a hydrogen atmosphere. After furnace cooling the specimens are again pressed to increase their density and hardness. A 10% tin bronze mixed with 10% graphite, heated to 880° and subjected to 3000 kilograms per square centimeter before heating and 2460 kilograms after heating, has a Brinell hardness of 35 and a compressive strength of 24.2 kilograms per square millimeter. Satisfactory results are also obtained with a crude copper containing tin 5.51, lead 7.18 and zinc 1.81%. Brass and aluminum bronze give poor results because of the small temperature interval between the solidus and the liquidus.

Thermal Treatment of Noble Metal Alloys
German Patent 581,259

Alloys of gold, platinum of palladium with up to 2% of silicon and up to 10%

of another metal, e.g., iron, nickel, cobalt, molybdenum, tungsten, tantalum, chromium, or copper, are heated to above 750°, chilled, and then heated for some time at 300–700°. Alloys of gold with silicon up to 2 and platinum or palladium up to 10% may be treated similarly. The treatment raises the hardness of the alloys.

Flux for Use in Soldering Metals
U. S. Patent 1,927,355

Venice turpentine 1, paraffin oil 1, resin 1 and balsam copaiba 3 parts are used together.

Finishing Die Cast Metal

In polishing zinc base die cast work the first operation consists of removing the parting fins by the use of felt or other suitable polishing wheels which are glued with 120–180 emery. The next operation is buffing with a loose or stitched cloth wheel, using tripoli as the buffing medium. Next, wash in gasoline, dry in sawdust, and color on a loose cloth wheel, using a small amount of a lime composition as the coloring medium.

The die cast work should be cleaned in a mild alkaline solution made of trisodium phosphate 2 ounces, carbonate soda 2 ounces, water 1 gallon. Use at boiling temperature, either as a plain cleaner or electric cleaner with direct current. Rinse in clean cold water, pickle in 1% solution of 48% hydrofluoric acid and water. Rinse in clean cold water, and nickel plate in the following solution:

Double Nickel Salts	10 oz.
Sodium Chloride	7 oz.
Boric Acid	2 oz.
Sodium Citrate	1 oz.
Sodium Sulphate (Anhydrous)	4 oz.
Water	1 gal.

Temperature of both should be 80° F.; pH, 6.4 to 6.8. Strike with a high current density for a few seconds, then reduce current to 8 to 10 amperes per square foot for five minutes. Rinse in clean cold water and bronze plate in the following solution:

Copper Cyanide	4	oz.
Zinc Cyanide	½	oz.
Sodium Cyanide	5	oz.
Sodium Carbonate	2	oz.
Rochelle Salts	2	oz.
Water	1	gal.

Temperature, 95° F.; cathode current density, 2 to 3 amperes per square foot.

Controlling Metal Pickling Baths
U. S. Patent 1,932,015

Diorthotolylthiourea	2.0 lb.
Evaporated Cellulose Pulp Waste Liquor	3.0 lb.
Sodium Chloride	5.0 lb.
Sodium Carbonate (Anhydrous)	0.5 lb.

This aggregate quickly dissolves and is effectively diffused in the acid solution, forming a bath ready for pickling steel. As the bath is reinforced with acid from time to time to maintain its strength, additional amounts of aggregate are added in proportion of about 1 pound of aggregate to every 100 pounds of sulfuric acid employed. The above is formed into a briquette, which will sink immediately even though the bath is covered with foam and the admixture is immediately attacked by the bath and the elements thereof blasted apart by the disrupting gas and quickly and completely dispersed thereby through the bath in a wet condition. The results are that the exact quantities of materials may be added at any time, the materials are uniformly and quickly placed in condition for functioning, no materials are lost through being removed with the work and no time is lost through delaying the pickling operations.

Cutting Tool Alloy
U. S. Patent 1,937,185

A method of producing a hard tantalum alloy comprises, degasifying tantalum carbide powder at above 1600° C. in a vacuum, mixing from 85% to 97% of the degasified tantalum carbide powder with iron 15% to 3% of powdered nickel and heating the mixture in vacuo to from 1350° C. to 1400° C.

Hard Tantalum Alloy for Cutting Tools
U. S. Patent 1,937,185

Tantalum carbide powder is degasified in a vacuum at a temperature above 1600° C. then mixed with 3–15% powdered nickel, and the mixture is heated in a vacuum to 1350–1400°.

Deoxidation Slags
French Patent 752,508

Slags used for deoxidizing metals have

their fluidity increased by adding 4–30% of aluminum oxide, boric acid, cryolite, magnesium oxide, barium oxide. An example contains silicon dioxide 45–65, aluminum oxide 4–25, magnesium oxide 4–25 and calcium oxide 4–25%. Titanium oxide increases fluidity and heat and electrical conductivity. An example contains titanium oxide 70 and calcium oxide 30%.

Alloys

	Lead	Tin	Antimony	Copper	Others
Condenser Foil	90	9.25	0.75		
Lead Tape	95	0.5	4.5		
Shot Lead	99–100				Arsenic less than 1
Tea Lead	98	2			
Brittania Metal	0–9	94–70	3.7–15	1.8–5	Zinc 0–5
Pewter	2–20	90–73	0–8	0–2	
Tinsel	40	60			
Ounce Metal ("85 and three-5's")	5	5		85	Zinc 5
Machine Brass	1.7–2.6			88–62	Zinc 10–35

Fusible Alloys

By using combinations of two or more of the metals, lead, tin, cadmium and bismuth, alloys which will melt at low temperatures may be made. These alloys find considerable use for such purposes as electric fuses, automatic sprinkler systems, and boiler plugs.

Fusible Metals (in Per Cent)

	Melts About Deg. C	Lead	Bismuth	Tin	Cadmium	Others
Anatomical Alloy	60	17	53.5	19		Mercury 10.5
Bismuth Solder		25–40	40–25	20–50		
D'Arcet		25	50	25		
Eutectic	94	32	52.5	15.5		
Eutectic	91.5	40.2	51.6		8.1	
Eutectic	145	32		50	18	
Eutectic	181	37		63		
Eutectic	125	42	58			
Guthries		19.36	47.38	19.97	13.29	
Lichtenberg's	91.5	30	50	20		
Lipowitz	70	26.7	50	13.3	10	
Newton's		31.25	50	18.75		
Onion's (see Lichtenberg's)						
Rose's		28	50	22		
Rose's		35	35	30		
Wood's	71	25	50	12.5	12.5	

Type Metals (in Per Cent)

	Lead	Tin	Antimony	Copp
Electrotype Metal	93	3	4	
Electrotype Metal	96		4	
Linotype Metal	85	3	12	
Linotype Metal (English)	83	5	12	
Monotype Metal	87–84	3–4	10–12	
Stereotype Metal	82–67	3–17	12–23	0–0
Type Metal (Small Type)	70	10	18	

Hard Lead Alloy Compositions (in Per Cent)

	Lead	Antimony	Other Metals
Antimonial Lead (Hard Lead).................	100–75	0–25	....
Battery Plates	93	7	¼ Tin
Bullets, Shrapnel	88	12	....
Cable Sheath Alloy............................	99	1	....
Collapsible Tubes	100–96	0–4	....
Pattern Alloy	87	13	....

Bearing Bronzes

Bearing bronzes are in wide use for a multitude of purposes, such as the backing of Babbitt-lined bearings and for uniform unlined bearings. These bronzes generally consist of copper alloyed with smaller amounts of tin.

Bearing Bronzes (in Per Cent)

	Lead	Copper	Tin	Others
Journal Bearing Backs:				
Baldwin	15–22	68–77	5–7	
American Locomotive	18	75	6	less than 3 Impurities
N. Y., N. H. & H. C., B. & Q. Norfolk & Western	16–24	67–77	5–7	less than 4 Impurities
Illinois Central	15–20	70–75	5–7	less than 4 Impurities
Pennsylvania	15–22	72.5 minimum	5–7	less than 2 Zinc less than 1 other Impurity
Phosphor Bronze:				
Unlined Bearings	10	80	10	less than 1 Phosphorus
Pennsylvania	8–11	77–82	9–11	Phosphorus
Baldwin	10–15	77¼–82¼	7–10	Phosphorus
American Locomotive	15 maximum	78	7 minimum	Phosphorus
N. Y., N. H. & H.........	9–11	79–81	9–11	Phosphorus
C., B. & Q..............	9–11	77–81	9–11	Phosphorus 0.4–1
Illinois Central	10–15	75–80	6–10	
Norfolk & Western.......	8–11	77–82	9–11	Phosphorus 0.5–1
Driving Journal Bearings:				
P. R. R................	23½–26½	63½–71½	4–6	
Others (Nominal)	10	80	10	
Ordinary Machinery				
Bearings	10	80	10	
Hall-Scott Motor Bearings..	25	75		

Lead Base White Metal Bearing Alloys (in Per Cent)

	Lead	Tin	Antimony	Copper	Others
Journal Bearing Linings...	85–88	3–5	8–10		Tin and Antimony 12–14
Piston Packing (P. R. R.).	70.5–75.5			18.5–21.5	Nickel 5–7
Piston Packing	73–76	12–14	10–15		
B. & O. Thin Linings.....	96–94	0.5–1.5	3–5		
B. & O. Thick Linings.....	86	3–5	10–12		
Chesapeake & Ohio........	91.5	1.5	7		
Dandelion Metal (Lining Crossheads)	72	10	18		
Light Duty Bearings......	90–80	0–10	10–17	0–1	
Frary Metal	98				Barium plus Calcium 2
French Auto Bearings.....	75	10	15		
Metallic Packing	100–70	0–17	0–13		
Atlantic Coast Line.......	85		15		
Pennsylvania R. R........	87		13		

Alloys of Precious Metals
U. S. Patent 1,946,231
An alloy which may be annealed and/or hardened by heat treatment consisting of about 70% to 95% of gold, 1 to 25% of platinum, and 0.05 to 5.0% of iron.

Acid Resisting Alloy
U. S. Patent 1,930,956
A copper base alloy containing 1½% to 5% silicon and one-tenth of 1% chromium the balance being commercially pure copper.

Alloy (For Lead Shot)
U. S. Patent 1,900,182
An alloy which is non-poisonous to animals comprises lead 95.0–96.5, antimony 3.0–4.0, phosphor-tin 0.1–1.0%.

Alloy
U. S. Patent 1,945,653
An alloy comprising from 5–10 per cent of nickel, .5 to 1.5 per cent of beryllium and chromium in substantial amounts up to a total of 25 per cent with a balance substantially iron; the said alloy being easily rolled and very heat resistant.

Bearing Alloy
German Patent 584,020
Alloys for making bearings contain copper 52–68, lead 23–38, silicon 2–4, nickel 5–8 and iron 0.5–1.5%. The alloys may be prepared by adding lead to an alloy of copper, silicon, nickel and iron heated to 1150°, oxidation of the lead being avoided by covering the surface of the alloy with glass, wood charcoal, silicon dioxide, etc.

Bearing Alloy
U. S. Patent 1,937,465
The method of producing sintered bodies such as bearings, bushings, or the like, comprises forming a mixture of powdered copper about 84%, powdered tin about 10% and powdered metallic iron about 5%, forming a briquette from the mixture, and sintering the briquette to produce a relatively soft sintered matrix having hard points of unalloyed metallic iron therein.

Self-Lubricating Bearing Alloy
German Patent 581,425
A finely divided mixture of:

Copper	85%
Tin	9%
Zinc	2–3%
Graphite	4%

is subjected to high pressure (1,000–3,500 kilograms per square centimeter) from opposite sides. It is then heated in an auto-clave at 750–800° C. for 10–20 minutes to form a porous alloy, which may be dipped in lubricating oil before or after shaping.

Bimetal Thermostat
U. S. Patent 1,902,589
A high-expansion alloy having the δ to γ change eliminated from the range 0–500° comprises: Manganese 10–22 (16), silicon 0.01–050, carbon 0.10–0.40, nickel 1.0–10–0 (3) %, iron remainder; and the low-expansion alloy; manganese less than 1.0, carbon less than 1.0, nickel 26–45 (28), cobalt 2–19 (16) %, iron remainder.

Carburizing Box
U. S. Patent 1,933,900
A carburizing box is composed of a ferrous alloy embodying the following constituents in the proportions specified; carbon .05–1.2%, manganese 2–10%, silicon .25–5%, chromium 20–45%, copper over 5–15%, aluminium 1–5%, the balance being substantially iron.

Corrosion Resistant Alloy
U. S. Patent 1,945,679
A corrosion resistant metal article comprises an alloy containing the following constituents in about the proportions by weight of from 1 to 20 per cent iron, 10 to 20 per cent of at least one metal of the chromium group from 5 to 12 per cent of copper and from 0.5 to 1.5 per cent of beryllium with a balance substantially nickel, said alloy having the hardness and elasticity produced by bringing the alloy into a state of solid solution by heating from about ½ to 1 hour at temperature above 900° C. but below the melting point of the alloy, followed by quenching and by age-hardening at temperatures ranging from about 250° to 600° C.

Improving Cobalt-Molybdenum Alloys
U. S. Patent 1,949,313

A process for obtaining alloys of high hardness consists in forming an alloy of 10 to 35 per cent of molybdenum and the balance substantially cobalt and heating said alloy to a temperature above 1000° C. and below the melting point of the alloy, then rapidly cooling said alloy and subsequently annealing it at temperatures between 500° and 900° C.

Electrode Alloy
Canadian Patent 337,987

An alloy for use as cathode material in a television tube contains nickel 70, cobalt 20, iron 8 and titanium 2%. These proportions may be varied and zirconium or thorium may be substituted for the titanium.

Electron Emitting Element
U. S. Patent 1,943,027

An electron emitter, homogeneous throughout, made of an alloy characterized by ready electronic emissivity, ductility and resistance to heat and corrosion comprises a solid solution of calcium and nickel, calcium constituting from .04% to .10% of the total and the balance consisting essentially of nickel, and approximately 2% manganese.

Hard Alloys
French Patent 749,190

Alloys for tools, etc., are composed principally of tungsten carbide with one or more metals of the iron group in amount up to 20%, and mixed crystals formed by carbides or carbides and nitrides in amount of 2–50%. The mixed crystals may be composed of titanium carbide-tantalum carbide or titanium carbide-columbium carbide.

Heat Treatment of Light Alloys
U. S. Patent 1,946,545

A method of increasing the elongation and resistance against intergranular corrosion embrittlement of duralumin and alloys commonly known as alloys of the duralumin type, which are of such a nature that their strength can be increased by a heat treating process including a period of aging, which comprises subjecting these alloys, during the period of aging and

for not less than nine hours, to a static pressure which will not result in any essential change of the shape or dimension of the material and which shall be not less than 600 pounds per square inch.

Welding Alloy
British Patent 386,836

A welding bar for depositing cutting edges on metal cutting, tools, etc., is made of an alloy of cobalt 40–50, tungsten 18–20, chromium 25–35 and iron 1–5%.

Welding Rod Alloys
U. S. Patent 1,952,842

Alloys useful for welding rods consisting of copper, iron, nickel, and zinc in approximately the following ranges of proportions: copper 42 to 64.5%, iron 0.25 to 3%, nickel 0.25 to 5%, zinc the balance, but in all instances the sum of the copper, iron and nickel being from 50 to 65%.

Arc-Welding Rod
U. S. Patent 1,921,528

A welding rod is formed of an alloy of iron together with carbon from a trace to 0.6 and silicon 1.5–4.0%.

Welding Rod
U. S. Patent 1,903,952

A suitable alloy, highly resistant to abrasion, consists of cobalt 20–65, chromium 15–45, tungsten 2–35, carbon 0.25–3.50, manganese not more than 0.90, and silicon not more than 2.0%; the ratio silicon to manganese should be not less than 0.4:1.

Welding Rod
U. S. Patent 1,926,090

A rod suitable for welding or resurfacing railroad rails consists of a base metal core such as nickel steel carrying a homogeneous coating of materials such as ferrochrome 5.12, "manganese titanium" 1.75, calcium molybdate 0.33 and fire clay 2 parts.

Weld-Rod Coating
U. S. Patent 1,903,620

The coating comprises kaolin 7, sodium silicate 41, glass 22, ferromanganese 22, sodium carbonate 6, and water not more than 2 parts.

Welding Composition
U. S. Patent 1,926,412

For welding so-called white metals such as alloys containing lead, antimony, tin and copper, a composition is used comprising copper 5, antimony 5 and zinc 90%.

Thermite for Welding

Powdered Magnetic Iron
Oxide 1 part
Powdered Aluminum 2 parts

These are thoroughly mixed and ignited with a strip of magnesium.

Welding Electrode and Flux
U. S. Patent 1,946,958

An electrode for arc welding comprises a base rod consisting predominantly of aluminum and a flux or coating applied uniformly to said rod, said flux comprising from 55% to 65% of potassium chloride, from 2% to 5% of sodium fluoride to cause the flux to adhere to the base rod without causing cracks in the flux during the welding operation, from 22% to 25% of lithium fluoride, and from 10% to 14% of sodium aluminum fluoride so that the melting point of the flux will be substantially the same or slightly higher than the melting point of the base rod.

Arc-Welding Fluxes
British Patent 394,610

A flux for arc welding contains ferromanganese, talc, feldspar and liquid sodium silicate. The mixture preferably consists of ferromanganese of carbon content less than 1.5% 10, talc 10, feldspar 30 and sodium silicate of specific gravity 1.36 24 parts. It may also contain cellulosic material, e.g., cotton yarn, cloth or gauze or braid.

Welding Flux
U. S. Patent 1,923,375

A flux suitable for use in arc welding of aluminum and aluminum alloys is formed from sodium-aluminum fluoride 3–11, carbon not more than about 2, sodium fluoride not more than about 4, sodium chloride about 8–20, potassium chloride about 5–12 and magnesium oxide not more than about 2 parts.

Hardening Zinc Alloys
French Patent 753,035

Alloys containing zinc up to 37, nickel at least 2, aluminum at least 0.5% and copper the rest are hardened by heating to 800–900°, cooling rapidly and reheating to 300–600°.

Die Cast Zinc Alloy
Formula No. 1

Nickel 0.1–10%
Zinc 99.9–90%

Formula No. 2

Manganese 0.05–10%
Zinc 99.95–90%

Purification of Mercury

Mercury is made the anode at 110 volts direct current with platinum electrodes in a stoppered bottle, first in 10% sulphuric acid and then in 5% sodium chloride. Finally in 10% sulphuric acid the mercury is made cathode. The mercury loss is from 2 to 4%.

Annealing of Duralumin Wire

The best results are obtained by heating to 370–400° for 2–3 hours, cooling in the furnace to 250–70° and then cooling in air. This gives a tensile strength of 22–3 kilograms per square millimeter and an elongation of 17–8%.

Clean Tungsten Wire for Sealing into Pyrex Glass Tubes

Heat wire to bright redness, plunge into granular or stick sodium nitrite (not nitrate). Remove promptly, as the nitrite will not only remove oxide, but will wear away the wire rapidly; plunge in water.

After this cleaning process, wire should be sheathed with a close-fitting pyrex glass sleeve of very thin stock, and the whole heated in blast flame.

Tough Tungsten Carbide Alloy
U. S. Patent 1,924,384

A cast alloy is formed comprising tungsten carbide together with nickel 2.5–18 and copper 2.5–18%.

Spark Plug Electrode
U. S. Patent 1,953,228

A spark plug electrode having improved corrosion and erosion resistance to condensed acid products of combustion of gasoline comprises a ferrous base alloy containing 6% to 18% chromium and 2% to 10% aluminum.

Spark Plug Electrode
U. S. Patent 1,953,229

A spark plug electrode adapted to resist corrosion and having improved resistance to erosion by the electric arc when the gas surrounding the electrode contains lead, lead compounds, halogen compounds or sulfur comprises a ferrous base alloy containing 6% to 18% chromium, 8% to 30% nickel and not over 4% silicon.

Oxidation Resistant Bimetallic Material Suitable for Bimetal Thermostatic Devices
U. S. Patent 1,929,655

A high-expansion component containing iron together with chromium 5.7, nickel 22, silicon 0.2 and manganese 0.7% and carbon about 0.47% is used with a low-expansion component containing iron together with chromium 17, silicon and manganese each 1, carbon 0.1–0.45% and a trace of nickel.

Bimetal Thermostat
U. S. Patent 1,947,065

A thermostatic element comprises a pair of cooperating metallic members having different temperature coefficients of expansion, the member having the higher temperature coefficient of expansion being an alloy including from 20% to 40% nickel, 60% to 25% copper and 20% to 35% zinc and the member having the lower temperature coefficient of expansion being an iron alloy including from 30% to 39% nickel, from 2% to 10% cobalt and traces of manganese and carbon.

Permanent Magnet
U. S. Patent 1,947,274

An age hardened magnet consists substantially of about 6% to 15% aluminum, about 20% to 30% nickel with the remainder iron.

(Foundry) Moulding Composition
U. S. Patent 1,902,419

The use of 1–10% of a hygroscopic chlorine-free halide salt, e.g., calcium chloride, in moulding compositions comprising a finely-divided base material, e.g., hard carbon or bank sand, a plasticizer, e.g., bentonite (2–10%), and sufficient water to render the whole mouldable reduces the tendency of the mould to dry out before the casting operation.

Manufacture of Boron
German Patent 593,425

A fused mixture of magnesium borate (or boron oxide and magnesium oxide) with a fluoride is electrolyzed. A suitable mixture contains boron oxide 140, magnesium oxide 40 and magnesium fluoride 62 parts, and may be treated at 1100° and 10–15 volts.

Non-Metallic Permanent Foundry Mold
U. S. Patent 1,935,362

A non-metallic permanent mold is composed of three parts of chrome ore, by weight, one part of fused silica and one part of bentonite.

Foundry Moulding Sands
British Patent 401,217

A binder for moulding sand, especially regenerated sand, comprises a mixture (3:1) of coal dust and the dried and pulverized product from sulphite-cellulose liquor.

Preparation and Etching of Galvanized Wire for Metallographic Examination

The following treatment was found to give very distinct structures under the microscope: mounting of sample in ebonite; polishing with magnesium oxide suspended in denatured alcohol, etching by means of a diluted Kourbatoff reagent (1 volume of 4% nitric acid in acetic anhydride is mixed just prior to use with 10 volumes of a mixture of equal volumes of methyl alcohol, ethyl alcohol and iso-amyl alcohol).

Rust "Solvent"

Phosphoric Acid	26 %
Sulphuric Acid	2 %
Lactic Acid	53 %
Water	18.9%
Inhibitor	.1%

Rust-Proofing Oil
U. S. Patetnt 1,943,808

"White" Mineral Oil	92–98%
Naphthalene	8–2 %

Warm and stir until dissolved. This oil is non-staining and may be used on needles and other metallic surfaces contacting fabrics.

Production of Manganin

Melt electrolytic copper and nickel in a closed graphite crucible with a covering of charcoal; after melting add manganese preheated to 300–400°; complete melting as quickly as possible, and cast at 1080–1100°. Before rolling hold for 1 hour at 800–850° in a neutral or slightly oxidizing atmosphere. For drawing anneal 90 minutes at 700–750° and pickle in a solution of 10% sulphuric acid and 2% potassium dichromate at 60–70°.

Torsion Spring for Clocks, Watches, etc.

U. S. Patent 1,931,251

A torsion-spring combining high elastic limit, ductility and a substantially-uniform modulus of elasticity comprising the following elements within the ranges given:

Nickel	34.50 to 37.50%
Chromium	11.00 to 13.00%
Manganese	0.60 to 1.00%
Silicon	0.25 to 0.90%
Carbon	0.07 to 0.35%

Iron forming substantially all of the balance.

Foamite Corrosion Inhibitor

U. S. Patent 1,948,039

An acid foam solution comprises an aqueous solution of aluminum sulfate or alum, and from 1% to 10% by volume of molasses to prevent the corrosion of an iron container by said aqueous solution.

Preventing Corrosion in Radiators

U. S. Patent 1,940,041

Take thirty-six grains of borax, thirty grains of sodium salicylate and seven grains of sodium nitrite, and dissolve same in one quart of water. In this solution submerge bright pieces of iron, copper or brass, on the surfaces of which has been melted soft solder. This is for the purpose of having the solder in contact with a dissimilar metal. Next, heat the solution from time to time, to a rather high temperature, so as to approximate the conditions the metals would undergo if same were contacted in the cooling system of an automobile. On examination some time afterwards, it will be observed that the water remains clear and free of all rust, and with the exception of the cuprous metals, which are somewhat tarnished, both the iron and soft solder retain their original brightness. This same effect is noticed when the solution contains forty per cent alcohol, but should the solution contain an equivalent quantity of glycerine instead of alcohol, then no evidence of corrosion is apparent, even with the cuprous metals. If to this aqueous or aqueous alcoholic solution, a reducing agent be added, such as sugar of milk and the above experiment be repeated, then it will be observed that the oxidation of the cuprous metals is so retarded as to be almost negligible. This will be apparent even though the solution may contain only about twenty-four grains of the sugar.

Prevention of Brine Corrosion

It is generally conceded that the sodium bichromate treatment of refrigerating brines offers the most efficient and economical method of combating corrosion. The following recommendations are based on an exhaustive study of the subject, both in the laboratory and in actual plant practice. The bichromate should be used in approximately the following proportions:

125 pounds per 1,000 cubic feet of calcium brine.

200 pounds per 1,000 cubic feet of sodium brine.

These quantities should afford protection against corrosion providing the other properties of the brine are satisfactory.

The bichromate may be hung in a bag in the brine at a point of rapid circulation, but it is better to dissolve the bichromate in a small quantity of warm water and pour the solution slowly into the circulating brine.

If the original brine is neutral, bichromate of soda will produce an acid condition. The addition of 35 pound of 76% commercial flake caustic soda for each 100 pounds of bichromat used will correct this acid condition Less caustic should be used if the original brine is alkaline. The pH value as determined by an approved type of pH comparator, should be used to determine the exact dosage. If the pH of the brine is too high add bichromate if it is too low add caustic soda.

The optimum pH value for a refrigerating brine is pH 7.0 to pH 8.0. The brines should not be allowed to remain more acid than pH 6.8 nor more alkaline than pH 8.5.

To maintain the necessary chroma concentration in refrigerating brine the bichromate of soda may be added when strengthening the brine with calcium chloride of salt. The following tables give the proportions needed:

For Calcium Chloride Brine

Final Density of Brine Specific Gravity	Deg. Salometer	Lbs. of Sodium Bichromate per 100 lbs. Calcium Chloride
1.160	80	.695
1.169	84	.656
1.179	88	.621
1.188	92	.587
1.198	96	.556
1.208	100	.528
1.218	104	.502
1.229	108	.478
1.239	112	.455
1.250	116	.435

For Sodium Chloride Brine

Final Density of Brine Specific Gravity	Deg. Salometer	Lbs. of Sodium Bichromate per 100 lbs. Sodium Chloride
1.118	60	1.7
1.126	64	1.675
1.134	68	1.57
1.142	71.7	1.47
1.150	75.2	1.39
1.158	79.1	1.32
1.166	82.8	1.24
1.175	86.8	1.18

Finishing Die Castings

Best results in plating aluminum die casting are secured when the surface of the casting has been roughened, not by sand-blasting or other mechanical means, but by an etching which dissolves out selectively a part of the alloy used in the casting and leaves an under-cut surface to which the subsequent plating adheres much more firmly than to a commercially smooth surface. The etching solution is composed of three parts—concentrated nitric acid (specific gravity 1.24) and one part of 60 per cent hydrofluoric acid. A dip of 10 to 30 seconds in this solution is sufficient to provide the desired anchorage for the plate. Prior to etching, the castings are dipped for about 30 seconds in a hot alkaline cleaner containing one ounce per U. S. gallon each of sodium carbonate and tri-sodium phosphate. This solution attacks the metal lightly and causes a surface gassing which helps to clean mechanically as

well as chemically in much the same way as electrolytic cleaning.

Nickel is the preferred primary coat, as the "anchors" formed are stronger than those of a softer metal. Subsequent coatings of almost any of the common plating metals may be applied. Many nickel baths have been used, but one of the two following is most generally employed:

Formula No. 1

	Per Gal.
Nickel Sulphate	19. oz.
Magnesium Sulphate	10 oz.
Ammonium Chloride	2 oz.
Boric Acid	2 oz.

Formula No. 2

Nickel Sulphate	16 oz.
Anhydrous Sodium Sulphate	26 oz.
Ammonium Chloride	2 oz.
Boric Acid	2 oz.

In either case the temperature of the solution should be 90 to 95° F. and a current density of 15 amperes per square foot is advised. A pH value of 5.8 to 6.0 (colorimetric) gives good results. In general, the lower the pH the better will the solution "throw" into recesses.

Aluminum die castings are prepared for plating in much the same way as those of the zinc-base type so far as grinding, polishing, buffing and coloring are concerned. These operations are followed by the alkaline dip and acid etch, referred to above, by a clear water rinse and by a 30-second dip in a water solution containing 37 ounces of nickel chloride and 2 ounces of hydrochloric acid (specific gravity 1.18) per U. S. gallon. Plating time depends, of course, upon the thickness of plate required. Thirty minutes to 1 hour is usually sufficient in the nickel-plating solution if a current density of 15 to 20 amperes per square foot of cathode surface is maintained. When properly plated the coating should adhere firmly to the metal, even when the latter is slightly deformed. Nickel is always recommended as the primary coat, but chromium, copper, brass or other metals may be plated over the nickel.

Although some die castings of tin-, lead- and copper-base types are employed, they are rarely plated except on occasional decorative articles, as all such alloys are highly resistant to atmospheric corrosion or are not affected beyond slight surface tarnishing or discoloration.

Some zinc-base and many aluminum-base die castings are finished by chemical dipping processes. There are many solutions for forming black deposits on zinc, but most of these are not very adherent, and often they have to · be "fixed" or coated with lacquer or wax. One such solution contains white arsenic, 2 ounces; sodium cyanide, 2.5 ounces; and caustic soda, 5 ounces per U. S. gallon of water. This gives a gunmetal or black finish, depending on the length of dip and the temperature. Deposits are formed with the castings immersed and with constant agitation in the solution, which is usually held between 180 and 200° F., until the desired deposit is formed uniformly on the surface. After washing in hot water and drying fixing is done, as a rule, by dipping in thin transparent lacquer. Solutions of antimony and of copper are used sometimes to secure similar effects.

Cleaning Die Castings

An efficient cleaning bath for die castings can be made up with 3½ to 5 ounces sodium metasilicate per gallon of water. In the majority of cases 3½ to 4 ounces are sufficient. The cleaning solution should be boiling or nearly so, and the parts cleaned cathodically for 2 to 3 minutes. The die castings will be slightly darkened which is a good sign the work is clean. The addition of powdered rosin equal to 1 per cent of the sodium metasilicate (0.05 ounce per gallon) is beneficial but not necessary.

In cleaning anodically, the zinc is easily attacked and will blacken parts in 2 minutes time which cannot be removed in acid dip. No nickel deposit will hold to such a surface. After cleaning, the acid dip for die castings should be strong enough to etch deep into the surface. In order to obtain a maximum adherence of nickel use a 1 normal (3½%) muriatic acid dip. This is easily prepared by taking commercial concentrated acid and diluting with water to one-tenth its original strength. The die castings should be left in acid dip for at least 30 seconds which darkens them considerably, but is an indication that the surface is etched deep and will hold nickel so that it cannot be worked off by bending. The chromium plate always holds to a die casting prepared in this manner.

PAPER

Paper Sizing
British Patent 386,018

700 parts of resin are melted with 200 parts of turpentine oil of high boiling point, e.g., that known commercially as Yarmor pine oil, and are stirred into a solution of a caseinate-resin soap emulsifying medium in 1,000 parts of water preferably using a steam jet or high-speed rotary stirrer. The resin soap is made by adding 100 parts of commercial caseinate powder to 80 parts of powdered resin and a 10 per cent solution of ammonia is added with mild heating to the mixture which is kneaded till saponification is completed and the mixture of swollen ammonium caseinate and ammonium resinate has reached its highest viscosity.

Hygroscopic Sizes in Paper Making

Wrinkling of sized paper when it becomes very dry, is due to a greater shrinkage coefficient in the size than in the paper. To prevent wrinkling the hygroscopicity of the size must be increased by adding enough glycerol to equalize the shrinkage.

Sizing Paper Pulp
British Patent 381,633

Powdered size consisting of rosin saponified with an alkali is added to stock in the beater. In an example 100 pounds of commercial rosin is cooked with 40 pounds of water and 18 pounds soda ash or with sodium, potassium, or ammonium hydroxide, sodium silicate, potassium carbonate, etc. The product may be atomized in a current of hot air. Alum may be added as a fixative.

Paper Pulp
Canadian Patent 335,222

Pulp is produced by circulating a dilute aqueous solution of ammonia over wood chips in the proportion of solution to chips of 5:1 by weight for about 45 minutes at 100–110°. The ammonia solution is removed without reduction of pressure. About 2 parts by weight of a relatively more dilute solution of ammonia per part of wood chips is circulated over the wood chips for 15 minutes at 70°. All but about 0.85 part of the solution is removed and the wood chips are cooked with 4.4 parts of sulphurous acid of 5.5% concentration.

Sulphite Pulp
Canadian Patent 334,891

Wood chips, etc., are cooked in the usual manner for 5–11 hours, until a temperature of 120–55° is reached. The liquor is drained off without substantial lowering of temperature and pressure. A second-stage cooking liquor preheated to the transfer temperature is forced into the digester. The liquor has a higher relative acidity than the first-stage liquor and preferably has a lower concentration of total sulphur dioxide. Cooking is continued until the fiber is completely liberated. The pressure is kept constantly at 75 pounds. The temperature is raised gradually from the transfer temperature to 145–55°, which is reached in 3–6 hours.

Chemical Pulp
Canadian Patent 334,879

Manila, sisal, ramie or similar bast or leaf fibrous material is steeped in water containing 5–8% chlorine, based on the weight of the dry fiber. After the chlorine has been practically consumed, for which purpose a 30 minute period is ample, the material is washed free of residual chlorine and reaction products. The washed material is digested at 300–335° F. in a plain solution of sodium sulphite, which need not be of greater strength than 2%. After 3 hours' cooking the material has been resolved into a pulp of excellent strength and other physical characteristics. The pulp is washed and bleached with a liquor containing as little as 2–3% calcium hypochlorite based on the weight of dry pulp. The cooking liquor may advantageously contain sodium carbonate and (or) sodium sulphide in substantial amount.

Bleaching Wood Pulp
Canadian Patent 334,467

Alkali-cooked wood pulp is treated with water containing 4–8% chlorine, based on pulp, until the chlorine is substantially consumed. The chlorinated pulp is digested for 1–4 hours at 40–70° in an alkaline liquor containing 2–4% alkali, calculated as sodium hydroxide, to remove most of the ligneous reaction products from the pulp while substantially preserving the a-cellulose contents.

Cooking Raw Manila Fiber
U. S. Patent 1,921,539

Raw manila fiber is cooked for about 4 hours at a temperature of about 170° in a liquor containing about 2–3% sodium carbonate and about 2% sodium sulphide, to obtain a strong, long-fibered, easily bleachable pulp suitable for making white paper of high grades.

De-inking Printed Paper
U. S. Patent 1,925,372

For separating the cellulosic fibers of filled and printed paper, the paper is beaten in a solution containing sodium hydroxide, sodium metasilicate or sodium phosphate in such proportion as to give the solution a pH of 9.0–12.6 until the paper has been substantially disintegrated, and an emulsifying agent such as a soap or sulphonated oil is then added to the resulting aqueous suspension and the beating is continued for a short time. The material is filtered through a sieve fine enough to retain the cellulosic fibers but coarse enough to pass the fillers and ink particles.

Manufacture of Parchment-Like Paper
British Patent 397,550

Unsized paper made from high-α-cellulose sulphite pulp is impregnated (at 140–160° for 1–3 minutes) with a molten, partly polymerized glyptal resin plasticised with, e.g., tricresylphophate, and the resin is then completely polymerized, e.g.,at 130° for 6 hours.

Fluid for Making Safety Paper
U. S. Patent 1,939,378

This compound may be made in any quantity by dissolving one grain of tannic acid in each gallon of water and adding 2½ to 5½ cubic centimeters of glacial acetic acid; to this add one-tenth of a gallon of decolorized iodine. Ordinary paper treated by this process is transmuted into safety paper which discolors and sets ink written on its surfaces when eradicators are used to remove the ink. If only 2 cubic centimeters of acetic acid are used the inkset will be thrown out and will not appear; the same thing will happen if 6 cubic centimeters or more of acetic acid are used. Between 2½ and 5½ the different colors require different amounts to get the best results. Different grades of paper require special formulas to be worked out between the two extremes specified.

Asphalt Paper
U. S. Patent 1,940,431

A papermaking procedure designed for the production of a mulch paper using an asphalt dispersion product may be practised substantially as follows. A suitable stock, such as kraft pulp, may be beaten out to the desired degree of hydration, say, from one to four hours, whereupon the desired amount of Montan wax-asphalt dispersion may be added thereto, say from 10% to 20%, based on the weight of dry pulp. After the dispersion has been uniformly disseminated throughout the pulp, about 4% to 6% alum, based on the weight of dry pulp, may be added to effect a fixation of the dispersed particles on the fibers. When the sized pulp is then run out on a paper machine, the white water removed during papermaking operations is substantially clear, showing that substantially all the size has been retained by the fibers. Moreover, because of the fine particle size, no gumming of the wires or felts is encountered during the papermaking operation. The resulting paper is admirably adapted for use as a mulch, being highly waterproof because of the uniform distribution of waterproofing material therethrough. Rather than having a black color, the paper has a tan shade, which indicates that the asphalt has been distributed therethrough as very fine particles and has blended with the color of the fiber.

In preparing dispersions of Montan wax and paraffin, a mixture consisting of 15% or more of crude Montan wax may be employed. The mixture is melted and heated to a temperature of about 220° F which temperature is materially above the melting point of the paraffin (130 F.). To about 50 parts of the Montan paraffin wax mixture may be added with vigorous agitation 50 parts of a solution of caustic soda of ¼% to ½% strength and at 95° to 100° F. The resulting dispersion is of a creamy consistency and is sufficiently fluid to be sucked up with a pipette. The final dispersion, in a case when a ½% caustic solution is used, may contain as high as 50% of the alkali originally used in free condition. Preferably the resulting paraffin dispersion, while at a temperature above that of the melting point of the paraffin, is suddenly chilled, as the fine particles of paraffin have a tendency to agglomerate and coalese while in the liquid state, if allowed to cool slowly. The resulting dispersion is, as in the case of the asphalt dispersion, an excellent waterproofing composition for felts, papers, yarns, textiles, fabrics, or the like. It is adapted for use on the calender rolls of a paper machine, to impart a high gloss or finish to the paper, under which conditions it may be used at very low concentration and maintain its stability. It is also suitable for the sizing of paper pulp in the beater engine and for the so-called tub-sizing of paper to render it waterproof and to impart thereto the appearance and characteristics of waxed papers produced by dipping in molten paraffin. In sizing papers after they have been formed by passing through a bath of the paraffin dispersion, the paper may be either in waterleaf condition or in partially sized conditions, in which latter case when passed through the paraffin wax dispersion too great an absorption of paraffin wax will not take place.

Tracing Paper
U. S. Patent 1,946,338

A lacquer is prepared according to the following illustrative formula:

Formula No. 1

Nitrocotton (¼″ or ½″)	50 parts
Tricresyl Phosphate	30 parts
Butyl Acetate	150 parts
Toluene	100 parts

For the production of tracing paper, the above lacquer is sprayed on one or both surfaces of, for example, an onion-skin typewriter paper in sufficient quantity to produce a relatively thin film thereon, on drying. After drying of the film, the paper is ready for use.

As a further illustration, for example, a lacquer is prepared according to the following illustrative formula:

Formula No. 2

Nitrocellulose (½″ or ¼″)	10%
Dibutyl Phthalate	5%
Butyl Acetate	20%
Ethyl Acetate	6%
Toluene	45%
Butanol	6%
Ester Gum	8%

The above lacquer may be sprayed on one or both surfaces of the paper, or the paper dipped therein. Alternatively, the paper, having been dipped in the lacquer according to the above formula, or one containing nitrocellulose and a plasticizer, is sprayed with a solution of nitrocellulose, which may contain a small amount of plasticizer or resin, to form a thin film on the surface thereof, which will give a surface free from greasiness, a characteristic of films including plasticizers, in any large amount. If desired, the surface of the paper not intended to receive ink may be coated more heavily than the ink-receiving surface for the purpose of increased waterproofing effect, it being noted that the film on the ink-receiving surface should be relatively thin, since its adaptability for desirably receiving ink will be decreased with excessive thickness of film.

Tests for Transparent Papers

Type	Behavior	Smell	Ash
Viscose	Burns slowly	Burnt paper	Small quantity
Cupra	Burns slowly	Burnt paper	Small quantity
Acetate	Burns only in flame. Fuses	Acrid	Blistered carbon residue
Nitro	Burns rapidly		
Gelatin	Fuses	Burnt feathers	Black fused mass

The action of the ash to litmus is acid in each of the synthetic types of film; alkaline in the case of gelatin.

Solubility tests may be applied as follows:—Acetate films are soluble in acetone and glacial acetic acid, while the nitrocellulose variety is soluble in an ether-alcohol mixture. A hot 10% solution of caustic soda rapidly dissolves gelatin films.

Among the most interesting are the reactions of these films to certain dyeing

reagents, and for this purpose small quantities of the following three may be kept on hand:—
(1) Zinc chloride-iodine solution.
(2) Diphenylamine-sulphuric acid.
(3) A solution of 0.4 gram thodamine B extra and 1.0 gram diamine pure blue FF per litre of water.

In (1) viscose and cupra foils dye bluish violet, acetate dyes yellow, while gelatine dyes yellowish brown.

In (2) all except nitro-cellulose foils are unaffected, this particular type, however, dyeing blue to bluish red.

In (3), which is a special reagent suggested by A. V. Schlutter, we have a very interesting means of determining whether a film is of the viscose or of the cupra variety. According to this author, if the sample of film is placed at room temperature in reagent 3 for one minute, being afterwards washed for ¼ min. and dried at 80–100° C., the viscose film dyes a shade between violet-rose and blue-violet. The cupra film dyes a pure blue.

The particular shade between violet-rose and blue-violet, to which the viscose film dyes, often forms a guide to the origin of the particular film, provided up-to-date standard dyeings of known brands are kept for purposes of comparison.

Water and Greaseproof Solution for Coating Paper

Latex (60%)	50 cc.
Paraffin Emulsion (50%)	30 cc.
Sodium Silicate	5 cc.
Water	30 cc.

The paraffin emulsion is diluted with the water and the silicate stirred in. The latex is added last and stirred until homogeneous.

The solution may be applied to the paper either by a brush or through the use of a spray gun.

Paper Coating
U. S. Patent 1,746,888

Fifteen pounds of shellac are dispersed in three gallons of water containing 3 pounds of borax, 100 pounds of bentonite are carefully mixed with it. Latex containing 1% ammonia is added sufficient to produce 10 pounds of rubber solids in the mixture. Paper coated with this compound is said to have great strength and flexing properties, and high resistance to ink adsorption.

Paper Coating, Carnauba Emulsion

Dry Castile Soap	12 lb.
Water	25 gal.
Boil till dissolved, then add	
Camauba Wax	75 lb.

Boil three hours; cool to 100° F. and add 3 pints 26° ammonia. Make up to 75 gallons with water. This emulsion is added to the casein coating mixture.

Thinning Clay

Taking a leaf from the potters' book, paper-makers are beginning to find that the use of a small amount of silicate in their clay is a great advantage. It makes the clay mixes much more fluid than they would be with the same amount of water without silicate. This makes the clay easier to handle through pipes and screens. Another important use is in connection with the use of clay in coating paper, where the addition of a small amount of silicate means that considerably less water may be used. The amount of silicate is around 1% of the weight of the clay, though it is worth while to test different batches of clay to find out what amount gives the best result in each case. A little too much diminishes the result obtained. The "S" Brand of "A-Syrup" Sodium Silicate will be found satisfactory for this.

Waterproof Solution for Paper

Latex (60%)	35 g.
Water	35 g.
Glue Solution (10%)	10 g.
Hydrowax Cream (50%)	40 g.

The glue solution is prepared in the usual way by soaking and heat. The hydrowax cream is dissolved in the water the glue added and when homogeneous the latex is stirred in.

The solution may be applied either by brush or spray and is specially suited for Kraft paper.

Waterproofed Paper

A sheet of paper or paper board may be waterproofed by treating with amber petrolatum and pressing between rollers. The paper is not oily or greasy. Evidence of the treatment is not a parent and the paper is very water repellent.

Waterproofing ''Cellophane''

Nitrocellulose films are coated with the following composition:

Nitrocellulose	31.5	parts
Ethyl Acetate	315.0	parts
Castor Oil	70.5	parts
Ethyl Ether	315.0	parts
Ceresin	0.315	part
Petroleum (Refined)	1.575	parts
Wool Fat	1.575	parts
Ester Gum	0.315	part

Waterproofing Solution for Blueprints

To waterproof blueprints and give them a sheen and greater legibility, rub them with a soft cloth that has been dampened with a solution of rosin, 50 grains, paraffin, 100 grains, and turpentine, 1 ounce.

Waterproofing Paper and Fibreboard

The following composition and method of application will render uncalendered paper, fibreboard and similar porous material waterproof and proof against the passage of penetration of water.

Paraffin (Melting Point about 130° F.)	22.5%
Trihydroxyethylamine Stearate	3.0%
Water	74.5%

The paraffine or a wax is melted and the stearate added to same. The water is then heated to nearly boiling and then vigorously agitated with a suitable mechanical stirring device, while the above mixture of melted wax and emulsifier is slowly added. This mixture is cooled while it is stirred.

The paper or fibreboard is coated on the side which is to be in contact with water. This is then quickly heated to the melting point of the wax, which then coalesces into a continuous film that does not soak into the paper which is preferentially wetted by the water. This method works most effectively on paper pulp moulded containers and possesses the advantages of being much cheaper than dipping in melted paraffine as only about a tenth as much paraffine is needed. In addition, the outside of the container is not greasy, and can be printed upon after treatment which is not the case when treated with melted wax.

Oil and Greaseproofing Paper and Fibreboard

This solution applied by brush, spray, or dipping will leave a thin film which is impervious to oils and greases. Applied to paper or fibre containers, it will enable them to retain oils and greases. All the following ingredients are by weight:

Starch (preferably Cassava)	6.6%
Caustic Soda (76% Na_2O)	0.1%
Glycerol	2.0%
Sugar	0.6%
Water	90.5%
Sodium Salicylate	0.2%

The caustic soda is dissolved in the water and the starch is then made into a thick paste by adding a portion of this solution. This paste is then added to the water. This mixture is placed in a water jacket and heated to about 85° C. until all the starch granules have broken and the temperature maintained for about half hour longer. The other substances are then added and thoroughly mixed and the composition is completed and ready for application. A smaller water content may be used if applied hot and a thicker coating will result. Two coats will result in a very considerable resistance to oil penetration.

Acid Casein Sizing
U. S. Patent 1,347,845

For use where a size or adhesive having an acid reaction is desired.

Sodium Fluoborate	15 parts
Casein	85 parts
Water	400 parts

Mix thoroughly and heat.

Moisture Grease-proof Paper
U. S. Patent 1,895,527

Multi-ply box board with hard sized liners, and preferably containing an inner layer of asphalt, is coated on one side first with a solution containing, e.g., 75 volumes of 38% aqueous sodium silicate, 25 volumes of glycerin, and 1–2% of alkali (sodium carbonate), and then, after drying, with (paraffin) wax. Such a board is moisture and grease-proof and suitable for foodstuffs.

Moistureproof Paper
French Patent 754,614

The sheet, printed or unprinted, is first passed through a bath of some wax dissolved in a suitable liquid hydrocarbon or mixture of hydrocarbons, with the concentration of the solution kept at 10 per cent or below. The solvent is removed at 50° C., and the sheet is subsequently coated on one or either side with nitrocellulose containing a natural or artificial resin, and then subjected again to 50° C., for removal of the solvent. The permeability of the paper to moisture is always expressed in milligrams of water per hour, per square centimeter, at 24° C. The wax alone has practically no effect on moistureproofing. Thus a paper showing an initial permeability of 0.56mg/hr./cm,[2] after impregnation with a 10 per cent solution of paraffin (melting at 150° C.) in naphtha, showed a permeability of 0.57. The effect of a subsequent treatment with 16 grams nitrocellulose in an admixture of 8 grams of dibutylphthalate 4.8 g. of ester gum, and 84 grams of a mixture of 20 per cent ethyl acetate, 30 per cent alcohol, and 50 per cent toluene, showed a permeability of only 0.0031. The waxes used are (primarily) paraffin, beeswax, candellila wax, and spermaceti. Wax solvents (primarily) toluene, naphtha, and other petroleum fractions. The plasticizers, (primarily) dibutylphthalate, diethylphthalate, tributyl phosphate, triphenylphosphate, castor oil, o-cresyl toluenesulphonate, etc. Among the resins used (mainly) the ester gums, the cumarone and indene resins, the phenolformaldehyde resins, and sandarac. The nitrocellulose solvents include suitable mixtures chosen from the following: ethyl acetate, ethyl alcohol, acetone, toluene, benzene, and naphtha.

Washable Wall-Paper

U. S. Patent 1,955,626

I. Paper with a finish of approximately

No. 1 Dry Unbleached Spruce	25%
No. 1 Wet Unbleached Spruce	10%
No. 1 Ground Wood	65%
Clay	5%
Size	2%
Alum	4%

is coated in manufacture with potentially a water-resistant coating, such as casein and formaldehyde and clay, or rosin and starch and clay, or linseed oil and starch and clay. A typical coating solution is as follows:

Clay Slip

Clay	750 lb.
Water	750 lb.

Casein Solution

Water (Heated to 90° F.)	525 lb.
Casein	105 lb.

Add 9 pounds, 8 ounces of ammonia water (sp. gr. 0.90) to casein solution and 3 pounds of borax which has been dissolved in 2 gallons of water. Heat this mixture to 120° F. Then add 7 pounds of pine oil (used to prevent foaming) and 4 pounds of formaldehyde. (40% solution).

The finished casein solution is then added to and mixed with the clay slip solution.

Sufficient coating of this mixture is then applied to the paper web on the paper machine to give the finished paper resistance to fading. This amounts to approximately 10 pounds per 24x36—500 sheet ream basis.

II. In the preferred methods, designs or patterns are printed on the dry paper with water ink.

A typical ink formula is as follows:—

Casein Solution for Binder

Color paste (formula varied according to particular pigment used):

Water	9 lb. 10 oz.
Helio Fast Red RLD Paste	11 lb. 6 oz.
Clay	21 lb.
Casein Solution (Formula Given Above)	40 lb.
Pine Oil	80 cc.

Casein Solution

Mix thoroughly and soak for twenty minutes:

Water at 65° C.	54 lb.
Casein (30 mesh California)	20 lb.

Add a solution of:

Water (at 80° C.	12 lb.
Borax	3 lb.

Add a solution of:

Water (at 60° C.)	4½ lb.
Soda Ash	½ lb.

The temperature of the casein solution should now be 55 to 60° C. Hold at this temperature for at least one-half hour, or until perfectly "smooth." Cool to about 30° C. (not higher than 35° C.).

Add a solution of:

Cold Water	5 lb. 6 oz.
Hexamethylenetetramine	1 lb. 9 oz.

III. The third step involves treating the printed paper with a fixing or hardening agent. A typical formula would be:

Dissolve 3 pounds alum (papermaker's) in 10 gallons warm water, add 1.5 pounds (680 cc.) formaldehyde (40% solution).

This solution is preferably applied by means of a roll, although other method of uniformly wetting the surface of the web may be used.

At the conclusion of this step the pape is dried.

A glossy surface may be obtained b applying an emulsion of wax to th printed surface.

Paper Waxing
U. S. Patent 1,953,085
Formula No. 1

Paraffin Wax	33 parts
Silicate of Soda	4 parts
Alum	2 parts
Glue	1 part
Water	60 parts

This formula is stated in parts by weight.

The silicate of soda used in the above formula is a commercial form of syrupy consistency containing about 50 per cent of water. This silicate of soda syrup is mixed with 65 per cent of the total water. The remaining 35 per cent of the total water is used to dissolve the alum. The glue is added to the diluted silicate of soda. The wax composition is prepared by first melting the wax, adding the silicate of soda solution containing the glue and agitating, finally adding the alum solution with agitation. The temperature at which the composition is prepared is about 170° F.

Formula No. 2

Paraffin Wax	32.5 parts
Montan Wax	0.85 part
Silicate of Soda	5 parts
Alum	2.75 parts
Glue	0.4 part
Water	58.5 parts

This formula is stated in parts by weight.

In this case, as in Formula No. 1, the silicate of soda specified is a syrup containing about 50 per cent of water. In like manner this syrup is diluted with 65 per cent of the total water and the alum is dissolved in the remainder of the water employed.

The glue is dissolved in the silicate of soda solution. It is desirable to add 1 or per cent of phenol, based on the amount of glue, in order to preserve the latter. The waxes are melted together and the silicate of soda solution containing the glue is added thereto with thorough agitation, the temperature being about 170° F. The alum solution is then added and agitation continued for a short time.

Formula No. 3

Paraffin wax	31 parts
Montan wax	7 parts
Silicate of soda	2.65 parts
Alum	1.35 parts
Water	58 parts

This formula is stated in parts by weight.

The silicate of soda syrup is diluted with a part of the water and the alum is diluted with the remainder, somewhat similar to the manner set forth in formulas 1 and 2. The waxes are melted together and the silicate of soda solution is added thereto with vigorous agitation at a working temperature of 170° F. When the incorporation is thoroughly effected the alum solution is introduced. A very fine wax dispersion results.

Formula No. 4

Paraffin wax	37 parts
Quick lime	1.7 parts
Alum	4.6 parts
Water	56.7 parts

This formula is stated in parts by weight.

The quick lime is treated with about 65 per cent of the total water and the alum is dissolved in the remaining water. The wax is melted and placed in an agitator and the milk of the lime is added. After stirring vigorously for about one minute the alum solution is added and agitation continued for two minutes, when the wax composition is ready for use. As in the foregoing, it is recommended that the working temperature in the preparation of the wax composition be about 170° F.

Dilute the wax composition (which comes from the agitator containing approximately 30 per cent of waxy materials) until it contains about 5 per cent of waxy material and add it in the diluted state to the beater through a screen of about 60 mesh. This insures a better dispersion through the stock in the beater. When using waxes containing rosin soaps and free rosin it is necessary to closely control the acidity of the stock after the size is added to assure proper coagulation of the size on the fibres, this acidity having a value corresponding to pH of about 4.5. Rosin size, furthermore, should be preferably added to stock that is slightly alkaline. When the wax composition involved in the present invention is used the acidity of the paper stock, after said composition is added, is not at all critical, and may vary from an acidity corresponding to a pH value of 4.5 to 7, without affecting the retention of the wax by the fibre. The wax composition can be added to the pulp if the pulp is acid, that is, corresponding to a pH value of from 5 to 6.5. In this case it is not necessary to add any further precipitant such as alum. This allows of a considerable saving of alum. The stock being less acid increases the life of the various parts of the system such as pumps, piping and wires.

Mounting Azo Paper on Metal

Aluminum with a matte finish, and thick enough to resist the curling action of the paper (about 1/32 inch), is commonly used. Cut the plates to size and round the corners. Clean in caustic soda solution, to remove grease and increase the grain. The paper can be mounted directly on the aluminum, or the metal can first be sprayed on both sides with a matte-black baking japan, used as directed by the manufacturer. If art work is to be done on the paper, Azo Grade E, double weight, will be found very suitable.

For mounting, use a stiff paste made of water and a very good casein cement. The paste must be worked smooth, and brushed on the plate, being worked into the surface with a stiff brush. Do not put paste on the back of the paper. The paper must be dry. Lay it on the prepared plate in a dark room. Cover with a sheet of plain paper, and squeegee by hand until smooth contact is secured. Put plates under a letterpress. Cover each one with paper, and allow to remain until the paste has set. The mounted sheets are used just like unmounted paper. In the art room the surface works like a good, hard-surfaced, semi-matte drawing paper.

Bleaching Solar Bromide Paper
(When Used for Making Pen Drawings)
Stock Solution No. 1

| Water | 32 | oz. |
| Potassium Permanganate | 1¾ | oz. |

Stock Solution No. 2

| Cold Water | 32 | oz. |
| Sulphuric Acid, C.P. (pure concentrated) | 1 | fl. oz. |

For use, take Stock Solution No. 1, 1 part, Stock Solution No. 2, 2 parts, water

64 parts. When the paper has been sufficiently reduced, immerse in a plain hypo solution, or in a fresh acid fixing bath for a few minutes, to remove yellow stain, after which wash thoroughly.

The best means of dissolving the permanganate crystals in Solution No. 1 is a small volume of hot water (about 180° F., 82° C.). Add the permanganate and shake or stir vigorously until the crystals are completely dissolved. Then dilute to volume with cold water. When preparing Stock Solution No. 2, always add the sulphuric acid to the water slowly, while stirring, and never the water to the acid; otherwise the solution may boil and spatter the acid on the hands or face, causing serious burns.

Blueprint Bleach

Gum Arabic	3 g.
Soda Ash	2 g.
Water	1 oz.

Add a very small amount of potassium ferrocyanide to make the bleach blue white, rather than yellow white.

"Magic" Writing Pad

Cardboard is coated with the following composition and covered with a sheet of waxed or oiled paper. When the latter is "written" on with a stylus or pencil the writing appears. When this sheet is lifted away from the coated cardboard the writing disappears.

Beeswax	4	parts
Venice Turpentine	9	parts
Lard	4	parts
China Clay	3½	parts
Carbon Black	1	part
Mineral Oil	2	parts

The consistency may be varied by varying the proportions of liquid in the formula.

PHARMACEUTICAL AND PROPRIETARY PREPARATIONS

Lotion Vegetale au Seringa

Tincture of Vanilla	200 g.
Tincture of Cantharides	20 g.
Terpineol	10 g.
Oil of Rose Geranium	1 g.
Oil of Ylang-Ylang	1 g.
Heliotropin	2 g.
Alcohol	2000 cc.
Water	80 cc.

Psoriasis Lotion

Picis Pini	170 parts
Liquor Picis Carbonis	690 parts
Olive Oil	190 parts
Bitumen Sulphonatum	60 parts
Bismuth Subnitrate	175 parts
Water enough to make	3785 parts

Rub the pine tar with the liquor picis carbonis and then slowly work in the olive oil. Rub up the bitumen sulphonatum with some of the water, add the bismuth subnitrate to this mixture and incorporate this aqueous mixture with the first mix above. Triturate well till thoroughly mixed and bring up to required amount with additional water.

Calamine Lotion

Formula No. 1

Calamine	16 g.
Zinc Oxide	30 g.
Glycerin	45 g.
Oxyquinoline Sulphate	4 g.
Distilled Water	450 cc.

Dissolve the oxyquinoline sulphate in the water. Mix the dry materials with glycerin, add the oxyquinoline solution.

Formula No. 2

Zinc Oxide	20 g.
Calamine	100 g.
Zinc Sulphocarbolate	30 g.
Tincture of Orris	50 cc.
Glycerin	50 cc.
Violet Synthetic	1 cc.
Water to produce	1000 cc.

Blackhead Lotion

Oil Bitter Almonds	3	drams
Oil Liquor Potassa	1½	drams

Tincture Benzoin Compound	½	dram
Rose Water to make	4	oz.

Apply at night. Wash with warm water in the morning.

Blackhead Remover

Triethanolamine	1 part
Glycerin	5 parts
Alcohol	35 parts
Water	60 parts

Acne Lotion

Triethanolamine	2 parts
Potassium Carbonate	1 part
Witch Hazel	65 parts
Rose Water	26 parts

In above, the triethanolamine penetrates and emulsifies sebaceous matter in pores.

Bust Developers

Artificial nourishment of skin to counteract atrophy of bust, by rubbing in skin creams has been looked upon with disfavor, as it was doubtful that such preparations could be effective. However, it is now claimed that since stable and highly active hormone extracts are available, such creams are procurable of sufficient effectiveness to warrant their recommendation. For example, 20 parts lanolin are mixed with 10 parts coconut oil, 10 parts stearin, 120 parts olive oil, 4 parts lecithin, 2 parts cholesterin, 60 parts water, 0.4 part paraoxybenzoic acid ester and 1 part sodium benzoate.

Antiseptic Ointment

The value of ointments seems to depend as much on the type of base used as on the antiseptic constituent according to recent test.

Petrolatum	73%
Anhydrous Lanolin	25%
Phenol U.S.P.	2%

Variations of more than ½% will materially lower the antiseptic value. Grind together.

Antiseptic Cream

Carbolic Acid	5	grains
Camphor	10	grains
Lanolin (Anhydrous)	1	oz.
Paraffin Wax (Soft)	1½	oz.
Cocoa Butter	½	oz.

Melt the lanolin, paraffin, and cocoa butter together. Rub together the carbolic acid and camphor, and when they are liquid, add to the first mixture. Stir until cold.

Wound Antiseptic
U. S. Patent 1,924,169

Meta-cresol	1	part
Camphor	3	parts
Oil of Pine Needles	0.5	part

Antiseptic Lotion

The following lotion has been found suitable for use after shaving and for minor cuts and skin infections:

70% ethyl alcohol (a commercial rubbing alcohol) is saturated at its boiling point with crystalline boric acid. The solution is allowed to cool and then filtered from the excess boric acid.

Methyleugenol Antiseptic Lotion

Methyleugenol	4 fl. oz.
Alcoholic Soap	14 fl. oz.
Distilled Water	20 fl. oz.

Two teaspoonfuls diluted to a pint may be used as a soluble antiseptic lotion.

Obstetric Antiseptic

Laboratory experiments indicate that the commonly used antiseptics when diluted sufficiently to become non-irritating are much less efficacious than non-irritating solutions of chlorothymol in 20 per cent alcohol and 10 per cent glycerin The clinical results in the 164 cases in which chlorothymol was used were better than those following the use of mercurochrome and iodine in a control series of 164 cases in spite of the fact that a larger number of the various factors which might cause morbidity were noted in the chlorothymol group.

Preservative for Hæmoglobin

Alcohol and glycerin are mostly used for this purpose. A maximum proportion of glycerin is 25%. An ideal preservative is also claimed in esters of parahydroxybenzoic acid. About 10% alcohol and 10% glycerin is a good mixture. Tests should, however, be made on a small sample of the hæmoglobin to determine maximum allowable alcohol content of preparation. One formula calls for 333.3 parts liquid hæmoglobin, 50 parts tincture of cacao and 135 parts simple syrup, 431 parts water. Preparation is more stable when small proportion of glycerin is added.

Antiseptic Flexible Collodion

Gun Cotton	4.0 g.
Alcohol	18.0 cc.
Tincture of Benzoin	9.0 cc.
Ether	75.0 cc.
Phenol	0.5 g.

Shake the gun cotton with the ether until a pulpy mass results. Then add the other ingredients and shake well.

Antiseptic

Brilliant Green	1%
Alcohol	40%
Water	59%

The above solution can replace tincture of iodine in most cases.

Toothache Remedy

Spirits of Peppermint	½ oz.
Tincture Ginger	½ oz.
Tincture Capsicum	½ oz.
Spirits Camphor	½ oz.
Spirits Chloroform	½ oz.
Ammonia Water	2 drams

Dental Pulp Protector
Part No. 1 (Liquid)

Eugenol

Part No. 2 (Solid)

Zinc Oxide	97.5%
Silver Proteinate	2.5%

Mix in quantities to make pasty before use.

Pyorrhea Astringent

Copper Sulphate	15%
Monsel's Solution	10%
Boric Acid	3%
Glycerin and Water	72%

Pyorrhea Surgical Cement
Part No. 1

Zinc Oxide	1 part
Rosin Powdered	1 part

Bolt through silk cloth separately and mix in above proportions.

About half hour before use mix with sufficient of following to give proper consistency:

Part No. 2

Eugenol	1 part
Mineral Oil	1 part

Local Dental Anæsthetic

Ethyl Amino Benzoate	1 part
Diethylene Glycol-Ethyl Ether	2 parts

Tooth Bleach
(Removing Nicotine Stains)

Formula No. 1

Sodium Bicarbonate	50 parts
Precipitated Chalk	50 parts
Flavor as desired.	

Formula No. 2

Lactic Acid	3 parts
Talc	30 parts
Flavor as desired.	

Taste Correctives

Lecithin tablets are sometimes objectionable because of taste. Ethyl malate, vanillin or lemon oil are best used for correcting this. A 2 to 5% solution of vanillin in lemon oil is suitable. Ethyl malate may be used alone or in mixtures with 7.5 parts of the ester with 100 parts of amyl valerianate and 892.5 parts of alcohol. Suitable mixture is 2 grams of vanillin dissolved in 10 grams of ethyl malate and 3 grams of lemon. Still another suitable mixture consists of 0.1% vanillin, 5% aromatic tincture and 0.2% ethyl malate. Quinine wine is treated with tincture of orange peel, tincture of cinnamon, tincture of cardamom seed and tincture aromatica. Also useful is addition of strong or weak alcoholic extract of cocoa powder.

Coloring Matter for Mouth Washes

A dilute ethyl alcohol extract of red sandal wood or/and cochineal is suitable for coloring mouth washes containing about 60% hydrogen peroxide. Reddish, non-poisonous food colors may also be used, such as raspberry red. Deep color is obtained even when the color is added in small proportions. The mouth wash prepared in this manner does not flocculate. 10 to 12 grams of coloring matter are used per 100 kilograms of the mouth wash.

Chlorine Antiseptic

Sodium Hypochlorite	4.05 g.
Sodium Chloride	2.50 g.
Calcium Hydroxide	0.14 g.
Calcium Chloride	0.65 g.

Use full strength or diluted with one or two parts of water for application to mucous membranes or infected bone. For mouth wash use 10 drops per half glass of water.

Antiseptic Mouth Wash (Powdered Form)

Sodium Perborate, N.F.	90 parts
Corn Starch	3 parts
Color and Flavor to Suit.	

Vincent's Infection Mouthwash

Chloramine U.S.P.	5%
Sodium Bicarbonate	5%
Eucalyptol	2%
Saccharin	1%
Salt	87%

Dissolve 1⁄3 teaspoonful in glass of water.

Antiseptic Solution (Double Strength)

Boric Acid	25	g.
Thymol	1	g.
Eucalyptol	5	cc.
Methyl Salicylate	1.2	cc.
Oil Thyme	0.3	cc.
Menthol	1	g.
Sodium Salicylate	1.2	g.
Sodium Benzoate	6	g.
Alcohol	300	cc.
Water to make	1000	cc.

Mouth Wash

Acid Tannic	1 dram
Spirits Lavender Compound	1 oz.
Water	3 oz.

Mouth Wash

Oil Eucalyptus	10	drops
Oil Wintergreen	10	drops
Menthol	10	grains
Thymol	10	grains
Boric Acid	1⁄2	oz.
Alcohol	4 1⁄2	oz.
Water	16	oz.

Mouth Wash, Swedish (Amykos)

Boric Acid	50 parts
Tincture of Cloves	25 parts
Borax	5 parts
Water	4000 parts

Mouth Wash

Salol	5	parts
Oil of Peppermint	1	part
Oil of Clove	0.04	part
Oil of Fennel	0.04	part
Saccharin	0.004	part
Alcohol	190	parts

Mouth Wash (Quinosol)

Quinosol	30 parts
Glycerin	100 parts
Rose Water	900 parts
Carmine	Sufficient

Mouth Wash (Thymol)

Thymol	0.3 g.
Alcohol	160 g.
Rose Geranium Oil	15 drops
Calamus Oil	10 drops
Glycerin	120 g.
Venetian Soap	16 g.
Sassafras Oil	15 drops
Eucalyptus Oil	6 drops
Pine Needle Oil	40 drops
Distilled Water	700 g.

Mouth Wash (Peppermint)

Powdered Angelica Root	25 parts
Powdered Anise Seed	30 parts
Powdered Cinnamon	6 parts
Powdered Nutmeg	3 parts
Powdered Cloves	10 parts
Alcohol (90%)	1000 parts
Vanillin	1 part
Peppermint Oil	8 parts
Tincture of Cochineal	Sufficient

Mouth Wash (Salol)

Salol	8 parts
Spearmint Oil	2 parts
Oil of Cloves	1 part
Oil of Cinnamon	1 part
Oil of Star Anise	1 part
Alcohol, to make	400 parts

Mouth Wash (Lactic Acid)

Lactic Acid	40 parts
Cochineal	1 part
Oil of Peppermint	30 parts
Oil of Cloves	3 parts
Oil of Cinnamon	6 parts
Distilled Water	400 parts
Alcohol	1600 parts

Mouth Wash (Peroxide)

Solution of Hydrogen Dioxide	250 g.
Peppermint Oil	1 drop
Ponceau R R	0.01 g.

Mouth Wash (Phenol)

Phenol	4 drams
Camphor	1 oz.
Chloroform	2 oz.
Oil of Cajeput, to make	4 oz.

Triturate the phenol with the camphor; add the chloroform and then the oil of cajeput.

Wadsworth's Oral Antiseptic

Sodium Chloride	2 g.
Sodium Bicarbonate	0.5 g.
Glycerin	15 cc.
Oil of Gaultheria	0.1 cc.
Alcohol	100 cc.
Distilled Water, to make	200 cc.

For use, dilute with an equal part of warm water.

Wadsworth's Solution (Gargle)

Sodium Chloride	8 g.
Sodium Bicarbonate	2.4 g.
Glycerin	420 g.
Alcohol	300 g.
Menthol	0.24 g.
Thymol	0.24 g.
Methyl Salicylate	0.70 cc.
Oil of Cinnamon	0.50 cc.
Oil of Eucalyptus	0.30 cc.
Tincture of Cudbear	16 cc.
Tincture of Krameria	8 cc.
Distilled Water, to make	1000 cc.

Iodo-Phenol Gargle

Liquefied Phenol	2 oz.
Tincture of Iodine	16 oz.
Glycerite of Tannic Acid	54 oz.
Glycerin	54 oz.
Aromatic Elixir, to make	1 gal.

Mix the ingredients in the order given. Dilute one part to sixteen parts of water to use.

Solution for Painting Throat

Iodine	2 grains
Potassium Iodide	5 grains
Oil of Cinnamon	1 drop
Glycerin, to make	1 oz.

Dissolve potassium iodide in the smallest amount of water necessary.

Add iodine; then add glycerin.

For painting throat, use small cotton swab dipped in above solution.

Sore Throat Relief

Mix tincture of ferric chloride, 30 grams, alcohol 30 grams, and potassium chlorate, 60 grams, with sufficient water to make 250 cubic centimeters.

Hard Candy Cough Drops

Granulated Sugar	100 lb.
Water	4 gal.
*Flavor	12 oz.

*Flavor Formula

Eucalyptol	30 cc.
Oil Peppermint	40 cc.

Menthol 20 cc.
Oil Bergamot 20 cc.
Oil Anise 20 cc.

Melt the sugar down with the water in a steam kettle. Cover to steam down sides of the kettle. Heat to about 160° C. or to hard crack. Cool down to about 135° C., add the flavor, mix in well. Roll and cut into drops.

Agar and Paraffin Confection

Alkanna Root 8 grains
White Soft Paraffin 5 oz.
Agar 2½ oz.
Citric Acid 40 grains
Oil of Lemon 16 minims
Sucrose to 8 oz.

Digest the alkanna root in the melted soft paraffin on a water-bath for fifteen minutes, strain on to the agar, citric acid, and sucrose, previously powdered and mixed in a mortar. When cold add the oil of lemon.

Dose: 1 to 4 drams.

Menthol Rock Candy

Sugar (crystal A) 20 lb.
Cream of Tartar 2 drams
Menthol 4 drams
Water Sufficient

Put the sugar in a suitable container; add the cream of tartar and 5 pints of water. Set the container on the fire and stir the batch until it comes to a boil. With a little water wash down the sides of the container, and cook the batch to 340° F. Pour the mass on an oiled slab; let it cool a little, and work in the menthol by folding the mass over and over. Work the mass into a long round strip; pull into sticks and cut into pieces.

Old-Fashioned Cough Candy

Canada Snake-Root 1 oz.
Pectoral Species, N.F. 2 oz.
Sugar 12 lb.
Molasses 8 oz.
Oil of Wintergreen 10 minims
Oil of Sassafras 10 minims
Oil of Anise 10 minims
Water Sufficient

Make a decoction of the herbs with 4 pints of water, and strain. Cook the sugar, the molasses and a little water until it forms a homogeneous mass; slowly stir in the decoction, and cook the batch to 310° F. Pour the mass on an oiled slab; and as it cools incorporate the oils. Finally cut into drops.

Peppermint Cough Candy

White Sugar 7 lb.
Tartaric Acid 0.5 oz.
Oil of Anise 2 drams
Oil of Peppermint 1 dram
Water 3 pints
Saffron Color Sufficient

Boil to crack (about 252° F.) and pull. The pulling process makes the candy look like satin. It is formed into rods and cut into cushion-shaped pieces with scissors.

Hoarhound Drops

Sugar 20 lb.
Hoarhound 2 oz.
Water 8 pints
Cream of Tartar 90 grains

Boil the hoarhound with 3 pints of water until reduced to one pint, and squeeze through muslin. Cook the sugar with three pints of water and the cream of tartar to 335° F.; add slowly the infusion of hoarhound and cook the batch to 340° F. Pour out on an oiled slab, fold over the edges as it cools and when cool enough stamp into tablets.

Glegg's Nasal Lotion

Sodium Chloride 6 drams
Sodium Sulphate 2 drams
Sodium Phosphate 2 drams
Sugar 14 drams
Thymol 3 grains
Menthol 3 grains
Water, to make 6 oz.

Menthol Inhalant

Menthol U.S.P. 10 parts
Oil Lavender Water Special 5 parts
Oil Guaiac Wood 5 parts
Alcohol 80 parts

Mix and dissolve.

Lee's Antiseptic Inhalation

Creosote 2 drams
Phenol 2 drams
Tincture of Iodine 1 dram
Spirit of Ether 1 dram
Spirit of Chloroform 2 drams

Fuller's Inhalation

Menthol 20 grains
Guaiacol 30 minims
Camphorated Tincture of Opium 2 oz.
Compound Tincture of Benzoin, to make 4 oz.

Thymol Inhalant

Thymol	10 grains
Menthol	30 grains
Oil of Lavender	20 minims
Oil of Eucalyptus	10 minims
Strong Solution of Ammonia	10 minims
Alcohol (90 per cent)	to 1 fl. oz.

A few drops sprinkled on a handkerchief or respirator and intermittently inhaled through the nose.

Ephedrine Glycanth

Ephedrine hydrochloride	9 grains
Eucalyptol	1 minim
Distilled water	Sufficient
Glycanth, to make	2 oz.

Dissolve the ephedrine hydrochloride in a small quantity of distilled water, and mix with the other ingredients.

Ephedrine Oil Solution

Ephedrine Alkaloid	1 cc.
Camphor	0.20 cc.
Menthol	0.20 cc.
Methyl Salicylate	0.05 cc.
Oil Cinnamon	0.05 cc.
Liq. Petrolatum (Light) to make	100 cc.

Triturate the camphor and menthol to a liquid, add the methyl salicylate and oil of cinnamon, then the ephedrine alkaloid and liquid petrolatum, light, to make 100 cubic centimeters. The solution may be heated to not over 40° C. to facilitate solution.

Glycerin and Tragacanth Base; Water-Soluble Base

Tragacanth	26 grains
Starch	25 grains
Distilled water	1 fl. oz.
Glycerin	1 fl. oz.

Mix the starch and cold distilled water in a dish and boil. Rub the tragacanth and glycerin in a mortar, add the contents of the dish and mix thoroughly. Dispense in a collapsible tube.
Uses: As a base for ephedrine hydrochloride, or 0.1 per cent of the following may be incorporated: camphor, eucalyptol, menthol, methyl salicylate, oil of cinnamon, and thymol, for application for the nostrils.

Golden Eye Water

A solution of sulphate of hydrastia, two grains to the ounce, forms an excellent eye wash for ordinary inflamed and granulated lids.

Rheumatism Liniment

Menthol	1 lb.
Camphor Liniment	6 lb.
Olive Oil	5 lb.
Methyl Salicylate	20 lb.
Alcohol	4 lb.

Dissolve the menthol in the camphor liniment. Add oil, methyl and alcohol. Filter.

Anti-Rheumatic Bath Salts

Sodium Sulphate	20 g.
Sodium Chloride	20 g.
Sodium Bicarbonate	80 g.

For one bath.

Rheumatic Salts and Powders

Formula No. 1

Magnesium Sulphate	53%
Rochelle Salt	15%
Bicarbonate of Soda	15%
Sodium Chloride	2%
Sodium Sulphate	15%

Mix.

Formula No. 2

Sodium Sulphate	52%
Sodium Bicarbonate	25%
Sodium Chloride	15%
Potassium Sulphate	8%

Mix.

Formula No. 3

Magnesium Citrate	12%
Powdered Rhubarb	12%
Sodium Bicarbonate	25%
Sodium Carbonate	25%
Powdered Sugar	25¾%
Oil Peppermint	¼%

Mix.

Mustard Liniment

Oil of Mustard	35 cc.
Camphor	50 g.
Menthol	20 g.
Olive Oil	150 cc.
Alcohol to make	1000 cc.

Dissolve camphor and menthol in alcohol, add the oils and filter.

White or Camphor Liniment

Camphor Powdered	4¾ lb.
Turpentine	26 lb.
Blendene	10 lb.

Mix the above until dissolved. Put in a pot fitted with a high speed agitator and while mixing rapidly add the following which has been dissolved previously:

Ammonium Chloride	7½ lb.
Stronger Ammonia Water	3¾ lb.
Water	28 gal.

Continue mixing for about 15 minutes and bottle.

Thompson's Liniment

Menthol	0.5 oz.
Camphor	0.5 oz.
Oil of Turpentine	0.5 oz.
Oil of Eucalyptus	0.5 oz.
Chloroform	1 oz.
Tincture of Capsicum	1 oz.
Methyl Salicylate	1 oz.
Liquid Petrolatum	1 oz.

Mott's Anodyne Liniment

Chloroform	15 cc.
Tincture of Aconite	15 cc.
Tincture of Iodine	15 cc.
Ammonia Water	15 cc.
Soap Liniment, to make	120 cc.

Kerosene Liniment

Camphor	1 gram
Oil of Peppermint	0.5 cc.
Oil of Wintergreen	0.5 cc.
Oil of Cloves	0.2 cc.
Oil of Cassia	4 cc.
Oil of Cottonseed	8 cc.
Oil of Cajeput	8 cc.
Oil of Turpentine	4 cc.
Kerosene	72 cc.

Camphorated Soap Liniment

Stearic Acid	50 g.
Water	100 cc.
Sodium Carbonate (Monohydrated)	10 g.
Thymol	25 g.
Oil Rosemary	6 cc.
Oil Lavender	4 cc.
Strong Ammonia Solution	50 cc.
Alcohol to make	1000 cc.

Heat the water and dissolve the sodium carbonate in it, add 200 cubic centimeters of alcohol and the stearic acid and continue heating until saponification takes place and effervescence ceases. Dissolve oils in 500 cubic centimeters of alcohol and stir into soap solution. Add rest of alcohol. Filter.

Belladonna Liniment

Chloroform	2 lb.
Tincture Belladonna	2 lb.
Tincture Opium	4 lb.
Tincture Aconite	2 lb.

Soft Soap	1 lb.
Glycerin	2 lb.
Alcohol	32 lb.
Ammonia	4 lb.
Water	1 lb.
Oil Cedar Leaf	4 oz.
Oil Anise	4 oz.
Oil Sassafras	4 oz.

Mix tinctures with alcohol, add oils and chloroform. Mix soft soap and water, stir into above solution. Add ammonia and agitate. Filter.

Athletic Liniment

Oil Sassafras	15 cc.
Oil Peppermint	20 cc.
Chloroform	400 cc.
Methyl Salicylate	175 cc.
Soap Liniment to make	3785 cc.

Mix in order given.

Muscular Liniment

Tincture Capsicum	12 oz.
Spirits of Camphor	10 oz.
Ether	10 oz.
Sodium Chloride	2 oz.
Alcohol	22 oz.
Oil of Turpentine	1 oz.
Ammonia	2 oz.
Ammonium Chloride	5 oz.
Water	35 oz.

Dissolve sodium and ammonium chlorides in water. Add the capsicum, camphor to the alcohol. Mix the two solutions and add the turpentine and ether. Filter.

Potassium Iodide Liniment

Glycerin	15 oz.
Potassium Iodide	4 oz.
Water	5 oz.
Coconut Oil	12 oz.
Oleic Acid	4 oz.
Caustic Potash	5 oz.
Water	20 oz.

Make the soap first by dissolving the caustic in water with heat. Melt and add the oil. Continue heat until saponified. Add glycerin. Dissolve potassium iodide in rest of water, add to the soap solution and stir in for fifteen minutes while continuing heat.

Beta Naphthol Liniment

Ether	5 lb.
Precipitated Sulphur	2 lb.
Beta Naphthol	6 lb.
Soap Liniment	20 lb.
Glycerin	5 lb.
Alcohol	70 lb.

Dissolve sulphur in ether. Stir into soap liniment. Add glycerin, betanaphthol and alcohol. Filter.

Analgesic Balm Liniment

Wintergreen Oil	60 cc.
Menthol	19 g.
Saponin	1 g.
Hydrous Lanolin	65 g.
Water	210 cc.
Chloroform	10 g.

Dissolve the saponin in water. Mix wintergreen and menthol, add chloroform. Bring the two mixtures together and stir in the lanolin.

Glycerin Liniment

Soap Liniment	8 lb.
Oil Cajeput	1 lb.
Camphor	1 lb.
Glycerin	24 lb.
Rose Water	16 lb.
Alcohol	3 lb.

Dissolve camphor in alcohol, add cajeput. Stir in glycerin, rose water and soap liniment.

Croton Oil Liniment

Croton Oil	18 oz.
Calamus Oil	72 oz.
Alcohol	70 oz.

Add oils to alcohol and filter.

Stimulating Liniment

Peanut Oil	30 oz.
Oleic Acid	3 oz.
Eucalyptus Oil	18 oz.
Turpentine	16 oz.
Camphor	4 oz.
Ammonia	40 oz.

Dissolve camphor in turpentine, add peanut and eucalyptus oils and oleic acid. Then add ammonia and agitate thoroughly.

Anodyne Liniment

Chloral Hydrate	2 g.
Thymol	2 g.
Chloroform	4 cc.
Ether	3 cc.
Tincture Opium	1 cc.
Soap Liniment	120 cc.
Oil of Wintergreen	1 cc.

Dissolve thymol in chloroform and the chloral hydrate in the ether. Mix the two solutions and add to soap liniment. Then add tincture of opium and wintergreen.

Aconite Liniment

Tincture Aconite	5 g.
Chloroform	5 g.

Pain Killer Liniment

Capsicum Powder	500 oz.
Turpentine (Approximately)	1650 oz.
Menthol	80 oz.
Rape Seed Oil	240 oz.
Oil Rosemary	40 oz.
Oil Thyme	30 oz.
Oil Lavender	30 oz.
Oil Wintergreen	60 oz.

Percolate the capsicum with the turpentine so as to obtain about 2000 ounces of percolate. Add rest of ingredients to this, mix thoroughly and filter.

Household Liniment

Tincture of Capsicum	120 cc.
Spirits of Camphor	100 cc.
Ammonia	110 cc.
Chloroform	85 cc.
Alcohol	120 cc.
Oil of Cedar Leaf	15 cc.
Oil of Hemlock	12 cc.
Oil of Sassafras	15 cc.
Oil of Rosemary	17 cc.

Mix the ingredients thoroughly and filter.

General Liniment

Camphor	2 oz.
Myrrh	2 oz.
Guaiacum	1 oz.
Capsicum	2 oz.
Oil of Sassafras	1 oz.
Oil of Wintergreen	1 oz.
Oil of Sandalwood	1 oz.

Macerate ten days and filter.

Athletes' Rub

Alcohol	100 oz.
Witch Hazel	50 oz.
Methyl Salicylate	2 oz.

Mix and filter.

Chest Rubs

Formula No. 1

Benzoated Lard	400 g.
Olive Oil	100 g.
Mace Oil	36 g.
Eucalyptus Oil	10 g.
Tincture of Benzoin	15 g.

Melt the lard, add the olive oil and the mace. Add eucalyptus and tincture of benzoin.

Formula No. 2

White Petrolatum	400 g.
Heavy Mineral Oil	200 g.
Eucalyptus Oil	50 g.
Menthol	10 g.
Oil of Turpentin	10 g.

Dissolve the menthol in the eucalyptus. Melt the petrolatum and then add the mineral oil. Pour in the eucalyptus and stir. Finally add turpentine.

Formula No. 3

White Petrolatum	500 g.
Hydrous Lanolin	50 g.
Soap Liniment U.S.P.	100 g.
Tincture of Aconite	50 g.
Chloroform	25 g.

Melt the lanolin and the petrolatum. Add the soap liniment and then the aconite and chloroform.

Formula No. 4

Vanishing Cream	600 g.
Hydrous Lanolin	100 g.
Mustard, Ground	150 g.
Camphor	20 g.
Methyl Salicylate	30 g.

Dissolve the camphor in the methyl salicylate. Mix the vanishing cream and the lanolin. Rub in the mustard. Add the methyl salicylate.

Formula No. 5

Menthol	10 g.
Chloral Hydrate	100 g.
Camphor	100 g.
Anhydrous Lanolin	300 g.
Petrolatum	490 g.

Melt the petrolatum and the lanolin. Rub up the menthol, chloral hydrate and camphor until liquefied. Meanwhile allow the fats to become cool (120° F.) and then mix in the active ingredients.

Formula No. 6

Menthol	10 g.
Camphor	10 g.
Eucalyptus Oil	18 g.
White Petrolatum	400 g.
Methyl Salicylate	15 g.

Add the menthol to the methyl salicylate and when dissolved add camphor. Bring temperature of petrolatum to about 120° F., add the eucalyptus and the mixture of active agents. Stir well.

Formula No. 7

Formaldehyde	0.25 g.
Creosote	10 g.
Turpentine	20 g.
Menthol	20 g.
Eucalyptus Oil	20 g.
Hydrous Lanolin	300 g.

| White Petrolatum | 579.75 g. |
| Croton Oil | 50 g. |

Dissolve the menthol in the eucalyptus oil. Bring the temperature of the petrolatum to 120° F.

Formula No. 8

Benzoated lard	400 g.
Olive Oil	100 g.
Mace Oil	36 g.
Eucalyptus Oil	10 g.
Tincture Benzoin	15 g.

Melt the lard, add the olive oil and the mace. Add eucalyptus and tincture of benzoin.

Formula No. 9

White Petrolatum	400 g.
Heavy Mineral Oil	200 g.
Eucalyptus Oil	50 g.
Menthol	10 g.
Oil of Turpentine	10 g.

Dissolve the menthol in the eucalyptus. Melt the petrolatum and then add the mineral oil. Poor in the eucalyptus and stir. Finally add turpentine.

Formula No. 10

White Petrolatum	500 g.
Lanolin, Hydrous	50 g.
Soap Liniment U.S.P.	100 g.
Tincture Aconite	50 g.
Chloroform	25 g.

Melt the lanolin and the petrolatum. Add the soap liniment and then the aconite and chloroform.

Formula No. 11

Goose Grease	500 g.
Lanolin, Hydrous	200 g.
Petrolatum	200 g.
Camphor	25 g.
Oil of Turpentine	50 g.
Oil of Lavender	15 g.

Melt the petrolatum. Stir in the goose grease and the lanolin. Mix the camphor with a little of the melted fats until dissolved then add to the rest of the batch. Add the turpentine and the lavender.

Formula No. 12

Petrolatum, White	500 g.
Oil of Mustard	30 g.
Camphor	10 g.
Croton Oil	200 g.

Dissolve the camphor in the croton oil, stir it into the petrolatum and finally add the oil of mustard.

Formula No. 13

Cold Cream	600 g.
Pulverized Mustard	200 g.
Lanolin, Hydrous	200 g.

Make the cold cream in the usual way. Melt and add the lanolin. Mix in the mustard.

Formula No. 14

Vanishing Cream	600 g.
Lanolin, Hydrous	100 g.
Mustard, Ground	150 g.
Camphor	20 g.
Methyl Salicylate	30 g.

Dissolve the camphor in the methyl salicylate. Mix the vanishing cream and the lanolin. Rub in the mustard. Add the methyl salicylate.

Formula No. 15

Menthol	10 g.
Chloral Hydrate	100 g.
Camphor	100 g.
Lanolin Anhydrous	300 g.
Petrolatum	490 g.

Melt the petrolatum and the lanolin. Rub up the menthol, chloral hydrate and camphor until liquefied. Meanwhile allow the fats to become cool (120° F.) and then mix in the active ingredients.

Formula No. 16

Menthol	10 g.
Camphor	10 g.
Eucalyptus Oil	18 g.
Petrolatum, White	400 g.
Methyl Salicylate	15 g.

Add the menthol to the methyl salicylate and when dissolved add camphor. Bring temperature of petrolatum to about 120° F., add the eucalyptus and the mixture of active agents. Stir well.

Formula No. 17

Thymol	10 g.
Menthol	20 g.
Eucalyptus Oil	20 g.
Cubeb Oil	20 g.
Petrolatum	730 g.
Mineral Oil	200 g.

Dissolve the menthol and thymol in the oils. Melt the petrolatum, add the mineral oil and then mix in the thymol menthol solution.

Formula No. 18

Oil of Pine Needles	40 g.
Oil of Turpentine	30 g.
Camphor	20 g.
Menthol	20 g.
Hydrous Lanolin	200 g.
Petrolatum	690 g.

Rub up camphor and menthol until liquefied. Add oil of pine and turpentine. Bring the temperature of the petrolatum and lanolin to 120° F. and stir in the foregoing mixture.

Formula No. 19

Formaldehyde	0.25 g.
Creosote	10.00 g.
Turpentine	20.00 g.
Menthol	20.00 g.
Eucalyptus Oil	20.00 g.
Hydrous Lanolin	300.00 g.
White Petrolatum	579.75 g.
Croton Oil	50.00 g.

Dissolve the menthol in the eucalyptus oil. Bring the temperature of the petrolatum to 120° F., add the hydrous lanolin, croton oil, creosote, turpentine, eucalyptus mixture, and formaldehyde in the order given, mixing well after each addition.

Formula No. 20

Petrolatum	200 g.
Mineral Oil, Light	300 g.
Spermaceti	100 g.
Stearic Acid	40 g.
Menthol	10 g.
Camphor	10 g.
Eucalyptol	40 g.

Rub the menthol and camphor together until liquefied and add to the eucalyptus. Melt the stearic acid, spermaceti and petrolatum, add the mineral oil and mix. When cooled down to 120° F. add the camphor-phenol solution.

Sore Muscle Liniment

Olive Oil	60 cc.
Methyl Salicylate	30 cc.

Mix the two ingredients thoroughly and keep in a well-stoppered bottle until desired. Apply externally.

"Chest Rub" Salve

Vaseline brown, 1 pound; paraffin wax, 1 ounce; oil eucalyptus oil, 2 ounces; menthol crystals, ½ ounce; oil cassia, ⅛ ounce; turpentine, ½ ounce; carbolic acid, ⅛ ounce. Melt the vaseline and paraffin wax together, then add the menthol crystals and stir till dissolved. Remove from fire, and while cooling add the oils, turpentine and acid. Pour into one-ounce tin boxes when it begins to thicken.

Burn and Insect Bite Cream
U. S. Patent 1,947,568

The ingredients of the composition are as follows: benzocaine, aluminum acetate, zinc oxide, cottonseed oil, lime water, chloramine, lanolin, acid stearic, white solid petrolatum (so-called white vaseline), liquid petrolatum, white wax, borax, triethanolamine and benzoic acid with perfume.

The first step in manufacturing the product is to melt the lanolin anhydrous, acid stearic, white petrolatum, white wax and liquid petrolatum. In this mixture is dissolved the benzocaine and benzoic acid powder. In this first step it is preferable to do the melting at a temperature of 50° to 60° C.

In the second step the aluminum acetate powder, zinc oxide powder and cottonseed oil are rubbed in a mortar to a smooth paste. The triethanolamine is dissolved in lime water and all are mixed together. This mixture is passed through a colloid mill producing a saponified fine product.

The third step comprises adding the mixture formed by the first step to the emulsion formed by the second step.

The fourth step comprises dissolving borax powder in lime water and in this mixture dissolving chloramine. The mixture of borax, lime water and chloramine is then added to the mixture formed by the third step and the complete mixture is stirred until effervescence ceases and the cream is emulsified after which a small amount of perfume may be added.

In actual practice the following proportion of ingredients have been found to work out excellently for the first step: lanolin anhydrous 18 pounds, acid stearic 5 pounds, white petrolatum solid 3 pounds, white wax 5 pounds, liquid petrolatum 10 pounds, benzocaine 1 pound, and acid benzoic powder 1 pound.

The following proportions of ingredients have been found desirable for the second step: aluminum acetate powder 1 pound, zinc oxide powder 5 pounds, cottonseed oil 25 pounds, triethanolamine 1 pound, and lime water 20 pints.

For the fourth step, one pound of borax powder, 5 pints of lime water and 2 pounds of chloramine U.S.P., have been found satisfactory.

The aluminum acetate in the composition acts as an astringent and a mild antiseptic to relieve soreness, reduce inflammation and stimulate healing. The benzocaine acts as a soothing local anesthetic and is non-toxic and non-irritating. The zinc oxide also acts as a mild antiseptic and a sedative for ulcerations and minor wounds. The oil with the addition of the chloramine and acid benzoic acts as antiseptic fluid for easing pain and soreness in treating minor burns and wounds.

Mosquito Cream

Glyceryl Monostearate	11.0 parts
Oil of Cedar Leaf	4.0 parts
Oil of Pennyroyal	4.0 parts
Linalyl Acetate	3.0 parts
Gasoline	5.0 parts
Menthol	0.5 part
Phenol	2.0 parts
Glycerin	5.0 parts
Water	65.5 parts

Put the glyceryl monostearate into the water, add the glycerin and bring the mixture to the boiling point with constant stirring. Keep stirring and when the temperature drops to about 45° C. add the rest of the ingredients which have been mixed together.

Insect Repellent Creams

Formula No. 1

Beeswax	3–4 parts
Oil Citronella	15 parts
Spirits of Camphor	8 parts
Cedar Wood Oil	8 parts
White Petrolatum	60 parts

Melt petrolatum and beeswax, then add other constituents and stir until smooth.

Formula No. 2

A

Castor Oil	15.00 parts
Pennyroyal Oil	3.75 parts
Oil of Pine Tar	3.75 parts
Oil of Camphor	7.50 parts
Oil Citronella	3.75 parts

B

Beeswax	6.625 parts
Lanolin (Anhydrous) Petrolatum	} 59.625 parts

Melt (B) and then add above oils (A).

Insect Bite Lotion

Epsom Salt	20 parts
Camphor	3 parts
Menthol	1 part
Glycerin	1 part
Alcohol	25 parts
Water	50 parts

Dissolve the camphor and menthol in the alcohol. Add the glycerin. Dissolve the salt in the water. Then add the alcohol solution gradually to the water solution. Mix well.

Burn Salve ("Beebe" Salve)

Pure Mutton Tallow	12 oz.
Olive Oil	2 oz.
Phenol	4 drams

Antiseptic Astringent Ointment

Phenol	34 parts
Sulphonated Bitumen	70 parts
Salicylic Acid	65 parts
Hydrous Wool Fat	400 parts
Petrolatum to make	1000 parts

Mix the materials thoroughly in order given, melting the fats and triturating till cool.

Casein Ointment

14 parts dry casein are mixed with 0.43 parts of a mixture of 4 parts of potassium hydroxide, and one part of sodium hydroxide. Final mixture contains 7 parts glycerin, 21 parts petrolatum, 56.57 parts water and one part preservative as well. Ichthyol may also be added. If this substance is used in making the ointment, the other ingredients must be previously diluted with water since ichthyol thickens the mixture. A casein ointment containing zinc oxide is made from 10 parts of the oxide, 10 parts of petrolatum and 80 parts casein ointment. Boric acid may also be used in making the ointment, which then consists of 5 parts boric acid, 5 parts petrolatum and 80 parts casein ointment base.

Dermatologist's White Drying Salve

Make salve of:

Petrolatum, White	2 oz.
Zinc Oxide Ointment	2 oz.
Powdered Gum Camphor	2 oz.
Phenol Crystals	2 oz.

Mix this with a salve of:

Corn Starch	4 oz.

with amount necessary of white petrolatum.

Metol Dermatitis Ointment

Many photographers show an idiosyncrasy to metol, which causes dermatitis of their fingers. Other organic chemicals used in photography may cause this, and the following ointment is recommended:

Lanolin	10 g.
Soft Paraffin	15 g.
Ichthyol	5 g.
Boric Acid	15 g.

Rubber gloves should be worn if at all practicable.

Pile Ointment

Bismuth Iodogallate	5 parts
Zinc Oxide	6 parts
Resorcin	1 part
Balsam of Peru	2 parts
Soft Paraffin, Yellow	to 100 parts
Hydrous Wool Fat	

Some prefer a larger proportion of Soft Paraffin.

Glycerin Suppositories

One gram of sodium bicarbonate is treated with 33 parts glycerin and heated until most of carbon dioxide has been evolved. Then 3 grams stearin are gradually added with agitation and mixture is heated over small flame with continued agitation, until it becomes homogeneous.

Lassar Paste

Lard	50 g.
Zinc Oxide	25 g.
Starch	25 g.
Salicylic Acid	3 g.

Mineral Jelly

Technical White Oil	45%
Ceresine	15%
Refined Paraffin (135° F. m.p.)	20%
Snow White Petrolatum	20%

Moorhaf's Bone Wax

Iodoform	20 parts
Spermaceti	40 parts
Sesame Oil	40 parts

Heat the ingredients slowly to 100° C. and allow the mixture to cool while stirring. For use as a bone plug it is heated to 50° C.

Butylphenol Preservative
U. S. Patent 1,887,662

Butyl alcohol (140 parts) and concentrated hydrochloric acid (28 parts) are added slowly to phenol (85 parts) and zinc chloride (150 parts) at 130–140°. The fraction of the product of boiling point 220–265° contains much ortho-secondary-butylphenol, and has a phenol coefficient of 36–40.

Ointment Base

Hexadecyl alcohol 4, adeps lanae 10, white vaseline 86 can take up more than its own weight of water; besides, it can never turn rancid.

Salves for Abscesses

Effective preparations for abscesses have been proposed. For example 7.5 parts gum elemi, 7.5 parts larch turpentine, 75 parts lard, ten parts lanolin and 25 drops liquid carbolic acid. A few preparations contain merely salve base and medicament. Thus 75 parts paraffin unguentum are mixed with 25 parts bismuth subnitrate; 90 parts of mixture of paraffin unguentum and glycerin unguentum are mixed with ten parts of iodoform.

A more complex mixture consists of half part boric acid, 4 parts zinc oxide, 4 parts talc, 0.15 part iodoform, 0.15 part balsam Peru, 35 parts unguentum diachylonum, 35 parts petrolatum and enough prepared suet to make 100 parts.

In another are 2.5 parts boric acid, 0.5 part pulverized salicylic acid, 5 parts zinc oxide, 0.8 part bismuth subgallate, 0.8 part lead subacetate solution, 10 parts anhydrous lanolin and yellow beeswax to make 50 parts.

Stick Salve

A base for producing salves in pencil form is as follows: 10 parts coconut oil, 30 parts white beeswax, 22 parts lard and 35 parts lanolin. The base should be carefully melted and agitated and the pulverized medicament thoroughly mixed in the melted mass before it congeals.

Sulphur Salve

Highly active sulphur salves can be made with precipitated sulphur, when the latter is made into a salve paste with paraffin oil and peanut oil. The preparation contains 150 parts of precipitated sulphur, 50 parts of paraffin oil and 50 parts of peanut oil. Thereafter 20 parts of white beeswax and 20 parts of spermaceti are added, and finally 100 parts of anhydrous lanolin and 100 parts more of the peanut oil. This mixture is melted on a water bath and 60 parts of water are added. The mixture is well agitated while on the bath and the agitation is continued until cool.

Poison Ivy Lotion

Oil Sassafras	½ oz.
Solution Boroglycerine	¼ oz.
Glycerin	1½ oz.

Poison Ivy Lotion

Ferric Chloride	3 parts
Glycerin	1 part
Alcohol	96 parts

Poison Ivy Remedies

Formula No. 1
Dusting Powder

Colloidal Sulphur	50%
Sodium Hyposulphite (Powder)	5%
Lycopodium (or Colloidal Clay)	35%
Calcium Stearate	10%

Mix. Particularly useful when lesions are weeping profusely.

Formula No. 2
Lotion

Colloidal Sulphur	20%
Sodium Hyposulphite	5%
Chloral Hydrate	2%
Benzocaine	2%
Glycerin	10%
Tincture Benzoin	5%
Water	56%

Rub up sulfur and glycerin. Dissolve sodium hyposulphite in the water, add chlorate, benzocaine, bezoin and sulphur mixture.

Formula No. 3
Lotion

Ammonium Oleate	5.0%
White Mineral Oil	10.5%
Menthol	0.1%
De-Odorized Kerosene	77.4%
Eugenol	5.0%
Carbon Tetrachloride	2.0%

Dissolve the menthol in the eugenol, add the mineral oil and the carbon tetrachloride and then dissolve the ammonium oleate in the mixture. Add this to the kerosene.

Formula No. 4
Lotion

Silver Proteinate	0.5%
Potassium Permanganate	0.2%
Basic Aluminum Acetate	5.0%
Triethanolamine	2.0%
Glycerin	7.0%
Tincture Benzoin	5.0%
Camphor	0.5%
Lime Water	79.8%

Dissolve camphor in tincture of benzoin. Dissolve the rest of the ingredients in the lime water and add the benzoin solution.

Formula No. 5
Lotion

Phenol	0.5%
Liquor Lead Subacetate	30.0%

Tincture of Opium	7.0%
Tincture of Benzoin	5.0%
Rose Water	57.5%

Mix all ingredients with the water.

Formula No. 6
Lotion

Picric Acid (20% Solution)	0.5%
Benzocaine	1 %
Tincture of Benzoin	7 %
Tincture of Arnica	3 %
Carbitol	10 %
Witch Hazel	78.5%

Dissolve picric acid in witch hazel. Mix the tinctures, add the carbitol and the benzocaine. Mix both solutions. Note: as picric acid is explosive under certain conditions handle carefully and avoid heat.

Formula No. 7
Lotion

Green Soap	25.0%
Potassium Permanganate	1.0%
Phenol	0.5%
Alcohol	10.0%
Glycerin	5.0%
Colloidal Sulphur	5.0%
Camphor Water	53.5%

Soak green soap in camphor water in which the permanganate has been dissolved, with occasional stirring until it dissolves. Put up and add sulphur and glycerin. Finally add alcohol and phenol.

For those who wish to make ointments any of the above ingredients made with a suitable vehicle such as short fibered petrolatum, lanolin, mineral oil mixed. Or straight cold cream may be used.

Poison Oak Remedies

Solutions of sodium thiosulphate or potassium permanganate are frequently used for skin irritations from poison oak. A remedy that often works well is a 5% aqueous solution of ferric chloride, mixed with an equal volume of ethyl alcohol.

Lotion for Hives or Prickly Heat

Menthol	2 g.
Alcohol	3 oz.
Sodium Bicarbonate	10 g.
Witch Hazel	3 oz.
Water to make	1 pint

Dissolve menthol in alcohol, add sodium bicarbonate and witch hazel. When dissolved add the water, stirring vigorously. Should not be used near the eyes, delicate skin, cuts, etc.

Magnesium Sulphate Solution for Boils

This is prepared by dissolving 40 ounces of crystalline magnesium sulphate in 30 ounces of boiling water and 10 fluid ounces of glycerin, and sterilizing in the autoclave. This is recommended for use after the central slough has separated and healthy granulating surface is visible. It is applied on lint wrung loosely out of the solution.

Porous Plaster
British Patent 402,101

A plaster consists of a mixture of pulverized mineral anhydrite, an accelerator, e.g., potassium sulphate, zinc sulphate, or aluminum sulphate, separately or together, gas-producing ingredients, and Portland cement. A suitable mixture is anhydrite 650, aluminum sulphate 28, cement 9, potassium sulphate 6, and chalk 3 parts.

Zinc Oxide Plasters
Formula No. 1

Indiarubber (Cut Fine)	20 g.
Dammar	11 g.
Resin	8 g.
Zinc Oxide	30 g.
Wool Fat	30 g.
Petroleum Benzine	148 g.

Mix the rubber with 120 grams of petroleum benzine and set aside, frequently turning the flask, for about three weeks, until a uniform, colloidal solution is produced. Dissolve the dammar and resin in 20 grams of petroleum benzine and strain. Mix the zinc oxide, previously dried at 100° C., with 8 grams of petroleum benzine to form a thick paste, and then incorporate the wool fat. The paste, the solution of resins and the rubber solution are then mixed in a flask by rolling the latter, whereupon the mass is set aside for a few hours before spreading.

Formula No. 2

Zinc Oxide	1 part
Mineral Oil	7 parts
Caoutchouc	2 parts
Benzine	10 parts

Dissolve the caoutchouc in the benzine and add the zinc oxide made into a cream with the liquid paraffin.

Healing Plaster

Take two parts of beeswax, four parts of pine tar, and four parts of rosin. Melt these materials together, when almost cool mold into sticks.

In its use, it is found best to heat the wax and apply a thin coat on a piece of muslin and place over the injury.

Athlete's Foot Relief

Dissolve one-tenth of a gram of basic fuchsine in 10 grams, and add to it 225 cubic centimeters of water, phenol, 5 grams, boric acid, 1 gram, acetone, 5 grams, and resorcinol, 10 grams. Paint the affected parts with this solution, letting it dry in before putting on stockings or shoes.

"Athlete's Foot" Treatment

An anesthetic ointment is liberally applied to all exposed areas. Considerable patience is necessary in the selection of the proper ointment for each case. The following alone or in various combinations have proved satisfactory: Ethylaminobenzoate ointment, 10 per cent; nupercaine ointment, 1 per cent; camphor-phenol ointment, 1 per cent of each. Following the application of the ointment, a protective gauze dressing is applied. Dressings may be changed once or twice a day, depending on the severity of the case.

"Athlete's Foot" Solution

1.3 grams of iodine crystals, 1.9 grams of potassium iodide, 1.9 grams of salicylic acid, 3.8 grams of boric acid and enough 50 per cent alcohol to make 59.1 grams. This preparation is applied as a paint once or twice a day.

Remedy for "Athlete's" Foot

"Hong - Kong Foot," "Athlete's Foot," ringworm and similar infections are usually stopped with one application of a 5% solution of iodine and of phenol in 95% alcohol.

"Athlete's Foot" Ointment

Benzoic Acid	50 g.
Salicylic Acid	30 g.
Raw Linseed Oil	400 cc.
Paraffin	150 g.
Petrolatum to make	1000 g.

Dissolve the acids in the warmed linseed oil. Mix this solution with the melted paraffin and petrolatum and stir l cool.

Foot Relief

Tannic acid, 30 grams; powdered alum, grams; tincture iodine, 2 cubic cen-

timeters; grain alcohol, 60 cubic centimeters; water, to make 500 cubic centimeters. Dissolve first two in water, cool and add rest. Let stand 1 day, then filter.

Use 1 tablespoonful per quart of water. Bathe the feet every night in this warm solution. Gently massage the feet while in the water. Do not use any soap. Let the feet dry without wiping.

Foot Powder

Salicylic Acid	1	dram
Powdered Zinc Oleate	1½	oz.
Starch Powder	3	oz.

Mix.

Dusting Powder (for Feet)

Boric Acid	4 parts
Magnesium Carbonate	4 parts
Salicylic Acid	1 part
Talcum	30 parts

Baby Dusting Powder

Benzocaine	2%
Picric Acid (20%)	4%
Tincture of Benzoin	5%
Boric Acid	5%
Talc	74%
Purified Kaolin	10%

This powder is intended for badly chafed skin, diaper scalds, prickly heat and sunburn.

The picric acid is mixed with the kaolin. This is thoroughly dried and sifted. The benzocaine and tincture of benzoin are mixed with a little talc until absorbed. Then all ingredients are mixed together thoroughly and sifted.

Surgical Powder

Procain Hydrochloride	3 parts
Balsam of Peru	10 parts
Thymol Iodide	3 parts
Eucalyptol	1 part
Willow Charcoal	5 parts
Kaolin	78 parts

Corn Application

Acid Salicylic	90 g.
Ext. Cannabis Indicis	10 g.
Collodium	1 oz.

Corn "Cure"

An effective formula is 20 parts each of 25% 40-second nitro-cotton in ethyl acetate, specially denatured alcohol No. 1, and ethyl acetate; 10 parts of sulfuric ether; 12 parts salicylic acid; 1

part each of chlorbutanol, castor oil and gum camphor and 15 parts toluol to make 100. This solution may be tinted an amber or green and coloring is best accomplished by an oil soluble color.

Acetone Collodion

Pyroxylin	5 g.
Oil of Cloves	2 g.
Amyl Acetate	20 g.
Benzol	20 g.
Acetone, to make	100 g.

Haarlem Oil

Heat four parts of linseed oil and one part of sulphur in an iron vessel to a temperature of 165° C., stirring all the time until the mixture drops off the stirrer with a glassy appearance. Remove from the fire and add fifteen parts of oil of turpentine, mix thoroughly and filter.

Anti-Acid

Cerium Oxalate	3 drams
Bismuth Subnitrate	½ oz.
Magnesium Calcined	1 oz.

A teaspoonful 3 times a day after meals.

Keely Institute Tonic (Liquor "Cure")

Gold and Sodium Chloride	12	grains
Strychnine Nitrate	1	grain
Atropine Sulphate	0.125	grain
Ammonium Chloride	6	grains
Aloin	1	grain
Hydrastine	2	grains
Glycerin	1	oz.
Compound Fluid Extract of Cinchona	3	oz.
Fluid Extract of Coca	1	oz.
Distilled Water	1	oz.

Diagnostic Bismuth Liquid

Bismuth Subcarbonate	120 g.
Acacia (Powdered)	20 g.
Tragacanth (Powdered)	5 g.
Syrup	150 cc.
Orange Flower Water	25 cc.
Distilled Water	350 cc.

Rub the powders till well mixed. Work the syrup into this till smooth mixture results and incorporate the orange flower water and distilled water.

White Pine Expectorant Syrup

Fluidextract White Pine Bark	1 fl. oz.
Fluidextract Wild Cherry	1 fl. oz.
Compound Tincture Opium	2 fl. oz.
Chloroform	2 drams
Syrup Tolu	8 fl. oz.
Glycerin, to make	1 pint

Mix as given and filter clear.

Elixir Phenobarbital

Phenobarbital	21 parts
Alcohol	170 parts
Tincture Persionis	125 parts
Aromatic Elixir U.S.P. to make	3785 parts

Dissolve the phenobarbital in the alcohol, add the tincture of persionis and then add enough U.S.P. aromatic elixir to make required amount. Agitate well and filter clear.

Elixir of Amidopyrine

Amidopyrine	83 g.
Aromatic Elixir USP	500 cc.
Syrup USP	500 cc.
Blood Orange Certified Dye Solution (1 g. in 30 cc.)	1 cc.

Dissolve the amidopyrine in the aromatic elixir add the simple syrup and dye. Mix thoroughly and filter clear if needed.

"Feminine Hygiene" Jelly

Gum Tragacanth	80 g.
Boric Acid	55 g.
Water	1200 cc.
Glycerin C.P.	60 cc.
Lactic Acid (85%)	13 cc.

Dissolve boric acid in 500 cubic centimeters of boiling water. Take up the tragacanth in balance of water; add other ingredients and stir until gel results.

Female Hygiene Suppositories

Cocoa Butter	233 g.
Salicylic Acid	15 g.
Boric Acid	66 g.
Quinine Bisulphate	21 g.

Melt the cocoa butter and dissolve in it the salicylic acid. Allow to cool until it is the consistency of soft butter, then work in the other ingredients. Compress into cone shaped suppositories weighing about 3 grams each. The suppositories can be moulded in plaster of Paris moulds and placed on ice to chill for removal if desired. However, when moulding it is necessary to heat the material until it will just pour into the mould stirring every few minutes to prevent the insoluble material from settling to the bottom of the pouring vessel.

If the suppositories are to be cast in plaster of Paris moulds, the mould should be painted with a very thin coat of shellac, afterwards giving several coats of best grade of linoleum lacquer by spraying on. The mould should be dusted with talc before using each time.

Lubricating Jelly

Irish Moss	10	oz.
Water	12½	pints

Allow to macerate 12 hours then add

Glycerine	1¼	pints
Sodium Benzoate	1½	drams

Heat for 15 minutes. Strain—then add

Quince Seed	2½	oz.
Water	3 pints 12	oz.

Allow to stand 24 hours. Strain and add to above. Then add

Eucalyptol	1½	drams
Sodium Benzoate	1½	drams

Douche Powder

Alum (Powdered)	150 g.
Boric Acid (Powdered)	750 g.
Liquefied Phenol	50 cc.
Methyl Salicylate	30 cc.
Corn Starch to make	1000 g.

Rub the alum and boric acid till thoroughly mixed. Add the liquefied phenol and methyl salicylate and work into powder evenly. Add enough starch to make 1000 grams.

Cocoa Butter Substitute

A mixture of refined coconut oil and palm kernel oil is melted together and allowed to cool slowly to form large crystals of "stearine." The mass is then wrapped in cloth, about 10 pounds in each cloth, stacked on a hydraulic press with iron plates between each layer and pressure applied slowly. The "oleine" is pressed out and the "stearine" remains in the cloth. This is neutralized, washed and deodorized and then cast into blocks weighing about 15 pounds.

Rectal Antiseptic

Tannic Acid	1.3	parts
Glycerin	15	parts
Water	105	parts

Rectal Anaesthetic and Analgesic
U. S. Patent 1,941,220
Formula No. 1

Ether	2 fl. oz.
Liquid Petrolatum	1 fl. oz.
Sulphonated Olein	0.1% by weight

Formula No. 2

Ether	2 fl. oz.
Olive Oil	2 fl. oz.
Sulphonated Olive Oil	0.2% by weight

Formula No. 3

A rectal analgesic especially suitable for obstetrical purposes as given below:

Ether	2½	fl. oz.
Quinine	20	grains
Alcohol	45	drops
Liquid Petrolatum to make	4	fl. oz.
Sulphonated Olein	.01% by weight	

Dyspepsia Mixtures
Formula No. 1

Bismuth Subnitrate	7.5%
Compound Tincture Gentian	5 %
Tincture Nux Vomica	5 %
Syrup of Ginger	20 %
Essence of Pepsin	62.5%

Rub up bismuth with the ginger syrup. Mix rest of ingredients and add the bismuth.

Formula No. 2

Dilute Hydrochloric Acid	2.5%
Tincture Capsicum	1.5%
Tincture Calamba	45 %
Essence Pepsin	51 %

Mix the tinctures and the essence. Add the acid.

Formula No. 3

Sodium Bicarbonate	3%
Aromatic Spirits Ammonia	3%
Tincture of Ginger	12%
Compound Infusion Gentian	82%

Dissolve the bicarbonate in the gentian. Add the ginger and the ammonia.

Formula No. 4

Pepsin	4.00%
Tartaric Acid	.12%
Glycerin	18.00%
Sherry Wine	77.88%

Dissolve the pepsin in the wine. Add the tartaric and then the glycerin.

Formula No. 5

Bismuth Subnitrate	.2 %
Magnesium Carbonate	.3 %
Solution Potassium Hydroxide	.1 %
Dilute Hydrocyanic Acid	.03%

Tincture of Ginger	.1 %
Peppermint Water	99.27%

Rub up bismuth and magnesium with some peppermint water. Add solution of potassium hydroxide. Mix with rest of peppermint water. Add tincture of ginger and acid.

Formula No. 6

Sodium Bicarbonate	3.5%
Tincture Nux Vomica	2.5%
Compound Tincture Gentian	94 %

Dissolve bicarbonate in gentian. Add nux vomica.

Formula No. 7

Sodium Sulphocarbolate	2 %
Calcium Carbonate	1.5%
Bismuth Carbonate	2 %
Pepsin	.5%
Tincture Rhubarb	3 %
Glycerin	3 %
Peppermint Water	88 %

Rub up calcium and bismuth carbonate with glycerin and rhubarb. Dissolve pepsin and sodium sulfocarbolate in peppermint water. Add the first mixture.

Formula No. 8

Syrup of Pepsin	15%
Tincture Nux Vomica	2%
Syrup of Ginger	7%
Glycerin	4%
Chloroform Water	72%

Mix the first four ingredients, add to the last.

Formula No. 9

Sodium Bicarbonate	15%
Tincture Nux Vomica	2%
Essence Peppermint	10%
Tincture Rhubarb	2%
Chloroform Water	71%

Dissolve the bicarbonate in the water, mix and add rest of ingredients.

Formula No. 10

Pepsin	.5%
Sodium Bicarbonate	4 %
Bismuth Carbonate	4 %
Aromatic Spirits Ammonia	6 %
Spirits of Chloroform	2 %
Tincture Ginger	6 %
Tincture Cardamom	4 %
Tincture Rhubarb	8 %
Infusion of Calamba	65.5%

Dissolve pepsin in calamba, add sodium bicarbonate. Rub up bismuth with the tinctures and add. Then add ammonia and chloroform.

Formula No. 11—Tablets

Activated Carbon	5 grain

Add binder, granulate and press.

Formula No. 12—Pills (5 Grain)

Pulverized Rhubarb	30%
Sodium Bicarbonate	30%
Extract of Gentian	40%

Add binder, mass in a mix and roll.

Formula No. 13—Tablets (3 Grain)

Pepsin	33%
Pancreatin	33%
Diastase	34%

Add binder, granulate and press.

Formula No. 14—Tablets (6 Grain)

Bismuth Carbonate	50%
Pulverized Rhubarb	17%
Sodium Bicarbonate	33%

Mix, granulate and press.

Formula No. 15—Tablets (5 Grain)

Bismuth carbonate	50%
Pepsin	20%
Pancreatin	20%
Diastase	10%

Mix, granulate and press.

Formula No. 16—Powders (1 Dram)

Pepsin	18%
Bismuth Carbonate	33%
Sodium Bicarbonate	16%
Magnesium Carbonate	33%

Wine of Cod Liver Extract

Formula No. 1

Gaduol	256 grains
Guaiacol	64 minims
Creosote	128 minims
Eucalyptol	30 minims
Extract of Malt	6 oz.
Alcohol	4 oz.
Syrup	8 oz.
Diluted Hydrocyanic Acid	1 dram
Compound Syrup of Hypophosphites	6 oz.
Fuller's Earth	2 oz.
Wine, to make	4 pints

Mix the gaduol with 2 ounces of alcohol and triturate with the fuller's earth add the syrup and 2 pints of the wine Set aside for several days, shaking occasionally, then filter. Add the extract of malt and the syrup of hypophosphites, let stand for 24 hours and again filter. Add the guaiacol, the creosote and the eucalyptol dissolved in 2 ounces of alcohol add the diluted hydrocyanic acid and enough wine to make 4 pints.

Formula No. 2

Gaduol	2.5	pints
Oil of Orange Peel	9	oz.
Fuller's Earth	10	lb.
Port Wine	17	gal.

Compound Tincture of		
Gentian	5	gal.
Fluid Extract of Wild		
Cherry	2.5	gal.
Glycerin	5	gal.
Extract of Malt	4	gal.
Compound Syrup of		
Hypophosphites	6.34	gal.
Fluid Extract of Licorice	1.25	gal.
Caramel	2.5	pints
Alcohol	2.5	gal.

This formula is for a 40-gallon batch.

Acriflavine Emulsion

Acriflavine	1 g.
Wool Fat	50 g.
Distilled Water	25 cc.
Liquid Paraffin	925 cc.

Dissolve the acriflavine in the distilled water, heated to about 90°, and add the solution to the anhydrous lanoline in a large warmed mortar. Triturate until a thick cream is produced, then add gradually the liquid paraffin.

Acriflavine Oleated

Acriflavine	17.5 grains
Potassium Acetate	17.5 grains
Distilled Water	Sufficient
Oleic Acid	4 fl. oz.

Dissolve the acriflavine in 1½ fluid ounces of distilled water in a small separating funnel. Add the potassium acetate and dissolve. Shake with the 3 fluid ounces of oleic acid. Separate the aqueous layer, and shake this with the remainder of the liquids by shaking with 5 grammes of anhydrous sodium sulphate. Allow to stand until clear, or after a few minutes filter through paper.

Agar-Agar and Paraffin Emulsions

To obtain the best results, triturate sodium bicarbonate 10 grains, powdered tragacanth 20 and powdered acacia 60 grains to a smooth paste with glycerol 6 grams. Dissolve agar 17.5 grains in boiling water 2 fluid ounces, add this while hot to the mixture in the mortar while stirring. Then add liquid paraffin ounces, 2–3 drams at a time. Triturate until cold, allow to stand for 1 hour, triturate again for 5 minutes and add water to complete 6 ounces. If an emulsifying apparatus worked by hand is available, a thick emulsion is obtained with agar 15, sodium bicarbonate 6 grains, liquid paraffin 3 ounces, tragacanth 5, acacia 5 grains, glycerol 2 grams, water to make 6 ounces.

Cacao Emulsion of Castor Oil

Castor Oil	2 oz.
Powdered Acacia	6 drams
Oil of Peppermint	3 minims
Powdered Tragacanth	16 grains
Saccharin	4 grains
Glycerin	3 drams
Cacao	2 drams
Water, to make	4 oz.

A mixture of the cacao in the glycerin and water is made by boiling for five minutes. A mucilage of the acacia is made with 4 drams of the cacao mixture and the oils are gradually incorporated by trituration until emulsified. Then the remainder of the cacao mixture is added.

Cod Liver Oil Emulsion

75 grams gum arabic powder is dissolved in 900 cubic centimeters water. This solution is placed in a tank fitted with mechanical stirrer, and 600 cubic centimeters cod liver oil are added. Stirring is continued for about a quarter of an hour, when a good primary emulsion is made. Further ingredients, such as oil of bitter almonds, hypophosphites, glycerophosphates, etc., are added, and the primary emulsion is fed to a De Laval homogenizer. Homogenization is carried out at a pressure of 3,500 pounds per square inch, when an absolutely stable and even emulsion is obtained of a suitable viscosity for pouring from a bottle.

White Oil Emulsion

Agar	0.7%
Gum Acacia	2.1%
Tragacanth	0.3%
White Mineral Oil	50.0%
Sodium Benzoate	0.2%
Flavoring	0.7%
Water	46 %

Aromatizing Mineral Oil

Acetic acid is useful for acidifying mineral oils, since addition of this acid does not spoil the product in any way. Other organic acids may also be used for same purpose, such as citric, tartaric and lactic acids. The first step in the process of mixing acid with mineral oil is to make a saturated solution of the acid in ether. Thus 100 parts of ether will dissolve 1.06 parts of citric acid, acetic acid in all proportions, 0.61 part of tartaric acid in all proportions, 0.61 part of tartaric acid and lactic acid in all proportions. Mineral oil is miscible with ether in all proportions. After the acid solution is made, mineral oil is added. The mixture is then shaken and ether removed by evaporation on water bath. This process requires considerable care and must be pushed as far as possible to remove greatest part of ether without causing mineral oil to become turbid due to separation of acid. It is also claimed that mineral oil may be mixed directly, 85% strength formic acid with the mineral oil or 1.21 parts of lactic acid may be used. The mineral oil will acquire a more pleasant sour taste due to this treatment. The shaken oil is allowed to stand, various suspended substances including excess acid falling to bottom and then clear oil is poured off. It is also claimed that slight warming of oil facilitates mixing of the acid with it. The quantity of acid retained by oil is very small, but sufficient to give it a distinctive taste while not having any physiological action.

Indigestion Mixture

Bismuth Subcarbonate	1 lb
Powdered Rhubarb	1 lb.
Sodium Bicarbonate	4 lb.
Magnesium Carbonate	2 lb.
Peppermint Oil	2 oz.

Mix the peppermint oil with the magnesium carbonate. Add other powders and mix in ball mill for one hour.
Take one-half teaspoonful in one-half glass water.

Indigestion Powder

Pepsin	48 grains
Pancreatin	48 grains
Ginger	48 grains
Sodium Bicarbonate	48 grains
Bismuth Subgallate	152 grains
Milk Sugar	1 oz.

Seasick Remedy

3–5 grains sodium nitrate taken every two hours will aid or partly or entirely prevent seasickness.

Constipation Remedies

Formula No. 1

Magnesium Sulphate	6.5 %
Sodium Sulphate	6.5 %
Potassium Sulphate	.06%
Sodium Bicarbonate	.52%
Sodium Chloride	.42%
Water	36 %
Honey	50 %

Dissolve the several ingredients in the water, add the honey.

Formula No. 2

Castor Oil	15 %
Glycerin	15 %
Tincture Auranti	5 %
Tincture Senega	.8%
Cinnamon Water	64.2%

Mix the last four and add the first. Dispense with a ''shake-well'' label on it. A reasonably stable preparation can be made by adding a small percentage of an emulsifying agent such as gum acacia.

Formula No. 3

Fluid Extract Cascara Sagrada	25%
Tincture Nux Vomica	15%
Tincture Belladonna	5%
Glycerin	30%
Cinnamon Water	25%

Mix the first three add to a mixture of the last two.

Formula No. 4

Syrup of Senna	30%
Potassium Bitartrate	5%
Sulphur Precipitated	4%
Iron Subcarbonate	4%
Honey	30%
Water	27%

Dissolve the bitartrate in the water, add the iron and sulphur, the syrups and the honey. Place ''shake-well'' label on this product.

Formula No. 5

Pulverized Rhubarb	1.75%
Sodium Bicarbonate	4 %
Pulverized Ipecac	.25%
Tincture Nux Vomica	5 %
Peppermint Water	89 %

Dissolve the bicarbonate in the water. Add the rhubarb and ipecac, the nux vomica. ''Shake-well'' label.

Formula No. 6

Pulverized Rhubarb	3.5 %
Sodium Bicarbonate	7 %
Pulverized Ipecac	.25%
Tincture Nux Vomica	5 %
Fluid Extract Cascara Sagrada	7 %
Peppermint Water	77.25%

This is prepared in the same way as Formula No. 5.

Compound Cascara Mixture

Formula No. 1 (King's Hospital, London)

Fluid Extract of Cascara Sagrada	20 minims
Ammonium Carbonate	2 grains
Tincture of Belladona	10 minims

Tincture of Nux Vomica   5 minims
Glycerin   10 minims
Water, to make   1 oz.

Formula No. 2 (London Hospital, London)

Magnesium Sulphate   1 dram
Glycerin   1 dram
Fluid Extract of Cascara
    Sagrada   1 dram
Fluid Extract of Licorice   1 dram
Tincture of Hyoscyamus   20 minims
Tincture of Nux Vomica   5 minims
Compound Decoction of
    Aloes, to make   1 oz.

Formula No. 3 (St. Mary's Hospital, London)

Fluid Extract of Cascara
    Sagrada   1   dram
Fluid Extract of
    Licorice   ½ dram
Sodium Sulphate   1   dram
Ammonia Water   5   minims
Water, to make   1   oz.

Formula No. 4 (St. Thomas' Hospital, London)

Fluid Extract of Cascara
    Sagrada   20 minims
Fluid Extract of Licorice   30 minims
Tincture of Belladona   5 minims
Tincture of Nux Vomica   5 minims
Aromatic Spirit of
    Ammonia   20 minims
Chloroform Water, to make 1 oz.

Formula No. 5 (Samaritan Hospital, London)

Fluid Extract of Cascara
    Sagrada   20 minims
Tincture of Nux Vomica   5 minims
Tincture of Belladonna   4 minims
Aromatic Spirit of
    Ammonia   10 minims
Choloroform   1 minim
Water, to make   1 oz.

Formula No. 6 (Women's Hospital London)

Fluid Extract of Cascara
    Sagrada   20 minims
Aromatic Spirit of
    Ammonia   10 minims
Fluid Extract of Licorice   15 minims
Peppermint Water, to
    make   1 oz.

Saline Laxative

Sodium Sulphate Crystals   5   lb.
Sodium Phosphate   2½ lb.
Distilled Water   5   gal.

Effervescent Saline Laxative

Sodium Chloride (Dried)   14 oz.
Sodium Sulphate (Dried)   27 oz.

Sodium Phosphate (Dried)   29 oz.
Sodium Bicarbonate   18 oz.
Acid Citric   2 oz.
Acid Tartaric   10 oz.

Hofer's Laxative Tea

Senna   10 g.
German Chamomille   1 g.
Acacia Flowers   1 g.
Poppy Petals   1 g.
Lamium Flowers   1 g.
Clove   1 g.

Tuma's Laxative Chocolate Syrup

Sugar   10 oz.
Cocoa Powder   2 oz.
Magnesium Sulphate   2 oz.
Water   8 fl. oz

Mix the dry ingredients thoroughly and then add cold water and mix again until entirely smooth paste results. Heat to boiling point with constant stirring.

Mineral Water Crystals

Epsom Salts   29   lb.
Sodium Sulphate   2   lb.
Ferrous Sulphate   ½ lb.
Sodium Bicarbonate   30   lb.
Potassium Bicarbonate   10   lb.
Calcium Sulphate   ¼ lb.
Ammonium Chloride   2   lb.

All of the above should be powdered; mixed and sift twice through a 20 mesh screen. All containers should be dry and air-tight.

Compound Sarsaparilla (Blood Purifier)

Formula No. 1

Fluid Extract Sarsaparilla   125   cc.
Fluid Extract Yellow Dock   125   cc.
Fluid Extract Tarazacum   125   cc.
Fluid Extract Senna   60   cc.
Fluid Extract Podophyllum   30   cc.
Oil Fennel   1.5 cc.
Oil Wintergreen   1.5 cc.
Alcohol   125   cc.
Syrup   250   cc.
Glycerin   200   cc.
Potassium Iodide   30   g.
Water, to make   2000   cc.

Mix, set aside a few days, filter, and put up in 6-ounce and 12-ounce bottles.

Formula No. 2

Fluid Extract Sarsaparilla   3 fl. oz.
Fluid Extract Stillingia   3 fl. oz.
Fluid Extract Yellow Dock   2 fl. oz.
Fluid Extract Colocynth   2 fl. oz.
Potassium Iodide   1 dram
Iron Iodide   10 grains

| Sugar | 1 oz. |
| Distilled water | 2 fl. oz. |

Mix the fluid extracts, and dissolve the potassium iodide in the mixture; add the water containing the ferrous iodide and sugar in solution; shake well, and filter.

The above mixtures are usually given in doses of a tablespoonful three times a day.

Diuretics

The following prescriptions yield preparations probably as palatable as may be. All of these contain 1 gram of the drug per teaspoonful, which must be well diluted with water before being taken.

Formula No. 1

Ammonium Chloride	30 g.
Anise Water	30 cc.
Syrup of Glycyrrhiza, to make	120 cc.

Formula No. 2

| Ammonium Nitrate | 30 g. |
| Syrup of Glycyrrhiza, to make | 120 cc. |

Formula No. 3

Urea	30 g.
Acacia Powder	12 g.
to make	120 cc.
to make	120 cc.

Formula No. 4

Potassium Citrate	1 dram
Potassium Bicarbonate	3 oz.
Infusum Uva Ursi	3 oz.
Tincture Hyoscami	1 dram

The usual dosage is a tablespoonful in water, 3 times daily.

Gelatin Capsules

U. S. Patent 1,898,507

A mixture of gelatin (400 parts), glycerin (100 parts), petroleum (0.015–0.02 part), and gum benzoin (0.03–0.04 part) is formed into capsules and hardened with 3–5% formaldehyde, ethyl alchol, and glycerin.

Pill Binder

A binder for making pills is as follows: 3 parts glycerin ointment and 3 parts kaolin; or, 5 parts magnesium carbonate, 0.3 parts gum tragacanth and 10 drops of glycerin. This is useful in making pills containing valerianated menthol, quinine hydrochloride, digitalis leaves and other substances.

Casings for Drugs

Drugs that have to pass through the stomach unchanged but which must dissolve readily in juices of intestines are provided with suitable casings. For example the drug mixture may first be mixed with the pill forming substance or filler and then the pills are dipped into a solution of 5% nitrocellulose and 5% olive oil in equal parts of an alcohol-ether mixture. A rapidly hardening skin is obtained over the pills. Pills may be coated with a thin sugar-gelatin layer and then dipped into a solution of 5% cellulose acetate and 5% castor oil in acetone. Capsules may also be made from mixture of 5% nitrocellulose and 5% lecithin dissolved in equal parts of alcohol and ether. Capsules are then filled with the drug. They will not dissolve in the stomach but will readily dissolve in intestines.

Coating of Tablets

For coating 5000 tablets of reduced iron (approximately 2500 grams) heat the tablets to 30–35° with warm air while slowly rotating the mixing vessel. Add four 20-cubic centimeter portions of 66⅔% sugar solution, drying the surfaces of the starch-sugar solution (250 grams starch–750 grams sugar solution) rapidly in 25-cubic centimeter increments, and as soon as the tablets have been uniformly covered pass in warm air until the material is dry. Add the remaining ⅔ of the starch-sugar solution more slowly and after the addition of each ⅓ carefully dry the tablets with warm air, then cool with cold air before adding the next portion. Glazing is done by slowly adding 250 grams of a dilute sugar solution and slowly drying after the addition of each increment of solution. Polishing is accomplished by first adding approximately 8 cubic centimeters 35% sugar solution and rotating till dry, then more sugar solution until the surfaces are left moist. After drying, add a mixture of wax, olive oil and fat together with talc and continue the rotation.

Shellac Coating for Tablets

Shellac coatings for tablets, pills and capsules are made in such manner that shellac does not dissolve in stomach and thus release medicaments, but passes through unchanged and first dissolves in intestinal juices. Furthermore, coating is made in such manner that it has no dangerous physiological action on human system. This is accomplished by dissolving shellac in a mixture of equal parts of ammonia and alcohol. Shellac apparently undergoes decomposition when dissolved in ammonia. The alcohol in mixture is added merely to hasten evaporation and prevent penetration of coating into tablet, pill or like. Pure bleached shellac in flake form free from arsenic must be used. A 25% solution is preferred.

Cocaine Solutions, Tablets, and Ampoules

Formula No. 1—French Codex:
Bonain's Anaesthetic Mixture

Cocaine hydrochloride	1 g.
Menthol	1 g.
Phenol	1 g.

Place the cocaine hydrochloride, menthol, and phenol in a small wide-mouthed flask and shake together until liquefied.

Formula No. 2—Abraham's Solution

Cocaine Hydrochloride	3	parts
Suprarenin Hydrochloride 1:100	5	parts
Solution Potassium Sulphate (2 per cent)	2.5	parts
Solution Acid Carbolic (0.5 per cent)	100	parts

Three per cent cocaine has been found as effective for surface anaesthesia as the 20 per cent solution.

Formula No. 3—(Central Throat Hospital)

Cocaine Hydrochloride	1 grain
Boric Acid Solution or Sodium Chloride Solution to	110 minims

Preparation of Nitroglycerin Tablets

To 90 parts lactose and 5 parts powdered agar-agar add a 10% alcoholic solution of nitroglycerin and dilute alcohol until the product can be granulated to go through a No. 15 sieve. Dry the powder, mix with 5 parts of talc and pass through a No. 10 sieve; then compress it into tablets. Nitroglycerin is lost during drying. Also some nitroglycerin is lost during storage.

Compressed Tablets of Acetylsalicylic Acid ("Aspirin")

Acetylsalicylic Acid	1620 g.
Gum Acacia (Powdered)	100 g.
Corn Starch (Dried)	200 g.
Talc (Purified)	50 g.
50% Alcohol	Sufficient

Mix the acetylsalicylic acid with the acacia, moisten the mix with enough dilute alcohol to granulate through a No. 20 sieve. Dry the granulation.

Mix the starch and purified talc with the granulation and compress with medium pressure into 5000 tablets.

Compressed Tablets of Phenacetin and Caffeine

Phenacetin	1620 g.
Caffeine	648 g.
Acacia (Powdered)	100 g.
Corn Starch (Dried)	300 g.
Talc (Purified)	75 g.

Mix the caffeine and phenacetin thoroughly, add the acacia. Granulate with distilled water through a No. 20 sieve. Dry the granulation, mix with starch and Talc and then compress into 10,000 tablets.

Rhubarb and Soda Tablets

Rhubarb (Finely Powdered)	1944 g.
Sodium Bicarbonate	1296 g.
Acacia (Powdered)	170 g.
50% Emulsion Petrolatum	Sufficient

Mix rhubarb, acacia and sodium bicarbonate well. Dampen with just enough 50% emulsion of petrolatum to granulate through a No. 20 sieve and compress into 10,000 tablets.

Salol Tablets

Phenyl Salicylate (Powdered)	3240 g.
Sugar (Powdered)	162 g.
Starch (Dried)	250 g.
Talc (Purified)	75 g.

Mix the salol, sugar well and dampen with distilled water to granulate through a No. 20 sieve. Dry the granulation, mix with starch and purified talc and compress into 10,000 tablets.

Acetanilid Tablets

Acetanilid	648 g.
Gelatin (20% Solution)	Sufficient
Starch (Dried)	648 g.
Talc (Purified)	200 g.

Granulate the acetanilid with the 20% gelatin solution through No. 20 sieve. Dry the granulation, mix with the starch and talc and then compress into 10,000 tablets.

Anti-Neuralgic Tablets

A good formula for an anti-neuralgic tablet consists of 20 parts codeine hydrochloride, 500 parts phenacetin, 6 parts white gelatin, 54 parts water,

500 parts acetylsalicylic acid, 40 parts talc and 144 parts starch. First, codeine hydrochloride and phenacetin are mixed together. Then this mixture is worked up with hot solution of gelatin in water in mixing or kneading machine. Moist mass is passed through screen and granulated material is dried in boxes in hot air drying oven. After superficial drying has taken place, mass is screened again and then dried at 50° C. until entirely dry. Mass is shaken occasionally during drying process. Acetylsalicylic acid ingredient is first dried in cold air drying chamber and then mixed with dried mixture obtained above. Talc and starch are also mixed in without pressure being applied to form uniform mass. Tablets are ob-,ained from this mixture at average pressure, weighing 0.6 and 1.2 grams each.

Aromatic Formaldehyde Tablets

1,600 grams white sugar, 60 grams coconut oil, 30 grams gum arabic, 20 grams citric acid, 30 grams paraformaldehyde, 285.5 grams lactose, 0.5 grams orange flower oil, 1.0 gram peppermint oil, 40 cc. ether and sufficient water. Powdered sugar is screened and worked up with an emulsion of coconut oil and gum arabic which is prepared on a water bath. About 45 grams of hot water are used. After this mixture has been prepared, enough water is added to obtain a crumbly, moist mass. This product is screened and the granulated mass is dried superficially only and screened again and then dried thoroughly at 40 to 50° C. in a drying chamber. Mixture of the citric acid, paraformaldehyde and lactose is prepared without pressure and screened. It is then poured over the aforementioned preparation and the mixture is filled into 5 liter wide-necked bottles provided with stoppers. The mass is then impregnated with a solution of essential oils in ether, mixed by careful shaking and allowed to stand for 24 hours. The mass is then spread out on paper, the ether is allowed to evaporate and powder is made into one gram tablets by pressing under high pressure.

Formaldehyde Tablets

To obtain a tablet which contains a vegetable slime as well as formaldehyde, careful selection must be made of the former substance so that the most suitable is used. Marshmallow root powder has been found to be most stable and hence most suitable for this purpose. The composition contains 20 parts of para-formaldehyde, one part of menthol, 100 parts of marshmallow root, 879 parts of powdered sugar, all of which is moistened with 10 parts of stearin dissolved in 45 parts of 95% ethyl alcohol. The mass is granulated and dried. A similar composition contains 7.5 parts of para-formaldehyde, 7.5 parts of sugar, 0.5 part of orange oil, 0.5 part of peppermint oil, 0.2 part of menthol, all the ingredients being well-mixed, and then incorporated with 1,000 parts of sugar, 10 parts of citric acid, 75 parts of marshmallow root (sometimes 100 parts is used), 25 parts of talc (very finely granulated). The mixture is thoroughly mixed and then made into tablets. Quince slime is also used in making the formaldehyde tablets. This substance is almost odorless. Agar-agar solution is also useful.

A slimy but loosely constituted tablet mixture is obtained by using dry pectin powder. The latter may be used alone or in admixture with dried wheat flour in the ratio of 1 : 2. Also gladderwrack (fucus vesiculosus) root finely pulverized may be added in small quantity to the mixture. About 1 to 2% of carragheen moss is also useful. However, the addition of gum arabic, gum tragacanth or dextrin is useless. The trouble is that tablets which are made from mixtures containing these substances become hard in course of time and dissolve only with great difficulty making the effective ingredient, paraformaldehyde, slow to act. Only a slight pressure should be used in making these tablets in the regular tablet machines.

Grippe Tablets

0.15 part acetylsalicylic acid, 0.15 part phenacetin and 0.15 part salipyrin. Another mixture contains 0.2 part salicylic acid, 0.15 part caffeine, 0.05 part potassium iodide and 0.1 part sodium thiosulfate. Another mixture contains 0.2 part acetylsalicylic acid, 0.05 part caffeine, and 0.05 part ipecac. Quinine salicylate is used in another mixture in proportion of 0.2 part mixed with 0.1 part caffeine, 0.1 part phenacetin and 0.1 part sodium chlorate. Another formula is 0.05 part phenacetin, 0.1 part salipyrin and 0.15 part amidopyrine. Cocoa, sugar and vanillin may be used to improve taste and increase volume of all these preparations.

Migraine Pencil

Stearic Acid	70 parts
Menthol	30 parts

Melt together on water bath and cast in molds.

Witch Hazel Foam

Stearin (Granulated)	100 g.
Sodium Carbonate	5 g.
Glycerin	15 g.
Distilled Extract Witch Hazel	500 g.
Distilled Water to make	1000 g.

Mix the sodium carbonate and glycerin with 500 grams of water, and heat on a water-bath until the salt is dissolved. Add the stearin and continue heating and stirring until saponification is complete. Remove from heat and when the mixture has cooled to about 160° to 175° F., add the extract of witch hazel slowly, with constant stirring, and continue stirring vigorously until cold.

Stabilizing Hydrogen Peroxide

Formula No. 1

The addition of 0.1% barbituric acid stabilizes hydrogen peroxide solutions.

Formula No. 2

Canadian Patent 337,601

A hydrogen peroxide solution is stabilized by adjusting the acidity to pH 2–6 and adding an amount of sodium stannite equivalent to 5–100 milligrams of tin per liter of solution and 0.01–0.2 grams of sodium pyrophosphate per liter of solution.

Formula No. 3

Tests run for some time show that amounts as small as 0.15% of methyl para-oxybenzoate will preserve peroxide solutions. Controls showed a loss of almost 15 times more than those so preserved. This is indeed a help in the formulation of peroxide lotions for bleaching, etc.

Spirit of Mercurochrome

2 parts mercurochrome, 54 parts 90% alcohol, 10 parts acetone, and 34 parts distilled water.

Solidified Iodine

Solidified iodine may be made from the tincture with from 5 per cent to 10 per cent of sodium stearate dissolved by heat and poured into suitable moulds. It must be kept fairly airtight.

Iodized Carbon as Protector Against Mercury-Vapor Poisoning

Iodized carbon is an activated carbon impregnated with iodine. The most favorable amount of iodine is about 5%. The carbon either can be used in gas masks or can be sprinkled directly upon open areas or in cracks contaminated with mercury. Experiments are reported which demonstrate the great effectivenss of this preparation.

Liver Concentrate

U. S. Patent 1,895,977

Small pieces of fresh liver are covered with salt and, after keeping at 10–15° for 40 hours, the liquid extract thus obtained is separated and evaporated to dryness.

PHOTOGRAPHY

Photographic Prints on Wood

Beat thoroughly the whites of one or two eggs, and let stand for one or two hours. In this interval the albumen will be reduced to a liquid form. Mix it with powdered English oxide of zinc, and place a smooth coating on the surface of the wood, making it as thin as possible. After the surface is thoroughly dry, sensitize it with a solution of silver nitrate (Eastman Hydrometer test about 50). The sensitized surface should be thoroughly dried in a dark place, after which it will be ready for printing, in sunlight or under strong artificial light. After being printed, it should be fixed with a weak solution of plain hypo for about five or ten minutes, and then washed.

Cockle-Proofing Film
U. S. Patent 1,947,160

The method of cockle-proofing photographic film of cellulose acetate containing moisture repelling agents so that the combustion retarding properties of said film will not be materially diminished by said method, comprises thoroughly and carefully drying the film, and subsequently applying to both sides thereof a relatively thin coating resulting from a solution containing by weight not more than 10% nitro-cellulose, approximately 1% camphor, approximately 10% each of acetone, butyl alcohol, and mono ethyl-ether ethylene glycol, and approximately 60% methyl alcohol, and subsequently re-drying the thus coated film.

Cellulose Acetate (Flexible Films)
U. S. Patent 1,896,145

Cellulose acetate (30–35% acetate) is dissolved in a mixture of 3 parts of ethylene chloride and 1 part of a monohydric aliphatic alcohol containing less than six carbon atoms, e.g., methyl alcohol.

Abrasion-Resisting Film
U. S. Patent 1,888,952

Unseasoned cellulose ester film, after removal of a portion of the solvent, is impregnated with a colloidal solution of carnauba or similar wax in alcohol (aqueous ammonia 2 parts, ethyl alcohol 1 part), both film and solution being heated to approximately the melting point of the wax; the treated film is finally dried or seasoned.

Preventing Loss of Clarity in Prints
British Patent 401,961

To minimize the loss of clarity of the prints during drying, they are treated with a toning solution in such concentration or for so short a time that no visible toning is produced. In examples (1) a solution containing 40 grams of ammonium thiocyanate and 40 cubic centimeters 1% auric chloride solution in 1 liter of water and (2) 0.1 gram sodium selenate in 1 liter fixing solution are used.

Making Reverse Prints on Metal

Make the print with halftone enamel. Develop in the regular way. Then flow with cold-top, whirl in the usual way, and expose to light, giving the usual time for printing. Develop the cold-top as an ordinary cold-top print. Wash this print under water. Then dry the print, and place it in an acetic acid stripping bath to lift the halftone enamel. Develop with cotton. Dry and proceed with etching.

Cold-Top
Solution No. 1

Bring 160 ounces water to boil; then add 3½ ounces ammonium carbonate. Stir in 1 pound orange shellac (high grade). Boil for ½ hour.

Solution No. 2

Dissolve ¾ ounce ammonium bichromate in 8 ounces water. Add 4 fluid ounces concentrated ammonia (28%).

Thoroughly mix Solutions No. 1 and No. 2.

Developing Solution

Alcohol *	1	gal.
Malachite Green Crystals	¼	oz.
Water	8	oz.

* 95% denatured alcohol, free from benzol or oils.

Modified Gaslight Photo Prints

Soft and detailed prints may be produced from very contrasty negatives in the following manner:

Prepare the stock solution:

Copper Sulphate	½	g.
Potassium Persulphate	1	g.
Nitric Acid	2	cc.
Water	800	cc.

The paper is exposed in the usual manner and for the correct time. Before development the exposed paper is plunged in the above bath for one minute, rinsed for a moment under the tap and then developed.

With an increase of bath strength five times it may be used to develop negatives of brilliant subjects or sunsets.

Transparent Projection Screens

Method No. 1

If varnished silk be given a coat of a flat varnish either by spraying or dipping, it will make an excellent screen for film or slide projection. In using this screen the light is projected from the rear through the screen to the eyes of the audience and consequently much greater illumination, relatively, is obtained than in the case of the opaque screen.

Method No. 2

Sheets of dental rubber also make fine transparent projection screens. To be most effective the rubber should be stretched on a suitable frame

Gelatine Dark Room Screens

Tartrazine	8 g.
8% Gelatine Solution	1000 cc.

This gives a yellow.

Rose Benzol	3 g.
8% Gelatine Solution	1000 cc.

This gives an orange.

To experiment with other colors, the following dyes are suggested.

Eosin Y
Amacid Yellow T Ex (Tartrazine)
Amacid Yellow M (Metanil)
Amacid Yellow S (Naphthol)
Amacid Pink B (Xylene Red)
Amacid Green 2G (Light Green)
Amacid Green N Conc. (Naphthol Green)
Fuchsine
Amacid Violet 3B (Conc.)
Rhodamine B
Victoria Blue B Conc.
Erythrosine
Safframine Y Spec.

Duplicate Halftone Screens

Prepare the following solution: Beat up the albumen of one egg and dilute with 24 ounces of distilled water to which add about eight ounces of photo-engravers liquid glue. In a double boiler bring this to the boiling point and then allow to cool. When cold bring this to a specific gravity of 1.25. Filter carefully.

Make a solution of ammonium bichromate to test 1.25 (specific gravity).

To every 500 cubic centimeters of the clarified glue solution add 125 cubic centimeters of the bichromate solution and again filter.

Thoroughly clean a plate glass, polish and dry. Then pour a pool of the glue solution in the center of the plate balancing on the fingers of one hand. Gradually "coax" the solution over the surface of the plate and then invert it on the whirler and revolve at about 60 revolutions per minute over gentle heat until quite dry. This must be done under a yellow light.

Put this dry gelatine plate in contact with a master screen and expose to a 35 ampere arc lamp at four feet distance for about one minute. This is approximate; exact time can be found by trial.

After exposure put the plate in a tray with warm water at about 80° F. and allow to remain for a minute or two, then gently rock the tray until the unexposed parts develop clean. Rinse under the cold tap and then put the plate in a strong solution of tannic acid. Exact strength is immaterial even to saturation.

The plate is then washed under the tap and flowed with silver solution previously prepared as follows:

Distilled Water	1000 cc.

to which is added silver nitrate to make a registered solution of 40 grains to the ounce by argentometer test.

Decant 200 cubic centimeters of this silver solution and to the remaining 800 cubic centimeters add ammonium hydroxide gradually and in small amounts with constant stirring until the silver is redissolved and the solution is again clear. Now carefully add some of the reserved silver solution until a very slight turbidity appears in the bulk solution.

The application of this silver solution to the plate that has been bathed in the tannic acid bath will give the image a mahogany color. Wash the plate again under the tap and then treat with a weak Farmer solution to clear away any possible veil after which wash again and repeat the bathing and washing until full density is attained. Dry the plate away from dust.

A thin cover glass will serve to protect the surface in use and this may or may not be sealed with Canada Balsam.

If a master screen cannot be obtained for making the print, a fairly satisfactory substitute may be prepared by photography using a contrast dry plate in contact with a sealed screen in the camera exposing against a piece of white paper with a small diaphragm in the lens. Develop with Hydros developer. If carefully made the dots in this photographic master will be sharp, square and wholly clean.

Photographic Flash Lamp
U. S. Patent 1,915,591

A sealed glass bulb contains 0.1 gram of aluminum foil in an atmosphere of oxygen at a pressure of 100 millimeters of mercury.

Dark Room Lamp Filter

A waterproofed lamp is immersed in a bottle filled with fluids made up as follows:

Media for Use With Orthochromatic Plates

Formula No. 1

Ferric Chloride	30	g.
Potassium Thiocyanate	0.2	g.
Water	500	cc.

Formula No. 2

Potassium Dichromate	25	g.
Crystal Violet	0.1	g.
Water	500	cc.

Formula No. 3

Cobaltous Sulphate, Heptahydrate	60	g.
Triethanolamine	5	cc.
Water	500	cc.

Any one of the above three solutions will transmit red light only.

Media for Use With Printing Papers

Formula No. 1

Potassium Dichromate	25	g.
Water	500	cc.

Formula No. 2

Water in dark amber bottle.

Media for Use With Panchromatic Plates

Formula No. 1

Ferric Chloride	30	g.
Nickelous Chloride Hexahydrate	100	g.
Chromic Chloride Hexahydrate	50	g.
Cupric Chloride Dihydrate	50	g.
Water to make	500	cc.

Formula No. 2

Tartrazine	10	g.
Naphthol Green	3	g.
Carmine Blue	5	g.

Panchromatic plates are sensitive to red light, but are not affected by dull green radiation. It is difficult to find light filtering media which will absorb the blue and all of the red radiation and transmit green light of sufficient intensity for manipulative purposes. Formula No. 1 passes light rich in green rays, but they also contain a small amount of red in them, hence the plate should not be exposed over long periods of time to the rays from the lamp. Formula No. 2 is free from this defect, but the dyes are not as readily available as the inorganic components of the first medium.

Hardening of Negatives in the Tropics

Negatives can be processed in the tropics without the use of formalin or cooling, by the following procedure: the metol-hydroquinine-borax developer is used which suppresses softening. Development is followed by rinsing for 1 minute in running water or in a large tank (60 by 60 by 60 centimeters). Next, the film is immersed for 1 minute or longer in a hardening bath consisting of a 10% solution of potassium alum. After a rinse of $\frac{1}{2}$-1 minute, it is fixed in the usual acid-fixing bath. It is then rinsed in running water and washed for 10 and 30 minutes respectively in 2 large tanks (as above).

Printing and Enlarging of Hard Negatives

The negative is printed onto a rapid chlorobromide paper and developed in the following fine-grained slow-acting developer: (1) water 1500 cubic centimeters, sodium sulphite heptahydrate 55 grams, glycine 3 grams, hydroquinone 10 grams, potassium bromide 1.5 grams. (2) Water

1000 cubic centimeters, potassium carbonate 50 grams. Equal parts of 1 and 2 are taken and diluted with 2–3 times the amount of water. The print is then fixed and washed. Except for very hard negatives, the result is said to be satisfactory. For very hard negatives, the print is placed for 5 minutes in a 1% solution of sodium sulphide, washed, and bleached in a potassium ferricyanide-potassium bromide solution until only a faint silver sulphide image is visible. The print is now placed, by daylight, in the original developer until the image has the desired strength. It is then fixed again. If only a very slight intensification of the silver sulphide is necessary, this may be done by printing out the image by exposure to daylight instead of redeveloping.

Making a Positive from a Negative
First Development
Developer:

Water	1000	cc.
Metol	3.25	g.
Sodium Sulphite (Anhydrous)	25	g.
Hydroquinone	1	g.
Potassium Bromide	1.5	g.
Ammonia (Sp. Gr. 0.91)	7.5	cc.

For stock solution omit ammonia. When ready for use, add from 2½ to 3 minims of ammonia to each ounce of developer. Development time 3 minutes.

Rinsing the Plate
Rinse the developed plate for about 1 minute in running water.

Reversing Bath
Stock Solution:

Water	32	oz.
Potassium Bichromate	1	oz.
Concentrated Sulphuric Acid 3¼ oz.		

For use take 10 parts water and to this add 1 part stock solution. Solution may be used until it becomes too muddy or of a greenish tint. After the plate has been in the reversing bath for 1 minute, turn on a strong artificial light or take the tray into *diffused* daylight. As soon as the blackened silver is completely eaten out, wash the plate in running water for 2 minutes.

Redevelopment
Place the plate in the original developing bath or other metol-hydroquinone plate or paper developer and leave until it is absolutely black (2 minutes). This can be done in bright light. Rinse for ½ minute in water and dry immediately.

Note: If the original developer is used for redevelopment it must not be used for the first development again.

Collodion Wet Plate Negatives
Cleanliness is essential to the successful working of all photographic operations, and particularly so when wet collodion is used. The darkroom must be kept clean and well ventilated, if trouble is to be avoided. There should also be plenty of space in which to work. A crowded darkroom is an obstacle to satisfactory results.

Negative glass, whether old or new, should be soaked in lye solution made as follows:

Crude Caustic Soda	2 lb.
Water	1 gal.

New glass should remain in the lye for half an hour. Glass carrying old negatives should be allowed to soak until the old film is loosened. The film is then scrubbed off.

The glass is next placed in this solution:

Nitric Acid	1 qt.
Water	1 gal.

Let the glass soak in the acid for a few hours, or for a full day if possible. It should then be scrubbed in running water with a soft scrubbing brush or pad of felt, and well rinsed. If the glass is not properly cleaned, fog and streaks will inevitably result.

While still wet, the glass is flowed with the following albumen solution:

Dried Albumen	70 grains
Ammonia	10 minims
Water	1 qt.

The solution should be filtered before being used. The wet plate is flowed once with it and drained, then flowed and drained again, and put to dry in a dust-free place.

If dried albumen is not at hand, the white of a strictly fresh egg may be used in place of the 70 grains of albumen in the above solution.

If preferred, the following solution may be substituted for the albumen solution:

Gelatin	150 grains
Glacial Acetic Acid	150 minims
Water	½ gal.

Or rubber solution, diluted with benzol, may be used:

| Rubber Solution | 1 oz. |
| Pure Benzol | 10 oz. |

. This rubber solution can oly be poured on after the glass is dry.

If it is preferred not to apply any first coat or ''substratum'' to the plates, they should be allowed to dry after scrubbing and rinsing, and should be sprinkled with this solution:

Iodine	200 grains
Denatured Alcohol (Free from	
Grease)	½ gal.

They are then polished off with a piece of lintless rag or chamois leather. Glass that is not substratumed should be edged with rubber solution applied with a small brush, which is better than absorbent cotton.

Eastman Complete Collodion is supplied in two containers, one holding the plain collodion and the other the iodizer. The reason for this is that an iodized collodion gradually gets slower as it ages, but the plain collodion and iodizer separately will keep indefinitely if the bottles are kept well corked. Collodion not in use should be kept in a cool place, otherwise the corks may be blown out.

Add the contents of the two-ounce bottle of iodizer to half a gallon of plain collodion and the collodion is complete and ready for use. *Do not add anything else.* The collodion works best after it has been iodized a few hours, but it can, if necessary, be used at once, and will keep for months without losing very much speed. However, as it does eventually become slower, it is recommended that no more be iodized than will be used in about two weeks.

Since Eastman Collodion is, in the course of its manufacture, filtered under pressure through paper half an inch thick, there is no need to refilter it. But if it is thought desirable to filter it again *after iodizing*, use a small filter loosely fitted with Filter Cotton. This speeds up the operation and minimizes evaporation. The collodion solution will not work well if it is too warm. The best temperature is beween 65° and 70° Fahrenheit. If you are compelled to work in very warm weather, it is advisable to place the solution on ice. In doing halftone work during hot, humid weather, it will be found almost impossible to get a sharp, clean dot unless the collodion is cooled in this way.

This collodion is equally good for line and halftone work. It is a mistake to suppose that different collodions are required for these two types of work. A halftone screen negative is exactly the same as a line negative inasmuch as the black halftone dots must be as dense as possible and the clear spaces quite transparent. If the collodion is good for line work, it is also good for screen negative making, and vice versa. However, if you have both old and new collodion, use the old for line work and the new for halftone, because the new collodion is faster.

If you are using a freshly iodized collodion for line work, be careful not to over-expose. To get the best results with fine line work, use the collodion after it has been iodized for some time.

Before coating a plate, it should first be carefully dusted with a camel's hair brush kept especially for the purpose. If the brush is not clean it may leave more dust on the plate than it takes off. Never leave this brush lying on the bench, but keep it hanging up.

Hold the plate horizontally, either on a holder or with one corner resting on the first two fingers of the left hand and held down by the thumb.

The collodion bottle is taken in the right hand and the collodion is poured in a pool on the upper part of the plate, the right top corner being covered during the pouring. Then the plate is gently tilted so that the left top corner is next covered, then the left bottom corner, and finally the excess of collodion is drained off at the right bottom corner.

During these operations, the bulk of the collodion should be kept well in the middle of the plate, which is only tilted sufficiently for the collodion to flow to the edge, but not over it and off the plate. Be sure at all times that the edge of the collodion is kept flowing *forward*, and that no flow-back occurs; otherwise the collodion will thicken and cause a mark in the negative.

The draining of the excess collodion should be done very gradually into a bottle, the plate being tilted only slightly at first and then gradually brought up to a vertical position. While this tilting is going on, the plate must also be rocked laterally by free movement of the arm, so that the collodion flows down toward the edges of the plate, and then to the corner, rather than *immediately* toward the corner. Ribbed markings are nearly always caused by hasty and improper draining of the collodion. As the draining nears completion and the position of the plate becomes more nearly vertical, the bottle should be moved with the plate, not scraped against it.

Care should be taken in the first place to pour out a quantity of collodion that is just sufficient for the size of the plate being used. Any excess over the correct amount will make it necessary to tilt the plate too quickly in draining and will produce streaks, besides wasting the collodion that drips over the sides of the plate.

Care in the methods of coating applies just as much to rubber solution and stripping collodion as to coating the collodion for the negative. Streaks visible in the negative after stripping are usually caused by careless coating. Sometimes this also leads to the splitting of the image, especially in halftone negatives.

When the film of collodion has set—a condition which is ascertained by touching one of the thickened edges with the finger—the plate is immersed at once in the silver sensitizing bath. It is especially important to do this very promptly in hot weather, because if it is delayed the end of the plate first flowed will dry and fail to sensitize.

The silver bath is the most important solution used in the wet plate process, and great care should be exercised in its preparation and use. Cleanliness is essential. Keep all dust and contaminating substances away from the bath. Have the darkroom clean, thoroughly ventilated and free from dust, so that the bath may be left uncovered without any danger. This has the advantage of allowing the collodion solvents in a used bath to evaporate to some extent. Use the purest silver nitrate possible to procure.

For halftone work, make up the bath according to the following formula:

Silver Nitrate	10 oz.
Distilled Water	3 qt. 14 oz.
Potassium Iodide	5 grains

If this bath is tested with a silver-solution hydrometer, sometimes called an argentometer, it will be found to register 40 (i.e., the number of grains of silver per ounce of solution), which is the strength generally preferred for halftone work. *The bath should not be stronger than 40.*

The hydrometer should be checked occasionally to make certain that its scale is correct. To do this, make up a new solution, using a known number of grains per ounce of distilled water, and place the hydrometer in it. If it does not register the right amount, a correction should be made in the future for the error on the scale.

For line work a weaker bath is advisable. If, instead of the amount of water given in the formula, one gallon is used, the resulting bath will contain 35 grains of silver to the ounce, which is a good strength for all-around work.

A common practice, to which there is no objection, is to omit the iodide in the above formula and to leave the collodionized plate in the bath for a few hours.

If distilled water cannot be obtained, tap water may be used. In this case the bath should, after dissolving, be set out in the sunlight until the cloudy precipitate is settled.

By setting any silver bath in the sun, all organic matter which it may contain will be reduced as a black precipitate. It is therefore advisable to make enough solution for two baths, so that one may always be out in the light while the other is in use.

When ready to use the bath, filter it and add one-sixth dram of pure nitric acid—just enough to make the bath acid, but not too strongly acid. A bath can be tested for this with litmus paper. It should turn blue litmus paper gradually red. If it turns red litmus paper blue, then acid must be added.

Be sure to have a big enough vessel for the silver bath, and enough solution to allow some movement, while in the bath, to the largest plate which you will have to sensitize; otherwise streaks are bound to occur. Keep the bath cool, because if it becomes too warm it is likely to cause fog. In very warm weather it is advisable to place the bath in a tank of water cooled with ice. Be careful, also, not to have the bath contaminated by fumes. The fumes from burning coal gas are almost certain to cause black spots.

A bath made as directed, and used for collodion, will keep indefinitely. When it becomes over-iodized, pour it into a clean, clear glass bottle containing about one-third water. This will precipitate the excess of iodides. Then test it with the hydrometer, adding silver nitrate until the hydrometer reads 40 (or 35, if that is the strength to be used). Neutralize acidity by putting in a crystal of carbonate of soda. Then set the bath in the light until all the precipitate has settled, filter, re-acidify with pure nitric acid, and it is ready for use. When silver nitrate is used, neutralization with the soda may be omitted, unless the bath has been over-acidified, and then it will be unnecessary to re-acidify the bath.

If the bath becomes overloaded with alcohol from the collodion plates sensitized in it, put it into an evaporating dish and boil it down. (This overloading happens more than twice as fast with a dipping bath as when a tray is used.) When nothing else is wrong with the bath, only about one-third need be boiled away, but if the bath is giving fog because of the presence of other organic impurities, it should be boiled down until it has formed a pasty mass and has become liquid again. It is then allowed to cool. The resulting fused silver nitrate is dissolved in distilled water, and this is poured into the quantity of water required to bring the bath to its original volume. It is then tested with the hydrometer, brought up to strength with silver nitrate and set in the sun for a few days. After being filtered and acidified, the bath is again ready for use.

If negatives are not right, seek other causes before deciding that the bath is at fault. Here are a few simple queries that often help to determine the source of the trouble: (1) Is the glass quite clean? (2) Is the air of the darkroom fresh and free from dust? (3) Is the developer right? (4) Is the temperature right? (5) If the darkroom was allowed to become cold during the night, has it (and the solution) had time to warm up? Suspect the *composition* of the bath *last*, and in any case avoid the addition of various chemicals sometimes recommended. They are never necessary, since any bath, if it is out of order, can be put right by sunning or by boiling down.

When a collodionized plate is put into the silver bath, the silver nitrate combines with the iodides in the collodion and forms silver iodide. This gives a creamy film, and sometimes operators think that the whiter the appearance of the plate in the bath, the better their negatives will be. This is not so.

There is one certain proportion of iodides which, if it is present in the collodion, will give the densest image. More iodides will not give any better negative, and therefore the extra iodides, although they make the plate look whiter in the bath, have no good effect whatever. On the contrary, they consume the silver nitrate, with the result that the bath becomes over-iodized more quickly than is necessary. When a bath has become over-iodized, silver iodide will precipitate on the plate as a sort of sandy deposit and these particles wash off in developing, leaving a granular negative covered with pinholes.

Pinholes also occur when a fresh bath *containing no iodide at all* is used. They are caused by the iodide being eaten out of the collodion film to satisfy the affinity of the silver bath for iodide. This defect, however, will at most be evident only in the first two or three plates.

The plate must be immersed in the bath solution in one even movement, so that the solution covers the plate without a stop; otherwise a ''bath-mark'' line will show in the negative. If a dipping bath is used, the plate is moved gently as soon as it is placed in the solution (or, if a tray is used the tray is rocked). This movement is repeated occasionally during sensitizing, which should be completed in about 3 minutes. If the bath is alcoholic, leave the plate in for about one minute after all greasiness has disappeared.

Before withdrawing the plate from the solution, be sure that your hands are clean, so that no dirt will adhere to the plate or get into the bath. Then take out the plate, allowing as much solution to drain back into the bath as possible. Set the plate on a clean piece of blotting paper and wipe the back with blotting paper or a lintless rag.

Provide the dark slide of the plate holder with a strip of blotting paper on which the plate may be set to absorb the silver drainings. The latter are corrosive and soon ruin woodwork with which they come into contact. Plate holders should be wiped carefully with a damp rag at the end of the day's work, and all parts which silver drainings are liable to touch should be coated with shellac once a week.

The sensitized plate should be placed in the camera immediately, and exposure started; otherwise some of its sensitivity will be lost. In warm weather a wet plate will begin to dry in from six to eight minutes. Because of this drying, an exposure can under no circumstances be longer than fifteen minutes.

After exposure the plate is developed, a suitable developer being the following:

Ferrous Sulphate	4 oz.
Glacial Acetic Acid	4 oz.
Water	½ gal.

More acid may be added, in hot weather, up to another two or three ounces, if there is any tendency to fog. If the developer solution is old and plates tend to repel it, a small quantity of denatured alcohol may be added.

Pour enough developer on the plate to cover it without any stop, or it will leave a mark. Do not use any more developer than is necessary, however, because just as little silver nitrate solution as possible should be washed off, the strength of the image depending upon the amount of silver left on the plate.

The exposure, already made, should be such that a good image is obtained with a development of from 20 to 30 seconds. If the development veils the plate, over-exposure is indicated. On the other hand, forcing development causes grain in the clear spaces of the negative. This and over-intensification will cause "dry effect" and possibly film splitting, especially if the glass was not thoroughly clean.

Do not go to the trouble of looking at negatives with a magnifying glass during development, because if the exposure is incorrect you cannot make it right in development.

After the plate has been developed, the developer must be thoroughly washed out of the collodion, otherwise it will cause a stain on the negative. Wash the latter at least 20 seconds under a good stream of water, and longer if any tendency to stain appears.

The following solution of potassium cyanide or sodium cyanide is flowed over the plate to fix the image:

Potassium or Sodium Cyanide 2 oz.
Water   1 qt.

If a tray is used the plate should be immersed in the solution for an interval twice as long as it takes for the white silver iodide to disappear. If a plate is left too long in the cyanide, the fine detail will tend to be dissolved away.

Both potassium cyanide and sodium cyanide are *deadly poisons,* and bottles containing solutions for either should be conspicuously labelled "poison."

After fixing the image, wash the plate thoroughly.

The commonest intensifier is copper bromide and silver.

First make these two solutions:

Solution No. 1

Copper Sulphate   12   oz.
Water to make   ½ gal.

Solution No. 2

Potassium Bromide   6   oz.
Water to make   ½ gal.

When dissolved, mix the two solutions and the bath is ready. The plate is placed in the solution until it is bleached white through and through. It is then washed in running water for about one minute. While under the tap, the plate should be moved to and fro, so that the stream of water will not continually strike the same spot.

Next the plate is well drained, and then blackened by pouring over it the following solution:

Silver Nitrate   1 oz.
Water to make   12 oz.

Freshly made silver solution for intensification will often give streaks, but the addition of a few drops of nitric acid will insure its working smoothly.

The washing, before blackening, must not be too prolonged or the plate will not blacken. On the other hand, if it is not washed long enough, silver bromide will be precipitated by the silver nitrate solution.

If the negative is not sufficiently dense, it may be intensified a second time by proceeding as before.

The silver solution should be of the strength given above, as the use of a weaker silver solution for intensification will give trouble in reducing. The plate may reduce suddenly and unevenly.

To get the utmost intensification the following intensifier can be used:

Lead Nitrate   3   oz.
Potassium Ferricyanide 3   oz.
Glacial Acetic Acid   3   oz.
Water to make   ½ gal.

The plate is placed in this until the color is evenly yellow through and through. It is then washed thoroughly and flowed with a weak nitric acid solution (1 part nitric acid to 30 parts water), rinsed again and blackened with this solution:

Sodium Sulphide   4   oz.
Water to make   ½ gal.

Ammonium sulphide may be used instead of sodium sulphide if there is no objection to the unpleasant odor. The plate is once more rinsed and again flowed with weak acid. Thereafter it is flowed with gum arabic or weak glue solution, to protect the film. Never use a sulphide in the same room in which the silver sensitizing bath is located.

Any reduction necessary in a negative that is to be intensified by the lead ferricyanide method just described must be done prior to intensification. It cannot be done afterward.

Sometimes, for fine line work, this mercury intensifier is used:

Mercuric Chloride   5   oz.
Ammonium Chloride   3   oz.
Water to make   ½ gal.

The bleaching of a wet plate negative in this solution is slow, but it may be hastened by warming the solution. After thorough washing, the blackening is done with the following:

| Ammonia Water | 3 | oz. |
| Water to make | 1 | qt. |

In this case also, any reduction necessary should be done before the intensification.

The usual method of reduction is to bleach the plate, after intensification, in the following solution:

Potassium Iodide	2	oz.
Iodine Resublimed	1	oz.
Water to make	½	gal.

After bleaching it is flowed with a weak solution of cyanide:

| Potassium or Sodium Cyanide | 1 | oz. |
| Water to make | ½ | gal. |

The negative must be carefully watched during this operation as it is very easy to over-reduce and so ruin it.

After the reduction is complete the plate is rinsed, and then blackened by flowing it with the following:

| Sodium Sulphide | 4 | oz. |
| Water to make | ½ | gal. |

If there is any sign of yellow stain, it is removed by flowing with weak nitric acid solution.

The wet plate negative is now complete. If it is not to be reversed, it should be coated, when dry, with engravers' hard varnish.

In varnishing a negative, be sure to have both varnish and negative fairly warm, and of even temperature.

If it is necessary to reverse the wet plate negative, or if it is desired to transfer a number of negatives to a large piece of glass in order to make one print instead of a number of prints, take the following steps.

After thorough drying, and when cool, the negative is flowed with rubber solution.

When the rubber is dry, the plate is coated with stripping collodion, which is plain collodion containing a small quantity of castor oil. It is made thin in order that it will flow easily and dry quickly. It forms a very tough and very flexible coating or film.

Read again the instructions previously given for coating collodion. They apply just as much to rubber solution and stripping collodion.

When the film of stripping collodion is dry, it is cut around with a knife, and the plate is put into the following solution:

| Glacial Acetic Acid | 2 | oz. |
| Water | 1 | qt. |

When the film begins to lift, remove the plate from the bath, lift up the film by one corner with a pocket knife and transfer it to the position required, turning it over if reversal is desired.

To be quite sure that the film is not damaged or stretched in stripping, it may be handled on paper. Thoroughly wet a piece of thin paper so that all stretch is taken out, and bring it into perfect contact with the negative by means of a rubber roller or other squeegee. Lift one corner of the paper and with it a corner of the film, which may be started with the point of a pocket knife. Draw the paper and film off together. For reversing, the film is now transferred to another piece of paper, the second paper and the film being trimmed with scissors and then laid on glass, preferably moistened with a little gum water. If the negative is not to be reversed, the transfer to the second piece of paper is of course omitted.

When the film has been placed in position on the glass, it can be squeegeed down evenly by stroking with a few pieces 3″ x 3″ blotting paper, or with a piece of velvet rubber. If the latter is used, keep it in water when not in use, so that it will remain soft and pliable.

Blisters may be caused in stripping by pouring the stripping collodion on when the plate is too warm, by heating the plate too much before the collodion has set, or by air bells in the collodion.

Be sure to have the negative quite dry before applying rubber or stripping collodion.

Negatives from which many prints are to be made should be protected. A good plan is to cover them with a thin transparent sheet that can be fastened to the corners of the negative with film cement.

Negatives which are to be preserved for future use should be stripped on to celluloid sheet. This is light, unbreakable, and requires little space for storage. Negatives transferred to it are instantly available at any time and can be reprinted just as easily as though they were on the original glass.

Causes and Correction of Wet Plate Defects

The following difficulties are sometimes met with in wet collodion work. Others have been dealt with in the foregoing text.

Mottling and Streaks. These defects are sometimes caused by not rocking or moving the plate sufficiently during sensitizing. Streaks and lines may also be caused by holding the negative vertically when developing, or by flowing the developer in one direction only.

Metallic Streaks. In very cold and dry weather streaks are caused by collodion made with solvents that are too dry. Try adding a few drops of water.

Diagonal Ribbing or Streaks. This condition is caused by rocking the plate through too narrow an angle when coating it, or by not rocking it long enough. The plate should be coated with just about the right quantity of collodion, the latter should be flowed slowly and evenly over the plate, and most of the excess should be allowed to drain off while the plate is almost horizontal. If the draining is too rapid at first, the coating is very likely to become thin at the top and thick at the bottom, and will give uneven negatives. If, in addition, it is brought too quickly to a vertical position, and rocked rapidly through a narrow angle (instead of slowly through a full right angle) the film will almost inevitably show ribbed markings. If the collodion is too thick, it should be thinned before iodizing, or Eastman Special Iodizer should be used. Always keep collodion tightly corked; otherwise it will thicken by evaporation.

Other Streaks may be caused by removing the plate from the bath too soon, especially if the bath has become loaded with alcohol and is slow in its action. Scum on the bath, bubbles, or dust in the collodion will also cause streaks, as will dirty glass. Stopping the even flow of the bath over the plate when sensitizing, or the even flow of the developer when developing, will give the same trouble.

Fog, due to a reduction of metallic silver on parts of the negative which should be clear, may arise from a variety of causes. It is shown on the negative by a gradual or, in some cases, by a sudden darkening all over the film, or it may only spread over a small portion. Some of the causes are:

1. An insufficiently acid state of the silver bath (one of the most frequent causes of fog). If the bath is suspected, test it with blue litmus paper. If it is in the condition mentioned, the litmus paper will only slightly discolor. Add, drop by drop, a 10 per cent pure nitric acid solution to the bath solution until it turns the blue litmus paper a decided red color. Note, however, that too much acid prevents the bath from sensitizing at all.

2. Light making its way to the plate through cracks in the bellows of the camera; or internal reflections. Cracks can be located by placing one's head in the back of the camera and covering the space between the camera and the head with a focusing cloth, thus excluding all light. When the eyes have become accustomed to the darkness, any cracks present will become apparent.

3. Unsafe darkroom illumination. Coat and sensitize a plate in the shade, expose half to the lamp suspected, keep the other half covered, and develop. If the light is unsafe, the exposed part will show fog. The yellow light employed in a wet plate darkroom may be bright, but must not contain any blue, violet or ultra-violet light.

4. Fumes of some chemicals, turpentine and fresh paint; also smoke. Ammonia and ammonium sulphide fumes sometimes make their way into the darkroom and cause fog; likewise fumes from burning gas or electric arcs. Keep the darkroom well ventilated and, if possible, prevent any fumes from entering.

5. Dirty glass, new glass that has not been sufficiently soaked in acid, imperfectly cleaned dry plate glass, a dirty scrubbing brush or cloth. Clean these by soaking them in soda solution, and rinse them in clean, fresh water. Do not use soap. Fog arising from this cause can generally be detected by examining the back of the plate, because a peculiar iridescence will be visible.

6. Decomposed albumen solution used for substratum. Some albumen substitutes or dried albumens are not suitable for substratum and always cause spots and fog.

7. Impure solvents employed in making rubber edging solution will show a fog which begins at the edge of the plate and gradually spreads toward the center.

8. Aprons upon which sodium sulphide or ammonia has been spilled will, especially in hot weather, cause a fog to appear on that side of the negative that has been held near the body. Fingers contaminated with these chemicals will give fog in the portions of the plate near which they have been.

9. Some collodions when freshly iodized have a tendency to give fog. Should this be the case, add a few drops of a 10 per cent solution of iodine in alcohol until the collodion assumes an orange color, or set the collodion aside to ripen. An iodized collodion of a light

yellow color is more liable to fog than one of a darker orange or red color. The former is, however, the faster working of the two.

10. A developer lacking in acetic acid, or one that is too warm or too strong.

11. A new bath will sometimes give a surface fog, which can be removed with a pad of absorbent cotton applied with care to the surface of the film while water is running upon it. If fog from this source continues, add a little more acid to the developer.

12. The use of impure chemicals will cause fog.

13. Bath container unsuitable. Use glass, porcelain or hard rubber.

14. Condensation of moisture on screen, due to lower temperature in the gallery than in the darkroom.

15. One of the worst causes of fog, because usually unsuspected, is insufficient ventilation in the darkroom.

Uneven Negatives may be caused:

1. By uneven coating.

2. By placing the plates in the silver bath after the collodion has become too dry. This produces uneven sensitivity and may also cause the film to contract and split on drying. The moving of the plate or rocking of the bath should not be in one direction only, especially if it contains much alcohol, as this may produce streaks.

3. By not allowing the developer to flow evenly over the plate, or pouring it on one spot only, or allowing it to remain on one part longer than another. Keep the plate moving gently, and see that distribution of the developer remains even all over the plate.

4. By leaving plates in too strong a fixing solution.

5. By constantly pouring the reducer on one side of the film.

Grain on the Negative. This is caused by too prolonged development. It is noticeable in cases of under-exposure, where the temptation is to force development. A developer with too high an iron sulphate content will also cause grain.

Scum on Negative. This is usually due to impurities. The glass may not be properly clean; the silver bath or the dark slide may be dirty. If scum is evenly distributed all over the surface it may be due to high temperature in the darkroom, to insufficient ventilation or to an insufficiently acid developer.

Collodion Found Slow. If the collodion has been iodized for a very long time it will be slower than when freshly iodized. Therefore do not iodize too much at a time. Before making up your

mind that it is too slow, be sure that it has had a fair trial. Be certain that your lights are right, your lens is clean, your developer correct.

Collodion Too Soft. This is sometimes found under conditions of high temperature and high humidity. Some operators find that collodion resumes its quality if it is cooled by being placed on ice.

Black Spots and Comets. These may be caused by: (1) dust falling on the plate before or after albumenizing, or while being coated; (2) dust already in the collodion; (3) decomposed or unsuitable albumen; (4) unfiltered developer; (5) dried particles of collodion, especially from mouths of dirty collodion bottles; (6) dry plate developer dust floating about the room; (7) mixing of chemicals in the sensitizing room; (8) fumes from coal gas being burned nearby; (9) poor ventilation in the darkroom; (10) sulphur from a hard rubber dipper (coat the dipper with shellac); (11) wood or composition baths used as containers for the silver bath; (12) building operations in the neighborhood (use muslin over the open windows).

See that the shutter of the plate holder does not grind; if it does, it may cause dust to settle on the plate and give rise to spots.

Fine Black Lines are caused by scratches in glass which has not been sufficiently cleaned. They can usually be removed by turning over the negative film and wiping with a piece of wet cotton. But to prevent this defect see that the glass is thoroughly cleaned.

Small Transparent Spots (Pinholes). These may be caused by an over-iodized bath. A somewhat less definite spot may result from an under-iodized bath. Hypo dust will cause a transparent spot. Keep solid hypo out of the room, and do not let hypo solution drip on the bench or floor, where it may crystallize and then be reduced to dust.

A Brown Stain When Blackening is Done With Sulphide. This is generally due to insufficient washing between developing and fixing, but it may also be caused by using a stale and exhausted fixing bath.

A Blue Stain after bleaching with lead is due to insufficient washing after development.

An Opalescent Stain is produced by insufficient fixing.

Film Cracking. This is most commonly due to stale, thick collodion. It may also be caused by using rubber solution which is so thin as not to be impervious to the stripping collodion applied afterward. In this case, and especially if the coating of rubber is at all streaky, the upper collodion will penetrate to the lower and dissolve it, thereby causing cracks.

Splitting of Negatives is due to excessive intensification. Fine cracks after intensification with lead may be caused by collodion that is too thick. Thin out the collodion, give full exposure and short development. See that the substratum is not too weak. If the collodion is allowed to become too dry before sensitizing it may crack.

Splitting may occur if the solvents in the collodion are too dry.

Films Refusing to Strip. This results from using dirty glass or impure rubber solution. More acetic acid than usual must be used to soak old negatives, and sometimes a very stale negative can not be stripped at all.

Films Not Sticking After Stripping. This is due to grease on the glass, or to impurities in the stripping collodion.

"Oyster-Shell" Markings. Generally these are due to one of the following: A dirty plate holder (the plate holder should be wiped out every day with a wet cloth, and once every month it should be coated inside with shellac). Dirty or soaked blotting paper. Insufficient draining when removing the plate from the silver bath. Oyster-shell markings from this cause are very liable to appear when a new bath is being used. If the old bath is available, add a little of it to the new one. By draining the plate longer and rocking it in the meantime, the markings can usually be avoided.

Blisters or Bubbles when stripping are due to putting the rubber solution or stripping collodion on the plate when too warm.

Fuzzy Halftone Dots. These are due either to incorrect relation between the screen separation and the stop size, or insufficient exposure.

Plate Too Blue on Taking out of Bath. Be sure that the proportion of iodizer is correct. When thinning iodized collodion with solvent, be sure to add more iodizer also. If, with this amount of iodizer, you think the plate is too blue, try an exposure. You will find that you can get no more density even if you put in more iodizer.

Silver Solution Fails to Blacken Negative after bleaching in copper bromide. Possible reasons are a silver nitrate solution that is too weak or insufficiently acid, or an image that has been washed too long after bleaching.

"Dry Effect." This term is used to designate a relief appearance of the image on the negative after it is dry, making it difficult to strip and difficult to print. It is due entirely to excessive intensification. If sufficient exposure is given so that only a normal intensification is needed, this trouble will not occur.

Ground Glass Effect or graininess is caused by under-exposing and forcing development. The remedy is to give more exposure and to shorten development.

Irregular Clear Marks. These occur when the plate is immersed in the silver bath before the collodion has set, the water of the bath precipitating the cotton.

Sharp Line Across Plate. This is a "bath mark," which is caused when the first flow of the sensitizing solution does not completely flood the plate, or by a stoppage of the developer when it is flowing over the plate.

Conclusion

In making wet plate negatives it is well to remember that prevention is better than cure. The foregoing list of defects may seem formidable, but there will seldom, if ever, be any defects if the methods described earlier in this section are followed. Too much emphasis cannot be placed on the necessity of practicing cleanliness. Let the darkroom have frequent doses of fresh air, and daylight if possible.

Wet Plate Iodine Solution

Potassium Iodide	340 g.
Iodine	170 g.
Water	1900 cc.

Wet Plate Albumen Solution

Egg Albumen	14.2 g.
Water	2660 cc.
Potassium Chrome Alum (Saturated Solution)	10 cc.
Formaldehyde	15 cc.

Wet Plate Cyanide Solution

Sodium cyanide in water to saturated solution.

Wet Plate Sulphide Solution

Sodium sulphide in water to saturated solution.

Wet Plate Developer

Ferrous Sulphate	2041 g.
Sulphuric Acid	15 cc.
Water	4 gal.

Photostat Developer

Metol	13.9 g.
Sodium Sulphite	204 g.
Hydroquinone	55.3 g.
Sodium Carbonate	306 g.
Potassium Bromide	9.3 g.

Photostat Fixer

Hypo	900 g.
Sodium Sulphite	115 g.
Ammonia Alum	85 g.
Citric Acid	60 g.
Water to make 1½ gallons.	

Fountain Solution

Magnesium Nitrate	1½ oz.
Water	1 qt.
Baumé	25°

Development of Images in Bluish Tones on Silver Chloride Papers

By the use of a developer saturated with nitrobenzimidazole (a sparingly soluble substance) or by addition of benzotriazole to the developer (20 cubic centimeters of 1% aqueous solution) to 1 liter of developer, chloride papers may be made to yield bluish black images of slightly increased contrast, and free from fog. The benzotriazole may alternatively be added to the emulsion or to the protective gelatin supercoating.

Quick Developer

Appreciable decrease in the size of the grain can be obtained with the following formula, though it does not give as pronounced a decrease as the formula without hydroquinone; water 1,000, génol 4, hydroquinone 2, anhydrous sodium sulphate 65, borax crystals 7.5; the time of developing is 3 minutes at 18°. In order to obtain maximum decrease in grain size, developing should not be prolonged beyond the indicated time, otherwise the size of the grain gradually increases, and becomes about the same as with ordinary developers when the time is doubled. With fine-grained emulsions the grain size obtained is about the same as with coarse-grained emulsions.

"High-Temperature" Photographic Developer

U. S. Patent 1,933,789

Formula No. 1

Para-Aminophenol	7 g.
Sodium Carbonate	50 g.

Sodium Sulphite	50 g.
Trioxymethylene	2 to 10 g.
Water	to 1 l.

Formula No. 2

Monomethyl Paraminophenol Sulphate	5 g.
Sodium Sulphite	75 g.
Sodium Carbonate	25 g.
Potassium Bromide	1.5 g.
Formalin (40%)	5 to 25 cc.
Water	to 1 l.

Formula No. 3

Pyrocatechol	7 g.
Sodium Carbonate	50 g.
Sodium Sulphite	50 g.
Formalin (40%)	5 to 25 cc.
Water	to 1 l.

Universal Developer

The following three-solution developer is suggested:

Solution A. Water, 1 l.; sodium sulphate heptahydrate, 140 g.; metol, 14 g.

Solution B. Water, 1 l.; sodium sulphate heptahydrate, 100 g.; hydroquinone, 17 g.

Solution C. Water, 1 l.; sodium carbonate decahydrate, 150 g.

Hydros Developer

Especially satisfactory for contrast negatives and will desensitize all dry plates, orthochromatic or panchromatic.

Stock Mixture

Sodium Hydrosulphite	100 g.
Sodium Bisulphite	165 g.
Potassium Bromide	35 g.

These dry powders should be thoroughly mixed and kept in a sealed container to be used as needed. The hydrosulphite should be 80% and the bisulphite 60% sulphurous content.

To Develop

Dissolve 12 grams of the hydros mixture in 88 cubic centimeters of water. Development starts slowly but after the image appears it builds up rapidly.

Formulas for Developing Dry Plates

Avoirdupois Weights (437½ Grains to the Ounce) are Used in All Formulas

Pyro-Soda Developer

A

Pure Water	16 oz.
Bisulphite of Soda	75 gr.
Pyrogallic Acid	1 oz.

B

Pure Water	16 oz.
Dry Sulphite of Soda	2 oz.

(Which will test 60° by hydrometer.)

If negatives are too yellow use more sulphite; if too grey use less.

C

Pure Water	16 oz.
Dry Carbonate of Soda	1 oz.

(Which will test 30° by hydrometer.)

Mix for immediate use in the following proportions:

For Tray

A	1 oz.
B	1 oz.
C	1 oz.
Water	8 oz.

For Tank

A	2½ oz.
B	3½ oz.
C	2½ oz.
Water	58 oz.
Saturated Solution of Bromide of Potassium	5 drops

Temperature 65° F.   Time 20 minutes.

In summer the developer should be used cooler (about 60° F.), or with more water.

In winter it should be used warmer (about 70° F.), or with less water.

Less water hastens development, whereas more water slows development.

Pyro-Acetone Developer
Stock Pyro Solution

Pure Water	16 oz.
Bisulphite of Soda	25 gr.
Pyrogallic Acid	1 oz.
Dry Sulphite of Soda	2¾ oz.

For Tray Take

Stock Pyro Solution	1 oz.
Water	14 oz.
Acetone	3 drams

For Tank Take

Stock Pyro Solution	1¼ oz.
Water	58 oz.
Acetone	5 drams

(Temperature 70°.   Time 30 minutes.)

Developer for Line Work
(Black and White)

Also for lantern slides, or wherever increased *contrast* is desired.

A

Water	32 oz.
Hydrochinon	1½ oz.
Sodium Sulphite (Dry)	1 oz.
Sulphuric Acid	60 minims

B

Water	32 oz.
Sodium Carbonate (Dry)	1 oz.
Potassium Carbonate	3 oz.
Potassium Bromide	120 grains
Sodium Sulphite (Dry)	3 oz.

To develop in either tray or tank, take equal parts A and B.

Develop for from 6 to 10 minutes; temperature 65° F., according to exposure and density desired.

Hydrochinon-Metol Developer
A

Pure Water	25 oz.
Metol (or Substitutes)	30 grains
Hydrochinon	90 grains
Dry Sulphite of Soda	1 oz.

B

Pure Water	25 oz.
Dry Carbonate of Soda	½ oz.

(Which will test 10° by hydrometer.)

For use, mix A and B equal parts. Can be used repeatedly.

Above Developer in Single Solution

Equal parts A and B keep well when mixed. The above ingredients may be put together in one solution. With fresh developer it may be necessary to add to each ounce 1 drop of bromide of potassium solution (containing 1 part of bromide potassium to 10 parts of water).

Note.—This is a very fine and desirable developer. It should not be used too old or too much diluted, as it is then liable to produce peculiar streaks and blotches.

Pyro-Metol Developer
A

Water	30 oz.
Metol (or Substitutes)	1 oz.
Bisulphite Soda	75 grains
Pyrogallic Acid	½ oz.

B

Pure Water	30 oz.
Dry Sulphite of Soda	4 oz.

(Which will test 64° by hydrometer.)

C

Pure Water	30 oz.
Dry Carbonate of Soda	4 oz.

(Which will test 64° by hydrometer.)

For use, take:

A	½ oz.
B	½ oz.
C	½ oz.
Water	10 to 15 oz.

(According to density desired.)

B and C may be mixed together and keep well in one solution, which should be diluted for use with from 6 to 10 parts water.

Developer for Transparencies
(Lantern Slides)
Hydrochinon and Pyrocatechin
A

Pure Water	32 oz.
Dry Sulphite of Soda	6 oz.
Pyrocatechin	240 grains
Hydrochinon	240 grains
Bromide of Potassium	120 grains

B

Water	32 oz.
Caustic Potash	240 grains

For use: Mix equal parts A and B.

This formula is excellent for producing clear, brilliant transparencies and slides.

Film Hardening

The following method is a certain preventive of frilling or blistering on any make or brand of plates.

Water	10 oz.
Chrome Alum (Granular)	1 oz.

This solution may be kept in either tank or tray.

Immediately after development and without previous washing or rinsing, immerse the plate in the chrome alum solution (as above) and allow it to remain therein for from 15 to 20 seconds. Then again without washing or rinsing, pass it directly into the regular fixing bath. After fixation is completed the film will be found to be extremely tough and practically insoluble; so much so, that it may be washed in water of high temperature without injury. The chrome alum solution may be used over indefinitely.

Acid Fixing and Hardening Bath
A

*Water (1 gal.)	128 oz.
Hyposulphite of Soda	32 oz.

(Which will test about 80° by hydrometer.)

B (See Note Below)

Water	32 oz.
Dry Sulphite of Soda	3 oz.

(Which will test 45° by hydrometer.)

Sulphuric Acid C. P.	½ oz.
Powdered Chrome Alum	2 oz.

Note.—Be sure to mix solution B exactly in given proportions and order.

Always pour B into A while stirring well. If this is not done precipitation will take place.

During the cold season one-half the quantity of Solution B is sufficient for full quantity of Solution A.

B can also be prepared as follows:

Water	32 oz.
Potassium Metabisulphite	3 oz.
Powdered Chrome Alum	2 oz.

This bath remains clear after frequent use, does not discolor the negatives and hardens the film to such a degree that the negatives can be washed in tepid water and dried by gentle artificial heat if necessary. To insure permanency, freedom from stain and perfect hardening, plates should be left in the bath ten to twenty minutes after the bromide of silver appears to have been dissolved.

If the bath becomes exhausted by continued use, replace it by a new one.

* For quick fixing of postal plates add to A solution

Ammonium Chloride	4 oz.

Plain Fixing Bath

Water	32 oz.
Hyposulphite of Soda	8 oz.

(Which will test about 80° by hydrometer.)

Do not use the bath when it becomes discolored, it must be made fresh each day.

Formula for Continuous Pyro Tank Developer
For 3½ Gallon Tank

Water (Hot)	1 gal.
Sodium Bisulphite	7½ oz.

For 1 Gallon Tank

Water (Hot)	½ gal.
Sodium Bisulphite	2½ oz.

(Meta-bisulphite must not be substituted.)

For 3½ Gallon Tank

Sodium Sulphite	7½ oz.

For 1 Gollon Tank

Sodium Sulphite	2½ oz.

After the above is dissolved boil five minutes then add:

For 3½ Gallon Tank

Water (Cold)	2½ gal.
Pyro	6 oz.
Sodium Carbonate	10 oz.

For 1 Gallon Tank

Water (Cold)	½ gal.
Pyro	2 oz.
Sodium Carbonate	3½ oz.

The developer is then ready for use, and can be used continuously until exhausted. To keep it in proper working condition and up to its original strength and bulk, add equal parts of the following ("A" and "B") solutions after each day's development.

Solution "A"

Water (Hot)	64 oz.
Sodium Bisulphite	2½ oz.

(Meta-bisulphite must not be substituted.)

Sodium Sulphite	2½ oz.

Boil five minutes, and when cool add:

Pyro	2 oz.

Solution "B"

Water (Cold)	64 oz.
Sodium Carbonate	4 oz.

The most frequent trouble in tank development is too much dilution. A definite amount of silver must be converted and this cannot be accomplished by water.

The color of the negative, when the developer is new, is a little more neutral than negatives developed in tray formula, but after two or three days' use, it gives a beautiful tone, and remains constant if proper attention is given to the bath. Toward the fifth or sixth week, if the bath has been overworked, it may impart a little reddish tone to the negative, but if the bath has been properly taken care of, this will not occur.

The best way is to calculate how much tray solution should be used for the quantity of negatives to be developed, and add an equivalent amount of stock solution to the bath, every day.

The factor is about the same as the tray formula.

The average time of development for normal exposures is six (6) minutes at 65 degrees F., although a minute more or less makes little or no difference in the printing quality. If a shorter time of development is desired, increase the temperature of the developer, bearing in mind, however, that too high a temperature will produce fog. Our tests show that to develop normally exposed plates from 5 to 6 minutes, a temperature of 65 to 70 degrees is required. Prolonged development increases density without increasing contrast.

To obtain the best results and a uniform quality it is highly important that the bath be kept at a uniform temperature.

To regulate the temperature of the developer we find it most convenient to place a rubber tank containing developer in a wooden box, or "jacket," allowing about an inch and a half (1½) space around the tank, also the bottom; this space to be filled with water, either cold or warm, according to needs.

There is no necessity to boil the sulphite and the bisulphite in the whole quantity of water. It may be done as follows:

To make 3½ gallon bath: boil 7½ ounces each of sulphite and bisulphite in not less than 1 gallon of water, cool and pour in tank; add 2½ gallons of cold water, then dissolve the pyro first and then the carbonate and when the temperature is 65° the bath is ready for use. Each time before development add enough stock solution as above suggested.

If this bath is left at rest 3 or 4 days, it may refuse to work, and, if so, take out one-third of the solution and add one-third new. If, instead of leaving the bath in the tank it is put away in bottles, this fault will not occur.

The tank must be covered when not in use to prevent oxidation, and we find by placing a sheet of glass with a weight on top, that it answers the purpose very well.

This developer is not intended for use in the "closed" metal tanks. We recommend the "open" hard rubber tank, same as used in the "Core" system, and which we have used in our experiments. These can be obtained at any dealer in photo supplies.

In our experiments we find the 3½ gallon developer, on account of its greater bulk, to have a longer life than the 1 gallon developer.

For "closed metal tanks" this formula diluted with from ⅓ to equal quantities of water, depending upon the time of development required, can be successfully used.

Closed Metal Tanks
(Tank Development)

While we do not find that this method is economical of either time or developer, we are satisfied that good negatives can be made by it. However, we believe a few words about the care and use of closed metal tanks will be opportune.

First: never use the tank or cage to fix the plates in.

Second: thoroughly clean the tank and cage at least once a week with hydrochloric (muriatic) acid, 1 ounce and water 10 ounces, rinsing the tank out well and observing that the solution has access to all parts of the cage, special attention being paid to the underside of the corrugations holding the plates.

Third: use either the pyro-soda tank formula or the pyro-acetone—the last we consider the best closed metal tank formula ever published.

These formulas can be varied to suit local water conditions or requirements, by remembering that if in the pyro-soda formula the negatives are too yellow, more sulphite solution should be used, and if they are not yellow enough, the quantity of sulphite solution should be reduced.

If the negatives are too dense and the time and temperature are correct, use less of the carbonate solution. If the negatives are too thin use more of the carbonate solution. Leave the quantity of pyro solution always as indicated in the formula.

If with the pyro-acetone formula the negatives are too thin and the time and temperature correct, increase both the A and B solution, or increase the time.

If the negatives are too dense decrease both the A and B solutions, or decrease the time.

Do not vary the proportions of the A and B solutions if the negatives are too yellow, but make a fresh acid fixing and hardening bath.

While it is best to maintain a constant temperature, yet in summer it is safe to start at a temperature of 65° F., for a half-hour development, provided the temperature does not rise higher than 75° F. This gives an average of 70° F. Reverse the tank at least every 5 minutes (and more often if convenient), and be methodical regarding temperature and time.

There is a tendency to flatness in most tank developed negatives (due to the diluted form in which the developer is used) which is not found in similarly lighted and timed negatives developed in the more energetic tray developer.

Do not develop black and white ground negatives together, as the black grounds require less development, and the white grounds more, and do not expect the same quality of negative from a normally timed and an undertimed plate.

Should you be obliged to use a higher or lower temperature than given in the formula, remember that for each degree of higher temperature the time should be shortened one minute, and for each degree lower temperature the time should be increased one minute.

By attention to this the tank formulas can be used anywhere between 55° F. and 75° F.

Finally: take care of your tank. Keep it clean. Use a correct thermometer, and when you have once settled on satisfactory proportions for your developer, temperature and time, use it the same every day. Reverse the tank as often as directed, and fix the negatives thoroughly.

Note.—There is a point beyond which the developer cannot be safely diluted without the production of peculiar streaks and blotches on the negative.

Reducing Solutions

The printing qualities of negatives which are too dense can be improved by submitting them wholly or in part to the action of the following solutions:

Make a saturated solution of permanganate potassium (as stock solution) and use as follows:

For general reduction—

Water	10 oz.
Stock Solution	1 dram
Sulphuric Acid	1 dram

For local reduction use only—

Water	2 to 4 oz.

Before applying this solution the negative must be washed well to free it from hypo, thus avoiding stains and streaks.

If the negative should show any stain, immerse in fresh 1 to 4 hypo solution.

For negatives which are too dense all over, and lack contrast, due to over-exposure and over-development, use the following:

A

Water	16 oz.
Hyposulphite of Soda	1 oz.

B

Water	16 oz.
Red Prussiate of Potassium	1 oz.

As this B solution is affected by light, the bottle containing it should be of amber color or wrapped in opaque paper and kept in the dark when not in use.

Mix for immediate use:

A	8 oz.
B	1 oz.

The negative can be placed in the solution directly after fixing. If a dry negative is to be reduced, it must be soaked in water for a few minutes before applying the reducer. To avoid streaks, always rinse the negative before holding it up for examination. As soon as sufficiently reduced wash thoroughly. This reducer increases contrast in the negative.

For negatives in which the shadow detail is correct, but whose lights are too dense, due to under-exposure combined with prolonged development, make the following solution:

Water	10 oz.
Ammonium Persulphate	1 oz.
Dry Sulphite of Soda	90 grains
Sulphuric Acid, C. P.	90 minims

For use, take:

Stock Solution	1 oz.
Water	4 to 9 oz.

The negative must be washed well to remove all traces of hypo.

Dry negatives should be soaked for about 10 minutes.

Keep the dish in motion during reduction.

After sufficiently reduced, place for 15 minutes in fresh 1 to 4 solution of hypo.

Intensifying Negatives

Prepare the following solution, which will keep indefinitely.

Solution No. 1

Water	16 oz.
Bichloride of Mercury	¼ oz.
Bromide of Potassium	¼ oz.

Solution No. 2

Number 2 should be freshly mixed:

Water	8 oz.
Dry Sulphite of Soda	1 oz.

(Which will test 60° by hydrometer.)

After the negative is well fixed and washed, immerse in No. 1 until it has become thoroughly whitened, and after rinsing carefully place it in No. 2, leaving it there until entirely cleared. If sufficient intensification has not been gained, wash for ten minutes, repeat the operation and finally wash well. If after intensification the negative is too dense it may be reduced by placing it for a few seconds in water 16 ounces, hypo 1 ounce.

If the negative has not been thoroughly fixed and washed before intensification, stains will ensue.

Failures and Their Remedies

Weak negatives with clear shadows: Under development, too cold or too weak developer.

Weak negatives with plenty of detail in the shadows. Over-exposure, lighted too flatly, or too weak developer (use less water).

Strong negatives, blocky lights and heavy shadows with too much contrast: Under-exposure or lighted too harshly or too strong developer containing too much alkali. Add more water and use less carbonate of soda (or potassium), as an excess of alkali blocks the lights.

Too much intensity: Developer too warm or too strong, or development carried too far. Negatives dried in warm, sultry air assume more intensity than when dried in a cool place with draft.

Light fog: Caused by over-exposure, white light entering dark room, unsafe light for developing, unclean lenses, reflections from shining edges of diaphragm, interior of tube and camera, extraneous light falling upon the lens from top or side when not properly shielded, light entering plate-holder when drawing or replacing the slide, or entering between the plate holder and the back of the camera.

Chemical fog: Caused by old plates, unclean trays, prolonged development, developer too warm or too strong, containing too much alkali without restrainer.

Some developers for example, hydrochinon, metol, etc., when quite fresh need the addition of some old developer (which has been used), or a few drops of bromide of potassium solution, to work perfectly clear.

If chemical fog occurs, use the developer cooler and add a few drops of bromide of potassium solution, containing 1 ounce bromide of potassium to 10 ounces of water.

Yellow and brown stains: Caused by decomposed pyro solution, insufficient or decomposed sodium sulphite, too small a quantity of developer so that the plate is not sufficiently covered, or taking the plate out of the developer too frequently for examination; prolonged development or using the developer too warm, or too strong, insufficient washing of the negative after developing and before placing in the fixing bath, also by plain hypo solution, which by continued use, has assumed a dark color. An addition of chrome alum acid solution to the fixing bath will prevent discoloration of the negative, provided same is allowed to remain in the bath a sufficient length of time to secure perfect fixing.

To remove yellow stain: Try gently rubbing the negative while still wet, with a tuft of cotton; if this does not remove the stain, wash well and place in iron clearing solution as follows:

Water	20 oz.
Clean Crystals of Sulphate of Iron	3 oz.
Sulphuric Acid C. P.	1 oz.
Powdered Alum	1 oz.

Let the negative remain in this solution until stain is removed, then wash well.

To remove chemical fog: After the negative is fixed, rinse and place in a weak solution of red prussiate reducer, using double the quantity of water. As soon as the fog is removed, wash well.

Iridescence (metallic appearance) of the surface can be removed by rubbing the negative with a tuft of cotton while the negative is wet, or by treating as for chemical fog. Caused by dirty trays, forced development and old fixing bath.

To remove silver stain from negative: Immerse in the red prussiate clearing solution as recommended for the removal of chemical fog.

Frilling, softening of the film, film leaving the glass—caused by too high temperature. When this occurs cool the developer and use the fixing and hardening bath.

For reticulation (puckering of the film) leave the negative in the acid fixing bath for half an hour to harden the film completely throughout.

In hot climates use the "acetone" developer with the acid fixing bath.

Cloudy, wavy appearance of negative, caused by not using a sufficient quantity of developer to cover the plate, or by not sufficiently rocking the developing tray.

Mottled appearance of negative is caused by precipitation from the fixing bath containing alum, or if the solution is old or turbid, or allowing plate to lie in the developer without rocking.

Crystallization of the negative and fading of image. Imperfect elimination of the hypo.

Sandy surface of negative, making it difficult to retouch, may be easily removed by using iron clearing solution.

Peculiar streaks and blotches closely resembling brush marks, finger marks and insensitive spots, and appearing as though the plate had been scrubbed with a dirty or greasy brush or improperly cleaned, are caused by the uneven action of the developer, if it is too old or too much diluted.

This trouble is more liable to occur if hydrochinon is used in combination with eikonogen or metol; and can be prevented by using the developer more concentrated, or by a radical change to a different developer.

Spots: A book might be written on "spots" and then not cover the entire subject, but by far the greater part of such a volume would be rendered useless if cleanliness were observed in the dark room, in the camera and its fittings.

Round transparent spots: Air bubbles in the developer. Can be avoided by going over the plate with a tuft of cotton while developing.

Transparent spots of irregular shape. Caused by dust. Keep the camera and plate holder free from dust. Grasp the plate by two edges and tap edge gently on table before placing in the plate holder. Floor and shelves of the dark room should be wiped frequently with a wet cloth.

Transparent spots are also caused by splinters of glass, jarred off the edges of plates when several are developed together in one tray.

Transparent angular spots of irregular size. Caused by scum forming on the surface of the developer, or by scale or sediment from dirty bottles and trays.

Fine transparent lines: Using too stiff a brush in dusting off plates.

Brown or purple spots are caused by dry pyro when undissolved, or floating in the air and settling on the plate. Can be removed by using iron clearing solution or by placing the plate in water to which a few drops of muriatic acid have been added.

Small opaque spots caused by iron scale in the water. Tie several thicknesses of cotton cloth over the faucet. Also caused by sediment in the fixing bath.

Blisters or frilling may occur if the fixing bath is too strong. Remedy, add more water; also caused by sudden changes of temperature between the different solutions and the wash water.

Negatives washed under a stream of water pumped at a high pressure may show numerous small blisters. This can only be avoided by using the water for washing negatives after it has stood long enough for the air to escape, or these same small blisters can be caused by insufficient washing between the developer and the acid fixing bath, the combination of the alkaline developer and the acid in the fixing bath forms a gas in the film which raises minute blisters.

Small, sharply defined opaque spots have been caused by changing plates in the dark room when allowing water to run from the faucet. The surface of the plate became spattered either with clear water or by impurities from the bottom of the sink, which afterwards dried while awaiting development. This causes spots varying in size, character and intensity.

Numerous parallel vertical lines are produced by acid in an old fixing bath cutting the pyro stain out in streaks when precipitating. The remedy would be to use a new acid hypo bath.

Small semi-transparent spots occurring in tank development are due to air in the water on account of high pumping pressure. This causes minute air-bells to adhere to the surface of the plate during the first few minutes of development. The remedy is to allow the water to stand long enough for the air to escape. Also the tank should be shaken to break adhering air-bells.

Double spots: Semi-developed. Caused by placing plates face to back after exposure, and letting them remain so until time to develop; particularly liable to happen during warm humid weather. In some instances quite large patches of the plate will refuse to develop entirely, owing to the almost optical contact of the film of one plate with the slightly moist back of another.

Remedy—Always repack plates after exposure, face to face, in original boxes.

Spots of almost any shape or size can be caused by brushing the plates with a dirty dusting brush.

Removing Stain Caused by Proof Paper Sticking to Wet Negative

Mix 6 ounces alcohol with 6 ounces ether. In this solution soak the plates about one minute and then in the reducer (5% solution of cyanide of potassium). Do not keep negatives over one minute in the reducer.

Antidote for Metol Poisoning

With certain individuals there is a serious drawback to the use of metol, as it produces an irritating skin trouble, leaving the fingers very sensitive and tender. Therefore it is advisable to wear rubber gloves when using any developer containing metol.

As a cure for metol poisoning, the following is recommended:

Take first Rochelle salts to open the bowels, and next day use Swift's Special Specific (S.S.S.) according to directions for Scrofula. Attention should be paid to the general health, any tendency toward dyspepsia and constipation should be at once checked.

An Italian authority gives the following prescription for an ointment:

Ichthyol	1 part
Lanoline	2 parts
Pure Vaseline	3 parts
Boracic Acid	4 parts

The use of the irritating cause should be at once suspended.

Pyro stains on fingers may be removed by rubbing with a large crystal of citric acid directly after development, before the fingers have been dried.

Plate lifters (thimbles with hook), also India rubber finger tips are useful for the protection of fingers from stain and metol injury.

Negative Varnish

Best Grain Alcohol	20 oz.
Crushed Dark Shellac	1 oz.

Shake occasionally for several days until dissolved (without heat). Allow to settle then decant carefully from the settlings, and add 2 drams of oil of lavender. The negatives must be warmed slightly before varnishing.

Retouching Dope

Powdered Rosin	120 grains
Oil of Turpentine	2 oz.

Moisten a clean cotton rag, and rub over the parts of the negative to be retouched.

Ground Glass Varnish

Gum Sandarac	2¼ oz.
Gum Mastic	½ oz.
Ether	24 oz.
Benzol	6 to 18 oz.

The quantity of benzol determines the nature of the matt obtained.

Dead Black Varnish

Alcohol	8 oz.
Lamp Black	½ oz.
Liquid Shellac	1 oz.

Blackening Brass Diaphragms
A.

Nitrate of Copper	¼ oz.
Water	1 oz.

B.

Nitrate of Silver	¼ oz.
Water	1 oz.

Mix A and B, clean the brass thoroughly and dip in the solution for a moment only. Allow to dry and then heat almost red hot.

To Photograph Polished Surfaces

Dab the surface of the article lightly with a ball of soft putty. All such articles should be photographed against a black ground, the light falling only from one direction, and all reflections stopped with black screens. Hollow cut glass may be filled with ink or a black aniline water soluble dye. Bright parts of machinery may be painted as follows: Make a thin cream of white lead and turpentine, gray with lampblack and add one-sixth part of gold size (which can be bought in any paint store). This paint can be removed with cotton waste wet with turpentine or benzine.

How to Remove the Film Quickly from Old Negatives

Water	20 oz.
Fluoride of Soda	60 grains
Sulphuric Acid	60 minims

Use a deep rubber tray, immerse the old dry negative. In about one minute you can begin to roll up the film at one corner, and may then quickly strip the film from the glass.

Stripping Film from Cracked Negatives

Place the cracked or broken negative face up, in water 8 ounces; formalin ½ ounce, for from 5 to 10 minutes, then remove, rinse briefly in water and set up to dry. This hardens the film and prevents distortion by uneven swelling during the stripping and mounting.

In a separate tray dissolve 40 grains of sodium fluoride in 4 ounces of water and have ready in a measure, 1 ounce of hydrochloric acid 10 per cent solution. Cut through the dry film on the negative with a sharp knife about ⅛ inch from the edge around the plate and immerse it in the fluoride solution after adding the acid.

Rock the tray and in a very few minutes the film will begin to loosen. Strip off the edges and roll the film away from the glass with the finger until it floats free. Remove the glass, pour off the solution and wash by allowing one or more changes of clear water, then slide into the tray a fixed out and washed undeveloped plate, face up, and under the film. Remove from the tray and with the fingers fit the film in position, touch lightly with blotting paper and rear up to dry.

When completely dry, the break will be invisible, or nearly so, depending upon how nicely the edges were matched. Take plenty of time in mounting, the film is tough and perfectly insoluble.

Both the sodium fluoride and the hydrochloric acid may be kept in separate stock solutions and used as required, and a single trial will assuredly prove the value of the method.

Developers Recommended for Process Film

A.

Solution No. 1

	Metric	Avoirdupois
Hot Water	500 cc.	16 oz.
Hydroquinone	25 g.	350 grains
Potassium Metabisulphite	25 g.	350 grains
Potassium Bromide	25 g.	350 grains
Cold Water to make	1 l.	32 oz.

Solution No. 2

	Metric	Avoirdupois
Cold Water	500 cc.	16 oz.
Potassium Hydroxide (Caustic Potash-stick)	50 g.	1 oz. 260 grains
Cold Water to make	1 l.	32 oz.

Mix equal parts of solutions 1 and 2 immediately before use. Shake bottles well before using. Develop within 3 minutes at 65° F.

B.

	Metric	Avoirdupois
Hot Water	500 cc.	16 oz.
Metol	5 g.	75 grains
Sodium Sulphite (Anhydrous)	38 g.	1 oz. 120 grains
Hydroquinone	6 g.	90 grains
Potassium Carbonate	38 g.	1 oz. 120 grains
Potassium Bromide	3 g.	45 grains
Water to make	1 l.	32 oz.

For good contrast this formula is used *without dilution*. Develop about 3 to 4 minutes at 65° F. If less contrast is desired, dilute with an equal volume of water.

Developers Recommended for Commercial Films

For Normal Negatives

	Metric	Avoirdupois
Water (Hot 125° F.)	500 cc.	16 oz.
Metol	1 g.	15 grains
Sodium Sulphite (Anhydrous)	15 g.	½ oz.
Hydroquinone	2 g.	30 grains

Potassium Carbonate	15 g.	½ oz.
Potassium Bromide	1 g.	15 grains
Cold water to make	1 l.	32 oz.

Time of development 5 minutes at 650° F.

For Contrasty Negatives

	Metric	Avoirdupois
Water (Hot 125° F.)	500 cc.	16 oz.
Metol	5 g.	75 grains
Sodium Sulphite (Anhydrous)	40 g.	1 oz. 145 grains
Hydroquinone	6 g.	90 grains
Potassium Carbonate	40 g.	1 oz. 145 grains
Potassium Bromide	3 g.	45 grains
Cold water to make	1 l.	32 oz.

Time of development 5 minutes at 65° F.

Rodinal Developer

Rodinal diluted 1 to 15 up to 1 to 40.
Dilution 1 to 15 yields contrasty negatives.

Dilution 1 to 40 gives negatives of most delicate gradation.

Rodinal has the advantage of great modulation by the simple means of dilution with water. In order to control development with Rodinal in high concentration and to attain more brilliant results, we recommend the addition of 1¼ ounces of a 10% solution of potassium bromide to each 32 ounces of diluted developer.

Development should be at 65° F.

Glycin Developer

	Metric	Avoirdupois
Water (Hot 125° F.)	500 cc.	16 oz.
Glycin	15 g.	½ oz.
Potassium Carbonate	15 g.	½ oz.
Sodium Sulphite (Anhydrous)	20 g.	290 grains
Potassium Bromide	½ g.	8 grains
Cold water to make	1 l.	32 oz.

Use full strength and develop 5 to 10 minutes at 65° F.

Developer for Printon Film

The following special formula is recommended for best results:

	Metric	Avoirdupois
Water (Hot)	500 cc.	16 oz.
Metol	5 g.	15 grains
Sodium Sulphite (Anhydrous)	40 g.	1 oz. 145 grains
Hydroquinone	6 g.	90 grains
Potassium Carbonate	40 g.	1 oz. 145 grains
Potassium Bromide	6 g.	90 grains
Water to make	1 l.	32 oz.

Dissolve chemicals in the order given and use solution at full strength at a temperature of 65° F. Time of development is 2 to 3 minutes, but it can be extended up to 5 minutes without risk of fog.

Pyro is unsuitable for use in developing Printon Film because of its effect

on the anti-halation coating. Uneven brown stains may result.

Developers for Reprolith Film
Paraformaldehyde Developer

This is the standard developer in quite general use for material of this kind. Reprolith works very satisfactorily with this formula, which permits development from 2 to 3 minutes at 65–70° Fahrenheit. In the case of Reprolith Orthochromatic film, development should be fully 3 minutes to bring out the superior orthochromatic quality of this product.

	Metric	Avoirdupois
Water (about 120° F.)	2000 cc.	64 oz.
Sodium Sulphite (Anhydrous)	120 g.	4 oz.
Trioxymethylene (Para-Formaldehyde)	30 g.	1 oz.
Potassium Metabisulphite	10 g.	150 grains
Boric Acid	30 g.	1 oz.
Hydroquinone	90 g.	3 oz.
Potassium Bromide	6 g.	90 grains
Water to make	4000 cc.	1 gal.

Reprolith Developer

This formula provides a developer which can be compounded in two solutions which have excellent keeping quality when stocked in separate well-stoppered bottles, a convenience appreciated by many workers. The development time with this formula should not exceed 2 minutes, in which time the standard quality of both Reprolith regular and Reprolith Orthochromatic will be accurately obtained.

Solution No. 1

	Metric	Avoirdupois
Water	3000 cc.	96 oz.
Sodium Sulphite (Anhydrous)	120 g.	4 oz.
Hydroquinone	90 g.	3 oz.
Boric Acid	30 g.	1 oz.

Solution No. 2

	Metric	Avoirdupois
Water	1000 cc.	32 oz.
Potassium Metabisulphite	10 g.	140 grains
Formaldehyde (40%)	100 cc.	3 oz.

For use add one part Solution No. 2 to three parts Solution No. 1 (or all of Solution No. 2 to all of Solution No. 1) and develop 1½ to 2 minutes at 65° Fahrenheit.

Fixing Baths
Acid Fixing Bath

	Metric	Avoirdupois
Hypo	290 g.	9½ oz.
Potassium Metabisulphite	25 g.	¾ oz.
Water	1 l.	32 oz.

The Metabisulphite should be added only when the hypo solution is cool, not when it is hot.

Hardening Fixing Bath
Solution No. 1

	Metric	Avoirdupois
Water	1 l.	32 oz.
Hypo	240 g.	8 oz.

Solution No. 2

	Metric	Avoirdupois
Water (about 125° F.)	75 cc.	2½ oz.
Sodium Sulphite (Anhydrous)	15 g.	½ oz.
Acetic Acid 28%	45 cc.	1½ oz.
Potassium Alum	15 g.	½ oz.

To make 28% Acetic Acid from Glacial Acid, dilute 3 parts Glacial with eight parts of water.

Dissolve chemicals thoroughly in order given. Cool Solution No. 2 after mixing and add it slowly with constant stirring to Solution No. 1.

A fresh bath should be prepared frequently, as the gelatin-coated backs of the films are likely to become stained in an old or discolored fixing solution.

Special Fixing Bath for Printon and Reprolith Films

Accuracy in registration for multi-color work being of prime importance, we recommend for use in such cases a fixing bath without hardener, as follows:

	Metric	Avoirdupois
Water	1 l.	32 oz.
Hypo	485 g.	16 oz.
Potassium Metabisulphite	75 g.	2½ oz.

In case this bath should lose its acidity by frequent use, giving the film a yellowish stain, add more potassium metabisulphite to restore the acidity of the solution.

Monckhoven's Intensifier for Reproduction Films

Gives very great intensification and contrast.

Solution A

	Metric	Avoirdupois
Potassium Bromide	23 g.	¾ oz.
Mercuric Chloride	23 g.	¾ oz.
Water	1 l.	32 oz.

Solution B

	Metric	Avoirdupois
Potassium Cyanide	23 g.	¾ oz.
Silver Nitrate	23 g.	¾ oz.
Water	1 l.	32 oz.

The silver and the cyanide are dissolved in separate lots of water, and the former added to the latter until a permanent precipitate is produced. The mixture is allowed to stand 15 minutes, and after filtering, forms Solution B.

Place the negative in A until bleached through; then rinse and place in Solution B. If intensification is carried too far, the negative may be reduced with a weak solution of hypo.

Because of the deadly poisonous character of this intensifier, it should be used with care and bottles containing it should be suitably marked.

Mercuric Iodide Intensifier

	Metric	Avoirdupois
2% Mercuric Chloride	100 cc.	10 oz.
10% Potassium Iodide	25 cc.	2.5 oz.
10% Hypo	40 cc.	4 oz.
Water	200 to 300 cc.	20 to 30 oz.

Part of the mercury solution is added to the water and then part of the Iodide solution, continuing until all the mercury and iodide is added to the water.

When solution is clear, add the hypo. Use full strength.

Farmer's Reducer for Reproduction Films

In case of overexposure or overdevelopment, this well-known reducer can be used effectively for clearing. It is easily compounded by making up first a 1:4 solution of plain hypo—for example, 8 ounces of hypo dissolved in 32 ounces of water—and adding to this just enough potassium ferricyanide to turn the solution to a lemon-yellow color. Most workers prepare the ferricyanide as a 10% solution in advance, for use as needed; others shake a little of the powder directly into the plain hypo solution. The lemon-yellow color disappears with use of the reducer, but may be restored by adding more ferricyanide. The stronger the color, the stronger the reducing action, and vice versa. If the reducer is used too strong its action is not so easy to control. The film may be immersed in the reducer solution, after being soaked in water to assure even action, or, in cases where only local reduction is desired, the reducer may be applied to the moist film with a tuft of cotton, with rinsing during inspection and afterwards.

Fine Grain Developers

Paraphenylenediamine	10 g.
Sodium Sulphite (Anhydrous)	60 g.
Water to	1000 cc.

The temperature should be about 65° F., and development is complete in about 30 minutes, the image being light-brown but sufficiently contrasty. Paraphenylenediamine, according to the temperature at which development is to take place, requires an exposure of 4x to 6x normal. Addition of alkali to the developer shortens the time; an addition of a small quantity of metol produces a comparatively large grain. Glycin when added produces a fine grain, and it requires a 2x to 3x normal exposure in order to achieve results comparable with metol-borax. The following developer gives good contrast, not obtainable with paraphenylenediamine alone, when used to develop ultra-rapid films:

Sodium Sulphite (anhydrous) 30 g.
Water   420 cc.
Paraphenylenediamine   4 g.
Glycin   4 g.

The above formula is specially useful for 35 millimeter panchromatic ciné films. For super pan films 1 gram more each of paraphenylenediamine and glycin are to be added. Development is complete in thirty minutes, but can be speeded up by keeping the film in motion. The image appears brownish by transmitted light.

Many developers for fine-grain work contain borax, and the following formulae have been published:

(1) is Gevaert fine-grain developer—ten to fifteen minutes at 65° F.; (2) is Kodak borax developer D76 for contrast, and (3) is for softer negatives. The development time is ten to twenty minutes at 65° F.

	(1)	(2)	(3)
Metol	61 grains	30 grains	40 grains
Sodium Sulphite (Anhydrous)	3 oz.	3½ oz.	3½ oz.
Hydroquinone	30 grains	75 grains	40 grains
Borax	30 grains	30 grains	30 grains
Water	32 oz.	32 oz.	32 oz.

Some fine-grain developers do not contain borax, and the physical factors must then receive careful attention. Agfa fine-grain developer 14 is as follows:

Metol   60 gr.
Sodium Sulphite, anhyd.   2¾ oz.
Sodium Carbonate, anhyd.   15 gr.
Water   32 oz.

Time for development is 10 to 20 minutes at 65° F.

Fine Grain Developer

The following formulae have been tested on miniature camera film of all makes.

Formula No. 1

Paraphenylenediamine   5.0 g.
Glycin   3.0 g.
Sodium Sulphite (Anhydrous) 37.5 g.
Water (Distilled)   500 cc.

Use the developing times given below for all films except Eastman Panatomic and Du Pont Quota Speed (Micropan). For these films use half the developing times indicated.

Formula No. 2

A simple formula which seems to be good on all films is:

Paraphenylenediamine   5.0 g.
Sodium Sulphite (Anhydrous) 37.5 g.
Water (Distilled)   500 cc.

When fresh, this developer gives flat, rather weak negatives. If it is made up and kept in a corked, brown bottle for ten or twelve weeks, its quality is greatly improved, and it will be found to give a vigorous black negative with a very fine grain and plenty of detail.

For the above two formulae the normal developing time is 30 minutes at 65° Fahrenheit. Films processed in these developers should be given twice the exposure given films processed in developers containing sodium carbonate or borax.

Formula No. 3

A very good, though rather troublesome, method for fine grain development follows:

Solution No. 1
Metol   0.5 g.
Sodium Sulphite (Anhydrous) 75.0 g.
Water (Distilled)   500 cc.

Solution No. 2
Paraphenylenediamine   5.0 g.
Sodium Sulphite (Anhydrous) 30.0 g.
Water (Distilled)   500 cc.

Develop in solution No. 1 *without much agitation* for 8 minutes at 20° F. Rinse with water, and develop in solution No. 2 for 20 minutes at 70° F. with frequent agitation of the solution.

A hardener to be used, if desired, before development is the following:

Trioxymethylene   3.0 g.
Sodium Sulphite (Anhydrous) 100.0 g.
Water (Distilled)   1000 cc.

The film is immersed in this for about two minutes before development. If preferred, the Trioxymethylene may be incorporated in the developer to the ratio of 3 grams Trioxymethylene per 100 grams of sodium sulphite (Anhydrous) present in the developer.

Fine Grain Technique in General

Paraphenylenediamine is a deep penetrating developer, and the film should be *thoroughly washed* between development and fixation.

Developer. wash-water, and fixer should all be at the same temperature. Paraphenylenediamine developers w o r k equally well at all temperatures between 60 and 70° Fahrenheit, but sudden temperature changes cause strains in the gelatin and increase grain size.

Film should be dried in a dust-free atmosphere.

Do not use stale fixing solution. Make it up fresh just before using, and throw it out before it is a week old. Many fine negatives have been spoiled for the sake of a few cents' worth of hypo.

Agitate the developer occasionally while the film is in it.

Rapid Fine-Grain Development

Good results are obtained with the following formula: elon 5 grams, p-phenylenediamine 10 grams, sodium sulphite 60 grams, sodium phosphate 3.5 grams, potassium bromide, 1 gram, water to 1 liter. Time of developing 7 minutes at 18°. The resultant negatives are thin. If such developers are used for fine-grained emulsions (Micro Lumiere), there is no diminution in graininess, and in all cases the final image is formed after the silver halide is dissolved by the developer, which then acts as a physical developer.

Phenylenediamine Fine-Grain Developer

The graininess of very fast negative materials is considerably reduced by doubling the normal time of exposure and developing for 20–30 minutes in the following developer: phenylenediamine (base) 10 grams, sodium sulphite (anhydrous) 60 grams, water 1000 cubic centimeters. The addition of sodium hydroxide to this developer increases the rate of development, but the graininess also increases.

Photo-Engraving Process
U. S. Patent 1,928,899

In forming a relief printing plate, an image of the copy to subsequently formed in relief on the plate is first printed by the action of light. The plate is then subjected to a solution formed of metol 120 grains, sodium sulphite 4 ounces, paradihydroxy-benzene 400 grains, solution, potassium bromide 40 grains and water 62 ounces to reduce those portions of the silver bromide affected by light to a metallic state and leave the remaining portions in a gelatinous state. The plate is then subjected to a relief bath solution formed from cupric sulphate 384, potassium bromide 336, potassium dichromate 12, chromic oxide 3.5 grains and water 16 ounces, then immersed in water heated to about 240°, removed from the heated water and dried to produce the desired relief surface.

Photo-Engraving

A process of forming a relief printing plate by first printing thereon by the action of light an image of the copy to be subsequently formed in relief on plate then subjecting the plate to the following solution: mono-ethyl-paramide-phenol-sulphate 120 grains, sodium sulphite 4 ounces, para-di-hydroxy-benzene 400 grains, sodium carbonate 2 ounces, bromide potassium 40 grains, and water 62 ounces, this solution due to its action on the silver bromide on the plate reduces those portions affected by light to a metallic state and leaves those portions of the plate not affected by the light in their gelatinous state, then subjecting the plate to the action of the following relief bath solution: cupric sulphate 384 grains, potassium bromide 336 grains, potassium bichromate 12 grains, chromic acid 3½ grains, and water 16 ounces, then immersing the plate in water heated to approximately 75° F. causing the entire gelatinous surface of the plate to swell to a predetermined thickness then removing the plate from the heated water and allowing it to dry thereby causing the gelatin around the relief portions of the plate to return to its normal unswollen condition.

Etching Fluids
Formula No.

Copper Sulphate	16.00%
Zinc Sulphate	1.20%
Sodium Chloride	14.30%
Water	68.50%

This etching fluid is slower than acid but will reproduce fine lines and will not blister varnish or wax. It can be used on iron also.

Formula No. 2

Nitric Acid, Concentrated	66%
Acetic Acid	34%

Formula No. 3 (For Deeper Work)

Hydrochloric Acid	
(Concentrated)	10.90%
Water	87 %
Potassium Chlorate	2.1 %

Formula No. 4 (For Brass)

Potassium Chlorate	2%
Water	95%
Nitric Acid	3%

Formula No. 5 (For Copper, Non-Corrosive)

Iron Chloride	6%
Water	30%
Alcohol	36%
Sodium Chloride	28%

Etching Copper

To etch copper use a solution of sodium bichromate and sulphuric acid. Dissolve 12 ounces of sodium bichromate in one gallon water, then add 8 fluid ounces sulphuric acid.

Etching on Steel

In attempting to etch on steel and to attain sufficient depth, the chief difficulty lies in finding an acid resist that will adhere to the steel. Either zinc or copper sensitizer can be used. For etching, use the usual solution of ferric chloride, 35° to 38° Baumé, to which is added a small amount of concentrated hydrochloric (muriatic) acid. (Nitric acid penetrates the enamel.)

Etching Stainless Steel

Stainless steel is etched with a 25% solution of muriatic acid and water. The steel is placed in the electrolyte as the anode. Use 6 to 8 volts reverse current with lead cathodes.

Etching Acid for High Speed Steel

Dilute Hydrochloric Acid	
(1–1)	350 cc.
Concentrated Nitric Acid	175 cc.
Molybdic Acid (85%)	70 cc.
Sodium Chloride (C.P.)	14 cc.

These mixed together and used on a rubber stamp, or used with the letters cut in paraffin act as a very good etching reagent for high speed steel.

Counter-Etch

Glacial Acetic Acid	6 oz.
Water	1 gal.

Copper Etching Sensitizer

Photo-Engraving	Metric	Avoir-dupois
Glue	360 g.	12 oz.
Ammonium Bichromate	52.5 g.	1¾ oz.
Dried Egg Albumen	7.5 g.	¼ oz.
Water to make	1 l.	32 oz.

Dissolve the bichromate in 8 ounces of the water, and the albumen in 4 ounces. Filter the albumen solution. Mix the glue with the remaining 20 ounces of water and add, successively, the albumen and bichromate solutions, stirring very thoroughly to insure complete mixing. If desired, 2 ounces of albusol may be substituted for the ¼ ounce of dried albumen in the above formula.

The film on the metal should be a thin one, but not so thin as to show interference colors. Therefore, it may be advisable to add more or less water, according to the film obtained under the conditions of whirling.

Photogravure Sensitizer

No sensitizing formula can be exclusively recommended, since a slight modification will affect the contrast of the final result. However, the following is a good general formula:

	Metric	Avoir-dupois
Potassium Bichromate	97.5 g.	3¼ oz.
Water	4 l.	1 gal.
Ammonia (Concentrated 28%)	7.5 cc.	¼ fl. oz.

Photogravure Spirit Sensitizer

A solution of ammonium bichromate is prepared, of three times the strength required. This should be neutralized by ammonia if the sensitive material is to be kept. Twice its volume of methylated spirits or denatured alcohol should then be added. *Keep in the dark.*

The sheet of tissue to be sensitized is then brushed with the solution, the strokes being made to cross each other, in order to avoid any irregularity.

Silk Screen Sensitizers

These should be made in a dark room and should at all times be kept away from direct light rays.

Formula A—8 fluid ounces of hot distilled water; one-half fluid ounce of strong (28%) ammonia water; 5 grams of C. P. glycerine; 10 grams of ammonium bichromate; 1 gram of chromic acid; 4 fluid ounces of photoengraver's glue.

While the above is melting together, be careful that temperature at no time reached the boiling point. Put this liquid in amber bottles, and, when ready to sensitize the screen, prepare the sensitizer as follows: The white of two eggs is to be stirred into the above liquid and allowed to settle twelve hours. Then strain the mixture and it is ready for use.

Formula B, which is simpler to make, is as follows: 100 cc. of Process Glue is mixed with 4 grams of C. P. glycerin and then with a solution of 6 grams of ammonium bichromate in 60 cc. of distilled water, and enough more water added to make 200 cc. total volume. This solution should be strained through a strainer made of several layers of muslin or fine cheesecloth wrung out in water, and 5 cc. of strong (28%) ammonia water added. It is then ready for use and will keep for two or three weeks.

Caution: This work must all be done in the dark room.

Ultra-Violet Sensitization

Photographic plates can be made sensitive to short-wave ultra-violet light by immersion in a 1% alcoholic solution of citric acid or an alkali metal salicylate. The plate is withdrawn immediately from the solution and dried in less than a minute by waving in the air before putting in the holder. After exposure the plate can be developed without previous washing and without special precautions. Plates thus sensitized can be kept, as the layer left after the evaporation of the alcohol is very adherent. Citric acid and its salts have no action on the emulsion or the developer. The sharpness of the image is not diminished.

Zinc Sensitizer

Line work on zinc is usually printed with bichromated albumen. The following is a satisfactory formula:

	Metric	Avoir-dupois
Albumen (White of Egg)	64 cc.	2 fl. oz.
Photo-Engraving Glue	15 cc.	½ oz.
Ammonium Bichromate	15 g.	½ oz.
Water to make	1 l.	32 oz.

A few drops of concentrated ammonia (28%) may be added. Dried, powdered egg albumen or prepared liquid albumen may be used in the place of the fresh albumen.

After exposure the print is rolled up with the thinnest possible coating of photo-transfer ink, and is then developed by swabbing with a piece of wet cotton while the plate is held in a tray of water, or under the tap.

Glue Sensitizer

Photo Engraver's Glue	6 oz.
Ammonium Dichromate	½ oz.
Water	32 oz.

Photographic Glue

Ammonium Dichromate (Saturated Solution)	2 oz.
Water	4 oz.
Ammonia (26°)	8 drops

Glue-Gum Sensitizer

Water	12 oz.
Glue	1½ oz.
Gum Arabic (Powdered)	¼ oz.
Ammonium Dichromate	¼ oz.

Violet Stain for Sensitized Coatings

Methyl Violet	125 gm.
95% Alcohol	2500 cc.

Use 3 ounces per gallon water.

Lithographic Formulae

Formula No. 1—Albumen Sensitizer

Egg Albumen	20 g.
Water	100 cc.
Ammonium Dichromate (Saturated)	15 cc.

Formula No. 2—Gum Sensitizer

	No. 1	No. 2
Gum Arabic	30 g.	600 cc. (Sp. Gr. 1.100)
Water	100 cc.	
Ammonium Dichromate	15 cc.	50 g. in 100 cc. water to Sp. Gr. 1.045

Formula No. 3—Casein Sensitizer

Casein	100 g.
Ammonium Hydroxide (28%)	3 oz.

Water	300 cc.
Ammonium Dichromate	75 g.
Water	100 cc.

Wall's Rapid Bichromate Bath

Suitable for the gum bichromate process and for the various oil processes:

Ammonium Bichromate	56 g.
Glycerine	20 cc.
Copper Sulphate	2 g.
Distilled Water	1000 cc.

Tests show this to be four times faster than the usual sensitizing baths.

This may be used with glue or gums, with pigments or other additions.

Photographic Intensifier

Mercuric iodide 2 grams, potassium iodide 2 grams, hypo 2 grams, water 100 cubic centimeters. Dissolve together in very little warm but not hot water and then make up to bulk. The solution may be used repeatedly until exhausted. The intensification is continuous but can be stopped at will. The negatives after washing may be dried and used but are not completely permanent. To render them so they should be immersed in a 1% sodium sulphide solution until the image has been wholly changed, as viewed from the back, from gray to brown-black. This sometimes, but not always, reduces down slightly. Intensification, if unsatisfactory, can be removed entirely by treating the plate with 20% hypo solution. This must be done before any after-treatment is applied.

Intensification with Mercuric Iodide

Solutions for mercuric iodide intensification are usually prepared by dissolving mercuric iodide in a potassium iodide solution containing a trace of sodium thiosulphate. The intensified images obtained with this solution are not stable and turn yellow rapidly. Good results are obtained with the following formula: mercuric iodide 10 grams, potassium iodide 20 grams, hypo 20 grams, water to make 1 liter.

Mercuric Iodide Intensifier

Mercuric Iodide	1 g.
Sodium Sulphite	10 g.
Water	100 cc.

Soak negative until sufficiently dark. Redevelop and fix.

Mercuric Chloride Intensifier

Soak negative 10 minutes in water and then in saturated solution mercuric chloride until white through back. Wash in running water 10–15 minutes and rede-

velop 5 minutes in solution of ammonium hydroxide C.P. concentrated 1 ounce per gallon water.

Composition Lithographic Rollers

Hide Glue	30 parts
Glycerine	65 parts
Water	60 parts
Borax	1 part

This is a very cheap composition but gives satisfactory performance.

Protective Lithographers Hand Cream

Lanolin (anhydrous)	20 parts
Paraffin Wax	6 parts
Paraffin Oil	7 parts
Glycol Stearate	12 parts
Water	55 parts

Melt lanolin, glycol stearate, and wax; add oil; then add water while agitating rapidly until cold. This gives protection against water solutions of moderately concentrated acids.

Cleaning Photographic Lenses

Water	3 oz.
Alcohol	1 oz.
Nitric Acid	3 drops

After dusting the lens, rub with an old clean cotton cloth dipped in this solution and polish with a dry piece of the same cloth. Protect the lens by a dead black hood of sufficient depth, to prevent any direct light from the sun, skylight, or windows from striking its surface.

Photographic Plate Backing

Caramel	12 oz.
Lamp Black	4 oz.
Denatured Alcohol	16 oz.
Water	8 oz.

Apply to the glass side of the plate with a flat brush.

Photographic Stay-Flat Formula

	Avoirdupois	Metric
Gelatin	2 oz.	60 g.
Golden Syrup (Karo)	2 oz.	60 g.
Glycerin	2 oz.	60 g.
Chrome Alum	16 grains	1.1 g.
Water to make	32 oz.	1 l.

Mix the syrup and glycerin in 24 ounces of water, and soak the gelatin in this mixture for ½ hour, warming it up to 120° F. Dissolve the alum in two ounces of water. Add this to the mixture and bring the total to 32 ounces with water. Strain through cheesecloth. One ounce will cover 100 square inches. The mixture sets in 24 hours. All of it must be used at one pouring, and it cannot be remelted.

Anti-Halation Layers

Anti-halation layers for photographic films are composed of a colored colloidal layer insoluble in water but soluble in akaline photographic baths, a synthetic resin containing an hydroxyl or carboxyl group capable of forming a square salt being used as binder for the dye. Thus, a solution containing resin (from salicyclic acid and paraldehyde) 10, fuchsin 1 and iso-butyl alcohol 100 grams is applied to the back of a cellulose acetate film to give a layer of 1–2 μ.

Matting Glass
Solution No. 1

Sodium Fluoride	60 g.
Potassium Sulphate	12 g.
Water	500 cc.

Solution No. 2

Zinc Chloride	14 grains
Hydrochloric Acid	65 cc.
Water	500 cc.

Positive Collodion

Alcohol	2000 cc.
Ether	2000 cc.
Negative Cotton	42 g.

Paint Out Gum

Gum Arabic Solution (Sp. Gr. 1.100)	32 oz.
Chrome Alum	1½ oz.
Green Dye, to suit	

Wet Plate Copper Solution

Copper Sulphate	1021 g.
Potassium Bromide	510 g.
Water	4 gal.

Ground Glass Finish

Gum Sandarac	5.8 g.
Gum Mastic	1.3 g.
Ether	60 cc.
Benzol	30–45 cc.

Bitumen Varnish for Stopping Out

Bitumen	2 lb.
Turpentine	1 pint
Benzol	1 pint

Photographic Cold Glue

Soak 6000 grams of bone glue in 3000 cubic centimeters of cold water until the glue is swollen. Heat until dissolved Add 300 cubic centimeters of nitric acid and then add the following mixture:

Water	1700 cc.
Calcium Chloride	300 g.
Oil of Cloves	60 cc.

PLASTICS

Graphite Plastics
Blocks

Cerelose	14 %
Pitch	5 %
Creosote	0.5%
Furfurol	0.5%
Ceylon Flat Lump Graphite	28 %
Mexican Graphite	25 %
Alabama Graphite	20 %
Borax	3 %
Greenbaum Clay	2 %
Bentonite	2 %

This is dry molded under pressure and calcined at 2000° F. (approximately).

Plastic Composition
British Patent 383,324

A composition for making an acoustic horn is made from fine sand 55, crushed or rolled natural rock, asbestos fiber and wood flour, each 15, and ground magnesite 45 pounds, the whole being mixed with 6 gallons (22° Bé.) magnesium chloride solution. A finishing composition is made by mixing the above quantities of magnesite and magnesium chloride with 25 pounds each of the other materials.

Plastic (Plaster) Composition
U. S. Patent 1,926,543

A non-cracking plaster is composed of clay 6, paper pulp 2 parts, and sawdust 1 part, manufactured wet and dried for shipment.

Plasticizer
U. S. Patent 1,946,202
Formula No. 1

160 grams of phthalic anhydride and 60 grams of ethylene glycol are heated to 180° C. and then cooled. 20 grams of ethylene glycol and 5 cc. of concentrated sulphuric acid are added and the mixture is then heated in an open flash to 115–125° C. for three hours. The product is cooled, washed with sodium hydroxide solution and with water, and is then dried by heating to 180° in an open dish. The resultant product is apparently liquid in character but is so viscous that it can not be poured except when warm. It is dark brown in color and is difficultly soluble in benzol.

Formula No. 2

74 grams of phthalic anhydride and 20 grams of ethylene glycol are heated together at 150° C. for one-half hour. 5 cc. of concentrated sulphuric acid, 11 grams of glycol, and 50 grams of benzol are then added and the mixture is heated to distill off the benzol. As the benzol vapor distills off, it carries with it, as a constant-boiling mixture, the water formed during the reaction. Heating is continued for three hours, benzol being added from time to time as needed. In this manner, the reaction temperature is maintained at 100° C. or less. The product is washed with alkali and with water and is dried in vacuo at 150° C. It is a viscous oily liquid easily soluble in benzol.

In place of the sulphuric acid used as a catalyst in the examples above, other esterification catalysts such as hydrochloric acid or phosphoric acid may be used. In place of benzol, other inert liquids forming constant-boiling mixtures with water, for example—carbon tetrachloride—may be employed.

Inorganic Plastic Bases
French Patent 752,452

Powdered substances or mixtures which are hardened by water are transformed by anhydrous liquids into homogeneous plastic masses, and kept out of contact with moisture until the time of use. Thus, zinc oxide 45, zinc sulphate 20, monoacetin 15 and dichlorohydrin 20 parts are mixed, or Portland cement 72.5 and an alcoholic solution of zinc chloride 27.5 parts.

Increasing Water Content of Plastics
French Patent 747,728

The amount of water which may be added to compositions of cellulose derivatives and organic liquids for making rayon, films, plastics, etc. is increased by the addition of 0.5–2% of aniline or the ethanolamines.

Phenol-Aldehyde Plastic
U. S. Patent 1,892,409

Finely-divided wood (100 parts) is digested with phenol or a homologue thereof (29), dilute ammonium hydroxide (density 0.9) (12), and water (800 parts) for 4 hours at 200–300 pounds per square inch. Then 72 parts of 40% formaldehyde (or other aldehyde) are added and the mixture is refluxed for 1 hour. The resulting product when dried and moulded under heat (150°) and pressure (1000 pounds per square inch) is strong, water-resistant, and resembles lignum vitae in many respects.

Plastic "Wood" and Wood Fiber
U. S. Patent 1,947,438

Formula No. 1

Nitrocellulose		100	parts
Dibutyl Phthalate	15 to	25	parts
Organic Filler	175 to	250	parts
Gum or Resin	60 to	65	parts
Castor Oil	20 to	35	parts
Volatile Solvent	400 to	640	parts

Formula No. 2

Pyroxylin	100	parts
Dibutyl Phthalate	15	parts
Wood Flour	250	parts
Ester Gum	65	parts
Castor Oil	35	parts
Volatile Solvent (Preferably Comprising 70% Acetone and 30% Denatured Ethyl Alcohol	640	parts

Formula No. 3

Cellulose Acetate (Preferably of Variety Soluble in Acetone)	100	parts
Triphenyl Phosphate	5	parts
Paraethyltoluenesulphonamid	15	parts
Dibutyl Tartrate	5	parts
Dimethyl Phthalate	35	parts
Ester Gum	35	parts
Wood Flour	250	parts
Volatile Solvent (Preferably Comprising 85% Methylene Chloride and 15% Methyl Alcohol)	600	parts

Formula No. 4

Cellulose Derivative Substantially Non-Volatile Modifier in	100	parts
Liquid Form	35 to 60	parts
Resin	60 to 65	parts
Comminuted Organic Material	175 to 250	parts
Volatile Solvent	400 to 640	parts

A stabilizer such as urea preferably in proportions ranging from .4 to 1 per cent is generally added.

The above ingredients may be mixed and processed by any means known in the art. For instance, all the ingredients may be combined in a kneader and after thorough mixing, the mass is ready for use. To prevent premature evaporation of volatile solvent, the wood filler is preferably stored and shipped in air-tight containers.

By soaking nitrocellulose, substantially non-volatile liquid modifiers, gums or resins, stabilizer and 50% of the volatile solvents in an air-tight container for a period from six to twenty-four hours a satisfactory product is produced. At the end of the soaking period the organic filling material is wet with the balance of the solvent and then all of the ingredients are thoroughly mixed in a kneader. After thorough mixing, the mass is kept in air-tight containers until used.

Before evaporation of the volatile solvents, the mass has a plastic consistency similar to that of putty.

Vinyl Resin Plastic

Twenty parts of zinc oxide are dissolved in 1,000 parts palmitic acid, the water resulting from formation of zinc palmitate is expelled, and the product introduced into an autoclave where it is mechanically stirred with a 2:1 mixture of acetylene and nitrogen under 15 to 20 atmospheres pressure at 160° C. After about five hours the acetylene absorption comes to an end, and the reaction product is then dissolved in one-half its volume of warm benzine, the solution filtered from zinc palmitate, and the filtrate fractionally distilled, when vinyl palmitate is isolated in 80 to 90 per cent of the theoretical yield. It boils at 165° C. under 2 millimeters pressure.

To illustrate the production of a vinyl phthalate, 500 parts monoethyl phthalate, in which 15 parts zinc oxide are dissolved by warming, are introduced into an autoclave, followed by nitrogen and acetylene at 5 and 10 atmospheres pressure respectively. The stirrer is started up and the temperature raised to 180° C., more compressed acetylene being introduced as required. Ethyl vinyl phthalate is formed as an oil, which readily polymerizes in the presence of a little benzoyl peroxide when kept at 95° to 100° C. The polymer itself can be hardened by heating at 100° to 120° C. Purification of the crude product as formed in the autoclave can be effected by passage over charcoal or by preliminary dissolution in benzine.

Plasticized Pitch Compound
U. S. Patent 1,911,131

Pitch (200 pounds) is mixed with asphalt (10–80 pounds) and a phenol-aldehyde condensation product (5–30 pounds) and the mixture is heated to about 230° until the whole forms an elastic solid.

Fillers for Asphaltic Composition Materials

The following is a partial list of filler materials which are employed in asphalt compositions, each giving different results. Usually several of them are used in combination.

Asbestos Fibers	Wood Flour
Tripoli	Celite
Saw Dust	Cotton Flocks
Talc	China Clay
Portland Cement	Plaster of Paris

"Thiokol" Type Plastic
U. S. Patent 1,950,744

Method No. 1

An aqueous solution of calcium polysulphide is prepared, having a density of 31° Bé., the empirical formula of which is, by analysis, CaS_{47}. Magnesium hydroxide is formed in this solution, preferably by the addition of a solution of sodium hydroxide, followed by a solution of magnesium chloride or sulphate, or other suitable compound. The amount of magnesium hydroxide may vary within quite wide limits say from 2 to 50 grams per thousand cubic centimeters of the polysulphide solution. Preferably from 5 to 10 grams per liter provides a desirable form of latex-like dispersion of the final product best adapted for handling. Thus, to each 1000 cubic centimeters of the above calcium polysulphide solution, there may be added 50 cubic centimeters of a solution containing 290 grams sodium hydroxide per liter and 100 cubic centimeters of a solution of magnesium chloride containing the chemical equivalent of 105 grams of magnesium hydroxide per liter, these proportions providing a slight excess of sodium hydroxide. A semi-gelatinous dispersion of magnesium hydroxide in the polysulphide solution results. If desired, the magnesium hydroxide suspension may be formed separately and added to the polysulphide solution, but preferably the suspension is formed in the solution itself, as before described.

The desired olefin-dihalide, for example, ethylene dichloride, is now added to the polysulphide solution containing dispersed magnesium hydroxide, the proportion of dihalide added being very slightly less than the equivalent required for reaction. Thus, with the solution above described, 157 cubic centimeters of ethylene dichloride per liter of original calcium polysulphide solution is used. To avoid excessive heating, the olefin-dihalide is added in small portions with vigorous agitation. If desired, the polysulphide solution may first be slightly warmed, say 80 to 100° F., to accelerate the beginning of the reaction. The addition of ethylene dichloride is controlled so that the temperature does not rise excessively, say to above 150 to 175° F. After the dihalide has been completely added, the reaction may be completed by further warming the mixture to, say, 175 to 190° F. for a short period, until complete disappearance of the olefin-dihalide is indicated by loss of its odor. The olefin-polysulphide plastic forms in this reaction mixture as a fluid dispersion, latex-like in character, which settles out of the reaction mixture. The supernatant solution may be removed by decantation and the latex-like dispersion may be readily washed with water and settled, the wash water being decanted. This is repeated until the soluble salts and other undesirable constituents present are completely removed. This latex may then be coagulated or compounded and coagulated, as set forth below.

Method No. 2

A stabilized sodium polysulphide solution of empirical formula Na_2S_4, and of density 37.0° Bé. at 60° F. is formed by dissolving sulphur in a solution of sodium hydroxide in the requisite proportions and heating the mixture to atmospheric pressure boiling point for a substantial period of time, say 20 hours, as disclosed in U. S. Patent Application Serial No. 369,912. Dispersed magnesium hydroxide is formed in this polysulphide solution in the same manner and suitably in the same proportions as set forth in connection with the first example above. The mixture is then heated to 120° F., and slightly less than equivalent quantity of ethylene dichloride is added in small portions at a time with vigorous agitation, while holding the temperature of the mixture below about 175° F. The amount of ethylene dichloride so added is about 208 grams per liter of the original polysulphide solution. The latex of olefin-polysulphide plastic forms and is settled and separated from the other products of reaction by decantation or centrifuging, thoroughly washed as in the first example and is then ready for coagulating or compounding and coagulating.

The latex-like dispersion of the olefin-polysulphide plastic produced in accordance with the method of the present invention may be coagulated by acidifying it with any suitable acid, such as hydrochloric or sulphuric acid, or organic acids such as acetic acid, formic acid, or the like; and on acidulation, the olefin-polysulphide plastic coagulates to form an elastic, spongy mass, from yellow to white or nearly white in color. Its characteristics may be improved by incorporating into it, suitably on a rubber mill, any desirable proportion of a metallic oxide such as litharge, zinc oxide, magnesium oxide or other compounds of the class of sulphur carriers well known in the rubber industry. Various inert materials, such as those of the type used in compounding rubber, for example, fibers, wood flour, carbon black, glue, asbestos, and the like, may be compounded with the latex either with or without the sulphur carrier. The compound may then be suitably heated to effect its stabilization, the time of heating varying with the temperature employed. Temperatures from 80° C. upwards may be used. Thus, by heating in autoclave under 40 pounds steam pressure for 40 minutes to 1 hour, a completely stable and homogeneous product may be secured. Instead of first coagulating the latex and incorporating a sulphur carrier and other compounding ingredients into the coagulum, such materials may be incorporated in the latex before coagulation. Thus, from 1 to 10% of litharge or zinc oxide may be thoroughly mixed with a thick latex, the mixture then heated (as it does not undergo coagulation on heating) under the same conditions as set forth in connection with the compound above, and the heated and stabilized mixture then coagulated. Other compounding ingredients may be incorporated in the latex for such heating and coagulation, or if desired, after the stabilized latex-sulphur carrier mixture has been coagulated, additional compounding ingredients may be milled into the coagulum in the ordinary manner.

The resulting product has the characteristics of a soft unvulcanized rubber, does not flow under pressure, is not separable by organic solvents; it is not affected or swelled by organic rubber solvents, including carbon disulphide, and has dielectric qualities superior to those of rubber. Furthermore, it completely resists the action of oils, salt water, and the like, and has an extremely low power loss when used as a dielectric.

The latex of the olefin-polysulphide plastic is stable and may be preserved and stored as such, or may be treated with a sulphur carrier, as hereinbefore set forth, and the stabilized latex preserved and stored for further use. If the latex is stored for periods of a week or more, it may undergo a change somewhat analogous to polymerization. The addition of an acid then produces the subsidence of a granular product from the liquid. The latex can, however, be restored to its original condition by subjecting the dispersion, before adding the acid, to temperatures around the boiling point of water for a few minutes, the addition of the proper amount of acid will then produce the massive coagulum.

Waterproof Cellulose Ester Wrappings
British Patent 399,191

(1) A mixture of cellulose acetate (85–90 parts), cellulose stearate (10–15 parts), tricresyl phosphate and diethyl phthalate, is dissolved in a mixture of acetone and benzene, and the solution cast on to a film wheel, band or the like. (2) A mixture of cellulose acetate (90–95 parts), cellulose palmitate (5–10 parts), diacetone alcohol and diethyl phthalate is dissolved in acetone-benzene mixture and cast on a suitable surface. (3) 90 parts of cellulose acetate-propionate dissolved in acetone are mixed with 10 parts of cellulose stearate-laurate dissolved in benzene, incorporated with triphenyl phosphate and diethyl phthalate, and the solution employed in the production of films.

Phonograph Record
U. S. Patent 1,946,597

A thermo-plastic composition for molding phonograph records comprises about 77% of the polymerized vinyl ester of acetic acid, about 20% of amorphous silica, about .25% of dibuty phthalate and about 2.75% of coloring matter.

Molding Powder
U. S. Patent 1,899,542

The powders comprise 100 parts of pyroxylin (containing 10.8–11.4% nitrogen), 75–80 parts of tricresylphosphate or other suitable plasticiser, and 300–350 parts of calcium sulphate, preferably hydrated, together with coloring pigments or dyes, and sufficient solvent (ethyl alcohol, methyl alcohol, acetone, ethyl acetate, methyl-ethyl-ketone, or mixtures thereof) to allow the ingredients to be masticated together.

Molding Hat-Blocks
British Patent 404,404

A hat-block is molded from a mixture of magnesite 1, magnesium chloride 1 and cork 5 parts, and the mold is lined with material consisting of plaster of paris 1 and plastic wood 2 parts.

Sulphur Treatment of Shellac

The best results are obtained by heating shellac at 180° with 3–4% sulphur until a drop of shellac on cooling has a typical green color. Water and abrasion resistance, elasticity, adhesion and molding properties are markedly improved.

White Metal Ornamental Molding Compound

Sulphur	98.5%
Aluminum Metal	1.5%

Melt the sulphur at a temperature of 260° F. Add aluminum powder and cast in mold.

PLATING

Cleaning Brass Prior to Nickel and Chromium Plating

4 ounces sodium metasilicate per gallon of water is satisfactory. A sulphuric acid dip should be used at a concentration of ½–1 Normal.

This alloy may be cleaned either anodically or cathodically but in order to get 100% efficient work it is necessary to clean properly. The best method is to clean anodically, and in the majority of cases the work is clean in 1 minute, but it can be left in the cleaning solution for indefinite periods without fear of etching. After cleaning the parts should be hand scrubbed. If the anodic method is not convenient brass may be cleaned cathodically, but for no longer than 20 seconds; the parts should then be hand scrubbed. The adherence of metals to be plated over brass is much better when cleaning is accomplished anodically.

If a nickel plate or nickel and chrome plate cannot be made to adhere to brass which has been cleaned electrolytically, there is still one sure way to accomplish it, and that is to clean brass without any current. By soaking the brass parts in 4 ounces sodium metasilicate per gallon for 1 to 2 minutes, and then scrubbing, the surface will be in a condition to take nickel and chrome, and the adherence is far greater than when cleaned electrolytically, either cathodically or anodically.

Chromium Plating
U. S. Patent 1,953,484

The method of electroplating a wire with a coating of chromium which will not flake upon deformation of said wire when plated comprises forming a solution of chromic acid (25%), sulphuric acid (2.5 to 4.5%) and electrolyzing for only a limited interval of time substantially no greater than that required to deposit approximately 3 to 4 per cent of the original chromium content of the solution, the temperature of the solution being maintained between about 20 and 30° C. and the cathode current density being maintained between about 8.5 amperes per square inch and 33.0 amperes per square inch.

Chromium Plating Printing Plates

The solution used at the Bureau of Standards:

Chromic Acid   33   oz. per gal.
Chromium Sulphate   0.4 oz. per gal.
Chromium
  Carbonate   1.0 oz. per gal.

Temperature—113° F.; amperage—100 amp/sq. ft. on steel plates; 200 amp/sq. ft. on nickeled plates.

Very good ventilation is required as a large amount of spray is formed, which is injurious to the nose.

Combined Chromium-Nickel Plating
U. S. Patent 1,948,145

Separate anodes of chromium and of nickel are immersed in a bath which contains nickel and chromium. The chromium and nickel are deposited simultaneously on a cathode which may be a steel plate by applying an electrical current in the usual way.

Preferably the chromium anode in this case will be somewhat smaller than the nickel anode while the voltage employed will be approximately 6 volts at about 25 amperes per square foot of area of the cathode. The nickel and chromium are deposited simultaneously and in desired proportions so as to provide a very satisfactory plating of nickel and chromium alloy.

In another way desirable results are obtained by the following: An anode of chromium and nickel alloy is employed in a solution containing chromium and nickel with a voltage and amperage as above stated. 25% chromium and 75% nickel provide a suitable anode. While from 32 ounces to 96 ounces of nickel sulphate, 16 to 32 ounces chromic carbonate and 2 ounces of boric acid per gallon of water from a very suitable solution or bath for functioning in the electro-depositing operation.

Chromic sulphate and sodium carbonate in proper proportions in conjunction with the nickel sulphate and boric acid in water provide a suitable solution, it being one feature to employ a solution having chromium and nickel as a constituent thereof.

By employing the current specified in connection with the bath and anodes described the deposit of chromium and nickel alloy is rapid and of uniform thickness and luster. Uniformity of deposit is particularly desired where large areas are to be plated, but by proceeding as outlined it has been found possible to provide large flat plates for instance having coatings of alloy which are uniform in thickness and which do not tend to peel or crack.

The alloy deposit is more brilliant than that of a deposit of either nickel or chromium alone, it has the combined characteristics of both, is not likely to rust and has a very desirable white appearance.

"Stripping" Chromium Plate

Chromium can be stripped in a hydrochloric acid solution or in the following solution using reverse current:

Sodium Cyanide	12 oz.
Sodium Hydroxide	4 oz.
Water	1 gal.

"Stripping" chromium from nickel:

Immerse at room temperature in following solution:

Concentrated Muriatic Acid (By Volume)	1 part
Water (By Volume)	1 part

or make the article the anode in any alkaline cleaner or solution.

"Stripping" copper or brass from nickel or steel:

Make article anode in following solution:

Sodium Cyanide	50 g.
Sodium Hydroxide	50 g.
Water	1 l.

Brightening Dip for Cadmium Electrodeposits

U. S. Patent 1,816,837

After thoroughly rinsing from the plating solution immerse in one of the following solutions for about a minute:

Formula No. 1

Concentrated Nitric Acid (By Volume)	1 part
Water (By Volume)	99 parts

Formula No. 2

Chromic Acid	150 g.
Sodium Sulphate (Anhydrous)	8 g.
Water	1 l.

Solution No. 2 is recommended.

"Copper Plating" (Of Glass Mirrors)
U. S. Patent 1,890,094

A protective varnish for the reflecting surface of silvered mirrors comprises shellac 1 part, ethyl alcohol 9, leafy copper about 7 parts.

Copper Plating Bath

The harmful effect of antimony in a copper plating bath is avoided by use of cream of tartar as follows:

Sulphuric Acid	110 g.
Copper	35 g.
Antimony	0.2 g.
Cream of Tartar	4 g.

The above for one liter solution.

Plating Copper on Cast Iron and Die Cast Metal

Formula for cyanide copper solution:

Copper Cyanide	3½ oz.
Sodium Cyanide	4½ oz.
Carbonate of Soda	2 oz.
Hyposulphite of Soda	1/64 oz.
Water	1 gal.

Operate solution at 110° F., using 1 volt pressure.

Brightener for Acid Copper Plating Solutions

To secure brighter deposits from acid copper plating solutions add phenolsulfonic acid in the ratio of about 1 gram to each liter of solution. The solution will work better after being operated for a short period.

"Stripping" Copper from Silver

Dip article in following:

Chromic Acid	4 lb.
Concentrated Sulphuric Acid	7 oz.
Water	1 gal.

"Stripping" Nickel from Copper or Silver

Make the article the anode in following (stated in parts by volume):

Concentrated Muriatic Acid	1 part
Water	24 parts

"Stripping" Nickel from Aluminum

Make the article the anode in a 10% (by volume) sulphuric acid at room temperature at a current density of 25 to 50 amperes at a per square foot.

"Stripping" Nickel from Steel

Make the article the anode in a 53° Bé. sulfuric acid solution. Use a low current density. Avoid the presence of arsenic.

"Stripping" Silver from Nickel Silver

Immerse the article in the following (stated in parts by volume):

Concentrated Sulphuric Acid	19 parts
Concentrated Nitric Acid	1 part
Temperature	180° F.

Electrodepositing Indium and Indium and Silver

U. S. Patent 1,935,630

A process of electro-depositing indium comprises the steps of immersing an article to be plated as the cathode in a prepared bath, bath comprising 1 gallon of water, 15 ounces potassium cyanide, and 7 ounces glycine, or equivalent thereof selected from the group: cyanuric acid, abietic acid, adipic acid, acetyl m-aminobenzoic acid, acetyl p-aminobenzoic acid, acetaldehyde sodium bisulphite, m-aminobenzoic acid, immersing an anode of indium in said bath, and passing a current of about .035 amperes and upwards per square inch at the cathode to deposit indium on said article.

Indium Plating Bath

A typical cyanide indium plating bath as follows:

Indium Metal (as the Double Cyanide)	5	oz.
Potassium Cyanide (or the Equivalent Amount of Sodium Cyanide)	9	oz.
Sugar	2½	oz.
Water to make	1	gal.

Indium plate as it comes from the bath is soft, uniform, and gray. It can be easily diffused into the base metal and thereby hardened. This procedure both hardens and stabilizes the surface so treated; that is, the surface is very resistant to oxidation and tarnish.

Polishing Silver Plating

In producing a high color on silver plated work, the kind of wheel used the speed at which it is operated, and polishing medium used are important factors.

The first operation is done by using a muslin buff of 84 x 92 weave and hard rouge. The wheel, if 8 or 10 inches in diameter, should be operated at 2800 to 3000 revolutions per minute. The coloring operation is done on a soft muslin wheel, 64 x 68 weave, operated at 3600 revolutions per minute, using a soft silver rouge.

It requires considerable experience to be able to produce a high finish on silver plated work.

Silver Plating Die Cast Lead Objects

For a cheap finish the work should be polished, then cleaned and plated in a cyanide copper solution until completely covered with copper, and then plated in the acid copper soltuion to produce a heavy deposit of copper. They are then bright dipped, flashed in a brass solution, silver plated, and lacquered. Formula for silver solution for this class of work:

Silver Chloride	3 oz.
Sodium Cyanide	5 oz.
Carbonate Soda	2 oz.
Water	1 gal.

A deposit of 4 or 5 minutes, using 1 volt, is all that is required for this class of work.

Silver Plating

Silver Cyanide (Troy)	4 oz.
Potassium Cyanide (Avoir.)	9 oz.
Potassium Carbonate (Avoir.)	8 oz.
Water	1 gal.

Maintain a "free cyanide" content of 4 ounces per gallon (which automatically results using above quantities). To secure a brighter deposit add a very small amount of the above solution saturated with carbondisulfide and aged for several months. Avoid adding an excess of brightener. Operate the above solution at 80° F., at a cathodic current density of 8 to 10 amperes per square foot with agitation. Use an anode area approximately equal to the cathode area. Before entering the plating solutions the articles should be cleaned and flashed with silver as described in Volume 1. The formula given above is believed to be an improvement over most sodium silver plating solutions.

Gold Plating Without Electricity

For gold plating by the immersion process, use the following formula:

Fulminate of Gold	4 pennyweight
Yellow Prussiate of Potash	12 oz.
Carbonate Soda	24 oz.
Caustic Soda	¼ oz.
Water	1 gal.

The solution should be kept at the boiling temperature in a cast iron tank for an hour or so, and then cooled to 180° F. before using.

Klondike "Gold" Plating

Solution is made up of 3 gallons of water, 2½ pounds sodium carbonate and 8 ounces sal ammoniac. It is operated at 180° F., with anodes of Oreide metal (90% copper, 10% zinc) and some steel anodes. Correct operation is important and this may require working form a while without results. Addition of an ounce of potassium or sodium cyanide is recommended to clear the solution if it turns green.

Gold Plating Base from Gold Filings

First a magnet is passed through the filings to remove iron and steel. Then they are burned in a frying pan to remove paper, dust, etc. Then they are treated with nitric acid, to remove all the base metal that can be reached. After being washed they are treated with aqua regia, which dissolves the gold and the remaining base metals, (except silver, which is instead converted into a heavy scum of silver chloride). These treatments produce unpleasant fumes.

The solution, which contains dissolved gold, etc., is filtered, and the gold is precipitated by any one of several substances, possibly ferrous sulphate of oxalic acid; this yields a powder which is fairly pure gold when washed.

The careful worker will dissolve this gold a second time, and precipitate it again, to insure high quality. These latter steps also produce fumes.

The gold is now ready to be converted into fulminate. It is once more dissolved in aqua regia, and the solution is slowly evaporated down to a syrup. Next, water is added (preferably distilled water), and finally, a carefully adjusted amount of ammonia.

The ammonia converts the dissolved gold into the brown so-called fulminate.

Zinc-Cadmium Plating

Good results may be obtained by inexperienced platers, or where control equipment is limited.

Cleaning Solution—1½ lb. sodium carbonate and 2 oz. lye per gal. water.

Pickle—½ lb. sulphuric acid per gal. water and cool.

Bright Dip—Add ½ pound of sulphuric acid to 5 pints water. Cool and add 8 ounces nitric acid and ¼ ounce hydrochloric acid.

Plating Solution

1. Dissolve in two gallons of boiling water:

Sodium Cyanide	4 lb.	2 oz.
Sodium Hydroxide	5 oz.	

2. Make paste of even consistency with water from:

Cadmium Oxide	1 lb.	2 oz.
Zinc Cyanide	5 oz.	
Glucose	½ oz.	

Add No. 2 to No. 1. (Cadmium oxide will probably require several hours to dissolve.) ALWAYS USE GREATEST CARE WITH CYANIDES! Make up to 5 gallons with water.

Anode—½ Zinc ½ Cadmium.

Current—25-30 amperes per square foot 5-6 minutes. Buffing results in polish nearly equal to chromium plate.

Iron Plate on Brass

The following formula for an iron solution will produce good results upon brass or bronze castings:

Ferrous Chloride	16 oz.
Calcium Chloride	16 oz.
Water	1 gal.

The solution should be operated at 150° to 175° F., with a cathode current density of 20 amperes per square foot. Use wrought iron andoes. The solution will require the constant addition of a small quantity of hydrochloric acid to properly maintain the pH.

"Stripping" Nickel Plating

Probably the best and cheapest way to remove defective deposits of nickel from iron or steel parts is to use reverse current with a sulfuric acid strip. The surface of the metal will be somewhat roughened, especially where the deposit has peeled, and a polishing operation will be necessary before plating.

In preparing the strip, 66° Baumé sulfuric acid should be diluted with water to 45° Baumé. In the water that is used to dilute the acid dissolve enough copper or nickel sulfate so that each gallon of the strip will contain 3 or 4 of either salt. This will prevent the strip from producing a pitted or rough surface to a great extent.

Use a lead lined tank or stoneware crock with lead anodes and 6 to 8 volts.

Salve for Nickel Plater's Itch

This annoying malady is caused mainly by uncleanliness and is very irritating especially if allowed to continue for some time without some form of treatment. As the first requisite is cleanliness, it is advisable to bathe the body freely, and also to take some form of cathartic. A tablespoonful of Rochelle salts dissolved in a glass of warm water, taken each morning before breakfast is beneficial.

The afflicted parts of the body should receive the following treatment:

Formula No. 1

Sodium Bicarbonate	4 oz.
Potassium Chlorate	2 oz.
Sodium Borate	2 oz.

Dissolve a tablespoonful of No. 1 in 2 quarts of warm water and bathe the afflicted parts, bathing for about 15 minutes, keeping the solution as warm as you can stand it. Then dry with a warm towel. Do not rub, but allow the towel to absorb the moisture. Rubbing will irritate and may spread the infection. Then immediately apply the following by rubbing lightly:

Formula No. 2

Lanolin	4 oz.
Olive Oil (sterilized)	1 oz.
Thymol Iodide	1 dram
Starch	Sufficient

This ointment will penetrate very readily and give almost instant relief. If treatment is used regularly once or twice a day, it will soon cure this ailment which is so prevalent among the help that is employed in the plating departments.

Plating Aluminum with Nickel

After degreasing with organic solvent, the oxide film is best removed by pickling in a hot acid solution of ferric chloride (containing 50 cubic centimeters of ferric chloride solution, specific gravity 1.42 and 50 cubic centi-meters of hydrochloric acid, specific gravity 1.18, per liter). The nickel-plating bath may contain (a) nickelous sulphate heptahydrate, 125 grams per liter, ammonium sulphate 20 grams per liter, sodium sulphate 30 grams per liter, magnesium sulphate heptahydrate, 30 grams per liter, sodium chloride 5 grams per liter; or (b), for rapid plating, nickelous sulphate heptahydrate 400 to 450 grams per liter, nickelous chloride hexahydrate 22 grams per liter, boric acid, 22 grams per liter, with a little nickelous nitrate to prevent pitting. Bath (a) has a pH of 5.5 to 6.5 and is operated at 35° to 45° with 2 to 4 amperes per square decimeter, while (b) has a pH of 5.3 to 5.7, and is operated at 45° to 50° with a current density of 3 to 10 amperes per square decimeter. Nickel-plated aluminum can readily be soldered with tin to give strong joints.

Rhenium and Molybdenum Plating

Rhenium deposits are recommended where hardness is of importance. It is best deposited from a sulphate bath containing 1 gram of rhenium per liter and requiring 4–6 volts; it develops a noxious mist. Molybdenum is deposited from a solution of 10 grams ammonium molybdate and 10–20 grams of ammonium nitrate in 1 liter water, with 0.2–0.3 amperes per square decimeter at about 2 volts; a fine black deposit is obtained.

Black Nickel Plating on Steel Tubing

The steel tube is polished on a belt polishing machine with No. 240 belt to produce an even surface. It is then copper plated in a warm cyanide copper solution for 30 to 45 minutes to obtain a heavy copper deposit. It is then run through the belt polishing machine again, using a finer belt than the one used on the steel; a No. 320 will produce the finish.

The work is plated in the regular nickel solution for about 3 or 4 minutes, and then plated in the following black nickel solution:

Double Nickel Salts	8 oz.
Sodium Sulphocyanide	2 oz.
Zinc Sulphate	1 oz.
Water	1 gal.

Use 1 volt; keep both at 80° F. and pH at 5.6. Plate work for 20 to 30 minutes, dry with alcohol and cold hardwood sawdust; coat with a lacquer that has a high gum content.

Nickel Plating Stainless Steel

To plate stainless steel it is necessary first to etch the surface very lightly to obtain adherence of the deposit. A dip for this purpose is made by using 1 pound of nickel chloride dissolved in ½ gallon of water; after the nickel chloride is dissolved add ½ gallon of hydrochloric acid.

The work is cleaned and immersed in this dip for half a minute, rinsed in water, then nickel plated.

Plating Stainless Steel

To produce an adherent electro-deposited metal coating on stainless steel, the passive film on the steel surface must be removed. The use of a pure, 6-normal, hydrochloric acid dip gives uncertain results, while the use of a 12-normal hydrochloric acid resulted in good adhesion. This second solution also contained 45 to 60 grams per liter of arsenic trioxide and 15 grams per liter of ferric chloride. The best adherence was obtained with a nickelous chloride dip containing 120 grams per liter of nickelous chloride hexahydrate in 6-normal hydrochloric acid, the stainless steel being processed as follows: electrolytic cleaner, rinse, nickelous chloride dip for 30 seconds, rinse, nickel plate from an ordinary nickelous sulphate—nickelous chloride plating solution for 8 minutes, followed by a chromium coating to obtain a lustrous non-tarnishing finish. Other metals, for example, gold, copper, brass, and bronze can be plated on the stainless steel after the nickelous chloride dip for the production of decorative finishes.

Electrodeposition of Platinum Metals

The current efficiency in the electrodeposition of platinum metals is at present still very poor and cannot be compared with those for nickel, copper, and silver which range between 90% and 100%. One of the palladium baths employs a solution of 4 to 10 grams of potassium palladonitrate (obtained from potassium palladonium chloride and potassium nitrate) in one liter of water to which is added 2 to 3 times the amount of sodium citrate. From 2 to 4 volts and a current density of 0.5 to 1.5 amperes per square decimeter are used at room temperature. For the deposition of rhodium alkali nitrates are employed or sodium sulphate in weakly acid solution with about 5% phosphorus pentoxide, oxalic acid or sulphuric acid.

Tin Plating
British Patent 395,377

In the use of tin plating baths containing 0.25–0.65 mole of alkali stannate per liter, and 1.6 moles of caustic alkali per mole of stannate, plus an additional amount of caustic alkali of 0.2–0.4 mole per liter. The alkali content is corrected when it has become too high by the addition of (a preferably weak) acid, for example acetic acid and at about the same time, an oxidizing agent, for example hydrogen peroxide, alkali metal peroxide, permanganate, persulphate, or perberoate. Thymolphthalein is preferably used as an indicator in the determination of the caustic alkali which includes that set free by hydrolysis of the stannate. The plating temperature is preferably 60 to 80 degrees and the current density 10 to 100 amperes per square foot at the cathode and up to 20 amperes per square foot at the anode.

Armor Plate

A steel alloy containing about 0,12–0.45 carbon, chromium 2–5% and molybdenum 0.15% to 1% with or without, up to 3% nickel, is cast, without subsequent rolling. The molybdenum may be partly replaced by 2 to 3 times as much tungsten. Heat treatment may comprise hardening and then reheating, preferably to not over 720°.

Tin Plating Copper Tubing
Formula for Alkaline Tin Solution

Sodium Stannate	12	oz.
Caustic Soda	1	oz.
Sodium Acetate	2	oz.
Sodium Perborate	⅛	oz.
Water	1	gal.

Anodes, Straits tin; ratio anode to cathode area, 3 to 1; cathode current density, 10 to 15 amperes per square foot; temperature, 140° F.; voltage, 3 to 4.

Work must be thoroughly clean before plating.

Tinning Pins

The copper film should be removed, and the work tumbled bright before placing in the tin solution. The copper can be removed by using a hot cyanide solution made of 6 ounces sodium cyanide to a gallon of water. They may be tumbled bright by placing in a bag made of 10 or 12 ounces duck with some fine rouge and Vienna lime, and rolled dry.

Procure a galvanized iron pan about 14 by 18 inches, 3 inches deep. Have some iron wire trays made to fit into the pan, and place the work in thin layers on the iron trays. Then cover with a perforated zinc sheet. The zinc sheet should be of ⅛ inch stock, and the pan may be filled with alternate layers of work and zinc sheets. They are kept boiling for 45 minutes in the solution given below; rinsed thoroughly in clean cold water; and dried in hardwood sawdust. If not bright enough, they may be tumbled with steel balls and soap water for a few minutes. Use the following solution:

Stannous Chloride	2 oz.
Sodium Sulphate	20 oz.
Water	1 gal.

In preparing solution, dissolve the chemicals in about ⅓ of the hot water, and then add this to the remaining ⅔ of the hot water which has first been placed in the galvanized iron pan. Then add 1 drop of sulfuric acid to each gallon of solution, and boil. The solution is used for only one batch of work, and it is prepared fresh each time.

Tinning Pins

Method No. 1

Make up a solution of stannous chloride 2 ounces, sodium sulfate 21 ounces. In preparing the solution dissolve the required amount of these salts in about one-third of the final volume of hot water, and add this solution to the remaining two-thirds of hot water which has been placed in the iron tank that contains the work to be plated. The articles to be tinned are placed in thin layers on iron trays covered with perforated zinc sheets in a galvanized iron tank. The work is left in the solution, which is kept near boiling temperature, for 45 minutes. The addition of a very small amount of hydrochloric acid (about 1 drop for each gallon of solution) will make the solution work faster.

Method No. 2

The other method requires a solution of ammonium alum, 2 ounces, cream tartar, 2 ounces, chloride tin, ½ ounce, water, 1 gallon.

A copper vessel is used with this method, with sheets of zinc placed across the bottom and on the sides.

The work is placed in thin layers on iron trays, and covered with perforated zinc sheets. The solution is used boiling hot, and the operation takes 30 to 45 minutes to tin plate the work.

Tinning Acid

Muriatic Acid	50 cc.
Zinc Metal	50 gm.
Saturated Solution of	
Salammoniac	87 cc.

After the acid shows no further action on the zinc, mix the salammoniac solution and muriatic solution together. When used on cast iron, steel or brass of any analysis, a very good heavy coating of tin can be obtained. It is quite hard to tin cast iron so that there are no defects in the tin surface. However, with this fluxing solution used on the cast iron two or three times during the tinning operation, a perfect job can be obtained.

Thermoelectric Element and Thermocouple

U. S. Patent 1,920,559

The couple consists of a 45:55 nickel-copper alloy and an alloy of antimony 60–63, zinc 34–37, and arsenic 1–4.5. A difference of 300° between the two junctions gives an electromotive force of 0.09 volt.

Thermoelectric Element

U. S. Patent 1,947,595

A negative thermoelectric element consists of 61 to 64% antimony, 35 to 37% zinc, and 0.5 to 2% beryllium.

Electrolytic Deposits on Aluminum

The metal must first be degreased in a 90° hot solution of 7.5–22.5 grams each of sodium phosphate and sodium carbonate per liter and, after rinsing in cold water, immersed in a 5% bath of hydrofluoric acid. The surface is then roughened in a sand blast, or better, in a pickling bath, to make the deposit adhere well. The nickel solution consists of 150 grams nickelous sulphate, 75 grams manganese sulphate, 15 grams of ammonium chloride, and 15 grams of boric acid per liter; temperature 35°; current density 1.5 amperes per square decimeter; pH 5.8–6. A minimum thickness of 0.0125 millimeter is recommended. Any other metal may be deposited on the nickel; zinc and chromium may be deposited directly on aluminum. Nickel-plated aluminum is not suitable for further working; heat treatment (1½ hours at 130°; then 3 hours at 315°) improves its behavior especially for articles which are exposed to high temperatures; gradually increase to 315° and hold this temperature for several hours.

Electrolytic Surface Treatment of Aluminum

British Patent 395,390

The metals are treated anodically in a bath containing an aluminum salt as the initial component. A suitable electrolyte for a soft coating is obtained by mixing 160 liters of dilute sulphuric acid containing 2½ parts of acid to 2 parts of water, with 20 liters of a 20% aluminum sulphate solution; a lower strength being used for a hard coating. The voltage may be 80 to 100 and the temperature 60 to 65° F. A small quantity, for example, 1%, of a water-dispersible albumenoid colloid or a carbohydrate or polysaccharide such as dextrin, gum tragacanth, gum acacia, agar, etc., is added for soft coatings. Organic dyes or metallic salts imparting color to the coatings may be added or the coated articles subjected to solutions thereof.

Plating on Aluminum Castings

Clean the article in following for about 1 minute:

Sodium Carbonate	2 oz.
Trisodium Phosphate	2 oz.
Water	1 gal.

Use above at 180° to 200° F.
Rinse in cold water.

Etch the article for 20 seconds in the following formula stated in parts by volume:

Concentrated Nitric Acid	3 parts
Hydrofluoric Acid (50%)	1 part

Mix above with a wooden stick in a lead lined container painted with the following formula, stated in parts by weight:

Beeswax	1 part
Paraffin	4 parts

Rinse in cold water.

Plate in solution such as the nickel solution given in volume 1 of this book.

Anodic Finishing of Aluminum

The chromic acid method for the anodic treatment of aluminum makes use of a 10% solution of chromic acid. The current is raised gradually to 45 volts, where it is kept long enough to produce an oxide of aluminum.

When the work is suspended in the bath, the points of contact must be positive. Aluminum racks or wire are used to support the work during treatment.

Anodic Treatment for Aluminum Bath

U. S. Patent 1,933,301

Ammonium Phosphate	10 %
Ammonium Vanadate	0.1%
Water	Balance

Anodic Oxidation of Aluminum

Aluminum or its alloys are cleaned in any of the standard cleaners to obtain a surface free of dirt, oil or grease. It may then be oxidized by either of the following methods:

Chromic Acid Method

Use a 3% solution of chromic acid free from sulphates or chloride. The aluminum object is made anode in this bath, increasing the voltage gradually from zero volts to 50 volts.

Sulphuric Acid Method

For hard coatings use a 10% solution. For soft coatings use stronger acid solutions up to 70%. Aluminum is made anode in these baths with an applied potential of 15 volts. The temperature of the bath should be between 15 and 30° C. It is desirable to dip the treated article, after a rinse in water, in dilute ammonia.

If desired the aluminum oxide formed may be dyed by immersing the article in a dye bath at 80° C. for about twenty minutes. The dyes used are generally of the alizarin group. Many oxide dye stuffs are also suitable.

Plating Without Electricity

The following mixed powders are rubbed with a wet cloth onto the surface to be coated.

Nickel

Double Nickel Salts	67%
Magnesium Powder	3%
Powdered Chalk	30%

Tin		Zinc	
Stannous Chloride	15%	Zinc Dust	45%
Ammonium Sulphate	15%	Ammonium Sulphate	15%
Magnesium Powder	3%	Magnesium Powder	3%
Powdered Chalk	67%	Powdered Chalk	37%

POLISHES, ABRASIVES, ETC.

Automobile Polish

Sulphonated Olive Oil (30%)	10%
Mineral Oil	30%
Pumice Powder	5%
Amyl Alcohol	10%
Carnauba Wax (Powdered)	5%
Water	40%

Mix the oils, wax and alcohol; add the water slowly with good stirring, mix in other ingredients.

Auto Polish

Kieselguhr	8	parts
Paraffin	8	parts
Methylated Spirit	2	parts
Glycerin	2	parts
Water	40	parts
Gum Tragacanth	¼	part

The glycerin can be omitted if desired, as, although it possesses qualities to recommend it, it has a tendency to leave a slightly sticky surface.

Polishing Emulsion for Automobiles and Furniture

Carnauba Wax	120	parts
Kerosene	50	parts
Stearic Acid	15	parts
Oleic Acid	3	parts
Benzaldehyde	6	parts
Triethanolamine	8	parts

Melt the carnauba wax, stearic acid and oleic acid. Remove from fire and add the kerosene and the benzaldehyde. Stir thoroughly. Now add a hot solution (80–85° C.) consisting of 240 parts of water and the amount of triethanolamine as given above. Stir well until a smooth emulsion is formed then add sufficient hot water to make 825 parts. Continue stirring until nearly cool then bottle. This polish will form a permanent emulsion which is stable at all temperatures.

Automobile Polishing Powder

Oleic Acid	2	lb.
White Mineral Oil	12	lb.
Carnauba Wax	2½	lb.
Yellow Beeswax	2	lb.
Water	28	lb.
Triethanolamine	13	oz.

Infusorial Earth	100	lb.
Gum Tragacanth (Powdered)	4	lb.

Take the carnauba wax, bees wax, oleic acid and white mineral oil and melt at gentle heat. In another vessel heat to near boiling point the water and the triethanolamine. Add the oil-wax mixture to the water mixture in a slow stream under steady stirring and retain at near boiling temperature until a smooth emulsion has been formed. Then remove from fire and continue stirring until nearly cold. Now add to the emulsion the infusorial earth, mix well and let dry until all the water has been expelled. Then add the powdered gum tragacanth, mix thoroughly and sift. 8 to 12 ounces of the powder are sufficient to make 1 gallon of polish by merely adding sufficient water.

Automobile Polish

White Mineral Oil	80	lb.
Deodorized Mineral Oil	9	lb.
Stearic Acid	9	lb.
Triethanolamine	4	lb.
Methanol	4	lb.
Water	120	lb.

Mix the white oil and mineral oil and add the stearic acid. Heat the mixture to about 140° F. at which time the stearic acid will dissolve to give a clear solution. In a separate container mix the triethanolamine, methanol and water and heat likewise to about 140° F. Then add the first mixture to this and stir vigorously until the emulsion is smooth. Continue to stir gently until cool.

Furniture Polish

This gives an oil polish of the highest type. After an hour it does not smear or "attract" dust.

Naphtha	10	parts
Thin Mineral Oil	72	parts
Perilla Oil	30	parts
Nelgin	7	parts
Water	88	parts

Soak Nelgin in water for a few hours.

Stir until dispersed. To this add oil mixture slowly with rapid stirring.

Polishing Cloth

Crude oleic acid, 1 pound; stearic acid, ½ ounce; vaseline, 1 ounce. Melt together, remove from fire and add cassia oil or methyl salicylate, or terpineol, ½ ounce. Cut good weight canton flannel into desired size (½ yard by 1 yard), dip in the mixture till thoroughly saturated, then run through a tight wringer. Fold and wrap in oiled paper.

Oil Polish for Furniture

This composition produces a high gloss on varnished and lacquered surfaces and at the same time cleans the surface.

This formula is stated in percentage by weight:

Spindle Oil	51%
Fatty Acids	12%
Pine Oil (Steam Distilled)	8%
Benzol	16%
Methanol	12%
Nitrobenzene	1%
Red Dye (Oil Soluble Type)	to color

The spindle oil is a light bodied lubricating oil of less than 100 seconds Saybolt at 100° Fahr. The fatty acids may consist of a mixture of oleic and stearic acids as obtained from cotton seed oil. Single pressed stearic acid alone serves the purpose. The fatty acids are melted and added to the spindle oil and this mixture added to the other ingredients.

Furniture Polish
Canadian Patent 339,547

Diamond paraffin oil 12, white gasoline 6, asphaltum varnish 4 quarts, lilac oil 6 fluid ounces and varnish 1 pint thoroughly mixed by agitation.

"Removing" Scratches from Furniture

Just take a pecan kernel (right from the shell as for eating), and rub it on the scratch. The scratch will disappear like magic. Pecan oil has no equal as a furniture polish. Grind up a few kernels (not too many, as Pecans are very full of oil), and rub them into a piece of cheesecloth, which is then rubbed over the furniture. This oil does not spread easily, and the furniture requires good rubbing all over. But it does not catch dust and is lasting. The polish is dull, that is, not a high shine. The cloth will last for years.

Furniture Finishing Oil

Raw Linseed Oil	10%
Paraffin Oil (100 seconds viscosity Saybolt)	60%
Turpentine	15%
Benzine	15%

Furniture, Lacquer Enamel and Polished Nickel or Chromium Wax Polish

Carnauba Wax	12%
Beeswax	10%
Benzine	40%
Turpentine	13%
Xylol	20%
Paraffin Oil	5%

Floor Wax, Liquid

Carnauba Wax	1	lb.
Paraffin Wax	¼	lb.
Raw Linseed Oil	½	pint
Turpentine	½	pint

Melt up together and heat to 85° C. Add a gasoline-kerosene mixture (1:2) or naphtha with thorough stirring to make a total volume of 2 gallons. Add 1 ounce of ammonium linoleate and stir thoroughly until cool. If the mix fails to emulsify, heat up again and stir well until cool.

Non-Slippery Floor Wax

One pound candelilla or carnauba wax; one-half pound Japan wax melted together by heat.

When melted, add gradually by stirring one gallon of light gravity paraffin oil. This mixture is to be mixed with four gallons more of the paraffin oil and one gallon of turpentine substitute (petroleum spirit). Kerosene or turpentine will do, but the odor will not be pleasant, except if some essential oil is added to hide the odor. The compound will not be slippery. This mixture can also be used as a reviver or polish for any varnished surface.

Homemade Floor Wax

Beeswax	¼	lb.
Paraffin	1	lb.
Raw Linseed Oil	¼	pint
Turpentine	1½	pints

Melt the beeswax and paraffin, add

linseed oil and turpentine, and stir mixture vigorously. Unfinished wood will be darkened somewhat by this wax as a result of the absorption of the linseed oil.

Floor Finish
Canadian Patent 339,024

A floor finish comprises as a base, linseed oil 313, tung oil 94, gloss oil 130, naphtha 365 gallons, paraffin wax 724, zinc stearate 50 lb. and 50% by volume of xylene.

Floor Sweeping Compound
For Polished Floors

Sifted pine sawdust, 1 bushel; clean sand, 1½ pecks; common salt, 1½ pecks; paraffin oil (floor oil), 4 quarts; paraffin wax, 8 ounces. Dissolve the paraffin oil and wax hot. Aniline color (oil-soluble) red or green, sufficient. Pine oil, 4 ounces; eucalyptus oil, 8 ounces. Saturate the sawdust with this, then stir in the salt and sand. For unfinished floors, same but use 5½ quarts of the paraffin oil.

Bright Drying Wax Polish
Formula No. 1

Carnauba Wax No. 1	10 g.
Ammonium Linolate Paste	3 g.
Borax	2 g.
Water	90 cc.
Turpentine	1 cc.
Pine Oil	¼ cc.

Heat the wax, turpentine and oil to 100° C. in a jacketed vessel fitted with a high speed stirrer. To this add slowly the paste, borax, and water previously heated to 100° C. Stir vigorously until emulsified.

Formula No. 2

Carnauba Wax No. 1	20	g.
Ammonium Linoleate Paste	8	g.
Ammonium Hydroxide	7½	cc.
Water	180	cc.
Ammonium Hydroxide	7½	cc.

Heat the wax to 100° C. in a jacketed vessel fitted with a high speed stirrer and to it add slowly the paste, water, and 7½ cubic centimeters of the ammonium hydroxide dissolved and heated to 80° C. As soon as emulsified add the balance of the ammonium hydroxide while stirring virorously.

Rubless Wax Polish

Hydromalin	138 lb.
Carnauba Wax No. 1 Yellow	250 lb.

Heat to 120–140° C. half hour. Cool to 100–105° C.

Add to the above slowly with stirring 280 pounds of water heated to 100° C. Stir to smooth paste then add slowly with good stirring 1500 pounds water. Keep as close to 100° C. as possible for 15 minutes. Cover and allow to cool to 100° in temperature. Filter and pack in glass or lacquered cans.

Easy Rubbing Wax Polish

Carnauba Wax	60	oz.
Dry Castile Soap	15¼	oz.
Oleic Acid	2¾	oz.
Caustic Soda	1½	oz.
Water to make	4	gals.

Heat the wax, soap, and acid to 100° C. and while stirring rapidly add the caustic dissolved in the water, heated to 80° C., slowly. A beautiful stable creamy emulsion results.

Wood-Preserving and Finishing Compound
U. S. Patent 1,907,796

The preservative comprises creosote oil 4 parts, ethyl alcohol 1 part, and turpentine 1 or 2 parts. If desired, a filler to act as finish may be incorporated, comprising white-lead 3 and wood-filler paste 4 parts.

Dustless Mop Oil

Paraffin Oil	2 parts
Turpentine	1 part
Oil of Cedar	Amount desired.

Stove Polish
Formula No. 1

Kerosene	38.7%
Asphaltum	14.3%
Lampblack	3.0%
Mexican Graphite	44.0%

Formula No. 2

Benzol	19 %
Kerosene	19.7%
Asphaltum	14.3%
Lampblack	3.0%
Mexican Graphite	44.0%

Window Polish

Precipitated Chalk	10 parts
Ground Quassia	1 part
Ammonium Carbonate	1 part

These should be ground together, and are used by moistening water as required, the result being to impart a smooth, high glaze.

"Italian" Powder for Polishing Marble

Alumina is used as a substitute for putty powder with good results. The product is prepared by complete precipitation of a saturated solution of alum with ammonium hydroxide (density 0.91), the mixture is allowed to stand 18 hours, the precipitate is filtered off, dried, heated, comminuted and sifted to a diameter of 2.3–4.7.

Colored Burnishing Clay

A typical formula for the preparation of red or blue clay to be brushed or sprayed on (like paint) is as follows:

Clay	1	lb.
Warm Water	1½	lb.
Cooked Glue	1	lb.
Denatured Alcohol	1	oz.

The cooked glue should be made up in the proportion of about ¼ pound of glue to ¾ pound of water, soaked overnight and cooked in a double boiler until dissolved. This formula may need to be changed slightly to put the material in the best condition for service at the time it is used. Experience and good judgment are essential here and in fact all along the line. Put the clay into a suitable container (for example, a gallon can), pour in warm water and mix thoroughly. Add the glue a little at a time, stirring it constantly. Add the alcohol last; this clay should be applied the same as the whiting, especially in reference to keeping it warm. A better burnish will be obtained if two coats of clay are applied. The red clay will be a dark brick red when it is sprayed on, and will dry to a dull rose color.

The blue will dry to a sort of pale flat blue. The color coat should dry thoroughly before the bronze is applied.

Sharpening Compositions
British Patent 397,529

An abrading composition for application to the rollers of a razor-blade-sharpening machine comprises finely divided corundum 70, ozokerite 7 and paraffin 23 parts. The corundum may be partly replaced by powder ferric oxide.

Silver Polishing Soap

Cocoanut Oil	25 kg.
Caustic Soda (36° Bé.)	13 kg.
Precipitated Chalk	10 kg.

The chalk is first added to the caustic soda solution and this is then saponified with the cocoanut oil.

Copper Polish

This gives excellent results with a minimum of labor.

Heavy Paraffin Oil	1	quart
Wood Alcohol (Methanol)	1	pint
Spirits of Camphor	½	pint
Turpentine	½	pint
Aqua Ammonia	¼	pint
Infusorial Earth (Fine Powder)	3½	lb.

Pour ammonia into the paraffin oil; then add wood alcohol and turpentine, and finally camphor.

Then, while stirring, add infusorial earth. Tripoli may be substituted for the infusorial earth.

Metal Polish for Chromium or Silver

Orthodichlorbenzol	400 cc.
Water	70 cc.
Strong Ammonia	1 cc.
Saturated Solution of Castile Soap	2 cc.
Levigated Alumina	35 g.

If carefully made a stable emulsion results.

For brass or copper polish substitute precipitated chalk for levigated alumina. If a more abrasive polish is desired substitute kieselguhr for precipitated chalk.

Brass Polish

Triethanolamine Oleate	1	lb.
Potassium Oleate	4	oz.
Naphtha	1	gal.
Ammonia (28° Bé.)	7	fl. oz.
Silica (300 Mesh)	3½	lb.

Mix thoroughly the silica and naphtha then stir in thoroughly the two oleates and finally the ammonia. This makes a superior metal polish which gives a very fine finish on brass. If carefully prepared, there will be no separation of the ingredients on long standing.

RESINS, GUMS AND WAXES

Identifications of Resins

A sample of the resin is dissolved in acetic anhydride in a test tube, and a drop of concentrated sulphuric acid allowed to run down the wall of the tube into the cold solution and the immediate color change noted. After standing twelve minutes and diluting with benzine or similar water-free solvent the color is again noted. The resulting characteristic color reactions are given in the following table:

Resin	Acetic Anhydride Solubility in	Color with Concentrated Sulphuric Acid Immediately	After 12 minutes (diluted)
Dammar	d. cloudy	light red	olive brown
Shellac	e. clear	yellow	yellow
Accroides	e. clear	dark red	reddish brown
Manila	e. clear	dark brown	olive brown
Rosin	e. clear	purplish red	brown
Gutta Percha	e. cloudy	red brown	reddish brown
Amberol	d. cloudy	purplish red	olive brown
Paramet Ester Gum	d. cloudy	purplish red	olive brown
Rezyl	e. clear	light orange	light olive

e. stands for "easily soluble" and d. for "difficultly soluble."

Rosin Resins

100 parts of a 25% solution of boron fluoride in technical crude cresol are added over a period of 4 hours with continued stirring and cooling to a mixture of 300 parts American rosin and 400 parts carbon tetrachloride. The temperature is maintained in the vicinity of 10° C., when a dark brown viscous mass is formed which is stirred at room temperature for a further 14 hours. After pouring the mass into water and treating with a current of steam a faint-yellow brittle resin is formed which melts at 105° C., as compared with 79° C. for the original rosin.

Advantage can still be taken of the reactivity of rosin with glycerine, since the acid groups of the rosin molecule are not completely reached in the above condensation with cresol. Thus 400 parts of the clear resinous product (softening point 105° C.) are esterified with 31 parts of glycerine at 250 to 260° C. Nor are the possibilities of raising the softening point exhausted by this operation, since condensation with 16 to 17 parts of paraformaldehyde for 3½ hours at 180 to 250° C. leads to the formation of a transparent resin with a softening point of 145°

C. This substance is difficulty soluble in cold alcohol, but dissolves readily in benzine and linseed oil.

Another variation of the process consists in starting from a glycerine ester gum (softening point of 83° C.) which is condensed with a phenol in the presence, as before, of boron fluoride. In this case 400 parts of the ester gum mixture with 450 parts carbon tetrachloride are treated with 100 to 120 parts of a 30% solution of boron fluoride in crude cresol. Further refinement is effected on the lines of the original process, and the final yellowish-white clear resinous product possesses a softening point of 98° C.

Instead of applying the paraformaldehyde condensation to the products of reaction between cresol-rosin product and glycerine the former can be directly condensed with the aldehyde, again yielding a product with a much higher softening point. For example, 390 parts of the condensation product of rosin with crude cresol are treated at 170° C. with 12 parts of 100% paraformaldehyde, several hours being taken over the addition of the aldehyde. The softening point of the new resin is 130° C.

Diene Resins
U. S. Patent 1,947,416

As an improvement in the manufacture of resins from phthalic anhydride, glycerin and fatty acids to form special resins soluble in coal tar hydrocarbons, the improved process which comprises distilling castor oil fatty acids under a reduced pressure of approximately 30 millimeters at temperature of about 260 to 270° C., collecting the distilled fatty acids thus obtained, mixing between 50 and 100 parts of said distilled fatty acids with about 100 parts of phthalic anhydride and 50 parts of glycerin, heating the mixture to between 180 and 230° C. until the mixture becomes clear continuing the heating until an elastic, light-colored, substantially non-tacky resin is obtained upon cooling and then cooling the mixture to obtain the resin, said resin being soluble in benzene, toluene and like coal tar oils.

Molding Powder Resin
U. S. Patent 1,949,831
Formula No. 1

300 grams of granulated white sugar are dissolved in an aluminum vessel in 250 cubic centimeters of 36% formalin solution at a sufficiently low temperature to prevent caramelization of the sugar. Heat is applied to the vessel and the temperature gradually raised. As the temperature increases there is a slow evolution of formaldehyde gas until the temperature reaches a point when there is a vigorous and very large increase of ebullition which continues for a few moments and then, despite further increase of the temperature, all evolution of gas ceases.

After the reaction between the sugar and the formalin has been completed, the temperature is raised still further and higher than the caramelization point of the sugar without discoloration or darkening of the fluid in the vessel. The purpose of this increase of temperature is to promote the expulsion of water vapor and is accompanied by boiling of the liquid. During this boiling 10 grams of hexamethylenetetramine is added as a hardening agent, the liquid being stirred sufficiently to get the hexamethylenetetramine into solution. After the hexamethylenetetramine is fully dissolved, the temperature is further increased with increased boiling. There is now added to the solution in small install-

ments 75 grams of urea. This is sprinkled gently on the surface of the reaction mass in the vessel in twelve equal installments of about 6 grams each, with constant agitation and taking care to see that each installment is fully dissolved before the next is added. This takes about half an hour and during this period the heat is maintained considerably above the boiling point of water and stirring is continuous. When the last of the urea has been dissolved the reaction mass is at once poured into molds.

Formula No. 2

The process described under Formula No. 1 is carried out in all respects as stated, except that the addition of the hexamethylenetetramine is omitted. After the addition of the urea, it is found that the reaction mass rapidly thickens till it can not be poured; but when it is quickly transferred into molds before it has become unpourable and allowed to harden therein it hardens to a substance differing from the substance made in accordance with Formula No. 1 in that neither further heating nor prolonged contact with the air causes it to harden fully, and, on being broken up into powder and, hot pressed into molds, it continues to retain its yielding quality though being clear, transparent and water-white.

The reaction mass produced in accordance with Formula No. 1 was in one instance poured into a mold, in another instance poured on a polished metal plate for drying in contact with the air, and in another instance extruded as a rod.

The reaction mass of Formula No. 1 poured into a mold was found to harden to a transparent, water-white substance which was at first somewhat yielding and resilient, but on exposure to the air, after it had hardened sufficiently to permit the removal of the mold, became progressively tougher and more nearly the consistency of hard rubber as time went by. When broken, the interior was found to be somewhat less stress resistant than the exposed surface, but on exposure of the fractured surface to the air this assumed the hard characteristic of the other exposed surfaces. It was found that the mass fully and accurately reproduced every configuration of the mold, including minute scratches or marks.

The reaction mass of Formula No. 1 poured on a polished metal plate rapidly hardened in the form of a thin sheet.

Synthetic Resin
U. S. Patent 1,938,642

One molecular weight of tri-cresyl phosphate is heated to a temperature of 600–640° F. under a reflux condenser. At this temperature three molecular weights of calcium oxide are added to the hot tri-cresyl phosphate, the heating being temporarily suspended. As the vigor of the resulting reaction subsides the temperature declines. When the temperature falls to 475–500° F. the heating is resumed and continued until the reaction product, when cooled, solidifies to form a synthetic resin hard at ordinary temperature.

Tri-phenyl phosphate may be substituted for the tri-cresyl phosphate and the other oxides mentioned may be substituted for the calcium oxide in the foregoing example. Small amounts of tri-oxymethylene, for example, may be added during the resinification; in amounts corresponding to any phenol or cresol liberated by the reaction. Drying oils such as linseed oil and tung oil and natural resins such as the varnish resins and rosins may be incorporated in the product during resinification. The addition of drying oils in amounts approximating 5–10% (by weight) on the synthetic resin or of natural resins in amounts approximating 5–20% on the synthetic resin, for example, adds to the toughness of the product.

The synthetic resin produced in accordance with the foregoing example is light amber in color and water-resistant. It may be used in thermo-plastic molding and in varnish, lacquer and impregnating compositions. Fibrous materials impregnated with this synthetic resin may be cured under heat and pressure to form materials having excellent electrical and mechanical properties.

Synthetic Resin
U. S. Patent 1,890,668

Propylene Glycol	76 parts
Phthalic Anhydride	148 parts

Heat together while refluxing for 2½ hours up to 290° C.
A soft pale straw colored resin results which is soluble in butyl and amyl acetates. It is compatible with nitrocellulose.

Trimethylene Glycol	76 parts
Phthalic Anhydride	148 parts

Heat as above to obtain a similar product, suitable for use in lacquers.

2–3 Butylene Glycol	100 parts
Phthalic Anhydride	148 parts

Heat as above to obtain a softer resin which is completely soluble in toluol as well as in the acetates and is compatible with nitrocellulose.

Synthetic Resin
U. S. Patent 1,897,977

Phthalic Anhydride	444 g.
Glycerol	184 g.

Heat to 250–240° C. with stirring. Reaction is complete when a spongy mass results. This resin is insoluble in most organic solvents.

Phthalic Anhydride	20 parts
Oleic Acid	16 parts
Glycerol	7.5 parts
Zinc Oxide	.5 parts

Heat slowly, with stirring, up to 260° C. when a soft gummy, opaque brown resin results. This is compatible with nitrocellulose, giving transparent flexible films.

Synthetic Resin

Citric Acid	96 parts
Triethylene Glycol	113 parts
Rosin	151 parts
Sodium Bisulphite	3 parts

Heat gradually with stirring to 220–240° C.

Synthetic Plastic Resin
U. S. Patent 1,950,516

A mixture comprising 10 grams flake sodium phenate, 5 grains methylene dichloride, 0.5 gram ammonium carbonate and water is heated under pressure, to a temperature of 160° to 190° C. for 6 hours. The reaction mixture is then acidified, followed by steam distillation, after which the reaction product is washed, dried and powdered.

The resinous, condensation product may then be molded by any suitable procedure after the addition of any preferred ingredients.

Synthetic Resin Plasticizers
U. S. Patent 1,940,092
Formula No. 1

150 parts of castor oil and 50 parts of colophony are heated at a pressure of 15 millimeters mercury gauge at 260° Centigrade for 6 hours. A clear pale yellow very viscous oil is obtained.

By employing from 100 to 150 parts

of colophony for each 150 parts of castor oil in the condensation, the almost neutral oil obtained is still more viscous and is readily miscible with paraffin oil. It is eminently suitable as a softening agent for celluloid condensation products of formaldehyde with urea or phenols, casein and the like.

Hydroxy stearic acid esters or other similar oils or waxes containing hydroxy groups may be employed instead of castor oil.

Formula No. 2

15 parts of ricinoleic acid are heated together with 13.5 parts of colophony for 10 hours at 250° Centigrade at a pressure of about 20 millimeters mercury gauge. While water is split off and the acid value is reduced, a brown, highly viscous oil is obtained which is soluble in most organic solvents with the exception of the lower members of the aliphatic alcohol series.

Formula No. 3

3.4 parts of dihydroxystearic acid ethyl ester are heated together with 3 parts of abietic acid at 260° Centigrade in vacuo until the acid value has been reduced to practically zero. A resinous product is obtained.

Synthetic Sugar Resin
U. S. Patent 1,949,832

Formula No. 1

300 grams of granulated white sugar is dissolved in a vessel in 250 cubic centimeters of 36% commercial formalin solution. Heat is applied to the vessel and the contents are stirred until the sugar has dissolved. The temperature is then gradually raised taking care not to permit caramelization of the sugar until a temperature is reached at which there is a large evolution of gas and finally, without temperature reduction, the liquid becomes quiet. There is then added 10 grams of hexamethylenetetramine, with sufficient stirring to get it into solution. Thereafter the heat is gradually raised to the neighborhood of 180° C. and there is then added to the solution in ten equal installments at intervals of about a minute with constant stirring and agitation, 300 grams phthalic anhydride. Each installment is sprinkled gently on the surface of the fluid with constant stirring and agitation. The addition of phthalic anhydride produces substantial bubbling which con-tinues for about a minute after the addition of each installment.

After all the phthalic anhydride has been gotten into solution, the temperature is reduced to 90° C. and maintained there for several hours. After the heating at this temperature, it is found that the water-white characteristic of the reaction mass in the vessel has not materially changed. The liquid is now poured into molds and permitted to cool. It is found, after cooling, to have set as a brilliant substantially water-white non-resilient substance which is somewhat brittle and breaks with a sharp fracture. It is soluble in spirit solvents such as acetone, but insoluble in water.

Formula No. 2

The process of Formula No. 1 is carried out to the point at which the phthalic anhydride has been gotten into solution.

The temperature is now increased from 180° C. to the neighborhood of 250° C. and maintained there for several hours. During this time the color of the solution steadily darkens and at the end of ten hours the liquid has assumed a dark brown or blackish color.

This is poured into molds and hardened on cooling to a shiny brownish-black substance which is insoluble in water but soluble in spirit solvents.

Formula No. 3

225 grams of granulated cane sugar is placed in an autoclave and gradually heated with occasional agitation until the sugar has melted to a light amber fluid. Without diminishing the temperature there is then added 100 grams of solid paraformaldehyde, and the contents of the autoclave are agitated until the solution is completed. The temperature is then gradually increased and the reaction between the sugar and the formaldehyde is permitted to proceed under self-generated pressure. When the temperature has reached about 200° C. the pressure is released and there is added 10 grams of hexamethylenetetramine. The liquid is sufficiently agitated to get the hexamethylenetetramine into solution. There is then added 200 grams of phthalic anhydride in ten installments or portions with constant agitation. The solution froths and bubbles noticeably and its color, already amber yellow, deepens somewhat.

When the phthalic anhydride has

been gotten into solution, the liquid is at once poured without further heating into molds where, on cooling, it is found to have hardened to a shiny brownish resinous substance which is brittle, breaks with a brilliant fracture, is insoluble in water, but soluble in spirit solvents.

Formula No. 4

The poured substance made in accordance with Formula No. 1 is, after cooling, broken up and dissolved in acetone. It is found to go into solution readily and the concentration is so adjusted as to produce a thick viscous paint-like solution. This, being substantially water-white, is found to be an excellent crystal varnish which is waterproof and has good covering properties.

Formula No. 5

The varnish made in accordance with Formula No. 4 is sprayed on a metal surface and then baked thereon at 300° C. for several hours. It is found to have darkened in color to a deep brownish black, and is hard, elastic, and tough. It is found to be durable even in thin application and is impervious to water.

Synthetic Terpene Resins
U. S. Patent 1,939,932

Using one mol pinene to two mols toluene within a suitable container or polymerizing vessel, and an activating compound such as powdered anhydrous aluminum chloride is added while the mixture in the vessel is being agitated. The aluminum chloride is preferably pulverized to give better contact and to increase the rate of and ease of solution. The activating compound is added in small quantities at a time while the contents of the polymerizing vessel are being agitated. It is found that the activity of the catalyst is directly proportional to the amount which goes into solution in the reaction mixture. It is therefore desirable that the solution should be as complete and as rapid as possible. For this reason rapid agitation during the addition of the catalyst and during the polymerization reaction is desirable. When 25 gallons of a mixture of active ingredients are to be treated within the polymerizing vessel, aluminum chloride may be added in quantities of approximately 6 to 8 ounces at a time. With the addition of the first batch of aluminum

chloride there is a rather active chemical reaction with a resultant rise in temperature. Where an atmospheric polymerizing vessel is used it is desirable that the temperature of the reaction be controlled so that it is not permitted to rise much above 40 degrees C. Otherwise the polymerizing vessel should be kept under pressure to prevent undue volatilization and loss of resin. Polymerization at higher temperatures, such as above 65° C., requires a special pressure vessel. Very satisfactory results are secured when the temperature is controlled between 25° and 35° C., this being readily accomplished by introducing the aluminum chloride in small amounts with proper agitation to prevent local overheating and by cooling the reaction mass by a suitable cooling jacket. This polymerizing reaction is preferably carried out in the absence of water. Water present during the reaction hydrolyzes the catalyst to form an acid which in turn affects reaction, resulting in a darker and quite different resin product.

Additional batches of aluminum chloride are added from time to time as may be done without unduly increasing the temperature of the reaction mass, and this is repeated with continuous agitation until no further temperature rise results. The amount of aluminum chloride used is also controlled in accordance with the materials being treated as this materially affects the yield and character of the resin. As the amount of catalysts used is increased, the yield of resin is found to increase, while the resin tends to be harder, lighter in color and lower in iodine number. The quantity of catalyst needed for producing the particular grade or quality of resin desired can be readily determined by tests for the particular ingredients being treated. Thus, when treating the above mixture in 25 gallon batches, satisfactory results are secured by the addition of about 10 pounds of catalyst, this being equivalent to approximately 5.0 grams of catalyst for 100 cc. of active ingredients.

Thermoplastic Resin Compound
U. S. Patent 1,953,951

Natural shellac, preferably purified, is suitably ground or powdered and mixed with preferably from about 30 per cent to about 50 per cent by weight of finely powdered zinc oxide, or with a suitable proportion of some other basic zinc compound, in the absence of water. The mixture is gently and uniformly heated, care being taken not to overheat locally, while the mixture is kept thoroughly stirred. Or the shellac may be melted and the desired quantity of dry zinc oxide or other dry zinc compound thoroughly mixed therewith. At a temperature slightly above 100° C., fluidity is rapidly developed due to the melting of the shellac, and the mass becomes a viscous liquid. At this stage the color is a light coffee brown, being simply that produced by mixing the dark brown of the shellac with the white of the zinc compound. Commercial zinc oxide, which is substantially anhydrous, may be used.

As the heating is continued the temperature gradually rises, and at 120–150° C. the color quickly changes to a characteristic dark pink, and gas begins to be evolved. These changes have been understood to indicate the development of a chemical reaction between the two materials. That such a reaction actually does take place is proved by the remarkable non-additive changes in the properties of the new resin, as described herein.

Ordinarily, heating of the mass is continued until a product is obtained having the desired softening range of temperature. That is to say, the reaction is a progressive one so far as properties such as softening range are concerned. If a low softening range is desired, heating is discontinued as soon as the reacting constituents have been brought into complete physical contact and enabled to form the desired chemical union, but if a maximum softening range is desired the heating is continued at above 200° C. for the necessary time. For example, if after being heated at about 150° C., for some minutes the mass is cooled, it solidifies at a temperature in the neighborhood of 130° C. to a hard, tough moderately insoluble resin, of homogeneous properties and differing markedly from shellac in those properties. Thermoplasticity without dry spongification has been developed, inasmuch as the resin can be repeatedly heated to softness and cooled to hardness without change in structure. It has developed considerable insolubility in organic solvents, those tried including alcohols such as methyl and ethyl alcohols, ethers such as ethyl ether, ketones such as acetone, hydrocarbons such as benzene, toluene and various petroleum distillates, terpenes such as turpentine, glycols such as

ethylene glycol, and carbon disulphide and carbon tetrachloride. Slight surface softening without dissolving has been noticed with certain solvents.

Physically, it is extremely hard and tough and has a high breaking strength for a non-fibrous moldable organic material. Its surface is vitreous and cannot be scratched with the fingernail. Its fracture is conchoidal and its internal structure homogeneous.

In this intermediate stage, therefore, the new resin shows all its definite unique properties. These may now be enhanced by further heat treatment, which causes the reaction to progress as indicated by the continued evolution of small amounts of gas. As heat is further applied, therefore, and the temperature slowly raised, the softening temperature range is raised, the product becomes tougher, and insolubility appears to be increased as evidenced by decreased softening in the presence of certain solvents, without, however, being accompanied by any change in color, or in the property of reversible thermoplasticity, until at a temperature of 280–310° C. decomposition commences, the color darkens, and an odor of charring is noticed. Even at such elevated temperatures, however, and after decomposition has obviously commenced, the mass is still reversibly thermoplastic.

When, as usual, the heating is stopped below the decomposition temperature, at 250° C., or thereabouts, the resin has acquired approximately its maximum properties of hardness, toughness, high temperature reversible thermoplasticity, and insolubility. Its physical appearance is not changed, however. The softening range at this stage is upwards of 200° C.

With bleached shellac the procedure is the same as with the natural shellac described above, and the product has the same properties except for the color which is a very light ivory. This ivory colored product may, if desired, have incorporated with it suitable coloring matter or colored fillers to obtain any desired colored product.

In use, it is obvious that the molding temperature of the resulting product must be adapted to the softening range of the resin used. The process of admixture with fillers of various kinds, and of incorporation into or with fabrics or fibers or the like, will be obvious.

Take 200 parts by weight of ground shellac, 70 parts by weight of dry zinc oxide and 48 parts of triethanolamine (about 15 per cent by weight), mix thoroughly, and while stirring the mixture constantly, heat it. At about 120° C., the resin reaction commences to take place, as evidenced by development of the characteristic color. Heating and stirring are continued considerably above that temperature, at 200° C. or higher, although the exact temperature is not important; and the resin and diluent are then easily and completely mixed and the mass becomes homogeneous. The mixture is allowed to cool at once to obtain a softened resin of the lowest softening range, which in this case is 100° C. to 150° C. But higher softening ranges can be obtained by continuing the heating, as described heretofore. Start with 85 parts by weight of the new resin and 15 parts of triethanolamine, heat them together and stir. As the resin softens, it may then be mixed with the liquid, rapidly and easily forming a homogeneous mixture that when cooled is in no essential way different from that prepared by conducting the resin reaction in the presence of the liquid as described above. Thus it appears that the function of the liquid is that of fluid diluent of the new resin.

When a still softer resin is desired a greater proportion of triethanolamine is used. Twenty-five per cent by weight, for example, gives a minimum softening range of 50° C. or thereabouts. Such a resin when cold is soft enough to be indented by pressure of the fingernail, and when in slab form and folded, or stretched, shows a noticeable resistance to permanent deformation, tending to resume its cast shape.

Similarly, using a 20 per cent proportion of ethylene glycol a homogeneous resin is obtained that has a softening range in the vicinity of 60° C.–100° C. This also exhibits the property of elasticity or resistance to permanent deformation to a noticeable extent.

The solubilities of such softened resins are different from those of the untreated resin, because of this treatment, as is to be expected. While they are insoluble in almost all organic solvents, they are softened by such liquids as alcohol and ethyl acetate; and if the percentage of softening agent is high, they may become soluble. For example, the softened resin made with 25 per cent of triethanolamine is soluble in alcohol, softened by ethyl acetate, benzene, carbon disulfide, and but little or not at all affected by acetone, carbon tetrachloride, ether, toluene, gasolene, or turpentine.

The solubility may be affected in a different way by high temperature reaction with an oil. This reaction is of the type that has been described in another application for patent (Boughton, reaction products of resins and certain other materials, Serial Number 556,297, filed August 10, 1931) and is not claimed as a novelty here except as a specific example not previously described.

Briefly, to carry out this reaction melt the pure resin and when smoothly melted add a pre-determined percentage of high temperature reactant, raise the temperature to 200° C.–300° C. and keep it there until the reaction is completed as indicated by quiescence. The product is then cooled. Specifically, the procedure is substantially as follows—100 parts by weight of the new resin are melted and the desired proportion in effective amounts (up to about 50 per cent) of an oil such as tung oil, linseed oil, cottonseed oil, etc., is added and the temperature maintained at 200° C.–280° C., while the mixture is stirred. There is no indication of solubility or reaction at first, but eventually large quantities of gas are rapidly evolved and complete homogeneity of the mixture is obtained. The product has modified properties, the most important one being solubility in turpentine and other solvents.

To produce one type of product melt seven to nine parts by weight of the new resin, and add one to three parts of oil, preferably linseed or China-wood oil (tung oil). At first the melted resin and oil show no signs of reaction or even mixing, but at the maximum temperature there is a sudden and marked gas evolution which soon thereafter quiets down, indicating completion of the reaction.

Such a reacted product when cooled is soluble in various solvents such as carbon tetrachloride, turpentine, etc., and in such solution shows properties characteristic of paint-making resins of high quality, namely, forms films that are glossy, tough and flexible.

Such high temperature reactions with consequent modification of properties may also be carried out with other substances as described in detail in the above mentioned application for patent, Serial Number 556,297.

We may further modify the properties of the new resin, as well as in specific cases cheapen its cost by preparing it in the presence of a proportion in effective amounts of another resin that is soluble or blendable therewith when the two are in solution or melted of the general type capable of forming a liquid of moderate to low viscosity when melted. For example, 200 parts of shellac, 200 parts of rosin, and 80 parts of zinc oxide by weight when heated and stirred react and/or mix to form a homogeneous, hard, tough, resinous mass with these desirable properties somewhat lessened, as compared with the pure new resin.

Similarly, take 200 parts of shellac, 200 parts of a meltable vinyl resin and 80 parts of zinc oxide, treat the mixture in the same manner and obtain a mixed resin of enhanced hardness.

The following compositions are illustrations of a wide range in proportions of suitable diluting resins that may be added to zinc-shellac compounds.

The proportions are by weight:

1. Shellac 50; zinc oxide 15, rosin 40.
2. Shellac 50; zinc oxide 15, copal 35.
3. Shellac 50; zinc oxide 15, coumarone resin 50.
4. Shellac 50; zinc oxide 15, chlorinated diphenyl resin 25.
5. Shellac 50; zinc oxide 15, teglac (unknown composition) 10.
6. Shellac 50; zinc oxide 15, chicle 5.

Other natural or synthetic resins which are blendable or soluble may also be used. The solubility of such a product in common organic solvents has been somewhat affected as indicated in the former example by extraction with some solvents (carbon disulfide, ether, toluene), and softening and extraction in others (alcohol, acetone, benzene, ethyl acetate, gasolene).

Refining of Rosin

One hundred parts of rosin by weight is heated for 24 hours with 25 parts by volume of 40 per cent formalin and 0.1 part by weight of hydrogen chloride. The condensation product is distilled under reduced pressure at 300°. The refined colophony rosin can be used advantageously in the manufacture of varnishes and paints.

Artificial Beeswax

This formula covers the process to be followed in the manufacture of artificial beeswax. This material is intended as a substitute for beeswax.

Carnauba Wax	½ lb.
Paraffin	9½ lb.
Ceres Yellow Color	1/16 lb.

Melt the carnauba wax and paraffin in separate vessels and pour the paraffin into the carnauba wax. Stir until thoroughly mixed. To the liquid material add the ceres yellow and continue the stirring until the mixture is perfectly homogeneous.

Cast in cakes of convenient size in metal molds. Lycopodium may be used to prevent the wax from sticking to the mold.

Dental Wax
U. S. Patent 1,933,907

A wax composition for dentists' use comprises beeswax 99 per cent and impalpable aluminum powder 1 per cent, admixed and incorporated into the said wax.

Electrotyper's Wax
Formula No. 1

Ozokerite (Green-Austrian)	100 lb.
Pitch	20 lb.
Beeswax	10 lb.

Formula No. 2

Ozokerite (Green Austrian)	50 lb.
Candelilla Wax	6½ lb.
Pitch	10 lb.
Beeswax	5 lb.

Modelling Waxes for Engravers
Formula No. 1 (Winter Wax)

Syrian Asphalt	15 g.
Gum Mastic	30 g.
Beeswax	40 g.

Formula No. 2 (Summer Wax)

Syrian Asphalt	60 g.
Gum Mastic	30 g.
Amber	30 g.
Beeswax	120 g.

Sculptors' Modelling Wax
Formula No. 1

Burgundy Pitch	1 g.
Beeswax	10 g.
Lard	1 g.
Venice Turpentine	1 g.

Formula No. 2

Burgundy Pitch	2 g.
Beeswax	16 g.
Lard	1 g.

The lard can be replaced by tallow.

Upholsterers' Wax

For rubbing the inside of the ticking cover before the down is put in so the feathers do not pierce the fabric.

Yellow Wax	50 g.
Venice Turpentine	8 g.
Burgundy Pitch	1 g.

Wax-Like Emulsifiable Materials
U. S. Patent 1,932,643

A wax-like material capable of combining with 5 times its quantity of water containing 2% sodium carbonate to form a colloidal mass may be formed of spermaceti 65, cetyl alcohol 25 and stearic acid 10%.

Hard White Wax
U. S. Patent 1,730,563

448 grams of stearic acid is heated to 130°–150° C. with 31.6 grams magnesium oxide for ½ hour. This mixture may be diluted with paraffin or other waxes to make a hard white high melting point mixture.

White Hard Wax

Heat one part cumar resin and 15 parts paraffin. When solidified on cooling this mixture will be white. Another process for getting similar results is the addition of aroclor No. 1268 and paraffin.

Slow Flowing Wax
(High Melting Point)

Prepare by mixing equal parts of methyl abietate and aluminum palmitate, stirring to a paste. Add ethylene dichloride in quantities sufficient to thoroughly dissolve with heat. To this paraffin is added by constant stirring. This turns into a gel which solidifies on cooling. On reheating the gel reappears in temperatures up to 350°F. By varying the portions of wax the characteristics may be changed.

Soft Elastic Wax

This wax has proved valuable in the chemical laboratory, especially for making joints which must be air-tight and somewhat flexible. It does not harden on exposure to air.

Yellow Beeswax	100 g.
Finely Powdered Red Mercuric Oxide	10 g.
Venice Turpentine	10 g.
Oil-Soluble Red Dye	3 g.

The above ingredients are melted together, stirred thoroughly to distribute the mercuric oxide and allowed to cool. The hardness of the wax may be altered by varying the amount of the Venice turpentine used. The dye is used only to give a pleasing color and may be omitted.

White Carnauba Wax

A white carnauba wax having a melting point of 80.8° C. can be obtained from "fat-gray" carnauba wax of melting point 82.1° C. by fractional crystallization. A by-product consisting of a wax-paraffin residue, melting point 60.2° C. is also obtained. Carnauba wax partially purified with the aid of paraffin is dissolved in 10 times its weight of benzine at 55° C. The benzine fraction used is that boiling between 80 and 130° C. Turbidity appears at 50° C. The mixture is allowed to cool with gentle stirring to 16°, when crystallized wax is filtered off. It is freed from solvent by steam distillization. A yellow wax-paraffin mixture is obtained from the filtrate as a by-product. Cooling cannot go much below 16°, as paraffin separates from benzine solution at about 12°.

Modelling Wax for Brass Foundries

Burgundy Pitch	5 g.
Beeswax	50 g.
Lard	5 g.
Venice Turpentine	5 g.

Beeswax Candles
U. S. Patent 1,960,994

A molded candle comprises a mixture of seventy parts of beeswax, twenty parts of stearic acid, ten parts of paraffin and one part of monoethyl ether of ethylene glycol.

Dripless, Rigid Candle
U. S. Patent 1,959,164
Formula No. 1

A modified glyptal resin is prepared by heating a mixture of 202 parts phthalic anhydride, 40.5 parts phthalide and 92 parts glycerin for 2-3 hours at 180° C. or until the desired degree of esterification has been reached. The product, dissolved in suitable solvents, is coated onto ordinary stearin or "tallow" candles of approximately 1¼ inches diameter by dipping or by painting with a brush or stick, and allowed to harden. The treated candles have a hard, glistening surface and do not feel greasy under the fingers. They burn easily and without drip, and show no tendency to stick together when packed together in a warm place.

The proportion of phthalide to phthalic anhydride may be varied in order to vary the hardness of the resinous coating. Thus, for example, a slightly softer resin is produced by using 60.5 parts of phthalide and 191 parts phthalic anhydride in the above preparation, and still greater plasticity is obtained by using 82 parts phthalide, 180 parts phthalic anhydride to 92 parts glycerin.

Corresponding amounts of other modifying agents may also be used, such as benzoic acid, cottonseed oil and other fat acids, etc. By this means, resinous coatings of any desired degree of hardness may be obtained.

Formula No. 2

A mixture of 90 parts by weight of glycerin, 195 parts phthalic anhydride and 18–24 parts rosin are heated to 170–185° C. until reaction ceases and a clear product is obtained. The resin so obtained is dissolved in acetone, amyl acetate or an alcohol and 50–75 parts nitrocellulose, cellulose acetate or cellulose ether are stirred in. The solution is thinned to the proper consistency, preferably using benzol, toluol or other cheap solvent, and is applied to wax or stearin candles by dipping, spraying or other coating methods. The coating, after drying in air, is hard and brilliant and produces a candle having a good appearance and excellent burning qualities.

The coating composition may be further plasticized by the addition of suitable amounts of high boiling esters, such as diethyl or dibutyl phthalate, or phthalide, substituted phthalides or hydrogenated phthalides may be used. Esters of keto aromatic acids, such as the methyl, ethyl or propyl esters of benzoyl benzoic or naphthoyl benzoic acid may also be used as plasticizers.

Formula No. 3

Beautiful color effects may be obtained by the incorporation of dyes or other coloring materials in the resinous coatings, either as such or in the form of a color lake:

A coating composition is prepared as in previous examples and dissolved in a suitable solvent or mixture of solvents. Amounts up to 1–3% of an aluminum,

iron, chromium, nickel or cobalt salt of a keto aromatic acid, e.g., aluminum benzoyl benzoate or zinc naphthoyl benzoate are added. A solution of a suitable lake-forming dye is then prepared, such as a solution of alizarin or other anthraquinone dye in alcohol, and is stirred into the solution containing the coating composition. An intense coloration is thus produced, consisting of a color lake in fine suspension throughout the solvent and giving a brilliant coating when applied to candles of any type by spraying, brushing or dipping. Such colored coatings are particularly useful in the manufacture of ornamental wax candles and the like.

A wide range or dyes may be applied in this manner, any lake-forming dye being suitable that is soluble in an organic solvent compatible with the coating to be applied. Further details of the process of preparing such color lakes are described in the co-pending application of Daniels & Jaeger, Serial No. 503,855, filed December 20, 1930, and any of the lake colors there described may be used with success in the present invention.

Formula No. 4

A phenol-formaldehyde condensation product is prepared, for example, by heating the components in the ratio of 1 mol formaldehyde to about 2 mols phenol, with or without the use of accelerators, until the product separates into layers. The resinous layer is collected and dried and dissolved in amyl acetate, acetone, fusel oil or mixtures of these, with or without the addition of benzol, toluol, or other cheap hydrocarbon solvents. 6–8% diethyl or dibutyl phthalate are added as plasticizer and the composition is applied to a candle in the usual manner. After drying and hardening a glistening black coating is produced, which adheres well and serves to protect the candle against damage during shipment and to produce a better candle flame.

If desired, cellulose plastics such as nitro-cellulose, cellulose acetate, cellulose ethers, etc., may be dissolved in the solvent in amounts up to 25–50% of the phenol-formaldehyde resin.

Improved Candle
U. S. Patent 1,950,814

A fused mixture of 97 parts of coarsely crystalline, hard paraffin wax (melting point 53.6° C.) and 3 parts of octodecyl alcohol is cast into candles.

The candles obtained are entirely homogeneous, have a smooth opaque appearance and a milky white color and are free from cracks, whereas candles prepared from the paraffin wax alone are transparent and show numerous cracks.

Colored-Flame Candles

German Patent 565,250

Candles which burn with a colored flame are prepared by incorporating in the candle mass about 1–2% of a suitable detonator, for example, copper acetylide and the picrates, azides, and fulminates of strontium, barium, lithium, and sodium. A small difference between the wick temperature and the detonation temperature of the added substance may be compensated by addition of an oxidizing agent. A larger difference may be counteracted by incorporating another detonator which does not color the flame and has a lower detonating temperature. The process is particularly intended for candles normally burning with a colorless flame, for example, candles of ethylurethane, dimethyl oxalate, or ethyl oxamate.

RUBBER

Rubber Compounding

Crude or raw rubber is a commercial product which is available in a large number of different varieties. The manufacturer selects the grade most suitable, depending upon the properties sought in the finished article. Raw rubber, however, is seldom used by itself and is usually mixed with various other ingredients which are added to change its properties in accordance with the purpose for which it is intended. Thus, compounded rubber may be slightly or heavily loaded, soft and elastic or hard and stiff, resistant to abrasion or readily abraded, high or low in tensile strength, good or bad in wearing properties, strong or weak in tear resistance and so on.

Crude rubber lends itself readily to being compounded with other materials. It must first be plasticized on a rubber mill or internal mixer and is then ready to receive the other ingredients which are to be added, and the rubber compounder has a wide choice of materials from which he can make his selection. These materials can be broadly classified in accordance with their functions, bearing in mind that some materials serve more than a single purpose and that a single compound does not require materials from each class with the single exception of vulcanizing ingredients which are absolutely essential in all rubber compounds. The various classes of ingredients are as follows:

Accelerators

The purpose of these materials is to increase the rate of vulcanization and to avoid the necessity of prolonged heating. They also smooth out curing variations which may exist in the crude rubber or reclaim. Their activity varies considerably. Accelerators are available in two classes, either organic or inorganic. The former class is growing rapidly and new types are coming into the market constantly. The latter class which includes litharge, lead carbonate, calcined magnesia, magnesium carbonate, antimony sulphide, etc., are still in use but not to any great extent. Boosters or sec-ondary accelerators are also sometimes used.

Accelerator Activators

Practically all accelerators prove most effective in the presence of certain metallic oxides notably zinc oxide and lead oxide or in the presence of organic acids such as stearic acid or oleic acid.

Anti-oxidants or Age Resisters

Vulcanized rubber tends to deteriorate on prolonged exposure to sunlight and air. It is the function of these materials to slow down oxidation and retard the deterioration of rubber goods upon aging.

Anti-Sun Materials or Sun Checking Agents

Sun checking is an inherent property of rubber. These materials afford sun protection by forming a protective film on the outside of the rubber.

Anti-Scorch Materials or Retarders

The purpose of these is to prevent premature curing during processing operations or while the rubber is kept in storage bins. They are chiefly used in connection with very fast curing compounds.

Colors or Pigments

In the early days of rubber manufacture inorganic colors were used for coloring rubber compounds. These have been very largely replaced in recent years by organic colors. Dyestuffs give much more brilliant colors in general than mineral pigments. Very small amounts of organic dyes are sufficient to give strong color to rubber stocks. Pure white goods are usually obtainable by using strong whitening pigments with high covering power.

Dispersing Agents

These materials serve as wetting agents for pigments causing better and more intimate distribution of the compounding materials with the raw rubber. In this way the homogeneity of the compounded stock is improved which in turn enhances the physical properties of the rubber article.

Fillers

Raw rubber to which no fillers are added is not suitable for many rubber products. Various materials or fillers are used to alter the physical properties of the raw rubber. In this way the tensile strength, stiffness, tear resistance, abrasive resistance, etc., of the rubber may be modified and controlled. The shape and size of the filler particles are important factors in the resulting properties of the compounded rubber. Where high physical qualities are not essential, the compounder may use inert fillers which serve merely as diluents, increasing the volume and lowering the cost of the rubber stock. If high physical properties are important, fillers of the reenforcing type should be used.

Reodorants

The odor of rubber articles is sometimes considered objectionable. In such cases reodorants may be used which counteract the existing odor and impart a pleasing scent to the rubber stock. The odor can be varied to suit.

Rubber Substitutes or Factice

Although not similar to raw rubber in their properties, these materials which are made by treating vegetable oils with sulphur or sulphur chloride are substituted for rubber in some types of articles. They aid in the processing of some rubber stocks.

Rubber Softeners

Prior to the incorporation of the filler materials, pure gum or crude rubber is worked on a hot rubber mill until it becomes soft and plastic. Various softeners are used which aid in the plasticizing of the rubber and keeps the mill temperature down, thereby preventing scorching and prevulcanization. They also aid in the dispersion of the fillers and in reducing the milling time and power consumption.

Stiffeners

It is sometimes found useful to stiffen up uncured rubber compounds. In such cases stiffeners of either organic or inorganic types are employed.

Vulcanizing Ingredients

When rubber is vulcanized its chemical and physical properties are completely altered from those which it possessed in its raw or unvulcanized condition. Vulcanization may be brought about either by the application of heat, which practice is employed in the manufacture of the great majority of rubber goods or in the cold way which method has limited application in connection with the vulcanization of comparatively thin layers of rubber. Vulcanization cannot be accomplished without the use of an agent which brings about this change. Several substances bring this about but sulphur is the principal vulcanizing agent. Selenium and tellurium are sometimes used in combination with sulphur as secondary vulcanizing agents.

Rubber Bands

Pale Crepe	100	parts
Stearic Acid	1	part
Antioxidant	0.50	part
Zinc Oxide	2	parts
"Tuads" (Tetramethyl-thiuram Disulphide)	0.50	part
"Captax" (Mercaptoben-zothiazole)	1	part
"Telloy" (Tellurium)	0.25	part
Sulphur	0.20	part

Cure: In steam 5 minutes at 30 pounds.

Surgeon's Gloves

Pale Crepe	100	parts
Stearic Acid	1	part
Antioxidant	1	part
Zinc Oxide	2	parts
"Captax" (Mercaptoben-zothiazole)	0.50	part
"Tuads" (Tetramethyl-thiuram Disulphide)	0.50	part
Sulphur	0.40	part

Cure: In press. 45 minutes at 20 pounds.

Water Bottle—Heat Resistant

Pale Crepe	100	parts
"Plastogen"	5	parts
Barytes	45	parts
Zinc Oxide	10	parts
"Kalite No. 1"	65	parts
"Rodo No. 10"	0.25	part
"Tuads" (Tetramethyl-thiuram Disulphide)	0.30	part
"Captax" (Mercaptoben-zothiazole)	1	part
"Telloy" (Tellurium)	0.50	part
Sulphur	0.50	part

Cure: In press. 4 to 5 minutes at 40 pounds.

Water Bottle

Smoked Sheets	100	parts
Stearic Acid	0.75	part
Antioxidant	1	part
Barytes	70	parts
Whiting (Gilders)	32.50	parts

Zinc Oxide	10	parts
Color		To suit
"Rodo No. 10"	0.25	part
"Altax" (Benzothiazyl Disulphide)	1	part
"Tuads" (Tetramethyl-thiuram Disulphide)	0.10	part
Sulphur	2.50	parts

Cure: In press. 6 minutes at 60 pounds.

Steam Hose

Smoked Sheets	100	parts
Whole Tire Reclaim	50	parts
Pine Tar	3	parts
Stearic Acid	3	parts
Antioxidant	3	parts
"Kalite No. 1"	100	parts
Zinc Oxide	5	parts
Clay	50	parts
"Tuads" (Tetramethyl-thiuram Disulphide)	0.50	part
"Telloy" (Tellurium)	1	part
"Altax" (Benzothiazyl Disulphide)	1	part
Sulphur	0.80	part

Cure: 10 minutes at 40 pounds steam.

Fire Hose

Pale Crepe	23	parts
Smoked Sheets	23	parts
Antioxidant	0.50	part
Zinc Oxide	20	parts
Gas Black	2	parts
"Kalite No. 1"	30	parts
"Captax" (Mercaptoben-zothiazol)	0.25	part
Sulphur	1.25	parts

Cure: In steam. 30 minutes at 247° F.

Medium Grade Hose Tube

Rolled Brown Crepe	10	parts
Whole Tire Reclaim	40	parts
Stearic Acid	0.25	part
Mineral Rubber	8	parts
Antioxidant	0.375	part
Whiting	36.875	parts
Zinc Oxide	3	parts
"Accelerator 808" (Bu-tyraldehyde-Aniline)	0.15	part
Sulphur	1.25	parts

Cure: 30 minutes at 274° F.

Medium Grade Hose Cover

Smoked Sheets	18	parts
Whole Tire Reclaim	22	parts
Stearic Acid	0.50	part
Paraffin	0.50	part
"Cumar CX"	3	parts

Clay	10	parts
Zinc Oxide	3	parts
Soft Carbon Black	14	parts
Whiting	17	parts
Aldehyde-Amine Type Accelerator	0.30	part
Sulphur	1.25	parts

Cure: 30 minutes at 40 pounds.

30% Wire Insulation

Smoked Sheets	100	parts
Stearic Acid	0.50	part
Paraffin	3	parts
"Oxynone"	0.125	part
"Feectol A" (Condensa-tion Product of Aniline and a Ketone)	1	part
Whiting	150	parts
Zinc Oxide	60	parts
"Ureka C"	0.56	part
Sulphur	3.25	parts

Cure: Approximately 60 minutes at 274° F.

30% Wire Insulation-Free Stipping

Smoked Sheets	50	parts
Pale Crepe	50	parts
Paraffin	3	parts
Zinc Oxide	100	parts
Whiting (Gilders)	100	parts
Antioxidant	2	parts
"Tuads" (Tetramethyl-thiuram Disulphide)	1.50	parts
"Telloy" (Tellurium)	0.50	part

Cure: In talc. 30 minutes at 274° F.

White Tire Sidewall

Pale Crepe	100	parts
Stearic Acid	1	part
Zinc Oxide	10	parts
Clay	35	parts
Lithopone	70	parts
"Tuads" (Tetramethyl-thiuram Disulphide)	0.25	part
"Altax" (Benzothiazyl Disulphide)	1	part
"Telloy" (Tellurium)	0.50	part
Sulphur	0.50	part

Cure: Approximately 35 minutes at 35 pounds steam.

Hard Rubber Sponge

Rubber	20	parts
Tube Reclaim	10	parts
Mineral Rubber	10	parts
Sodium Bicarbonate	5	parts
Kerosene	2	parts
Whiting	15	parts
Alum	0.75	part
Lime	1	part

"Ureka C" 0.1875 part
"Guantal" 0.125 part
Sulphur 8 parts
Cure: 90 minutes at 60 pounds open steam.

Black Stamp Pad-Sponge Rubber

Pale Crepe 100 parts
Stearic Acid 10 parts
Petrolatum 10 parts
Antioxidant 1 part
Sodium Bicarbonate 10 parts
Whiting 90 parts
Carbon Black 2 parts
"Thionex" (Tetramethyl-
  thiuram Monosulphide) 0.15 part
Sulphur 3.50 parts
Cure: 15 minutes at 40 pounds steam.

Black Molded Sole

Smoked Sheets 24 parts
While Tire Reclaim 25 parts
"Laurex" 1 part
Mineral Rubber 8 parts
Antioxidant 0.75 part
Zinc Oxide 2 parts
Carbon Black 36.50 parts
"Beutene" (An Aldehy-
  de-Amine Condensation
  Product of Butyl Al-
  dehyde and Aniline) 0.375 part
Sulphur 1.375 parts
Cure: 16 minutes at 60 pounds.

Bathing Cap

Pale Crepe 100 parts
"Plastogen" 8 parts
Stearic Acid 1 part
"Rodo No. 10" 0.10 part
Zinc Oxide 5 parts
Color To suit
"Zimate" (Zinc Dimethyl-
  dithio-Carbamate) 0.05 part
"Captax" (Mercapto-
  benzothiazole) 1 part
Sulphur 2.25 parts
Cure: 5 to 10 minutes at 50 pounds steam.

Heat Resistant Packing

Smoked Sheets 50 parts
Pale Crepe 50 parts
Stearic Acid 2 parts
Antioxidant 2 parts
Zinc Oxide 20 parts
Clay 100 parts
"Pyrax" 100 parts
Red Oxide 2 parts
"Tuads" (Tetramethyl-
  thiuram Disulphide) 4 parts
Cure: 20 minutes at 50 pounds.

Rubber Printing Rolls

Pale Crepe 100 parts
"Cumar RH" 15 parts
Antioxidant 1 part
Thiocarbanilide 0.60 part
Anhydroformaldehyde-
  paratoluidene 1 part
Zinc Oxide 5 parts
Red Oxide 3 parts
Sulphur 9 parts
Cure: 2 hours rise to 260° F., 4 hours at 260° F. and cooled for ½ hour to 1¼ hours depending on size of rolls.

Hard Rubber High-Grade Battery Container

Smoked Sheets 25 parts
Whole Tire Reclaim 150 parts
Mineral Rubber 20 parts
Petrolatum 5 parts
"Accelerator 833"
  (An Aldehyde-Amine
  Condensation Product
  of Butyraldehyde and
  Monobutyl Amine) 0.25 part
Silica 140 parts
Carbon Black 4 parts
Whiting 40 parts
Sulphur 42 parts
Cure: 17 minutes at 90 pounds steam in mold equipped with heated sides and cores.

Auto Topping

Smoked Sheet 100 parts
Mineral Rubber 8 parts
Antioxidant 1.50 parts
"Kalite No. 1" 100 parts
Zinc Oxide 5 parts
Litharge 5 parts
Soft Carbon Black 10 parts
Clay 30 parts
"Captax" (Mercapto-
  benzothiazole) 0.80 part
Sulphur 3 parts
Cure: 60 minutes rise to 255° F. and 45 minutes at 255° F.

Rubber Clothing Stock

Pale Crepe 100 parts
Petrolatum 1 part
Zinc Oxide 10 parts
Lithopone 75 parts
Whiting 63 parts
Color As desired
"Thionex" (Tetramethyl-
  thiuram Monosulphide) 1 part
Sulphur 1 part
Cure: In air. 60 minute rise to 245° F. and 45 minutes at 245° F.

Tire Tread

Smoked Sheets	100	parts
Pine Tar	3.50	parts
"Laurex"	1.25	parts
Antioxidant	1	part
Carbon Black	45	parts
Zinc Oxide	5	parts
"Beutene"		
(An Aldehyde - Amine Condensation Product of Butyl Aldehyde and Aniline)	1.125	parts
Sulphur	3.25	parts

Cure: 60 minutes at 40 pounds.

Tire Tread

Smoked Sheets	50	parts
Pale Crepe	50	parts
Pine Tar	2	parts
Stearic Acid	4	parts
Antioxidant	2	parts
Gas Black	40	parts
Zinc Oxide	5	parts
"Tuads" (Tetramethyl-thiuram Disulphide)	0.25	part
"Altax" (Benzothiazyl Disulphide)	1	part
"Vandex" (Selenium)	0.50	part
Sulphur	1	part

Cure: 45 minutes at 35 pounds.

Red Rubber Tubing

Pale Crepe	100	parts
Petrolatum	4	parts
Antioxidant	1	part
"Altax" (Benzothiazyl Disulphide)	1.25	parts
Zinc Oxide	12	parts
Clay	40	parts
"Kalite No. 1"	225	parts
Red Oxide	12	parts
Sulphur	3	parts

Cure: In talc. 30 minutes at 20 pounds.

Tire Friction

Smoked Sheets	60	parts
Amber Crepe	40	parts
"Cumar Resin"	1	part
Mineral Rubber	5	parts
Stearic Acid	0.50	part
Pine Tar	3	parts
Antioxidant	1	part
Zinc Oxide	10	parts
"Accelerator 808" (Bu-tyraldehyde Aniline)	0.6875	part
Sulphur	3.25	parts

Cure: 45 minutes at 281° F.

Black Heel

Smoked Sheets	37	parts
Whole Tire Reclaim	100	parts
Mineral Rubber	7	parts
Antioxidant	0.50	part
Whiting (Natural)	20	parts
Clay	35	parts
Zinc Oxide	4	parts
"Monex" (Tetramethyl-thiuram Monosulphide)	0.20	part
Sulphur	3	parts

Cure: 8 minutes at 40 pounds.

Inner Tube

Smoked Sheets	100	parts
Antioxidant	3	parts
Blanc Fixe	40	parts
Zinc Oxide	6	parts
"Beutene"		
(An Aldehyde - Amine Condensation Product of Butyl Aldehyde and Aniline)	0.75	part
Diphenylguanidine	0.75	part
Sulphur	2.25	parts

Cure: 6 minutes at 60 pounds.

Bus and Truck Inner Tube

Smoked Sheets	50	parts
Pale Crepe	50	parts
"Plastogen"	4	parts
Stearic Acid	1	part
Antioxidant	1	part
Zinc Oxide	5	parts
Soft Carbon Black	40	parts
"Tuads" (Tetramethyl-thiuram Disulphide)	0.10	part
"Altax" (Benzothiazyl Disulphide)	0.50	part
"Captax" (Mercapto-benzothiazole	0.50	part
Sulphur	1	part

Cure: 5 minutes at 55 pounds.

Tan Heel Compound

Smoked Sheets	5	%
Inner Tube Reclaimed	54	%
Clay	30	%
Zinc Oxide	1½	%
Red Iron Oxide	4	%
Diphenylguanidine	½	%
Mineral Rubber	3	%
Sulphur	2	%

Cure: 7 minutes at 60 pounds steam.

Cheap Matting

Whole Tire Reclaimed	33	%
Whiting	33¾	%
Mineral Rubber	30	%

Engine Oil	2	%
Sulphur	1	%
Diphenylguanidine	¼	%

Cure: 30 minutes at 30 pounds steam.

Tan Sole Stock

Smoked Sheets	20	%
Red Tube Reclaimed	20	%
Mineral Rubber	5	%
Zinc Oxide	5	%
Stearic Acid	2	%
Mercaptobenzothiazole (Captax)	½	%
(Antioxidant) Phenyl-beta naphthylamine	¾	%
Clay	8	%
Paraffin	1	%
Magnesium Carbonate	36½	%
Sulphur	1¼	%

Cure: 11 minutes at 60 pounds steam.

High-Grade Translucent Rubber

Pale Crepe	100	parts
Light Magnesium Carbonate	25	parts
Sulphur	3	parts
Zinc Stearate	2	parts
Benzothiazyl Disulphide (Altax)	0.8	part
Tetramethylthiuram Disulphide (Tuads)	0.2	part

Cure: 30 minutes at 30 pounds steam.
In the above formula the light magnesium carbonate having practically the same index of refraction as rubber gives the translucent compound.

Transparent Rubber

Concentrated Latex (60% Solids)	150	parts
Sulphur	2.5	parts
Zinc Diethyldithiocarbamate	0.5	part
Mineral Oil	5	parts

Rubber to Resist Boiling Water

Pale Crepe Rubber	100	parts
Zinc Oxide	5	parts
Stearic Acid	1.5	parts
Tetramethylthiuram Disulphide	3	parts
Mercaptobenzothiazole	0.5	part

Cure: 10 minutes at 298° F.

Bath and Kneeling Mats

Pale Crepe Rubber	100	parts
Whiting	60	parts
Zinc Oxide	5	parts
Stearic Acid	10	parts
Petrolatum	10	parts
Sulphur	3	parts
Phenyl-beta-naphthylamine	1	part
Color		To suit
Tetramethylthiuram Monosulphide	0.1875	part
Sodium Bicarbonate	10	parts

Cure: 20 minutes at 287° F.

Sponge Rubber for Abrasion

Pale Crepe Rubber	100	parts
Carbon Black	36	parts
Zinc Oxide	5	parts
Stearic Acid	11	parts
Petrolatum	11	parts
Sulphur	3.25	parts
Phenyl-Alpha-Naphthylamine	1.5	parts
Tetra Methylthiuram Monosulphide	0.2	part
Sodium Bicarbonate	10	parts

Cure: 15 minutes at 40 pounds.

Sponge Rubber Compounds
Sponge Balls

Pale Crepe Rubber	100	lb.
Whiting	65	lb.
Zinc Oxide	5	lb.
Petrolatum	8	lb.
Stearic Acid	8	lb.
Sulphur	3	lb.
Phenyl-Beta-Naphthylamine	1	lb.
Color to suit.		

Mill above for 1 hour per day for 2 consecutive days, then add:

Tetramethylthiuram Monosulphide	0.125	lb.
Sodium Bicarbonate	5	lb.

During third milling on third day, let batch stand overnight before vulcanizing 40 minutes at 300° F.

Colored Rubber
U. S. Patent 1,912,939

This is a process for making washing-proof dyeings of products manufactured from aqueous rubber dispersions:

Formula No. 1

100 grams of crepe rubber are intimately mixed on the rolls with 50 grams of du Pont rubber red RL (powder). The mixture thus obtained is dispersed in 700 grams of benzol, to which had been added 22 grams of ox gall (reckoned as dry substance). Twenty grams of this dispersion are then added to a mixture consisting of 130 grams of a

concentrated rubber latex containing 75% total solids, 4 grams of zinc oxide, 2 grams of sulphur, 1 gram of the accelerator Thionex (tetramethyl thiurammonosulphide). The resulting mixture is then ready for use, for example, as coating composition.

Formula No. 2

A mixture obtained by incorporating on the masticator 50 grams of ultramarine into 50 grams of crepe rubber is dispersed in 900 grams of paraffin oil. Thirty-five grams of this dispersion are then added to the following mixture: 130 grams of a concentrated rubber latex containing 75% total solids, 2 grams of zinc oxide, 1 gram of sulphur, 1 gram of accelerator 833 (aldehyde-amine). With the resulting mixture coatings, toys, bathing caps, etc., which are washing-proof, can be produced.

Coloring Rubber Articles

An oil soluble dye is dissolved in benzol or other volatile hydrocarbon. The rubber is soaked in this until slightly swollen. It is then removed and dried. The depth of color depends on the concentration of the dye and the time of immersion.

Colored Rubber Goods
U. S. Patent 1,895,088

A rubber compound is prepared containing approximately 100 parts by weight of rubber, 50 parts whiting (about 15% by volume), 5 parts zinc oxide, 3 parts sulphur, ½ part organic accelerator, and, if a soft stock is desired, from 5 to 10 parts of a light-colored neutral mineral oil. After a thorough mixing and mastication the compound is calendered to a suitable thickness for the articles to be manufactured. Blanks of the calendered stock are then molded and vulcanized, say in the form of bathing caps. At this stage the caps are all of a somewhat translucent white color and are completely manufactured except that the color is lacking.

Solutions of oil-soluble dyes of various colors in xylene in a concentration of 5 to 10% are prepared. One group of bathing caps is dipped bodily into the dye solutions, drained, and allowed to dry to give solid colors. If no one of the dyes represents the exact shade of color desired, 2 or more dyes may be mixed or the caps may be dipped successively in several dye solutions. In the course of one or two days the migration

of the dye is substantially complete, and the color is uniform throughout the mass of rubber.

Another group of caps is colored by covering portions of the rubber with stencils or masks and painting or spraying dye solutions on the exposed areas. After a short time, when the first dye solution is substantially absorbed by the rubber, a different stencil or mask may be applied, and a different dye solution is applied in the same manner. If the areas covered by the 2 dyes overlap, 3 colors are obtained with only 2 dyes, the overlapping area presenting the composite color of the 2 dyes. If desired, a ground color may be imparted to the rubber by dipping in another dye, either before or after the treatment described above. The variety of patterns and color combinations obtainable is practically endless, being limited only by the ingenuity and skill of the operator.

Mottled effects may be obtained by applying irregular blotches of various dye solutions to the rubber.

The method of this invention is advantageous in that it permits the rubber goods manufacturer to mix, handle, shape, and even vulcanize a single, colorless base stock and to color individual articles made from this stock with a large number of different colors, either in solid colors or in an innumerable variety of patterns, and even permits him to color the individual parts of the same article with any number of different hues.

Yellow Rubber
U. S. Patent 1,897,129
Formula No. 1

A rubber mixture is prepared from 100 parts of crepe rubber, 160 parts of calcium carbonate, 50 parts of kaolin, 2.5 parts of petrolatum, 5 parts of zinc white, 0.15 part of diphenyl-guanidine, 1 part of mercaptobenzothiazol disulphide, 1.5 parts of stearic acid, 3 parts of sulphur, and 3 parts of the azo dyestuff obtainable by coupling 1 molecular proportion of tetrazotized 3.3'-dichlor-4.4'-diaminodiphenyl with 2 molecular proportions of acetoacetic acid m-xylidine. The mixture is vulcanized for 12 minutes at a steam pressure of 3½ atmospheres. The resulting bright yellow vulcanizate may serve as a floor covering.

Formula No. 2

A mixture is prepared from 100 parts of crepe rubber, 2.5 parts of sulphur,

0.35 part of thiuram, 5 parts of zinc white, 0.6 part of ozocerite, 0.5 part of stearic acid, and 2 parts of the azo dyestuff obtainable by coupling 1 molecular proportion of tetrazotized 3.3'-dichlor-4.4'-diaminodiphenyl with 2 molecular proportions of acetoacetic acid anilide. The mass is vulcanized for 15 minutes at a pressure of 2 atmospheres. The resulting product, a beautiful yellow, may be used for bathing caps.

Colors—Inorganic Rubber

Black:	
Lampblack	10–30%
Bone Black	10–30%
Blue:	
Ultramarine	20–40%
Brown:	
Sienna	10–25%
Mapico	10–25%
Green:	
Chrome Green	20–30%
Red:	
Antimony Reds	10–30%
Iron Oxide	10–20%
White:	
Titanium Dioxide	25–75%
Lithopone	25–75%
Zinc Oxide	5–100%
Zinc Sulphide	5–200%
Yellow:	
Chrome Yellow	10–20%

Dipped Rubber Goods

For making rubber balloons, finger cots, etc., dip form (which may be test tube, etc.) in rubber solution composed of 1½ pounds rubber dissolved in 1 gallon naphtha. About 6 dips are required to secure proper thickness. After solvent has dried from dipped form, immerse it in 2% sulphur monochloride, in benzol or carbon tetrachloride solution for several seconds. Dust with starch and remove article from form.

Compound for Making Rubber Articles by Dipping

Concentrated Latex (60% Solids)	170	parts
Sulphur	4	parts
Zinc Oxide	0.5	part
Zinc Dimethyl Dithiocarbamate	0.5	part

Rubber Coating
British Patent 397,270

A coating composition for producing a smooth mat finish on articles, e.g., on fabric or rubber surfaces consists of a flocculent precipitate of rubber, obtained from an aqueous dispersion thereof, admixed with starch. 100 parts of 60% latex are diluted to 5% solution containing 120 parts sodium silicate and then a 5% solution containing 96 parts aluminum sulphate are stirred in to form the precipitate which is drained on a filter plate to a mass containing 15–16% solids and starch, equal in weight to that of the solids, is ground in until a smooth cream is obtained which may be diluted as required. Vulcanizing, compounding, etc., ingredients may be added.

Oil Proof Rubber Coating
U. S. Patent 1,907,231

A mixture is made of rubber 40–50, fillers 20–40, fibers 10–20, animal glue 2–5, glycerin 1–2, and a vulcanization accelerator 0.5–1.0%, with sulphur 5–12% of the rubber.

Rubber Belting Preservative

Shellac	1	quart
Denatured Alcohol	1	pint
Ammonia (Household)	1½	quarts
Water	3	quarts

Rubber Printing Plate
U. S. Patent 1,902,048

A composition for printing plates consists of commercial rubber cement 3 pounds, carbon tetrachloride 2 pounds, benzol 2 pounds, talcum 4 pounds, carbon black ½ ounce.

Glossy Finish on Rubber

For producing gloss on molded rubber products, paint mold with saturated solution of brown sugar in water.

Accelerators—Inorganic Rubber

Lime (Hydrated)	2–10%
Litharge (Commercial)	2–10%
Magnesia (Calcined)	2–8 %
Magnesium Carbonate	5–15%

Binders, Fibrous Rubber

Cotton Flock	10–60%
Rayon Flock	10–60%

Reenforcers for Rubber

Carbon Black	20–45%
Clay	20–60%

"Solvents" for Rubber
Gasolene
Naphtha
Carbon Bisulphide
Turpentine
Carbon Tetrachloride
Benzol

Softeners for Rubber

Factice	5–30%
Burgundy Pitch	2–10%
Cumar	2–10%
Palm Oil	1–5 %
Petrolatum	1–5 %
Pine Tar	1–5 %
Rosin Oil	1–5 %
Castor Oil	1–10%
Cottonseed Oil	1–5 %

Chlorinated Rubber
British Patent 400,898

Chlorine is introduced into solutions of rubber at low concentration (2% or less), and low viscosity; and the resulting rubber chloride solution is slowly poured into hot water, which may contain substances to raise the boiling point, in which is immersed a disintegrating apparatus, whereby the solvent is evaporated and the chloride reduced to granular state. Prior to introduction into the water, the solution may be reduced in viscosity by standing, and the addition of alkali, or exposure to light. For example, 35 kilograms of raw rubber is dissolved in 1700 kilograms of ethylene bromide, allowed to stand to decrease the viscosity, then chlorine is passed through, the solution is alowed to stand further for several days or treated with caustic alkali or exposed to light for several hours and run into a vessel containing 8 parts of glycerol and 1 part of water, heated to 135–140°, in which an immersed propeller runs at full speed.

Synthetic Rubber
Formula No. 1
British Patent 318,115

Two hundred parts butadiene, 90 milk, 4 glue, 4 methyl cellulose, 2 sodium butyl-α-naphthalenesulphonate, 1 s o d i u m isopropyl-β-naphthalenesulphonate, 4 castor oil, and 1 urea-hydrogen peroxide compound are emulsified and heated for 7 days at 50° to 55° C. A while polymer, stable in air, is obtained in quantitative yield. Without the urea-peroxide compound the yield under similar conditions is less than 20 per cent.

Formula No. 2
British Patent 312,201

Four hundred parts isoprene by volume are emulsified with 500 parts water, 15 ammonium oleate, 10 trisodium phosphate, 5 parts 30 per cent hydrogen peroxide solution, 25 parts 5 per cent solution of glue. After standing for 190 hours at room temperature there is obtained a viscous, homogeneous latex which can be coagulated to a plastic and elastic rubber.

Formula No. 3
German Patent 532,211

One hundred fifty parts butadiene by weight and 15 parts hexachloroethane are emulsified in a solution of 15 parts sodium stearate in 150 parts water. At ordinary or slightly increased temperature a substantially quantative yield of synthetic rubber is obtained in 5 days. Without the hexachloroethane the yield is only 45 per cent and cannot be substantially increased by prolonging the duration of the process.

Formula No. 4
French Patent 709,637

Eight and three-tenths parts by weight butadiene, 2.52 methylmethylene ethyl ketone, and 8 of a 3 per cent solution of the hydrochloride of diethylamino-ethoxyoleylanilide, after agitation at 60° C. for several days, give a quantitative yield of a product which becomes soft and plastic on the mill and produces a good vulcanizate.

Cleaning of Moulds (Used in Rubber-Curing)
U. S. Patent 1,891,197

The mould is filled with caustic soda solution (15%) and heated until the pressure reaches 110 pounds per square inch. The mould is then emptied by releasing the pressure from the bottom.

Fireproof Rubber Curing Solution

Sulphur Chloride	2⅛	oz.
Carbon Tetrachloride	1	gal.

Fire Resisting Rubber
British Patent 404,691

The proportion of selenium necessary for fire-proofing rubber compounds can be reduced if carbonates, e.g., magnesium carbonate, and softeners, e.g., fatty acids, are also added. The following composition is indicated: rubber 25–40, selenium 2–6, carbonates 45–55, softeners 2–6%.

Inflated Balls
British Patent 397,209

The inflation pressure of tennis balls and other permanently sealed objects is maintained or rehabilitated by a continuous generation of gas at an appropriate rate inside the object by embodying therein a substance or substances generating gas. One or more of the gas-generating substances is preferably incorporated in the material of the walls or in a mixture applied to the inner surface of said walls. A suitable mixture is rubber 100, mercaptobenzothiazole 1.3, sulphur 3, zinc oxide 5, stearic acid 1 and aluminum carbide 30 parts. Such a mixture can be vulcanized and in the presence of water generates methane at a suitable rate. Alternatively the gas-generating substance (s) may be supported in gelatin-glycerol jelly, the rate of generation being controlled by the hydrogen-ion concentration. Suitable gas-generating substances are aluminum, zinc, barium peroxide and manganese dioxide, urea and oxalic acid with a solution of sodium nitrite, hydrogen peroxide, etc.

Toy Balloon Base
Formula

60% Latex	170	g.
Colloidal Sulphur (23.5% Solids)	7.5	g.
Colloidal Zinc Oxide (21.6%)	15	g.
Colloidal Zinc Sulphide (20.0%)	50	g.
Antioxidant	1	g.
No. 552 Accelerator	0.5	g.
Zimate Accelerator	0.5	g.
Aniline Color	2	g.

All pigments and colors employed must be in colloidal form, finely ground in water solution. Such solutions or pastes are now available on the open market.

The pigments, colors, etc., are all ground together in a mortar and then carefully stirred into the solution stirring for at least half an hour.

Such solutions must be kept tightly sealed when not in use and strained before using.

High Percentage Caoutchouc Solutions
U. S. Patent 1,909,219

A concentrated solution of low viscosity, adapted for processes such as spreading or dipping, is obtained from a homogeneous mixture of rubber (100 parts) and a rubber solvent (400–600) by adding nitric acid, e.g., of 50% concentration (1–3 parts) and stirring.

"Imitation Rubber"
U. S. Patent 1,938,015

A vegetable oil such as perilla oil is heated to about 250° and mixed with 5–10% of lead carbonate. The mixture is gradually heated to about 300° and the heating is continued until the lead carbonate decomposes, liberating carbon dioxide and forming a lead salt with the oil. The product thus obtained may be used for coating metals, wood or cloth, etc.

Rubber Substitute
French Patent 751,798

A rubber substitute contains approximately strong glue 100, salicylic acid 0.2, gum tragacanth 25, wood flour 435, water 200, glycerol 125, castor oil 18, colza oil 11, fatty soap 6, suet 3, resinate treated with formaldehyde 3, solution of resin (e.g., bakelite in naphtha) 5, potassium dichromate 3 and potassium alum 6 parts.

Rubber Cements

Rubber cement is ordinarily understood to be a solution of rubber, in gasoline, benzene, carbon tetrachloride or other organic solvent with or without accessory ingredients. Within recent years, however, rubber latex and preparations made from it have come to be used for many of the same purposes as the above-mentioned cement and are sometimes called rubber cements. These products will be treated briefly, but to avoid confusion they will be designated as latex cements, and the term rubber cement will be used to mean solutions of rubber in gasoline or other non-aqueous solvents.

Both rubber and latex cements, as the names imply, are used as adhesives, but they are also employed extensively for other purposes such as for binding and sealing compounds, for applying rubber coatings to various materials, and for making products such as gloves and toy balloons by the dipping process.

I. Ingredients of Rubber Cements

The essential ingredients of a rubber cement are rubber and solvent. To these may be added sulphur and accessory ingredients for vulcanization, antioxidants, pigments, and other materials as required for special purposes.

---

1. Rubber

The rubber commonly used for cements is the ordinary raw, or crude rubber of commerce, which comes to the market chiefly in the form of crepe or ribbed smoked sheets. "Number 1 thin latex crepe" is used where a light colored cement is desired—otherwise either crepe or smoked sheets may be employed. Some specifications for cements call for certain kinds of wild rubber, especially for "fine up-river Para rubber." There is, however, little evidence that such cements are superior for general use to cements made from the crepe or smoked sheets which are produced on plantations. A percentage of a tacky African rubber is sometimes used with other rubber to increase the tackiness of the resulting cement. Particular attention is directed to the fact that the rubber used for cements must be *raw*, or as it is called by the trade, *crude* rubber. Manufactured rubber articles such as inner tubes, gloves, rubber bands, and the like have been vulcanized and are insoluble in all rubber solvents. There is no practical means of making a satisfactory cement from vulcanized rubber.

2. Solvents

The common solvents used in making rubber cement are gasoline, benzol and carbon tetrachloride.

Gasoline is the standard solvent for rubber so far as American practice is concerned. It is employed in grades having different rates of evaporation depending on the type of cement to be made. For cement to be used for adhesive purposes one authority recommends a grade boiling in the range 65° to 128° C. (149° to 262° F.). A grade boiling in the range 67° to 140° C. (152° to 284° F.) is recommended for cement that is to be spread on fabric. Gasolines of a lower initial boiling point than 65° C. are looked upon with disfavor for cement purposes because of lesser solubility of the rubber, larger losses by volatilization in making and handling the cement, and "blushing" of the cement when applied as the result of condensation of moisture. Gasolines containing high boiling or difficultly volatile fractions are objectionable because cements made from them dry slowly and may be mechanically weak on account of

the presence of residual oils. Most motor gasolines are unsatisfactory in this regard for use as rubber solvents. Some high test and aviation gasolines may be used as rubber solvents with satisfactory results, but in general it will be found advantageous to employ a grade of gasoline designed for solvent purposes.

Benzol is generally regarded as being a better solvent for rubber than gasoline, in that it softens and dissolves rubber more readily than gasoline and gives a cement that is smoother and of more uniform consistency. Benzol is a definite chemical substance having a boiling point of 80° C. and constant physical properties. When used in a cement it has the advantage that it evaporates at a uniform and rapid rate, and that it leaves no oily residue to weaken the rubber film obtained from the cement. Impure benzols contain a higher boiling homolog, toluol, as the chief impurity. This decreases the rate of evaporation somewhat but it is not possible to secure with benzol the same range of volatility that can be obtained with gasolines.

Carbon tetrachloride is similar to benzol as a solvent for rubber. It is a definitely chemical substance and when pure has the boiling point 76° C. The outstanding advantage of carbon tetrachloride over both benzol and gasoline is the fact that it is non-flammable.

Both benzol and carbon tetrachloride would be more widely used in industry than is now the case were it not for the fact that *they are toxic when inhaled as vapor or brought into contact with the skin.* The susceptibility to poisoning from these substances varies greatly with the individual but *neither should be used without adequate ventilation and reasonable care to avoid contact of solvent or cement with the skin.*

3. "Solvent Activators"

A solvent activator is a substance which increases the fluidity of solutions of rubber. Although rubber and solvent can be blended in any desired proportion, such mixtures become too stiff for practical use as cements with relatively low percentages of rubber. A solvent activator will increase the amount of rubber that can be incorporated in a cement which is being made to a given consistency, or will increase the fluidity of a cement containing a given percentage of rubber.

4. Vulcanizing Agents

Rubber cements may be vulcanized either with sulphur, or with certain organic compounds containing sulphur, or with sulphur chloride.

Any form of powdered sulphur is satisfactory for use in rubber cements. Flowers of sulphur and ground roll sulphur are the forms most commonly available.

Sulphur is seldom used alone as a vulcanizing agent on account of the relatively long time and high temperature needed for vulcanization. Certain complex organic substances known as accelerators are commonly used to speed up the vulcanizing process. These will be discussed more fully in a subsequent section in which formulas are given.

Accelerators usually require the addition of zinc oxide to function properly. Zinc oxide for this purpose should be the dry powder, and not the linseed oil paste used in paints. The addition of stearic acid may also be advantageous.

Certain complex organic compounds containing sulphur function both as accelerators and vulcanizing agents. These will be described in connection with particular formulas.

Sulphur chloride, more properly termed sulphur mono-chloride, is a yellow, ill-smelling, corrosive liquid capable of vulcanizing rubber quickly in the cold. As a vulcanizing agent for cements and thin layers of rubber it is commonly employed as a 2 per cent solution in anhydrous carbon tetrachloride or carbon bisulphide.

5. Antioxidants

Antioxidants are substances added to rubber to minimize the rate of deterioration. It is often advantageous to incorporate them in cements that are to be used for work for which a reasonable degree of permanence is desired.

II. Ordinary or Non-Vulcanizing Cements

The simplest rubber cement consists of crude rubber in a solvent. The ease with which the rubber will dissolve and the quality of the resulting cement depend upon the amount of milling or "breakdown"* which the rubber receives prior

---

* Crude rubber as received commercially in a tough, gristly material and when in this condition it is difficult to incorporate other ingredients with it. The first process in mixing a rubber compound is usually to "break-down" or soften the rubber by working it between steel rolls on a rubber mill or between the blades of a mixer. The rolls or blades are arranged so that they may be heated or cooled.

to being placed in the solvent. If the rubber is not milled at all, it is very reluctant to go into solution and the resulting cement is non-homogeneous and will contain a low percentage of rubber (although quite stiff). Rubber which is milled a great deal, particularly on cold rolls, will dissolve very readily. The cement which results, however, is poor from the standpoint of adhesive qualities. It has a consistency somewhat like molasses with very little "tack." The best cement is produced when the rubber is given a small amount of hot milling just sufficient so that a reasonable amount, say 10%, may be put in solution.

This solubility of rubber may be changed as indicated in section I—3 with a solvent activator. By the use of about 2% of Bondogen in gasoline, it is possible to make a cement containing 5% of rubber which has not been milled at all. However, the most important use for the activator is not for dissolving unmilled rubber but (1) to facilitate putting milled rubber in solution, (2) to increase the amount of rubber which can be put into a solution of given consistency, and (3) to decrease the consistency of a cement of any given rubber content.

Small amounts of cement may be made by placing 8 to 10% of milled rubber in a solvent and allowing it to swell for several hours. Occasional shaking or stirring after that period will yield a uniform cement in two or three days. Mechanical stirring will hasten the process.

Non-vulcanizing cements are used where a very strong bond is necessary. They are not suitable for use where the parts may become hot, as the cement will soften.

III. Vulcanizing Cements

With a vulcanizing cement a permanent bond can be made between most rubber surfaces and between certain other surfaces. Two types are described (1) heat vulcanizing cements and (2) cold curing cements vulcanized with sulphur chloride.

1. Heat Vulcanizing Cements

The basic constituents of heat vulcanizing cements are rubber, sulphur, an accelerator and usually an accelerator activator such as zinc oxide. The accelerator is used to increase the rate of vulcanization and avoid the necessity for prolonged heating. The time required for vulcanization is greatly affected by the amount of sulphur, the kind of accelerator used and the temperature employed. To illustrate the wide range of vulcanizing conditions, a cement containing simply rubber and sulphur would vulcanize in about 2 hours at a temperature of 288° F. The addition of certain accelerators may reduce the required time to as little as one or two minutes. In general, however, when very active accelerators are used, it is more practicable to use a little longer time than this for vulcanizing at a lower temperature. Cements can be made which will vulcanize in fifteen minutes or less in boiling water.

If a cement will vulcanize at a sufficiently low temperature, it is often referred to as self-vulcanizing. This is not an exact term, however, as the temperature has a great influence on the time of vulcanization and a cement which will vulcanize in a few days in summer might not vulcanize at all at winter temperatures.

Heat vulcanizing cements are applied by spreading one or more coats of cement on the parts to be united, allowing the solvent to evaporate, pressing the parts together, and holding them under pressure while being heated at the vulcanizing temperature. After the cement is vulcanized, it is not dissolved by the solvents which dissolve unvulcanized rubber and is little affected by moderate temperatures.

2. Cold Curing Cements Vulcanized with Sulphur Chloride

If sulphur mono-chloride is in contact with rubber, vulcanization takes place in a very short period of time. A common way of using sulphur chloride for cementing is to give the parts to be united one or more coats of a non-vulcanizing cement. After the solvent has evaporated, a solution of 2 per cent of sulphur chloride in a solvent is applied, the parts quickly put together and held under pressure for a few minutes. No heat is necessary.

The use of sulphur chloride in the manufacture of dipped goods is described in section V—4.

A rapid vulcanizing cement may be also made by adding with care a *very small* percentage of sulphur chloride directly to a non-vulcanizing cement. The amount added determines the time required for vulcanization. An example of such a cement is as follows:

Dissolve 10 ounces of rubber in 100 fluid ounces of benzol. Add 5 fluid ounces of a 2 per cent solution of sulphur chloride in benzol and immediately mix thoroughly.

This cement must be used soon after being mixed as it will ''jell'' or vulcanize and become worthless in about an hour.

In vulcanizing with sulphur chloride a more careful control of the procedure is necessary than with other types of cements.

3. Typical Formulas

Following are some typical formulas for rubber cements together with information on their vulcanizing characteristics:

Cement No. 1 (Vulcanizing)

	Parts by Weight
Rubber	100
Sulphur	4
Zinc Oxide	5
Zinc Stearate	1
Diphenylguanidine	1
Benzol (or other solvent)	1000

A film of rubber deposited from this cement will vulcanize in 30 minutes at 288° F. Other accelerators may be substituted for diphenylguanidine such as diorthotolylguanidine or mercaptobenzothiazole.

Other materials are often added to cements as for instance, stearic acid as an added activator, antioxidants to improve aging qualities and mineral fillers for specific purposes.

Cement No. 2 (Self-Vulcanizing)

Ingredients	Parts by Weight	
	A	B
Rubber	50	50
Sulphur	4	—
Zinc Stearate	2	—
Antioxidant	1	—
''Zimate''	—	1
''Captax''	—	1
Benzol (or other solvent)	500	500

This cement is made in two parts (A and B) in order to prevent premature vulcanization. Before using, mix together equal parts of A and B. The mixture of A and B can ordinarily be kept for several days but it tends to vulcanize or ''jell'' and becomes worthless if kept longer.

A film of rubber from this cement will vulcanize in boiling water in about 15 minutes. It will self-vulcanize in time at room temperature.

Cement No. 3 (Self-Vulcanizing)

Ingredients	Parts by Weight	
	A	B
Rubber	50	50
Sulphur	4	—

Zinc Oxide	2	—
Zinc Stearate	1	—
Antioxidant	1	—
Accelerator 552	—	1.5
Benzol (or other solvent)	500	500

This cement is made in two parts and used the same as cement No. 2. Its vulcanizing characteristics are similar to cement No. 2.

Cement No. 4 (Self-Vulcanizing)

Ingredients	Parts by Weight	
	A	B
Rubber	50	50
Zinc Oxide	5	—
"Tetrone A"	—	2
Benzol (or other solvent)	500	500

With this cement active sulphur is contained in the accelerator and no additional sulphur is necessary.

The cement is made in two parts which are used the same as cement No. 2. Its vulcanizing characteristics are similar to cements Nos. 2 and 3.

Cement No. 5 (Self-Vulcanizing)

Ingredients	Parts by Weight	
	A	B
Rubber	50	50
Sulphur	5	—
Zinc Stearate	2	—
Antioxidant	1	—
"Captax"	—	2
"Tetrone A"	—	2
Benzol (or other solvent)	500	500

This cement contains an excess of sulphur and a combination of accelerators in order to produce very rapid vulcanization. It is made in two parts and used the same as cements Nos. 2, 3 and 4. A film of rubber from this cement will vulcanize in boiling water in about 7 minutes. It will self-vulcanize at room temperature.

4. Making Rubber Cements

The method of making a rubber cement consists of first making a rubber stock by mixing rubber with the various ingredients on a rubber mill or in an internal mixer. In the case of self-vulcanizing cements such as those listed, the two parts should be mixed separately. After mixing, the rubber stocks are put in solution by first allowing them to swell in a solvent for several hours and then stirring or shaking until a uniform mixture is obtained.

It is possible to mix cements containing small amounts of fillers more or less satisfactorily by placing the fillers with the rubber directly in a solvent. Better dispersion is obtained, however, by mixing rubber compounds as described.

After a cement is made, it should so far as possible be protected from light and also, of course, from exposure to the air. A cement which is exposed to light deteriorates quite rapidly and decreases greatly in consistency due to oxidation of the rubber.

IV. Latex Cements

Rubber latex is the milk-like product obtained by tapping the rubber tree, and consists essentially of a dispersion of extremely small rubber particles in a watery medium. It usually has a higher rubber content than rubber cement but is much less viscous than the cement. As received in this country it contains about 3% by weight of concentrated aqueous ammonia as a preservative. Rubber latex finds many of its applications by reason of the fact that it has an aqueous medium and hence gives off no flammable or toxic fumes on drying. Artificial latex can be made from either crude or reclaimed rubber, and is used industrially to some extent. Rubber latex is a cement itself and may be used in the same manner as a non-vulcanizing cement by applying it and allowing the water to evaporate. It ordinarily contains 30–40% of rubber while the usual solvent cement contains about 10% so that the amount of rubber left from solution will be proportionately greater for latex.

It is possible to make a vulcanizing cement from latex by adding sulphur and an accelerator. All added materials, however, must be soluble in water or properly dispersed in it as a fine suspension.

Formula No. 6 (Latex Cement)

	Parts by Weight
Rubber Latex (33% rubber content)	300
Zinc Oxide	1
"Zimate"	1
"Captax"	1
Sulphur	1
Antioxidant	1

Before adding the ingredients to the latex they should be ground for 12 hours or more in a ball mill with water. Less sulphur and zinc oxide are needed than in a corresponding rubber cement.

Care must be observed in adding ingredients to rubber latex so as not to cause the rubber to coagulate.

A rubber film deposited from this latex cement vulcanizes readily in boiling water. It is stated that it will vulcanize in 15 minutes in an oven at 175° F.

While normal latex contains 30 to 40 per cent of rubber, it has been found possible to remove a portion of the water and by adding certain materials to prevent coagulation, to produce a latex containing as much as 75 per cent of rubber.

There has come into the market recently a vulcanized latex. This consists of ordinary latex in which the rubber particles have been vulcanized by chemical means. Simple evaporation of the water from this material leaves a vulcanized rubber.

This product requires no mixing or compounding and needs no heating or other treatment after application. From the standpoint of simplicity and convenience it leaves little to be desired.

There are many patents relating to rubber latex and anyone utilizing this material should give consideration to the patent situation.

V. Uses of Rubber and Latex Cements

1. Adhesives

(a) Rubber to Rubber.

One important use for cement as an adhesive is for a bond between rubber surfaces, as in the manufacture of many rubber products such as boots and shoes, hospital supplies, tires and tubes to some extent, and many hand-made articles. Also the use of rubber cement as an adhesive between rubber surfaces is important in repair work as for instance in the repair of tires and tubes.

When using cement for repair purposes, the preparation of the surfaces to be cemented is quite as important as the cement used. In putting a patch on an inner tube, the portions to be cemented should be well roughened using coarse sandpaper or a metal scraper. They should then be given two or more thin coats of cement allowing each to dry. The best patches are made from vulcanized sheet rubber coated on one side with a layer of unvulcanized rubber—the unvulcanized side being placed next to the tube. The patch should be held on the tube with a clamp and the whole heated to cause vulcanization. The same general procedure is followed with other types of repairs. If a self-vulcanizing cement is used and it is not feasible to heat the article, the parts should be clamped together and held under pressure until vulcanization takes place. In case of a patch on an inner tube, pressure can be applied by inflating the tube in the casing.

(b) Rubber to Textiles.

Good adhesion can be obtained between rubber and cotton and between rubber and some other textiles. It is common practice to first "friction" the fabric, or in other words, force rubber into the meshes of the fabric. This is done on a friction calender. A substitute for frictioning consists of running the fabric through a bath of rubber or latex cement.

Either of these methods leaves the meshes of the fabric filled with rubber so that with further coating the adhesion is really between rubber surfaces and the process may be carried out the same as a rubber to rubber adhesion.

(c) Rubber to Leather.

Adhesion between rubber and leather cannot be made so easily as between rubber and cotton because of the character of the leather and because leather cannot be heated to the usual vulcanizing temperatures.

It is quite common to use cements which will self-vulcanize without heating above room temperature. The desirability of making the leather surface to be cemented very rough and fuzzy before applying the cement cannot be emphasized too much.

Rubber soles for cementing to leather soles have become popular in the past few years. In one type, the sole is coated when received with a layer of tacky rubber and the cement which accompanies it appears to be non-vulcanizing. In another type, the bond is made by the equivalent of a two part vulcanizing cement. The sole is coated when received with one part of cement as for instance, formula 5A in section III (the solvent of course having evaporated). The cement accompanying the sole is equivalent to formula 5B. The two parts (5A and 5B) come into contact in putting on the sole and vulcanization gradually takes place.

(d) Rubber to Metals.

In general, rubber does not adhere easily to metals, but by using certain combinations of metal and rubber, satisfactory adhesion can be secured. Rubber does adhere fairly well to brass and in particular to brass of a composition of approximately 75% copper and 25% zinc.

Cement No. 7 (Rubber to Brass)

	Parts by weight
Rubber	100
Antioxidant	1
Stearic Acid	1
Zinc Oxide	5
Carbon Black	40
Sulphur	5
''Vandex''	0.5
''Captax''	1
Solvent	1000

This cement requires about 30 minutes at 288° F. for vulcanization.

If adhesion is desired between rubber and steel, the steel may first be plated with brass and then treated the same as though it were solid brass.

Another method of securing adhesion between steel and rubber is by the use of a thermoprene cement. This material is of a different class from others described in this circular. It is derived from rubber by the action of certain reagents such as phenol sulphonic acid. The basic patents on this cement are held by the B. F. Goodrich Company of Akron, Ohio. The cement which they produce for uniting rubber to metals is designated by the trade-name of ''Vulcalock.''

There are several other methods for securing adhesion between rubber and metals and many patents covering special materials and processes.

2. Binders

A rubber or latex cement is often used as a binder for other materials. For instance, many packings made principally of asbestos fiber contain rubber as a binder; some brake lining compositions employ rubber for this purpose; and in the shoe industry it is used as a binder for felt and cotton insoles.

3. Sealing Compounds

One of the large uses for latex cement is as a sealing material in the manufacture of ''tin'' cans. The cement is applied to the surfaces of the metal and is forced into the joint during the rolling process used for making the can.

4. The Manufacture of Dipped Goods

In making dipped goods, a form of wood, glass, porcelain, aluminum or a similar material is dipped into a rubber or a latex cement, lifted out and the adhering solution allowed to dry. This operation is repeated until the rubber coating reaches the desired thickness.

If a non-vulcanizing rubber or latex cement is used, the product may be vulcanized by dipping it into a 2% solution of sulphur chloride. (Carbon bi-sulphide is considered the best solvent for this purpose.) The time required for vulcanization is from about 30 to 45 seconds. After vulcanization the product should be dipped in an ammonium hydroxide solution to neutralize the excess of sulphur chloride and check the vulcanizing before it has gone too far.

If a vulcanizing cement is used, the product may be vulcanized with heat. In general, products which are vulcanized with heat are apt to have better aging characteristics than those vulcanized with sulphur chloride.

The process of making dipped goods, although simple in principle, requires great care and attention to details in order to obtain a satisfactory product. The cement must be uniformly smooth and of the proper consistency. Dipping must be done with care, the drying process carried out at the proper rate and precautions taken to avoid any dirt or rust.

5. Rubberizing Fabrics

Rubber-coating of fabrics for use in the manufacture of products such as raincoats, balloons, sheeting, etc., is an important part of the rubber industry. The rubberizing is usually done by the spreading process. The untreated cloth is run from a roll into a spreading machine where rubber cement is distributed evenly over the surface by means of a spreader knife. The cloth is then passed over a hot plate to evaporate the solvent and then is re-rolled. It is passed through the machine as many times as may be necessary to obtain the desired thickness of rubber. If a non-vulcanizing cement is used, vulcanization is produced by passing the cloth through a chamber containing sulphur chloride vapor, after which it is neutralized to prevent ''tendering'' of the fabric. If a vulcanizing cement is used, the cloth may be vulcanized with heat by hanging in a heated chamber or by rolling it between liners and vulcanizing in steam.

If vulcanized latex is used for spreading, no vulcanizing treatment is necessary. This makes the process well adapted to use for silk, rayon, etc., to avoid damage to the fabric or injury to delicate colors.

VI. Production of Rubber Cement on a Small Scale

Although various types of rubber cements can be purchased ready mixed, there are circumstances under which the individual may wish to prepare cements in lots ranging from a few ounces to a few gallons at a time. The first difficulty likely to be encountered is in the purchase of ingredients. These are available in only a few of the larger cities, and the dealers for the most part are accustomed to selling to manufacturers in large lots, rather than at retail in small lots.

The second difficulty which may be met is in the milling or "breaking down" of the rubber. This requires the use of a rubber mill or internal mixer. As indicated in section II, this difficulty may be partly overcome with a solvent activator. It is not known whether there are any firms which regularly engage in the business of supplying milled, or "broken down" crude rubber in small lots, but many small firms are willing to mill occasional lots of rubber for a reasonable consideration. Some users obtain a sufficient supply of milled rubber to last for several months and make up cement in small lots as required.

The third difficulty which may be met is in the mixing of the cement. As indicated elsewhere this can be done by hand, but if large or frequent batches are required, hand mixing is apt to be tedious and time-consuming. Mechanical mixing can be accomplished by various laboratory mixing, stirring, or shaking devices, or by small dough mixers or churns which are designed along the same lines as the large scale commercial cement mixing equipment.

To offset these difficulties, there may be certain advantages in making rubber cement on a small scale. The uses of rubber cement are so multitudinous that there may be purposes which cannot be served as well by standard commercial cements as by a cement made up for the particular job. Furthermore, there may be cases where the characteristics desired for a cement cannot be specified in advance, but where it is necessary to "cut and try" until a product of optimum characteristics is obtained. Any development process such as this can often be best served if the investigator has in hand the necessary ingredients for a cement and can blend them in various proportions until the desired results are achieved. In still other cases it may be desirable to make rubber cement on a small scale for the reason that the cost of ingredients for making a cement may be much less than the price of the same product, ready-mixed, when purchased at retail.

Vulcanizing Agent
U. S. Patent 1,860,320

A 50–50 mixture of selenium sulphide and bentonite in powdered form is heated to 125 to 150° C., the sulphite fusing and being taken up by the clay in some sort of adsorbed state. In this form the sulphur is very active, and the compound is said to be a more active vulcanizing agent, which can be used with or without accelerators.

Low Temperature Vulcanization

Dilute the fresh latex to 3 pounds of dry rubber per gallon and treat with 1 part of sodium bisulphite, added in the form of the 5% solution, to 200 parts of rubber. The vulcanizing ingredients are 3 pounds of zinc oxide, 1½ pounds of sulphur, 0.75 pound of accelerator (zinc or sodium diethyldithiocarbonate) and ½ pound of coloring matter. The wetting ingredient is 1¼ liters of saponin solution dissolved in 1½ gallons of water. The coagulant is 3¼ pounds of alum dissolved in 3¼ gallons of water. Thoroughly grind the vulcanizing ingredients into a cream with the solution of wetting agent, and slowly add the cream to the latex with vigorous stirring. Add the alum solution and stir until thickening occurs. After one hour the thick paste forms a coagulum when stirred. It is preferable to machine the coagulum within a few hours but it can be left overnight. The rubber is made into lace crepe and air-dried within 3 to 4 days. Dry rolling is done 12 to 14 days after preparation, and the rubber is made into laminated sheets by the usual Ceylon sole crepe method. The product, suitable for table mats, bath mats, etc., resists heat and sunlight as well and absorbs water less readily than raw rubber. Tests showed that alum was a better coagulant than formic acid for this type of mixing and that sodium bisulphite assisted vulcanization and improved the color of the product.

Eliminating Air-Bags in Vulcanization of Tires
British Patent 397,508

The use of air-bags in vulcanizing tires is obviated by applying a layer of gas-impermeable composition to the surface to be subjected to gas pressure. A suitable composition is gelatin 22.9, water 35.1, phenol 7.8, alcohol 18.0 and glycerol 15.3 parts; this is spread as a thin film and allowed to dry before vulcanizing

Mould Dipping Latex
U. S. Patent 1,814,473

Emulsion.—Especially for mould dipping. To 100 cubic centimeters of 35% latex are added 3.5 grams of bentonite in the form of a 12% aqueous dispersion. This mixture forms a gel with thixotropic properties, thus causing thicker layers to be deposited on the dipping mould.

Joint Filler
U. S. Patent 1,803,178

Bentonite	95–70%
Oil	4–25%
Latex	1–5 %

Bentonite Latex Dispersions

In making emulsions, one part of bentonite is mixed with 7½ to 8 parts of water. 12 parts of this suspension is used with 70 parts of latex. With Para or other grades of crude or reclaimed rubber or mixtures, the compound is heated to 300 or 400° F., under pressure of 50 to 200 pounds and stirred. Another latex emulsion is made by making a water emulsion of latex containing about 35% rubber solids. It is then mixed so that the final batch contains 18 pounds bentonite, 100 pounds water, 10 pounds 35% rubber latex emulsion.

A vulcanizable cement is made with six parts sulphur, one part bentonite, and 20 parts of water. This is mixed with 30 parts of latex.

Rubber isomers are dispersed by grinding with water and bentonite in a colloid mill. Paper coating is made by dispersing 15 pounds of shellac in three gallons of water containing three pounds borax. With this, 100 pounds of clay is mixed and to this mixture rubber latex containing 1% of ammonia is added to produce ten pounds of rubber solids in the mixture.

Coagulants for Rubber Latex

The following materials can be employed as coagulants for rubber latex, each producing slightly different results. The crystalline substances should be first dissolved in water before being added to the latex.

Petroleum Naphtha
Sodium Chloride
Formic Acid
Aluminum Potassium Sulphate
Aluminum Sulphate
Sulphuric Acid

Fillers for Rubber Latex Compositions

The following materials can be employed as fillers for rubber latex compositions each giving different results.

Celite Snow Floss
Powdered Asbestos Fiber
Tripoli
Wood Flour
Cotton Flocks
China Clay

TEXTILES AND FIBERS

Flour and Starch Sizing Preservatives

When chloride of magnesium is used as a deliquescent substance, chloride of zinc should be used as the preservative. In pure sizing, salicylic acid, carbolic acid, boric acid, sulphate of copper may be used. The following table shows the amounts of the various antiseptics required in pure sizing. The amounts are based on the total weights of starch or flour:

	Wheat Flour Percent	Other Starches Percent
Zinc Chloride	8.0	6.0
Salicylic Acid	0.5	0.3
Cresylic and Carbolic Acids	1.0	0.5
Sulphate of Copper	0.5	0.3
Boric Acid	3.0	2.0
Silicofluoride of Soda	3.0	2.0

Flexible Textile Size

Formula No. 1

Glyceryl Phthalate	10 parts
Soda Ash	2 parts
Water	50 parts

Heat and stir until dissolved.

Formula No. 2

Glyceryl Phthalate	10 parts
Ammonia 26°	3 parts
Water	100 parts

Tracing Cloth

U. S. Patent 1,934,824

Formula No. 1

40 parts by weight of nitrocellulose and 10 parts by weight of tricresyl phosphate are dissolved in 1000 parts of a suitable lacquer solvent mixture in the usual way and to this solution 100 cubic centimeters of a suspension of about 30% starch, as well as 60 cubic centimeters of a 6% colloidal calcium carbonate solution, are added. This mixture is applied to cellulose hydrate film.

Formula No. 2

To a solution of 40 parts by weight of nitrocellulose and 10 parts by weight of a softening agent in 950 parts of a known solvent 80 parts of a resin, for instance dewaxed dammar, are added and this mixture is applied to the film.

This new tracing material is much superior to prior art materials due to its higher permeability to photo-chemical rays. India ink drawings may be reproduced by the usual processes of exposure to light much more quickly than by means of the usual tracing material. Furthermore, lines which do not reproduce well, such as lead pencil or chalk lines, and which only give blurred outlines with the usual tracing materials, give clear, sharp outlines when the new material is used. This constitutes an important innovation since, in industry, for the sake of simplicity about 80% of all drawings are made with a lead pencil. The new material is, furthermore, superior to nitrocellulose in being less flammable, and is superior to gelatine in having higher mechanical strength.

The new material is not damaged by drawing pens and erasures can be made well on it.

Vulcanized Fiber

U. S. Patent 1,935,692

Formula No. 1

The manufacture of a specifically heavy, very horn-like fiber is carried out by means of a zinc chloride lye containing 70% by weight of zinc chloride and 1.5% by weight of calcium chloride. The specific gravity of the lye is 1.975 at a temperature of 20° C. and the temperature of working is, for example, 60°C. In this manner a very horn-like, easily workable vulcanized fiber is produced of a specific gravity exceeding 1.4.

Formula No. 2

From a porous and absorbent paper a vulcanized fiber is produced by means of a zinc chloride lye containing 71% by weight of zinc chloride and 0.6% by weight of calcium chloride. The specific gravity of the lye is 1.980 at a temperature of 20° C. and the working is effected at a temperature of 60° C. The specific gravity of the fiber produced is 1.25–1.35. it is tough and can be easily punched.

Formula No. 3

From a porous and absorbent raw material a vulcanized fiber is produced by means of a zinc chloride lye containing 70.5% by weight of zinc chloride and 3% by weight of ammonium chloride. The specific gravity of the lye is 1.975 at a temperature of 20° C. and the working is effected at about 55° C. A light and flexible fiber is obtained, possessing a specific gravity of about 1.2.

Dressing, Sizing and Softening Oil
U. S. Patent 1,946,332

In order to obtain a fatty acid emulsion suitable for the textile industry melt together at a temperature of 56–60° C. equal parts of olein and dried phosphatides which have a soya bean oil content of 30% (and are obtained by extraction with 90 parts of benzol and 10 parts of 96% alcohol). Then by adding an equal quantity of water, keeping the temperature at 50–60° C. and adding small quantities of dilute soda lye, the mixture is emulsified under vigorous stirring. When required for use, this emulsion is diluted with further quantities of water in the desired manner.

Pre-shrinking
U. S. Patent 1,959,406

A pre-shrinking bath for cotton textile materials comprises an aqueous solution containing 20 parts of water, 2 parts of ammonium alum and 1 part of sodium bisulphite.

Textile Gloss Oil
Formula No. 1

Paraffin Wax	20 parts
Lanette Wax Sx	10 parts
Water	70 parts

Heat together and stir until dispersed.

Formula No. 2

Paraffin Wax	20 parts
Petrolatum White	2 parts
Glycosterin	10 parts
Water	180 parts

Heat together and stir until dispersed.

Plush Softening Liquid

Olive Oil	50 parts
Lanette Wax Sx	10 parts
Water	43 parts

Manufacture of Thin-Boiling Starch
U. S. Patent 1,871,027

Dried starch containing 1.5–2% of water is treated with 0.02–0.06% of dry chlorine in an oil- or steam-jacketed mixing vessel at 105–120° until the desired modification is attained, usually in 4–5 hours.

Oil Cloth
U. S. Patent 1,956,343

The following are examples of compositions for producing an oil cloth with only one application. They may be put on by hand or machine, but should not be pressed into the cloth.

Formula No. 1

Casein Paste	2 lb.
Heavy Boiled Linseed Oil	1 lb.
Dry China Clay	4 lb.

Formula No. 2

Casein or Glue Paste	1 lb.
Latex	1 lb.
Dry China Clay	2 lb.
Dry Silex	2 lb.
Varnish	¼ lb.

Formula No. 3

Casein or Glue Paste	1 lb.
Heavy Boiled Linseed Oil	1 lb.
Dry China Clay	2 lb.
Dry Silex	2 lb.

Formula No. 4

Casein Paste	1 lb.
Heavy Boiled Linseed Oil	½ lb.
China Clay	3 lb.
Flour or Starch Mixture	1 lb.

Formula No. 5

Casein Paste	2 lb.
Flour or Starch Mixture	2 lb.
Heavy Boiled Drying Oil	1 lb.
China Clay	5 lb.

Formula No. 6

Casein Paste	2 lb.
Flour or Starch Mixture	1 lb.
Heavy Boiled Drying Oil	1 lb.
China Clay	4 lb.

Formula No. 7

Casein Paste	1½ lb.
Heavy Boiled Drying Oil	½ lb.
China Clay	4 lb.
Varnish	¼ lb.

Formula No. 8

Rubber Latex	1 lb.
Casein Paste	½ lb.
Heavy Boiled Drying Oil	¼ lb.
China Clay	2 lb.

Formula No. 9

Casein Paste	1	lb.
Rubber Latex	1½	lb.
China Clay	4	lb.
Varnish	½	lb.

Formula No. 10

Rubber Latex	1	lb.
Casein Paste	½	lb.
China Clay	4	lb.
Varnish	½	lb.

The casein paste mentioned in the above examples preferably consists of a mixture of casein, ammonia, and water, about 1 pound of casein to about 7 pounds of water and enough aqua ammonia to dissolve the casein. One pint (12 ounces) of ordinary household ammonia is sufficient. When the more concentrated commercial aqua ammonia is used, a smaller amount is sufficient.

Washable Shade Cloth
U. S. Patent 1,951,119
Formula No. 1

To a web of sheeting weighing 2.7 ounces per square yard and having a thread count of 68x68 is applied on one side from 0.2–0.8, preferably 0.4, ounces of a coating composition as follows, parts being by weight:

Cellulose Ester	25–60 parts
Softener	15–30 parts
Pigment	25–40 parts

The preferred composition is as follows:

Cellulose Ester	41 parts
Softener	20 parts
Pigment	39 parts

The weights given above and throughout the specification include only the so-called "solids" of the coating composition, that is, they do not include solvents and diluents which may or may not be added, as desired, to obtain a coating composition of the desired viscosity. This coating composition may be applied in one or several coats and by any known means, the conventional doctor knife being the preferred form of apparatus for this coating operation.

A like amount of the coating composition is applied to the other side of the web, which is then, while the coatings thus applied are still in a plastic state, run through a plate press of the type usually employed for embossing, or through pressure rolls. The plates in the press may be smooth or may have a shallow engraving upon them, such as a "Skiver" grain, or the like. The pressure exercised by the plate press or the pressure rolls may be varied widely, but is preferably about 575 pounds per square inch.

After the pressure treatment, the same coating composition as above, or a similar one, is then applied to both sides of the coated web in sufficient thickness to give a total coating composition weight of about 1.5 ounces per square yard on each side. This second coating may also be applied in any conventional manner, the doctor knife being preferred.

A shade cloth thus prepared is of the so-called translucent type and is of greatly improved appearance as compared with shade cloth as heretofore made, in which no pressure treatment is given to the cloth prior to application of the final coat of the coating composition. Due to the pressure treatment, the cloth is appreciably improved with respect to the uniformity of translucency and clearness when viewed by transmitted light, two properties of the greatest importance when the appearance of shade cloth is considered.

Formula No. 2

Shade cloth was prepared according to the method set forth in Formula No. 1, except that the coating composition had the following formula:

Cellulose ester	22–40 parts
Softener	15–30 parts
Pigment	35–40 parts

The preferred composition in this case was:

Cellulose ester	31 parts
Softener	23 parts
Pigment	46 parts

The shade cloth prepared according to this example is similar to that obtained proceeding as in Formula No. 1, except that the shade cloth is of the so-called opaque type, although it is to be understood that the shade cloth is by no means absolutely opaque.

The particular cellulose ester coating composition used in the present method, both before and after the pressure treatment, may be varied widely and need not differ from such compositions heretofore used in this art. Generally, cellulose nitrate is the preferred cellulose ester, although cellulose acetate or other esters may be used. A wide variety of softeners is available for these compositions, as is well known in the art. Among those most suitable are raw castor oil, blown castor oil, blown cottonseed or rapeseed oil, among the vegetable oil softeners, and the so-called solvent softeners such as dibutyl phthalate, dibutyl tartrate, ethyl tartrate, and the like. Obviously, the choice of pigments is extremely wide and includes those commonly used in the industry, or combinations of several pigments, depending upon the translucency and color of the shade cloth desired.

The improvement obtained in shade cloth material by the present method is quite unexpected, as the pressure treatment on the partially coated shade cloth not only securely anchors the coating composition in the cloth and thus improves the durability of the material, but also improves the appearance of the shade cloth to a remarkable degree, increasing the uniformity of the translucency of the cloth beyond expectation, and also giving the cloth greater clearness when viewed by transmitted light.

Impregnating Shoes, Gloves and Garments against Tetraethyl Lead

U. S. Patent 1,933,704

One pound of sodium stearate is mixed with four pounds of commercial glycerin and the mixture heated until the soap is dissolved. The mixture is then cooled to approximately 80° C. and one pound of ethyl alcohol is added to the solution. Leather gloves or other articles to be treated are dipped into this mixture while the temperature of the solution is maintained at around 80° C. and are kept immersed for approximately one minute. The gloves or other articles are then suspended from a suitable device or rack where they are allowed to drain and cool. The articles treated in this way become saturated with the composition and will remain impervious to tetra ethyl lead, and other alkyl lead compounds until they have been completely worn out.

Other material, such as fibrous wearing apparel, and cotton fabrics, may be similarly treated. It will be understood, of course, that the proportions of the ingredients forming the composition may vary within wide limits.

Artificial Crêping of Thread Goods

Impregnate thread with citric acid or lead acetate solution before twisting. Then treat with barium hydroxide solution. Concentration of above solution should be about 4%; this being determined by the degree of twist in the thread.

Radioactive Artificial Fibers

British Patent 392,377

Radioactive substances are mixed with the spinning solution. Up to 0.02%

radium sulphate may be added for producing weak radiators and up to 0.2% for strong radiators.

Lusterizing Black Sisal

Liquor	10	gal.
Gelatine Size	2	lb.
Logwood Extract	2	lb.
Fustic Extract	½	lb.
Pyrolignite of Iron	½	lb.

Work the goods in this bath for 30 minutes, allow to drain off, and brush with suitable brushing machines until dry. If the fiber is not to be polished, 8 ounces whitening per 10 gallons liquor are added to the bath of pyrolignite of iron.

Removing Shine from Cloth

U. S. Patent 1,942,523

A solution of ammonium-acetate in alcohol is rubbed on the fabric. On drying the shine disappears.

Delustering Fabric

Canadian Patent 338,583

Relatively permanent subdued luster is produced on fabric containing yarns of cellulose acetate by treating the fabric in a bath that acts as a swelling agent for the cellulose acetate, which bath contains a soluble salt of an alkaline earth metal and then treating the fabric with a reagent to precipitate the alkaline earth metal in the form of an opaque and substantially insoluble salt. *E.g.*, a fabric consisting solely of yarns of cellulose acetate is treated for 1 hour at 70° in a 25° Tw. solution of barium thiocyanate. It is then treated for 20 minutes in the cold in a 1% solution of sulphuric acid, after which it is rinsed and dried.

Tarnishproof Cloth

U. S. Patent 1,933,302

Take any suitable type of a textile fabric designed for the wrapping or packing material, preferably a pile or a napped fabric such as velveteen or duvetyn and treat it in any suitable manner with a suitable bath, preferably containing a cadmium salt in partial or entire solution therein. While any one of the more or less soluble types of cadmium salts may be employed, preferably employ one of the more soluble salts such as cadmium acetate. Pass the cloth through a bath containing the cadmium acetate in solution, or, if desired, the solution may be merely poured or sprayed over the fabric or the fabric may be dipped therein.

Satisfactory results are gotten by passing the material through a bath made up of one-half pound of cadmium acetate to a gallon of water which, however, may be either strengthened or further diluted if desired. The application of the tarnish proofing cadmium salt is preferably made after the usual dyeing operation and hence does not in any manner affect the type of dye which may be put upon the fabrics. In addition, as sulphur attacks the fabric cadmium sulphide is formed, a light yellow and insoluble substance which does not materially affect even the lightest shades of dye. This is quite an advantage over other sulphides such as silver or lead which tend to strongly discolor any dyed fabric as the sulphide forms in use.

Fire-Proofing Cloth

Paint or soak in a mixture of 50 grams of boric acid, and 60 grams of borax dissolved in 1000 cubic centimeters of water.

Flameproofing Fabric Solutions
Formula No. 1

Ammonium Sulphate	2	lb.
Ammonium Chloride	4	lb.
Water	3	gal.

Formula No. 2

Sodium Tungstate	4	lb.
Di-ammonium Phosphate	1	lb.
Water	2	gal.

Formula No. 3

Boric Acid	1¼	lb.
Crystallized Borax	1½	lb.
Water	3	gal.

Formula No. 4

Ammonium Chloride	3	lb.
Borax	3	lb.
Vinegar	1	pint
Water	3	gal.

Formula No. 5

Di-ammonium Phosphate	1	lb.
Ammonium Chloride	2	lb.
Water	1½	gal.

Formula No. 6

Treat cloth with 2 pounds of di-ammonium phosphate per gallon of water, then treat with a solution of 1½ pounds of alum per gallon of water.

Fireproofing Textiles
U. S. Patent 1,961,108

Saturate the fibrous material with a dilute solution of sodium stannate and then dry it. The sodium stannate should be made slightly alkaline with sodium hydroxide to prevent precipitation.

The dried textile is then immersed in any of the following solutions:

Solution No. 1

Ferric Sulphate Crystals	20 g.
Water	80 g.

Solution No. 2

Vanadyl Nitrate	5–10 g.
Water	90–95 g.

Nitric Acid sufficient to acidify.

The fabric is taken out and dried.

When prevention of afterglow is desired the fabric should previously be treated with earthy pigment such as umber, sienna, prussian blue, chrome yellow, etc., in conjunction with a binder.

If the material to be treated is an unbleached cotton fabric, it should first be soaked in a 1% solution of a "wetting-out" agent to aid penetration.

Moisture-Proofing Clothing

Clothing may be made moisture repellant in the course of dry cleaning by the addition to the dry cleaning solvent of 1 per cent aluminum palmitate (or certain other metallic soaps such as magnesium stearate or oleate which have been used successfully for this purpose). The addition of rosin or paraffin is preferred by various cleaners since it seems to prevent any further deposit of white film on the fabric. The proportion of these latter materials is quite small and varies, depending upon shop conditions and temperature of the solvents used. It is very difficult to put metallic soaps into solution, and is therefore far preferable to make about a 10 per cent solution in cleaners' naphtha by heating the soap and naphtha together in a kettle for several hours and pouring the resulting solution into the system. After it is allowed to cool, it will set to gel, but may be redissolved very easily on reheating.

Carbonizing Wool in Cotton Mixture

Some kinds of burnt out embroideries which consist partly of pure cotton and partly also of artificial silk and cotton, are prepared on a ground of wool or cotton. The ground is then usually carbonized before the dyeing, that is to say, removed so that the actual embroidery alone remains standing out.

For cotton embroidery, a wool ground is usually used, and is carbonized by a hot treatment or by boiling for 20 to 30 minutes with caustic soda lye or 3°-5° Tw. The embroidery is then rinsed thoroughly, soured off and dried, the destroyed wool then being removed by heating.

Protecting Silk Stockings Against "Runs"

Silk stockings may be protected against runs by washing them as usual in soap, squeezing them as dry as possible, and afterwards rinsing them in ½ to 1 per cent alum solution. It is quite immaterial whether the aluminum salt used is potash or ammonia alum or aluminum sulphate.

Rubberizing Jute Sacks

Latex	130	parts
Whiting	150	parts
Zinc Oxide	3	parts
Sulphur	2	parts
Anchoracel 2. P.	0.2	parts
Casein Solution (10%)	25	parts
Water, about	35	parts

This formula is stated in parts by weight.

The first step is to make the casein solution; 100 parts by weight of crude acid casein are homogeneously distributed in 900 parts of water and mixed with 15 parts of ammonia 26°; 2 parts of preserving agent should be added. The simplest manner of making the mix is first to add the casein solution to the latex, then the water, and lastly to mix the fillers in gradually by means of a suitable mixing mill. The mix should finally be passed through a color mill.

The mix is spread on the fabric by means of a spreading machine. The fabric should be as smooth and as free of impurities as possible before being treated. The more even and smooth the surface is, the smaller will be the quantities of spreading mix used. The thickness of the spread will naturally be dependent upon the purpose, whether the sacks are required to be merely dustproof or whether they must be more or less waterproof. Generally speaking, it is found desirable to keep costs down by spreading once only. This entails the application of about 2½ ounces of dry substance per square yard. The material must be calendered after each spread in order to secure the penetration of the mix into the pores of the fabric and to press the jute fibers flat against the fabric. The thickness of the spread can be controlled by the position and sharpness of the knife and by the speed of spreading.

It is sometimes desirable to impregnate the jute fabric right through and not merely to rubberize its surface. The best method of impregnation is to pass the jute through a trough containing the mix, to press off the superfluous mix, and then to dry it.

Rubberizing Textiles

The material to be treated, whether it is in the form of individual fibers or as a fabric, must satisfy several conditions to insure success. Mineral acids have a tendency to promote premature coagulation, and may break down the rubber. The removal of such acids from the fiber may often mean neutralization with either soda or ammonia. This applies particularly when dealing with wool because of the tenacious manner in which the fiber retains acid. Clean scouring, the basis of all good finishing, definitely assists in the success of rubberizing for residual grease prevents good adhesion and may be responsible for the setting up of harmful oxidation influences. Provided soft water is always used, residual alkali need not, however, present difficulties. The use of water of either permanent or temporary hardness may result in deposits on the fiber of certain lime and magnesia salts which prevent the best results being obtained. The cloth, for reasons explained above, should also be free from copper and the salts of iron and manganese. Special care should, therefore, be taken when dealing with material dyed with aniline black, because the salts of copper and manganese are often present. Colored goods must be dyed before rubberizing.

When treating fabrics, the side to be rubberized should have as smooth a surface as possible. It is obvious, then, that certain weaves give the best results, although stockinette fabrics are often treated with good results. In some cases smart cutting is helpful in achieving the required surfaces.

It is worthy of note that of all the filling agents it is possible to use for textiles, latex is the only one to impart what is a most desirable property, namely, resilience. Some hair fibers possess this characteristic, but these when rubberized acquire other valuable properties. The power of recovery of rubberized fabrics is a most remarkable and valuable feature.

Fabrics doubled with latex have many additional properties. It is not surprising, therefore, that such cloths are finding an increasing use in the production of e.g., shoe linings, boot tops, wind jackets, printers' blankets, motor car hoods, miners' clothing and balloon fabrics. In this case a standard doubling mix consists of:

Concentrated Latex		
(e.g., Revertex)	100	parts
Whiting	50	parts
Zinc Oxide	3	parts
Sulphur	2	parts
Du Pont Accelerator 552	0.2	part
Casein Solution (10%)	5	parts
Water, about	15	parts

French chalk or barytes may be substituted for whiting, which serves to promote permeability to air as well as to cheapen the spreading mix. If more weight or a better handle is required the quantity can be increased, in which case it may be advisable to increase the quantities of water and casein solution. Curing is effected by the addition of zinc oxide, sulphur, and accelerators under heat. If Du Pont Accelerator 552 is taken, it is sufficient to run the fabric over a heating plate or cylinder heated to about 100° C.–120° C. Vulcanization of the mix sets in and continues at ordinary room temperatures.

The quality of the material, and the degree of waterproofness, determine the amount of mix to be spread. The doubling mix will adhere to untreated material better than to a smooth dressed cloth. Too pronounced penetration of the mass into the fabric may be prevented by using a sharp spreading knife, set at an acute angle to the direction in which the fabric is travelling. The fabric should also pass under the knife as quickly as possible and without friction.

With light weight fabrics it frequently happens that a light colored mix is visible through the new material. This has the effect of causing an apparent change in shade. In such cases suitable coloring agents should be added to the mix. When dealing with pure white cloths, it is possible to avoid discoloration by displacing 10 parts of whiting in the above mentioned mix, and substituting 20 parts of titanium dioxide along with a further 5 parts of 10% casein solution. Here the addition of a water soluble blue is also helpful to tint the mix slightly.

Where good adhesion only is required, as in shoe fabrics, the cloths may be doubled in one operation by the aid of a combined spreading and doubling machine. It is also possible to use a calender having a spreading knife in front and drying apparatus behind. One fabric, after passing under a pulley, passes a spreading knife without counter pressure, and joins the second length at the doubling roller. The doubled fabric is then dried and rolled up. Speaking generally, when about 1 ounce of the mix is used for 1 square yard the material is sufficiently permeable to air. Penetration during calendering must be avoided, and the mix must not have time to dry out before doubling. For this reason it is recommended that the spreading knife and doubling rollers be set at 18" to 1 yard apart. When uniting the cloths, the highest possible pressure should be exerted to insure equally strong, adherence to both cloths.

Where waterproof doublings, or heavier cloths are treated, it is preferable to carry out the process in more than one operation. Here the whiting may be reduced or even omitted altogether. Each length of cloth is coated separately on the spreading machine, either one or several times. To prevent the subsequent applications penetrating into the base of the fabric a sharp spreading knife is preferable. A blunt knife may be used for subsequent coats, however. Three to four ounces of the mix per square yard should give good waterproof qualities. The spreads of each length of cloth must be sufficiently dry before they are doubled on the calender to avoid penetration.

There is an increasing demand for artificial velvet to serve a variety of purposes, e.g., for curtains and upholstery. Of the several methods of production the simplest is the use of a concentrated rubber latex dispersion. Practically any cloth is suitable for the fabric base, but, of course, certain weaves are better than others. It is not unusual to give the fabric a preliminary light dressing of latex on one side. This promotes a smooth surface in the case of cloths which lack the required smoothness, and also prevents excessive shrinkage. Where a light weight fabric forms the base, or a waterproof product is required, a thorough initial coating is required. A representative formula for a suitable compound is as follows:

Concentrated Latex		
(e.g., Revertex)	100	parts
Sulphur	3	parts
Zinc Oxide	5	parts
10% Casein Solution	5	parts
Captax	0.5	part
Zimate	0.5	part

Modifications of the above may be necessary to suit particular cases. The initial coating is effected by a spreading machine, during which penetration of light-weight fabrics is avoided by stretching and not exerting counter pressure. The spreading knife should always be fairly sharp, and set in the opposite direction to that in which the cloth is running.

The mix is then diluted with water, and poured into the trough through which the fabric is running on a rubber roller. The level of the mix is adjusted so that it just reaches the surface of the fabric. The artificial velvet base then goes forward to the machine which applies the dust on the still moist surface. Rotating sieves, preferably hexagonal in shape and of 1 millimeter gauge, carry the dusting media. They should cover the same area. As the amount of dust they release varies with the amount they contain, both sieves should be in operation when they are full or nearly empty, and only one operating when they are half empty, and, therefore, releasing most dust. About 5/6 times as much dust is shaken on the base as is finally required. Meanwhile, the fabric is beaten quickly and regularly from underneath with flat instruments to insure that each individual hair assumes as prependicular a position on the cloth as possible. The material is then left for about 5 minutes on a hot plate, or pressed over drums heated by approximately 15 pounds of steam to dry it.

Brushing with soft cylindrical brushes takes place when the artificial velvet is quite cold. The superfluous dust removed by this process may be recovered and used again. Vulcanization, the final stage, is best carried out in a hanging position and heating for 20–30 minutes at a temperature of 140° C. to 150°C. Where production on a large scale is practiced drying cupboards are used, but it is also possible to use a tenter frame.

Different effects may be obtained by varying the dusting media, silk, artificial silk, wool, and cotton dust are all used. The depth of pile may also be varied, and it is possible to emboss these artificial velvets. A similar process is used to produce cloths hardly distinguishable from moquettes and suede.

A further development using latex is in the production of carpets. In the axminster type the tuffs are loosely held together by the back of the carpet, and kept in place by a size of starch or glue. When a suitable latex compound is used in place of the starch or glue a much more pliable product results. The compound is applied to the back of the carpet by means of a roller dipping in a bath of the mix. A doctor knife removes the excess. The compound is dried

off and vulcanized by passing the carpet through a heated chamber. A superior finish to that usually obtained by back sizing results, and the product is more resilient, stiffer, less susceptible to wear, and the pile more firmly bound in the foundation. Yet another type of carpet, in which the pile is not woven with the fabric but formed from a card web of suitable hair material, is also being made. The web passes over a fluted roller and the bases are treated with a latex mix while in the crimped form. This is then made to come in contact with a simple fabric base also treated with latex. In this way the pile is made to adhere to the fabric, and after vulcanization by moderate heat, brushing and cropping give a lofty and serviceable carpet.

Of equal importance is the use of latex in the felt industry. Doubled woolen felts are chiefly used in the manufacture of slippers, and a great advantage accruing to the user of latex is that excellent adhesion is obtained with the minimum of material. Thin layers have the added advantage of giving permeability to air, although by using a thicker layer complete waterproof qualities will result. When coarse materials such as molten (swan skin, felt padding) which absorb large quantities of mix are used, it is clear that the spreading method entails a larger consumption of mix than should be necessary to obtain good adhesion. In such cases both materials should be sprayed with a mix by means of spraying guns, and then passed through a calender. The nozzles should be situated in such a way that the cloth is sprayed immediately before it passes through the calender. It is also possible to have a battery of stationary guns.

The single sided rubberizing of felts is carried out in the same manner as the rubberizing of fabrics. The spreading machine is the most suitable as it enables a thicker spread to be obtained with comparatively few spreadings. As this method is only suitable for long lengths of felt however, short lengths may be sprayed. It is possible to emboss these rubberized felts. This may be done before the surface has completely dried. The felts should be pressed under a hydraulic press in which the pattern is marked out by means of press sheets. Vulcanization should be effected simultaneously with the pressing operation. Less pressure is required when the surface is still damp, and a smaller pressure affects the thickness of the felts in a lesser degree. The treated felts increase in thickness again during the drying operation. A better final product is obtained by the use of a press in which vulcanization can be effected at the same time, although the operation may be carried out in a press calender.

In the rubberizing of smooth or embossed felts for use as non-skid backings, a low filled mix may be used as the addition of fillers beyond certain point has an adverse effect on the non-skid properties. Although the addition of either oils or resins enhances the non-skid properties, their use is not advisable because they affect the aging properties of the rubber. This is of paramount importance for the very thin layer which is applied is very susceptible to oxidation influences.

Matt Printing on Rayon
British Patent No. 309,194

White or colored matt effects are possible upon yarns or fabrics composed of the esters or ethers of cellulose, or its transformation products, by printing thereon a suitable thickened solution containing urea or a derivative thereof. Depending upon the quantity of urea or its derivative used, more or less opaque effects are obtained. By adding a suitable dye to the printing paste, the matt effects can be dyed or shaded. For example, a white fabric is printed with a paste containing 200 to 300 grams of urea, 400 grams of 65 in 1000 tragacanth paste, and 400 to 500 cc. of water. After drying and steaming for 5 to 10 minutes, it is washed and dried again. A milky matt effect on a lustrous ground is obtained.

British Patent No. 328,978

The application of preparations containing mono- or di-methylolurea to textiles, either locally, or generally and then removed in portions by a solvent. In printing pastes these preparations give clean prints, or without dyes they give luster and damask effects. On mercerizing the printed goods the methylolureas protect the fabric and crepe effects are obtained. For example, a matt effect on acetate silk fabric is obtained by padding with an aqueous paste containing 15 per cent of dimethylolurea and 45 per cent of gum arabic, and drying. Glossy effects on a matt ground are obtained by printing the above padded and dried goods with a paste containing 800 parts of methylcellulose and 200 parts of triethanolamine, steaming, and rinsing.

Printing Bronze Powders on Textiles
Formula No. 1

Sericose (or Similar Compound)	125 g.
Phenol (90%)	350 g.
Methylated Spirit	300 g.
Commercial Acetone	250 g.

Formula No. 2

Sericose	125 g.
Phenol (90%)	350 g.
Commercial Acetone	525 g.

The bronze powders are added in suitable proportions.

The sericose dissolves easily in the cold; more rapidly if heated on a water bath.

The greater part of the solvents is driven off during the passage of the printed goods through the drying chambers connected with the printing machines, and partial fixation of the metallic powders takes place at this stage. But a certain amount of phenol is still retained by the cloth, and can only be removed by a short steaming in a rapid ager—an operation which completes the precipitation of insoluble sericose, and finally fixes the bronze and, at the same time, frees the goods from the odor of phenol.

Colors made as above are distinguished by their great fastness. Boiling soap is without action on them; and if they are printed along with ordinary steam colors, or with discharges on dyed grounds, they withstand all the after-treatments demanded by these styles (steaming, fixing in tartar-emetic, soaping, etc.). They work well in the printing machine, do not stick in the engraving, and keep almost indefinitely provided the solvents are not acid.

Metallic Finish Cellulose Acetate Thread ("Bayco" Yarn)

The following compositions are printed by means of rolls on leather, fabric and metallic fabric.

Formula No. 1

Malachite Green	1%
Cellulose Acetate	5%
Mono Methyl Glycol Ether	30%
Benzol	32%
Acetone	32%

Formula No. 2

Rhodamine B. Extra	¾%
Cellulose Acetate	6½%
Ethyl Lactate	18 %
Mono Methyl Glycol Ether	15 %
Benzol	29⅞%
Acetone	29⅞%

Formula No. 3

Supramine Yellow R.	1½%
Cellulose Acetate	6½%
Mono Methyl Glycol Ether	36 %
Benzol	28 %
Acetone	28 %

Formula No. 4

Alizarine Rubinole R.	½%
Cellulose Acetate	6¾%
Mono-Methyl Glycol Ether	33 %
Benzol	29⅞%
Acetone	29⅞%

Formula No. 5

Fanal Blue R.	4%
Cellulose Acetate	6%
Mono-Ethyl Glycol Ether Acetate	30%
Benzol	30%
Acetone	30%

Formula No. 6

Helio Fast Red R. L.	5 %
Cellulose Acetate	7 %
Mono-Ethyl Glycol Ether Acetate	33 %
Benzol	27½%
Acetone	27½%

Logwood Textile Printing

Formula No. 1

Logwood blacks as used to-day in many printworks are made up on the following lines:

Printing Black Standard

Water	4½ gal.
Hematin (crystals)	42 lb.
Acetic Acid (80%)	1½ gal.

To this mixture add:

Bichromate of Potassium	16 lb.
Water	3 gal.
Sulphuric Acid (168° Tw.)	½ gal.

Then add:

Iron Liquor (24° Tw.)	1 gal.
Bisulphite of Sodium (72° Tw.)	3 gal

Make up to 11 gallons Black Standard. From this standard the printing color is made up as follows:

Water	9 gal.
Printers' Starch	12½ lb.
British Gum	7 lb.
Olive Oil	½ gal.
Acetic Acid	½ gal.
Printing Black Standard	6 gal.

Make this up to 12½ gallons Printing Color.

Formula No. 2

Another black in common use to-day is made up as follows:

Black for Printing

Starch Paste	5 gal.
Logwood Extract (48° Tw.)	½ gal.
Acetic Acid (40%)	½ gal.
Persian Berry Extract	⅛ gal.
Chromium Acetate Solution	1 gal.

In this case the Persian berry extract is added to turn the rather bluish cast of the black, in order to give it density and depth of tone. The recipes that have been given are typical of those in use in practically every printworks. There are other blacks on the market fixed by a chrome mordant. Many of these possess advantages over logwood in fastness properties, etc., but their price is against them and for ordinary styles in which fastness is not paramount, logwood blacks are used.

Black Cloth Printing Paste

Starch	150 g.
British Gum	80 g.
Water	528 g.
Chlorate of Soda	40 g.
Oil (Cotton-seed)	35 g.
Methyl Violet (to lighten the color)	5 g.

Boil, cool and add:

Aniline Salt (Hydrochloride)	95 g.
Aniline Oil	17 g.
Sulphide of Copper (30% paste)	50 g.

Print on white cloth, dry, and pass for three minutes through the rapid ager at not above 65° C. If desired, the development of the black may be more slowly but quite as well effected by hanging the printed goods in an aging room for 10-24 hours. This method is adopted when the black is printed along with aluminium or iron mordants, and it develops very well at the temperature of a cool chamber 55°-60° C. Ordinary goods, after aging or steaming for three minutes, are afterwards chromed in a hot solution of bichromate of potash (1 per cent).

Printing of Vat Colors on Cotton and Rayon

Color	20 %
Sodium Sulphoxalate Formaldehyde	7.5%
Potassium Carbonate	8 %
Glycerine	4 %
Turpentine	1 %
British Gum (2 lbs. per gallon)	59.5%

Reduce Color Strength with

Sodium Sulphoxalate For-
maldehyde 3%
Potassium Carbonate 8%
Glycerine 4%
Turpentine 1%
British Gum (2 lbs. per gallon) 84%

Print on cloth, dry, steam in aging chamber for four minutes at 215° F., oxidize in mild peroxide, perborate, chrome and acetic acid or any such oxidizers, rinse, soap, rinse and dry.

The above colors are the fastest colors used on cotton and rayon. Used on shirtings, dress goods, etc.

Printing of Basic Colors on Cotton, Rayon Silk, etc.

Basic Color 4%
Acetic Acid 56% 10%
Water 15%

Heat to dissolve and add

Gum Tragacanth Solution
(6 ozs. per gal.) 53%
Tannin Acetic Solution
(4 pounds Tannic Acid dissolved in 1 gallon 28% acetic acid) 18%
Reduce color strength with
Gum Tragacanth Solution
(6 ozs. per gal.) 90%
Acetic Acid 28% 10%

Print on cloth, dry, steam in a continuous steamer for 1½ hours, run through the following bath.

Tartar Emetic or Antimony
Salt 1 %
Sodium Carbonate ⅛%

Then given a light soaping, rinse and dry. Basic colors are extremely bright but inferior in fastness to light and washing. They are used mainly for printing draperies where fastness is not essential.

Acid Colors on Silk and Wool

Color 4%
Glycerine 10%
Water 15%

Paste color with glycerine, add water and heat to dissolve

Gum Tragacanth Solution
(6 ozs. per gal.) 69%
Oxalic Acid 2%

Reduce color strength with

Gum Tragacanth
(6 ozs. per gal.) 93%
Oxalic Acid 2%
Glycerine 5%

Print on cloth, steam in continuous

steamer for 1½ hours, run through cold water then soap lightly, rinse and dry. These colors are used extensively on silk and wool because of their extreme fastness. They are used on dress goods, ties, etc.

Mordant Colors on Cotton, Rayon, Silk and Wool

Dry Color or 20% Paste 4%
Water 15%
56% Acetic Acid 3%
Heat to dissolve
Gum Tragacanth Solution
(6 ozs. per gal.) 68%
Chromium Acetate Solution
(10%) 10%
Reduce color strength with
Chromium Acetate Solution
(10%) 5%
Acetic Acid 56% 2%
Gum Tragacanth Solution
(6 ozs. per gal.) 93%

Print on cloth, age 1 hour in a continuous steamer, run through cold water, light soaping, rinse and dry. Used as an intermediate class of colors as they possess good all around fastness properties. However they are somewhat duller in shade than basics and not so fast as vats. They are used on fast color draperies to quite an extent.

Pigment Colors on Cotton, Rayon, etc.

Dry Color 20%
Glycerine 4%
Gum Tragacanth
(6 ozs. per gal.) 56%
Blood Albumen Solution (40%) 20%
Reduce
Glycerine 4%
Gum Tragacanth
(6 ozs. per gal.) 96%

Print on cloth, steam 4 minutes at 215° F., soap well, rinse and dry. This class of colors is used largely on awnings. The colors are not fast to rubbing but otherwise are fairly fast. They give the goods a harsh handle.

Resists for Acid Colors on Wool

Practically all resists to acid colors on wool are variations of the same basic idea. Method 1 is suggested for previously dyed wool.

Method No. 1

Scour and dye skeins as usual. Work for 10 minutes in a cold bath containing

25% sulfuric acid. Wash and immerse in a second bath containing 25% tannic acid. Bring to a boil in one hour and keep at a boil for an additional two hours. Cool down 15 minutes and add 20% Tartar Emetic. Bring to a boil and hold at a boil for 45 minutes. Cool to 100° F. and add 4% tin crystals. Remove skeins after 15 minutes and wash thoroughly.

Method No. 2 is somewhat similar to the above but will not felt the wool as much.

Method No. 2

Immerse the skeins in 40% tannic acid at 180° F. The skeins are worked half an hour and are then left in the cooling liquor overnight. The next day wring the skeins out evenly and immerse in a cold bath containing 20% Tartar Emetic for half an hour. Then rinse lightly and immerse in a fresh cold bath containing 4% tin crystals for 30 minutes. It is usually necessary to add about 1% hydrochloric acid to the tin crystal bath to keep it clear. A final rinse completes the process.

Method No. 3 is similar to No. 1, but requires less time to accomplish.

Method No. 3

Enter the skeins into a bath containing 30% tannic acid and 5% acetic acid. Bring slowly to a boil and boil two hours. 15% acetic acid is gradually added while boiling. Then transfer to a second bath containing 4% tin crystals and 6% formic acid. Treat for one hour at 195° F. A final wash completes the resist.

Method No. 4 is the shortest resist process. It is not so completely effective as the others but will answer the purpose in most cases.

Method No. 4

Work the skeins 15 minutes at a boil in 10% tannic acid. Wring out evenly and immerse in a cold bath containing 5% Tartar Emetic for 30 minutes. Wring or squeeze out evenly again and immerse cold in a third bath containing 5% tin crystals for 30 minutes. A wash completes the process.

If it is desired to resist only part of the skein, the tub used for resisting is partially filled and the skeins suspended from sticks so that only the portion to be resisted is immersed in the liquor. All percentages given above are based on the weight of woolen material being treated. The treatments can be used both on dyed wool and on undyed skeins.

Wool Discharge

This line is still dyed almost exclusively with acid colors. The weak link in this group lies in the absence of a vivid level dyeing blue which can be discharged to a permanent white. Up to now such an article does not exist.

Every printer is acquainted with the unsatisfactory appearance of materials dyed blue which are not permanently discharged. Because of the deficient fastness of the dyeings to perspiration, washing and pleating, there still remains much to be desired. However, the Chrome Fast Dyestuffs are destined to fulfill a requirement long desired.

In their total fastnesses, they are naturally considerably superior to the acid dyestuffs. The following procedure is recommended for their use:

Dyeing: Enter into the dyebath at 60–70° C. with addition of 10% Glauber salt and 2–3% Acetic Acid (30%) or 1% Formic Acid (85%). Bring to boil slowly. Boil for ½ hour; then add 3–5% Acetic Acid or 1–2% Formic Acid or 1–2% Sulphuric Acid and boil for additional ½ hour. The steam is then turned off and Potassium Bichromate or Sodium Chromate is added and the shade is developed by boiling ½ hour longer.

In general, the Erio Chrome Dyestuffs require relatively small quantities of Bichromate; half the amount of dyestuff used. A 2% dyeing therefore requires 1%, and a 3% dyeing requires 1½ Bichromate.

The 1½% Bichromate should be the maximum amount necessary as otherwise the dyeings are ''over-chromed.'' Therefore this limit of 1½% is also sufficient for shades which have to be dyed heavier than 3%.

The Erio Chrome Dyestuffs are distinguished by their purity of shade, level dyeing qualities, penetration and by their remarkable fastness properties. In addition, these dyes are fast to steaming, i.e., they do not change their shade after steaming. The following dyestuffs are suitable for discharging:

Erio Chrome Yellow 6 G
Erio Chrome Yellow S Extra
Erio Chrome Orange R
Erio Chrome Phosphine RR
Erio Chrome Red PE
Erio Chrome Brilliant Violet B
Erio Chrome Cyanine R
Erio Chrome Azurole B
Erio Chrome Azurole G
Chrome Discharge Blue HF
Chrome Discharge Black HFN

The following formula is recommended for discharging:

White Discharge

Hydrosulphite FD Conc.	200 g.
Water	120 g.
Glycerine	50 g.
Gum Water 1:1	250 g.
Sodium Citrate	80 g.
Zinc Oxide 1:1	200 g.
Egg Albumen 1:1	100 g.

The citrate serves the purpose of neutralizing the effect of chrome oxide on the fiber, thereby obtaining a better white. The white could be improved further by the addition of ultramarine to the discharge paste.

After printing the material is dried slightly, steamed 5 minutes in the Mather-Platt (with saturated steam), then rinsed.

Attractive multicolored discharges are obtained with

Chinoline Yellow
Diphenyl Chlor. Yellow FF Supra
Xanthaurin
Acid Red XG
Wool Pure Blue FFB
Gallazol Cyanole B
Nigrosine Yellowish

by using the following formula:

Dyestuff	25–50 g.
Glycerine	50 g.
Water	320 g.
British Gum 1:1	250 g.
Hydrosulphite FD Conc.	150 g.
Sodium Citrate	80 g.
Zinc Oxide Powder	100 g.

The foregoing ingredients are ground thoroughly into as fine a paste as possible.

Wool Pure Blue FFD is dissolved with addition of Resorcine in the proportion of 5 parts dyestuff to 3 parts of Resorcine.

After printing, the material is steamed 5 minutes in the Mather-Platt.

For more permanent fixation of the multicolored discharge, an additional half hour steaming is suggested, after which the material is rinsed and dried.

Gas-Impermeable Coating

A suitable fabric is first coated with the following composition, six coats being applied.

Latex (40%)	10 lb.
Caustic Soda Solution (20%)	1 lb.
Water	14 lb.

It is then coated with six or more coats of the following:

Latex (40%)	25 lb.
Viscose (7%)	550 lb.
Glycerol	20 lb.
Water	45 lb.

The viscose being regenerated by dipping the fabric in a bath of the following composition:

Glycerol	18 parts
Alcohol	48 parts
Boric Acid	6 parts
Water	54 parts

Elastic Aluminum Coating
Russian Patent 840

Cloth is coated with two layers of:

Aluminum Powder	1–3 parts
Gasoline	3.5 parts
Rubber Cement	1 part

The cloth should first be rubberized.

Retting Flax
U. S. Patent 1,951,793

About 2% of the weight of the flax of saponified linseed oil is used as an auxiliary retting bath ingredient to facilitate retting.

Mercerizing Penetrant
German Patent 593,048

The wetting and penetrating capacity of mercerizing lyes is improved by addition of a phenol and an organic base. Thus, a mixture containing tar cresols 90 and aniline 10% may be added to a lye of 32° Bé. in the proportion of 1.17% by weight.

Wetting Out Agents for Mercerization
British Patent 354,946

Sulfuric Ester of n-Amyl Alcohol	85%
1-3 Butylene Glycol Ethyl Ether	15%

British Patent 279,784

Commercial Xylenol	80%
Napthenic Acid	20%

British Patent 365,323

Cresol	92%
Camphor	8%

British Patent 360,472

Cresol	85%
Pinacone	15%

"Wetting-Out" Agent
French Patent 751,422

Sodium Sulfo-Naphthenate	97%
Ammonium Ichthyolsulfonate	3%

Preparation of Alpha Pulp

Cook the wood by the acid sulphite process so as to have the lignin removed incompletely.

Method No. 1

The pulp is chlorinated so as to convert the lignin into a form soluble in alkali. Wash, then cook with 10% of its weight of sodium hydroxide at 15 lbs. steam pressure. Wash, Bleach.

Method No. 2

The pulp can also be treated with ½% to 2% of alkali at a pressure of 2–3 atmospheres for 3–6 hours.

Method No. 3

A third method is bleach a sulphite pulp of density of 5% with a 1% chlorine solution, wash, and squeeze to a density of 15%. This pulp is cooked at 200° F. for 6 hours with lime water containing 7% reactive calcium oxide. It is then washed and again bleached yielding an alpha pulp of 90–94%

Method No. 4

Kraft pulp is mixed with water to a strength of 8%. This is bleached with 14% Bleaching powder at 100° F. It is then acidified to a pH. of 3 and washed. It is then treated with a 6% caustic soda solution at 125° F. for 1 hour. Washed, acidified with sulphur dioxide and re-bleached with 4% bleach solution, washed and dried.

Method No. 5

Pulp from straw, jute, hemp or bamboo is cut into small pieces and cooked in a digester with a solution of caustic soda containing 12–14 grams per liter using 8% to 12% of the soda solution based on the weight of material, for several hours. It is then drained, washed and placed in a kneader to rub the fibres together. It is again washed and chlorinated for 10 minutes in a closed vessel by means of a strong chlorine water solution. It is neutralized with lime, washed and centrifuged. Water is added to bring the pulp to a consistency of 6–8% and then bleached in a beater at 35° C. with hypochlorite. Washed and dried.

Method No. 6

Pulp from wood is placed in an autoclave filled with a 1–2% caustic soda solution at 75° C. Cooked under a hydraulic pressure of 80 lbs. per square inch for 6–7 hours; washed and chlorinated in a closed chamber with a mixture of moist air and 25–50% chlorine gas for 10 hours at 20° C. The pulp is washed with ½% to 1% caustic soda solution, and the process repeated.

Conditions in Rayon Manufacturing

Note. The name of the pulp is the company name and is important because of the difference in pulps.

Pulp No. 1. is 210–pound Rayonier.
Pulp No. 2. is 160–pound Kipawa and 48½–pound Hercules.
Pulp No. 3 is 214–pound Kipawa.
Pulp No. 4 is 100–pound Kipawa.

Pulp No.	1	2	3	4
Steeping				
Percentage Caustic Soda	18%	18%	18%	17.9%
Per Cent Hemi-Cellulose	0.5	0.5	0.6	0.5
Time of Steep	50 min.	50 min.	50 min.	50 min.
Moisture in Pulp	6.5%	6.5%	4.84%	4.84%
Pressed to Weight	640 lb.	670 lb.	660 lb.	314 lb.
Shredding Time	2 hr.	2 hr.	3 hr.	2 hr.
Shredding Temperature	22° C.	22.5° C.	21° C.	20° C.
Aging of Alkali Cellulose				
Time	72 hr.	72 hr.	80 hr.	72 hr.
Temperature	20.5° C.	21.75° C.	18.5° C.	19.5° C.
Xanthation				
Weight of Carbon Bisulphide	72 lb.	74.5 lb.	74.5 lb.	74 lb.
Time	2 hr.	2 hr.	2.25 hr.	2 hr.
Temperature	21°–28° C.	21°–25° C.	20°–27° C.	21°–26° C.
Dissolving Solution				
Dissolve to 6.5% Caustic Soda and 7.6% Cellulose				
Time of Dissolving to Viscose	7 hr.	7 hr.	7.75 hr.	5 hr.
Time for Ripened Viscose	79 hr.	79 hr.	84 hr.	95 hr.
Temperature	18° C.	18° C.	17.8° C.	17° C.
Viscosity of Viscose	28–31	25–30	34	21
Spinning Index	3.5	3.5	3.5	3.5

The viscosity is the time required for a ⅛ inch steel ball to drop through 20 centimeters of viscose.

The spinning index is the percentage of salt needed to coagulate 1 drop of viscose.

The strength of the spin bath is the same for all the pulps and contains:

Sulphuric Acid	9 %
Sodium Sulphate	18 %
Glucose	4 %
Zinc Sulphate	0.6%
Water	Balance

Rayon Spinning Solution

Cellulose Acetate	27 %
Acetone	65.8%
Xylol	3.6%
Diacetone Alcohol	3.6%

Desulphuring Rayon

Treat with 1% sodium sulphide and 1% glucose solution at 60° C. or with 3% sodium hyposulphite at 70° C. Time of treatment 6–20–30 minutes.

Sizing for Rayon

Rosin	10–15 parts
Olein	2–10 parts
Triethanolamine	7–20 parts
Stearine	7–20 parts
Water	1000 parts

Sizing is done at 50° C. for 15 minutes. Then raise temperature to 60° C.

Desizing or washing is done with ½% soap solution at 60° C. for one hour.

Rayon Size

Lanette Wax Sx	10%
Water	90%

Heat together and stir until dispersed.

(Permanently) Lustreless Rayon
U. S. Patent 1,899,725

Viscose is spun into a bath containing 5–7% of sulphuric acid and 1.5–4.5% of nickelous sulphate at 30–45°.

Delustering Rayon
British Patent 393,985

Rayon comprising cellulose acetate or other acyl, alkyl or aralkyl derivative of cellulose is delustered by treatment at an elevated temperature with an emulsion of less than 5% concentration, of pine oil, i.e., an oil in which terpineols are the main constituents. A powerful wetting and swelling action is exerted by the oil which assists the penetration of dyes. Among examples 10 pounds of cellulose acetate of benzyl-cellulose rayon is delustered by treating 0.5–0.75 hours at 80–90° in a bath containing in 20 gallons of water 2 pounds of an emulsion of pine oil 25, Turkey-red oil 25 and water 50 parts.

Rayon Delustering

Olive Oil Soap (low titre)	10 parts
White Glue	10 parts
Water	100 parts

Dissolve by warming and stirring; add

Paraffin Wax

Heat and stir till emulsified.

Take 2½ parts of above emulsion and add to it

Water	100 parts

Stir and add

Infusorial earth	2.27–4.54 parts

Delustre Rayon with this at 38° C.

Water proofing (Acetate) Rayon
Formula No. 1

Paraffin wax	5 g.
Soap	5 g.
Ammonium Hydroxide (0.880)	30 cc.
Water	970 cc.

Heat to 55° C. and stir vigorously until a uniform emulsion results. Soap the rayon in this emulsion at 45° C. for 15 minutes; remove; centrifuge and treat for 10 minutes at 20° C. with a 5° Tw. solution of aluminum acetate.

Formula No. 2

Treat Rayon at 30° C. with following:

Tin Chloride	3 parts
Ammonium Thiocyanate	1 part

Water: smallest possible amount.
Time: 3 hours.

Rinse and treat with a 5% soap-carnauba wax emulsion at 45° C. Wash well and dry.

Making Rayon Hosiery ''Run-Proof''
U. S. Patent 1,929,705

Alum	70.25%
Salt	50%
Boric Acid	27.00%
Gum Arabic	2.00%
Powdered Casein	.25%

This formula is stated in percentages by weight.

A mixture of substantially these proportions is dissolved in lukewarm water in the proportion of one teaspoonful to an ordinary table glass full for each pair of hosiery whose treatment is desired. After the powder has been thoroughly mingled in the water, the hosiery or other garment is immersed therein and left for a period of about thirty minutes, after which it is removed and allowed to dry, and then washed with clean water and mild soap flakes. Precaution should be taken against the treatment in the same solution at the same time of garments of more than one color because of the tendency of some dyes to fade and run before the dye-setting influence of the solution here offered has had an opportunity to work.

Strengthening Composition for Treating Silk and Rayon Fabrics

U. S. Patent 1,929,705

An aqueous mixture suitable for impregnating hosiery fabrics in order to strengthen them is formed from alum 70 and boric acid 27 parts to which may be added about 3 parts of gum arabic with a smaller proportion of salt and casein.

Opaque or Matt Artificial Silk

U. S. Patent 1,902,529

A solution of a wax (Japan wax, ozokerite) in an organic solvent (benzol) is emulsified with viscose before filtration, deaeration, and spinning; suitable proportions are viscose 98.7, Japan wax 0.4, benzol 0.9%.

Scouring Tussah Black Silk

The fabric is given 8 turns (about 30 minutes) at 160° F. in a bath containing 8 pounds of olive soap and 2 pounds of trisodium phosphate per 450 gallons of water. The scour is followed by a thorough wash which consists of 6 separate rinses. The first rinse is a 30-minute bath containing 3 pounds of sal soda. The next 3 rinses are run for 15 minutes each in lukewarm water. The fifth rinse consists of two ends in cold water. A sour for ten minutes at 120° F. in a bath containing 1 gallon of 28 per cent acetic acid completes the scour.

Weighting of Natural and Artificial Silk Fibers

U. S. Patents 1,902,777 and 1,902,778

Silk weighted in the normal way with stannic chloride-disodium hydrogen phosphate is immersed in a suspension of lead chloride containing 15% of tenth-normal sodium acetate, 10% of concentrated pylnaphthalene-disulphonic acid for 30 minutes at 65° and fixed in a separate bath of 10% aqueous ammonium bicarbonate. Formic acid or chloracetic acid may be used (in conjunction with sodium phosphate in the lead chloride bath) either as a pre-treatment or, in the lead chloride bath, as a substitute for phosphoric acid.

In an alternative method separate baths of lead chloride dissolved by sodium thiosulphate, and of ammonium bicarbonate, ammonium hydroxide, or sodium phosphate are used.

Weighting Silk

U. S. Patent 1,955,440

Textile fibers, such as silk, after the preliminary boil-off treatment, are immersed or otherwise subjected to an aqueous solution containing 14% to 30% of a tin salt such as tin tetrachloride. The silk is permitted to remain in this solution until it is completely penetrated by and saturated with the tin salt. It is then removed and the surplus tin salt removed by centrifuging or whizzing. The whizzed material is then treated with water to hydrolyze the tin salt. The material containing the hydrolyzed tin salt is then subjected to a solution containing a phosphate, whereby the tin salt is fixed and the weighting obtained. If desired, the treatments with the tin salt and the phosphating solution may be repeated to secure the desired weighting.

The following examples illustrate several specific phosphating solutions which have given satisfactory results:

Formula No. 1

Sodium Chloride	40 grams per liter
Disodium Phosphate	75 grams per liter

Formula No. 2

Sodium Sulphate (Decahydrate)	60 grams per liter
Disodium Phosphate	75 grams per liter

Formula No. 3

Sodium Nitrate	30 grams per liter
Disodium Phosphate	75 grams per liter

Immunizing Textiles to Direct Dyes

U. S. Patent 1,895,298

Bleached cotton yarn is boiled in a 12% solution of caustic soda for one-quarter of an hour and then centrifuged while hot until it has an increased weight of about 50%. The alkali-treated material is then treated with a toluol solution containing 22% of para-toluene phosphoric acid and 5% of isopro-sulphochloride at 18 deg. C. for 1¼ hours, whereupon the yarn cannot be dyed with direct dyes. It is claimed that in the case of mercerized cotton this special alkali treatment increases the whiteness and luster. In the case of viscose rayon the luster becomes decreased when the concentration of caustic soda attains 10%.

A New Rapid Bleaching Process
British Patent 401,199

It is possible to avoid the prolonged caustic kier boils which are usually given to cotton yarns and fabrics when being bleached in preparation for dyeing. It appears that a process in which the cotton material is successively treated with an active chlorine liquor and one containing hydrogen peroxide is capable of yielding results hitherto considered possible only by means of one or two boils with 2% caustic soda, each of 5 to 8 hours duration, followed by chemicking.

The bleaching process recommended consists of impregnating the material with a liquor containing sodium hypochlorite (4 grams of active chlorine per liter) and then almost immediately entering it in a liquor consisting of the following:

Hydrogen Peroxide (40%)	7	l.
Caustic Soda	3	kg.
Water Glass (36° Bé.)	7½	l.
Water	2000	l.

and maintained at 90 to 95° C. The cotton material is worked in this liquor for about 4 hours and is then rinsed and preferably soured with dilute formic acid (0.05%).

It is found that by the time that the cotton is entered into the peroxide solution about 75% of the chlorine present in it has been rendered inactive by chemical reaction with the impurities present. The remaining 25% of active chlorine reacts with the hydrogen peroxide and with consequent liberation of oxygen.

The essential point to notice is that this combination of chlorine-peroxide treatment results in an excellent white color while it ensures that the goods have a soft handle and good absorbency.

Bleaching Pile Fabrics

While tussah silk pile is used mainly for black fabrics, there is a demand for fairly light shades at times. Due to the variation in shade of the tussah fiber obtained from different sources and also due to the brown or tan color of natural tussah, bleaching is necessary for practically all shades excepting black and other dark shades which will hide the natural brown of the wild silk fiber. Sodium hydrosulphite is often used alone as a means of obtaining a "half" bleach. For obtaining a full bleach on tussah the best bleaching agent is probably hydrogen peroxide or the latter combined in successive treatments with sodium hydrosulphite. Permanganate bleaches are claimed to give good results on tussah silk.

Tussah silk is more difficult to prepare for bleaching than most natural fibers due to its content of lime compounds in addition to the silk gum. Practically every mill has its own special way of preparing tussah.

Certain types of cultivated tussah silk may be prepared for bleaching simply by boiling for one to two hours in a 1 to 2 per cent olive soap bath. This is followed by a thorough rinse in soft water or in water softened by the addition of sodium carbonate. Other types of wild silk are prepared by boiling in a 1.5 per cent soda ash solution for 30 minutes followed by a boil in 1.5 per cent soap for one hour. The latter boil-off is completed by a thorough rinse in soft water. Still a third method which has proven to be effective is to use a continuous washer. The first box is filled with hot water at 200°–205° F. The second box is filled with a 2 per cent solution of soda ash at 200 to 205° F. This is followed by two boxes of warm water. Box 5 holds a one per cent solution of commercial hydrochloric acid at 100° F. The last two boxes are filled with cold water. Rotating brushes are used before each nip to lay the pile flat. The cloth is plaited off on a flat truck as it leaves the continuous washer. From here it is transferred to a rope dyeing machine and is run for 15 minutes in a .25% soda ash solution at 205° F. Enough previously dissolved olive soap is added to make a .75% solution in which the scour is continued for an additional 45 minutes at 205° F. A rinse in water containing one pound of soda ash per 100 gallons follows the scour and a final warm rinse completes the preparation for bleaching. The actual bleach varies in different plants almost as much as the scour prior to bleaching.

Tussah silk will withstand hydrogen

peroxide treatment at much higher concentration and temperatures than mohair or real silk. It is usually bleached at 180° to 190° F., although some firms bleach this fiber at a temperature closely approaching a boil. The bath is made up using either sodium peroxide or concentrated solutions of hydrogen peroxide. In any case the concentration of the peroxide bath is rarely below one volume or over three volumes of available oxygen. The time necessary varies with the temperature of the bleach liquor; the concentration of hydrogen peroxide, and the purity of white desired. The best "white" obtainable by the use of hydrogen peroxide is somewhat yellowish when compared to the pure white of bleached cotton, but is much better than the best white obtainable by the use of sodium hydrosulphite alone.

The bath is made up with highly concentrated solutions of hydrogen peroxide by adding the peroxide to the water and then slowly adding sodium silicate until the desired alkalinity is obtained. The bath is thoroughly mixed after each addition of silicate by using a wooden hoe.

Where sodium peroxide is used rather than concentrated solutions of hydrogen peroxide, the bath is made up by adding the necessary amount of sulphuric acid to the cold water. Then the sodium peroxide is slowly sifted into the acidulated water until the bath is almost neutral. This point is determined by the use of litmus paper. One ounce of sodium silicate per 50 gallons is then added while stirring constantly to keep any iron which may be in the water in a colloidal state. The proportion of sodium peroxide to sulphuric acid is 1.4 pounds of sulphuric acid to one pound of sodium peroxide for an approximately neutral bath.

Peroxide baths should always be accurately controlled. A sample is drawn off before starting the bleach and 10 cubic centimeter portions are titrated rapidly for alkalinity and hydrogen peroxide content. One practical bleacher starts a tussah bleach with a bath that requires between 28 and 30 cubic centimeters of tenth normal potassium permanganate for peroxide content and 2.3 to 2.5 cubic centimeters of tenth normal sulphuric acid for alkalinity per 10 cubic centimeters. In the latter bath the cloth is run three hours at 190° F. The bleach obtainable with hydrosulphite on tussah is at best a light cream color. However, this half bleach suf-

fices for a large number of mode shades and any light shades having a brownish cast. The bleach is accomplished by first preparing the cloth by any of the methods outlined above. As a rule, the less pure a white is desired the simpler is the method of preparation. The cloth would therefore probably be prepared for a half bleach by a simple soda ash scour followed by a soap scour. The actual bleach consists of simply immersing the cloth in a .5 per cent solution of sodium hydrosulphite at a temperature of 75° F. to 90° F. for a period of 2 to 3 hours. To obtain level bleaching the fabric is turned over once or twice during the bleach. In some cases the cloth is allowed to remain in the hydrosulphite liquor overnight. Following the bleach the cloth is rinsed well, hydro-extracted, and transferred to the dyeing vessel.

Bleaching Straw

A preliminary purification consists of treating the straw with a dilute solution of caustic soda at a temperature not exceeding about 60° C.; at higher temperatures the fiber is adversely affected. This is followed by a thorough washing, and the straw is then steeped for about 12 hours in a warm bath (about 50° C.) containing oxalic acid, potassium carbonate, and sodium bisulphite. Afterwards the straw, which is now pale yellow, is washed and bleached in a hydrogen peroxide bath. It is found that a 6-volume liquor is satisfactory but it is essential that it also contain oxalic acid and a stabilizing agent such as sodium pyrophosphate.

The bleaching process occupies about five days if a really good white color is desired. During the first day in which the fibrous material is steeped, the hydrogen peroxide solution should be maintained at about 50° C. On the second day the temperature should be raised to 60° C. and during the remaining three days the liquor should be at about 70° C. The straw is then washed and finally whitened by steeping for about 12 hours in a bath at 60° C. containing 1 to 2 grams of Blankit I (I. G.) per litre, followed by washing, and treatment with a liquor containing sodium bisulphite and oxalic acid.

After such a prolonged treatment the straw is obtained with a good but slightly greenish white color and is then ready for dyeing.

Bleaching of Manila Hemp

This is differently carried out, and the following is a recommended method: The raw material is first treated for one hour at 60° C. in a bath containing 2 grams of Blankit I per litre, then washed, and entered into a bath at 50 to 55° C. containing hydrogen peroxide and sodium phosphate; about 12 hours' immersion in this bleach liquor is necessary, and it should contain about 0.3 to 1% of hydrogen peroxide. Afterwards the hemp should be well washed in slightly acidified water, since this assists the removal of the slightly yellow color which the hemp retains when withdrawn from the bleach liquor. If a very clear pure white is required then the bleached hemp must be further treated with a dilute solution of Blankit I as in the first treatment.

Before dyeing any of these straw materials it is essential that they be well boiled out in water since this softens them and renders them more easily penetrated by the dye liquor subsequently applied. The acid dyes are most suitable but basic and direct colors are also applied. The basic dyes suffer from the disadvantage of rushing on the fibrous material and coloring it unevenly or of giving dyeings which have a bronzy appearance. Vat and sulphur dyes are not generally used for straw and hemp.

Discharging a Direct Color

Print on the following paste:

Sodium Sulphoxalate Formaldehyde   15%
Starch Tragacanth Thickening 85%

Run through ager, light soaping, rinse and dry. In many cases the addition of 4 per cent anthraquinone paste is quite beneficial. Most direct colors discharge to a good white but a careful selection of colors will eliminate any possibility of poor discharges.

Dyeing Silk with Aniline Black

A bath is prepared as follows: (a) 3 pounds of aniline are mixed with 3 pounds of hydrochloric acid, 30° Tw., diluted with 1 gallon of water; (b) 6 pounds of yellow prussiate (ferrocyanide) of potash are dissolved in 4 gallons of water; (c) 5 pounds of chlorate of potash are dissolved in 5 gallons of water. The three solutions are mixed together, and the silk is worked in the mixture for about half an hour; it is then taken out, wrung lightly, and placed in an aging chamber heated to about 150° F., and left for about fifteen to twenty minutes; then it is introduced into a steaming chest at a pressure of 5 to 10 pounds, and steamed for half an hour. It will now have acquired a fairly good black color, which is more fully developed by passing into a bath a bichromate of potash of a strength of about ½ pound to the gallon. It is then rinsed in water and brightened in the usual way.

Dyeing Assistant
U. S. Patent 1,946,079

2.7 parts of a crude sulfonated oil mixture in the form of free acids are mixed with 1 part of a mixture of di- and tri-ethanolamine.

In dyeing Ponsol Blue GD double paste, Schultz No. 842, on cotton yarn in the package machine, using

Color   5%
Sodium Hydroxide   2½%
Hydrosulphite   3%

The addition of 1% of the product formed by combining 1 mole of sulphoricinoleic acid and 1 mole tri-ethanolamine containing diethanolamine, results in a brighter, more level dyeing.

Detection of Various Colors on Cloth

Para Red. Boil with 1% cupric sulphate solution for one minute. Rinse with running cold water. A permanent brown shade indicates para red.

Indigo. Spot with concentrated nitric acid which will leave a yellow spot with a greenish rim if indigo is present.

Sulphur Colors. Heat a piece of the cloth in a beaker which contains equal weights stannous chloride and hydrochloric acid 20% in water. Place over the beaker a piece of moist lead acetate paper. If the paper blackens appreciably sulphur colors have been used.

Direct Colors. Heat in a test tube with a dilute soap solution at 140° F. Direct colors will bleed very badly. Deazotized and developed colors will not bleed to a very great extent. Both of these classes of colors will strip quite readily with caustic soda and sodium hydrosulphite at 150° F. This is another test that distinguishes both classes from vat colors.

Vat Colors. Heat in a solution of 3% caustic soda and 3% sodium hydrosulphite at 150° F. for a minute. Vat colors will reduce to an entirely different color from the original shade and the original color may be brought back by rinsing and giving a mild oxidizing treatment. Sulphur colors also reduce to another shade from the original but this class can readily be distinguished by the above mentioned test for this class of colors.

Naphthol Colors. In concentrated sulphuric acid this class of colors will give an entirely different shade from the original. The shade of the color in the concentrated acid is very brilliant. This class of colors also strips with caustic soda and sodium hydrosulphite. After stripping there is left a yellow coloration that is the naphthol itself.

Basic Colors. This class of colors is easily recognized by its brilliancy of shade and fugitiveness to light. Another test other than placing in a fadeometer to determine its light fastness is as follows:

Dissolve the color and cloth with concentrated sulphuric acid. Dilute with water and add a mixture of tannic acid and acetate of soda to the diluted mixture. If the dyestuff precipitates basic colors have been used.

Mordant Colors or Chrome Colors. Dissolve the color and cloth with concentrated nitric acid, dilute and test for chrome and aluminum.

Mercerization Test for Cotton. Place a known piece of mercerized cloth as well as a known unmercerized piece together with the sample to be tested in the following solution for 1 minute:

Iodine (Crystals)	10 g.
Saturated Potassium Iodide Solution	100 cc.

Then enter into running cold water. The unmercerized piece will wash out while the mercerized piece will not.

Iron Test on Cloth. Spot the cloth with a dilute solution of hydrochloric acid. Then spot with a weak solution of potassium ferrocyanide. A blue coloration is an indication that iron is present.

Leuco-Ester Vat Dyeing

British Patent 260,638

Formula No. 1

In the dyeing of woolen cloth, printed with a wax reserve as for ordinary batik, the bath consists (on the liter basis) of

Leuco-Ester Salt of 5-7-6'- Trichlor-Indigo	10 g.
pasted up with	
Warm Water	950 cc.
Ethylene Thio-diglycol	10 cc.
Sodium Benzyl-Sulphanilate	10 g.
Potassium Carbonate	10 g.
Formaldehyde Sulphoxylate (or Hydrosulphite)	10 g.

After treatment for about one hour in this bath in the cold, the goods are washed, and the dyestuff developed in a bath containing 20 grams of potassium persulphate and 20 cubic centimeters of concentrated sulphuric acid per liter, at about 35° C., followed by washing and drying. The wax reserve is then removed in any of the usual ways (benzine, benzene, etc.) and for final finishing soaped warm and washed. In this way, white effects upon a violet ground are produced.

Formula No. 2

Bleached woolen goods, or weighted or unweighted silk, are printed (on a machine with deeply engraved rolls capable of being heated during the printing) with a reserve of the following general composition:

Beeswax	50 g.
Colophony	650 g.
Spermaceti	30 g.
Melted Mutton Tallow	15 g.
Paraffine	25 g.
Oil of Turpentine	230 g.

After printing, the goods are dyed as in Formula No. 1, rinsed, and developed, well washed, dried, and the reserve removed in any suitable manner.

Formula No. 3

Ground dyeing—the goods are dyed as in Formula No. 1, in the cold, in any ordinary dyeing machine, jig, etc., and oxidized as before.

Formula No. 4

Spun or woven material (unweighted or weighted silk or chlored wool) the latter unprinted or printed with a wax reserve, are dyed in a bath of about the following composition:

Leuco-Ester Salt of 6-6'- Diethoxy-Thio-Indigo	10 g.
Formaldehyde Sulphoxylate	10 g.
Triethanolamine	10–20 g.
Water	960 g.

as before described. The dyestuff is then developed either as in Formula No. 1, with potassium persulphate, or with potassium chromate and sulphuric acid.

Formula No. 5

The materials mentioned above are dyed in a bath such as the following:

Leuco-Ester Salt of 5–7–6'-
Trichlor-Indigo	5 g.
Water	955 g.
Formaldehyde Sulphoxylate	10 g.
Triethanolamine	10–20 g.
Sodium Nitrite	10 g.

All details as in the previous cases.

Vat Dyeing Piece-Goods

For very pale shades where little actual dyestuff is used, it is often the practice to work from a freshly made standard of the vat dye.

The following represent very simple recipes used to produce a pale ecru shade:

2–5 pieces. Weight, 210 pounds. Length, 120 yards. Mercerized poplins. Shade, ecru 4.

Yellow Standard

Indanthren Yellow 3R	8 oz.
Hydrosulphite (Powdered)	12 oz.
Caustic Soda (76° Tw.)	1 gill
Make up to 1 gallon.	

Brown Standard

Caledon Brown RS	8 oz.
Hydrosulphite (Powdered)	12 oz.
Caustic Soda (76° Tw.)	1 gill

Make up to 1 gallon.

Brown Standard	5 pt.
Yellow Standard (divided)	3 pt.
Hydrosulphite	1 lb.
Caustic Soda (76° Tw.)	2 pt.
Oleine	2 pt.
Nekal BX Dry	1 lb.

in approximately 30 gallons; 4 ends; usual "finish-off." Dyeing temperature, 50° C.

The cloth is run on the dye jig through a weak warm soda ash bath, and is then ready for dyeing. It seems quite a common practice to give the goods two ends through hydrosulphite and caustic soda at 50° C., before entering the actual dye liquor. The idea lying behind this method is that so-called "air-bubbles" will otherwise interfere with a level dyeing. Whether this be the case or not, it has never been found that the extra two ends are beneficial. The standards may be strained before use, and while this is certainly a precaution against color specking, it is not an essential.

The color is divided equally over the first two ends, taking special care that sufficient time has been allowed between adding the standard to the dye jig and commencing running. In the case illustrated probably two to three minutes is quite long enough to ensure obtaining a proper vat. The process is again repeated before commencing the second end. Once the color and hydrosulfite have been added, special care must be taken in seeing that the temperature is not on any account allowed to rise above 50° C., and attention must be given to this point.

A common fault is to turn on the steam, take a reading and find that the temperature has risen above that advocated for dyeing, and then allow to cool down. The damage has already been done with some dyestuffs, and they may partially have decomposed, likewise the hydrosulfite. This is frequently the cause of "out-of-solutions," and a mistake in this direction cannot often be passed off. Two ends are then given in the cooling liquor, probably in the case of such a pale shade, without the addition of any Glauber's salt. During the fourth passage through the liquor, the batches are stopped for a moment, and the jigger cuts a small pattern out for shading purposes.

Mention may be made here of the also important point that once vat dyeing is commenced, the goods must never be left at a standstill between the ends. The strings or cords should be untied immediately and the batches allowed to swing round. This does not mean to say that one can leave the batches for long in this state—once started, vat dyeing must be carried right through to a finish.

1–4 pieces. Weight, 130 pounds. Length, 90 yards. Mercerized Brocades. Shade, Nigger.

Indanthren Brown BR	12 lb.
Paradone Yellow G	1 lb.
Hydrosulphite	2 lb.– 2 lb.
Caustic Soda (76° Tw.)	3 pints
Nekal BX Dry	1 lb.
Glauber's Salt	5 lb.– 5 lb.

in approximately 30 gallons; 8 ends; usual "finish-off." Dyeing temperature, 50° C.

2–3 pieces. Weight, 210 pounds. Length, 100 yards. Artificial Silk Brocades Shade, Blue 210.

Caledon Blue RCS	15	lb.
Caledon Violet RS	1	lb.
Hydrosulphite	1½ lb.– 1½ lb.	
Caustic Soda (76° Tw.)	4	pints
Nekal BX Dry	1	lb.

in approximately 30 gallons; 6 ends; usual "finish-off." Dyeing temperature, 45° C.

The principle is essentially the same as before, but we are dealing with a large weight of dyestuffs, and difficulties may very easily arise. The following are some of the main points to watch:

(1) Relation of liquor to weight of material.

(2) Relation of liquor to weight of dyestuff.

(3) Correct vatting of the dyestuff.

(4) Obtaining complete exhaustion.

(5) Obtaining shade required without various additions of color.

Twenty-five gallons of water per batch of 120 pounds serves as a useful guide, but much depends upon the nature of the cloth being dealt with. Obviously a batch of a low plain cloth will not appear anything like as bulky as the same weight of a very spongy heavily brocaded cloth, and less liquor will probably be needed. It is far better to have too much liquor than too little. Shortage of water is the main cause of batches appearing "endy" and too much liquor will tend to retard exhaustion. A happy medium must be chosen.

Dispersion Agents for Insoluble Dyes

U. S. Patent 1,959,352

Formula No. 1

500 parts of naphthalene are mixed with 500 parts of sulphuric acid monohydrate, and heated slowly to 160° C., and stirred at this temperature for 10 hours. The mixture is then allowed to cool to 60° C., and 200 parts of water are run in so slowly, that the temperature does not rise above 85° C. At this temperature 200 parts of resin (colophony) are slowly added, and when all is in and the frothing has subsided, the mixture is slowly heated to 100° C., and kept at this temperature for 6 to 8 hours. The batch is then cooled, diluted with 300 parts of water and partly neutralized with 200 parts of 40% caustic soda solution. A quantity of insoluble resinous matter agglomerates and is removed. A little salt is then added, when the product, hereinafter referred to as resino-naphthalene sulphonic acid, separates. The precipitate is filtered, washed with 10% brine, and dried. It is thus obtained as a light grey powder, easily and rapidly soluble in water. It has the further advantage of being readily obtained in a fine state of subdivision.

Formula No. 2

500 parts benzene are heated with 1100 parts of 96% sulphuric acid and refluxed with stirring for 8 hours. The mixture is cooled to 50° C. and 50 parts of finely powdered resin are added in small portions and the temperature slowly raised to 100° C. and kept thereabouts for 6–8 hours. The batch is then poured on to 2500 parts of crushed ice and the resulting liquid filtered from a little resinous matter.

After allowing to cool the acid is partly neutralized with 1000 parts 30% caustic soda solution and about 500 parts salt added whereby the resino-benzene sulphonic acid separates out and after a time is filtered off, washed with brine and dried.

Formula No. 3

500 parts phenol are sulphonated by heating with 1400 parts of sulphuric acid at 110° C. until a test portion is soluble in cold water. After cooling to 70–80° C. 50 parts of powdered resin are added and the mixture stirred on the water bath for two hours. The resulting mixture is then poured on to 3000 parts of ice, filtered and partly neutralized with 1000 parts of 30% caustic soda solution; 500 parts of salt added and stirred until cool, when the resino-phenol sulphonate separates as a grey powdery precipitate which is filtered, washed and dried.

Though specific reference has been made only to the use of resins in the process and to the introduction of sulphonic groups prior to or concurrently with the reaction between the resin and the aromatic compound it will be readily appreciated that the resins may be replaced by substances derived therefrom, for example products of hydrolysis, e.g., resin acids, or in some cases products of esterification, and that sulphonation may take place subsequently to the interaction between the resin or substance derived therefrom and the aromatic compound.

Other sulphonating agents may be employed in place of sulphuric acid for example, fuming sulphuric acid or chlorsulphonic acid.

Formula No. 4

To obtain a blue violet shade on 100 lb. of cellulose acetate woven fabric, 1 lb. of 1-amino-4-methylaminoanthraquinone is ground with 3 lb. of the dry sodium salt of resino-naphthalene sulphonic acid and well stirred into 10 gallons of boiling water. The dyestuff dispersion thus obtained is strained into a bath of 300 gallons of water in a suitable dyeing machine and the goods entered at 25–30° C. The temperature is

raised over half an hour to 75° C. and
the goods worked at this temperature for
one hour. The goods are then carefully
lifted, rinsed and dried or otherwise
treated as requisite.

Formula No. 5

To obtain a red shade on 10 lb. of
cellulose acetate yarn in hank form, 1½
ounces of 1-methylaminoanthraquinone
are ground with 6 ounces of resino-
naphthalene sodium sulphonate, and sufficient water to make a thin cream. This
is well stirred into 30 gallons of water
at 30–35° C. and the goods entered. The
temperature is raised very slowly to 75–
80° C. and the goods worked until the
desired shade is attained. The hanks are
then lifted, rinsed and dried, or treated
in any other desired manner.

Artificial Wool
U. S. Patent 1,903,828

Jute is treated successively with solutions of sodium phosphate (hot), 4%
sodium bisulphite (hot), 18% sodium
hydroxide (cold), and 3% sodium bisulphite (hot), the fibers being washed with
water between each treatment. It is then
bleached by treatment with dilute hydrochloric acid, washing with water, and
immersion in an aqueous solution of
potassium permanganate.

MISCELLANEOUS

Artificial Pumice Stone

Briquettes made by mixing 9.64 parts of a mixture of clay (6.67), kaolin (16.67), felspar (20), washed chalk (6.66), and water (50) with 64.25 parts of quartz sand and 26.11 parts of water are fired.

To Increase the Toughness of Plaster of Paris Products

The toughness of plaster of paris art objects can be greatly increased by dissolving in the water with which the plaster is to be mixed, from ¼ to 4 ounces of white or yellow dextrine or gum acacia.

Synthetic Precious Stone
U. S. Patent 1,952,255

A composition of matter adapted to produce when fused a synthetic spinel having a clear green daylight color and with respect to its appearance in artificial light, the characteristics of an alexandrite, consists of a basic mixture of about 85 parts by weight of alumina and about 15 parts by weight of magnesia, and incorporated in said basic mixture approximately one part by weight of chromium oxide and approximately 0.06 part by weight of metallic cobalt.

Synthetic Precious Stone
U. S. Patent 1,952,256

A composition of matter adapted to produce when fused a synthetic spinel having a violet color, consists of a basic mixture of aluminum oxide and magnesium oxide approximately in the proportion of 85 parts of aluminum oxide to 15 parts of magnesium oxide and about 1.5 per cent by weight of iron and a quantity of cobalt ranging between one thousandth and one tenth part by weight of the quantity of iron present, incorporated in said basic mixture.

Heat-Evolving Compositions
Austrian Patent 135,336

Compositions which evolve heat when moistened with water are prepared by mixing a powdered metal with a metal sulfide and a suitable electrolyte. A typical composition contains iron 92, ferrous sulphide or cupric sulphide 3, and cupric sulphate 5%. The compositions may contain inert additions, e.g., inorganic colloids or wood meal, and may be used for medical or cosmetic purposes, etc.

Metallic (Copper) Carbon Brushes
British Patent 396,250

Copper powder (300-mesh) is agitated with 0.25% aqueous solution of silver nitrate to coat the particles with a film of silver, and the product after drying is mixed with graphite and pressed into shape.

Eggs of Pharaoh's Serpents

Take sixty-four parts of mercuric nitrate in solution and add thirty-six parts of potassium thiocyanate. The resulting precipitate of mercuric thiocyanate should be filtered off, and washed three times with distilled water. The residue then should be placed in a warm place to dry.

To make the powder into pellets, gum tragacanth serves as the binder. For every pound of the powder an ounce of gum is required. This is softened by soaking in hot water to form a paste. The dried precipitate is gradually mixed into the paste by constant and thorough stirring. Add a little water if necessary, so as to present a somewhat dry pill mass, from which pellets of the desired size can be made by hand. Place on a piece of plate glass and allow to dry. They are then ready for use. On ignition they burn forming a voluminous ash, in the form of snake like tubes— the so-called ''Pharaoh's Serpents.''

''Soap Type Eraser''

Corn Oil	4 lb. 6 oz.
Light Magnesium Oxide	2½ oz.
Sulphur Chloride	16 oz.
Ground Pumice (Milled to 200 mesh)	1 lb. 8 oz.

Add the magnesium oxide and the pumice to the corn oil, and stir until the solids are uniformly distributed throughout the mass. Now add slowly, and with vigorous agitation, 8 ounces of sulphur chloride at such a rate that the temperature does not rise above 90° F. The mixture is now allowed to stand for 12 hours, whereupon the balance of the sulphur chloride may be added somewhat more rapidly than the initial quantity, but with equally vigorous agitation.

Upon the addition of the second quantity of sulphur chloride the mass thickens somewhat and is poured into wooden channels, open at the top, and having a cross section approximately one inch square. The mixture is allowed to stand for several hours and is removed from the molds when solidified, cut into three inch lengths, imprinted and packed.

Removing Plaster Casts

Plaster casts may be cut readily, if hydrogen peroxide is applied along the line one wishes to open, and then continually applied as the cutting progresses. In place of hydrogen peroxide, vinegar or ordinary acetic acid may be used in the same way.

To Remove Plaster from Hands

Rub either a little sugar syrup or a little moistened sugar into the hands. The plaster will disintegrate and wash off easily.

Easter Egg Dyes
Blue

Marine Blue	1	dram
Citric Acid	10	drams
Dextrin	2	oz.

Chocolate Brown

Vesuvin	1	oz.
Citric Acid	10	drams
Dextrin	1	oz.

Green

Brilliant Green	0.5	oz.
Citric Acid	5	drams
Dextrin	2	oz.

Orange

Azo Orange	2.5	drams
Citric Acid	5	drams
Dextrin	2.5	oz.

Rose

Eosin	75	grains
Dextrin	3	oz.

Violet

Methyl Violet	1	dram
Citric Acid	5	drams
Dextrin	2.5	oz.

Yellow

Naphthol Yellow	0.5	oz.
Citric Acid	10	drams
Dextrin	2.5	oz.

Red

Diamond Fuchsin	1	dram
Citric Acid	5	drams
Dextrin	2.5	oz.

One-twentieth of each of the above formulae, dissolved in a half pint of boiling water, is sufficient to color a dozen eggs. After the eggs have been immersed in the solution and allowed to dry, they should be polished with a little olive oil.

Removing Odors from Cutlery

Odors of garlic, onions, etc., may be removed from cutlery by washing with dilute hydrochloric acid.

Canary Bird Food

Yolk of Eggs (Dried)	2 parts
Poppy Heads (Coarse Powder)	1 part
Cuttlefish Bone (Coarse Powder)	1 part
Granulated Sugar	2 parts
Soda Crackers (Powdered)	8 parts

Radiator Scale-Remover

Use 5 ounces of tri-sodium phosphate to 5 gallons of water, and place in radiator. Run motor slowly for ten minutes. Draw off the water, and flush out with hose.

Metallic Dental Filling Composition
U. S. Patent 1,935,266

A dental filling mass consisting of a mixture of about 50% of dental cement and about 50% of a powdered chromium-nickel or steel alloy neutral to chemical agents, shows high durability.

Composition for Making Dental Casts
Austrian Patent 135,683

The composition comprises chamotte about 75, alabaster about 20 and a gypsum cement about 5%.

To Remove Teeth from Vulcanite Dentures

Boil the dentures in glycerine until the latter smokes. The teeth which will come away clean and not discolored are placed back in the glycerine for tempering. When they are cool, wash them with soap and water.

Cartridge with Primary and Secondary Charges
U. S. Patent 1,930,765

A detonating cap comprising a container and primary and secondary charges of explosive material stratified in the container; said primary charge consisting of a mixture of about 0.45 gram of trinitrotoluene and from 0.03 gram to 0.10 gram of lead azide; and said secondary charge consisting of a mixture of about 0.40 gram of trinitrotoluene and about 0.10 gram of lead azide.

Ammunition Primer

Mercury Fulminate	40%
Barium Nitrate	30%
Antimony Sulphide	20%
Zirconium Metal Powder	4–10%

Blasting Fuses
British Patent 404,335

The electric igniter of a blasting fuse is secured in a detonator casing, e.g., lined with fireproofed paper, by a composition comprising Colorado (montan) wax (I) mixed with talc, pumice or similar hardening substance and having no sulphur or other substances ignitable by the heat from the explosion. A suitable mixture comprises iodine 70, talc 20 and asphaltum or pitch 10%. The fuse may be sealed by a mass of similar composition, e.g., comprising pitch (Trinidad Lake) 38, iodine 24.5, aluminum silicate (kaolin) 18, pumice stone (ground) 12.5 and mineral jelly 7%.

Explosive Composition
U. S. Patent 1,891,500

Ammonium chlorate 54, barium nitrate 29.5, aluminum powder 1.5, aluminum granules 9, resin 3 parts, and, if desired, paraffin wax 2 parts and brown wax 1 part.

Explosive
U. S. Patent 1,895,144

An explosive which is milder than the gelatine type of product, but equally water-resistant, is composed of (e.g.) nitroglycerin 17, nitrotoluene 3.0, nitrocotton 0.2, ammonium nitrate 64.8, sodium nitrate 8.0, oat hulls 2.0 and ivory meal 5.0%.

Explosives
British Patent 399,553

Peat of a bulk density not exceeding 0.22 grams per cubic centimeter under 25 pounds per square inch pressure is used as a carbonaceous ingredient in blasting explosives. It may be slightly charred before use. In an example, the explosive contains a fully nitrated mixture of 4 parts glycerol and 1 part ethylene glycol 14, ammonium nitrate 44.5, sodium chloride 11, sodium nitrate 13, peat containing 5% moisture 17, resin 0.25 and diammonium hydrogen phosphate 0.25%.

Explosive
U. S. Patent 1,923,327

Sodium Chlorate (50–100 mesh)	30 parts
Sodium Nitrate (8–18 mesh)	55 parts
Wood Flour	5 parts
Calcium Carbonate	1 part

Dinitrotoluene sufficient for binding at 50° C.
This sets hard on cooling.

Yellow Smoke Composition
U. S. Patent 1,920,254

Potassium Dichromate	66%
Bismuth Subnitrate	20%
Magnesium	14%

Explosive
Canadian Patent 339,449

An explosive composition comprises nitroglycerin, nitrocellulose, or more oxygen-carrying compounds, absorbent materials for the liquid explosive, and 0.1–2.0% of dicyanodiamide; e.g., nitroglycerine 34%, nitrocellulose 1%, sodium nitrate 52%, wood pulp, 8%, starch 3%, sulphur 1.6%, and dicyanodiamide 0.4%.

Explosive
Canadian Patent 339,467

An explosive composition comprises nitroglycerin 85%, nitrocotton 1%, wood pulp 13%, and chalk 1%, and is capable of being subsequently combined with additional ingredients to form various types of nitroglycerin high explosives.

Explosive
Canadian Patent 339,468

A blasting gelatin comprises nitroglycerin 86–92%, nitrocellulose and kieselguhr 3–6%.

Dynamite
Canadian Patent 339,433

A dynamite of low density comprises nitroglycerin, sodium nitrate, ammonium nitrate, and the pith of cornstalks, e.g., nitroglycerin 15%, ammonium nitrate

58%, sodium nitrate 9%, and cornstalk pith 18%. Part of the cornstalk pith may be replaced by wood pulp.

Pyrotechnic Sparklers
Austrian Patent 136,232

For the manufacture of fireworks giving a colored spray effect, a mixture of a combustible substance, for example, a resin, fat, cellulose or nitrocellulose, with flame-coloring salts such as nitrates, chlorides, oxides, carbonates, acetates of barium, strontium, calcium, thorium, copper, potassium, or sodium, is worked into granules with the aid of a nonaqueous binder, e.g., a nitrocellulose lacquer. The granules are coated with a metal, e.g., copper, by known electrical or mechanical means and are then attached to a combustible core formed, e.g., from a mixture of dextrin 10%, aluminum 5%, iron 30,% and barium nitrate 55%.

Pyrotechnic ''Sparklers''
U. S. Patent 1,936,221

Barium Nitrate	85 parts
Strontium Carbonate	60 parts
Kryolith	40 parts
Potassium Chlorate	225 parts
Dextrine	30 parts
Shellac	55 parts

In preparing the composition, the several ingredients in the proportions given above are thoroughly mixed together, and sufficient water is added thereto to form a thick paste or plastic material. The sticks are dipped into this plastic material and then removed causing a layer of the material to adhere to the sticks. While the layer is still plastic, the granules or particles are sprayed thereon as by shaking sieves containing the granules above the sticks and while turning or rotating these sticks which are preferably held in horizontal positions during the spraying operation, so as to thereby cover the entire surface of the layer with these granules.

As the layer of composition dries out it hardens, thereby firmly uniting the granules to the surface portion of this layer. The dried composition is relatively hard and dense and is non-hygroscopic and keeps well under variations of climatic conditions. This composition is not subject to spontaneous combustion and will remain properly operative for the purposes intended for an indefinite period.

The granules are preferably composed of rapid burning metallic alloys although pure metals such as magnesium or aluminum in some instances may be used. Alloys of magnesium and aluminum with metals of higher melting points are preferred to the pure metals, however, because such alloys, having higher melting points than these pure metals themselves, do not tend to melt too rapidly while in contact with the burning composition but instead are caused by the generated gases to fly off in solid form and thereafter burn with explosive violence, as desired. Alloys of magnesium and copper or aluminum and copper have been found highly satisfactory for this purpose.

In use, the firework is held by the handle portion of stick or rod and the composition at the outer tip of the rod is ignited causing this composition to burn with an attractive colored flame. As soon as the composition commences to burn, the granules in the proximity of the flame are caused to be driven off from the firework at relatively high speeds in all directions. These granules glow somewhat as they move through the air, thereby creating the effect of a multiplicity of streamers moving rapidly away from the firework. After these granules have traveled to varying distances from the firework, they burst into flame or explode, thereby producing beautiful little lights or whiffs of flame of varying colors, depending upon the composition or compositions.

Economical Permanent Antifreeze

Commercial diethylene glycol is in many ways a suitable and economical antifreeze for use in automobile cooling systems. The pure substance is a colorless and practically odorless liquid of specific gravity 1.1185 boiling at 244.5° C. and melting at −6.5° C.

The following table gives the approximate freezing points of its aqueous solutions together with those of glycerine and ethylene glycol for comparison.

Substance	Freezing Point of Solution in Degrees F. (Concentrations in Percent by Volume)				
	10%	20%	30%	40%	50%
Distilled Glycerine	+29	+21	+12	0	−15
Diethylene Glycol	+28	+18	+8	−5	−20
Ethylene Glycol	+26	+16	+3	−11	−31

It is evident that volume for volume diethylene glycol is more efficient than glycerine and less efficient than ethylene glycol. The difference in price more than compensates for the greater efficiency of ethylene glycol.

Diethylene glycol has another advantage over glycerine. The latter substance shows a tendency toward the formation of allyl derivatives on decomposition and these might be expected to polymerize with the formation of gums. Diethylene glycol, on the other hand, appears to be free from this tendency. The following table gives the specific gravities of aqueous solutions of diethylene glycol together with those of glycerine and ethylene glycol solutions. This information should facilitate the testing of radiator solutions with a hydrometer.

Substance	Specific Gravities at 60° F.				
	10%	20%	30%	40%	50%
Glycerine	1.029	1.057	1.085	1.112	1.140
Diethylene Glycol	1.019	1.034	1.049	1.063	1.078
Ethylene Glycol	1.016	1.031	1.045	1.058	1.070

With respect to the use of this antifreeze three other points may be noted. 1. Like the other permanent antifreezes diethylene glycol has a tendency to seep through leaky connections so all loose connections should be tightened before using the antifreeze. It is very advantageous to seal hose connections with white lead when replacements are made. 2. Lacquer finishes should be protected from hot solutions of diethylene glycol. However, as these solutions boil above the boiling point of water the engine can operate at even the normal summer temprature without danger of boiling the antifreeze solution. 3. The solutions of diethylene glycol show considerable tendency to supercool and may not begin to freeze until much lower temperatures than those noted. Furthermore, the freezing points given although approximate are conservative in that they represent the point at which freezing begins and at these temperatures only a mush is formed. Complete solidfication of these solutions would occur only at much lower temperatures. Diethylene glycol has been used in air conditioning equipment and has been tested by actual use in automobile radiators. As an illustration of the significance of the points just mentioned a car has been operated throughout the winter with a 40% diethylene glycol solution and although the temperatures fell at times well below —10° F. no indication of freezing was noticed.

Non-Corrosive Antifreeze Liquid

U. S. Patent 1,911,195

Methyl alcohol 15–100% is mixed with 0.01–3.0% of borax.

Antifreeze Solution of Same Gravity as Water

This solution contains (in percentage by volume):

Water	51.6%
Ethyl Alcohol (95% grain alcohol)	34.3%
Glycerin	14.1%

The freezing point of this solution is from minus 10° to minus 20° F.

Antifreeze Solution

U. S. Patent 1,955,296

An antifreeze composition comprises an aqueous solution made up by adding to water a mixture of 35% by volume of methyl alcohol and 65% by volume of isopropyl alcohol, such composition substantially duplicating the specific gravity volatility, and freezing point of an aqueous solution containing the same percentage of ethyl alcohol.

Antifreeze Composition for Radiators

U. S. Patent 1,902,287

A jelly is made by boiling together sugar 3 pounds, salt 6 pounds, copperas 1 ounce, hydrochloric acid 2 ounces, heavy oil 1 pound, and water 14 quarts; the product is added to boiling water.

Antifreezing Compounds

U. S. Patent 1,906,972

To prevent the formation of ice-crystals in cooling-brines, 10–45 parts of urea or condensation products thereof with polyhydric alcohols are added per 100 parts of solution.

Warning-Odorant for Refrigeration

U. S. Patent 1,905,817

An alkyl mercaptan (e.g., 0.5–2.0% of ethyl mercaptan) is added to a hydrocarbon or halogenated hydrocarbon refrigerant.

Gas Absorbent (for Removing Odors from Foodstuff Refrigerators)

U. S. Patent 1,922,416

This is a mixture of sour cherry-wood charcoal 45%, coconut-shell charcoal 25%, boxwood charcoal 20% and meta-formaldehyde 10%. This absorbent converts alkylamines into hexamethylene tetramine and reacts with volatile organic sulphur compounds produced in the decomposition of foods.

Anti-Fogging Compounds

The purpose of an anti-fogging compound is to cause the moisture to spread over the surface in an even film. The most common uses for anti-fogging compounds are on automobile windshields and electric cars in rainy or foggy weather, and on the interior of automobile windshields, electric cars and store display windows where vision is impaired by condensation of moisture on the glass in the form of droplets. A further use is for spectacles which in cold weather become coated with moisture due to condensation when going inside of warm buildings, after having been outside. Spectacles also become coated with moisture in summer due to perspiration.

All the various compounds used for these purposes are listed below. Only the best and cheapest for the various uses are considered. Any compound listed will work for every anti-fogging use listed above but is not always ideally suited.

For Spectacles

Soap	70%
Glycerine C.P.	30%

Soaps with fillers of any kind should be avoided. A soap of cocoanut oil base is quite satisfactory. Prepared soaps such as Rose Glycerine and Kirk's Flake are also equally good. The amount of glycerine used is arbitrary, varying with different soaps and with different water content. The ideal consistency is a stiff paste. To use, apply a small amount of the paste to a soft cloth and rub on both sides of the spectacles, then polish until thoroughly dry.

For Automobile Windshields

Soap	50%
Glycerine C.P.	25%
Sulphonated Castor Oil	25%

The soap employed should answer the same specifications as for the above spectacle compound. The ideal consistency is a soft paste. To use, apply a small quantity to the outside and inside of the windshield and thoroughly polish the surface until no free compound remains. The windshield wiper should not be used in combination with the compound.

For Electric Car Windows

Soap	10%
Glycerine C.P.	85%
Sulphonated Castor Oil	5%

The formula is compounded in the same manner as for automobile windshields above. To use, apply to the glass both inside and out with a soft cloth and polish until the glass is clear. It is not necessary that all the compound be removed in the polishing.

For Display Windows

As soon as considerable moisture has condensed on the glass wipe from top to bottom in vertical strokes with plain unbleached muslin which has been prepared as specified below. The rag must be used folded. The glass will remain clear as long as it remains wet.

Asphaltum	1 gal.
Yellow Beeswax	1 lb.
Pine Tar	1 lb.
Gasoline or Naphtha	5 gal.

Melt up the asphaltum, yellow beeswax and pine tar together and stir until evenly mixed. Remove from fire and slowly pour the melted mixture into the solvent with rapid stirring. The cloth is dipped in this mixture until saturated and wrung out and allowed to dry over night.

To Prevent Formation of Ice on Auto Windshields

Water can be prevented from freezing on windshields by a liberal application of C.P. Glycerine. The application has to be renewed every half hour or so depending on the amount of water striking the glass.

Waterproofing Composition
British Patent 382,073

A composition for waterproofing tent material, awnings, etc., consists of alum 2, lead acetate 2.5 lb., paraffin wax 6.5, hard soap 4, gum tragacanth 4 ounces, and water 67 pounds.

Water Softening Compound
U. S. Patent 1,952,408

A water softening compound comprises the following ingredients in substantially the proportions named, sodium carbonate 62.5%, trisodium phosphate 30%, calcium chloride 5% and sodium chloride 2.5%.

Low "Conductivity" Water

Dissolve the following:

Potassium Hydroxide	700 g.
Potassium Permanganate	45 g.
Distilled Water	3.5 l.

Add the above solution to distilled water in the ratio of 250 cubic centimeters of the above to 14 liters of distilled water and redistill. Discard the first 3 liters of distillate. As a container in redistilling, use block tin or quartz. Hard Pyrex may also be used if necessary.

Composition for Treating Water Used in Air Conditioning
U. S. Patent 1,921,137

A composition for treating water to prevent corrosion and congestion of the apparatus is formed of water 4-10%, sodium hydroxide 0.5-2.0%, sodium dichromate 0.5-2.0%, sodium carbonate 1-5%, disodium hydrogen phosphate 0.5-2%, sodium silicate 75 to 90%, and tannin 0.006%.

Boiler Water Treatment

After extensive studies of feed water and scales from 20 ships located at widely separated stations, and comparative trials of the colloidal, coating, electrolytic and chemical methods of feed water treatment at the U. S. Naval Experiment Station, a new Navy formula was evolved. This consists of anhydrous disodium hydrogen phosphate 47%, soda ash 44% and cornstarch 9%.

Water "Wettable" Sulfur
U. S. Patent 1,939,403

Take a pound of tannic acid, 4 pounds of dry sulphite waste liquor or 8 pounds of a sulphite waste liquor containing 50 per cent solids, add 0.75 ton of water and a ton of sulphur and run the mixture through some sort of milling device to form a sulphur pulp. This sulphur pulp contains from 0.05 to 0.2 per cent of the emulsifying agent, based on the sulphur present. The sulphur can then be allowed to settle out. It may be washed while wet and will still retain its miscibility with water. With the dispersing agent washed out, the wet pulp of sulphur and water is still miscible with water and with wet paper pulp.

Friction Element
U. S. Patent 1,899,239

A brake lining is composed of black graphitic clay 45, zirconium oxide 25, felspar 15, agalmatolite 5, magnesite 5, kaolin 5%.

Brake and Bearing Materials
British Patent 391,155

Porous metal articles, e.g., bearings, bushings, etc., are made from briquets, molded from an intimate mixture of powdered iron and a volatile lubricant and sintered by an electric current in a reducing or inert atmosphere, e.g., hydrogen, helium, illuminating gas. In an example sponge iron, produced by reducing ferric oxide, is mixed with 2-5% stearic acid, preferably dissolved in ethyl ether or other volatile solvent which is allowed to volatilize before briquetting. Palmitic, oleic or other fatty acid or lubricating oil may be used in place of stearic acid.

Brake Fluid
U. S. Patent 1,943,813

A brake fluid composition is composed of 50% castor oil and 50% ethyl lactate with a small quantity of an organic base. This composition has a comparatively high boiling point, a low freezing point, is very stable and has no deleterious effect upon rubber or metals. The organic base which may preferably be triethanolamine is added to neutralize the free fatty acid (namely, ricinoleic acid), which is present in castor oil.

In preparing the solution, first neutralize a fatty acid by adding a suitable quantity of triethanolamine to the oil. Add a slight excess of the base, inasmuch as the excess has no injurious effect. The amount depends upon the amount of acid present in the oil, but by way of example it may be stated that to one liter of oil having 3% fatty acid, from 30 to 35% of triethanolamine should be added slowly and with constant stirring. Thereafter, the solvent is added to the solution also with constant stirring.

Hydraulic Brake Fluid
U. S. Patent 1,949,775

Castor oil 35, alcohol 45 and a toluene-sulfonamide ester such as an ethyl ester mixture 20 parts are used together.

Tire Puncture-Proofing Composition
Canadian Patent 336,531

A puncture-proofing composition for pneumatic tubes consists of gum tragacanth 8 pounds dissolved in 14 gallons of water, containing 16 ounces of short threads of cotton and plasticized with sugar 142 pounds, salicylic acid 8 ounces, carbolic acid 4 ounces, oil of tar 4 ounces, alcohol 7 gallons, magnesium carbonate 45 pounds, and magnesium silicate 27 pounds.

Non-Skidding Compound
U. S. Patent 1,943,917

Resin and alcohol of 96 vol. % in equal parts are mixed with heating. Of this solution small portions of equal volume are applied to the surface of the tires either while the vehicle is stationary or moving. The dilution with water is effected in a suitable way, for example by means of the humidity of the road, or by applying water to the tread of the wheels while the vehicle is stationary. In this way the resin is instantly liberated as a viscous, sticky material which makes any skidding of the tires impossible.

Furthermore, a solution can be employed composed for example of 30% of resin, 30% of wood pitch and 40% of ethyl alcohol of 96 vol. % heated together under a reflux condenser. This solution, after the removal if required of non-dissolved portions can be used as anti-skid material.

The effect of the sticky material as produced on the tires not being a permanent one, the application of the solution must be repeated when required. Such application can, for example, be effected by atomizing the solution onto the surfaces of the tires, the solution being distributed from the driver's seat from a container under pressure through pipelines.

Composition for Inhibiting Corrosion in Automobile Cooling Systems, Etc.
U. S. Patent 1,925,672

Yellow sodium chromate 20, pale paraffin oil 15, sulfonated red oil 50, liquid soap 2 and water 8 parts are used together.

Preventing Scale Formation in Boilers
U. S. Patent 1,927,027

Colloidal metallic iron is added to the boiler water in the proportion of about 6 grams to a boiler of 20,000 gallons daily capacity.

Prevention of Corrosion of Petroleum Pipe-Lines

Coating with a mixture of clay and asphalt, melting point 80° C. (1:1) is recommended.

Catalyst for Oxidizing Carbon Monoxide in Exhaust Gases
U. S. Patent 1,903,803

The catalyst comprises an alloy of lead 10, manganese 25, and copper 65% having an oxidized surface.

(Palladium) Catalyst
U. S. Patent 1,907,710

A palladium catalyst suitable for hydrogenation or dehydrogenation purposes is formed by electrolyzing palladium chloride solution at 0.005–0.015 amperes per square centimeter with the formation of a pebbly surface of palladium on a wire screen which is afterwards made the cathode in dilute sulphuric acid and heated in oxygen at 300–400° and in hydrogen at 200–350°.

Preservative Compound
U. S. Patent 1,937,813

The method of making a preservative compound consists in heating gelatine and wood creosote to a temperature of between 160° and 250° Centigrade to form vapors, cooling said vapors to form a liquid and further cooling to cause the liquid to form a colloidal mass.

Transparent (Gelatin) Sheet
U. S. Patent 1,893,172

Gelatin (3) is dispersed in water (12) and 50% phosphoric acid (0.036) is added, followed at 42° by 8% sulphonated castor oil (1) diluted with water (10), a blue dye (0.02) to counteract the yellow color, and 40% aqueous formaldehyde solution (0.066–0.26 part), and the mixture is spread in thin sheets on an endless belt.

Preserving Cadavers in the Tropics

The blood should be washed out of the blood vessels, if practicable, an hour or two after death. This is usually impracticable.

Into one femoral artery by a three-way cannula, equal parts of phenol U.S.P., glycerin and alcohol should be injected, 6 liters to each 150 pounds (68 kilograms). The injection should be done slowly by gravity pressure of 3½ or 4 feet for several hours (over night). Should any arteries be completely blocked through disease or impassable blood clots, the parts supplied by them can easily be distinguished from the parts into which the fluid has passed. Such parts may be further treated by the injection with a large hypodermic needle of a considerable amount of this preserving fluid directly into the tissues in several places. The femoral artery should be tied above and below the point of injection and about two days should be allowed for the fluid to penetrate the tissues thoroughly.

To inject arteries with a color mass (Souchon's method), a mixture should be used consisting of crimson aniline solution, 45 cubic centimeters; potassium antimony tartrate solution, 12 cubic centimeters; cornstarch (put through a sieve), 1 kilogram; hot (not boiling) water, 1 liter. The cornstarch and water are rubbed up in a mortar to make a thick cream. The crimson aniline solution is added and then the potassium antimony tartrate; the latter is to prevent the diffusion of the color mass through the walls of the smaller arteries. Force should not be used in injecting the color. It should be allowed to remain in the vessels for fifteen minutes and any excess permitted to run out freely.

Crimson aniline solution consists of crimson aniline crystals, 30 grams; alcohol, 30 cubic centimeters; water, 1 liter.

Potassium antimony tartrate solution is made by dissolving 4 grams of potassium antimony tartrate in 125 cubic centimeters of water.

After the color solution has been in the arteries for twenty-four hours, it sets. Cadavers may then be stored by immersion in tanks containing 3 per cent phenol in water. In this solution they will keep indefinitely, even in the tropics.

Bentonite Pastes and Suspensions

A paste of 17% bentonite and 83% water is the thickest mixture practical to make with a small mechanical stirrer and still be free from lumps. It will flow through a quarter inch diameter pipe when subject to ordinary water pressure. A paste of 10% bentonite and 90% water is the thickest one that will flow through a half inch inside diameter pipe under its own weight. In our test, the container was filled to a height of 13 inches above the entrance of the pipe.

4 grams of bentonite in 150 cubic centimeters of distilled water will carry in suspension 100 grams of 140 mesh silica flour.

Flocculating and Decreasing Swelling

Sodium chloride and many other soluble salts markedly decrease the swelling of bentonite, partially flocculate it and permit it to be broken down in water more easily. All acids cause flocculation.

Increasing Viscosity and Suspension Properties

Viscosity, suspending properties and gelling are greatly increased by adding 0.5% to 1¼% magnesium oxide or caustic calcined magnesite to the bentonite. However, when bentonite is so treated and allowed to stand in a thin solution, there will be recession and a layer of clear water will form at the top. This will not occur with straight untreated bentonite, although the whole suspension will be thinner.

Mixtures with Other Liquids

Bentonite can be readily mixed with lubricating oil, kerosene, gasolene, glycerine or alcohol without swelling. The following amounts will moisten bentonite to a very thick paste (using in each case, 100 grams of bentonite):

Alcohol	100 cc.
Glycerine	110 cc.
Kerosene	70 cc.

Thick gels can be made by adding water to the above mixtures, more easily than by mixing the bentonite and water directly.

Cork Composition for Gaskets

Glue Binder	1 part
Glycerine	15 parts
Granulated Cork	20 parts

Warm the glycerine and dissolve the glue in the hot liquid. Mix thoroughly and quickly the warm solution with the cork. Place in molds and compress to one-fifth of the original volume. Warm for eight hours at 120 degrees F.

Cork Composition Polishing Wheels

20/30 Granulated Cork	400 parts
Emery Dust	10 parts
Bentonite	10 parts

Mix thoroughly the above materials and add to a hot solution of

Glue Binder	20 parts
Glycerine	200 parts

When thoroughly mixed place in mold and compress to one-twelfth original volume. May be used as a wheel or as a block.

Non-Resonant Surfaces
British Patent 394,827

Non-resonant surfaces are obtained by treating the surface with a layer of adhesive and depositing pulverized vegetable or animal textile fibers thereon. A suitable adhesive comprises standard oil 25, calcium carbonate 45, synthetic resin 3, mineral oil 22.5, lead oxide 0.5 and siccative with manganese and cobalt 1%.

Composite Mica Sheet
U. S. Patent 1,953,950

Dissolve 1 pound shellac in 1 gallon alcohol; mix in thoroughly 114–204 grams zinc oxide. Paste sheets of mica together with above; press together and heat at 100° C. until dry. Press and mold at low temperature and pressure. Heat gradually up to 265° C. and increase pressure above 200 pounds per square inch. An insoluble reaction cement is formed and an excellent bond results.

Purification of Phosphoric Acid
U. S. Patent 1,894,289

Suspended matter is removed from crude phosphoric acid by addition of a substance that will precipitate sulphur and allow the mixture to settle. A suitable mixture is sodium thiosulphate 100 pounds, sodium pentasulphide 32 pounds and calcium oxide-sulphur solution (density 1.29) 17 gallons, per 1000 gallons of acid of density 1.26.

Stabilization of Chlorinated Solvents
British Patent 401,210

Carbon tetrachloride, tetrachloroethylene, chloroform and trichloroethylene are stabilized by the addition of, for example, 0.001–0.1% by volume of a mercaptan, for example, butyl mercaptan.

Stabilizing Chloroform and Other Chlorinated Solvents
U. S. Patent 1,925,602

Stabilization is effected by the addition of 0.0001–0.1% triethylamine by volume.

Calorific Powder
Belgian Patent 393,437

A mixture of potassium permanganate 10, silicon 2, iron 83 and carbon 3.7 on addition of a small amount of water generates heat, producing a maximum temperature of 100° C.

Purifying ''Mahogany Sulfonates''
U. S. Patent 1,930,488

Process of purifying impure salts of the group consisting of alkali metal salts and ammonium salts of the mahogany petroleum sulphonic acids derived from the treatment of a petroleum oil with sulphuric acid, which comprises dissolving such salts of the mahogany petroleum sulphonic acids in a solvent consisting predominantly of water and a mono hydroxyl alcohol of not exceeding 3 carbon atoms to the molecule, the water being present to the extent of from 25 to 75% by weight of the combined water and alcohol, thereby producing a mixture, agitating the said mixture with a petroleum hydrocarbon boiling predominantly below 300° C., permitting the said agitated materials to stratify, thereby producing a solvent layer and a hydrocarbon layer, separately collecting said solvent layer, and recovering the purified salts of the mahogany petroleum sulphonic acids from the said solvent layer.

Solution to Loosen Frozen Glass Stopcocks

Camphor	5 parts
Turpentine	95 parts

Microscopic Mounting Fuids

Dissolve 100 grams of styrax as supplied by perfumers in 200 cc. of benzol, chloroform, carbon tetrachloride or acetone and filter the viscous solution through filter paper to free it from impurities which are insoluble in any of these solvents. Place the filtered solution in an evaporating dish and heat over a low gas flame for two to four days, stirring from time to time by means of a thermometer, not allowing the temperature to rise above 140 to 145°

Centigrade. Heating on a sand bath is more suitable than a wire screen as the sand bath may be held to the proper temperature more easily. The process of evaporation is completed when the tip of a stick, such as a match stick, is immersed in the liquid, cooled to about 45° and the styrax which is now hard, can be chipped with the finger nail. If it does not chip, further evaporation is indicated. Now choose the solvent for the styrax in which it is to be used. Chloroform, benzol, xylene, benzol-methyl alcohol or benzol-ethyl alcohol are the most suitable solvents. Place the solvent you select in a thoroughly clean and dry glass stoppered bottle, using a bottle of about 8 oz. capacity, using a quantity of solvent about equal to the volume of the resulting evaporated styrax gum (this will be about 90 cc. of solvent). Before the styrax has become cold pour it into the solvent in the bottle and shake until the styrax is completely in solution. This dilution will very nearly approximate the consistency most adaptable for microscopical use. If desired, this dissolved styrax may again be filtered into one-half ounce bottles and tightly corked. In this way evaporation of solvent will be kept at a minimum.

Styrax-Piperine Mounting Medium

Dissolve a quantity of purified piperine in the above dissolved styrax solution, using about as much piperine as there is styrax in the first formula given.

Cystographic Medium
U. S. Patent 1,935,661

A 5–8% aqueous solution of sodium or potassium bismuth tartrate or citrate is used. These solutions are non-viscous, radiopaque, non-irritating and innocuous to the tissues which they contact.

Quick-Drying Canada Balsam

A quick-drying balsam can be prepared by heating ordinary Canada balsam for several hours until it has a decided orange color and becomes brittle when cold; the product is dissolved to saturation in xylol.

Neutralization of Tobacco Smoke

The total basicity of the smoke from a number of different tobaccos has been determined. The efficacies of various neutralizing media increase in the order: ferric chloride (3% solution), ferrous ammonium sulphate (6% solution in 60% aqueous solution of alcohol), lactic acid (A) (5% solution). A. may be employed in the form of pellets of wadding soaked in the solution and air-dried. The aroma of the tobacco remains unaltered.

Manufacture of Natural Sour Casein

In the manufacture of natural sour casein, skim milk of good quality is acidified to 0.25–0.30% acidity (as lactic acid) with whey containing the acid-producing organism. The organism is allowed to act on the milk for 4–6 hours, during which time the acidity increases to 0.64% or pH approximately 4.6. The milk is then heated rapidly to 120° F. to coagulate the casein, and the whey is drained off. The casein curd is washed twice with cold water, subjected to pressure to remove water, ground and dried.

Defecation of Cane Juice
U. S. Patent 1,887,879

Raw cane juice is treated with sodium (or other soluble) aluminate, e.g., a solution of 0.25–1 pound of sodium aluminate per 1000 gallons, and then limed, e.g., to pH 8.3–9, heated to 99–104° and clarified by subsidence.

Solid Lactic Acid
British Patent 395,990

An 80% aqueous solution of lactic acid is kneaded with an equal weight of a dry vegetable mucilage or mucilaginous substance, e.g., carob bean, carragheen, etc.

Lactic Acid from Molasses

Best results were obtained by fermentation of molasses by means of B. delbrucki with addition of 10% barley malt extract. The malt was added at 45° and this temperature maintained for 30 minutes; the mixture was then kept at 52° for 30 minutes and at 58° for 1 hour; heated to 62°, and filtered. Calcium carbonate was used to fix the lactic acid. The product was treated with sulphuric acid and filtered, and the lactic acid distilled with superheated steam. Laboratory tests gave a 25–30% lactic acid, that satisfies pharmacopoeial requirements. It is estimated that a 50–75% product could be obtained on the industrial scale.

Hematoxylin Stain

Hematoxylin (crystals)	4 g.
Alcohol (95 per cent)	25 cc.
Saturated Solution of Ammonia Alum	400 cc.

This solution is placed in an open dish, at a distance of 15 centimeters from a Cooper-Hewitt burner, operating at 140 volts and 3.3 amperes, for one hour. The solution is then filtered and to the filtrate is added:

Methyl Alcohol	100 cc.
Glycerine	100 cc.

This solution is placed under the Cooper-Hewitt burner at the same distance for two hours. The solution is then filtered and used for staining purposes.

Modified Hematoxylin Stain
Solution No. 1

Hematoxylin (White Crystal Preferred)	0.2	g.
Distilled Water	160	cc.

Dissolve by boiling.

Solution No. 2

Phosphotungstic Acid	10 g.
Distilled Water	100 cc.

Dissolve by boiling.

When cool, mix 80 cubic centimeters of Solution No. 1 with 20 cubic centimeters of Solution No. 2, allow the mixture to ripen from one to five months in a covered, not stoppered, bottle in the sun, if possible. The ripening can be hastened by the addition of a drop or two of hydrogen dioxide.

Pencil Sheath Composition
U. S. Patent 1,937,103

Methylcellulose Binder	2 %
Casein	17 %
Lime	2 %
Sodium Fluoride	2.5%
Turkey Red Oil	4 %
Wood Flour	54 %
Comminuted Rice Hulls	18.3%
Coloring Matter	0.2%

The alkyl cellulose is dissolved in four times its weight of boiling water, and the casein is mixed with about 125 parts of cool water. To the mixture of alkyl cellulose and casein is added the lime mixed with about 50 parts of water, and then the sodium fluoride mixed with about 50 parts of water is added. The whole mixture is thoroughly stirred until the mass is in plastic condition. The turkey red oil and the color are next thoroughly stirred into the mass, and then a mixture of the wood flour and the comminuted rice hulls is added and the mass stirred until the composition is homogeneous. In this condition the plastic composition can be extruded from the machines usually employed for the purpose of producing composition sheath pencils.

If the price of alkyl cellulose is too high for use in making the preferred form of pencil sheath, an excellent substitute pencil sheath, very slightly or not at all inferior in practical qualities to the preferred form, can be made from the following ingredients:

Casein	18 %
Lime	2 %
Sodium Fluoride	2.5%
Turkey Red Oil	5 %
Wood Flour	54 %
Rice Hulls	18.3%
Coloring Matter	0.2%

The foregoing composition can be compounded in substantially the same manner described for the previous composition.

Casein is a granular powder, it is therefore necessary to plasticize it so as to convert it into a bonding material. This can be done by mixing it with a basic hydroxide, such as lime or ammonia, or with a salt, such as sodium fluoride or sodium phosphate. If it is desired to substitute ammonia for the sodium fluoride or sodium phosphate, then it is desirable to add a preservative and strengthening agent, such as hexamethylenetetramine or phthalic anhydride. From one to two per cent of these preservative and strengthening agents is sufficient. The turkey red oil acts as a softening agent for the casein. Soap or rubber substitute, that is, sulfurized oil, may advantageously be substituted for the turkey red oil.

Temperature Indicators
British Patent 397,520

A substance of suitable melting point is made up with fillers into block or "pencil," which leaves a mark on the heated surface of an iron or calender at temperatures above the melting point. A composition containing anthracene (melting point 216°; 15 parts), calcium carbonate (15 parts) and carnauba wax (0.28 part) leaves a faint mark at 210° and a heavy mark at more than 220° The marks are readily wiped off.

Silver Copper Mirror
U. S. Patent 1,935,520
Solution No. 1

In a suitable glass or earthenware container preferably one that is capable of withstanding considerable heat without breaking, is placed 16 ounces of silver nitrate. To this silver nitrate is added 11 ounces of ammonia (26°), the ammonia being added slowly to prevent possible explosion and being stirred to insure complete dissolution of the silver nitrate in the ammonia. When all of the silver nitrate has been dissolved, 16 ounces of distilled water is added to the silver nitrate-ammonia solution, the solution being then cooled and filtered. To the filtered solution is added an additional 144 ounces of distilled water.

Solution No. 2

Copper Sulphate (Crystal)	1 lb.
Distilled Water	64 oz.

This Solution No. 2 is filtered and the filtered solution placed in a dark bottle where it is kept until used.

Solution No. 3

In a suitable glass or porcelain container is placed 64 ounces of distilled water to which is added two pounds of crystalline Rochelle salts. This salt solution is heated to the boiling point at which time 1 ounce silver nitrate dissolved in 4 ounces of distilled water is added. The Rochelle salt solution and the silver nitrate solution are then thoroughly mixed together after which the solution is again heated to the boiling point at which time 4 ounces of Solution No. 2 is added. After the addition of Solution No. 2 the solution is again boiled for at least ten minutes whereupon it is cooled and filtered. This filtered Solution No. 3 is then placed in a dark bottle in which it is kept ready for use.

Solution No. 4

Powdered Tartaric Acid	1 lb.
Distilled Water	48 oz.

This Solution No. 4 should be allowed to stand for at least one week after which it is filtered clear.

The final solution which is to be used upon the glass surface to be treated is prepared as follows:

Distilled Water	64 oz.
Solution No. 1	2 oz.
Solution No. 3	2 oz.
Solution No. 4	3 drams

Attention is here directed to the fact that in preparing this final solution, the solutions are added to the distilled water in the order given above and that Solution No. 4 is not added until after Solution No. 1 and Solution No. 3 have been thoroughly mixed together. Increasing the amount of Solution No. 4 which is added to the final action will slow up its action.

Before mixing together the numbered solutions in order to obtain the final solution, the glass or other surface to be coated is initially block polished or hand rubbed with rouge, after which it is well brushed with water. Following this water brushing operation, a weak solution of tin chloride is applied to the surface to be treated preferably by means of a felt block or bristle brush. The surface is then rinsed well with water and lightly brushed.

The glass so treated is then placed in a horizontal plane and accurately leveled with wedges, the surface to be coated being uppermost. Following the application of the tin chloride solution the surface to be treated must be kept wet until the final solution has been applied thereto. As much of the final solution is poured upon the leveled surface as the latter will hold without the solution running over the edges.

In a relatively short time (about 15 minutes) the first coating of silver copper alloy will have deposited out of the final solution and upon the glass. Thereupon the excess or undeposited solution is removed from the glass surface, preferably with a piece of chamois, the surface being then well brushed to obtain a clean metallic surface for a second coating. A second application of the final solution is then made. In about 10 minutes a second coating or film of silver copper alloy will have deposited out of solution upon the first coating, this second coating being also well brushed and then dried with the chamois. When the deposited film of metal shows no dark spots indicating the presence of moisture, a coating of shellac is applied and when this coating of shellac is thoroughly dry it is covered with a coating of paint.

The shellac for coating the silver copper alloy is preferably prepared by dissolving one pound of arsenic free shellac in 96 ounces of methanol or special denatured alcohol.

In some instances it may be desired to apply only a single coating of the silver copper alloy to the surface to be treated. In such instance the best results are obtained by decreasing the amount of distilled water in the final solution to 32 ounces.

Toning of Silver Pictures
U. S. Patent 1,899,972

A single-solution sepia toner, which gives reproducible depths of tone and does not affect the user's hands, comprises a solution containing ammonium thiocyanate 10, citric acid 1, and selenious acid 0.5 grams, in 100 cubic centimeters of water.

Luminous Neon Tube
U. S. Patent 1,918,012

By addition of 1% of krypton to the gas in an ordinary neon tube, light resembling that of daylight can be obtained when an electric discharge is passed, the blue of the krypton being complementary to the red of the neon.

Elastic Impression Compound
U. S. Patent 1,930,391

An elastic composition which is particularly applicable as a molding material to take casts or impressions of both animate and inanimate objects.

The elastic property of the said composition permits casts or impression molds made thereof to be distorted or stretched in effecting their removal from the object without breaking or marring the impression or the casting, and which will return to their original molded form after removal. It also has the advantage that a single casting may be made of an intricate object which would require it to be molded in parts if a stiff, unresilient material were used. Because of its elasticity it is particularly tough and will not fracture in ordinary use; it is therefore possible to obtain perfect casts of both animate and inanimate objects which have considerable undercut parts, without injuring the cast or requiring repair thereto after removal of the object.

In producing the material, the nitrogenous constituent, agar-agar and gelatine, are soaked in water separately for approximately four hours and then filtered through gauze and pressed to reduce the water content, after which they are mixed in a double boiler with glycerine and boiled for several hours until dissolved. The soapy body is then added to the boiling solution and dissolved therewith, after which resin ground to a powder and mixed with sodium borate is added to the hot solution and it is cooked until smooth.

The gum is reduced to a solution by the reaction therewith of sodium borate in the boiling operation and which combines with the resin to form an alkali solution thereof. This solution dissolves with the other ingredients in the solution to form the elastic impression composition.

The proportions of the various ingredients of the material may be varied in accordance with the use that is desired to be made of the material; for instance, if it is desired to use it on an inanimate object the material may be produced with gelatine and without a vegetable hydro colloid, but when it is for use on an animate object it is preferable that both vegetable and animal hydro coloids be used, as the vegetable hydro colloid agar-agar has a lower melting point than gelatine and the composition will therefore melt at a lower temperature, so that it can be applied directly to the human body without scalding effects.

The following proportions have been found to produce the composition having the properties hereinbefore described (parts by weight):

Formula No. 1

Soapy Body	50 parts
Sodium Borate	50 parts
Gum or Resin	50 parts
Animal and Vegetable Hydro Colloid	50 parts
Glycerine	100 parts

Water to be added to thin as desired.

Formula No. 2

Soapy Body	100 parts
Sodium Borate	50 parts
Gum or Resin	50 parts
Animal and Vegetable Hydro Colloid	200 parts
Glycerine	100 parts

Water to be added to thin as desired.

The composition melts at approximately 80° C. and may be applied to animate or inanimate objects in liquid form either by hand, plastic manipulation or with a brush at temperatures between 45° and 55° C. It solidifies quickly having elastic properties so that it can be removed from the object at approximately 33° C. without injuring the cast.

Bleaching Ostrich Feathers

In specially constructed glass troughs, made the length of an average ostrich feather, 15 or 20 of these feathers can be treated at a time. The bleaching fluid is made from a 30 per cent solution of hydrogen peroxide, with enough ammonia added to make it neutral; in other words when neutral, blue litmus paper will not turn red, and red will take a pale violet tinge. The previously cleansed feathers are entirely immersed in this bleaching bath, which may be diluted if desired. The trough is covered with a glass plate and put in a dark place. From time to time the feathers are stirred and turned, adding more hydrogen peroxide. This process requires 10 to 12 hours and if necessary should be repeated. After bleaching they are rinsed in distilled water or rain water, dried in the air, and kept in motion while drying.

To insure success in coloring feathers in delicate tints, they must be free from all impurities, and evenly white. It has been found of advantage to rub the quill of heavy ostrich plumes while still moist with carbonate of ammonia before the dyeing is begun.

Wood Floor Bleaching

Water used in the bleaching solutions raises the grain of the wood, consequently, after the surface has been allowed to thoroughly dry, it must be sand-papered to remove the raised wood fibers. A thin coat of shellac—about 2 lbs. of shellac gum to 1 gal. of denatured alcohol—brushed on and allowed to dry, makes these fibers easy to remove.

After bleaching, the surfaces may retain some of the chemicals; if so, immediate sponging with clean water will be necessary. A coat of ordinary vinegar will assist in neutralizing any traces of alkali left by the solutions, and will put the surfaces in suitable condition for the finishing process. At least 12 hours should be allowed for a thorough drying of the surface before finishing begins.

Many chemical solutions are used for bleaching; some are more effective on certain woods; others succeed better on other woods. Oxalic acid solutions are used by the majority of house painters. Before using any solution, clean and scrub the surface, using hot water to which soap and a small quantity of sal soda has been added. Use No. 2 or No. 3 steel wool for scrubbing, washing well with clean water and a sponge.

Make up a saturated solution, dissolving as much of the oxalic acid in a gallon of hot water as the water will take up. Apply the solution hot, using a flat wall brush, and let it dry on the surface. For bleaching dark sap stains and weather stains, about 8 oz. of oxalic acid to 2 qts. of water is sufficient. When the first application does not give the desired results in color, apply the

solution two or more times, as may be found necessary. For greasy surfaces, rub with denatured alcohol and let dry before the bleaching solution is applied.

Dissolved in water, chlorinated soda makes an effective bleach, especially if it is followed by a solution of peroxide of hydrogen. This solution is used as follows:

Solution No. 1

Sal Soda 5¼ oz.
Water 10 oz.
Dissolve.

Solution No. 2

Chloride of Lime 2½ oz.
Water 6 oz.
Mix well and allow to settle.

Pour the clear liquid of solution No. 2 in another container and a sediment will be found in the bottom of the first container. Add to this 6 oz. of water and stir well. Let settle and pour off the clear fluid as before. Add 1 or 2 oz. of water to the remaining sediments; stir well and strain into the second container through filter paper.

Mix solutions No. 1 and No. 2, and a clear liquid bleach of green color, and having a faint odor of chlorine and a strong alkaline taste, results. Use this solution hot and brush on with a flat wall brush; let dry and wash the surface with clean water.

For bleaching walnut, ordinary chloride of lime dissolved in water and brushed on the wood has been found satisfactory. It will work on many other woods too.

Permanganate of potash dissolved in water and used in varying strengths makes an excellent bleach. The wood will have a purple tint when dry. The solution is to be applied with a brush and, when dry, a second coating of a saturated solution of hyposulphite of soda in water will "fix" the tint. A 5% solution of oxalic acid in water has been found an effective second coat treatment over the permanganate of potash.

Hydrosulphite of soda, when used as a 10% solution in water, has been found to be a satisfactory bleach. One or two coats brushed on and allowed to dry thoroughly following each application, then washing with clear water, has proved to be the required treatment.

To Ebonize Veneer

For ebonizing veneer for inlays before they are laid: Dissolve 6 to 8 oz. of nigrosine to the gallon of alcohol, and place this solution in a tank large enough for dipping the veneers. Use steam for heat, and heat the solution to the boiling point. Leave the veneer in for about one hour; then remove the veneer and dry it, using a plate dryer if possible. Repeat this procedure until the veneer is properly colored clear through. This cycle of steaming and drying may have to be repeated four or five times, depending upon the thickness of the veneer. When the color is clear through and the veneers are dry, they are ready for use. Black inlay will harmonize effectively with practically any kind or color of veneers.

Some finishers claim they get an effective ebony black with a single dipping of the inlays by using a strong water solution of jet black nigrosine heated to a temperature under the boiling point. Inlays are seldom more than $\frac{1}{16}$ inch thick, and they are generally very narrow. A warm water solution of jet black nigrosine will penetrate deep and quickly —except possibly on some of the hardest of close-grained woods like maple, for example, which might require a second dipping or a longer period of immersion.

Preserving Cut Flowers

For those whose flowers are supplied from private gardens, advise cutting the flowers in the early morning or late evening when the stems are turgid. A sharp knife is recommended. The sharper the cut the less is the bruising of the conducting vessels and the greater the absorption of water. The elimination of ragged edges will lessen the chances of bacterial action.

The proper stage of the flower's development should be observed when cutting. Gladioli are best for cutting when the first floret is open; peonies, when the petals are unfolded; roses before the buds open; dahlias, when fully open; poppies, the night before they open.

Flowers after they are cut should be plunged stem-deep in water. All arranging should be postponed until after the stems have been thoroughly soaked.

Flowers should be kept in a humid room and never in sunshine. This reduces the evaporation to a minimum. It is well to keep them at 45° F. If they are kept cooler than that during the night, the lasting quality is improved. Containers which permit a free entrance of air through the top are recommended. Narrow-necked vases should be avoided. Stems should be cut each day with change of water. The aspirin treatment may be used with each change to prolong

the freshness of the flowers. In cutting stems, a slanting cut will prevent the ends from resting squarely on the bottom of the vase. All leaves which are submerged should be removed to prevent decomposition and fouling the water.

Wilted flowers may be revived by cutting their stems short, plunging them deep in water and storing in a cool dark place for ten hours or more. The so-called "hot water" treatment is also useful in restoring wilted flowers. Immerse the stems in hot water (not boiling) for half an hour, keeping them in the dark, and then change to cooler water. Usually several hours are required for restoration.

Coloring Natural Flowers

Freshly cut flowers are placed in tall jars so that stems are partially immersed in 1% solutions of the following dyes:

Light Green s.f.
Tartrazine
Brilliant Blue
Amaranth
Ponceau 3R

Coloring Organic Products
U. S. Patent 1,953,438
Formula No. 1

1 part of Kiton Fast Fellow 3 G (Schultz, Farbstofftabellen, 7th edition, No. 748) is dissolved in 9 parts of textile lecithin while stirring the mixture warmed on the water bath. The solution obtained may be mixed with 390 parts of an oil varnish, whereby a clear transparent solution is obtained.

Formula No. 2

1 part of Patent Blue V (Schultz, Farbstofftabellen, 7th edition, No. 826) is dissolved as prescribed in Formula 1 in 9 parts of textile lecithin. The deep bluish green mass thus obtained, may be used for dyeing a composition for candles consisting, for instance, of 60 parts of stearic acid and 40 parts of ceresin. There is thus obtained an intensely greenish-blue candle wax.

Formula No. 3

0.5 part of Acid Red XB (Schultz, Farbstofftabellen, 7th edition, No. 863) is heated to 90° C. together with 10 parts of lecithin until dissolved. The dye preparation may be applied for coloring an oil emulsion used for the treatment of leather.

Formula No. 4

1 part of Alizarine Viridin FF (powder) (Schultz, Farbstofftabellen, 7th edition, No. 1193) is dissolved in 15 parts of textile lecithin and the solution is mixed with a wax composition. By addition of a lake-forming compound, thus as, for instance, chromium naphthenate, the coloration may be brightened.

Formula No. 5

1 part of Alizarine Marron W (powder) (Schultz, Farbstofftabellen, 7th edition, No. 1159) is stirred on the water bath with 9 parts of textile lecithin until dissolved. The preparation may be used for coloring an oil varnish diluted with toluene. The intensity of the coloration may be increased by addition of aluminium oleate or chromium naphthenate.

Formula No. 6

1 part of Fanal Green LB (a complex salt of a basic dye) (Schultz, Farbstofftabellen, 7th edition, No. 765) is dissolved while being warmed in 15 parts of lecithin. The intensely yellowish green colored dye preparation may be used for coloring a cellulose ether varnish (containing, for instance, an aromatic hydrocarbon and an alcohol as solvents).

Formula No. 7

In a melt containing 60 parts of stearic acid and 5 parts of lecithin there are dissolved on the water bath 0.2 part of the azo dye obtained by diazotizing 3.5-dichloro-2-amino-1-hydroxybenzene and coupling with 1-phenyl-3-methyl-5-pyrazolone. The greenish yellow mass becomes reddish yellow on addition of aluminium chloride.

Formula No. 8

0.5 part of Alizarine RG (Schultz, Farbstofftabellen, 7th edition, No. 1154) and 10 parts of triethanolamine-monooleate are dissolved in 10 parts of textile lecithin. The dye solution may be added, for instance, to a cellulose laurate varnish dissolved in toluene. On addition of an aluminium-, copper-, iron- or nickel salt, the shade of the color may be varied in known manner.

Formula No. 9

Dyes which are insoluble in oils, such as linseed oil or paraffin oil, (for instance those mentioned in Formulas 2, 3 and 6) are dissolved in the manner described in lecithin. With the aid of the dye preparations thus obtainable, fatty acid esters, such as linseed oil, or benzine hydrocarbons, such as paraffin oil, may be homogeneously dyed without separation of the coloring matter.

Wax Coating for Fruits

U. S. Patent 1,943,468

Formula No. 1

65 parts of carnauba wax and 20 parts of oleic acid are heated to 95–100° C., and, when the wax is melted, the mass is thoroughly agitated or stirred to thoroughly mix the acid into the molten wax. In another vessel 10 parts of triethanolamine and 30 parts of water are mixed and heated to 95–100° C. With both mixtures at this temperature, the triethanolamine solution is added to the wax-oleic acid solution with rapid enough stirring to thoroughly mix the mass. As the last of the water solution is added the mass will thicken slightly and have the appearance of a clear jelly. Stirring is continued for 15 to 20 minutes, keeping the temperature at about 90–95° C. Then 35 pounds of paraffin wax, broken into small lumps, are added, and the mixture stirred for one-half hour, or until the paraffin is entirely melted and mixed through the jelly. Water heated to 90–95° C. is then added slowly with constant stirring; this clear transparent jelly dissolves in the hot water and produces an emulsion which has a decidedly bluish-white opalescence. Water is added until the total weight of the emulsion is 640 parts. The emulsion is then cooled rapidly.

This emulsion is better adapted for use in coating bananas, pineapples, and certain other fruits than for coating citrus fruits. It dries somewhat more slowly than does the emulsion to be described in Formula No. 2 below.

Formula No. 2

Another operable composition (emulsion) peculiarly adapted for use, in diluted form, where a high luster is desired on fruits, e.g., citrus fruits, may be prepared as follows:

125 pounds of carnauba wax is melted with 22 pounds of oleic acid at a temperature of about 85–90° C. A solution is made of 1 pound of caustic soda and 10 pounds of triethanolamine in 30 pounds of water and the solution is heated to 90° C. This solution is stirred into the wax-oleic acid mixture at the same temperature, and thoroughly mixed. A jelly-like mass is formed and to this

water, at 90° C., is added slowly until a total weight of 550 pounds is reached. A solution of 51 pounds of Pontianak gum in 350 pounds of water and about 6¼ to 6½ pounds of caustic potash is the amount required to just dissolve the gum and leave the solution only slightly alkaline. The weight is adjusted to 1085 pounds by addition of water, and the emulsion is then cooled. The product is a concentrated emulsion in a form favorable for shipping or storage.

The concentrated emulsion thereafter is suitably diluted with water for use as treating bath. In diluting the concentrated emulsion we find that water may be incorporated with it in the ratio of from 1 to 3 pounds per 1 pound of the concentrate. Fruits passed through a treating bath of this emulsion carry therefrom substantially no excess emulsion over and above a thin coating film thereof, and surface-dry so rapidly that they may be handled in the ordinary equipment (driers, conveyors, and the like) of a fruit packing plant without any material slowing down of the ordinary operations therein.

The specific emulsion above set forth may contain, dispersed or dissolved therein, a suitable amount of an appropriate sterilizing agent, e.g., borax, in an amount equivalent to 5–20% of the wax content of the composition (that is, from .5% to 5.0% of the emulsion). Or, one may in known manner substitute a relatively small amount of boric acid for a corresponding amount of the borax employed, e. g., use as the sterilizing agent component of the emulsion borax and boric acid in the proportions of 4.3% borax and 0.7% boric acid, by weight, based on the weight of the emulsion as used.

Transfer

U. S. Patent 1,941,697

Amberol (type F7)	150 parts
Carnauba Wax	84 parts
Ozokerite	32 parts
No. 6 Litho Varnish	35 parts
Blown Castor Oil	60 parts
Cobalt Drier	1 part
Butyl Carbitol	10 parts
Cadmium Selenide	215 parts

In order to prepare this composition for printing on the transfer base, the fusible elements are first melted and intermixed and the oils are then added, and the mass is thoroughly intermixed. The butyl carbitol is diethylene glycol mono

butyl ether and it has a boiling point of 220° C.

The temperature of the mixture is then lowered almost to the hardening point and the butyl carbitol is then added and the mass is again thoroughly mixed.

The cadmium selenide, which is the pigment utilized in the above-mentioned formula, is then added with thorough mixing.

The artificial resin above specified furnishes a base for the vehicle which is not easily fusible, and the addition of the carnauba wax and the ozokerite renders the composition more fusible. The addition of the oils increases the fusibility of the composition so that it has a melting point between 80° C.–120° C.

The No. 6 litho varnish slowly oxidizes so that the transfer marking hardens or sets upon the base of the transfer and this setting action increases with the age of the transfer while allowing the composition to be used for printing purposes when the composition is fused. The cobalt drier accelerates this ageing action. The butyl carbitol reduces the melting point of the composition and it disappears in a few days after the transfer marking is printed upon the paper base so as to increase the rapidity of the ageing action. Due to the relatively high boiling point of the butyl carbitol, said solvent is not completely driven off during the preparation of the mixture. However, said solvent evaporates within a few days after the transfer marking has been printed, as previously stated.

These transfers are ordinarily supplied to the consumer with the paper base wound into the form of a spiral form and rapid ageing is desirable in order to prevent the transfer markings from smearing.

The cadmium selenide is particularly advantageous because it is an opaque pigment which is very fine and very smooth. This pigment is very useful in producing a brilliant red marking on black hosiery. If a white marking is desired, titanium oxide can be used and if a yellow marking is desired, cadmium sulphide can be used.

Cadmium selenide has remarkable resistance to heat and it does not lose its brilliancy if it is overheated, so that it is especially suited for a transfer of this type. It can therefore be employed in the above-mentioned formula in a very large proportion and to secure a very intense color. Another advantage of using cadmium selenide is that it has low oil absorption.

The above formula can be varied. For

example, if greater penetration of the marking is desired, more wax can be used and if less fusibility is desired, less wax can be used.

Other ingredients may be added to increase the fusibility of the composition, such as ethyl abietate.

Extracting Carotene from Spinach
U. S. Patent 1,953,607

Raw, undried, fresh spinach is cooked in a sealed container from which the air has been exhausted. The cooked spinach or other material, preferably drained from excess water, is covered with a dilute alkaline solution at ordinary room temperature, such as a solution of dilute sodium hydroxide, or even the more expensive potassium hydrate, of approximately 3N concentration, although the exact concentration is immaterial. After standing a few hours or longer, preferably with exclusion of air, the cellulose becomes disintegrated and the chlorophyll hydrolyzes to form water soluble products which are easily separated. However, the carotene is not materially injured or changed by this alkali treatment.

The pasty mixture is next diluted with water, preferably about an equal volume, and the whole mass is gently stirred with chloroform, which extracts or dissolves the carotene as well as the xanthophyll, but no chlorophyll. Care is taken not to agitate the water and plant material while mixed with chloroform to such an extent that troublesome emulsions result.

The heavy and red solution of plant pigments in chloroform settles as a layer on standing and is easily drawn off. The watery plant material is again extracted with chloroform and the pigment solution thus formed is allowed to settle and is again drawn off and successive chloroform washings are continued until no worthwhile additional amounts of carotene are dissolved.

The several chloroform solutions are next mixed and are distilled to yield an orange colored mass of crude carotene, xanthophyll and some fatty material. Of course, this rich concentrate is a crude form of carotene, contaminated by other substances, but it is useful in this form and may be so used directly for animal feeding or for other purposes. However, if a pure or more pure product is desirable, the crude product is dissolved in petroleum ether in suitable quantities, the resulting solution of pigments being agitated with methyl alcohol of approximately 80% to 90% concentration. In the petroleum ether solution the methyl alcohol dissolves the xanthophyll but not the carotene and the mixture settles or stratifies by gravity with the methyl alcohol and its contained xanthophyll in one layer and the carotene in its petroleum ether solvent in another layer, and the two layers are readily separated.

By distillation, of course, a more pure carotene, now free of xanthophyll, may readily be separated, but to further purify it the petroleum ether solution, containing the carotene, has added thereto a suitable proportion of absolute ethyl alcohol, which coagulates and settles or precipitates certain fatty materials present in these plant substances. The fatty material is removed by filtration and the filtrate containing the carotene is concentrated by evaporation at reduced pressure, but to a point short of separation of carotene as a solid. The concentrated solution, preferably in an atmosphere of nitrogen, as in a flask, is then cooled and is held for an appreciable time at a suitable low temperature, for example, approximately 32° F., until the carotene crystallizes out as small dark red lustrous crystals whose solubility in the mixture is reduced by the presence of the alcohol.

Oiled Duck

Treat 11 ounce duck in continuous rolls in a vertical tower. It must be thoroughly penetrated by the mixture and coated on both sides.

Keep sufficient mixture in the treating vats so that the level of the liquid will be 3 inches above the top of the iron roll.

Pass the duck through one dip of boiled linseed oil thinned with benzine to a specific gravity of .875 at 15° C. Feed into the treating tank at regular intervals with boiled linseed oil thinned with benzine to a specific gravity .875 at 15° C.

Drying: Dry by giving the duck 3 passes up and down the tower at a rate of 22 inches per minute with a temperature of 105° C. (220° F.) in the tower.

Motion Picture Screen Coat (Very Good Detail)

To ½ pound stick glue melted in 1 gallon water, add ½ pound glycerine and stir in thoroughly 1 pound zinc oxide. Apply hot with large brush to stretched screen, and let dry before removing from stretcher. May be rolled without cracking or breaking.

Non-Poisonous Water Pipe-joint Seal

Thick spar varnish—1 part mixed thoroughly with equal amount linseed oil. Mix evenly 1 part graphite and 1 part slaked lime, and stir into above liquid. Will keep indefinitely in sealed container.

Dust Settler

Calcium chloride scattered generously on a dusty road, gathers moisture from the air sufficient to render the road practically dustless.

Spontaneous Combustion Reducer

Sulphur dust discourages spontaneous combustion when sprinkled on oily waste, paint soaked rags, etc.

PATENT LAWS ON CHEMICAL COMPOUNDS

Chemical Compounding Patents
Composition of Matter

While the Constitution of the United States makes no specific provision for the grant of a patent on a Chemical Composition or Composition of Matter, the Patent Statutes conferring such protection are based on Section 8 of Article I, which reads:

> "The Congress shall ·have power . . . to promote the progress of science and useful arts, by securing for limited timer. to authors and inventors the exclusive right to their respective writings and discoveries."

It is required, of course, that the subject matter of a patent shall fall within the statutory definition of patentable subject matter before a grant of an authorized legal monopoly may be had for the limited term of seventeen years; and under the Patent Laws

> "any new and useful art, machine, manufacture, or composition of matter, or any new and useful improvement thereof,"

may be the subject of a patent.

The particular phrase "Composition of Matter" is set forth in Walker on Patents to cover "all compositions of two or more substances. It includes, therefore, all composite articles, whether they be results of chemical union, or of mechanical mixture, or whether they be gases, fluids, powders or solids."

However, to be patentable, the composition must not be a mere aggregation of several ingredients not developing a different or additional property or properties which the several ingredients do not possess individually; but it must involve a true combination or be capable of a new use not obvious to those skilled in the art and remote from any known use.

Thus, an alloy has been held patentable for use as an electrical resistance where it had been known previously only as a non-magnetic material for making the parts of a watch. And a solidified fuel has been held patentable which consisted of nitrocellulose, methyl alcohol and ethyl alcohol, providing a spongy framework of combustible material (nitrocellulse) and combustible liquid (methyl and ethyl alcohol)—all of these ingredients cooperating to produce the product and all contributing in the production of heat in the intended use of the composition.

As an example of aggregation, there may be noted that a hair-treating or hair-food compound—where all except one of the ingredients had previously been used in compounds for the same purpose, and this additional ingredient was equivalent to a substance used before in the same relation—was held to involve nothing patentable because there was no chemical combination of substances nor any new function or effect brought out, but merely the combined functions and effect of the separate ingredients selected. Also, a remedy for treatment of lung disease has been held unpatentable which involved merely an aggregation of ingredients, the medicinal value of which had long been known.

Furthermore, a variation in the proportions of the elements of a composition does not constitute patentable novelty or invention, unless from the new proportions a new material or substance, or an old material or substance with new characteristics, or at least substantially enhanced qualities of utility, has been created.

In connection with the infringement of claims for a Composition of Matter, it may be noted, generally, that changes in the proportions of the ingredients of a patented composition (except where such proportions are of the essence of the invention), or additions thereto of other ingredients, which do not change the principal characteristics of the patented composition, do not avoid infringement; and the addition of an ingredient to the patented composition of such minor consequence and inconsequential effect as not to change the essential character thereof and to have it retain its character and function, does not avoid infringement; or, where such addition does not substantially destroy or alter the essential properties of the composition; but infringement may be avoided by the omission of an ingredient where no equivalent is substituted therefor. Infringement occurs only when the product has the same characteristics and is composed of the same or equivalent ingredients.

No Serial Number and/or date of filing is accorded an inventor, nor is an application complete, for an invention until there is met all of the requirements provided for in the Rules of Practice of the United States Patent Office, to wit: a Petition for the grant of a patent and duly signed by the inventor, or inventors; a similarly executed Specification (and, if required for the better understanding, a Drawing) describing the nature of the invention in such a manner that those skilled in the particular art can readily understand the same, together with Claims setting forth the features believed novel and patentable; and an Oath, affirmed or sworn to before an officer authorized to administer the same, and setting forth among other things that the applicant, or applicants, is the original, first and sole inventor (or are joint inventors) of the subject matter described and claimed, and that the same has not been published, nor been in public use or on sale for more than two years prior to the date of application.

However, it is frequently desirable to make some record prior to the filing of an application, which record may be useful as evidence of conception of the invention in the event of a subsequent interference proceeding in which the application may become involved. To this end, a statement describing the invention as fully as possible may be dated and signed by the inventor and witnessed by one or preferably several subscribing witnesses who fully understand the nature of the invention disclosed in the said statement, because the unsupported statements of the inventors are of no avail. Such statement is, also, preferably to be acknowledged before a Notary Public.

No provision is now in effect for filing Caveats in the United States Patent Office.

The grant of Letters Patent does not confer of itself the right of the patentee to make use of the invention covered thereby, but conveys merely the right, for a limited period, to exclude others from the manufacture, use and sale thereof. It may become necessary, therefore, to first secure a license under some controlling patent or to purchase an interest therein.

In the case of Letters Patent issued jointly to several patentees, as well as in the case of the owner of an undivided part of a patent, each party, in the absence of an agreement to the contrary, may proceed to dispose of his rights, grant licenses, and otherwise operate thereunder without first obtaining the consent of the other owner or owners, and without accounting to such for any returns therefrom.

In the case where patents are owned by corporations, or large manufacturers, difficulty may be encountered in securing a license to operate under the same, as the full rights are usually retained in such circumstances to secure a monopoly. When nothing has been done with respect to marketing the invention, it is generally possible to effect a satisfactory arrangement, as for a non-exclusive or an exclusive license, with royalty payments; and negotiations may be initiated, for example, through the attorney of record of the particular patent, or directly with the patentee if the address of the latter be readily available.

The purchaser of a patent right should endeavor to buy the patent outright, if possible; whereas, a seller should usually retain title to a patent and grant only licenses thereunder.

The foregoing sets forth only the general principles to which there may be exceptions which it is not possible to consider without unduly extending the length of this article; and, it is generally more satisfactory in matters of this nature to consult those expert therein.

FIRST AID FOR CHEMICAL INJURIES

General Treatment

1. Call a physician or an ambulance.
2. In the case of a corrosive poison, immediately discharge the contents of the mouth and wash out with large amounts of water.
3. Empty the stomach by means of an emetic or by use of the stomach tube, if some one trained in its use is present, unless later specific directions forbid this.
4. If the poison and its antidote are known, administer the indicated antidote. If poison or antidote are unknown, give repeated doses of the General Antidote, trying to induce vomiting after each dose.
5. Administer stimulants and combat collapse as previously described.

In poisoning due to the swallowing of a corrosive acid, the mouth should be washed out immediately with an alkaline solution, such as 5% sodium bicarbonate. If due to a corrosive alkali, a 5% boric or acetic acid solution should be used.

Emetics may be used to remove poison from the stomach in all cases except those in which a corrosive material is involved, which includes all mineral acids and strong alkalis. In the case of carbolic acid (phenol), an emetic or the stomach tube should be used only with extreme care, if at all.

Emetic—The most commonly used emetic is prepared by stirring one tablespoonful of powdered mustard in sufficient warm water to give a paste of the consistency of thick cream. This amount is one dose.

Tartar emetic, potassium antimony tartrate, may be given in doses of $\frac{1}{10}$th grain in 25 cc. of warm water.

Zinc sulphate may be given in doses of 15 grains in 25 cc. of warm water.

The *Physician* may administer $\frac{1}{10}$th grain Apomorphine Hydrochloride per hypo as an emetic.

In any case the dose is to be repeated after fifteen minutes if vomiting has not taken place.

General Antidote

Where an unknown poison is suspected, or where no specific antidote is known, administration of the General Antidote may be of value. This is prepared by mixing two parts of pulverized charcoal, one part magnesium oxide, and one part tannic acid. It is kept dry, and does not deteriorate. It is administered by stirring one heaping teaspoonful in a small glass of warm water. The charcoal absorbs many poisons, magnesium oxide neutralizes acids and precipitates many toxic materials, and the tannic acid neutralizes alkalies and also precipitates many toxic substances.

In cases of poisoning by corrosive materials, demulcent drinks are administered, to soothe the injured tissues. They may consist of raw eggs, directly from the shell, milk, a mixture of four eggs to one pint of milk, gruels of various grains, flour and water mixture, starch and water paste, crushed bananas, mucilage and water, oil and water emulsions, or any other available harmless material of similar character. They aid in preventing further action of the poison by lining the alimentary canal with a film of colloidal material.

In some cases, white of egg is administered to form an insoluble compound with the poison, as in the case of Denatured Alcohol, compounds of the heavy metals, etc. In such cases the white of one egg is stirred up with one quart of water, and as much of this mixture administered as may be possible. Many of this insoluble albuminous compounds so formed dissolve in excess of albumen, consequently the amount used must be kept low.

Acetone

1. Give emetic.
2. Stimulate; combat collapse.
3. Keep patient awake.

Acids, Mineral—The most quickly available remedy is the best. Neither the Stomach Tube nor Emetics should be used.

1. Give calcined magnesia, white magnesia, milk of magnesia, or Lime-water *immediately*, mixed in milk or any mucilaginous fluid that will act as a demulcent. Repeat the dose at short intervals until it may be inferred the poison is neutralized. Do not give carbonates as anti-

dotes for mineral acids. Oleaginous and mucilaginous fluids should be given freely, even as vehicles for antidotes. In the case of strong sulphuric acid, water, if given at all, should be given in large quantities, on account of the heat developed. Ice may be given to relieve thirst and pain.

2. Administer stimulants and combat collapse. An enema of strong coffee or of three to five grains caffein citrate is a valuable stimulant when a corrosive poison makes the administration of a stimulant by mouth difficult.

The *Physician* may administer Morphine Sulphate, per hypo, ⅛th to ¼th grain, to relieve pain; Atropine Sulphate, ¼₂₀th grain; Strychnine Sulphate, ⅟₆₀th grain; or Digitalin, ⅟₁₀₀th grain as stimulant.

Aldehydes (Formaldehyde)—Artificial Respiration or a Pulmotor may be required.

1. Repeatedly administer doses of one to two cc. of 1% ammonium hydroxide solution, to form harmless hexamethylene tetramine. Follow with an alkaline mineral water.

2. Demulcents such as raw egg, egg and milk, barley, water, etc.

3. Wash out the stomach using the stomach tube if necessary. The *Physician* may administer hypodermic of Strychnine or Digitalin as supportive.

Alkalies, Caustic—Neither the stomach tube nor powerful emetics should be used.

1. Administer a weak acid, such as 5% acetic acid, vinegar, or lemon juice, until it may be inferred the alkali has been neutralized. Give butter, olive or cotton seed oil, or other oils or fats to form soaps.

2. Assist vomiting by draughts of tepid water.

3. Stimulate and combat collapse.

4. Administer demulcents.

The *Physician* may administer Strychnine, Atropine, or Digitalin per hypo as stimulant.

Aniline and Derivatives—Artificial respiration or use of the pulmotor and the administration of oxygen may be necessary to overcome cyanosis.

1. Use emetic or stomach tube to empty stomach.

2. Administer 3% solution of acetic acid, or vinegar.

3. Loosen clothing, apply artificial respiration, or use Pulmotor and administer oxygen if necessary.

4. Administer stimulants and combat collapse.

The *Physician* may administer Atropine, Strychnine, or Digitalin per hypo, as stimulant.

Antimony and Compounds

1. Emetic of zinc sulphate or use stomach tube to empty stomach.

2. Large draughts of strong tea, or doses of two grams of gallic of tannic acid in warm water, or doses of the General Antidote.

3. Demulcent drinks, milk, raw eggs, etc.

4. Stimulate and combat collapse.

The *Physician* may administer Strychnine or Atropine per hypo, as stimulant.

Barium and Compounds

1. Emetic, preferably mustard.

2. 10% aqueous solution of sodium sulphate.

3. Demulcents, egg, milk, oils, gruels, etc.

The *Physician* may administer, per hypo, Morphine to relieve pain, and Strychnine, Atrophine, or Digitalin, as stimulant.

Bromine Solutions

1. Starch and water paste.

2. Evacuate stomach—do not use stomach tube.

3. Give 10% solution sodium bicarbonate freely.

4. Give demulcents.

5. Stimulate and combat collapse.

The *Physician* may administer, per hypo, Apomorphine as emetic, if immediately available; and Morphine to relieve pain.

Chlorine Solutions—(Bleaching powder solution, Javelle water, etc.).

1. ¼th teaspoonful ammonia in wine glass of water.

2. Evacuate the stomach by cautious use of stomach tube, or by emetic of zinc sulphate or mustard, followed by large draughts of warm water.

3. Give demulcents, milk, egg, gruel, etc.

4. Give 20 grains sodium thiosulphate in ½ glass water.

The *Physician* may administer, per hypo: Apomorphine as emetic, if immediately available; Strychnine, Atrophine, or Digitalin as stimulant; and Morphine to relieve pain.

Chromium and Compounds

1. Give emetic.

2. Milk of magnesia or precipitated chalk and water.

3. Stimulate and combat collapse.

Physician may administer stimulants per hypo.

Copper and Compounds

1. White of egg in water, milk, much water.
2. 10% solution of sodium hydrogen phosphate.
3. Emetic.

The *Physician* may administer Morphine to relieve pain and stimulants per hypo.

Cyanides—Speed is Essential!!

If Swallowed:

1. 3% Hydrogen peroxide solution in large quantities or a freshly prepared mixture of equal volumes of 5% solution of ferrous sulphate and sodium carbonate, followed by immediately evacuation of the stomach.
2. Stimulate and combat collapse.
3. Artificial respiration and Pulmotor.
4. If patient survives, give emetic, and after vomiting, supportive of strong coffee with brandy or whiskey.

If Inhales: Inhale Chlorine from a weak water solution.

Fluorides

1. Give emetic
2. Milk of magnesia.

Hyrdogen Sulphide—inhaled

1. Inhale ammonia from 5% ammonium hydroxide, or inhale fresh air containing a small proportion of chlorine.
2. Administer milk, white of egg in water, olive or cotton-seed oil, etc.
3. Stimulate and combat collapse.
4. Use artificial respiration, Pulmotor if necessary.

Physician may administer stimulant per hypo.

Illuminating Gas

1. Remove to freshest air quickly available.
2. Call Pulmotor and practice artificial respiration until Pulmotor and crew arrive, or for at least one hour if no Pulmotor is available. Administration of oxygen containing 5% Carbon dioxide is often of value, as is the inhalation of ammonia or vapors of amyl nitrite.
3. Stimulate with brandy or whiskey, or ammonia.

Nitrobenzene and Compounds

1. Emetic, or stomach tube if necessary, large quantities of water to wash out the stomach.

2. 3% aqueous acetic acid or vinegar.
3. Stimulate and combat collapse.
4. Employ artificial respiration and Pulmotor if necessary, with administration of oxygen.

The *Physician* may stimulate by Atrophine, Strychnine, or Digitalin, per hypo.

Oxalic Acid and Oxalates—The stomach tube should be used only with great care, if at all.

1. Precipitated chalk suspended in water, or milk of magnesia.
2. Give emetic.
3. Stimulate and combat collapse.
4. Demulcents, raw egg, milk, gruels, etc.

The *Physician* may administer Morphine, per hypo to relieve pain.

Phosphorous

1. Administer Liquid Petrolatum, then evacuate the stomach with stomach tube.
2. Emetic, 3 to 5 grains copper sulphate in two teaspoonfuls of water, repeated every five to ten minutes until vomiting occurs.
3. Much water, and five to ten cc. of French (Bordeaux) turpentine, or turpentine which has been exposed to the air for a long time and contains large amounts of oxygen, but *no other* fatty oils. Avoid *milk*.
4. Give purgative of 15 grams of magnesium sulphate.

The *Physician* may administer Morphine, per hypo, to relieve pain.

Permanganates

1. Give emetic.
2. White of egg in water.
3. Solution of Hydrogen Peroxide, slightly acidified with acetic acid.

Pyridine

1. Give emetic.
2. Two grams of gallic or tannic acid in 50 cc. water.
3. Stimulate and combat collapse.
4. Use artificial respiration or Pulmotor if necessary.

The *Physician* may administer Atropine, Digitalin, or Strychnine per hypo, as stimulant.

Silver and Compounds

1. 25 per cent solution Sodium Chloride.
2. Emetic, or evacuate the stomach by use of stomach tube.
3. White of egg in water, milk, or egg and milk.

The *Physician* may administer, per hypo, Morphine to relieve pain; or stimulant of Atropine, Digitalin or Strychnine.

Tin and Compounds

1. Emetic or evacuation of stomach by use of stomach tube and large amounts of water.
2. Milk of magnesia, followed by soap and water, gum arabic suspension in water, oil and water emulsion, or other demulcent.
3. Stimulate and combat collapse.

The *Physician* may administer Morphine, per hypo, to relieve pain; and Atropine, Digitalin or Strychnine to stimulate.

Zinc and Compounds

1. Emetic, or wash out the stomach by use of stomach tube, using large amounts of water containing sodium bicarbonate.
2. White of egg in water, milk, or egg and milk.
3. Strong tea.
4. Stimulate and combat collapse.

The *Physician* may administer, per hypo, Morphine for relief of pain; and Atropine, Strychnine, or Digitalin, as stimulant.

First Aid for Burns

From Flames or Hot Objects—Dress with Burn Ointment, and, if serious or covering a large area, bandage over the dressing. Where tissues have been destroyed, use a dressing of a pad of moist picric acid gauze large enough to cover the injured area, and keep in place by a bandage. The entire dressing should be kept moist.

Dressing should be completely changed at least once a day.

Blisters forming from burns should be opened and drained by puncturing in at least two places near the edge and pressing out the liquid. The puncture may be made with a flame-sterilized needle or razor blade. When changing dressing, any blisters present should be drained again.

From Strong Acids, bromine, chlorine, phosphorus or other material of acid character, are washed first with large quantities of water, then with 5% of sodium bicarbonate or 5% ammonium hydroxide solution, then dressed and bandaged as above.

From Strong Alkalies, sodium peroxide, metallic sodium or potassium or other materials of alkaline nature, are washed first with large quantities of water, then with 5% boric or acetic acid solution or vinegar, then dressed or bandaged in like manner as for Burns from Flames or Hot Objects.

From Phenol are washed immediately with strong pure alcohol, then dressed and bandaged, if necessary.

From Phenylhydrazine are washed with three to five per cent acetic acid or vinegar, then dressed and bandaged, if necessary.

First Aid for Scalds

Blisters that have formed are to be opened and drained by puncturing in at least two places near the edge and pressing out the liquid. Then the injury is dressed and bandaged, using suitable neutralizing agents if due to acid or alkaline materials, as prescribed for Burns.

Cuts

1. Remove all foreign matter, such as pieces of glass, dirt, etc., then apply 3.5% Tincture of Iodine, taking care that the tincture reaches all crevices of the wound. Note that the U. S. P. Tincture is 7%, and should be diluted with an equal volume of water or 95% pure alcohol.
2. If the cut is slight or does not bleed copiously, bandage, placing a small pad of gauze directly over the wound, and bandaging tightly enough to stop the flow of blood. If a small cut does not stop bleeding from the pressure bandage alone, apply peroxide of hydrogen or a saturated solution of ferric chloride to coagulate the blood, then bandage with a pad of gauze over the wound.
3. If the cut is severe and bleeds copiously, apply a tourniquet to check the bleeding until the arrival of a physician. If the cut is in an artery, indicated by the blood being a bright scarlet and flowing in an intermittent stream, the tourniquet is to be placed between the cut and the heart. If in a vein, shown by dark, purplish blood, the tourniquet should be placed between the capillaries and the wound.

Under no conditions should a tourniquet be allowed to remain in place for more than two hours at a time.

Collapse

In the case of injury due to Burns, Cuts or Scalds, as well as in many cases of poisoning, attention must be given to combating shock and collapse which might follow. Stimulants such as hot, strong coffee, inhalation of ammonia or smelling salts, applications of cold packs to the head and warm or hot objects to the remainder of the body are of value. Brisk rubbing of the chest and extremities may be sufficient to restore a patient who has fainted. Administration of brandy or whiskey in doses of ten to fifteen cc. is of value, to prevent collapse. Inhalation of ethanol, acetic acid, or amyl nitrate vapors likewise acts as a stimulant.

Toxic Headaches due to the inhalation of vapors of various materials may be combated by removal to fresh air, administration of five to ten grains of Aspirin, and allowing the patient to rest for a time.

Unconsciousness due to electric shock is treated by the administration of stimulants as given above, and by artificial respiration, either manual or by use of the pulmotor using the drowning resuscitation methods.

TABLES

Conversion Factors

1. Grams per litre (g./l.) multiplied by 0.134=avoirdupois ounces per gallon (oz./gal.).

2. Avoirdupois ounces per gallon (oz./gal.) multiplied by 7.5=grams per litre (g./l.).

3. Grams per litre (g./l.) multiplied by 0.122=troy ounces per gallon (troy oz./gal.).

4. Troy ounces per gallon (troy oz./gal.) multiplied by 8.2=grams per litre (g./l.).

5. Grams per litre (g./l.) multiplied by 2.44=pennyweights per gallon (dwt./gal.)

6. Pennyweights per gallon (dwt./gal.) multiplied by 0.41=grams per litre (g./l.).

7. Amperes per square decimeter (amp./dm.²) multiplied by 9.29=amperes per square foot (amp./sq. ft.).

8. Amperes per square foot (amp./sq. ft.) multiplied by 0.108=amperes per square decimeter (amp./dm.²).

Thermometer Readings:

Degrees Centigrade × 1.8 + 32 = deg. Fahr.

$$\text{Degrees} \frac{\text{Fahrenheit} - 32}{1.8} = \text{deg. Cent.}$$

$$\text{Degrees} \frac{\text{Reamur} \times 9}{4} + 32 = \text{deg. Fahr.}$$

$$\text{Degrees} \frac{(\text{Fahrenheit} - 32)4}{9} = \text{deg. Reaumur.}$$

$$\text{Degrees} \frac{\text{Reamur} \times 5}{4} = \text{deg. Cent.}$$

$$\text{Degrees} \frac{\text{Centigrade} \times 4}{5} = \text{deg. Reaumur.}$$

SPECIFIC GRAVITY
WEIGHT REQUIRED TO MAKE A GALLON

	Specific Gravity	Pounds to Gallon
Litharge	9.3	77.5
Red-Lead	8.7 to 8.8	72.5
Orange Mineral (orange lead)	8.6 to 8.7	73.0
White-Lead	6.7	55.8
Basic Lead Sulphate	6.4	53.3
Chrome Yellow (medium)	6.0	50.0
Zinc Oxide (white zinc)	5.6	46.6
Basic Lead Chromate	6.8	56.6
English (mercury) Vermillion	8.2	68.3
Bright Red Oxide of Iron	4.9 to 5.26	42.0
Indian Red Oxide of Iron	5.26	43.8
Brown Oxide of Iron (Prince's)	3.2	26.6
Ultramarine	2.4	20.0
Prussian Blue	1.85	15.4
Chrome Green (blue tone)	4.44	37.0
Chrome Green (yellow tone)	4.0	33.0
Lithopone	4.25	35.4
Ochre	2.94	24.5
Barytes	4.35 to 4.46	35. to 37.0
Blanc Fixe	4.25	35.4
Gypsum (terra alba)	2.3	19.0
Asbestine (magnesium silicate	2.75	23.0
China Clay (aluminum silicate)	2.6 to 2.7	22.5
Whiting	2.65	22.0
Silica	2.65	22.0
Natural Graphite	2.1 to 2.4	18.0
Acheson's Graphite	2.2	18.3
Lampblack	1.85	15.4
Carbon Black	1.85	15.4
Keystone Filler (ground slate)	2.66	22.0
Titanox	4.3	35.8
Titanium Oxide	3.9 to 4.0	33.3
Drop Black	2.5	20.8

To this table the following data may be added: The weight of one gallon of paste made with

	Pounds
Red-Lead	44.8
White-Lead (heavy paste)	34.0
White-Lead (soft paste)	30.8
White Zinc	25.0
Chrome Yellow (medium)	24.0
Chrome Green	24.0
Venetian Red	19.0
French Ochre	15.0
Prussian Blue	10.0
Lampblack	9.1
Drop Black	11.7

WEIGHTS AND MEASURES
ENGLISH SYSTEM

Avoirdupois and Commercial Weights

16 drams, or 437.5 grains   =1 ounce, oz.
16 ounces, or 7000 grains   =1 pound, lb.

WEIGHTS AND MEASURES, ENGLISH SYSTEM—*Continued*

28 pounds	=1 quarter, qr.
4 quarters (English)	=1 hundredweight, cwt.—112 lbs.
20 hundredweight	=1 ton of 2240 lbs., gross or long ton
2000 pounds	=1 net, or short, ton
2204.6 pounds	=1 metric ton=1000 kilos

1 stone=14 pounds; 1 quintal=100 pounds

Troy Weights

24 grains	= 1 pennyweight, dwt.
20 pennyweights	= 1 ounce, oz. = 480 grains
12 ounces	= 1 pound, lb.= 5760 grains
1 carat	= 3.168 grains = 0.205 gram

Troy weight is used for weighing gold and silver. The grain is the same in Avoirdupois, Troy and Apothecaries' weights.

Apothecaries' Weights

20 grains =1 scruple

2 scruples=1 drachm, ϶=60 grains
8 drachms=1 ounce, ℥=480 grains
12 ounces =1 pound, lb.=5760 grains

Apothecaries' Measures

60 minims (min.)=1 fluid drachm (fl. dr.)
8 fluid drachms =1 fluid ounce (fl. oz.)
20 fluid ounces =1 pint (O) +
8 pints =1 gallon (C) +

Relations of Apothecaries' Measures to Weights
(All liquids to be measured at 62° Fahr.)

1 minim is the measure of	0.0115	grains of distilled water
1 fluid drachm " "	54.687	" " " "
1 fluid ounce " "	437.5	" " " "
1 pint " "	8750	" " " "
1 gallon " "	70000	" " " "

Linear Measure

12 inches=1 foot	4 poles	=1 chain
3 feet =1 yard	40 poles	=1 furlong
6 feet =1 fathom	8 furlongs=1 mile=1760 yards	
5½ yards =1 rod pole, or perch		

Square Measure

144 square inches=1 square foot
9 square feet =1 square yard
30.25 square yards or 272.5 sq. feet=1 square rod
160 square rods or 4840 sq. yards or 43560 sq. feet=1 acre
640 acres=1 square mile
An acre equals a square whose side is 208.7 feet

Cubic Measure

1728 cubic inches =1 cubic foot
27 cubic feet =1 cubic yard
1 cord of wood =a pile 4×4×8 feet=128 cubic feet
1 perch of masonry=16.5×1.5×1 foot=24.75 cubic feet

1 cubic inch of water at 62° Fahr. weighs 252.286		grains
" " " " " " "	0.57665	oz. (av.)
" " " " " " "	0.036041	lb.
1 cubic foot " " " " " "	996.458	oz. (av.)
" " " " " " "	62.2786	lb.
1 cubic yard " " " " " "	0.75068	tons

CAPACITY MEASURE
Liquid

4 gills =1 pint
2 pints =1 quart
4 quarts=1 gallon

CONVERSION OF THERMOMETER READINGS

F°	C°	F°	C°	F°	C°	F°	C°	F°	C°	F°	C°
−40	−40.00	30	−1.11	80	26.67	250	121.11	500	260.00	900	482.22
−38	−38.89	31	−0.56	81	27.22	255	123.89	505	262.78	910	487.78
−36	−37.78	32	0.00	82	27.78	260	126.67	510	265.56	920	493.33
−34	−36.67	33	0.56	83	28.33	265	129.44	515	268.33	930	498.89
−32	−35.56	34	1.11	84	28.89	270	132.22	520	271.11	940	504.44
−30	−34.44	35	1.67	85	29.44	275	135.00	525	273.89	950	510.00
−28	−33.33	36	2.22	86	30.00	280	137.78	530	276.67	960	515.56
−26	−32.22	37	2.78	87	30.56	285	140.55	535	279.44	970	521.11
−24	−31.11	38	3.33	88	31.11	290	143.33	540	282.22	980	526.67
−22	−30.00	39	3.89	89	31.67	295	146.11	545	285.00	990	532.22
−20	−28.89	40	4.44	90	32.22	300	148.89	550	287.78	1000	537.78
−18	−27.78	41	5.00	91	32.78	305	151.67	555	290.55	1050	565.56
−16	−26.67	42	5.56	92	33.33	310	154.44	560	293.33	1100	593.33
−14	−25.56	43	6.11	93	33.89	315	157.22	565	296.11	1150	621.11
−12	−24.44	44	6.67	94	39.44	320	160.00	570	298.89	1200	648.89
−10	−23.33	45	7.22	95	35.00	325	162.78	575	301.67	1250	676.67
− 8	−22.22	46	7.78	96	35.56	330	165.56	580	304.44	1300	704.44
− 6	−21.11	47	8.33	97	36.11	335	168.33	585	307.22	1350	732.22
− 4	−20.00	48	8.89	98	36.67	340	171.11	590	310.00	1400	760.00
− 2	−18.89	49	9.44	99	37.22	345	173.89	595	312.78	1450	787.78
0	−17.78	50	10.00	100	37.78	350	176.67	600	315.56	1500	815.56
1	−17.22	51	10.56	105	40.55	355	179.44	610	321.11	1550	843.33
2	−16.67	52	11.11	110	43.33	360	182.22	620	326.67	1600	871.11
3	−16.11	53	11.67	115	46.11	365	185.00	630	332.22	1650	898.89
4	−15.56	54	12.22	120	48.89	370	187.78	640	337.78	1700	926.67
5	−15.00	55	12.78	125	51.67	375	190.55	650	343.33	1750	954.44
6	−14.44	56	13.33	130	54.44	380	193.33	660	348.89	1800	982.22
7	−13.89	57	13.89	135	57.22	385	196.11	670	354.44	1850	1010.00
8	−13.33	58	14.44	140	60.00	390	198.89	680	360.00	1900	1037.78
9	−12.78	59	15.00	145	62.78	395	201.67	690	365.56	1950	1065.56
10	−12.22	60	15.56	150	65.56	400	204.44	700	371.11	2000	1093.33
11	−11.67	61	16.11	155	68.33	405	207.22	710	376.67	2050	1121.11
12	−11.11	62	16.67	160	71.11	410	210.00	720	382.22	2100	1148.89
13	−10.56	63	17.22	165	73.89	415	212.78	730	387.78	2150	1176.67
14	−10.00	64	17.78	170	76.67	420	215.56	740	393.33	2200	1204.44
15	− 9.44	65	18.33	175	79.44	425	218.33	750	398.89	2250	1232.22
16	− 8.89	66	18.89	180	82.22	430	221.11	760	404.44	2300	1260.00
17	− 8.33	67	19.44	185	85.00	435	223.89	770	410.00	2350	1287.78
18	− 7.78	68	20.00	190	87.78	440	226.67	780	415.56	2400	1315.56
19	− 7.22	69	20.56	195	90.55	445	229.44	790	421.11	2450	1343.33
20	− 6.67	70	21.11	200	93.33	450	232.22	800	426.67	2500	1371.11
21	− 6.11	71	21.67	205	96.11	455	235.00	810	432.22	2550	1398.89
22	− 5.56	72	22.22	210	98.89	460	237.78	820	437.78	2600	1426.67
23	− 5.00	73	22.78	215	101.67	465	240.55	830	443.33	2650	1454.44
24	− 4.44	74	23.33	220	104.44	470	243.33	840	448.89	2700	1482.22
25	− 3.89	75	23.89	225	107.22	475	246.11	850	454.44	2750	1510.00
26	− 3.33	76	24.44	230	110.00	480	248.89	860	460.00	2800	1537.78
27	− 2.78	77	25.00	235	112.78	485	251.67	870	465.56	2850	1565.56
28	− 2.22	78	25.56	240	115.56	490	254.44	880	471.11	2900	1593.33
29	− 1.67	79	26.11	245	118.33	495	257.22	890	476.67	2950	1621.11

EQUIVALENTS OF TWADDELL, BAUME AND SPECIFIC GRAVITY SCALES

Twaddell	Baumé	Specific Gravity	Twaddell	Baumé	Specific Gravity	Twaddell	Baumé	Specific Gravity	Twaddell	Baumé	Specific Gravity
0	0	1.000	44	26.0	1.220	88	44.1	1.440	131	57.1	1.655
1	0.7	1.005	45	26.4	1.225	89	44.4	1.445	132	57.4	1.660
2	1.4	1.010	46	26.9	1.230	90	44.8	1.450	133	57.7	1.665
3	2.1	1.015	47	27.4	1.235	91	45.1	1.455	134	57.9	1.670
4	2.7	1.020	48	27.9	1.240	92	45.4	1.460	135	58.2	1.675
5	3.4	1.025	49	28.4	1.245	93	45.8	1.465	136	58.4	1.680
6	4.1	1.030	50	28.8	1.250	94	46.1	1.470	137	58.7	1.685
7	4.7	1.035	51	29.3	1.255	95	46.4	1.475	138	58.9	1.690
8	5.4	1.040	52	29.7	1.260	96	46.8	1.480	139	59.2	1.695
9	6.0	1.045	53	30.2	1.265	97	47.1	1.485	140	59.5	1.700
10	6.7	1.050	54	30.6	1.270	98	47.4	1.490	141	59.7	1.705
11	7.4	1.055	55	31.1	1.275	99	47.8	1.495	142	60.0	1.710
12	8.0	1.060	56	31.5	1.280	100	48.1	1.500	143	60.2	1.715
13	8.7	1.065	57	32.0	1.285	101	48.4	1.505	144	60.4	1.720
14	9.4	1.070	58	32.4	1.290	102	48.7	1.510	145	60.6	1.725
15	10.0	1.075	59	32.8	1.295	103	49.0	1.515	146	60.9	1.730
16	10.6	1.080	60	33.3	1.300	104	49.4	1.520	147	61.1	1.735
17	11.2	1.085	61	33.7	1.305	105	49.7	1.525	148	61.4	1.740
18	11.9	1.090	62	34.2	1.310	106	50.0	1.530	149	61.6	1.745
19	12.4	1.095	63	34.6	1.315	107	50.3	1.535	150	61.8	1.750
20	13.0	1.100	64	35.0	1.320	108	50.6	1.540	151	62.1	1.755
21	13.6	1.105	65	35.4	1.325	109	50.9	1.545	152	62.3	1.760
22	14.2	1.110	66	35.8	1.330	110	51.2	1.550	153	62.5	1.765
23	14.9	1.115	67	36.2	1.335	111	51.5	1.555	154	62.8	1.770
24	15.4	1.120	68	36.6	1.340	112	51.8	1.560	155	63.0	1.775
25	16.0	1.125	69	37.0	1.345	113	52.1	1.565	156	63.2	1.780
26	16.5	1.130	70	37.4	1.350	114	52.4	1.570	157	63.5	1.785
27	17.1	1.135	71	37.8	1.355	115	52.7	1.575	158	63.7	1.790
28	17.7	1.140	72	38.2	1.360	116	53.0	1.580	159	64.0	1.795
29	18.3	1.145	73	38.6	1.365	117	53.3	1.585	160	64.2	1.800
30	18.8	1.150	74	39.0	1.370	118	53.6	1.590	161	64.4	1.805
31	19.3	1.155	75	39.4	1.375	119	53.9	1.595	162	64.6	1.810
32	19.8	1.160	76	39.8	1.380	120	54.1	1.600	163	64.8	1.815
33	20.3	1.165	77	40.1	1.385	121	54.4	1.605	164	65.0	1.820
24	20.9	1.170	78	40.5	1.390	122	54.7	1.610	165	65.2	1.825
35	21.4	1.175	79	40.8	1.395	123	55.0	1.615	166	65.5	1.830
36	22.0	1.180	80	41.2	1.400	124	55.2	1.620	167	65.7	1.835
37	22.5	1.185	81	41.6	1.405	125	55.5	1.625	168	65.9	1.840
38	23.0	1.190	82	42.0	1.410	126	55.8	1.630	169	66.1	1.845
39	23.5	1.195	83	42.3	1.415	127	56.0	1.635	170	66.3	1.850
40	24.0	1.200	84	42.7	1.420	128	56.3	1.640	171	66.5	1.855
41	24.5	1.205	85	43.1	1.425	129	56.6	1.645	172	66.7	1.860
42	25.0	1.210	36	43.4	1.430	130	56.9	1.650	173	67.0	1.865
43	25.5	1.215	87	43.8	1.435						

Relation of Capacity, Volume and Weight

1 pint   = 28.875 cubic inches
1 quart   = 57.75 cubic inches
1 gallon (U. S.)   = 231 cubic inches
1 gallon (English) = 277.274 cubic inches
7.4805 gallons   = 1 cubic foot
1 gallon water at 62° Fahr. weighs 8.3356 lbs.

Dry

2 pints =1 quart
8 quarts=1 peck
4 pecks =1 bushel
1 U. S. standard bushel (struck)=2150.42 cubic inches.
0.80356 U. S. bushels (struck)   =1 cubic foot

METRIC EQUIVALENTS

Linear Measure

1 centimeter=0.3937 in.
1 decimeter=3.937 in.=0.328 ft.
1 meter=39.37 in.=1.0936 yds.
1 decameter=1.9884 rods
1 kilometer=0.62137 miles
1 inch=2.54 centimeters
1 foot=3.048 decimeters
1 yard=0.9144 meters
1 rod=0.5029 decameters
1 mile=1.6093 kilometers
(The meter, as used in Europe, is 39.370432 inches.)

Square Measure

1 sq. centimeter=0.1550 sq. inches
1 sq. decimeter=0.1076 sq. feet
1 sq. meter=1.196 sq. yards
1 are=3.954 sq. rods
1 hectare=2.47 acres
1 sq. kilometer=0.386 sq. miles
1 sq. inch=6.452 sq. centimeters
1 sq. foot=9.2903 sq. decimeters
1 sq. yard=0.8361 sq. meters
1 sq. rod=0.2529 ares
1 acre=0.4047 hectares
1 sq. mile=2.59 sq. kilometers

Weights

1 decigram=0.003527 oz.=1.5432 grains
1 gram=0.03527 oz. avoir., or about 15½ troy grains

1 kilogram=2.2046 lbs. avoir.
1 metric ton=1.1023 English short tons
1 ounce avoir.=28.35 grams
1 pound avoir.=0.4536 kilograms
1 English short ton=0.9072 metric tons

Approximate Metric Equivalents

1 decimeter=4 inches
1 meter=1.1 yards
1 kilometer=⅝ of a mile
1 hectare=2½ acres
1 stere, or cu. meter=¼ of a cord
1 liter=1.06 qt. liquid, 0.9 qt. dry
1 hectoliter=2⅚ bushels
1 kilogram=2⅕ lbs.
1 metric ton=2200 lbs.

Comparison of Avoirdupois and Metric Weights

Grains	Drams	Oz. Av.	Lbs. Av.	Deniers	Grams
1.000	......	......	......	1.296	0.065
27.340	1.000	......	......	35.437	1.772
437.500	16.000	1.000	......	566.990	28.350
7000.000	256.000	16.000	1.000	9071.840	453.592
0.772	.....	....	.....	1.000	0.050
15.432	.....	0.03527	.....	20.000	1.000

pH Values of Chemicals

Solution Strength	Reagent	pH
1%	Commercial Olive Oil Soap (Neutral)	10.1 –10.3
1%	Commercial Olive Oil Soap (Neutral)	10.1 –10.3
1%	Commercial Olive Oil or Tallow Soap Containing 20% Soda Ash	10.75–10.88
1%	Commercial Olive Oil or Tallow Soap Containing 5% Caustic	12.0–12.2
½%	Commercial Olive Oil or Tallow Soap	10.0 –10.2
¼%	Commercial Olive Oil or Tallow Soap	9.9 –10.1
1%	Sulphonated Oils (Neutral)	6.0 –7.0
1%	Sulphonated Oils Containing Free Acid	Below 6.0
1%	Sulphonated Oils Containing Soap or Alkalies	Above 7.0
¼%	Trisodium Phosphate	12.3
¼%	Sodium Silicate	12.2
¼%	Sodium Carbonate	11.3
¼%	Sodium Sulphite	9.7
¼%	Disodium Phosphate	8.9
¼%	Borax	8.8
¼%	Monosodium Phosphate	5.0

pH Ranges of Common Indicators

	Useful pH Range
Thymol Blue	1.2–2.8
Bromphenol Green	2.8–4.6
Methyl Orange	3.1–4.4
Bromcresol Green	4.0–5.6
Methyl Red	4.4–6.0
Propyl Red	4.8–6.4
Brom Cresol Purple	5.2–6.8
Brom Thymol Blue	6.0–7.6
Phenol Red	6.8–8.4
Litmus	7.2–8.8
Cresol Red	7.2–8.8
Cresolphthalein	8.2–9.8
Phenolphthalein	8.6–10.2
Nitro Yellow	10.0–11.6
Alizarin Yellow R	10.1–12.1
Sulfo Orange	11.2–12.6

Melting Points of Resins, Etc.

Material	Melting Point ° C.
Amber	250–325
Benzoin	75–100
Copal (Zanzibar)	280
Copal (Congo)	220
Copal (Kauri)	165
Copal (Manila)	120
Cumarone	127–142
Dammar (Batavia)	100
Dammar (Singapore)	95
Dragon's Blood	120
Elemi	75–120
Ester Gum	120–140
Gilsonite	123° C.
Guiac	85–90
Indene	127–142
Mastic	105–120
Pontianak	135
Rosin (Colophony)	100–140
Sandarac	135–150
Shellac	120

* Melting Points of Common Waxes

Wax	Melting Point ° C.
Bayberry Wax	40–44
Beeswax White	67.2
Beeswax Yellow	61
Candelilla Wax	64–67
Carnauba Wax	85
Ceresine	74–80
Chinese Insect Wax	92.2
Cocoa Butter	21.5–27.3
Japan Wax	54.5–59.6
Montan Wax Refined	95–96
Myrtle Wax	47–48
Ozokerite	65–110
Paraffin	55–65° C.
Spermaceti	44–47.5
Tallow (Beef)	42.5–44

* Very often there is considerable difference between the melting and solidifying point. Natural and commercially adulterated articles will also show variations.

REFERENCES

Agr. Gaz. N.S. Wales
Allg. Oel. v. Gettzeitung
Amer. Dyestuff Reporter
Amer. Electrop. Society
Amer. Paint Jol.
Amer. Perfumer
Amer. Photography
Anal. Fis. Quim.
Ault & Wiborg Varnish Wks. Handbook

Bakers Review
Better Enameling
Brewers' Tech. Review
Brit. Jol. of Photography
Brit. Medical Jol.
Bull. Imp. Hyg. Lab.
Bulletin of Imperial Institute
Bull. Soc. Franc. Phot.

Camera
Camera (Luzern)
Chemical Abstracts
Chemical Analyst
Chemical Formulary, Vol. I.
Chemist Analyst
Chemist & Druggist
Chem. Zent.
Combustion
Cramer's Manual

Dansk. Tids. Farm
Devt. Part. Zeitung
Drug & Cosmetic Industry
Druggists Circular
Drugs, Oils, & Paints

Eastman Kodak Co.
Electric Journal

Farbe v. Lacke
Farben Zeitung
Fein Mechanic v. Prazision
Fettchen, Umscham
Fils & Tissus
Focus
Fruit Products Jol.

Glass Industry

Hawaiian Planters' Record
Hide & Leather

India Rubber World
Indian Lac Research Inst.
Industrial Chemist
Industrial Finishing

Jol. Amer. Dental Assn.
Jol. Amer. Medical Assn.
Jol. Chinese Chem. Soc.
Jol. Federation Curriers
Jol. Federation Light Leather Tanners
Jol. Ind. & Eng. Chemistry

Jol. Soc. Leather Trades
Jol. Soc. Rubber Ind. Japan

Keram Steklo
Khimstroi
Kozhevenna-Obuvnaya Prom.
Kunstdinger, Undlam

Lakokras, Ind.

Malayan Agric. Jol.
Manufacturing Chemist
Melliand
Metal Industry
Metall und Erz
Metallurg
Metallurgist
Metals & Alloys
Munic. Eng. San. Record

Nat'l Butter & Cheese Jol.
Nitrocellulose

Ober Flachen Tach.
Oil & Soap

Paper Trade Jol.
Parfum Mod.
Peinture, Pigments, Vernis
Phar. Acta Helva
Pharmaceutical Jol.
Phot. Abstracts
Phot. Ind.
Phot. Korr.
Photog. Kronik
Phot. Rev.
Photo Runschau
Phytopathology
Plater's Guide Book
Portland Cement Assn.
Practical Druggist
Practical Everyday Chemistry
Printing Industry
Prob. Edelmetalle
Proc. World Petroleum Congress

Rayon & Mell. Tex. Monthly
Refiner & Nat. Gas. Mfr.
Rev. Aluminum

Science
Soap Gazette & Perfumer
Solvent News
Sovet-Sakhar
Spirits
Synthetic & Applied Finishes

Textile Colorist
Textile Mfr.
Textile Recorder

U. S. Bureau of Mines

Zeit. Unters. Lebensm.

INDEX

For Chemical Advisors, Special Raw Materials, Equipment, Containers, etc., consult Supply Section at end of book.

For Chemical Advisors, Special Raw Materials, Equipment, Containers, etc., consult Supply
Section at end of book,

For Chemical Advisors, Special Raw Materials, Equipment, Containers, etc., consult Supply Section at end of book.

For Chemical Advisors, Special Raw Materials, Equipment, Containers, etc., consult Supply Section at end of book.

For Chemical Advisors, Special Raw Materials, Equipment, Containers, etc., consult Supply Section at end of book.

All formulae preceded by an asterisk (*) are covered by patents.

For Chemical Advisors, Special Raw Materials, Equipment, Containers, etc., consult Supply Section at end of book.

For Chemical Advisors, Special Raw Materials, Equipment, Containers, etc., consult Supply
Section at end of book.

For Chemical Advisors, Special Raw Materials, Equipment, Containers, etc., consult Supply Section at end of book.

For Chemical Advisors, Special Raw Materials, Equipment, Containers, etc., consult Supply Section at end of book.

For Chemical Advisors, Special Raw Materials, Equipment, Containers, etc., consult Supply Section at end of book.

For Chemical Advisors, Special Raw Materials, Equipment, Containers, etc., consult Supply Section at end of book.

TRADE NAMED CHEMICALS

During the past few years, the practice of marketing raw materials, under names which in themselves are not descriptive chemically of the products they represent, has become very prevalent. No modern book of formulae could justify its claims either to completeness or modernity without numerous formulae containing these so-called "Trade Names."

Without wishing to enter into any discussion regarding the justification of "Trade Names," the Editors recognize the tremendous service rendered to commercial chemistry by manufacturers of "Trade Name" products, both in the physical data supplied and the formulation suggested.

Deprived of the protection afforded their products by this system of nomenclature, these manufacturers would have been forced to stand helplessly by while the fruits of their labor were being filched from them by competitors who, unhampered by expenses of research, experimentation and promotion, would be able to produce something "just as good" at prices far below those of the original producers.

That these competitive products were "just as good" solely in the minds of the imitators would only be evidenced in costly experimental work on the part of the purchaser and, in the meantime irreparable damage would have been done, to the truly ethical product. It is obvious, of course, that under these circumstances, there would be no incentive for manufacturers to develop new materials.

Because of this, and also because the "Chemical Formulary" is primarily concerned with the physical results of compounding rather than with the chemistry involved, the Editors felt that the inclusion of formulae containing various trade name products would be of definite value to the producer of finished chemical materials. If they had been left out many ideas and processes would have been automatically eliminated.

As a further service a list of the better known "trade name" products is appended together with the suppliers of these materials. The number after each trade name refers to the supplier given below with the corresponding number.

TRADE NAMES

SUPPLIERS OF TRADE NAME CHEMICALS

1. Acheson Graphite Corp., Niagara Falls, N. Y.
2. Advance Solvents & Chem. Corp., New York City
3. American Aniline Products, Inc., New York City
4. American Chem. Prod. Co., Rochester, N. Y.
5. American Colloid Co., Chicago, Ill.
6. American Cyanamid & Chem. Co., New York City
7. Anchor Chem. Co., Manchester, England
8. Anderson Prichard Oil Corp., Oklahoma City, Okla.
9. Ansbacher-Siegle Corp., Rosebank, N. Y.
10. Archer-Daniels-Midland Co., Minneapolis, Minn.
11. Arkansas Co., New York City
12. Atlantic Refining Co., Phila., Pa.
13. Bakelite Corp., New York City
14. Baker, J. T. Chem. Co., Phillipsburg, N. J.
15. Barber Asphalt Co., Phila., Pa.
16. Barrett Co., New York City
17. Beck, Koller & Co., Detroit, Mich.
18. Bick & Co., Inc., Reading, Pa.
19. Binney & Smith, New York City
20. British Drug Houses, Ltd., London, England
21. Bud Aromatic Chem. Co., Inc., New York City
22. Buromin Corp., Pittsburgh, Pa.
23. Bush, W. J. & Co. Inc., New York City
24. Cabot, Godfrey L. Inc., Boston, Mass
25. Calco Chem. Co., Bound Brook, N. J.
26. Campbell, John & Co., New York City
27. Carbic Color & Chem. Co., New York City
28. Carbide & Carbon Chem. Corp., New York City
29. Carborundum Co., Niagara Falls, N. Y.
30. Casein Mfg. Co., New York City
31. Celluloid Corp., Newark, N. J.
32. Century Stearic Acid & Candle Wks., New York City
33. Champion Fibre Co., Canton, No. Car.
34. Chaplin-Bibbo, New York City
35. Chemical & Pigment Co., Inc., Scranton, Pa.
36. Chemical Solvents Inc., New York City
37. Chesebrough Mfg. Co., New York City
38. Ciba Co., Inc., New York City
39. Colgate-Palmolive-Peet Co., Jersey City, N. J.
40. Colledge, E. W., Inc., Cleveland, O.
41. Columbia Alkali Corp., New York City
42. Commercial Solvents Corp., Terre Haute, Ind.
43. Commonwealth Color & Chem. Co., Brooklyn, N. Y.
44. Corn Products Refining Co., New York City
45. Darco Sales Corp., New York City
46. Deep Rock Oil Corp., Chicago, Ill.
47. Dennis, Martin & Co., Newark, N. J.
48. Dodge & Olcott Co., New York City
49. Dow Chem. Co., Midland, Mich.
50. Ducas, B. P. Co., New York City
51. DuPont, E. I., de Nemours & Co., Wilmington, Del.
52. Eastman Kodak Co., Rochester, N. Y.
53. Economic Materials Co., Chicago, Ill.
54. Emery Industries, Inc., Cincinnati, O.
55. Felton Chem. Co., Brooklyn, N. Y.
56. Fezandié and Sperrlé, Inc., New York City
57. Fougera, E. & Co., New York City
58. Franco-Amer. Chem. Works, Carlstadt, N. J.

59. Fries Bros., New York City
60. Fritzchie Bros., New York City
61. Geigy Co. Inc., New York City
62. General Atlas Carbon Co., New York City
63. General Chemical Co., New York City
64. General Drug Co., New York City
65. General Dyestuffs Corp., New York City
66. General Electric Co., Schenectady, N. Y.
67. General Naval Stores Co., New York City
68. General Plastics Inc., No. Tonawanda, N. Y.
69. Givaudan-Delawanna, Inc., New York City
70. Glyco Products Co., Inc., New York City
71. Goldschmidt Corp., New York City
72. Goodyear Tire & Rubber Co., Akron, O.
73. Grasselli Chem. Co., Cleveland, O.
74. Greef, R. W. & Co., Inc., New York City
75. Hall, C. P. & Co., Akron, O.
76. Halowax Corp., New York City
77. Harshaw Chem. Co., Cleveland, O.
78. Heine & Co., New York City
79. Hercules Powder Co., Wilmington, Del.
80. Hooker Electro-Chem. Co., New York City
81. Hopkins, J. L. & Co., New York City
82. Industrial Chem. Sales Co., New York City
83. Innis, Speiden & Co., New York City
84. International Pulp Corp., New York City
85. Johns-Manville Corp., New York City
86. Jungmann & Co., New York City
87. Kay-Fries Chemicals, Inc., New York City
88. Kessler Chem. Corp., New York City
89. Koppers Products Co., Pittsburgh, Pa.
90. Krebs Pigment & Color Corp., Newark, N. J.
91. Lehn & Fink Corp., New York City
92. Lewis, John D., Inc., Providence, R. I.
93. Liquid Carbonic Corp., Chicago, Ill.
94. Lucidol Corp., Buffalo, N. Y.
95. Magnus, Mabee & Reynard, Inc., New York City
96. Mallinckrodt Chem. Works, New York City
97. Martin, Dennis Co., Newark, N. J.
98. Mathieson Alkali Co., New York City
99. McCormick & Co., Baltimore, Md.
100. Merck & Co. Inc., New York City
101. Monsanto Chem. Works, St. Louis, Mo.
102. Moore-Munger, New York City
103. Mutual Chem. Co. of Amer., Newark, N. J.
104. National Aluminate Corp., Chicago, Ill.
105. National Aniline & Chem. Co., Buffalo, N. Y.
106. National Oil Products Co., Harrison, N. J.
107. National Rosin Oil & Size Co., New York City
108. Naugatuck Chem. Co., New York City
109. Neville Co., Pittsburgh, Pa.
110. New Jersey Zinc Sales Co., New York City
111. Nulomoline Co., New York City
112. Nuodex Products, Inc., Newark, N. J.
113. Onyx Oil & Chem. Co., Passaic, N. J.
114. Papermakers' Chem. Corp., Wilmington, Del.
115. Paramet Chem. Corp., Long Island City, N. Y.
116. Penick, S. B. & Co., New York City
117. Penn. Alcohol Corp., Phila., Pa.
118. Penn. Salt Mfg. Co., Phila., Pa.
119. Pfaltz & Bauer, Inc., New York City
120. Phila. Quartz Co., Phila., Pa.
121. Plymouth Organic Labs., New York City
122. Pylam Products Co., New York City
123. Rauh, Robert Inc., Newark, N. J.

124. Reilly Tar & Chem. Corp., Indianapolis, Ind.
125. Resinous Prod. & Chem. Co., Phila., Pa.
126. Resinox Corp., New York City
127. Revertex Corp., New York City
128. Robeson & ro^ss Co., New York City
129. Rohm-Haas Chem. Co., Phila., Pa.
130. Royce Chem. Co., Carlton Hill, N. J.
131. Rubber Service Labs. Co., Akron, O.
132. Russia Cement Co., Gloucester, Mass.
133. Salomon, L. A. & Bro., New York City
134. Sandoz Chem. Works, New York City
135. Scholler Bros. Inc., Phila., Pa.
136. Schliemann Co. Inc., New York City
137. Scott, Bader & Co., London, England
138. Sceley & Co., New York City
139. Sharples Solvents Corp., Phila., Pa.
140. Shawinigan, Ltd., New York City
141. Sherwood Petroleum Co., Brooklyn, N. Y.
142. Silver, Geo. Import Co., New York City
143. Sonneborn, L. Sons, New York City
144. Southwark Mfg. Co., Camden, N. J.
145. Spencer-Kellogg Co., New York City
146. Stamford Rubber Supply Co., Stamford, Conn.
147. Stanco, Inc., New York City
148. Standard Oil Co. of Calif., San Francisco, Cal.
149. Standand Oil Co. of New Jersey, New York City
150. Stauffer Chem. Co., New York City
151. Stein-Hall & Co. Inc., New York City
152. Sun Oil Co., Phila., Pa.
153. Swann Chem. Corp., Birmingham, Ala.
154. Synfleur Scientific Labs., Monticello, N. Y.
155. Texas Mining & Smelting Co., Laredo, Texas
156. Thomas, Arthur H. Co., Phila., Pa.
157. Titanium Pigments Co., New York City
158. Uhlich, Paul Co., New York City
159. United Color & Pigment Co. Inc., Newark, N. J.
160. United States Gypsum Co., Chicago, Ill.
161. United States Industrial Chem. Co. Inc., New York City
162. Van-Ameringen Haebler, Inc., New York City
163. Vanderbilt, R. T. Co. Inc., New York City
164. Varcum Chem. Corp., Niagara Falls, N. Y.
165. Verley, Albert & Co., Chicago, Ill.
166. Victor Chem. Works, Chicago, Ill.
167. Virginia Smelting Co., W. Norfolk, Va.
168. Vultex Corp. of America, Cambridge, Mass.
169. Wallerstein Co., Inc., New York City
170. Welch, Holme & Clark Co., Inc., New York City
171. Whittaker, Clark & Daniels, Inc., New York City
172. Will & Baumer Candle Co., New York City
173. Wishnick-Tumpeer, Inc., New York City
174. Woburn Degreasing Co. of N. J., Harrison, N. J.
175. Wolf, Jacques & Co., Passaic, N. J.

WHERE TO BUY CHEMICALS

Abietic Acid
Hercules Powder Co., New York, N. Y.

Acetamide
Amer. Chemical Products Co., Rochester, N. Y.

Acetanilide
Heyden Chemical Corporation, New York, N. Y.

Acetic Acid
The Cleveland-Cliffs Iron Co., Cleveland, Ohio.

Acetic Anhydride
American-British Chemical Supplies, Inc., New York, N. Y.

Acetone
W. S. Gray Co., New York, N. Y.

Acetphenetidin
Merck & Co., Inc., Rahway, N. J.

Acetyl Salicylic Acid
Monsanto Chemical Co., St. Louis, Mo.

Acriflavine
Abbott Laboratories, North Chicago, Ill.

Agar
American Agar Co., Inc., San Diego, Calif.

Albumen
Stein, Hall & Co., Inc., New York, N. Y.

Alcohol, Denatured
Rogers & McClellan, Boston, Mass.
L. R. Van Allen & Co., Chicago, Ill.

Alcohol, Pure
U. S. Industrial Alcohol Co., New York, N. Y.

Alkanet
J. L. Hopkins & Co., New York, N. Y.

Almond Oil
Magnus, Mabee & Reynard, Inc., New York, N. Y.

Aloes
Peck & Velsor, New York, N. Y.

Aluminum
Aluminum Co. of America, Pittsburgh, Pa.

Alums
The Grasselli Chemical Co., Cleveland, Ohio.

Aluminum Acetate
Niacet Chemicals Corporation, Niagara Falls, N. Y.

Aluminum Bronze Powder
U. S. Bronze Powder Works, Inc., New York, N. Y.

Aluminum Chloride (Solution, Crystals and Anhydrous)
The Calco Chemical Co., Bound Brook, N. J.

Aluminum Stearate
Franks Chemical Products Co., Inc., Brooklyn, N. Y.

Ammonia
Nat'l. Ammonia Co., Inc., Phila., Pa.

Ammonium Bifluoride
The Harshaw Chemical Co., Cleveland, Ohio.

Ammonium Chloride
Pennsylvania Salt Mfg. Co., Inc., Phila., Pa.

Ammonium Linoleate
Glyco Products Co., Inc., New York, N. Y.

Ammonium Oleate
Glyco Products Co., Inc., New York, N. Y.

Ammonium Phosphate
Swann Chemical Co., New York, N. Y.

Ammonium Stearate
Glyco Products Co., Inc., New York, N. Y.

Amyl Acetate
Chemical Solvents, Inc., New York, N. Y.

Aniline Dyes
Experimenters Supply Co., New York, N. Y.

Aniline Oil
Dow Chemical Co., Midland, Michigan.

Antimony Chloride
Seldner & Enequist, Inc., B'klyn, N. Y.

Antimony Oxide
O. Hommel Co., Pittsburgh, Pa.

Asbestos
Powhatan Mining Corp., Woodlawn, Baltimore, Md.

Asphalt
The Barber Asphalt Co., Philadelphia, Pa.

Asphaltum
Allied Asphalt & Mineral Corp., New York, N. Y.

Barium Carbonate
Barium Reduction Corp., Charleston, W. Va.

Barium Sulphate
C. P. De Lore Co., St. Louis, Mo.

Barium Sulphide
Chicago Copper & Chemical Co., Blue Island, Ill.

Barytes
Bradley & Baker, 155 E. 44th St., New York, N. Y.
Nat'l. Pigments & Chemical Co., St. Louis, Mo.

Bayberry Wax
The W. H. Bowdlear Co., Syracuse, N. Y.

Beeswax
A. C. Drury & Co., Inc., Chicago, Ill.
Theodor Leonhard Wax Co., Inc., Haledon, Paterson, N. J.

Bentonite
Amer. Colloid Co., Chicago, Ill.
Silica Products Co., Kansas City, Mo.
The Wyodak Chemical Co., Cleveland, Ohio.

Benzidine
General Aniline Works, Inc., New York, N. Y.

Benzocaine
Abbott Laboratories, No. Chicago, Ill.

Benzoic Acid
Carus Chemical Co., Inc., La Salle, Ill.

Benzol
The Barrett Co., New York, N. Y.

Benzoyl Peroxide
Lucidol Corp., 203 Larkin St., Buffalo, N. Y.

Beryllium
Belmont Smelting & Refining Wks., Inc., 316 Belmont Ave., B'klyn, N. Y.

Beta Naphthol
The Calco Chemical Co., Bound Brook, N. J.

Bismuth Subnitrate
The New York Quinine & Chemical Wks., Inc., B'klyn, N. Y.

Blanc Fixe
Adolph Hurst & Co., Inc., New York, N. Y.
L. H. Butcher Co., Los Angeles, California.

Blood Albumen
Morningstar, Nicol, Inc., New York, N. Y.

Bone Ash
Denver Fire Clay Co., Denver, Colorado.

Bone Black
Binney & Smith Co., New York, N. Y.

Borax
American Potash & Chemical Corp., New York, N. Y.

Boric Acid
Borax Union, Inc., San Francisco, Calif.

Bromo-Fluorescein
New York Color & Chemical Co., Belleville, N. J.

Bronze Powder
B. F. Drakenfeld & Co., Inc., New York, N. Y.

Butyl Acetate
Commercial Solvents Corp., New York, N. Y.
Publicker, Inc., Phila., Pa.

Butyl Aldehyde
Commercial Solvents Corp., Terre Haute, Indiana.

Butyl Alcohol (Normal)
Publicker, Inc., Phila., Pa.

Butyric Ether
The Northwestern Chemical Co., Wauwatosa, Wisconsin.

Calcium Arsenate
Bowker Chemical Corp., New York City.
Chipman Chemical Co., Inc., Bound Brook, N. J.

Calcium Carbonate
Limestone Products Corp. of Amer., Newton, N. J.

Calcium Carbonate (Precipitated)
Merck & Co., Inc., Rahway, N. J.

Calcium Chloride
Michigan Alkali Co., New York, N. Y.
Saginaw Salt Products Co., Saginaw, Mich.

Calcium Chloride (Anhydrous)
Fales Chemical Co., Inc., Cornwall Landing, N. Y.

Calcium Phosphate
Provident Chemical Wks., St. Louis, Mo.

Calcium Sulphide (Luminous)
Amer. Luminous Products Co., Huntington Park, Calif.

Calcium Stearate
The Synthetic Products Co., Cleveland, Ohio.

Camphor
E. J. Barry, 54 Fulton St., New York, N. Y.

Camphor Oil
Magnus, Mabee & Reynard, Inc., New York, N. Y.

Candelilla Wax
Innis, Speiden & Co., Inc., New York, N. Y.

Caramel Color
Alex Fries & Bro., Cincinnati, Ohio.

Carbolic Oil
Reilly Tar & Chemical Corp., New York, N. Y.

Carbon, Activated
The Jennison-Wright Co., Toledo, Ohio.

Carbon Bisulphide
J. T. Baker Chemical Co., Phillipsburg, N. J.

Carbon Black
United Carbon Co., Charleston, W. Va.
Binney & Smith, New York, N. Y.

Carbon, Decolorizing
Darco Sales Corp., New York, N. Y.

Carbon Tetrachloride
Niagara Smelting Corp., Niagara Falls, N. Y.

Cardamom Seed
Newmann-Buslee & Wolfe, Inc., Chicago, Ill.

Carnauba Wax
Frank B. Ross Co., Inc., New York, N. Y.

Casein
The Casein Mfg. Co. of America, Inc., New York, N. Y.

Castor Oil
The Baker Castor Oil Co., New York, N. Y.

Celluloid
Celluloid Corp., New York, N. Y.

Celluloid Scrap
Moses Sereinsky Co., Indianapolis, Ind.

Cellulose Acetate
Celanese Corp. of America, New York, N. Y.

Ceresin Wax
Sherwood Petroleum Co., Inc., Bklyn., N. Y.

Cetyl Alcohol
Hummel Chemical Co., Inc., 90 West St., New York, N. Y.

Chalk, Precipitated
Charles B. Chrystal Co., Inc., New York, N. Y.

Charcoal
Chas. L. Read & Co., Inc., New York, N. Y.
Western Charcoal Co., Chicago, Ill.

China Wood Oil
Balfour, Guthrie & Co., Ltd., New York, N. Y.

Chlorine (Liquid)
Electro Bleaching Gas Co., 9 E. 41st St., New York, N. Y.

Chloroform
The Dow Chemical Co., Midland, Michigan.

Chlorophyll
Pylam Products Co., New York, N. Y.

Cholesterin
Digestive Ferments Co., Detroit, Mich.
Merck & Co., Inc., Rahway, N. J.

Chrome Green
Kentucky Color & Chem. Co., Louisville, Ky.

Chromic Acid
Mutual Chemical Co. of America, New York, N. Y.

Citral
Givaudan-Delawanna, Inc., New York, N. Y.

Citric Acid
Chas. Pfizer & Co., Inc., New York, N. Y.

Clay
Kentucky Clay Mining Co., Mayfield, Ky.
Olive Branch Minerals Co., Cairo, Illinois.

Cobalt Acetate
Fred L. Brooke Co., Chicago, Ill.

Cobalt Linoleate
The McGean Chemical Co., Cleveland, O.

Cocoa Butter
Alpha Lux Co., Inc., New York, N. Y.
Thomas J. Shields Co., New York, N. Y.

Coconut Butter
Procter & Gamble Co., Cincinnati, O.

Coconut Oil
Franklin Baker Co., Hoboken, N. J.

Coconut Oil Fatty Acid
Acme Oil Corp., Chicago, Illinois.

Cod Liver Oil
H. H. Rosenthal & Co., Inc., New York, N. Y.

Collodion
Charles Cooper & Co., New York, N. Y.

Colors, Dry
Holland Aniline Dye Co., Holland, Michigan.

Copper Carbonate
Chas. Cooper & Co., New York, N. Y.
Jungmann & Co., Inc., New York, N. Y.

Copper Cyanide
Charles Hardy, Inc., New York, N. Y.

Copper Oxides
The O. Hommel Co., Inc., 209 Fourth Ave., Pittsburgh, Pa.

Copper Sulphate
Barada & Page, Inc., Kansas City, Mo.

Corn Oil
American Maize Products Co., New York, N. Y.

Corn Sugar
Staley Sales Corp., Decatur, Ill.

Corn Syrup
Clinton Co., Clinton, Ia.
Corn Products Refining Co., New York, N. Y.

Cottonseed Oil (Crude)
Battleboro Oil Co., Battleboro, N. C.
Welch, Holme & Clark Co., New York, N. Y.

Cream of Tartar
  The Harshaw Chemical Co., Cleveland, Ohio.

Creosote
  Koppers Products Co., Pittsburgh, Pa.

Cresols
  Coopers Creek Chem. Co., W. Conshohocken, Pa.
  Reilly Tar & Chemical Corp., New York, N. Y.

Cresylic Acid
  The Barrett Co., New York, N. Y.

Dammar, Gum
  George H. Lincks, New York, N. Y.

Derris Root
  W. Benkert & Co., Inc., New York, N. Y.

Dextrins
  Morningstar, Nicol, Inc., New York, N. Y.

Diastase
  Takamine Laboratory, Inc., Clifton, N. J.

Dibutylphthalate
  The Kessler Chemical Corp., New York, N. Y.

Diethyleneglycol
  Carbide & Carbon Chemicals Corp., New York, N. Y.

Diethylphthalate
  Van Dyk & Co., Inc., Jersey City, N. J.

Di Glycol Oleate
  Glyco Products Co., Inc., New York, N. Y.

Di Glycol Stearate
  Glyco Products Co., Inc., New York, N. Y.

Diphenyl
  Swann Chemical Co., New York, N. Y.

Dyestuffs
  National Aniline & Chemical Co., Inc., New York, N. Y.

Egg Dried
  W. P. Pray, New York, N. Y.

Ephedrine
  Abbott Laboratories, No. Chicago, Illinois.

Epsom Salt
  General Chemical Co., New York, N. Y.

Ester Gum
  John D. Lewis, Inc., Providence, R. I.
  Paramet Chemical Corp., Long Island City, N. Y.

Ether
  Carbide & Carbon Chemicals Corp., New York, N. Y.

Ethyl Acetate
  Merrimac Chemical Co., Boston, Mass.

Ethylamine
  F. C. Bersworth Labs., Framingham, Mass.

Ethyl Lactate
  American Cyanamid & Chemical Corp., New York, N. Y.

Ethylene Diamine
  F. C. Bersworth Labs., Framingham, Mass.

Ethylene Dichloride
  Dow Chemical Co., Midland, Michigan.

Ethyleneglycol
Carbide & Carbon Chemicals Corp., New York, N. Y.

Fish Oil
Falk & Co., Pittsburgh, Pa.

Formaldehyde
Heyden Chemical Corp., New York, N. Y.

Fuller's Earth
L. A. Salmon & Bro., New York, N. Y.
Sinclair Refining Co., Olmstead, Illinois.

Geranium Lake
Interstate Color Co., Inc., New York, N. Y.
R. F. Revson Co., New York, N. Y.

Gilsonite
George H. Lincks, New York, N. Y.
Utah Gilsonite Co., St. Louis, Mo.

Ginseng
C. H. Lewis & Co., New York, N. Y.

Glandular Products
The Wilson Laboratories, Chicago, Illinois.

Glauber Salt
Iowa Soda Products Co., Council Bluffs, Ia.

Glue
Cudahy Packing Co., 221 North La Salle St., Chicago, Ill.

Glycerin
Colgate-Palmolive-Peet Co., Chicago, Ill.

Glyceryl Mono Stearate
Glyco Products Co., Inc., New York, N. Y.

Glyceryl Phthalate
Glyco Products Co., Inc., New York, N. Y.

Glyceryl Stearate
Glyco Products Co., Inc., New York, N. Y.

Glycol Oleate
Glyco Products Co., Inc., New York, N. Y.

Glycol Phthalate
Glyco Products Co., Inc., New York, N. Y.

Glycol Stearate
Glyco Products Co., Inc., New York, N. Y.

Gold Chloride
Mallinckrodt Chemical Works, St. Louis, Mo.

Graphite
Adolphe Hurst & Co., Inc., New York, N. Y.
Asbury Graphite Mills, Asbury, N. J.

Gum Arabic
T. M. Duche & Sons, New York, N. Y.

Gum Benzoin
Peek & Velsor, Inc., New York, N. Y.

Gum Copal
George H. Lincks, New York, N. Y.

Gum Dammar
Thurston & Braidich, New York, N. Y.

Gum Karaya
Frank-Vliet Co., Inc., New York, N. Y.

Gum, Locust Bean
  Geo. H. Lincks, New York, N. Y.

Gum Manila
  Stroock & Wittenberg Corp., New York, N. Y.

Gum Tragacanth
  E. Meer & Co., Inc., New York, N. Y.
  J. L. Hopkins & Co., New York, N. Y.

Hemlock Bark
  Tanners Supply Co., Grand Rapids, Michigan.

Henna Leaves
  S. B. Penick & Co., New York, N. Y.

Hexamethylenetetramine
  Heyden Chemical Corp., New York, N. Y.

Hydrochloric Acid
  General Chemical Co,. New York, N. Y.

Hydrogen Peroxide
  The Warner Chemical Co., New York, N. Y.

Hydroquinone
  Eastman Kodak Co., Rochester, N. Y.

Iodine
  New York Quinine & Chemical Wks., Inc., B'klyn., N. Y.

Irish Moss
  S. B. Penick & Co., 132 Nassau St., New York, N. Y.

Iron Oxide
  Binney & Smith Co., 41 E. 42d St., New York, N. Y.

Isopropyl Alcohol
  Carbide & Carbon Chemicals Corp., New York, N. Y.

Insect Wax, Chinese
  Frank B. Ross Co., Inc., New York, N. Y.

Japan Wax
  Smith & Nichols, Inc., New York, N. Y.

Laboratory Equipment
  Central Scientific Co., Chicago, Ill.
  Chicago Apparatus Co., Chicago, Ill.
  Eimer & Amend, New York, N. Y.
  Experimenter's Supply Co., New York, N. Y.
  Fisher Scientific Co., Pittsburgh, Pa.
  N. J. Laboratory Supply Co., Newark, N. J.
  Scientific Glass Apparatus Co., Bloomfield, N. J.

Lacquers
  Maas & Waldstein, Newark, N. J.

Lactic Acid
  Apex Chemical Co., Inc., New York, N. Y.

Lamp Black
  Binney & Smith Co., New York, N. Y.
  L. Martin Co., New York, N. Y.

Lanolin
  American Lanolin Corp., Lawrence, Mass.
  Merck & Co., Inc., Rahway, N. J.
  Pfaltz & Bauer, New York, N. Y.

Lead Arsenate
  Barada & Page, Inc., Kansas City, Mo.
  General Chemical Co., New York, N. Y.

Lecithin
  American Lecithin Corp., New York, N. Y

Lemon Oil
  D. W. Hutchinson & Co., Inc., New York, N. Y.

Licorice
  MacAndrews & Forbes Co., New York, N. Y.

Lime
  J. E. Baker Co., York, Pa.
  Chazy Marble Lime Co., Inc., Chazy, N. Y.

Linoleic Acid
  Glyco Products Co., Inc., New York, N. Y.

Linseed Oil
  Bisbee Linseed Co., Phila., Pa.

Litharge
  The Eagle-Picher Lead Co., Cincinnati, Ohio.

Lithopone
  Krebs Pigment & Color Corp., Newark, N. J.
  Marshall Dill Co., San Francisco, Cal.

Magnesium Carbonate
  Merck & Co., Inc., Rahway, N. J.

Magnesium Chloride
  Wishnick-Tumpeer, Inc., New York, N. Y.

Magnesium Powder
  Belmont Smelting & Refining Wks., Inc., B'klyn, N. Y.

Manganese Dioxide
  B. F. Drakenfeld & Co., Inc., New York, N. Y.

Menthol
  Chas. L. Huisking & Co., Inc., New York, N. Y.

Mercury
  Chas. L. Huisking & Co., Inc., New York, N. Y.
  George Uhe Co., New York, N. Y.

Methanol
  Wm. S. Gray & Co., New York, N. Y.

Methyl Salicylate
  Dow Chemical Co., Midland, Michigan.

Mica
  Southern Mica Co., Franklin, N. C.

Milk Sugar
  Mallinckrodt Chemical Wks., St. Louis, Mo.

Montan Wax
  Strahl & Pitsch, 141 Front St., New York, N. Y.

Naphtha
  Deep Rock Oil Corp., Chicago, Illinois.

Napthalene
  The Barrett Co., New York, N. Y.

Neatsfoot Oil
  National Oil Products Co., Harrison, N. J.

Nickel Sulphate
  The Harshaw Chemical Co., Cleveland, O.

Nicotine
  Tobacco By-Products & Chemical Corp., Louisville, Ky.

Nicotine Sulphate
  Lattimer-Goodwin Chemical Co., Grand Junction, Colorado.

Nitric Acid
  Monsanto Chemical Co., St. Louis, Mo.

Nitrocellulose
E. I. Du Pont de Nemours & Co., Inc., Parlin, N. J.

Ochres
Smith Chemical & Color Co., B'klyn, N. Y.

Oil, Citronella
D. W. Hutchinson & Co., Inc., New York, N. Y.

Oil, Mineral
Standard Oil Co. of California, San Francisco, Calif.

Oil, Olive
Leghorn Trading Co., Inc., New York, N. Y.

Orange Oil
Dodge & Olcott Co., New York, N. Y.

Ortho Dichlorbenzene
Hooker Electrochemical Co., New York, N. Y.

Oxalic Acid
Mutual Chemical Co. of America, New York, N. Y.

Oxygen
Cheney Chemical Co., Cleveland, Ohio

Ozokerite Wax
Strohmeyer & Arpe Co., New York, N. Y.

Palm Oil
Wishnick-Tumpeer, Inc., New York, N. Y.

Paraffin Oils
S. Schwabacher & Co., Inc., New York, N. Y.

Paraffin Wax
Oil States Petroleum Co., New York, N. Y.

Peppermint Oil
Magnus, Mabee & Reynard, Inc., New York, N. Y.
The Sparhawk Co., Sparkhill, N. Y.

Petrolatum
Pennsylvania Refining Co., Butler, Pa.

Petroleum Jelly
L. Sonneborn Sons, Inc., New York, N. Y.

Phenol
American-British Chemical Supplies, Inc., New York, N. Y.

Phosphoric Acid
Victor Chemical Works, Chicago, Illinois.

Pine Oil
General Naval Stores Co., Inc., New York, N. Y.

Pine Tar
Southern Pine Chem. Co., Jacksonville, Florida.

Pitch
Robert Rauh, Inc., 480 Frelinghuysen Ave., Newark, N. J.

Plaster of Paris
Whittaker, Clark & Daniels, Inc., New York, N. Y.

Potash, Caustic
Niagara Alkali Co., New York, N. Y.

Potassium Carbonate
Joseph Turner & Co., New York, N. Y.

Potassium Chlorate
Joseph Turner & Co., New York, N. Y.

Potassium Hydroxide
Merck & Co., Inc., Rahway, N. J.

Potassium Iodide
New York Quinine & Chemical Wks., Inc., B'klyn, N. Y.

Potassium Oleate
Glyco Products Co., Inc., New York, N. Y.
Carl F. Miller & Co., Seattle, Washington.

Potassium Permanganate
Carus Chemical Co., Inc., La Salle, Ill.

Potassium Silicate
Philadelphia Quartz Co., Philadelphia, Pa.

Pumice
Charles B. Chrystal Co., Inc., New York, N. Y.

Psyllium Seeds
Laxseed Co., New York, N. Y.

Pyrethrum Extract
McLaughlin, Gormley, King & Co., Minneapolis, Minn.

Pyrethrum
S. B. Penick & Co., New York, N. Y.

Quinine Bisulphate
R. W. Greeff & Co., Inc., New York, N. Y.

Red Oil
Century Stearic Acid Candle Wks., New York, N. Y.

Resins, Synthetic
Beck, Koller & Co., Inc., Detroit, Mich.

Rochelle Salts
Chas. Pfizer & Co., Inc., New York, N. Y.

Rosin
General Naval Stores Co., Inc., New York, N. Y.

Rosin Oil
National Rosin Oil & Size Co., New York, N. Y.

Rotenone
Thorocide, Inc., St. Louis, Mo.
Cyrus Ward & Co., Ltd., New York, N. Y.

Rubber Latex
Littlejohn & Co., Inc., New York, N. Y.

Saccharine
Heyden Chemical Corp., New York, N. Y.

Salicylic Acid
The Dow Chemical Co., Midland, Michigan.

Sal Soda
Church & Dwight Co., Inc., New York, N. Y.

Salt
Morton Salt Co., Chicago, Ill.

Saponin
Experimenters Supply Co., New York, N. Y.
Jungmann & Co., New York, N. Y.

Shellac
Wm. Zinsser & Co., New York, N. Y.

Siennas
Fezandié & Sperrlé, Inc., New York, N. Y.

Silver Nitrate
Eastman Kodak Co., Rochester, N. Y.

Soda Ash
Diamond Alkali Co., Pittsburgh, Pa.

Soda, Caustic
Mathieson Alkali Works, Inc., New York, N. Y.

Soda, Sal
Consolidated Chem. Sales Corp., Newark, N. J.

Sodium Aluminate
National Aluminate Corp., Chicago, Ill.

Sodium Arsenite
Harrison Mfg. Co., Rahway, N. J.

Sodium Benzoate
Hooker Electrochemical Co., New York, N. Y.

Sodium Bicarbonate
Church & Dwight Co., Inc., New York, N. Y.

Sodium Bisulphite
The Grasselli Chemical Co., Cleveland, O.

Sodium Carbonate
Solvay Sales Corporation, New York, N. Y.

Sodium Cyanide
E. I. Du Pont de Nemours & Co., Inc., Wilmington, Del.

Sodium Fluoride
American Cyanamid & Chemical Corp., New York, N. Y.

Sodium Hydroxide
Merck & Co., Inc., Rahway, N. J.

Sodium Hypochlorite
Delta Chemical Mfg. Co., Baltimore, Md.
Mathieson Alkali Wks., Inc., New York, N. Y.

Sodium Hypochlorite Liquid
Riverside Chemical Co., No. Tonawanda, N. Y.

Sodium Hyposulphite
The Grasselli Chemical Co., Cleveland, O.

Sodium Metasilicate
Philadelphia Quartz Co., Phila., Pa.

Sodium Perborate
E. I. Du Pont De Nemours & Co., Inc., Wilmington, Del.

Sodium Phosphate
Swann Chemical Co., New York, N. Y.

Sodium Resinate
Paper Makers Chem. Corp., Wilmington, Del.

Sodium Silicate
Mechling Bros. Chemical Co., Camden, N. J.
Philadelphia Quartz Co., Phila., Pa.
Standard Silicate Co., Pittsburgh, Pa.

Sodium Silico Fluoride
The Grasselli Co., Cleveland, Ohio.

Sodium Sulphide
Barium Reduction Corp., Charleston, W. Va.

Sodium Sulphite
Mechling Bros. Chemical Co., Camden, N. J.

Sorbitol
Atlas Powder Co., Wilmington, Del.

Soybean Oil
Spencer Kellogg & Sons Sales Corp., Buffalo, N. Y.
Arthur C. Trask Co., Chicago, Ill.

Sperm Oil
Cook Swan Co., Inc., New York, N. Y.

Spermaceti
Strahl & Pitsch, New York, N. Y.

Starch
Starch Products Co., New York, N. Y.

Stearic Acid
Century Stearic Acid Candle Wks., New York, N. Y.

Sulfonated Castor Oil
Burkart-Schier Chem. Co., Chattanooga, Tenn.

Sulfonated Olive Oil
Jacques Wolf & Co., Passaic, N. J.

Sulphur
Stauffer Chemical Co., of Texas, Freeport, Tex.

Sulphur Dioxide
Virginia Smelting Co., Boston, Mass.

Sulphuric Acid
Merrimac Chemical Co., Everett Sta., Boston, Mass.

Talc
Charles B. Chrystal Co., Inc., New York, N. Y.

Tallow
Welch, Holme & Clark Co., Inc., New York, N. Y.

Tannic Acid
John D. Lewis, Inc., Providence, R. I.

Tartaric Acid
R. W. Greeff & Co., Inc., New York, N. Y.

Tea Seed Oil
Lundt & Co., New York, N. Y.

Thallium Sulphate
Jungmann & Co., Inc., New York, N. Y.

Thymol
Sherka Chemical Co., Inc., Bloomfield, N. J.

Titanium Dioxide
R. T. Vanderbilt Co., New York, N. Y.

Toluol
Jones & Laughlin Steel Corp., Pittsburgh, Pa.

Triacetin
Niacet Chemicals Corp., Niagara Falls, N. Y.

Trichlorethylene
International Selling Corp., New York, N. Y.

Tricresyl Phosphate
R. W. Greeff & Co., Inc., New York, N. Y.

Triethanolamine
Experimenters Supply Co., (Small lots) New York, N. Y.
Carbide & Carbon Chem. Co., (Large lots) New York, N. Y.

Triethanolamine Oleate
Glyco Products Co., Inc., New York, N. Y.
Marshall Dill Co., San Francisco, Calif.

Triethanolamine Stearate
Glyco Products Co., Inc., New York, N. Y.
Carl F. Miller & Co., Seattle, Washington.

Triphenylphosphate
Monsanto Chemical Co., St. Louis, Mo.

Tripoli
  Tamms Silica Co., Chicago, Ill.

Turkey Red Oil
  National Oil Products Co., Inc., Harrison, N. J.

Turpentine
  Antwerp Naval Stores Co., Inc., Boston, Mass.
  General Naval Stores Co., New York, N. Y.

Turpentine (Venice)
  National Rosin Oil & Size Co., New York, N. Y.

Umbers
  Fezandié & Sperrlé, Inc., New York, N. Y.

Urea
  Sherka Chemical Co., Inc., Bloomfield, N. J.

Vanillin
  Seeley & Co., Inc., New York, N. Y.
  Van-Ameringen-Haebler, Inc., New York, N. Y.

Wax, Synthetic
  Glyco Products Co., Inc., New York, N. Y.

Whiting
  Columbia Alkali Corp., New York, N. Y.
  Limestone Products Corp. of Amer., Newton, N. J.

Witch Hazel Extract
  E. E. Dickinson Co., Essex, Conn.

Wood Flour
  D. H. Litter Co., New York, N. Y.
  Wood Flour, Inc., Manchester, N. H.

Xylol
  The Barrett Co., New York, N. Y.

Zinc
  Hegeler Zinc Co., Danville, Illinois.

Zinc Chloride
  Wishnick-Tumpeer, Inc., New York, N. Y.

Zinc Chromate
  E. M. & F. Waldo, Inc., Muirkirk, Md.

Zinc Oxide
  Merck & Co., Inc., Rahway, N. J.
  N. J. Zinc Co., New York, N. Y.

Zinc Stearate
  Merck & Co., Inc., Rahway, N. J.
  Wishnick-Tumpeer, Inc., New York, N. Y.

Zinc Sulphate
  W. R. Russell & Co., New York, N. Y.
  Virginia Smelting Co., West Norfolk, Va.

www.ingramcontent.com/pod-product-compliance
Lightning Source LLC
Chambersburg PA
CBHW060817170526
45158CB00001B/4